DNA in Supramolecular Chemistry and Nanotechnology

DNA in Supramolecular Chemistry and Nanotechnology

Edited by

EUGEN STULZ
University of Southampton, UK

GUIDO H. CLEVER
Georg-August University of Göttingen, Germany

WILEY Blackwell

This edition first published 2015

Registered Office
John Wiley & Sons, Ltd, The Atrium, Southern Gate, Chichester, West Sussex, PO19 8SQ, United Kingdom

For details of our global editorial offices, for customer services and for information about how to apply for permission to reuse the copyright material in this book please see our website at www.wiley.com.

Library of Congress Cataloging-in-Publication data applied for

ISBN: 9781118696866

A catalogue record for this book is available from the British Library.

Set in 10/12pt Times by SPi Global, Pondicherry, India

1 2015

Contents

List of Contributors

Philippe Barthélémy
Université Victor Segalen Bordeaux 2
UFR Sciences de la Vie
and INSERM U869
Bordeaux, France

Jacqueline K. Barton
Division of Chemistry and Chemical Engineering
California Institute of Technology
Pasadena
CA, USA

Nina Berova
Department of Chemistry
Columbia University
Havemeyer Hall
New York
NY, USA

Glenn A. Burley
Department of Pure and Applied Chemistry
University of Strathclyde
Glasgow, UK

Niklaas J. Buurma
Physical Organic Chemistry Centre
School of Chemistry
Cardiff University
Cardiff, UK

Jungkweon Choi
The Institute of Scientific and Industrial Research (SANKEN)
Osaka University
Ibaraki, Osaka, Japan

Jean-Louis Duprey
Graduate School of Science
Department of Chemistry
The University of Tokyo
Tokyo, Japan

Alessandro D'Urso
Department of Science
University of Catania
Catania, Sicily, Italy

George A. Ellestad
Department of Chemistry
Columbia University
Havemeyer Hall
New York
NY, USA

A. Erbe
Helmholtz-Zentrum Dresden-Rossendorf
Institute of Ion Beam Physics and Materials Research
Division Scaling Phenomena
Dresden, Germany

Maria Elena Fragalà
Department of Science
University of Catania
Catania, Sicily, Italy

Ariel L. Furst
Division of Chemistry and Chemical Engineering
California Institute of Technology
Pasadena
CA, USA

Sofia Gallo
Department of Chemistry
University of Zurich
Zurich, Switzerland

Alice Ghidini
Karolinska Institute
Department of Biosciences and Nutrition
Novum, Sweden

Arnaud Gissot
Université Victor Segalen Bordeaux 2
UFR Sciences de la Vie
and INSERM U869
Bordeaux, France

Andrea Greschner
Department of Chemistry
McGill University
Montreal, Quebec, Canada

Helmut Griesser
Institute for Organic Chemistry
University of Stuttgart
Stuttgart, Germany

Michael A. Grodick
Division of Chemistry and Chemical Engineering
California Institute of Technology
Pasadena
CA, USA

Anika Kern
Department of Organic and Bioinorganic Chemistry
Humboldt-University of Berlin
Berlin, Germany

Alfonso Latorre
Instituto Madrileño de Estudios Avanzados en Nanociencia (IMDEA Nanociencia) and CNB-CSIC-IMDEA Nanociencia Associated Unit "Unidad de Nanobiotecnología"
Madrid, Spain

Chun-Hua Lu
Institute of Chemistry
The Center for Nanoscience and Nanotechnology
The Hebrew University of Jerusalem
Jerusalem, Israel

Tetsuro Majima
The Institute of Scientific and Industrial Research (SANKEN)
Osaka University
Ibaraki, Osaka, Japan

Andriy Mokhir
Friedrich-Alexander-University Erlangen-Nürnberg
Department of Chemistry and Pharmacy
Germany

David Monchaud
ICMUB (Institut de Chimie Moleculaire Universite de Bourgogne)
CNRS UMR6302
Dijon, France

Jens Müller
Institute for Inorganic and Analytical Chemistry
University of Münster
Münster, Germany

Merita Murtola
Karolinska Institute
Department of Biosciences and Nutrition
Novum, Sweden
and Department of Chemistry
University of Turku
Turku, Finland

Anastasia Musiari
Department of Chemistry
University of Zurich
Zurich, Switzerland

Khalid Oumzil
Université Victor Segalen Bordeaux 2
UFR Sciences de la Vie
and INSERM U869
Bordeaux, France

Amit Patwa
Université Victor Segalen Bordeaux 2
UFR Sciences de la Vie
and INSERM U869
Bordeaux, France

Ana G. Petrovic
Department of Chemistry
Columbia University
Havemeyer Hall
and Department of Life Sciences
New York Institute of Technology (NYIT)
New York
NY, USA

Fang Pu
Laboratory of Chemical Biology and State Key
Laboratory of Rare Earth Resources Utilization
Changchun Institute of Applied Chemistry
Chinese Academy of Sciences
Changchun, China

Roberto Purrello
Department of Science
University of Catania
Catania, Sicily, Italy

Hanna Radecka
Department of Biosensors
Institute of Animal Reproduction and Food
Research of Polish Academy of Sciences
Olsztyn, Poland

Jerzy Radecki
Department of Biosensors
Institute of Animal Reproduction and Food
Research of Polish Academy of Sciences
Olsztyn, Poland

Jinsong Ren
Laboratory of Chemical Biology and State
Key Laboratory of Rare Earth Resources
Utilization
Changchun Institute of Applied Chemistry
Chinese Academy of Sciences
Changchun, China

Clemens Richert
Institute for Organic Chemistry
University of Stuttgart
Stuttgart, Germany

Ana Rioz-Martínez
Stratingh Institute for Chemistry
University of Groningen
Groningen, The Netherlands

Gerard Roelfes
Stratingh Institute for Chemistry
University of Groningen
Groningen, The Netherlands

Fiora Rosati
Department of Chemistry
McGill University
Montreal, Quebec, Canada

Magdalena Rowinska-Zyrek
Department of Chemistry
University of Zurich
Zurich, Switzerland

Alexander Schwenger
Institute for Organic Chemistry
University of Stuttgart
Stuttgart, Germany

Oliver Seitz
Department of Organic and Bioinorganic
Chemistry
Humboldt-University of Berlin
Berlin, Germany

Mitsuhiko Shionoya
Graduate School of Science
Department of Chemistry
The University of Tokyo
Tokyo, Japan

Roland K. O. Sigel
Department of Chemistry
University of Zurich
Zurich, Switzerland

Indranil Sinha
Institute for Inorganic and Analytical Chemistry
University of Münster
Münster, Germany

Hanadi Sleiman
Department of Chemistry
McGill University
Montreal, Quebec, Canada

Álvaro Somoza
Instituto Madrileño de Estudios Avanzados en Nanociencia (IMDEA Nanociencia) and CNB-CSIC-IMDEA Nanociencia Associated Unit "Unidad de Nanobiotecnología"
Madrid, Spain

Peter Strazewski
Institut de Chimie et Biochimie Moléculaires et Supramoléculaires (UMR 5246)
Université de Lyon
Lyon, France

Roger Strömberg
Karolinska Institute
Department of Biosciences and Nutrition
Novum
Sweden

Claudia Stubinitzky
Karlsruhe Institute of Technology (KIT)
Institute of Organic Chemistry-Chair II
Karlsruhe, Germany

Yusuke Takezawa
Graduate School of Science
Department of Chemistry
The University of Tokyo
Tokyo, Japan

Manuel A. Tamargo
Department of Chemistry
Columbia University
Havemeyer Hall
New York, NY
USA

Stefan Vogel
Department of Physics
Chemistry and Pharmacy
University of Southern Denmark
Biomolecular Nanoscale Engineering Center (BioNEC)
Denmark

Hans-Achim Wagenknecht
Karlsruhe Institute of Technology (KIT)
Institute of Organic Chemistry-Chair II
Karlsruhe, Germany

Fuan Wang
Institute of Chemistry
The Center for Nanoscience and Nanotechnology
The Hebrew University of Jerusalem
Jerusalem, Israel

Christian Wellner
Karlsruhe Institute of Technology (KIT)
Institute of Organic Chemistry-Chair II
Karlsruhe, Germany

Itamar Willner
Institute of Chemistry
The Center for Nanoscience and Nanotechnology
The Hebrew University of Jerusalem
Jerusalem, Israel

Kazushige Yamana
Department of Materials Science and Chemistry
University of Hyogo
Himeji, Japan

Preface

Notably, one of the most influential scientific achievements of the last century concerns the structural elucidation of DNA by Watson, Crick, Franklin and Wilkins in 1953 [1]. As was foreseen in their seminal papers, the structure of DNA had profound implications on molecular biology:

> "We wish to suggest a structure for the salt of deoxyribose nucleic acid (D.N.A.). This structure has novel features which are of considerable biological interest."

Since then, the role of DNA in biology has been accepted as that of the bearer of the genetic code, and many genomic sequences have been deciphered, including the human genome. However, the story is far from over, and many questions still remain to be answered, in particular, the details of how DNA works to fulfil its duties, including the sequence dependent local structure, protein binding to DNA, regulation of gene expression (including methylation and selective unwinding from histones), to name but a few. Also, although knowledge of the three-letter genetic code allows for the identification of genes, and comparison between species to find similarities and heritage, it also raises the question of the exact role of the non-coding regions of the genomes.

Not long after the huge biological importance of DNA had been understood, the fascinating features of this self-assembled biopolymer began to inspire synthetic chemists, analytical specialists, biophysicists and even computer programmers to devote their research efforts to oligonucleotide-based systems.

The basic principle of the workings of DNA is simple: the molecule forms a well-understood double helix through the complementary base pairing of two antiparallel strands. Yet there is far more to DNA than just this concept. DNA can act both as a rigid stick (duplex) with a persistence length of about 40–50 nm, and as a flexible glue (single strand), giving rise to a Lego-like building block system for the creation of architectures with nanometre precision. In addition, more complex tertiary structures such as loops, triple strands, quadruplexes and various junctions, many of which have been found *in vivo*, add to the countless possibilities of creating DNA structures with functions beyond those of just information storage and heredity. The plethora of non-duplex tertiary structures is even more pronounced in the RNA context.

By taking DNA out of its biological environment, new systems have emerged that are beginning to play a major role in materials science, electronics, diagnostics, medicinal chemistry and much more. The supramolecular aspects of DNA have seen significant advances in the past few decades, particularly with the inclusion of modified nucleotides.

Organic synthesis has and will remain a key aspect of DNA chemistry. The concept of phosphoramidite chemistry for automated solid-phase synthesis (SPS) of DNA, as first described by Caruthers in 1981 [2], has, on the whole, remained unaltered, as it has proven to be a most versatile approach. In principle, any diol can be used for the stepwise build-up of functional molecules, as the concept is not restricted to nucleosides, provided the other functional groups that are present are compatible with the strongly acidic, basic and oxidising environment in SPS, although milder conditions are also available. In addition, nucleosides can be chemically functionalised with almost any additional group, be it small organic molecules or large metal complexes. The combination of all accessible building blocks has lead to a system that is now being explored in almost all areas of science, as mentioned earlier.

The chapters in this book present an excellent collection of state-of-the-art reviews covering research that is currently being carried out. The chapters are devoted to the most important fields being studied:

1. (Non-) covalently modified DNA with novel functions
2. DNA wires and electron transport through DNA
3. Oligonucleotides in sensing and diagnostic applications
4. Conjugation of DNA with biomolecules and nanoparticles
5. Alternative DNA structures, switches and nanomachines

Each chapter is devoted to a specialised topic within the larger context of DNA chemistry, and will give the reader both an introduction to the field and the relevant technologies, and a more focussed description of the authors' own research and viewpoint. General conclusions and outlooks for the future are also given. We are in the very lucky position of having been able to attract many of the key players in the relevant areas to devote their time to writing what we believe is an outstanding collection of well written and up-to-date chapters. As the field is growing so rapidly, it would be beyond the scope of any textbook to include everyone currently working in this field, or to mention the excellent new ideas of every research group. This demonstrates that applied DNA technology has reached the level at which it is playing a key role across the board.

Nucleotide technology is a fast moving area, where research is advancing rapidly. However, these chapters will certainly remain a key collection for some years to come. Even though the systems will steadily improve, the basic technologies presented and discussed here will remain the same for the foreseeable future, as can be seen from the basic SPS principles. The book is intended to give the novice an introduction and a basic overview of all aspects of nucleotide technology, including strategies, concepts and visions. The experienced reader will have an up-to-date volume at hand where – through the amalgamation of very different ideas – novel applications and concepts may be conceived. We are therefore convinced that this collection will be very useful to scientists at all levels, and we hope that it will indeed trigger the emergence of new research.

It is generally difficult, if not impossible, to predict where the future will take us, as new results are being published on a daily basis, sometimes in unrelated fields where the impact on one's own research may not be immediately evident. Many of the approaches used today could not have been predicted ten years ago, as some of the key technologies were not available then (e.g. DNA origami). DNA technology will continue to be important at both the giving and receiving ends: DNA bio-nanotechnology will influence material, medicinal and biological sciences, whereas inventions and demands from those fields will have an impact on the direction of the development of DNA technology.

We give great thanks to all contributors to this book and hope that the reader will enjoy it as much as we did during the editorial stages.

Eugen Stulz
Guido H. Clever

References

[1] (a) R. E. Franklin and R. G. Gosling, Molecular configuration in sodium thymonucleate. *Nature* 1953, **171** (4356), 740–741; (b) J. D. Watson and F. H. C. Crick, Molecular structure of nucleic acids: A structure for deoxyribose nucleic acid. *Nature* 1953, **171** (4356), 737–738; (c) M. H. Wilkins, A. R. Stokes and H. R. Wilson, Molecular structure of nucleic acids: Molecular structure of deoxypentose nucleic acids. *Nature* 1953, **171** (4356), 738–740.

[2] (a) S. L. Beaucage and M. H. Caruthers, Deoxynucleoside phosphoramidites – A new class of key intermediates for deoxypolynucleotide synthesis. *Tetrahedron Lett.* 1981, **22** (20), 1859–1862; (b) M. D. Matteucci and M. H. and Caruthers, Nucleotide Chemistry. 4. Synthesis of deoxyoligonucleotides on a polymer support. *J. Am. Chem. Soc.* 1981, **103** (11), 3185–3191.

Part I

(Non-) Covalently Modified DNA with Novel Functions

1.1

DNA-Based Construction of Molecular Photonic Devices

Glenn A. Burley

Department of Pure and Applied Chemistry, University of Strathclyde, Glasgow, UK

1.1.1 Introduction

Controlling the spatial arrangement of photonic materials reproducibly and with nanoscale precision is of fundamental importance for the development of optoelectronic devices and sensors of the future. Over the past 30 years, industry has made phenomenal progress in the fabrication of optoelectronic circuits and devices with a high level of accuracy and reproducibility. 'Top down' nanolithography has been the major driver in these developments, producing devices and circuits with resolution levels ranging from tens [1] to hundreds of nanometres [2]. Top down photolithographic approaches such as Extreme Ultraviolet Lithography (EUV) have been the predominant methods used to fabricate optoelectronic devices with sub-50 nm resolution levels. One of the major drawbacks in the further development of higher resolution circuits fabricated using EUV is the rising cost of the equipment required to produce smaller devices with sub-22 nm resolution [3]. The more recent development of Nanoimprint Lithography (NIL), for example, can replicate high-resolution patterns as small as 2.4 nm and is one of the leading contenders for the fabrication of sub-22 nm circuitry [1a], yet technological hurdles such as the defectivity and process variability of the resultant device platforms requires further development [4].

As a consequence of the increasing technological as well as economic challenges involved in fabricating devices through purely lithographic approaches, alternative methods and strategies of fabrication are now being investigated from both a fundamental as well as an applied perspective [5]. Building circuits and devices from functional molecular building blocks, that is, a 'bottom up approach', is a particularly attractive method for achieving molecular-scale precision [6]. There is increasing interest in using supramolecular

DNA in Supramolecular Chemistry and Nanotechnology, First Edition.
Edited by Eugen Stulz and Guido H. Clever.

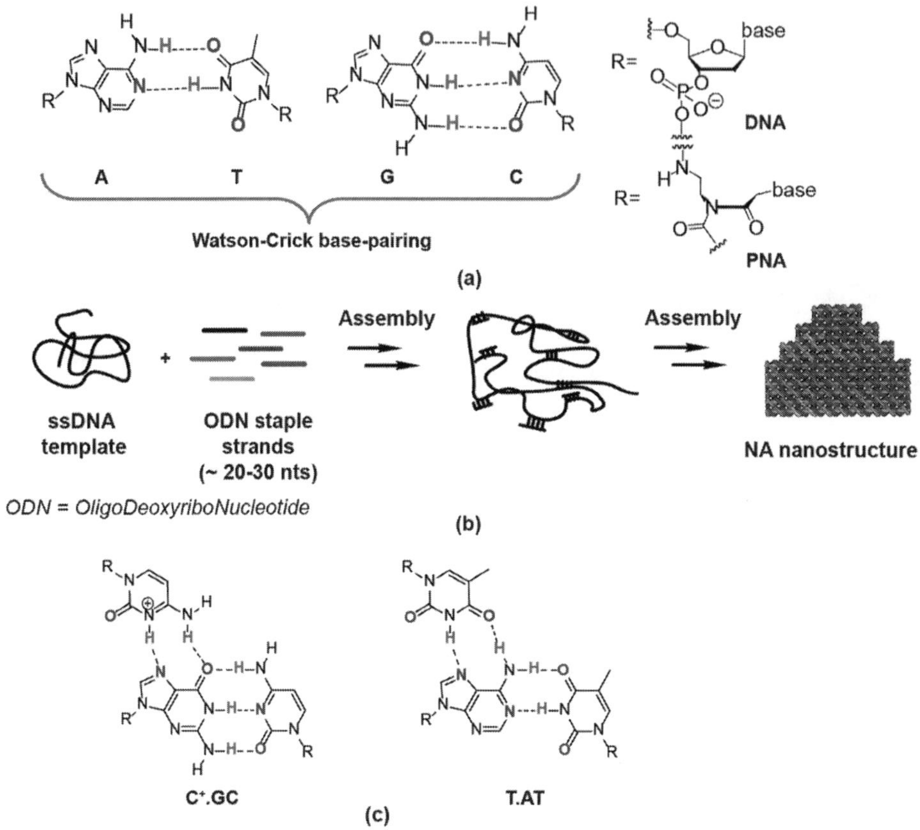

Figure 1.1.1 (a) Watson–Crick base-pairing is used in Nature to store genetic information and in DNA nanotechnology to direct the assembly of sophisticated multi-dimensional nanostructures. DNA analogues such as Peptide Nucleic Acids (PNA) have also been used to direct the assembly of DNA nanostructures. (b) Schematic representation of DNA origami. A single-stranded DNA template is weaved in two- and three-dimensional DNA nanostructures using a variety of oligodeoxyribonucleotide (ODN) staple strands. (c) Triplex Forming Oligonucleotides (TFOs) offer an alternative directing modality through the formation of triplex structures

assembly principles to form functional optoelectronic devices and sensors for device applications [7], yet despite a number of seminal advances in this area [8], a significant challenge still remains, that of fabricating precisely defined and error-free nanomaterials over micron-scale surface areas with complete 3D control and sub-nanometre resolution in a reproducible fashion *de novo* [9]. In contrast, Nature is astute at preparing micron-scale, self-assembled nanostructures via the use of a template-driven process to direct both the formation and the control of the growth of the overall nanostructure [10]. For example, the protein ferritin can be used as a template for the controlled biomineralisation of nanostructures [11]. Peptides can also be programmed to assemble in nanostructures and even act as templates for the assembly of non-natural functional materials; however, the ability to form bespoke functional materials is still restricted by our limited understanding of the rules that govern their self-assembly [10].

Of the biomacromolecules available in Nature, DNA molecules and their structural analogues have emerged as excellent templates to guide the synthesis [12] as well as the assembly of functional nanomaterials from the 'bottom up' (Figure 1.1.1a) [13]. By exploiting the predictable base-pairing rules of DNA and the high density of information embedded in its structure, DNA-programmed self-assembly can form sophisticated multi-dimensional assemblies ranging from 3D crystals [14], micron-scale 2D [15] and 3D [15b, 16] DNA nanostructures, as well as dynamic nanostructures [17], which can be reconfigured to release a therapeutic cargo in response to molecular cues [18].

The principal aim of this chapter is to highlight the recent developments in the use of DNA-programmed self-assembly to guide the construction of discrete photonic nanostructures. The advantages and disadvantages of using DNA-programmed self-assembly to construct arrays of organic fluorophores and proteins will be presented. The second half of the chapter will review efforts focusing on different modes of DNA-programmed self-assembly to fabricate optoelectronic circuits and light-harvesting complexes. For specific applications of DNA-directed assembly for the construction of supramolecular photosynthetic mimics, the reader is directed to a recent review by Albinsson, Hannestad and Börjesson [19]. DNA-programmed assembly of metallic and semiconductor nanoparticles is another rapidly expanding area of DNA nanotechnology. This has been the subject of recent reports and will not be discussed herein [13a, 13c, 20].

1.1.2 Using DNA as a template to construct discrete optoelectronic nanostructures

DNA is a unique self-assembling molecular system. This uniqueness arises from the inherent programmability of Watson–Crick base-pairing of Adenine (A) hydrogen-bonding with Thymine (T) and Guanine (G) hydrogen-bonding with Cytosine (C) [13b]. Both the programmability and flexibility of these pairing rules can be used to form a variety of structures ranging from simple duplexes through to more complex four-stranded Holliday junctions. Further enhancement of the stiffness of DNA nanostructures is also possible as a consequence of the development of double and triple cross-over motifs [13b]. The high level of programmability of DNA is also underpinned by the availability of pre-designed sequences – both short and long. This is a key aspect of DNA nanotechnology and sets it apart from other self-assembling biomolecules as both short and long DNA sequences can be prepared, and amplified to produce suitable amounts of the template for fundamental investigations. For example, solid-phase synthesis can produce modified oligodeoxyribonucleotides (ODNs) up to ~120 nucleotides in length [21], whereas longer DNA sequences of up to 20 kilobases in length can be prepared using the Polymerase Chain Reaction (PCR) [22].

Taken collectively, both the availability of material, the predictability of self-assembly rules and the more recent advent of computer software to facilitate the design of DNA nanostructures has spurred on the construction of sophisticated two- and three-dimensional nanostructures. One of the most successful exemplars of this has been 'DNA origami', which weaves a long single-stranded DNA template with the help of a series of shorter ODN strands (Figure 1.1.1b) [15a]. Since modified ODNs with a precise functionalisation pattern can be prepared by solid-phase synthesis, the insertion of non-natural functionality at precise locations in an origami-design DNA-programmed array can be realised [21, 23], and has been used with great effect to template a vast array of functional materials along a DNA nanostructure [21, 24].

Traditional strategies to construct DNA nanostructures have focused on utilizing Watson–Crick base pairing between two complementary DNA strands. A less investigated strategy to control the addressability of optical functionality is to exploit the topological features of higher order DNA structures (Figure 1.1.2a). With its 2 nm diameter, repetitive helicity of 3.4 nm, arrangement of base-pair 'bits' of information every 0.34 nm and its widespread occurrence in DNA nanostructures, B-type double-stranded DNA (dsDNA) offers an auxiliary mode to address optoelectronic materials. For example, the large surface area and solvent accessible major groove is the primary site for DNA-binding domains found in Transcription Factors [25]. Minor-groove binding small molecules such as Hoechst 33258 (**1**) and DNA-binding polyamides (PAs, **2**, Figure 1.1.2b) [26] offer an alternative mode of duplex DNA binding in the deep, hydrophobic minor groove [27].

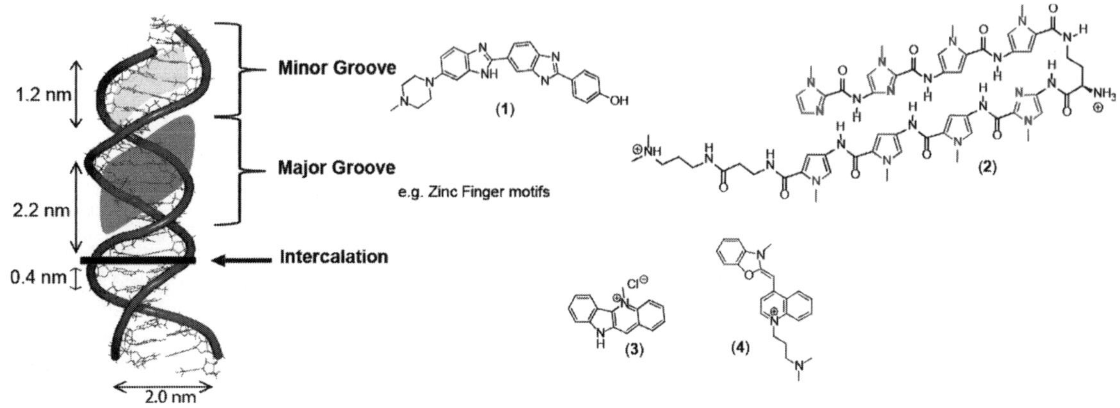

Figure 1.1.2 *(a) Structure of a B-type DNA duplex and the location of ligand binding. (b) Representative subset of DNA minor-groove binder (e.g.* ***1*** *and* ***2****) and DNA intercalators (e.g.* ***3*** *and* ***4****)*

Tethering functionality to specific sites on these molecules can therefore be used to direct optoelectronic materials to a specific dsDNA sequences within nanostructures.

Nucleic acid analogues such as Peptide Nucleic Acids (PNA) and Triplex Forming Oligonucleotides (TFOs) are another family of molecules that can bind to dsDNA in a sequence-selective fashion. PNA, for example, has a number of binding modes, ranging from strand invasion of dsDNA through to triplex formation (Figure 1.1.1a) [28]. The different PNA binding modes are influenced by the DNA target sequence, which allows one to develop binding strategies that are contextualised by sequence and the mode of binding. In contrast, TFOs form triplex structures with dsDNA via the recognition of the edges of Watson–Crick base-pairs protruding into the major groove (Figure 1.1.1c) [29].

Finally, more generic binding modes can also be exploited for binding to duplex DNA (Figure 1.1.2a). For example, the hydrophobic interior of dsDNA enables aromatic and positively charged molecules such as cryptolepine (**3**) and YO-PRO (**4**), which can intercalate between base-pairs (Figure 1.1.2b) [30]. Electrostatic interactions can also play an important role in templating functional materials. The highly charged anionic phosphodiester backbone can be used to template a wide range of cationic polymers [31] and polyamines [32] through electrostatic attraction, albeit in a non-sequence specific manner. Both of these modes do not possess the equivalent level of programmability of minor- or major-groove binders; however, they do provide the potential to interface with DNA nanostructures in a more generic fashion if programmability is not an essential requirement.

Currently, there are two major categories of DNA binding used to construct DNA-programmed assemblies and arrays:

i. Construction of arrays that utilise single-stranded DNA (ssDNA) as a template. These multi-chromophoric assemblies typically utilise modified ODNs and Watson–Crick base-pairing to direct the construction of higher order supramolecular arrays [33].
ii. Construction of arrays that utilise dsDNA as the template. This strategy exploits the topological characteristics of DNA duplexes to place optoelectronic materials in precise locations along a DNA architecture. These binding modes include the use of intercalators [34] and minor-groove binding ligands [35] to direct positional assembly within DNA nanostructures.

1.1.3 Assembly of photonic arrays based on the molecular recognition of single-stranded DNA templates

DNA-programmed photonic assemblies prepared by this strategy require the use of ODNs where the position of fluorophores at defined sites is controlled by the primary sequence of a complementary single-stranded DNA template. Although there are many reported examples of the use of Förster Resonance Energy Transfer (FRET) in DNA-based assemblies [36], this section will predominantly restrict itself to the discussion of systems comprising more than two types of fluorescent dyes (i.e. involving at least two energy-transfer steps). For a treatise of FRET in biological processes, including nucleic acids, the reader is directed to the review by Sapsford, Berti and Medintz [37].

Kawahara, Uchimaru and Murata reported the first use of a DNA template to produce a photonic wire in 1999 [38]. In this design, a 25-mer ODN 5′-modified with 6-carboxyfluorescein was used as the template. Sequential binding of two fluorophore-modified ODNs (4,7,2′,4′,5′,7′-hexachloro-6-carboxyfluorescein and 6-carboxy-X-rhodamine), whose sequence corresponded to a specific coding sequence within the template, were then added to produce the final photonic array (Figure 1.1.3a). Energy transfer was observed over 8 nm. Ohya *et al.* improved on this fundamental design through the construction of a longer photonic wire comprising three different fluorophores [39]. The absorption and emission characteristics of eosin (Eo), tetramethylrhodamine (TMR) and Texas Red (TR) were coupled with Förster Resonance Energy Transfer (FRET). FRET was observed over a maximum distance of 10 nm along a DNA duplex by exploiting a mixture of both hetero- and homo-FRET processes, starting from the initial excitation of the Eo dye and observing the emission of the TR dye. Two TMR 'jumper dyes' were used to transfer energy ultimately to the TR located at the end of a DNA duplex. By virtue of the narrow Stokes shift of TMR, energy transfer is achieved between the two TMR dyes by a homo-FRET process, whereas the first and last energy transfer steps occur via a uni-directional hetero-FRET process.

A similar DNA photonic model to that of Ohya *et al.* was utilised by Heilemann *et al.* but with crucial innovations [40]. A DNA template 5′-end-modified with a blue Rhodamine Green (RhG) strand formed the basis for the complementary binding of four ODNs outfitted with fluorophores, permitting an energy cascade to ensure energy transfer. Using this model, Heilemann *et al.* demonstrated the first DNA-based photonic wire where energy transfer proceeded in a truly unidirectional process over a distance of 13.6 nm and a spectral range of 200 nm. High energy transfer efficiencies (~90%) of this five colour system were also reported using a fully assembled photonic wire model when the interchromophore distance was confined to 3.4 nm or one helical turn of the DNA template. However, a limitation of this design is the need to assemble multiple components to construct the final photonic wire. This multi-component assembly results in poor overall FRET efficiencies (i.e. 10%). As a consequence of the poor yield of the final construct, energy transfer efficiencies were also poor (~15%) in bulk solution relative to the high energy transfer efficiencies observed in single-molecule examples where the photonic wire is fully assembled [41].

In order to address the structural heterogeneity of their original model, Heilemann *et al.* reported significant increases in energy transfer of a five colour photonic wire through the immobilisation of the DNA scaffold to a solid surface by biotin–streptavidin binding [42]. In this latest design a photonic assembly using a template strand comprising the injector RhG dye on the 5′-end and a biotin on the 3′-end was constructed. Three complementary strands incorporating the transmitting chromophores and a final emitting chromophore were then hybridised to form the final photonic wire assembly. Surface immobilisation of the DNA-based photonic wire was then achieved using streptavidin-coated surfaces. With this design, Heilemann *et al.* report homogeneous energy transfer with efficiencies of ~85% for a five colour photonic wire system. The key aspect for the increase in energy transfer is restriction of the conformational freedom of the fluorophores as a consequence of surface immobilisation. However, in order to increase the FRET efficiencies and improve the yield of the resulting array, future studies will require simplification of the complicated sequential assembly of multiple DNA strands.

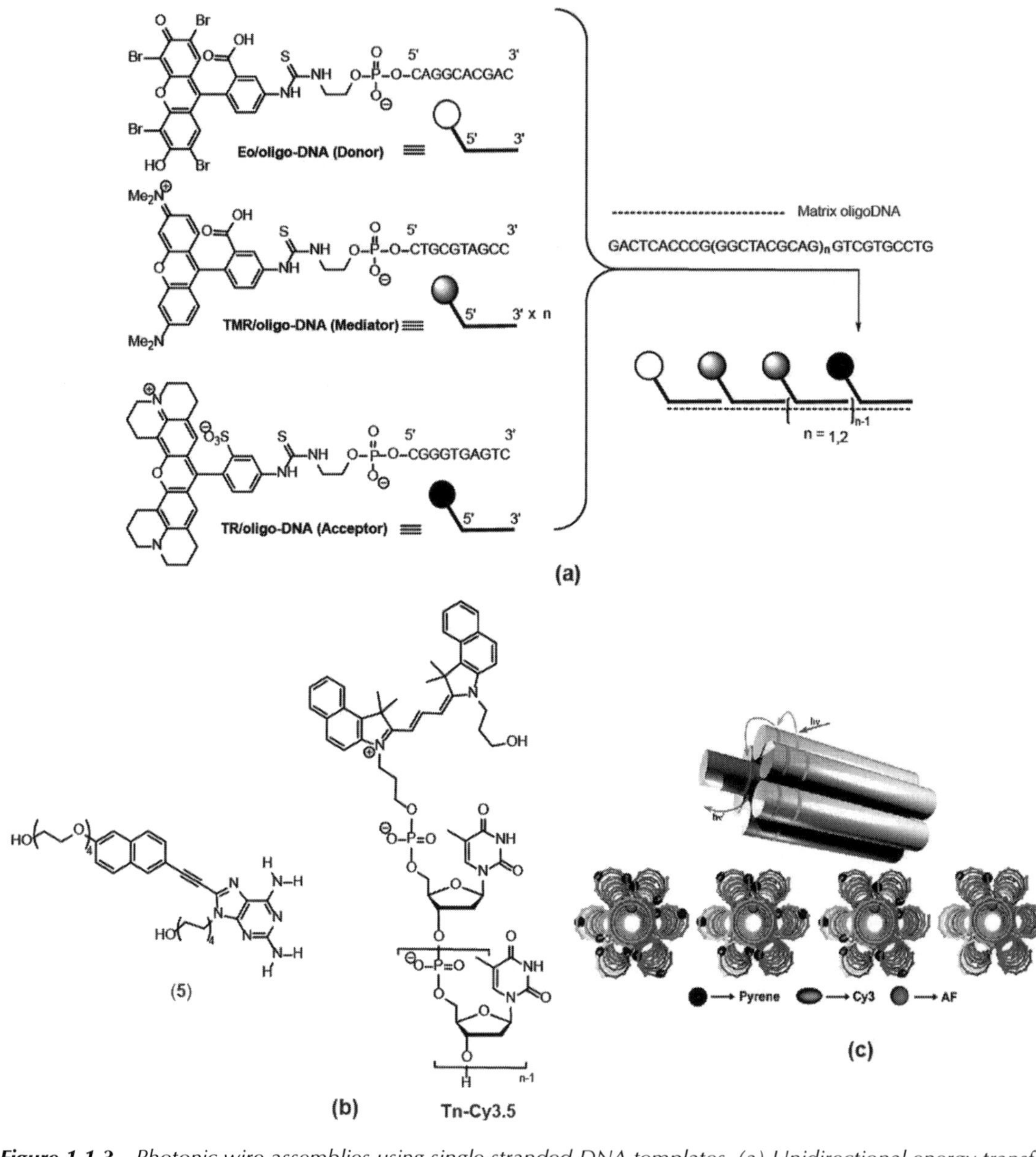

__Figure 1.1.3__ Photonic wire assemblies using single-stranded DNA templates. (a) Unidirectional energy transfer reported by Ohya et al. (adapted from [39]). This design utilises a single-stranded template and three different ODN end-modified with different transmitting fluorophores. Reprinted by permission of the publisher (Taylor & Francis Ltd, http://www.tandfonline.com). (b) Energy transfer was reported using single-stranded DNA to template the assembly of photoactive diaminopurine building blocks (__5__; Reprinted with permission from [48a], The Royal Society of Chemistry). (c) The formation of a DNA-programmed seven helix bundle artificial light harvesting complex. Reprinted with permission from [50]. Copyright 2011, American Chemical Society (See colour figure in colour plate section)

A photonic wire design first reported by Tong *et al.* [43], and later optimised by Vyawahare *et al.* [44], reduced the complexity of the multiple self-assembly processes required in the models of Ohya *et al.* and Heilemann *et al.* by reducing the construction of the photonic assembly to a single duplex forming process. In this model, photonic wire assemblies using two complementary DNA strands were prepared by the site specific incorporation of dyes at both ends of the ODN [6-FAM (6-carboxyfluorescein)] as the input and Cy5 as the output dye as well as internal positions [TAMRA (6-tetramethylrhodamine-5(6)-carboxamide)]. Using a simple hetero-FRET model of three fluorophores, Vyawahare *et al.* observed energy transfer efficiencies of 40% for this photonic wire assembly over 6.5 nm (19 base pairs). Increasing the number of intermediary energy transfer steps by the incorporation of TAMRA at ten base pair intervals, energy transfer efficiencies of 20% were observed in DNA-based photonic wires up to 13.6 nm in length (40 base pairs).

With each of the models aforementioned, the photonic wire assemblies involved the propagation of energy in a single dimension. In a recent series of studies, the Tinnefeld group reported the use of two-dimensional arrays to produce directional energy transfer [45]. Using a DNA origami approach [15a], these workers constructed a two-dimensional array and ODNs end-modified energy transmitting fluorophores. They demonstrated a directional two-step energy transfer based on the choice of ODN sequence binding to a defined sequence along the DNA template. Energy transfer rates of 25% from the blue injecting dye (ATTO488) through to the infrared dye (Alexa 750) were reported over a 9 nm distance. With further optimisation of DNA origami self-assembly techniques and the choice of photoactive materials, the multi-dimensional self-assembly techniques hold considerable potential for the design of future devices. Indeed Tinnefeld *et al.* have made inroads into this area by developing functional DNA-programmed 'nanopillars'. These highly rigid nanostructures can be immobilised to solid substrates and provide structural rigidity for possible multi-layered assembly of DNA origami tiles [46].

In these first generation models of DNA-templated fluorophore assemblies, the typical method of covalent attachment of fluorophores to ODNs used commercially available and flexible C6-linkers. This conformational flexibility allowed these fluorophores to adopt a range of unfavourable orientations and competing quenching pathways, which can have a significant impact on FRET efficiencies. In order to combat this, rigid linkers provide a feasible solution to reduce the conformational flexibility. Elegant examples of conformationally fixed assemblies of DNA-programmed chromophores have been reported and provide an additional level of sophistication to the construction of photonic wires and optical waveguides, however, the utility of these building blocks in the context of DNA-programmed photonic assemblies await comprehensive characterisation [21, 24e, 47].

Ruiz-Carretero *et al.* reported a different approach to the one-dimensional assembly of DNA-programmed photonic wires using single-stranded DNA templates [48]. In this approach, diaminopurine analogues incorporating a naphthalene energy donor were prepared and then assembled along a poly-T ODN sequence end-tagged with an energy reporter Cy3.5 dye (Tn-Cy3.5, Figure 1.1.3b). Cy3.5 emission was observed upon excitation of (**5**). This was the first reported approach of DNA-directed energy transfer that did not require a complementary DNA strand to pre-organise photonic components [48b, 49].

The exploitation of Watson–Crick base-pairing to construct a large array of light-harvesting modules was elegantly demonstrated by Dutta *et al.* [50]. They used a series of ODNs internally modified with chromophores using rigid linkers. An energy transfer gradient was set up using pyrene (Py) as the energy injector, cyanine 3 (Cy3) as the intermediate donor and finally the Alexa Fluor 647 (AF) dye as the ultimate acceptor. The key element of this design is the precise spatial positioning of these dyes within a seven-helix bundle (Figure 1.1.3c). Upon excitation of the Py injector, energy transfer was funnelled to the centre of the self-assembled DNA nanostructure, thus mimicking naturally-occurring light-harvesting complexes, such as those found in purple photosynthetic bacteria [51].

DNA-programmed light-harvesting complexes have also been prepared by the Haener group, where the Watson–Crick base-pairing of a DNA duplex was partially replaced by π-stacked phenanthrene and Py chromophores [52]. In this design, up to eight phenanthrene chromophores funnel energy towards a single Py chromophore embedded in a DNA duplex. The quantum yield obtained for this design was 41%, indicative that a highly efficient excitation energy transfer (EET) process had taken place.

The Haener group then extended this approach to a produce a three-dimensional light-harvesting complex. In this latest design a three-way junction (3WJ) was used to facilitate π-stacking of the light-collecting phenanthrene chromophores along one arm of a 3WJ [53]. Energy-collecting chromophores were pre-organised within close proximity to the light-collecting modules. EET was observed in a 3WJ exemplar, which incorporated a Py energy accepting chromophore. Replacement of the Py with a perylenediimide within the 3WJ design resulted in quenching of the fluorescence, whereas the use of a Cy dye resulted in fluorescence resonance energy transfer. This latest design demonstrates the potential to prepare more sophisticated DNA-programmed nanostructures beyond simple duplexes and to utilise other types of non-covalent interactions in order to construct sophisticated artificial light-harvesting assemblies.

1.1.4 Assembly of photonic arrays based on the molecular recognition of double-stranded DNA templates

Much less attention has been given to the assembly of photonic arrays by the molecular recognition of dsDNA templates. In this section, the progress made towards this goal will be summarised by categorising assemblies according to the binding mode utilized.

1.1.4.1 Intercalation

The utility of intercalators as an energy relay in a one-dimensional DNA-programmed photonic wire assembly was recently highlighted by Hannestad, Sandin and Albinsson [54]. These workers investigated energy transfer using DNA duplexes of up to 50 base pairs in length (~24 nm), by exploiting the homo-energy transfer capabilities of the oxazole yellow intercalating dye YO-PRO. YO-PRO exhibits attractive energy relay qualities for photonic wire applications, such as a narrow Stokes shift, enabling diffusive energy transfer along the duplex. Additionally, YO-PRO fluorescence is switched on only when bound to DNA, providing a readout of the binding. The basic design of the photonic wire system is highlighted in Figure 1.1.4a. Each end of the duplex contains a fluorophore on the 5′-end: one end with a blue injector [Pacific Blue, (PB)] and the other with a red reporter Cy3 dye. Upon the addition of increasing amounts of YO-PRO, energy transfer was observed from the PB injector through to the Cy3 reporter. They claim high end-to-end energy transfer efficiencies of up to 29% in a DNA wire of 50 base pairs in length or 24 nm. Moving on from simple DNA duplexes, Hannestad *et al.* have extended their one-dimensional photonic wire design to a more sophisticated two-dimensional DNA-programmed photonic network [55].

Looking beyond DNA-based photonic wire systems and towards constructing light-harvesting mimics, the Albinsson group has recently reported a functional DNA-programmed light-harvesting complex embedded within a lipid bilayer [56]. A porphyrin-linked uridine phosphoramidite was incorporated into a 39-mer ODN sequence by solid-phase synthesis. The porphyrin plays a dual role in this design. Firstly, it is an efficient acceptor for excitation energy transfer. Secondly, the hydrophobic porphyrin anchors the DNA complex within a lipid bilayer, thereby mimicking membrane-bound photosynthetic complexes found in Nature [51, 57]. Upon addition of YO-PRO to the DNA complex, a highly efficient energy transfer event was observed as a consequence of homo-FRET processes transferring energy along each YO-PRO and finally to the porphyrin chromophore.

Özhalici-Űnal and Armitage extended the utility of DNA-architectures to scaffold the assembly of fluorescent dyes in 3D DNA nanostructures [58]. They investigated the assembly of bis-intercalating YOYO

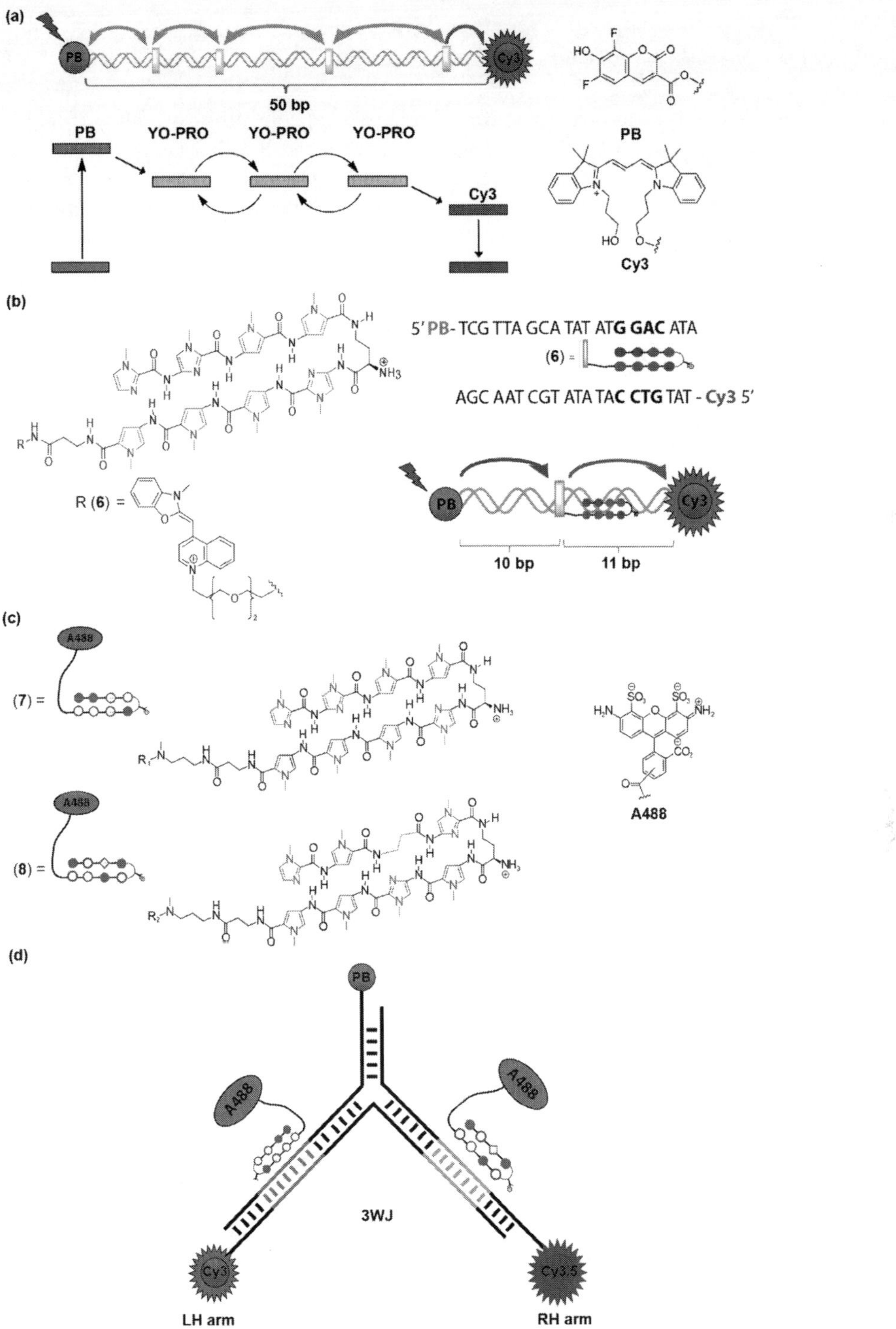

Figure 1.1.4 *Photonic wire assemblies guided by dsDNA templates. (a) DNA-programmed photonic wire assembly as reported by Hannestad, Sandin and Albinsson (adapted from [52]) utilises a double-stranded DNA template end modified by a blue injector (PB) and a red reporter (Cy3) dye. The intercalator dye YO-PRO (**4**) acts as an energy relay. Reprinted with permission from [52]. Copyright © 2012, WILEY-VCH Verlag GmbH & Co. KGaA, Weinheim. (b) Schematic representation of a DNA-based photonic wire assembly reported by Su et al. Reprinted with permission from [59]. Copyright © 2011, WILEY-VCH Verlag GmbH & Co. KGaA, Weinheim. (c) Structures of PAs (**7**) and (**8**) used to construct a DNA photonic array based on a 3WJ design. (d) Schematic representation of a three-dimensional photonic wire assembly based on a 3WJ. Reprinted with permission from [60] (See colour figure in colour plate section)*

dyes within a DNA-based tetrahedral scaffold and provided compelling evidence of intercalation of up to 24 YOYO molecules within the 3D scaffold. Using Cy3 end-labelled ODNs in the construction of the tetrahedral DNA-based scaffold, these workers reported highly efficient energy transfer from the YOYO to the Cy3 acceptor dye, thus opening the possibility of the development of three-dimensional light harvesting devices.

1.1.4.2 Minor-groove binding

Su *et al.* developed a highly efficient DNA-programmed photonic wire where the energy transfer efficiency is enhanced by the use of DNA-binding polyamides [59]. DNA-binding PAs bind to target sequences of six to ten base pairs in length within the minor groove of B-DNA [26]. Inspired by the work of Hannestad, Sandin and Albinsson (Figure 1.1.4a) [54, 55], a 21-mer DNA duplex incorporating a single PA target sequence 5′WWGGACW (where W = A/T) was designed. Both ends of DNA21 were modified with a PB injector and a Cy3 acceptor dye (Figure 1.1.4b). Using a PA-tethered YO (PAYO), Su *et al.* observed a threefold increase in energy transfer efficiency (DNA21@(**6**), 49%) with one equivalent of PAYO relative to a system lacking the PA (i.e. DNA21@(**6**), 15%). The generality of the approach was also demonstrated in DNA-based wire assemblies containing six PA binding sites with energy transfer observed over 27 nm.

Su *et al.* extended this work to programme the uni-directional transfer of excitation energy along a DNA three-way junction [60]. In contrast to the 3WJ design of Haener, these 3WJs incorporated two binding sites for PA (**7**) and (**8**), respectively (Figure 1.1.4c). Each PA was tethered to the fluorophore A488. Upon excitation of the PB injector, selective routing of light energy along the left-hand arm of the 3WJ was observed with the addition of (**7**). When PA (**8**) was added, light energy was routed along the right-hand arm of the 3WJ (Figure 1.1.4d). This study highlighted the first example that spatial and directional control of excitation energy can be achieved in a three-dimensional DNA-programmed nanostructure by using duplex DNA as a higher order addressable template within a DNA nanostructure.

Although minor-groove recognition is still an emergent concept in DNA nanotechnology, examples of the assembly of minor-groove binders in the presence of the double stranded DNA-templates have been reported. The Armitage group has reported the preparation of DNA-programmed supramolecular aggregates of cationic cyanine dyes both in one and two dimensions [61]. Cyanine dyes can self-assemble within A-T rich regions of the minor grooves, resulting in significant deviations in the optical properties of the resultant assemblies relative to their monomeric species. However, energy transfer investigations of these assemblies have not been reported to the best of our knowledge.

An innovative approach to prepare a DNA-programmed light harvesting complex using minor-groove binders was recently reported by Kumar and Duff [62]. DNA duplexes were used as templates for the assembly of the Hoechst 33258 minor-groove binder (**1**) with calf thymus DNA (ctDNA). Compound (**1**) is highly fluorescent when in complex with A-T rich regions of duplex DNA. A cationic version of the protein bovine serum albumin (BSA) was prepared and interfaced with the acceptor fluorophore Coumarin 540A (C540A). C540 has a high affinity for BSA resulting in a pre-organised protein scaffold adorned with acceptor molecules. Electrostatic attraction of the cationic BSA-C540 with the highly polyanionic ctDNA, results in the formation of a complex. Steady state fluorescence studies revealed that energy transfer was only observed in the presence of ctDNA, thus confirming the significance of Hoechst 33258 binding dsDNA to the scaffold, and bringing the donor (Hoechst 33258) and the acceptor (BSA-C540) functionalities within close enough proximity to permit energy transfer. Although these workers did not comment on the energy transfer efficiency, they observed >90% quenching of Hoechst 33258 emission.

1.1.5 Towards the construction of photonic devices

A series of fundamental exemplars of DNA-programmed photonic nanostructures have been highlighted; however, in order to gain widespread use one requires the flexibility to fabricate devices in the solid state as well as in solution. Research on the investigation of the optoelectronic properties of DNA thin films is a nascent field and one that is starting to gather momentum thanks to two key innovations:

i. The availability of large amounts of DNA in sufficiently high purity. Salmon sperm is a by-product of the fishing industry and is a cost-effective source of high molecular weight DNA [63].
ii. Techniques that render DNA templates soluble in organic solvents [64]. DNA in complex with the surfactant cetyl ammonium chloride (DNA-CTMA) is soluble in polar organic solvents enabling one to form DNA thin films derived from ethanol–chloroform solutions [63a, 64, 65].

With the availability of organic soluble DNA templates, a number of material science groups have now reported the use of DNA-CTMA thin films, including the preparation of organic emitting diodes [63e], nonlinear optics [66] and optical waveguides [67] derived from spin-coating organic solutions of DNA-CTMA. Ner *et al.* recently reported the utility of DNA as a scaffold for the fabrication of luminescent thin films [68]. These workers investigated the efficiency of FRET by doping an ethanol–chloroform (3:1) solution of CTMA-DNA with the fluorescent donor Coumarin 102 (Cm102) and the acceptor 4-(4-dimethylaminostyryl)-1-docosylpyridinium bromide (Hemi22). Both dyes are known to interact with DNA non-covalently: Cm102 is via intercalation, whereas Hemi22 is via minor-groove binding. Spin-coating of CTMA-DNA samples doped with varying ratios of Cm102 and Hemi22 revealed substantial energy transfer from Cm102 to Hemi22, even at low loadings of the Hemi22 acceptor. Furthermore, Ner *et al.* [68] demonstrated the significance of these findings from the observation that CTMA-DNA-Cm102-Hemi22 nanofibres convert UV-light into white light when fabricated into a light-emitting diode (Figure 1.1.5). The significance of the templating effect of the DNA scaffold was apparent with very low white luminescence observed in thin films formed in the absence of a CTMA-DNA scaffold. These findings demonstrate the potential of organic soluble CTMA-DNA to provide a matrix that controls both spatial arrangements of donor and acceptor molecules for device applications [69].

1.1.6 Outlook

Directing the assembly of optoelectronic materials using DNA offers a new bottom-up methodology that holds considerable potential to build functional assemblies for exploitation in the material science and biomedical arenas. Although by no means exhaustive, and open to interpretation, a summary of potential applications in which DNA-programmed self-assembly could directly have an impact on the photonics field is described below.

1.1.6.1 Optoelectronic circuits

The concept of templating of photonic components in DNA duplexes has already been applied in the fabrication of organic light emitting diodes [68], however, there is potential to move beyond this proof of concept study and utilise the programmability of DNA nanostructures to prepare discrete multi-dimensional nanostructures. Since the advent of DNA origami in 2006 [15a], structural DNA nanotechnology has rapidly progressed from mostly simple two-dimensional structures formed in moderate yields to the current state of

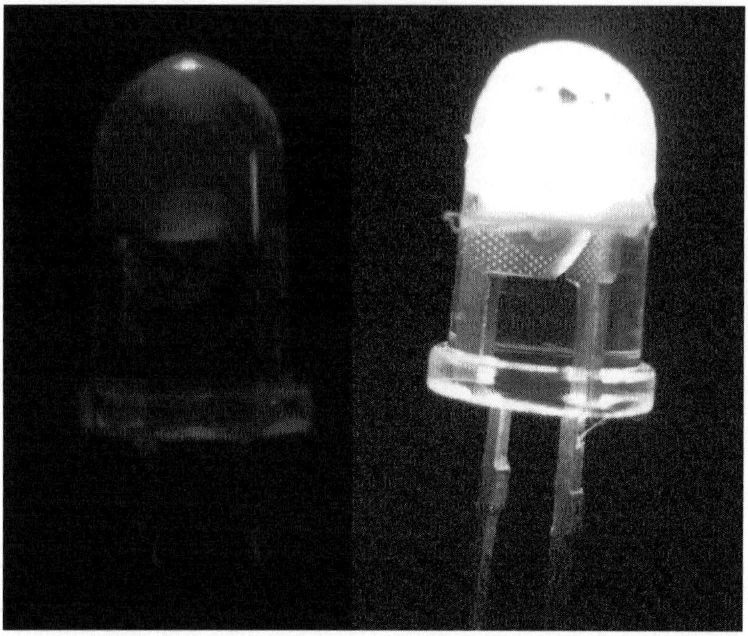

Figure 1.1.5 Photograph of a light emitting diode (LED) derived from DNA-CTMA thin films doped with Cm102 and Hemi22 (Reprinted with permission from [68]. Copyright © 2011, WILEY-VCH Verlag GmbH & Co. KGaA, Weinheim). Upon irradiation with 400 nm UV-light, the LED prepared from a dye doped DNA-CTMA thin film emits an intense white luminescence (right) compared with the LED lacking DNA-CTMA (left)

the art in which computer programs such as caDNAno [70] facilitate the design of high yielding and highly sophisticated three-dimensional structures [16a, 71].

1.1.6.2 Diagnostic platforms

The current research described here has been mainly confined to systems where the DNA architecture provides a static structural framework, but dynamic processes are also possible and could open up new opportunities for the investigation of biological processes interacting with the framework, such as nucleic acid–protein, small molecule–nucleic acid or indeed protein–protein interactions [72]. These systems are predicated on the precise arrangement of detection modules within the array, which in turn can detect a binding event. Indeed several groups have progressed towards this goal by using AFM as a diagnostic tool. Ke *et al.*, for example, have developed an RNA diagnostic platform using DNA origami-based methods to control the spatial arrangement of single stranded DNA sequences, which are used to detect specific RNA sequences [73]. Similarly, Subramanian *et al.* have reported a DNA origami-based method for the detection of Single Nucleotide Polymorphisms (SNPs) using AFM as the readout tool [74]. Interfacing optical outputs using FRET or surface plasmon resonance could potentially increase both the sensitivity and throughput of these pioneering detection platforms [75].

Although this chapter has focused on optoelectronics and diagnostics as applications earmarked for growth, there could indeed be many other uses that will also start to emerge. However, in order for DNA nanotechnology to be considered as a fabrication tool, the cost of preparing DNA nanostructures on a large scale needs to be addressed [13a]. This might involve developing methods to use cheaper sources of DNA to form DNA nanostructures, such as salmon sperm DNA [63c]. Other challenges include:

1. *Stability of photoactive materials*. The photo-instability of small-molecule fluorophores, especially fluorophores with absorptions in the red to far-red regions of the visible spectrum, prevents their widespread application in devices. Therefore, future devices will require the preparation of assemblies with stable photoactive materials, such as quantum dots, metal complexes or metallic nanoparticles [76]. The preparation of DNA-programmed quantum dot assemblies is not trivial [77] and further development to streamline the preparation of stabilised quantum dot-DNA nanostructures will be required.
2. *Compatibility of the DNA template*. UV-irradiation is known to damage DNA structures, producing a wide range of damaged bases, depurination as well as sugar adducts, which can alter the integrity of the double helix of DNA [78]. It is currently not known if DNA damage is exacerbated or even perturbed in DNA nanostructures, or if these alterations translate into differences in device performance. Another unanswered question is the role of the actual DNA nanostructure in mediating energy transfer events. Does the DNA nanostructure act merely as a passive scaffold or a conduit for energy transfer? The issue of whether DNA should be considered an electrical conductor, semiconductor or indeed an insulator remains contentious [79]. Therefore, a pertinent question that requires consideration is whether the *DNA primary sequence as well as the secondary and tertiary structure* of these assemblies facilitates energy/electron transfer processes and how such sequences/structures impact on device performance.
3. *Robustness of the device platform*. In order for these fundamental technologies to be of use in a device, one requires the reproducible preparation of each of the requisite components within a functional device platform [80]. At present, the preparation of DNA-programmed functional components and their integration into a device platform in a reproducible fashion is still a formidable challenge. Infrastructure is now required that can enhance the molecular robustness of fabrication and monitor quality control.

In summary, the development of DNA-programmed photonics is a vibrant and dynamic area of research. Fundamental discoveries in the design and construction of photonic assemblies and their integration into a device platform will undoubtedly be the next wave of exciting developments that could open up new opportunities for diverse applications in material science and biomedicine.

References

[1] (a) E. A. Costner, M. W. Lin, W.-L. Jen and C. G. Willson in *Nanoimprint Lithography Materials Development for Semiconductor Device Fabrication*, Annual Review of Materials Research,Vol. **39**: 155–180, DOI: 10.1146/annurev-matsci-082908-145336; (b) C. M. Kolodziej and H. D. Maynard, Electron-beam lithography for patterning biomolecules at the micron and nanometer scale. *Chemistry of Materials* 2012, **24**, 774–780.

[2] (a) B. D. Gates, Q. B. Xu, M. Stewart, D. Ryan, C. G. Willson and G. M. Whitesides, New approaches to nanofabrication: Molding, printing, and other techniques. *Chemical Reviews* 2005, **105**, 1171–1196; (b) S. R. Quake and A. Scherer, From micro- to nanofabrication with soft materials. *Science* 2000, **290**, 1536–1540.

[3] A. Vaglio Pret, R. Gronheid, T. R. Younkin, G. Winroth, J. J. Biafore, Y. Anno, K. Hoshiko and V. Constantoudis, Roughness and variability in EUV lithography: Who is to blame? (part 1) *Proc. SPIE*, 8679, Extreme Ultraviolet (EUV) Lithography IV, 867920 (April 1, 2013); DOI: 10.1117/12.2011584.

[4] A. V. Pret, P. Poliakov, R. Gronheid, P. Blomme, M. M. Corbalan, W. Dehaene, D. Verkest, J. Van Houdt and D. Bianchi, Linking EUV lithography line edge roughness and 16 nm NAND memory performance. *Microelectronic Engineering* 2012, **98**, 24–28.

[5] W. J. Parak, F. C. Simmel and A. W. Holleitner, Top-Down Versus Bottom-Up, in *Nanotechnology, Volume 1: Principles and Fundamentals*, Günter Schmid (Ed.). Wiley-VCH Verlag GmbH & C KGaA, Weinheim, 2008, ISBN: 978-3-527-31732-5, DOI: 10.1002/9783527628155.

[6] J.-M. Lehn, Toward self-organization and complex matter. *Science* 2002, **295**, 2400–2403.

[7] (a) Y. Yin and D. Talapin, The chemistry of functional nanomaterials. *Chemical Society Reviews* 2013, **42**, 2484–2487; (b) Y. Kim, J. Zhu, B. Yeom, M. Di Prima, X. Su, J.-G. Kim, S. J. Yoo, C. Uher and N. A. Kotov, Stretchable nanoparticle conductors with self-organized conductive pathways. *Nature* 2013, **500**, 59–77; (c) M. Heiss, Y. Fontana, A. Gustafsson, G. Wuest, C. Magen, D. D. O'Regan, J. W. Luo, B. Ketterer, S. Conesa-Boj, A. V. Kuhlmann, J. Houel, E. Russo-Averchi, J. R. Morante, M. Cantoni, N. Marzari, J. Arbiol, A. Zunger, R. J. Warburton and A. Fontcuberta i Morral, Self-assembled quantum dots in a nanowire system for quantum photonics. *Nature Materials* 2013, **12**, 439–444; (d) J. Yan, M. Bloom, S. C. Bae, E. Luijten and S. Granick, Linking synchronization to self-assembly using magnetic Janus colloids. *Nature* 2012, **491**, 578–581; (e) M. Naqshbandi, J. Canning, B. C. Gibson, M. M. Nash and M. J. Crossley, Room temperature self-assembly of mixed nanoparticles into photonic structures. *Nature Communications* 2012, **3**; (f) J. Vogelsang, T. Adachi, J. Brazard, D. A. V. Bout and P. F. Barbara, Self-assembly of highly ordered conjugated polymer aggregates with long-range energy transfer. *Nature Materials* 2011, **10**, 942–946; (g) H. Liu, J. Xu, Y. Li and Y. Li, Aggregate nanostructures of organic molecular materials. *Accounts of Chemical Research* 2010, **43**, 1496–1508; (h) L. Zang, Y. Che and J. S. Moore, One-dimensional self-assembly of planar pi-conjugated molecules: adaptable building blocks for organic nanodevices. *Accounts of Chemical Research* 2008, **41**, 1596–1608; (i) M. P. Stoykovich, M. Muller, S. O. Kim, H. H. Solak, E. W. Edwards, J. J. de Pablo and P. F. Nealey, Directed assembly of block copolymer blends into nonregular device-oriented structures. *Science* 2005, **308**, 1442–1446.

[8] (a) J. T. A. Jones, T. Hasell, X. Wu, J. Bacsa, K. E. Jelfs, M. Schmidtmann, S. Y. Chong, D. J. Adams, A. Trewin, F. Schiffman, F. Cora, B. Slater, A. Steiner, G. M. Day and A. I. Cooper, Modular and predictable assembly of porous organic molecular crystals. *Nature* 2011, **474**, 367–371; (b) Q. Chen, S. C. Bae and S. Granick, Directed self-assembly of a colloidal kagome lattice. *Nature* 2011, **469**, 381–384; (c) D. V. Talapin, E. V. Shevchenko, M. I. Bodnarchuk, X. Ye, J. Chen and C. B. Murray, Quasicrystalline order in self-assembled binary nanoparticle superlattices. *Nature* 2009, **461**, 964–967; (d) E. C. P. Smits, S. G. J. Mathijssen, P. A. van Hal, S. Setayesh, T. C. T. Geuns, K. A. H. A. Mutsaers, E. Cantatore, H. J. Wondergem, O. Werzer, R. Resel, M. Kemerink, S. Kirchmeyer, A. M. Muzafarov, S. A. Ponomarenko, B. de Boer, P. W. M. Blom and D. M. de Leeuw, Bottom-up organic integrated circuits. *Nature* 2008, **455**, 956–959; (e) E. V. Shevchenko, D. V. Talapin, N. A. Kotov, S. O'Brien and C. B. Murray, Structural diversity in binary nanoparticle superlattices. *Nature* 2006, **439**, 55–59; (f) Y. Lin, A. Boker, J. B. He, K. Sill, H. Q. Xiang, C. Abetz, X. F. Li, J. Wang, T. Emrick, S. Long, Q. Wang, A. Balazs and T. P. Russell, Self-directed self-assembly of nanoparticle/copolymer mixtures. *Nature* 2005, **434**, 55–59; (g) J. V. Barth, G. Costantini and K. Kern, Engineering atomic and molecular nanostructures at surfaces. *Nature* 2005, **437**, 671–679; (h) M. Kroutvar, Y. Ducommun, D. Heiss, M. Bichler, D. Schuh, G. Abstreiter and J. J. Finley, Optically programmable electron spin memory using semiconductor quantum dots. *Nature* 2004, **432**, 81–84; (i) D. Lingwood and K. Simons, Lipid rafts as a membrane-organizing principle. *Science* 2010, **327**, 46–50; (j) D. G. Gibson, J. I. Glass, C. Lartigue, V. N. Noskov, R.-Y. Chuang, M. A. Algire, G. A. Benders, M. G. Montague, L. Ma, M. M. Moodie, C. Merryman, S. Vashee, R. Krishnakumar, N. Assad-Garcia, C. Andrews-Pfannkoch, E. A. Denisova, L. Young, Z.-Q. Qi, T. H. Segall-Shapiro, C. H. Calvey, P. P. Parmar, C. A. Hutchison, III, H. O. Smith and J. C. Venter, Creation of a bacterial cell controlled by a chemically synthesized genome. *Science* 2010, **329**, 52–56; (k) J. A. Fan, C. Wu, K. Bao, J. Bao, R. Bardhan, N. J. Halas, V. N. Manoharan, P. Nordlander, G. Shvets and F. Capasso, Self-assembled plasmonic nanoparticle clusters. *Science* 2010, **328**, 1135–1138; (l) I. Bita, J. K. W. Yang, Y. S. Jung, C. A. Ross, E. L. Thomas and K. K. Berggren, Graphoepitaxy of self-assembled block copolymers on two-dimensional periodic patterned templates. *Science* 2008, **321**, 939–943; (m) Y. Yamamoto, T. Fukushima, Y. Suna, N. Ishii, A. Saeki, S. Seki, S. Tagawa, M. Taniguchi, T. Kawai and T. Aida, Photoconductive coaxial nanotubes of molecularly connected electron donor and acceptor layers. *Science* 2006, **314**, 1761–1764.

[9] (a) A. L. Boyle and D. N. Woolfson, *De novo* designed peptides for biological applications. *Chemical Society Reviews* 2011, **40**, 4295–4306; (b) S. A. Claridge, A. W. Castleman, Jr., S. N. Khanna, C. B. Murray, A. Sen and P. S. Weiss, Cluster-assembled Materials. *ACS Nano* 2009, **3**, 244–255; (c) C. B. Murray, C. R. Kagan and M. G. Bawendi, Synthesis and characterization of monodisperse nanocrystals and close-packed nanocrystal assemblies. *Annual Review of Materials Science* 2000, **30**, 545–610.

[10] J. Aizenberg, New nanofabrication strategies: Inspired by biomineralization. *MRS Bulletin* 2010, **35**, 323–330.

[11] (a) K. K. W. Wong and S. Mann, Biomimetic synthesis of cadmium sulfide-ferritin nanocomposites. *Advanced Materials* 1996, **8**, 928–932; (b) F. C. Meldrum, B. R. Heywood and S. Mann, Magnetoferritin - in vitro synthesis of a novel magnetic protein. *Science* 1992, **257**, 522–523; (c)F. C. Meldrum, V. J. Wade, D. L. Nimmo, B. R. Heywood and S. Mann, Synthesis of inorganic nanophase materials in supramolecular protein cages. *Nature* 1991, **349**, 684–687.

[12] L. Berti and G. A. Burley, Nucleic acid and nucleotide-mediated synthesis of inorganic nanoparticles. *Nature Nanotechnology* 2008, **3**, 81–87.

[13] (a) A. V. Pinheiro, D. Han, W. M. Shih and H. Yan, Challenges and opportunities for structural DNA nanotechnology. *Nature Nanotechnology* 2011, **6**, 763–772; (b) N. C. Seeman, Nanomaterials based on DNA. *Annual Review of Biochemistry,* 2010, **79**, 65–87; (c) H. Li, J. D. Carter and T. H. LaBean, Nanofabrication by DNA self-assembly. *Materials Today* 2009, **12**, 24–32.

[14] J. Zheng, J. J. Birktoft, Y. Chen, T. Wang, R. Sha, P. E. Constantinou, S. L. Ginell, C. Mao and N. C. Seeman, From molecular to macroscopic via the rational design of a self-assembled 3D DNA crystal. *Nature* 2009, **461**, 74–77.

[15] (a) P. W. K. Rothemund, Folding DNA to create nanoscale shapes and patterns. *Nature* 2006, **440**, 297–302; (b) R. P. Goodman, I. A. T. Schaap, C. F. Tardin, C. M. Erben, R. M. Berry, C. F. Schmidt and A. J. Turberfield, Rapid chiral assembly of rigid DNA building blocks for molecular nanofabrication. *Science* 2005, **310**, 1661–1665.

[16] (a) H. Dietz, S. M. Douglas and W. M. Shih, Folding DNA into twisted and curved nanoscale shapes. *Science* 2009, **325**, 725–730; (b) D. Han, S. Pal, J. Nangreave, Z. Deng, Y. Liu and H. Yan, DNA Origami with complex curvatures in three-dimensional space. *Science* 2011, **332**, 342–346.

[17] R. A. Muscat, J. Bath and A. J. Turberfield, A programmable molecular robot. *Nano Letters* 2011, **11**, 982–987.

[18] S. M. Douglas, I. Bachelet and G. M. Church, A logic-gated nanorobot for targeted transport of molecular payloads. *Science* 2012, **335**, 831–834.

[19] B. Albinsson, J. K. Hannestad and K. Boerjesson, Functionalized DNA nanostructures for light harvesting and charge separation. *Coordination Chemistry Reviews* 2012, **256**, 2399–2413.

[20] (a) S. J. Tan, M. J. Campolongo, D. Luo and W. Cheng, Building plasmonic nanostructures with DNA. *Nature Nanotechnology* 2011, **6**, 268–276; (b) Z.-G. Wang, C. Song and B. Ding, Functional DNA nanostructures for photonic and biomedical applications. *Small* 2013, **9**, 2210–2222; (c) U. Feldkamp and C. M. Niemeyer, Rational design of DNA nanoarchitectures. *Angewandte Chemie-International Edition* 2006, **45**, 1856–1876; (d) R. Freeman, J. Girsh and I. Willner, Nucleic acid/quantum dots (QDs) hybrid systems for optical and photoelectrochemical sensing. *ACS Applied Materials & Interfaces* 2013, **5**, 2815–2834.

[21] T. J. Bandy, A. Brewer, J. R. Burns, G. Marth, T. Nguyen and E. Stulz, DNA as supramolecular scaffold for functional molecules: progress in DNA nanotechnology. *Chemical Society Reviews* 2011, **40**, 138–148.

[22] H. L. Zhang, J. Chao, D. Pan, H. J. Liu, Q. Huang and C. H. Fan, Folding super-sized DNA origami with scaffold strands from long-range PCR. *Chemical Communications* 2012, **48**, 6405–6407.

[23] A. Heckel and M. Famulok, Building objects from nucleic acids for a nanometer world. *Biochimie* 2008, **90**, 1096–1107.

[24] (a) S. Jager, G. Rasched, H. Kornreich-Leshem, M. Engeser, O. Thum and M. Famulok, A versatile toolbox for variable DNA functionalization at high density. *Journal of the American Chemical Society* 2005, **127**, 15071–15082; (b) M. Fischler, U. Simon, H. Nir, Y. Eichen, G. A. Burley, J. Gierlich, P. M. E. Gramlich and T. Carell, Formation of bimetallic Ag-Au nanowires by metallization of artificial DNA duplexes. *Small* 2007, **3**, 1049–1055; (c) G. A. Burley, J. Gierlich, M. R. Mofid, H. Nir, S. Tal, Y. Eichen and T. Carell, Directed DNA metallization. *Journal of the American Chemical Society* 2006, **128**, 1398–1399; (d) R. J. Macfarlane, M. R. Jones, B. Lee, E. Auyeung and C. A. Mirkin, Topotactic interconversion of nanoparticle superlattices. *Science* 2013, **341**, 1222–1225; (e) V. L. Malinovskii, D. Wenger and R. Haener, Nucleic acid-guided assembly of aromatic chromophores. *Chemical Society Reviews* 2010, **39**, 410–422.

[25] J. E. Darnell, Transcription factors as targets for cancer therapy. *Nature Reviews Cancer* 2002, **2**, 740–749.

[26] P. B. Dervan and B. S. Edelson, Recognition of the DNA minor groove by pyrrole-imidazole polyamides. *Current Opinion in Structural Biology* 2003, **13**, 284–299.

[27] (a) C. Bailly and J. B. Chaires, Sequence-specific DNA minor groove binders. Design and synthesis of netropsin and distamycin analogues. Bioconjugate Chemistry 1998, **9**, 513–538; (b) M. Endo and H. Sugiyama, Chemical approaches to DNA nanotechnology. *Chembiochem* 2009, **10**, 2420–2443.

[28] P. E. Nielsen, Sequence-selective targeting of duplex DNA by peptide nucleic acids. *Current Opinion in Molecular Therapeutics* 2010, **12**, 184–191.

[29] (a) P. Simon, F. Cannata, J. P. Concordet and C. Giovannangeli, Targeting DNA with triplex-forming oligonucleotides to modify gene sequence. *Biochimie* 2008, **90**, 1109–1116; (b) D. A. Rusling, I. S. Nandhakumar, T. Brown and K. R. Fox, Triplex-directed recognition of a DNA nanostructure assembled by crossover strand exchange. *ACS Nano* 2012, **6**, 3604–3613.

[30] (a) J. N. Lisgarten, M. Coll, J. Portugal, C. W. Wright and J. Aymami, The antimalarial and cytotoxic drug cryptolepine intercalates into DNA at cytosine-cytosine sites. *Nature Structural Biology* 2002, **9**, 57–60; (b) W. C. Tse and D. L. Boger, Sequence-selective DNA recognition: natural products and nature's lessons. *Chemistry & Biology* 2004, **11**, 1607–1617.

[31] J. J. Green, R. Langer and D. G. Anderson, A combinatorial polymer library approach yields insight into nonviral gene delivery. *Accounts of Chemical Research* 2008, **41**, 749–759.

[32] (a) R. A. Casero and P. M. Woster, Recent advances in the development of polyamine analogues as antitumor agents, *Journal of Medicinal Chemistry* 2009, **52**, 4551–4573; (b) P. M. Cullis, R. E. Green, L. Merson-Davies and N. Travis, Probing the mechanism of transport and compartmentalisation of polyamines in mammalian cells. *Chemistry & Biology* 1999, **6**, 717–729.

[33] Y. Singh, P. Murat and E. Defrancq, Recent developments in oligonucleotide conjugation. *Chemical Society Reviews* 2010, **39**, 2054–2070.

[34] (a) Y. Ke, G. Bellot, N. V. Voigt, E. Fradkov and W. M. Shih, Two design strategies for enhancement of multilayer-DNA-origami folding: underwinding for specific intercalator rescue and staple-break positioning. *Chemical Science* 2012, **3**, 2587–2597; (b) A. A. Greschner, K. E. Bujold and H. F. Sleiman, Intercalators as molecular chaperones in DNA self-assembly. *Journal of the American Chemical Society* 2013.

[35] W. Su, V. Bonnard and G. A. Burley, DNA-templated photonic arrays and assemblies: Design principles and future opportunities. *Chemistry – A European Journal* 2011, **17**, 7982–7991.

[36] (a) C. Joo, H. Balci, Y. Ishitsuka, C. Buranachai and T. Ha in *Advances in Single-Molecule Fluorescence Methods for Molecular Biology, Annual Review of Biochemistry Vol.* **77**, 2008, pp. 51–76, DOI: 10.1146/annurev.biochem.77.070606.10154;(b) S. Preus and L. M. Wilhelmsson, Advances in quantitative FRET-based methods for studying nucleic acids. *Chembiochem* 2012, **13**, 1990–2001; (c) D. M. J. Lilley and T. J. Wilson, Fluorescence resonance energy transfer as a structural tool for nucleic acids. *Current Opinion in Chemical Biology* 2000, **4**, 507–517.

[37] K. E. Sapsford, L. Berti and I. L. Medintz, Materials for fluorescence resonance energy transfer analysis: Beyond traditional donor-acceptor combinations. *Angewandte Chemie-International Edition* 2006, **45**, 4562–4588.

[38] S.-i. Kawahara, T. Uchimaru and S. Murata, Sequential multistep energy transfer: enhancement of efficiency of long-range fluorescence resonance energy transfer. *Chemical Communications* 1999, **563–564**.

[39] Y. Ohya, K. Yabuki, M. Tokuyama and T. Ouchi, Construction and energy transfer behavior of sequential chromophore arrays on an oligo-DNA assembly. *Supramolecular Chemistry* 2003, **15**, 45–54.

[40] M. Heilemann, P. Tinnefeld, G. S. Mosteiro, M. G. Parajo, N. F. Van Hulst and M. Sauer, Multistep energy transfer in single molecular photonic wires. *Journal of the American Chemical Society* 2004, **126**, 6514–6515.

[41] G. Sanchez-Mosteiro, E. van Dijk, J. Hernando, M. Heilemann, P. Tinnefeld, M. Sauer, F. Koberlin, M. Patting, M. Wahl, R. Erdmann, N. F. van Hulst and M. F. Garcia-Parajo, DNA-based molecular wires: Multiple emission pathways of individual constructs. *Journal of Physical Chemistry B* 2006, **110**, 26349–26353.

[42] M. Heilemann, R. Kasper, P. Tinnefeld and M. Sauer, Dissecting and reducing the heterogeneity of excited-state energy transport in DNA-based photonic wires. *Journal of the American Chemical Society* 2006, **128**, 16864–16875.

[43] A. K. Tong, S. Jockusch, Z. Li, H.-R. Zhu, D. L. Akins, N. J. Turro and J. Ju, Triple fluorescence energy transfer in covalently trichromophore-labeled DNA. *Journal of the American Chemical Society* 2001, **123**, 12923–12924.

[44] S. Vyawahare, S. Eyal, K. D. Mathews and S. R. Quake, Nanometer-scale fluorescence resonance optical waveguides. *Nano Letters* 2004, **4**, 1035–1039.

[45] (a) I. H. Stein, C. Steinhauer and P. Tinnefeld, Single-molecule four-color FRET visualizes energy-transfer paths on DNA origami. *Journal of the American Chemical Society* 2011, **133**, 4193–4195; (b) J. Vogelsang, C. Steinhauer,

C. Forthmann, I. H. Stein, B. Person-Skegro, T. Cordes and P. Tinnefeld, Make them blink: Probes for super-resolution microscopy. *ChemPhysChem* 2010, **11**, 2475–2490.

[46] J. J. Schmied, C. Forthmann, E. Pibiri, B. Lalkens, P. Nickels, T. Liedl and P. Tinnefeld, DNA origami nanopillars as standards for three-dimensional superresolution microscopy. *Nano Letters* 2013, **13**, 781–785.

[47] (a) R. Varghese and H.-A. Wagenknecht, DNA as a supramolecular framework for the helical arrangements of chromophores: towards photoactive DNA-based nanomaterials. *Chemical Communications* 2009, 2615–2624; (b) Y. N. Teo, J. N. Wilson and E. T. Kool, Polyfluorophores on a DNA backbone: A multicolor set of labels excited at one wavelength. *Journal of the American Chemical Society* 2009, **131**, 3923–3933.

[48] (a) A. Ruiz-Carretero, P. G. A. Janssen, A. L. Stevens, M. Surin, L. M. Herz and A. P. H. J. Schenning, Directing energy transfer in discrete one-dimensional oligonucleotide-templated assemblies. *Chemical Communications* 2011, **47**, 884-886; (b) P. G. A. Janssen, S. Jabbari-Farouji, M. Surin, X. Vila, J. C. Gielen, T. F. A. de Greef, M. R. J. Vos, P. H. H. Bomans, N. Sommerdijk, P. C. M. Christianen, P. Leclere, R. Lazzaroni, P. van der Schoot, E. W. Meijer and A. Schenning, Insights into templated supramolecular polymerization: Binding of naphthalene derivatives to ssDNA templates of different lengths. *Journal of the American Chemical Society* 2009, **131**, 1222–1231.

[49] R. Iwaura, F. J. M. Hoeben, M. Masuda, A. Schenning, E. W. Meijer and T. Shimizu, Molecular-level helical stack of a nucleotide-appended oligo(p-phenylenevinylene) directed by supramolecular self-assembly with a complementary oligonucleotide as a template. *Journal of the American Chemical Society* 2006, **128**, 13298–13304.

[50] P. K. Dutta, R. Varghese, J. Nangreave, S. Lin, H. Yan and Y. Liu, DNA-directed artificial light-harvesting antenna. *Journal of the American Chemical Society* 2011, **133**, 11985-11993.

[51] R. J. Cogdell, A. Gall, J. Köhler, The architecture and function of the light–harvesting apparatus of purple bacteria: from single molecules to *in vivo* membranes. *Quarterly Reviews of Biophysics* 2006, **39**, 227–324.

[52] F. Garo and R. Haener, A DNA-based light-harvesting antenna. *Angewandte Chemie-International Edition* 2012, **51**, 916–919.

[53] M. Probst, S. M. Langenegger and R. Haener, A modular LHC built on the DNA three–way junction. *Chemical Communications* 2014, **50**, 159–161.

[54] J. K. Hannestad, P. Sandin and B. Albinsson, Self-assembled DNA photonic wire for long-range energy transfer. *Journal of the American Chemical Society* 2008, **130**, 15889–15895.

[55] J. K. Hannestad, S. R. Gerrard, T. Brown and B. Albinsson, Self-assembled DNA-based fluorescence waveguide with selectable output. *Small* 2011, **7**, 3178–3185.

[56] J. G. Woller, J. K. Hannestad and B. Albinsson, Self assembled nanoscale DNA-porphyrin complex for artificial light harvesting. *Journal of the American Chemical Society* 2013, **135**, 2759–2768.

[57] D. Gust, T. A. Moore and A. L. Moore, Mimicking photosynthetic solar energy transduction. *Accounts of Chemical Research* 2001, **34**, 40–48.

[58] H. Özhalici-Ünal and B. A. Armitage, Fluorescent DNA nanotags based on a self-assembled DNA tetrahedron. *ACS Nano* 2009, **3**, 425–433.

[59] W. Su, M. Schuster, C. R. Bagshaw, U. Rant and G. A. Burley, Site-specific assembly of DNA-based photonic wires by using programmable polyamides. *Angewandte Chemie International Edition* 2011, **50**, 2712–2715.

[60] W. Su, C. R. Bagshaw and G. A. Burley, Addressable and unidirectional energy transfer along a DNA three-way junction programmed by pyrrole-imidazole polyamides. *Scientific Reports* 2013, **3**, 1883.

[61] (a) A. L. Benvin, Y. Creeger, G. W. Fisher, B. Ballou, A. S. Waggoner and B. A. Armitage, Fluorescent DNA nanotags: Supramolecular fluorescent labels based on intercalating dye arrays assembled on nanostructured DNA templates. *Journal of the American Chemical Society* 2007, **129**, 2025–2034; (b) K. C. Hannah and B. A. Armitage, DNA-templated assembly of helical cyanine dye aggregates: A supramolecular chain polymerization. *Accounts of Chemical Research* 2004, **37**, 845–853.

[62] C. V. Kumar and M. R. Duff, DNA-based supramolecular artificial light harvesting complexes. *Journal of the American Chemical Society* 2009, **131**, 16024–16026.

[63] (a) K. Tanaka and Y. Okahata, A DNA-lipid complex in organic media and formation of an aligned cast film. *Journal of the American Chemical Society* 1996, **118**, 10679–10683; (b) W. Li, R. Jones, H. Spaeth and A. J. Steckl, Dose effects in electron beam irradiation of DNA-complex thin films. *Applied Physics Letters* 2010, **97**, 063702; (c) A. J. Steckl, DNA - a new material for photonics? *Nature Photonics* 2007, **1**, 3–5; (d) J. A. Hagen, W.-X. Li,

H. Spaeth, J. G. Grote and A. J. Steckl, Molecular beam deposition of DNA nanometer films. *Nano Letters* 2006, **7**, 133–137; (e) J. A. Hagen, W. Li, J. Steckl and J. G. Grote, Enhanced emission efficiency in organic light-emitting diodes using deoxyribonucleic acid complex as an electron blocking layer. *Applied Physics Letters* 2006, **88**, 171109.

[64] L. L. Wang, J. Yoshida and N. Ogata, Self-assembled supramolecular films derived from marine deoxyribonucleic acid (DNA)-cationic surfactant complexes: Large-scale preparation and optical and thermal properties. *Chemistry of Materials* 2001, **13**, 1273–1281.

[65] E. M. Heckman, J. A. Hagen, P. P. Yaney, J. G. Grote and F. K. Hopkins, Processing techniques for deoxyribonucleic acid: Biopolymer for photonics applications. *Applied Physics Letters* 2005, **87**.

[66] D. Wanapun, V. J. Hall, N. J. Begue, J. G. Grote and G. J. Simpson, DNA-based pPolymers as chiral templates for second-order nonlinear optical materials. *ChemPhysChem* 2009, **10**, 2674–2678.

[67] E. M. Heckman, J. G. Grote, F. K. Hopkins and P. P. Yaney, Performance of an electro-optic waveguide modulator fabricated using a deoxyribonucleic-acid-based biopolymer. *Applied Physics Letters* 2006, **89**.

[68] Y. Ner, J. G. Grote, J. A. Stuart and G. A. Sotzing, White luminescence from multiple-dye-doped electrospun DNA nanofibers by fluorescence resonance energy transfer. *Angewandte Chemie-International Edition* 2009, **48**, 5134–5138.

[69] D. Navarathne, Y. Ner, J. G. Grote and G. A. Sotzing, Three dye energy transfer cascade within DNA thin films. *Chemical Communications* 2011, **47**, 12125–12127.

[70] S. M. Douglas, A. H. Marblestone, S. Teerapittayanon, A. Vazquez, G. M. Church and W. M. Shih, Rapid prototyping of 3D DNA-origami shapes with caDNAno. *Nucleic Acids Research* 2009, **37**, 5001–5006.

[71] T. Liedl, B. Hogberg, J. Tytell, D. E. Ingber and W. M. Shih, Self-assembly of three-dimensional prestressed tensegrity structures from DNA. *Nature Nanotechnology* 2010, **5**, 520–524.

[72] (a) D. Y. Zhang and G. Seelig, Dynamic DNA nanotechnology using strand-displacement reactions. *Nature Chemistry* 2011, **3**, 103–113; (b) U. Rant, K. Arinaga, S. Scherer, E. Pringsheim, S. Fujita, N. Yokoyama, M. Tornow and G. Abstreiter, Switchable DNA interfaces for the highly sensitive detection of label-free DNA targets. *Proceedings of the National Academy of Sciences of the United States of America* 2007, **104**, 17364–17369.

[73] Y. Ke, S. Lindsay, Y. Chang, Y. Liu and H. Yan, Self-assembled water-soluble nucleic acid probe tiles for label-free RNA hybridization assays. *Science* 2008, **319**, 180–183.

[74] H. K. K. Subramanian, B. Chakraborty, R. Sha and N. C. Seeman, The label-free unambiguous detection and symbolic display of single nucleotide polymorphisms on DNA origami. *Nano Letters* 2011, **11**, 910–913.

[75] D. Graham and K. Faulds, Quantitative SERRS for DNA sequence analysis. *Chemical Society Reviews* 2008, **37**, 1042–1051.

[76] K. Boeneman, D. E. Prasuhn, J. B. Blanco-Canosa, P. E. Dawson, J. S. Melinger, M. Ancona, M. H. Stewart, K. Susumu, A. Huston and I. L. Medintz, Self-assembled quantum dot-sensitized multivalent DNA photonic wires. *Journal of the American Chemical Society* 2010, **132**, 18177–18190.

[77] (a) G. Tikhomirov, S. Hoogland, P. E. Lee, A. Fischer, E. H. Sargent and S. O. Kelley, DNA-based programming of quantum dot valency, self-assembly and luminescence. *Nature Nanotechnology* 2011, **6**, 485–490; (b) J. Farlow, D. Seo, K. E. Broaders, M. J. Taylor, Z. J. Gartner and Y. W. Jun, Formation of targeted monovalent quantum dots by steric exclusion. *Nature Methods* 2013, **10**, 1203–1205; (c) Y. G. Zhang, F. Lu, K. G. Yager, D. van der Lelie and O. Gang, A general strategy for the DNA-mediated self-assembly of functional nanoparticles into heterogeneous systems. *Nature Nanotechnology* 2013, **8**, 865–872; (d) C. M. Spillmann, M. G. Ancona, S. Buckhout-White, W. R. Algar, M. H. Stewart, K. Susumu, A. L. Huston, E. R. Goldman and I. L. Medintz, Achieving effective terminal exciton delivery in quantum dot antenna-sensitized multistep DNA photonic wires. *ACS Nano* 2013, **7**, 7101–7118; (e) C. Zhang, R. J. Macfarlane, K. L. Young, C. H. J. Choi, L. L. Hao, E. Auyeung, G. L. Liu, X. Z. Zhou and C. A. Mirkin, A general approach to DNA-programmable atom equivalents. *Nature Materials* 2013, **12**, 741–746.

[78] (a) C. J. Burrows and J. G. Muller, Oxidative nucleobase modifications leading to strand scission. *Chemical Reviews* 1998, **98**, 1109–1151; (b) J. R. Wagner and J. Cadet, Oxidation reactions of cytosine DNA components by hydroxyl radical and one-electron oxidants in aerated aqueous solutions. *Accounts of Chemical Research* 2010, **43**, 564–571; (c) J. Cadet, T. Douki and J.-L. Ravanat, Oxidatively generated damage to the guanine moiety of DNA: Mechanistic aspects and formation in cells. *Accounts of Chemical Research* 2008, **41**, 1075–1083.

[79] (a) X. F. Guo, A. A. Gorodetsky, J. Hone, J. K. Barton and C. Nuckolls, Conductivity of a single DNA duplex bridging a carbon nanotube gap. *Nature Nanotechnology* 2008, **3**, 163–167; (b) P. A. Sontz, N. B. Muren and J. K. Barton, DNA charge transport for sensing and signaling. *Accounts of Chemical Research* 2012, **45**, 1792–1800.

[80] J. R. Heath in *Molecular Electronics, Annual Review of Materials Research Vol.* **39**, 2009, 1–23, DOI: 10.1146/annurev-matsci-082908-145401.

1.2

π-Conjugated DNA Binders: Optoelectronics, Molecular Diagnostics and Therapeutics

Niklaas J. Buurma

Physical Organic Chemistry Centre, School of Chemistry, Cardiff University, Cardiff, UK

1.2.1 π-Conjugated compounds

π-Conjugated molecules constitute a very interesting class of compounds with versatile optoelectronic properties. For example, tuning π-conjugation can result in absorption of light across the UV–visible spectrum, in conductivity in π-conjugated polymers, in fluorescence and related energy transfer processes, and in useful redox properties. At the same time, molecular structures, and hence interaction patterns, solubility, and so on, can be optimised for applications of interest. The rise of π-conjugated compounds is, to a large extent, the result of concomitant developments in molecular electronics, organic synthesis and computational property prediction.

It is hard to exaggerate the wealth of progress in the field of molecular electronics over the past two decades. The development of (semi)conducting polymers has led to technologies such as OLEDs, OFETs, flexible displays and organic photovoltaics [1]. With myriad possible applications and the potential for relatively rapid impact, industry and academia alike have been driving this field forward.

Closely related to the advances in the application of molecular electronics are advances in the synthesis of π-conjugated molecular structures. Synthetic chemists can choose from reactions such as the Suzuki–Miyaura [2], Stille [3], Corriu–Kumada [4] and Mizoroki–Heck [5] cross-coupling reactions, which are now rapidly being complemented by C–H activation processes. Numerous relevant reactants are commercially available. The development of MIDA (*N*-methyliminodiacetic acid) esters as protective groups for boronic acids [6] and solid-phase strategies [7] even allow for iterative Suzuki cross-coupling reactions. As a result,

DNA in Supramolecular Chemistry and Nanotechnology, First Edition.
Edited by Eugen Stulz and Guido H. Clever.

viable procedures for the synthesis of mixed π-conjugated oligomers of, for example, thiophene, furan and pyrrole, are well known (see, e.g., references [8] and [9]).

At the same time, development of readily accessible and relatively user-friendly computational tools, such as time-dependent density functional theory (TD-DFT), allows reasonable predictions of UV–visible absorption and fluorescence spectra as well as the associated transition dipole moments [10]. Similarly, redox properties of molecules can be predicted to a reasonable extent [11].

As a result of these simultaneous developments, π-conjugated molecules have become a class of accessible functional components with predictable properties for (self-assembled) optoelectronic materials, biosensors and therapeutics, amongst others.

1.2.2 DNA binders for different applications

For π-conjugated DNA binders that have tunable structures and accessible properties, the desired properties for various applications need to be established. These properties strongly depend on the particular application, although there is considerable cross-over between the properties that are required for different applications, and we will explore these for π-conjugated DNA binders for molecular diagnostics, therapeutics and optoelectronics.

1.2.2.1 Molecular diagnostics

DNA is best known as the carrier of genetic information. As a result, genetic diseases and pathogens (for example) can be uniquely identified through the detection of an associated nucleic acid sequence. DNA has therefore provided a valuable target for sequence-selective biosensors for DNA, also called 'genosensors' (see also Chapter 3.1). Detecting RNA, and messenger RNA (mRNA) in particular, is also of significant interest as levels of mRNA are related to the actual expression levels of genetic information.

A frequently used approach for the selective detection of nucleic acids involves the use of a single-stranded so-called capture strand, which recognises its complementary target strand in solution. Typically, the single capture strand is either single-stranded DNA (ssDNA) or a single-stranded peptide nucleic acid (PNA). Sequence specificity in this type of assay is the result of the intrinsic sequence-selectivity of DNA base pairing. Figure 1.2.1 shows a cartoon representation of a genosensor involving a surface-immobilised capture strand (a) hybridising with a target strand (b).

The typical use of a π-conjugated compound in such biosensors is as a fluorescent and/or redox-active label or sensitiser. The various approaches to the detection can be broadly divided into two categories, namely labelled and label-free detection. The first category of sensing system involves covalent labelling of the potential target strands in the sample with the sensitiser. This labelling process involves potentially non-trivial sample preparation steps for the introduction of the label. The name of the second approach, that is 'label-free', is the slightly odd but generally accepted jargon for an assay in which the nucleic acid of interest does not need to be labelled covalently. Rather, the sensitiser is added as a separate species and binds non-covalently to the structure formed upon recognition of a target strand (event (c) in Figure 1.2.1). The approach where the sensitiser is added and binds non-covalently has obvious advantages because it requires little sample pre-treatment and has the potential to be very quick. If such an approach is combined with portable instrumentation, minimally trained clinical personnel at the point of care can use it.

An obvious area of application of π-conjugated DNA binders is as sensitisers in these label-free genosensors. Sensitisers can either generate an optical response (UV–visible or fluorescence) or an electrochemical response. Electrochemical genosensors are of particular interest because they are miniaturised more readily than optical sensors and typically require small sample volumes. Developments in electrochemical DNA

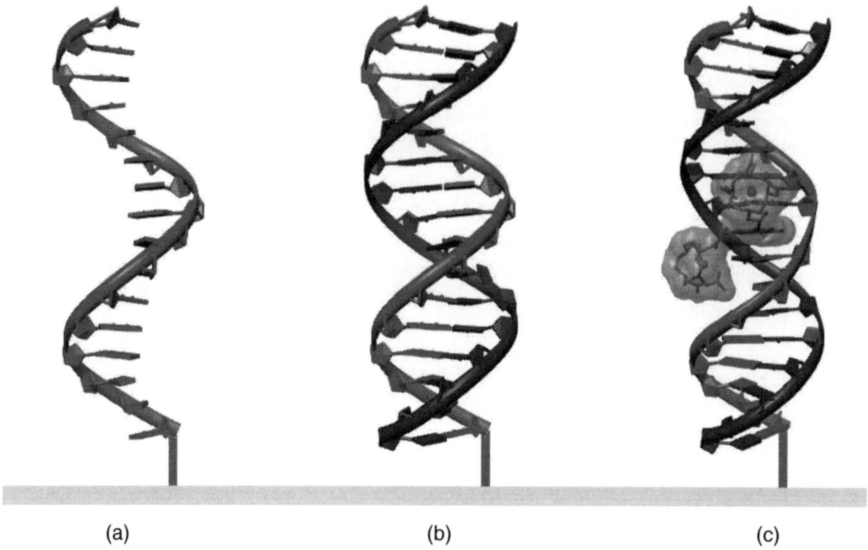

Figure 1.2.1 *Design of a genosensor sequence selectively detecting DNA involving a duplex-DNA binding sensitiser (See colour figure in colour plate section)*

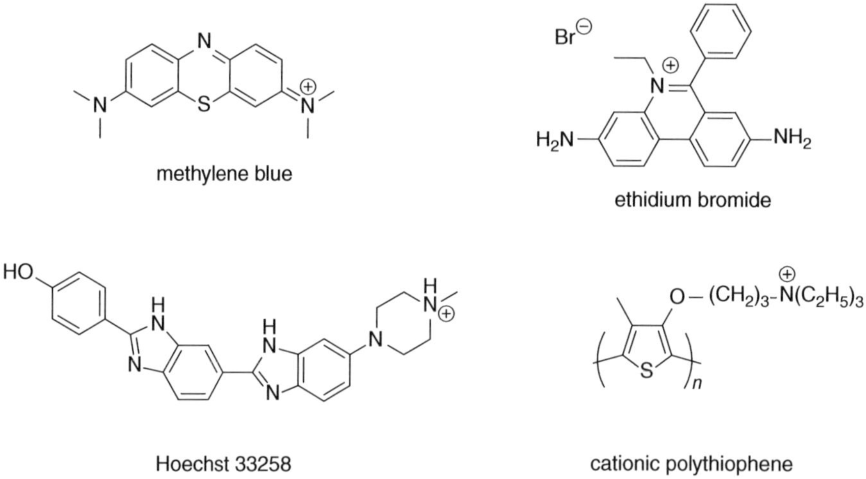

Scheme 1.2.1

biosensors have been reviewed in reference [12] and Chapter 3.1. The required properties of the sensitisers will be discussed briefly.

When discussing the general shape requirements of the π-conjugated molecules used as sensitisers, one needs to consider the binding modes of small-molecule binders with duplex DNA (other structures will be considered separately). The two main binding modes to duplex DNA are intercalation and minor-groove binding (*vide infra*). Intercalators are typically fairly rigid and flat structures, such as methylene blue and ethidium bromide (Scheme 1.2.1). Minor-groove binders are typically long and relatively flexible molecules, such as Hoechst 33258 (H33258) and cationic polythiophenes (Scheme 1.2.1).

For groove binders in particular, it is important that the molecule can follow the minor groove around the DNA. In practice this means that a curved conformation should be available to the molecule, although this is not essential (*vide infra*).

π-Conjugated compounds are often flat and fairly hydrophobic. As a result, stacking interactions between such compounds are likely and these may negatively affect the solubility. For aqueous solubility, conjugated scaffolds carrying charged groups are therefore of interest. Positively charged groups have the advantage that electrostatic interactions between the DNA binder and the anionic phosphate backbone provide an additional driving force for binding. However, the use of relatively hydrophobic positively charged DNA binders can also lead to extensive non-specific binding and precipitation as a result of charge cancellation between the DNA backbone phosphates and the bound ligand molecules. In addition, although charges may help solubilise the DNA binders, DNA binding of π-conjugated DNA binders is likely to be in competition with self-aggregation of the binder, as has been observed for the well-known DNA stain Hoechst 33258 (Scheme 1.2.1) for example [13].

The most sensitive sensor designs that rely on spectroscopic detection involve fluorescence. For this application, the sensitiser requires a UV–visible absorbance for excitation above 300 nm, to avoid competing absorbance of light by the DNA.

For redox-active sensitisers used in amperometric detection in combination with immobilised capture strands on gold electrodes, the available electrochemical window is from approximately –1 to +1 V. The lower limit is because DNA strands are often attached to gold electrodes through Au–S bonds, which are reductively broken below –1 V. The upper limit of the electrochemical window is the oxidation potential of the guanine base at neutral pH of 1.29 V [14]. A particularly popular redox sensitiser is methylene blue (see e.g. reference [15] and references therein), while the benzimidazole moiety is also known to provide useful redox properties, for example in H33258. If possible, redox-active sensitisers acting as redox catalysts are preferred. If such a catalyst achieves multiple turnovers, higher currents can be achieved because the electrochemistry is no longer limited by the stoichiometry of the added sensitiser, but rather by the amount of a potentially cheap sacrificial reactant, which does not need to bind to the nucleic acid structure on the electrode.

An alternative electrochemical technique that is amenable to biosensor design involves electrochemical impedance spectroscopy (EIS). In an EIS-based sensor DNA•DNA, or PNA•DNA hybridisation, is detected through changes in impedance of an electrode-immobilised layer of molecules resulting from the interaction with the target DNA strand [16]. Typically, this involves the use of a negatively charged redox couple that does not bind to the DNA, such as the $Fe(CN)_6^{3-}/Fe(CN)_6^{2-}$ couple. Strand hybridisation causes a small increase in the negative charge on the electrode (the increase is small because of counterion effects) and swelling of the electrode-immobilised layer. Both effects impede access of the $Fe(CN)_6^{3-}/Fe(CN)_6^{2-}$ couple to the electrode, resulting in a measurable increase in charge transfer resistance (R_{ct}) upon hybridisation.

Signal modulation in EIS-based genosensors can be achieved using intercalating molecules, resulting in further swelling of the electrode-immobilised layer and thus an increase in resistivity [17]. Addition of cationic DNA-binders reduces the local negative charge and may also lead to contraction of the nucleic acid structures on the electrode, thus decreasing the observed R_{ct}. A decrease in the observed signal may not seem ideal, but it can still be used for sensor validation purposes, in particular if the binding molecule is also redox active allowing simultaneous impedimetric and amperometric detection. Negatively charged DNA binders should in principle also give a detectable increase in the EIS signal, although electrostatic repulsion between the DNA and the binder would obviously make any binding weaker.

All the approaches described here exploit the natural sequence selectivity of single-stranded nucleic acids for each other. As a result, sequence selectivity of the binders is not required. However, if the binders are sequence-selective, this would provide a further opportunity for sensor self-validation. In general, sequence preferences may be helpful for sensor self-validation but they are not required. If a binder displays sequence preference, it is important to be aware of this because this preference may result in different sensitivities for different target sequences.

1.2.2.2 Therapeutics

DNA is an interesting therapeutic target because one can attempt to address diseases at source, be it by recognising a pathogen or diseased cell, or through modulation of transcription and subsequent translation of the genetic code. For example, one could target oncogenic transcription factors.

The development of π-conjugated DNA binders for use in biosensors may present the challenge of requiring specific structural and optoelectronic properties, but application in therapy provides a different set of challenges. In a biosensor, the DNA is effectively brought to the molecule. For therapeutic applications, however, the molecule must selectively find its target. The shape and solubility requirements on the molecule remain the same, but additional parameters such as target and sequence selectivity, toxicity of the parent compound and its metabolites (e.g. thiophene toxicity is related to bioactivation by liver enzymes, see reference [18]) and cell membrane permeability become critical.

Optoelectronic properties remain important for applications such as phototherapy where π-conjugated compounds need to display photoactivated reactivity. One of the more obvious possibilities is to use π-conjugated compounds as a source of singlet oxygen. In this respect, it is of interest that bi- and terthiophenes are singlet oxygen photosensitisers [19]. Alternatively, π-conjugated redox-active compounds can be envisaged to lead to therapeutics that employ sensitivity to the local redox potential in cells.

Regarding cell entry, significant work has been carried out on the modification of Dervan's hairpin polyamides (*vide infra*) in order to make these cell-permeable using alkene isosteres for the amide linkages found in DHPs. This has the added advantage of generating an extended range of π-conjugation along the compound [20].

Binders targeting higher order nucleic acid structures may also be of significant interest from a therapeutic point of view. For example, considerable effort is being made to develop binders for quadruplex DNA structures, and many currently known quadruplex DNA binders are π-conjugated molecules. It is hoped that these binders can stabilise telomeric DNA in its quadruplex fold, so that the telomerase enzyme cannot extend the telomeres in cancer cells. Blocking the elongation of telomeres should prevent cancer cells from becoming immortal.

1.2.2.3 Optoelectronics

Rather than being a target for a therapeutic or a sensor, nucleic acid structures can also be considered as a scaffold for the construction of multicomponent self-assembled nanoscale structures. This is essentially systems chemistry [21] in which the systems become more than the sum of their parts.

Of particular interest at the interface between systems chemistry and systems biology are non-covalently self-assembling systems consisting of bio(macro)molecules and purely synthetic chemical systems, such as π-conjugated molecules, displaying variable optoelectronic properties. Combining the versatility of the molecules of life with technology-defining molecular electronics in particular opens up a wealth of applications, including the programmable self-assembly of nanobioelectronic systems.

Self-assembly of complex multicomponent systems requires input of information into the system where the amount of information scales with the complexity of the system. An obvious choice for an organising molecule carrying complex instructions for use in aqueous solutions is DNA, because it codes for the controlled complexity of life with high information density and thus appears an excellent candidate in material science as well. Not surprisingly, many attempts have been made at harnessing the coding ability of DNA for the construction of (synthetic) multicomponent materials. These attempts have been fuelled by the generation of three-dimensional DNA-based structures as well as so-called 'DNA origami'.

DNA can be used for the programmable construction of three-dimensional structures, following the pioneering work by Seeman [22] (see also e.g. references [23] and [24]). DNA-origami, developed by Rothemund

[25], affords additional addressable structures (see e.g. references [23] and [26]). If (semi)conducting polymers can be made to interact sequence-selectively with DNA, these programmable three-dimensional structures of DNA can be 'wired' in a precisely predictable manner and three-dimensional nanocircuitry can be assembled, fulfilling one of the promises of DNA nanotechnology [27].

There are two main routes for using DNA for self-assembly and both routes have been the subjects of several reviews (see e.g. reference [28]). The first route covalently attaches molecular fragments with properties of interest to single-stranded oligonucleotides. The oligonucleotides recognise a single-stranded template through the intrinsic sequence selectivity of DNA duplex formation. This approach has, for example, recently been employed for the construction of a seven-fluorophore FRET cascade [29] in which energy is transferred unidirectionally through fluorescence resonance energy transfer (FRET). This approach is associated with good predictability and, typically, good aqueous solubility of the components. Drawbacks include the synthetic cost and the risk of oligonucleotide modifications affecting DNA duplex stability. The second route uses molecules with DNA-sequence selectivity involving either molecular fragments with sequence selectivity covalently attached to a molecular fragment with properties of interest, or sequence-selective duplex-DNA binders with intrinsic properties of interest. The approach separating sequence selectivity from properties is illustrated by two examples of sequence-selective Dervan hairpin polyamides (*vide infra*) covalently linked to fluorophores establishing a spectroscopic gradient [30] (see also Chapter 1.1 by Burley). These systems are inspiring, but further improvement might be possible through the removal of the flexible linker between the sequence-recognition unit and the fluorophores, potentially improving both the precision in orienting the fluorophores and also the fluorescence quantum yield, which is often enhanced by close interaction between nucleic acids and fluorophores as a result of reduced quenching by the solvent and molecular movement [31]. In addition, Dervan's hairpin polyamides are plagued by aggregation and solubility problems [32].

An approach utilising sequence-selective π-conjugated duplex-DNA binders with intrinsic properties of interest would benefit from the advantages of a more modular system. However, a disadvantage to this approach is that the identification of sequence-selective duplex-DNA binders with intrinsic properties of interest is non-trivial.

However, even without sequence recognition, DNA can assemble molecules. For example, Houlton and coworkers created DNA-templated π-conjugated structures [33]. Similar work was reviewed by Schenning and coworkers [28c], while DNA nanostructures for light harvesting and charge separation were discussed by Albinsson *et al.* [28b].

In a particularly elegant example of a DNA-templated self-assembled structure, an intercalator and a groove binder (*vide infra*) were bound simultaneously to duplex DNA, displaying FRET, as expected for the relative organisation of the dyes [34]. Thus, in essence, a nanoscale assembly was constructed that is observable through its optoelectronic properties. Similarly, FRET has been observed between two dyes where one dye selectively bound to duplex DNA and the other to quadruplex DNA, again illustrating a potential for directed assembly, although this assembly was intended for imaging purposes [35].

1.2.3 Targeting duplex DNA

DNA can be targeted in its double-helical form, but other nucleic acid structures are also available as targets. For example, interest in the recognition of various naturally occurring nucleotide-structures [36], such as G-quadruplexes [37], i-motif [38] and triplex DNA [39], has increased significantly.

In this section, selected π-conjugated binders targeting duplex DNA through minor-groove binding or through intercalation will be discussed first. Selected π-conjugated compounds targeting the various higher-order nucleic acid structures will be discussed in the following section. In both sections, a limited selection of representative compounds will be discussed.

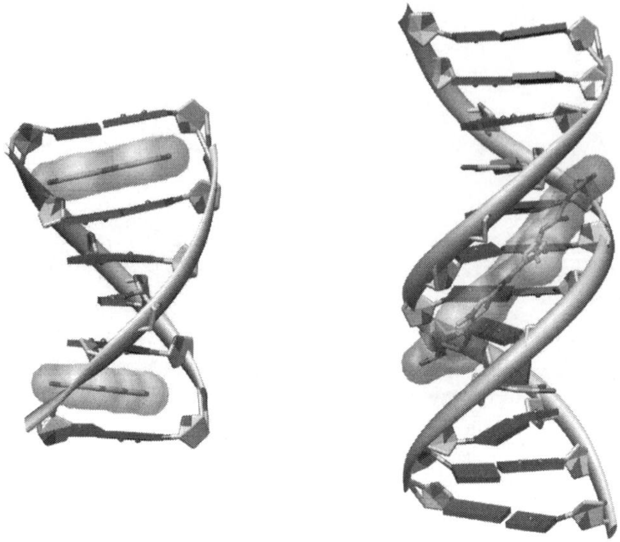

Figure 1.2.2 *Left, the intercalator proflavine (NDB ID: DD0103) and right, minor-groove binder netropsin (NDB ID: GDLB05 [40]) (images rendered using UCSF Chimera [41]) (See colour figure in colour plate section)*

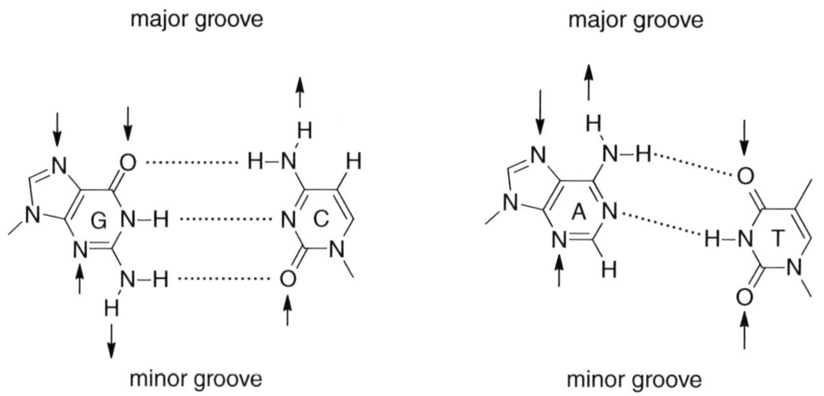

Scheme 1.2.2

There are three main ways in which a small molecule can bind non-covalently to DNA: intercalation, minor-groove binding and through electrostatic interactions (Figure 1.2.2).

One potential binding mode, namely major-groove binding, is missing from this list despite occurring in biological systems, for example in helix-turn-helix motifs in DNA-binding proteins. The reason for this omission is that minor-groove binding and intercalation can be driven by hydrophobic (including stacking) interactions, whereas major-groove recognition requires a multivalent array of hydrogen bonding interactions with the hydrogen bond donors and acceptors at the bottom of the major groove (Scheme 1.2.2), with less opportunity for hydrophobic interactions.

The multivalent array of hydrogen-bonding interactions available in the major groove allows excellent selectivity in the recognition processes, but synthetic systems reproducing this feat currently appear beyond our design capabilities.

The most frequently occurring binding modes for π-conjugated DNA binders, that is minor groove binding and intercalation, will be discussed in more detail.

Rigid π-conjugated molecules with extended surfaces tend to intercalate in between the DNA base pairs. This interaction is driven by hydrophobic stacking interactions. For this interaction to occur, the DNA needs to unwind so that the base step distance, as quantified by the so-called rise parameter, increases from a typical 3.3 to 7.0 Å (for a statistical analysis, see the Supplementary Information for reference [42]).

Minor-groove binders are typically longer molecules with flexibility allowing the molecule to follow the minor groove around the DNA. Important driving forces for minor-groove binders interacting with DNA are hydrophobic interactions with the walls of the minor groove, hydrogen bond formation with the sides of the base pairs at the bottom of the minor groove (Scheme 1.2.2) and removal of the so-called 'spine of hydration' when binding in A•T-rich sequences.

The distinction between the binding modes can be made using viscometry [43] and linear dichroism [44], and, employing empirical correlations, using circular dichroism spectroscopy [45] and isothermal titration calorimetry [46].

1.2.3.1 Examples of π-conjugated compounds interacting with double-stranded DNA – minor groove binders

A range of π-conjugated compounds binds in the minor groove of DNA; some of these display sequence selectivity, others do not. This section firstly reviews a series of compounds that bind without sequence selectivity, or at least with unknown sequence selectivity, and then a few compounds for which sequence selectivity has been achieved will be highlighted.

Conjugated polymers including cationic polythiophenes [47] and polypyrroles [48] are known to bind to DNA in aqueous solutions and their favourable electronic and spectroscopic properties render these compounds promising components of biosensors [47, 49]. The application of oligo- and polythiophenes as fluorescent probes for studies of biological events has been reviewed by Åslund *et al.* [50].

Cationic polythiophenes, for example, are used to detect single-stranded DNA (ssDNA) in a sequence-specific manner. This detection is achieved through combining a cationic conjugated polymer and a single-stranded capture probe resulting in a considerable red-shift of the UV–visible absorption spectrum of the cationic polythiophene. This red-shift is attributed to a combination of 'aligning' the cationic polythiophene and π-stacking between ssDNA-cationic polythiophene complexes [47d, 51]. Addition of the complementary single-stranded target sequence results in the formation of the complex of the duplex DNA with the cationic polythiophene and a concomitant blue-shift in the UV–visible absorption spectrum, although the resulting spectrum is still red-shifted compared with the free cationic polythiophene. Solid-support immobilised systems based on this approach have been successfully used to detect attomolar concentrations of an ssDNA sequence selectively [49a]. Similarly, a cationic polythiophene-based system has been used as a component of a sequence-selective electrochemical ssDNA sensor immobilised on a gold electrode [49b].

The interactions involved in the different complexes between cationic polythiophenes and both ssDNA and duplex DNA are sufficiently strong that they are difficult to quantify and the mode of interaction of the polymers with duplex DNA has not yet been proven unambiguously. In fact, even the stoichiometry of interaction has yet to be defined [52]. Nevertheless, the observation that interaction of cationic polythiophenes with duplex DNA is accompanied by a bathochromic shift indicates ordering of the polythiophene [47b, 47c]. In addition, an induced CD signal characteristic of a right-handed helical orientation of the polythiophene backbone appears upon interaction with duplex DNA [47c], suggesting that polythiophenes bind in either the major or minor groove. A marked change in fluorescence upon addition of the complementary single-stranded target suggests that π-stacked polymer aggregates are formed in the complex of ssDNA with cationic polythiophene and that these aggregates are broken up upon formation of duplex DNA [47d].

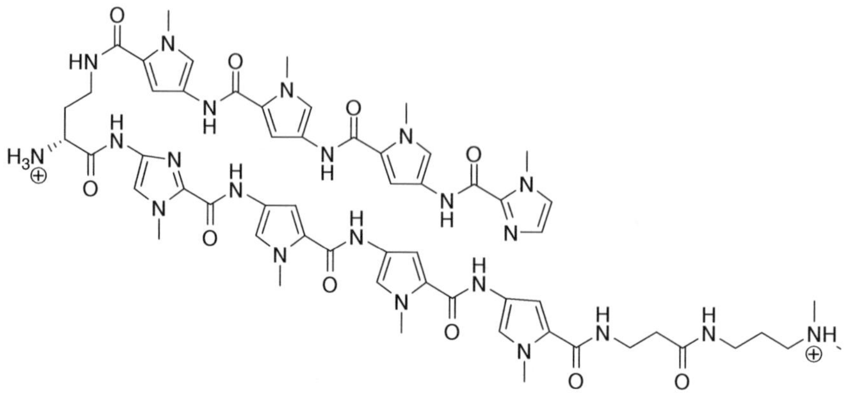

Scheme 1.2.3

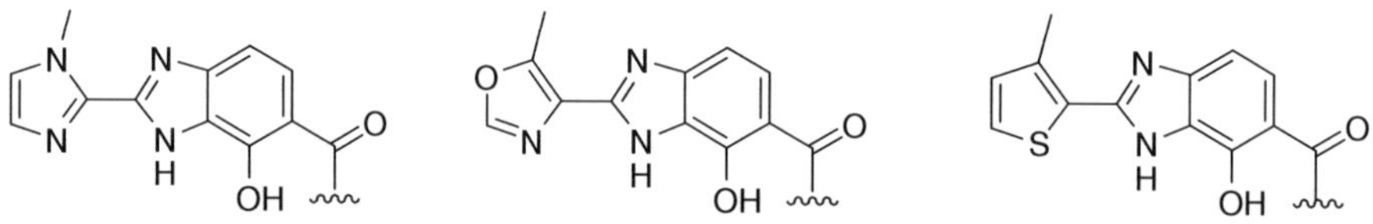

Scheme 1.2.4

In order to target duplex DNA directly, sequence selectivity of the π-conjugated DNA binder is required. The best-known modular system for sequence-specific recognition of duplex DNA is formed by Dervan's modular hairpin polyamides (DHPs; see the example in Scheme 1.2.3), in which effective π-conjugation is broken by the amide linkages.

DHPs are fully modular and can recognise up to approximately five consecutive base pairs, before a gradually increasing mismatch in the curvature of the molecule and the floor of the minor groove breaks the recognition sequence. The underlying cause for this gradually increasing mismatch probably lies in that the DHPs recognise duplex DNA at the single base pair level, whereas the overall curvature of DNA is determined on the tri- and tetranucleotide scale, as a result of the influence of base-pair stacking on the local DNA structure [53, 54]. As a result, sequence-recognition strategies targeting individual base pairs, whereas extremely elegant, are likely to be limited in the length of contiguous sequences that can be recognised. It is of interest that DHPs employ an intramolecular side-by-side binding mode, as this motif appears to be common, and is possibly required for DNA binders recognising sequences other than A•T-rich DNA.

A set of π-conjugated heterocycle dimers [55], which can be used as part of hairpin polyamides (Scheme 1.2.4), has also been developed.

The majority of current synthetic π-conjugated minor groove binders is A•T-selective. This preference is probably because of a combination of: (i) specific hydration effects in the minor groove of A•T-rich DNA, that is the so-called spine of hydration, making binding there thermodynamically particularly favourable and (ii) steric clashes resulting from the guanine NH_2 protruding into the minor groove and thus hindering binding at G•C-rich sequences.

Examples of A•T-selective π-conjugated minor-groove binders include Hoechst 33258 (Scheme 1.2.5) [56], a Hoechst 33258 analogue [57], DAPI [56, 58], furamidine DB75 (2,5-bis(4-amidinophenyl)furan) [58, 59], thiophene-based diamidine DB818 [59] and DB921 [60] (Scheme 1.2.5).

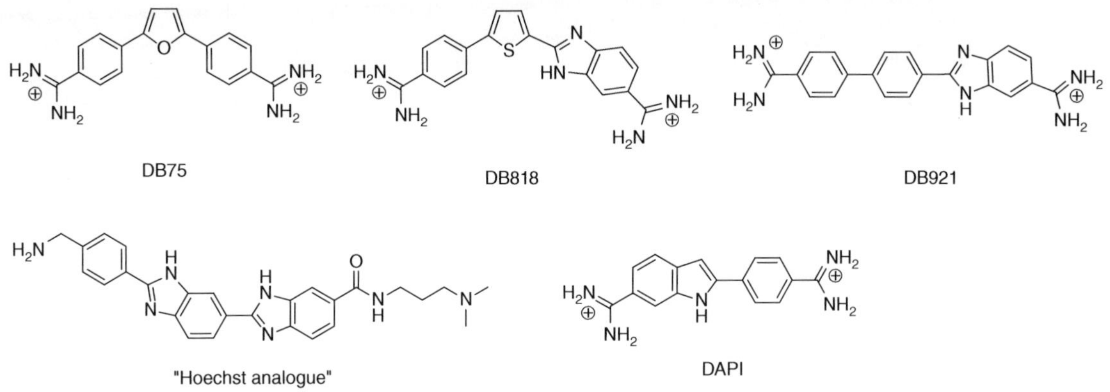

Scheme 1.2.5

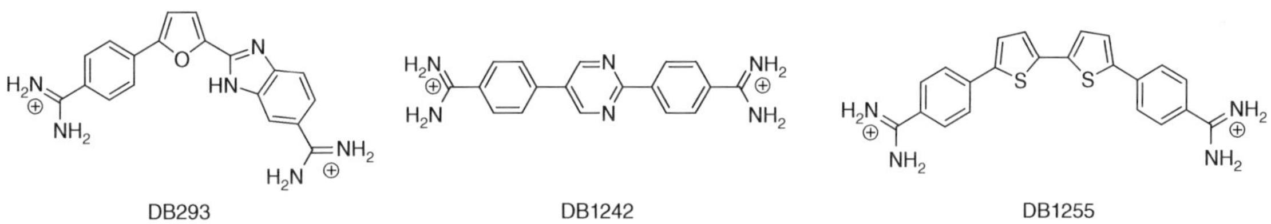

Scheme 1.2.6

It is of interest that thiophene-containing DB818 binds more strongly than DB293 (Scheme 1.2.6) to similar sequences. This observation was attributed to the angle between the substituents connected to a thiophene ring being ideal for the molecule to follow the curvature of the minor groove.

Nevertheless, minor-groove binders based on non-DHP motifs showing selectivity for sequences other than A•T have been developed. Prime examples of π-conjugated binders with selectivity for sequences other than A•T were developed in inspiring research by the team working with Boykin and Wilson who developed DB293 selective for ATGA [61], DB1242 selective for GCTCG [62] and DB1255, which is selective for ATGAT [63] (Scheme 1.2.6).

Like the DHPs, DB293 and DB1242 interact as dimers. DB1242 is of particular interest as it is a linear molecule that binds as an offset partially stacked dimer, where the dimer as a whole follows the curvature of the minor groove. An alternative binding mode is displayed by DB921 (Scheme 1.2.5), another linear compound, which follows the curvature of the minor groove by forming a water-mediated contact [60a]. The results from Boykin and Wilson's studies (for a review of selected heterocyclic diamidine minor-groove binders, see reference [64]) thus relax the importance of both ligand curvature and hydrogen bonding patterns in selective minor-groove binding. Boykin and Wilson's work further shows that appropriate binders can circumvent the steric problems associated with the guanine NH_2 protruding into the minor groove when binding to G•C-rich DNA, and that G•C-rich sequences are rather unforgiving towards less fitting ligands. Although it will make the search for π-conjugated binders with G•C selectivity harder than the search for A•T-selective binders, such an unforgiving nature should not be considered bad. In fact, a particular level of this unforgiving characteristic is essential for the selectivity for longer sequences because unforgiving sequences provide the opportunity to severely disrupt the binding of mismatched binders. As such, Boykin and Wilson's work shows that one can use the steric 'problems' associated with G•C base pairs to achieve selectivity.

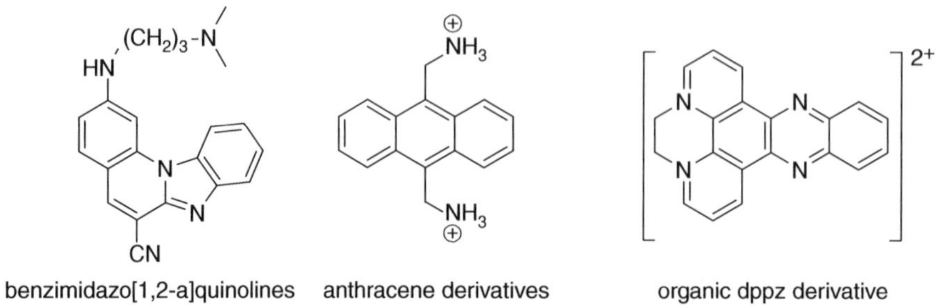

Scheme 1.2.7

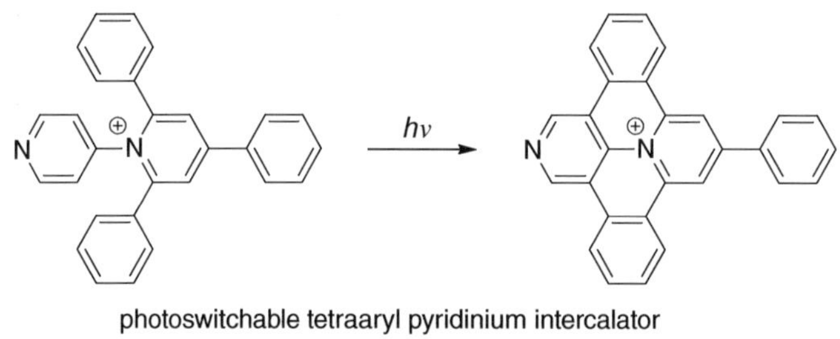

Scheme 1.2.8

1.2.3.2 Examples of π-conjugated DNA binders interacting with double-stranded DNA – intercalators

Many flat conjugated aromatic molecules intercalate between the base pairs of DNA. For example, methylene blue and ethidium bromide have already been mentioned. The opportunities for sequence selectivity for intercalators are significantly less than for groove binders because of the lack of specific hydrogen-bonding interactions, unless the intercalator is attached to a groove-binding moiety of course. Nevertheless, there seems to be a preference for intercalation to occur between G•C base pairs, possibly because of extended stacking interactions.

There are numerous further examples of π-conjugated intercalators, which include substituted pyrenes (e.g. reference [65]), dipyrido[3,2-a:2′,3′-c]phenazine (dppz) and its modifications [66] (for an example see Scheme 1.2.7) and metal complexes (see e.g. reference [67]), substituted anthraquinones [42], substituted benzimidazo[1,2-a]quinolines [68] (Scheme 1.2.7) and anthracene derivatives [69] (Scheme 1.2.7).

A very interesting photoswitchable intercalator (Scheme 1.2.8) based on a tetraaryl pyridinium compound only intercalates upon activation by light.

Overall, the compounds shown in this section represent the type of structures that lead to intercalation. There are many more π-conjugated intercalators, and even more will undoubtedly be discovered, so there is considerable scope to incorporate the optoelectronic properties required for the various applications.

1.2.4 Examples of π-conjugated compounds interacting with hybrid duplexes and higher order nucleic acid structures

Although duplex DNA may be the most obvious and best-explored target for π-conjugated DNA binders, higher order structures can be equally interesting. A particularly useful method for identifying structure selectivity is competition dialysis [70, 71], which has been used for a number of compounds, including a

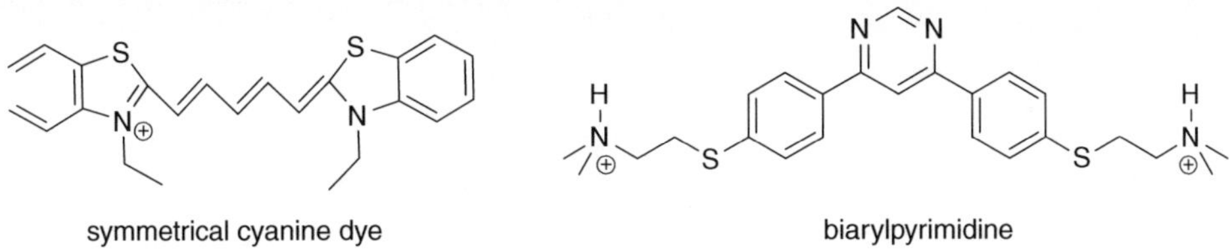

Scheme 1.2.9

series of aromatic diamidines [72]. Numerous examples of nucleic acid structure-selective binders can be found, for example, in reference [73].

1.2.4.1 Examples of π-conjugated compounds interacting with DNA•RNA and DNA•PNA hybrid duplexes

Several compounds interact with DNA•RNA hybrid duplexes and such compounds are of interest for use in sensors for mRNA. Binding to DNA•RNA hybrid duplexes has been reviewed by Shaw and Arya [74]. An example of a π-conjugated DNA•RNA binder is a biarylpyrimidine [75] (Scheme 1.2.9).

Binding to PNA•DNA duplexes is modulated in comparison with binding to DNA•DNA duplexes. In general, intercalators do not bind to PNA•DNA duplexes while groove binders may still bind but with modulated affinity and specificity [76]. An example of a PNA•DNA binder involves a symmetrical cyanine dye containing benzothiazole groups (Scheme 1.2.9), which binds with high affinity to PNA•DNA duplexes, a PNA•PNA duplex and a PNA•PNA•DNA triplex [77].

1.2.4.2 Examples of π-conjugated compounds interacting with higher order nucleic acid structures

Several compounds are also known to bind to higher order nucleic acid structures, such as triplex DNA, i-motifs and quadruplex DNA.

Neomycin conjugates [78] and 2-(2-naphthyl)quinoline [79] are examples of π-conjugated compounds showing selectivity for triplex structures. Similarly, coralyne binds strongly to triplex and ss-DNA and drives disproportionation of double-stranded poly(dA)•poly(dT) to form triplex poly(dT)•poly(dA)•poly(dT) and single-stranded poly(dA) [80].

Quadruplex structures have attracted a lot of interest. Distyrylpyridinium dyes [81], telomestatin [82], tetrasubstituted phenanthrolines [83], an acyclic oligoheteroaryl [84] and a cationic cyanine dye and berberine [85], a cyclic naphthalene diimide [86], several benzoindoloquinolines [87], ditercalinium [88], a bis(quinacridine) macocycle [89], are all known to selectively bind to quadruplex DNA (for examples of these compounds, see Scheme 1.2.10).

Similarly, many porphyrins bind to quadruplex DNA (for examples see reference [90]). For a review of fluorescent compounds binding to quadruplex DNA, see reference [91].

1.2.5 Conclusions

Many compounds with interesting optoelectronic properties resulting from π-conjugation interact with nucleic acid structures. These compounds can be used in optoelectronics, molecular diagnostics and therapeutics and, in many cases, properties that are useful for one application are also relevant for other

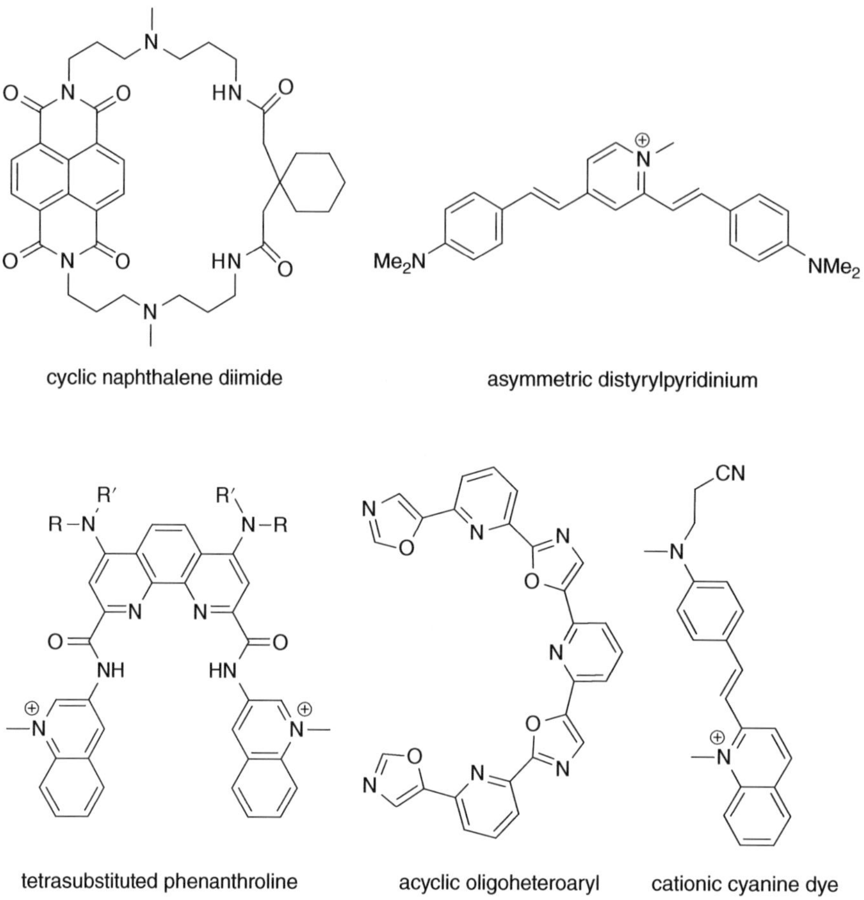

Scheme 1.2.10

applications. Although selectivity for nucleic acid structures has already been achieved, sequence selectivity for duplex DNA remains a challenge. The versatility of π-conjugated compounds in interactions with nucleic acids is illustrated by the sheer number of such compounds that are available, and this number is expected to continue to show a healthy growth.

References

[1] S. Kirchmeyer, K. Reuter, *J. Mater. Chem.* 2005, **15**, 2077–2088.
[2] N. Miyaura, A. Suzuki, *Chem. Rev.* 1995, **95**, 2457–2483.
[3] J. K. Stille, *Angew. Chem.-Int. Edit. Engl.* 1986, **25**, 508–523.
[4] (a) R. J. P. Corriu, J. P. Masse, *J. Chem. Soc., Chem. Commun.* 1972, 144; (b) K. Tamao, K. Sumitani, M. Kumada, *J. Am. Chem. Soc.* 1972, **94**, 4374–4376.
[5] (a) T. Mizoroki, K. Mori, A. Ozaki, *Bull. Chem. Soc. Jpn.* 1971, **44**, 581; (b) R. F. Heck, J. P. Nolley Jr, *J. Org. Chem.* 1972, **37**, 2320–2322.
[6] (a) E. P. Gillis, M. D. Burke, *J. Am. Chem. Soc.* 2007, **129**, 6716–6717; (b) D. M. Knapp, E. P. Gillis, M. D. Burke, *J. Am. Chem. Soc.* 2009, **131**, 6961–6963.
[7] M. S. Schiedel, C. A. Briehn, P. Bäuerle, *J. Organomet. Chem.* 2002, **653**, 200–208.

[8] L. Groenendaal, H. W. I. Peerlings, E. E. Havinga, J. A. J. M. Vekemans, E. W. Meijer, *Synth. Met.* 1995, **69**, 467–470.
[9] Y. Miyata, T. Nishinaga, K. Komatsu, *J. Org. Chem.* 2005, **70**, 1147–1153.
[10] (a) A. D. Laurent, D. Jacquemin, *Int. J. Quant. Chem.* 2013, **113**, 2019–2039; (b) J. Fabian, *Dyes Pigm.* 2010, **84**, 36–53.
[11] (a) A. P. Davis, A. J. Fry, J. Phys. *Chem. A* 2010, **114**, 12299–12304; (b) A. L. Speelman, J. G. Gillmore, J. Phys. *Chem. A* 2008, **112**, 5684–5690.
[12] A. Liu, K. Wang, S. Weng, Y. Lei, L. Lin, W. Chen, X. Lin, Y. Chen, *TrAC, Trends Anal. Chem.* 2012, **37**, 101–111.
[13] N. J. Buurma, I. Haq, *J. Mol. Biol.* 2008, **381**, 607–621.
[14] S. Steenken, S. V. Jovanovic, *J. Am. Chem. Soc.* 1997, **119**, 617–618.
[15] E. Farjami, L. Clima, K. V. Gothelf, E. E. Ferapontova, *Analyst* 2010, **135**, 1443–1448.
[16] S. D. Keighley, P. Li, P. Estrela, P. Migliorato, *Biosens. Bioelectron.* 2008, **23**, 1291–1297.
[17] E. M. Regan, A. J. Hallett, L. C. C. Wong, I. Q. Saeed, E. E. Langdon-Jones, N. J. Buurma, S. J. A. Pope, P. Estrela, *Electrochim. Acta* 2014, **128**, 10–15.
[18] A. Dreiem, F. Fonnum, *NeuroToxicology* 2004, **25**, 959–966.
[19] R. Boch, B. Mehta, T. Connolly, T. Durst, J. T. Arnason, R. W. Redmond, J. C. Scaiano, *J. Photochem. Photobiol. A: Chem.* 1996, **93**, 39–47.
[20] (a) C. J. Suckling, *J. Phys. Org. Chem.* 2008, **21**, 575–583; (b) N. G. Anthony, D. Breen, J. Clarke, G. Donoghue, A. J. Drummond, E. M. Ellis, C. G. Gemmell, J. J. Helesbeux, I. S. Hunter, A. I. Khalaf, S. P. Mackay, J. A. Parkinson, C. J. Suckling, R. D. Waigh, *J. Med. Chem.* 2007, **50**, 6116–6125; (c) N. G. Anthony, D. Breen, G. Donoghue, A. I. Khalaf, S. P. MacKay, J. A. Parkinson, C. J. Suckling, *Org. Biomol. Chem.* 2009, **7**, 1843–1850.
[21] R. F. Ludlow, S. Otto, *Chem. Soc. Rev.* 2008, **37**, 101–108.
[22] (a) J. Chen, N. C. Seeman, *Nature* 1991, **350**, 631–633; (b) N. C. Seeman, *Acc. Chem. Res.* 1997, **30**, 357–363.
[23] F. C. Simmel, *Angew. Chem., Int. Ed.* 2008, **47**, 5884–5887.
[24] R. P. Goodman, I. A. T. Schaap, C. F. Tardin, C. M. Erben, R. M. Berry, C. F. Schmidt, A. J. Turberfield, *Science* 2005, **310**, 1661–1665.
[25] P. W. K. Rothemund, *Nature* 2006, **440**, 297–302.
[26] I. H. Stein, C. Steinhauer, P. Tinnefeld, *J. Am. Chem. Soc.* 2011, **133**, 4193–4195.
[27] (a) N. C. Seeman, *Trends Biotechnol.* 1999, **17**, 437–443; (b) N. C. Seeman, *Chem. Biol.* 2003, **10**, 1151–1159; (c) N. C. Seeman, *Nature* 2003, **421**, 427–431; (d) Y. Osakada, K. Kawai, M. Fujitsuka, T. Majima, *Proc. Natl. Acad. Sci. U.S.A.* 2006, **103**, 18072–18076; (e) F. E. Alemdaroglu, A. Herrmann, *Org. Biomol. Chem.* 2007, **5**, 1311–1320.
[28] (a) Y. N. Teo, E. T. Kool, *Chem. Rev.* 2012, **112**, 4221–4245; (b) B. Albinsson, J. K. Hannestad, K. Börjesson, *Coord. Chem. Rev.* 2012, **256**, 2399–2413; (c) A. Ruiz-Carretero, P. G. A. Janssen, A. Kaeser, A. P. H. J. Schenning, *Chem. Commun.* 2011, **47**, 4340–4347.
[29] C. M. Spillmann, S. Buckhout-White, E. Oh, E. R. Goldman, M. G. Ancona, I. L. Medintz, *Chem. Commun.* 2014, **50**, 7246–7249.
[30] (a) W. Su, M. Schuster, C. R. Bagshaw, U. Rant, G. A. Burley, *Angew. Chem., Int. Ed.* 2011, **50**, 2712–2715; (b) W. Su, C. R. Bagshaw, G. A. Burley, *Sci. Rep.* 2013, **3**, article number 1883, DOI: 10.1038/srep01883.
[31] G. Cosa, K. S. Focsaneanu, J. R. N. McLean, J. P. McNamee, J. C. Scaiano, *Photochem. Photobiol.* 2001, **73**, 585–599.
[32] A. E. Hargrove, J. A. Raskatov, J. L. Meier, D. C. Montgomery, P. B. Dervan, *J. Med. Chem.* 2012, **55**, 5425–5432.
[33] S. M. D. Watson, A. R. Pike, J. Pate, A. Houlton, B. R. Horrocks, *Nanoscale* 2014, **6**, 4027–4037.
[34] D. Banerjee, S. K. Pal, *J. Phys. Chem. B* 2007, **111**, 5047–5052.
[35] Y. C. Chan, J. W. Chen, S. Y. Su, C. C. Chang, *Biosens. Bioelectron.* 2013, **47**, 566–573.
[36] D. E. Gilbert, J. Feigon, *Curr. Opin. Struct. Biol.* 1999, **9**, 305–314.
[37] W. Guschlbauer, J. F. Chantot, D. Thiele, *J. Biomol. Struct. Dyn.* 1990, **8**, 491–511.
[38] K. Gehring, J. L. Leroy, M. Gueron, *Nature* 1993, **363**, 561–565.
[39] E. Wang, J. Feigon, *Oxford Handbook of Nucleic Acid Structure* 1999, Oxford University Press, Oxford, pp. 355–388.
[40] M. L. Kopka, C. Yoon, D. Goodsell, P. Pjura, R. E. Dickerson, *J. Mol. Biol.* 1985, **183**, 553–563.

[41] E. F. Pettersen, T. D. Goddard, C. C. Huang, G. S. Couch, D. M. Greenblatt, E. C. Meng, T. E. Ferrin, *J. Comput. Chem.* 2004, **25**, 1605–1612.

[42] J. E. Jones, A. J. Amoroso, I. M. Dorin, G. Parigi, B. D. Ward, N. J. Buurma, S. J. A. Pope, *Chem. Commun.* 2011, **47**, 3374–3376.

[43] (a) L. S. Lerman, *J. Mol. Biol.* 1961, **3**, 18–30; (b) G. Cohen, H. Eisenberg, *Biopolymers* 1969, **8**, 45–55; (c) S. Satyanarayana, J. C. Dabrowiak, J. B. Chaires, *Biochemistry* 1993, **32**, 2573–2584; (d) D. Suh, J. B. Chaires, *Bioorg. Med. Chem.* 1995, **3**, 723–728.

[44] (a) E. Tuite, B. Norden, *J. Am. Chem. Soc.* 1994, **116**, 7548–7556; (b) W. D. Wilson, F. A. Tanious, D. Ding, A. Kumar, D. W. Boykin, P. Colson, C. Houssier, C. Bailly, *J. Am. Chem. Soc.* 1998, **120**, 10310–10321.

[45] N. C. Garbett, P. A. Ragazzon, J. B. Chaires, *Nat. Protoc.* 2007, **2**, 3166–3172.

[46] J. B. Chaires, *Arch. Biochem. Biophys.* 2006, **453**, 26–31.

[47] (a) K. Dore, S. Dubus, H. A. Ho, I. Levesque, M. Brunette, G. Corbeil, M. Boissinot, G. Boivin, M. G. Bergeron, D. Boudreau, M. Leclerc, *J. Am. Chem. Soc.* 2004, **126**, 4240–4244; (b) P. C. Ewbank, G. Nuding, H. Suenaga, R. D. McCullough, S. Shinkai, *Tetrahedron Lett.* 2001, **42**, 155–157; (c) H. A. Ho, M. Boissinot, M. G. Bergeron, G. Corbeil, K. Dore, D. Boudreau, M. Leclerc, *Angew. Chem., Int. Ed.* 2002, **41**, 1548–1551; (d) K. P. R. Nilsson, O. Inganas, *Nat. Mater.* 2003, **2**, 419–424; (e) H. A. Ho, A. Najari, M. Leclerc, *Acc. Chem. Res.* 2008, **41**, 168–178.

[48] H. Korri-Youssoufi, F. Garnier, P. Srivastava, P. Godillot, A. Yassar, *J. Am. Chem. Soc.* 1997, **119**, 7388–7389.

[49] (a) A. Najari, H. A. Ho, J. F. Gravel, P. Nobert, D. Boudreau, M. Leclerc, *Anal. Chem.* 2006, **78**, 7896–7899; (b) L. Y. Zhang, H. Sun, D. Li, S. Song, C. H. Fan, S. Wang, *Macromol. Rapid Commun.* 2008, **29**, 1626–1626.

[50] A. Åslund, K.P.R. Nilsson, P. Konradsson, *J. Chem. Biol.* 2009, **2**, 161–175.

[51] P. Björk, D. Thomsson, O. Mirzov, J. Wigenius, O. Inganäs, I. G. Scheblykin, *Small* 2009, **5**, 96–103.

[52] G. Barbarella, M. Melucci, G. Sotgiu, *Adv. Mater.* 2005, **17**, 1581–1593.

[53] (a) J. Farwer, M. J. Packer, C. A. Hunter, *In Silico Biol.* 2007, **7**, 595–600; (b) J. Farwer, M. J. Packer, C. A. Hunter, *Biopolymers* 2006, **81**, 51–61.

[54] P. B. Dervan, *Bioorg. Med. Chem.* 2001, **9**, 2215–2235.

[55] R. M. Doss, M. A. Marques, S. Foister, D. M. Chenoweth, P. B. Dervan, *J. Am. Chem. Soc.* 2006, **128**, 9074–9079.

[56] S. Y. Breusegem, R. M. Clegg, F. G. Loontiens, *J. Mol. Biol.* 2002, **315**, 1049–1061.

[57] C. Behrens, N. Harrit, P. E. Nielsen, *Bioconjugate Chem.* 2001, **12**, 1021–1027.

[58] K. Jansen, P. Lincoln, B. Nordén, *Biochemistry* 1993, **32**, 6605–6612.

[59] S. Mallena, M. P. H. Lee, C. Bailly, S. Neidle, A. Kumar, D. W. Boykin, W. D. Wilson, *J. Am. Chem. Soc.* 2004, **126**, 13659–13669.

[60] (a) Y. Liu, A. Kumar, S. Depauw, R. Nhili, M. H. David-Cordonnier, M. P. Lee, M. A. Ismail, A. A. Farahat, M. Say, S. Chackal-Catoen, A. Batista-Parra, S. Neidle, D. W. Boykin, W. D. Wilson, *J. Am. Chem. Soc.* 2011, **133**, 10171–10183; (b) Y. Miao, M. P. H. Lee, G. N. Parkinson, A. Batista-Parra, M. A. Ismail, S. Neidle, D. W. Boykin, W. D. Wilson, *Biochemistry* 2005, **44**, 14701–14708.

[61] C. Bailly, C. Tardy, L. Wang, B. Armitage, K. Hopkins, A. Kumar, G. B. Schuster, D. W. Boykin, W. D. Wilson, *Biochemistry* 2001, **40**, 9770–9779.

[62] M. Munde, M. A. Ismail, R. Arafa, P. Peixoto, C. J. Collar, Y. Liu, L. Hu, M. H. David-Cordonnier, A. Lansiaux, C. Bailly, D. W. Boykin, W. D. Wilson, *J. Am. Chem. Soc.* 2007, **129**, 13732–13743.

[63] (a) R. Nhili, P. Peixoto, S. Depauw, S. Flajollet, X. Dezitter, M. M. Munde, M. A. Ismail, A. Kumar, A. A. Farahat, C. E. Stephens, M. Duterque-Coquillaud, W. David Wilson, D. W. Boykin, M. H. David-Cordonnier, *Nucleic Acids Res.* 2013, **41**, 125–138; (b) M. Munde, A. Kumar, P. Peixoto, S. Depauw, M. A. Ismail, A. A. Farahat, A. Paul, M. V. Say, M. H. David-Cordonnier, D. W. Boykin, W. D. Wilson, *Biochemistry* 2014, **53**, 1218–1227.

[64] R. Nanjunda, W. D. Wilson, *Curr. Protoc. Nucleic Acid Chem.*, Issue SUPPL.51, December 2012, Article number 8.8, DOI: 10.1002/0471142700.nc0808s51.

[65] L. A. Mullice, R. H. Laye, L. P. Harding, N. J. Buurma, S. J. A. Pope, *New J. Chem.* 2008, **32**, 2140–2149.

[66] T. Phillips, I. Haq, A. Meijer, H. Adams, I. Soutar, L. Swanson, M. J. Sykes, J. A. Thomas, *Biochemistry* 2004, **43**, 13657–13665.

[67] C. Metcalfe, H. Adams, I. Haq, J. A. Thomas, *Chem. Commun.* 2003, 1152–1153.

[68] N. Perin, I. Martin-Kleiner, R. Nhili, W. Laine, M. H. David-Cordonnier, O. Vugrek, G. Karminski-Zamola, M. Kralj, M. Hranjec, *MedChemComm* 2013, **4**, 1537–1550.

[69] M. R. Duff Jr, V. K. Mudhivarthi, C. V. Kumar, *J. Phys. Chem. B* 2009, **113**, 1710–1721.
[70] (a) P. A. Ragazzon, N. C. Garbett, J. B. Chaires, *Methods* 2007, **42**, 173–182; (b) P. Ragazzon, J. B. Chaires, *Methods* 2007, **43**, 313–323.
[71] P. A. Holt, P. Ragazzon, L. Strekowski, J. B. Chaires, J. O. Trent, *Nucleic Acids Res.* 2009, **37**, 1280–1287.
[72] J. B. Chaires, J. Ren, D. Hamelberg, A. Kumar, V. Pandya, D. W. Boykin, W. D. Wilson, *J. Med. Chem.* 2004, **47**, 5729–5742.
[73] G. Song, J. Ren, *Chem. Commun.* 2010, **46**, 7283–7294.
[74] N. N. Shaw, D. P. Arya, *Biochimie* 2008, **90**, 1026–1039.
[75] R. T. Wheelhouse, N. C. Garbett, N. J. Buurma, J. B. Chaires, *Angew. Chem., Int. Ed.* 2010, **49**, 3207–3210.
[76] P. Wittung, S. K. Kim, O. Buchardt, P. Nielsen, B. Norden, *Nucleic Acids Res.* 1994, **22**, 5371–5377.
[77] J. O. Smith, D. A. Olson, B. A. Armitage, *J. Am. Chem. Soc.* 1999, **121**, 2686–2695.
[78] D. P. Arya, *Acc. Chem. Res.* 2011, **44**, 134–146.
[79] J. B. Chaires, J. Ren, M. Henary, O. Zegrocka, G. Reid Bishop, L. Strekowski, *J. Am. Chem. Soc.* 2003, **125**, 7272–7283.
[80] M. Polak, N. V. Hud, *Nucleic Acids Res.* 2002, **30**, 983–992.
[81] X. Xie, B. Choi, E. Largy, R. Guillot, A. Granzhan, M. P. Teulade-Fichou, *Chem. Eur. J.* 2013, **19**, 1214–1226.
[82] K. Shin-ya, K. Wierzba, K. Matsuo, T. Ohtani, Y. Yamada, K. Furihata, Y. Hayakawa, H. Seto, *J. Am. Chem. Soc.* 2001, **123**, 1262–1263.
[83] A. F. Larsen, M. C. Nielsen, T. Ulven, *Chem. Eur. J.* 2012, **18**, 10892–10902.
[84] F. Hamon, E. Largy, A. Guédin-Beaurepaire, M. Rouchon-Dagois, A. Sidibe, D. Monchaud, J. L. Mergny, J. F. Riou, C. H. Nguyen, M. P. Teulade-Fichou, *Angew. Chem., Int. Ed.* 2011, **50**, 8745–8749.
[85] A. Benz, V. Singh, T. U. Mayer, J. S. Hartig, *ChemBioChem* 2011, **12**, 1422–1426.
[86] Y. Esaki, M. M. Islam, S. Fujii, S. Sato, S. Takenaka, *Chem. Commun.* 2014, **50**, 5967–5969.
[87] P. Alberti, P. Schmitt, C. H. Nguyen, C. Rivalle, M. Hoarau, D. S. Grierson, J. L. Mergny, *Bioorg. Med. Chem. Lett.* 2002, **12**, 1071–1074.
[88] C. Carrasco, F. Rosu, V. Gabelica, C. Houssier, E. De Pauw, C. Garbay-Jaureguiberry, B. Roques, W. David Wilson, J. B. Chaires, M. J. Waring, C. Bailly, *ChemBioChem* 2002, **3**, 1235–1241.
[89] M. P. Teulade-Fichou, C. Carrasco, L. Guittat, C. Bailly, P. Alberti, J. L. Mergny, A. David, J. M. Lehn, W. D. Wilson, *J. Am. Chem. Soc.* 2003, **125**, 4732–4740.
[90] (a) J. Seenisamy, S. Bashyam, V. Gokhale, H. Vankayalapati, D. Sun, A. Siddiqui-Jain, N. Streiner, K. Shin-ya, E. White, W. D. Wilson, L. H. Hurley, *J. Am. Chem. Soc.* 2005, **127**, 2944–2959; (b) H. T. Le, M. C. Miller, R. Buscaglia, W. L. Dean, P. A. Holt, J. B. Chaires, J. O. Trent, *Org. Biomol. Chem.* 2012, **10**, 9393–9404.
[91] B. R. Vummidi, J. Alzeer, N. W. Luedtke, *ChemBioChem* 2013, **14**, 540–558.

1.3

Metal Ion- and Perylene Diimide-Mediated DNA Architectures

Christian Wellner, Claudia Stubinitzky, and Hans-Achim Wagenknecht

Karlsruhe Institute of Technology (KIT) Institute of Organic Chemistry-Chair II, Karlsruhe, Germany

1.3.1 Introduction

The role of DNA is increasing not only in biotechnology but also in nanotechnology. Nucleic acids are used as a scaffold to arrange molecules and materials into complex three-dimensional structures based on a number of unique properties: a simple four letter code (A–T, G–C), high fidelity of hybridization, known double helical geometry and dimensions, readily accessible through automated organic synthesis and biochemical tools to modify DNA [1].

Extended 2D structures based on crossover DNA junctions were pioneered by Seeman and coworkers [2]. More complex assemblies, which are termed DNA origami, have been designed by the groups of Winfree [3], Rothemund [4], Yan [5] and others [6]. Based on these key results, three-dimensional DNA assemblies have been reported, for example, by researchers working with Shih [7], Gothelf, and Kjems [8]. Such defined architectures are crucial when using nucleic acids in the field of nanoelectronics. However, nearly all of these complex self-assembled DNA nanostructures were based purely on unmodified nucleic acids. Self-assembling molecular electronic devices would require an adequate electronic conductance of DNA. We, and others, have shown that single excess electrons and positive charges can be transferred through short DNA parts [9]. However, it turns out that the ability of unmodified natural DNA to conduct electrons over a longer range is not sufficient to be able to use DNA in molecular electronic circuits. The reported ranges for the conductivity values differ enormously due to differences in the experimental settings and sequences studied [10]. Therefore, the current interest of research is focused on the improvement of the DNA electron transfer properties.

DNA in Supramolecular Chemistry and Nanotechnology, First Edition.
Edited by Eugen Stulz and Guido H. Clever.

Over the last decade we have worked out several synthetic protocols to modify DNA chemically to develop functional DNA architectures [11]. In particular, organic chromophores and metal ion ligands can be introduced to nucleic acids by three different approaches: (i) DNA base substitutes/surrogates, (ii) DNA base modifications, and (iii) DNA sugar modifications at the 2′-position. The first two types of modifications can be obtained mainly by the DNA building block approach whereas the last type is based on a post-synthetic methodology. In this chapter, we focus on (i) metal-ion complexes and (ii) π–π interacting perylene diimides, which are two important types of modifications able to introduce not only additional structural motifs to the purely hydrogen-bonded base pairing in DNA architectures, but also have the potential to improve the electron transfer abilities of DNA.

1.3.2 Metal ion complexes as DNA modifications: hydroquinoline and terpyridine

The spatial programmability and stability of nucleic acids can be tuned by transition metal complexes. These metal complexes show defined geometries that can be used to branch DNA oligonucleotides to form new DNA architectures. Metal–DNA conjugates can show an increased stability and/or rigidity and they can possess enhanced properties with respect to electronic, catalytic or photochemical properties. Furthermore, these conjugates can be reversible or switchable upon metal ion addition. These properties can be particularly promising for DNA nanoelectronics [1a].

Metal ions can interact with the DNA backbone or replace a hydrogen atom in the base pairs. The formation of metal mediated T–T base pairs upon complexation of Hg^{2+} ions has already been postulated by Katz in 1963 [12]. Other examples with divalent metal ions were shown by the Lee group [13]. This type of interaction is not base selective; hence the binding of metal ions to natural DNA cannot lead to higher order DNA assemblies. For this purpose it is crucial to achieve well-defined metal–DNA structures and here we will focus on the synthetic approaches.

While unmodified DNA synthesis is nowadays an automated procedure, the preparation of DNA–metal conjugates can still be challenging. Metal ions can bind non-specifically to the DNA phosphate backbone or to the DNA bases. Other challenges are the degradation of DNA by metal complexes, the decomposition of the metal complexes in aqueous solution or during solid-phase DNA synthesis, and the difficult characterization of those conjugates due to the intrinsic complexity.

Nucleosides bearing a metal-binding ligand are referred to as "ligandosides" in the literature. In principle, there are three categories for preparing these DNA–metal conjugates (Figure 1.3.1). The first is the substitution of a natural base pair by a metal-mediated base pair. The second is achieved by modifying the nucleoside with a ligand. In the third category, a metal binding ligand is tethered terminally to DNA, which forms a metal complex upon metal addition [1].

In 1999, Tanaka and Shionoya were among the first to report the *o*-phenylenediamine–Pd complex as an example of a ligand that potentially forms a metal complex inside a DNA double helix [14]. Important metal-mediated base pairs in DNA were published by Meggers and coworkers in 2000 [17]. This base pair comprised pyridine-2,6-dicarboxylate (dipic) as a planar tridentate ligand and pyridine (py) as the complementary single donor ligand. Upon addition of copper, the dipic–py base pair is selectively formed, which stabilizes the DNA duplex significantly.

These ligandoside complexes offer the possibility to discretely assemble metal ions along the DNA helix as linear DNA nanoarrays. Shionoya and Tanaka were the first to incorporate up to five consecutive copper hydroxypyridone base pairs in DNA. The copper ions are supposed to stack on top of each other in the center of the helix. The distance between the copper centers was estimated by electron paramagnetic resonance (EPR) spectroscopy to be 3.7 ± 0.1 Å, which comes close to the distance of 3.4 Å between natural base pairs in DNA [18].

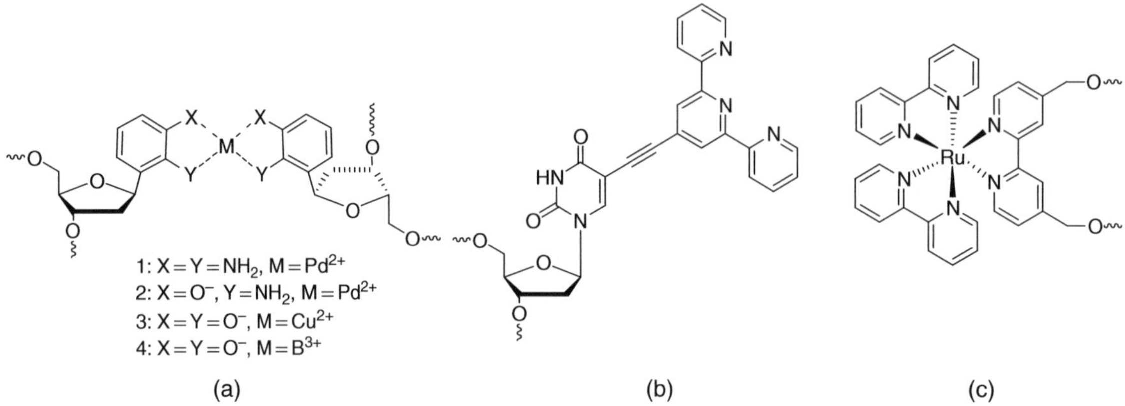

Figure 1.3.1 *Examples for each category of DNA–metal conjugates: (a) substitution of a natural base pair [14]; (b) ligand modified nucleobase (ethynyl terpyridine-modified 2′-deoxyuridine, tpy-dU [15]); (c) terminally tethered metal complex to DNA [16]*

Using adjacent base pairs with the *N,N′*-bis(salicylidene)ethylenediamine (salen) ligand the Carell group could also determine the Cu–Cu distance to be 3.7 Å. Tanaka *et al.* achieved basically a two letter metal ion based artificial genetic code. Two orthogonally different base pairs were capable of selectively binding Cu or Hg ions inside a DNA double helix so that the sequence of the metal ions can be programmed inside the duplex [19].

Recently, we reported a new copper-ion mediated base pair that was introduced for electron transfer studies in DNA [20]. The hydroxyquinoline ligand (Hq) was linked by its 6-position to the sugar moiety. This offers two major advantages, in contrast to the earlier published alternatives with attachment to the 7-position [21]. Firstly, it provides more structural similarity to natural pyrimidine nucleosides and thereby a hydrogen-bonded base pair can be expected if two of the Hq nucleosides are placed opposite to each other. Additionally, the synthetic steps to this C-nucleoside can be reduced significantly. DNA with non-metallated Hq base pairs show almost the same melting temperatures as the corresponding DNA duplexes with A–T base pairs. Moreover, the melting temperatures are increased tremendously upon copper insertion. To elucidate the potential of Hq–Hq base pairs as charge carriers for photoinduced energy transfer, the Hq base pair in double-stranded DNA was framed by two electron transfer probes. On the one hand, 2,7-diazapyrenium as a 2′-modification of uridine (Dp–U) is amenable to introducing oxidative hole transfer into DNA by photoexcitation at 339 nm (see Figure 1.3.2). Alternatively, 6-*N,N*-dimethyl aminopyrene attached to the 5-position of 2′-deoxyuridine (Ap–dU) represents the complementary electron hole acceptor. By analysis of fluorescence intensities and melting temperatures it can be assumed that photoinduced charge transfer across non-metallated Hq base pairs can occur and the efficiency is further increased in the presence of Cu(II) [22].

The combination of metal ion-mediated base pair and self-assembled higher order structures would potentially create a new class of material, combining the properties of DNA assemblies and transition metals. Metallated DNA-based three-dimensional cages were first reported by Sleiman and coworkers in 2009 [23]. Therein, DNA duplex triangles bearing three bis-2,9-diphenyl-1,10-phenanthroline (dpp) ligands were assembled using linking strands into a 3D prismatic cage [24] with a variety of transition metals. A possible application for 3D cages could be as stimuli-responsive host molecules for proteins or nanoparticles, as well as building blocks for even more extended metal–nucleic acid DNA architectures.

In order to assemble two or more oligonucleotides by metal ion-complex formation we applied 2,2′;6′,2″-terpyridine (tpy) as a ligand, which is known to form stable complexes with several metal ions [25]. The first

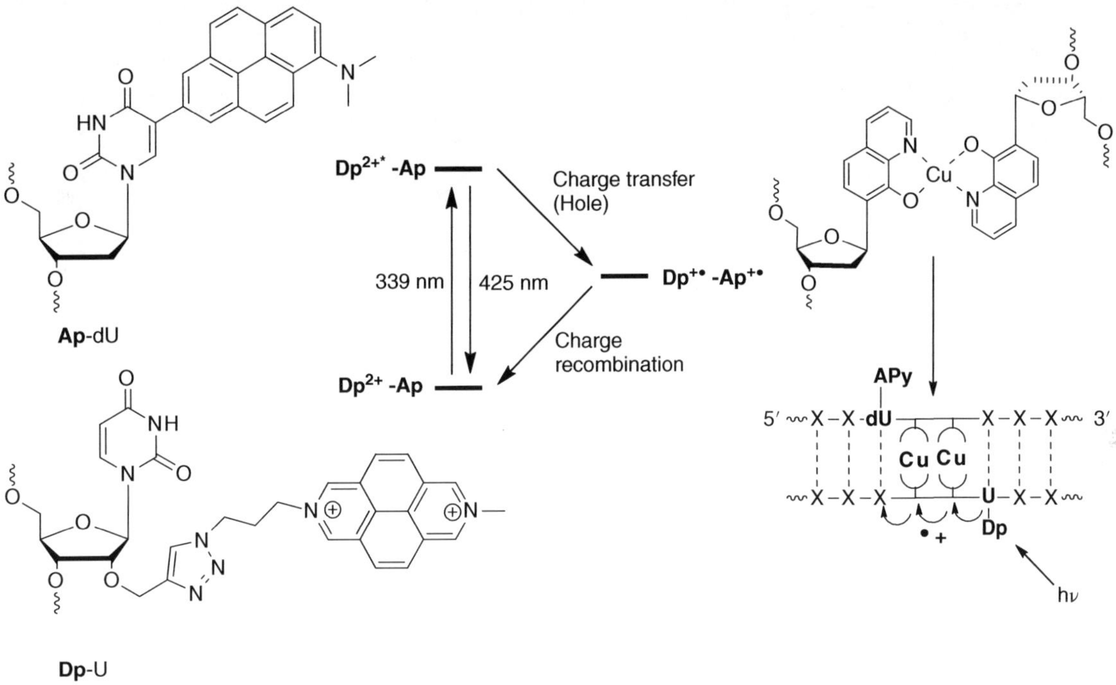

Figure 1.3.2 *Structures of the charge donor (Dp–U) and the charge acceptor (Ap–dU). The scheme shows the main indicated pathway between Dp and Ap including the metal-ion mediated way (right) [20]*

conjugates of tpy ligands with DNA were synthesized by McLaughlin *et al.* DNA-tethered bis(tpy)Ru(II) complexes were attached to a triethylene glycol linker and thereby hybridization at various complex ratios and linear arrays of varying lengths could be observed [26]. In a similar experiment, linear DNA nanoarrays were achieved by hybrid metal complex–DNA conjugates [27] with tpy units attached at the 3′-end by using the oxime chemistry.

Bis(tpy)Fe(II) complexes were applied as vertexes to assemble DNA triangles [28]. Allosteric control of oligonucleotide hybridization was performed with tpy–DNA conjugates that show a response behavior towards Fe^{2+} and Zn^{2+} ions. In the absence of a metal, an unmodified DNA strand is bound. After addition of one equivalent of metal, cyclization is initiated and dehybridization occurs. With an excess of the metal (in the case of Zn) hybridization is reestablished [29]. Duplex stability modulation was achieved by DNA duplexes carrying a tpy ligand at the N2′ position of 2′-amino-2′-deoxyuridine or its locked counterpart 2′-amino-LNA. The thermal stability of the tpy-modified DNA duplexes was affected upon addition of varying equivalents of transition metals.

In all of the examples described so far, a flexible linker was applied to tether the tpy ligand to the DNA. A linkage of two or more DNA oligonucleotides was only achieved by terminal labeling; a "side-on" connection by metal complexation cannot be found in the literature very often [30]. Regarding these observations, our idea was to synthesize a modified uridine (with the ligand tethered rigidly to provide the structural basis for an electronic coupling between the ligand and the base stack). This coupling is crucial for applications in the field of nanoelectronics. The correspondingly modified 2′-deoxyuridine (tpy-dU, see Figure 1.3.1) was synthesized according to Hocek *et al.* [31], but incorporated in DNA by solid-phase phosphoramidite chemistry [15]. 4′-Ethynyl-2,2′,6′,2″-terpyridine was tethered via Sonogashira cross-coupling to

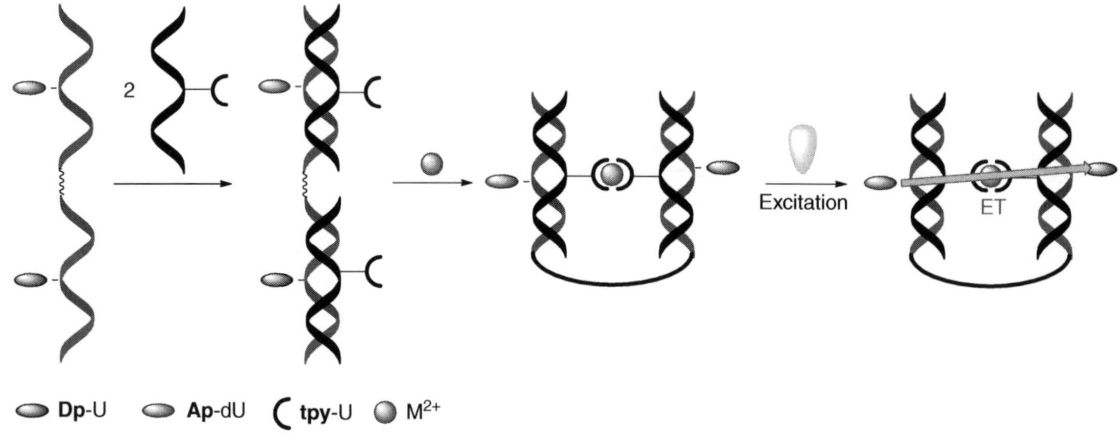

Figure 1.3.3 *The experimental setup for transferring an electron from one into another duplex over metal mediated tpy-dU units. For structures of Ap-dU and Dp-U see Figure 1.3.2 (See color figure in color plate section)*

5-iodo-2′-deoxyuridine [15, 30–32] and transformed into the phosphoramidite. This building block allows internal and terminal modification of oligonucleotides with tpy-dU.

The melting temperatures were investigated to show the influence of one or two tpy-dU units. With one internal tpy-dU, DNA duplexes are destabilized with all opposite bases (with a slight preference for A as the counter base). Remarkable results were obtained when two tpy-dU units were incorporated opposite to each other. It was shown that the hydrophobic interaction regains more hybridization energy than the destabilization due to incorporation of the tpy moiety. A similar observation was made with bipyridine pairs [33] and binaphthyl pairs [34].

Upon addition of metal ions (Cu^{2+}, Ni^{2+}, Zn^{2+}, Fe^{2+}) dimerization was induced with strands bearing only one internal or terminal tpy-dU unit and A as the opposite base. To establish whether larger DNA assemblies can be obtained, double strands containing two tpy-dU units either opposite to each other or at the termini, were analyzed by fluorescence quenching, T_m values, and PAGE analysis. Internally formed metal ion-mediated tpy-dU base pairs interfere with the formation of higher DNA assemblies. In the case of terminally labeled DNA, several gel bands of lower mobility could be observed especially with Ni^{2+} and Fe^{2+}, indicating the formation of metal ion-mediated DNA assemblies [15].

To investigate the applicability of the tpy-dU coordination for electron transfer between DNA strands, we designed DNA strands linked by tpy units to combine two DNA strands upon addition of metal. One of these strands contains Ap-dU as an additional modification and electron donor, while the other DNA strand bears Dp-U as an electron acceptor (see Figure 1.3.3). Our idea is to transfer and thereby translocate an injected charge from one DNA strand over the external tpy-dU unit into the other DNA double strand containing the electron acceptor.

Preliminary fluorescence data indicate that a hole transfer can be induced and an electron can be transferred over tpy-dU units into another DNA strand. However, to ensure these indications, time-resolved transient absorption studies have to follow.

1.3.3 Perylene diimide-based DNA architectures

Perylene diimides (PDIs), which have the advantages of strong extinction, high fluorescence quantum yields, high photochemical and thermal stability, are well established as pigments [35], fluorescence dyes [36], and conducting materials for organic electronics [37] (Figure 1.3.4). Moreover, the aggregation of the planar

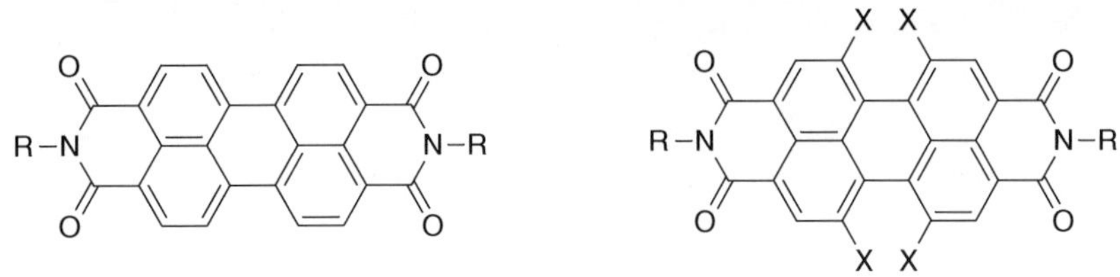

Figure 1.3.4 *Perylene diimides (PDIs) without (left) and with (right) substituents in the bay area (X = various electron withdrawing or donating groups) [45, 46]*

hydrophobic aromatic molecule has been extensively employed for molecular self-assembly [37p, 38]. The expanded quadrupolar π-system of this dye-class allows the construction of supramolecular architectures with outstanding photophysical properties [39]. The integration of PDI into oligonucleotide conjugates has been reported to yield intra- and intermolecular hydrophobic association of the chromophores in single strand [37j], duplex [37d, k, q, 40], hairpin [41], hairpin dimer structures [37k, l, n, 42], DNA triplexes [37l], DNA dumbbells [38h], as well as three-way junctions [43].

Water provides many advantages as a solvent for supramolecular chemistry compared with organic media. Primarily, the hydrophobic effect arises from the increasing interactions between the hydrophobic π-surfaces of the PDIs, hence the desired recognition and self-assembly processes leading to extended architectures may already occur in dilute solutions. For this reason tremendous opportunities can arise from the interaction of such supramolecular structures with biomolecules, for example, the DNA double helix, in water [44].

Two successful strategies have been developed to achieve soluble perylene diimide dyes, which provide the basis for the synthetic accessibility of this class of chromophores. The first one was worked out by Langhals and coworkers who introduced solubilizing substituents at the imide nitrogen [45]. The second synthetic strategy, developed by Seybold and co-workers, is the attachment of substituents at the so-called bay area [46].

Numerous PDI derivatives have been developed for application as fluorescent standards, in fluorescent light collectors, or as laser dyes due to their intense yellow–green photoluminescence [45]. Distinctive changes in the absorption and emission bands result if PDIs bear substituents in the bay area. In contrast, it has been demonstrated that the imide substituent has a minor influence on the absorption and emission properties [37p]. For quite some time it has been known that water-soluble PDIs undergo π-stacking interactions with DNA bases [45]. The chemical resistance, the particular fluorescence, and the self-assembling properties, as well as their electronic properties, have attracted interest for applications in DNA analytics and DNA-based nanotechnologies [37h–k, 47]. For this reason PDIs have been incorporated into DNA double strands as artificial DNA base surrogates [37i] at both internal (Figure 1.3.5a) and terminal positions (Figure 1.3.5b) [37q].

At temperatures below the melting temperatures (T_m) of the DNA duplexes (Figure 1.3.5) an excimer-type fluorescence of the PDI dimers occurs with a broad band at around 660 nm, which has also been observed in the nanoaggregates of PDIs [48]. PDI monomer fluorescence with well-defined signals (558/603 nm) is obtained at higher temperatures, which clearly shows that the DNA architecture is needed for PDI dimerization. Additionally, the UV–Vis spectra show two major bands at 506 and 545 nm, whose intensities support the occurrence of π–π excitonic interactions between the dyes [37q]. This motif was further used for the development of supramolecular DNA structures, which can be arranged through the aggregation of PDIs [44]. Accordingly, the synthesis of PDI DNA hairpins [41], hairpin dimers [37l], DNA dumbbells [38h], as well as DNA triplexes [37l], has been reported.

An oligonucleotide conjugate, which contains pure A–T base pair sequences connected by a PDI linker (Figure 1.3.6a), exists as a monomer in water (Figure 1.3.6b) but forms a hairpin dimer in the presence of NaCl (Figure 1.3.6c) [37k, l]. The monomer–dimer equilibrium is found to be dependent upon

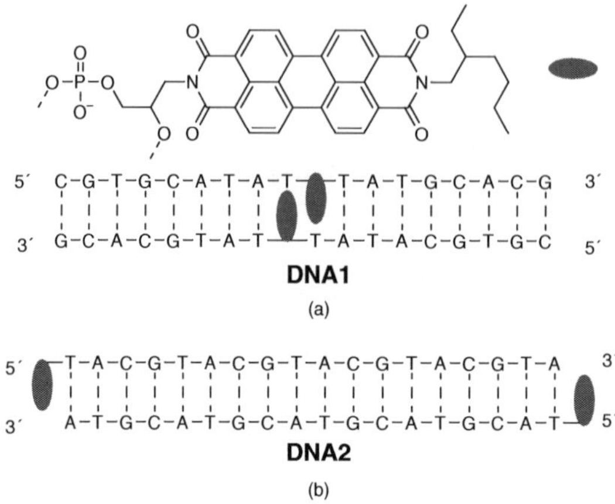

Figure 1.3.5 *PDI incorporated into DNA double strand at (a) internal and (b) terminal positions*

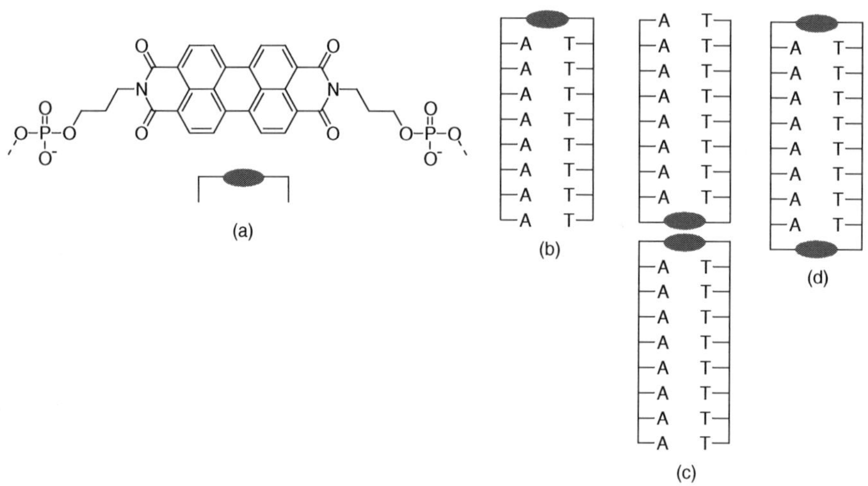

Figure 1.3.6 *Structures of (a) the PDI linker, (b) PDI hairpin, (c) PDI hairpin dimer, and (d) PDI dumbbell*

salt concentration and temperature, as well as the concentration of the conjugate. The dumbbell structures, containing PDI caps at both ends, are able to undergo end-to-end assembly (Figure 1.3.6d). AFM and cryo-TEM images showed branched fibers corresponding to the end-to-end assembly of ~10–30 monomers [38h].

Molecular beacons (MBs) are single stranded nucleic acids possessing a stem-and-loop structure, which are used for the specific detection of DNA targets [49]. The excimer-type fluorescence of PDI is shifted to a different wavelength and not simply quenched. For this reason PDI represents an attractive readout for MBs. Their modification with two PDI chromophores as DNA base substitutes (Figure 1.3.7) form PDI dimers in the interstrand mode, which shows a red-shifted excimer fluorescence. The hybridization with target oligonucleotides yields absorption and fluorescence changes typical for PDI monomers [50].

With respect to such bioanalytical applications, a major disadvantage of PDI as part of DNA is its low fluorescence quantum yield (less than 1%) [42]. To overcome this problem PDI has been combined with pyrene to obtain highly sensitive MBs (Figure 1.3.8). The stem containing pyrene and PDI can interact by

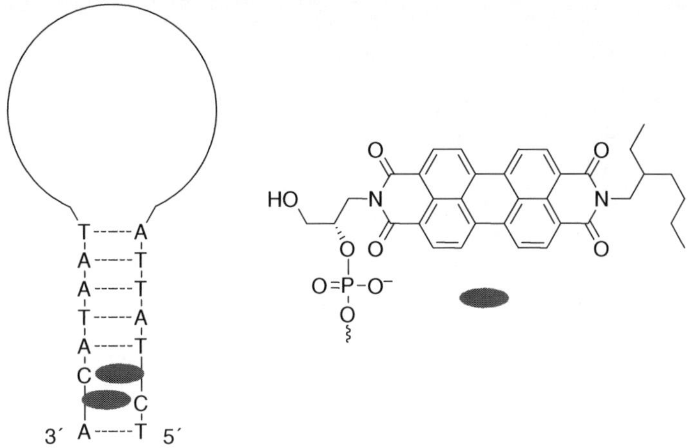

Figure 1.3.7 *MB with PDI as a dimeric internal DNA base substitution*

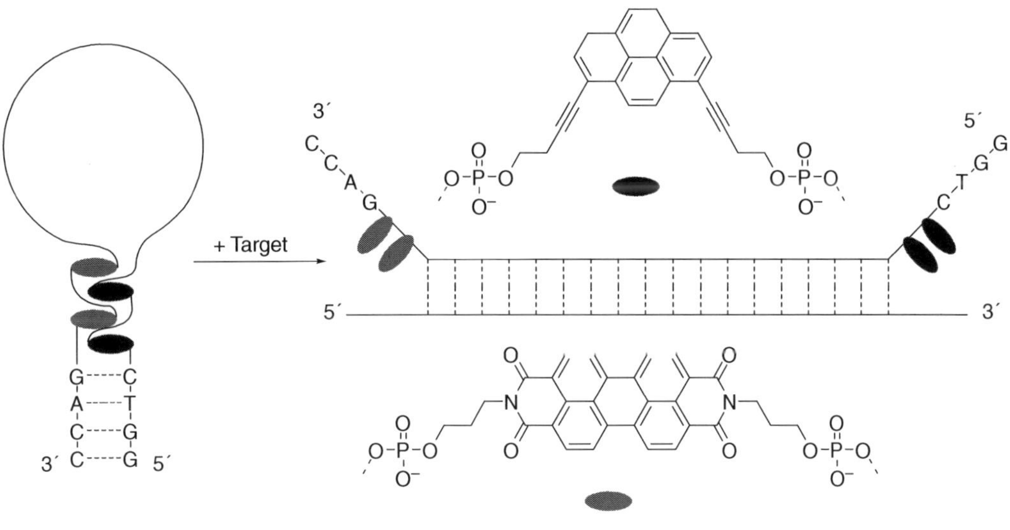

Figure 1.3.8 *Excimer controlled MB with PDI and pyrene modified DNA*

interstrand stacking, which yields a highly efficient quenching of the pyrene excimer fluorescence [37g, 51]. Upon hybridization with the target oligonucleotide, pyrene forms a strongly fluorescent excimer possessing an emission maximum around 520 nm with both a large extinction coefficient and high quantum yields [52].

An option to prevent fluorescence quenching of common PDI dyes in the vicinity of guanines has been the development of a green PDI derivate containing two *N*-pyrrolidinyl substituents in the bay area as a new building block for DNA modification (Figure 1.3.9) [37q, 42, 53]. The strongly electron-donating substituents reduce the electron deficiency and thus prevent fluorescence quenching by photoinduced charge transfer to guanines [53, 54].

In addition, the self-assembly of redox-active molecules into ordered arrays, applicable in rapid, long distance charge transport, is important for the development of functional nanomaterials for organic electronics [37r]. DNA shows a great promise as a structural scaffold for the helical arrangement of chromophores with

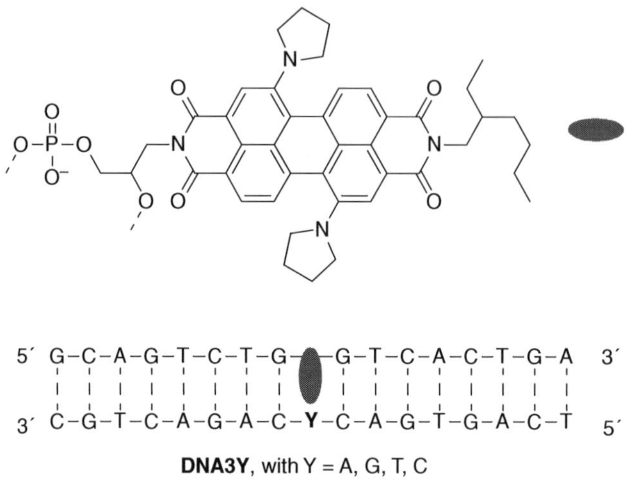

Figure 1.3.9 Double helix with a PDI derivative as DNA

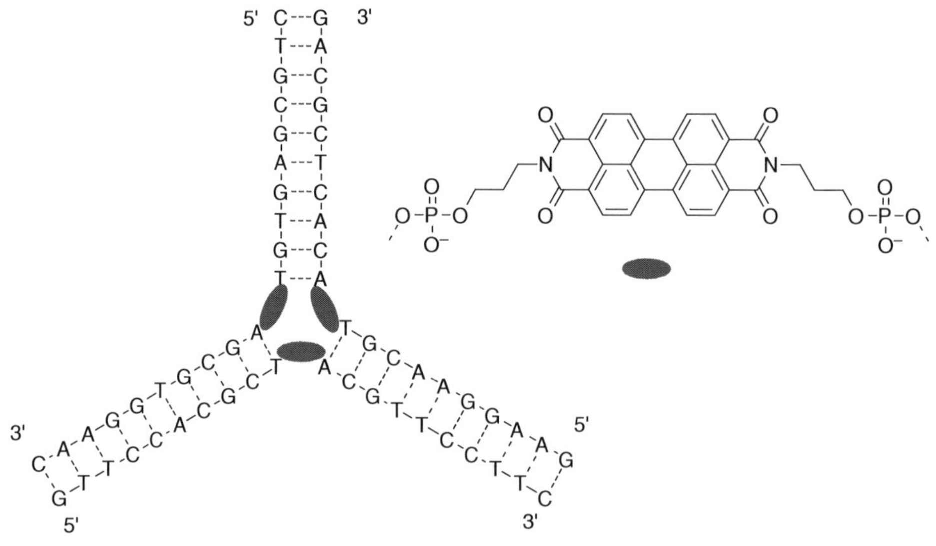

Figure 1.3.10 DNA three-way junctions containing PDI (a) at the 5′-end of the DNA strands or (b) in the branching point of DNA construct

defined distances and (semi)conducting materials [37r, 55]. The presence of PDI as an artificial chargeable island in DNA should enable electron transport via hopping processes. DNA hairpins containing PDI for intermolecular electron hopping experiments were synthesized. It has been demonstrated that electron hopping is possible among two to three π-stacked PDI moieties [37r]. Thus, PDI-modified DNA strands are interesting materials for the study of the conductivity of DNA nanoaggregates.

Furthermore, PDIs have been synthetically incorporated into DNA three-way and four-way junctions (3WJ and 4WJ, respectively). To this aim, two different strategies have been demonstrated [43]. 3WJs containing PDI in the branching point were designed by Häner and coworkers (Figure 1.3.10). In this case the tight aggregation of the PDI chromophores leads to the stabilization of the branched DNA construct [43b].

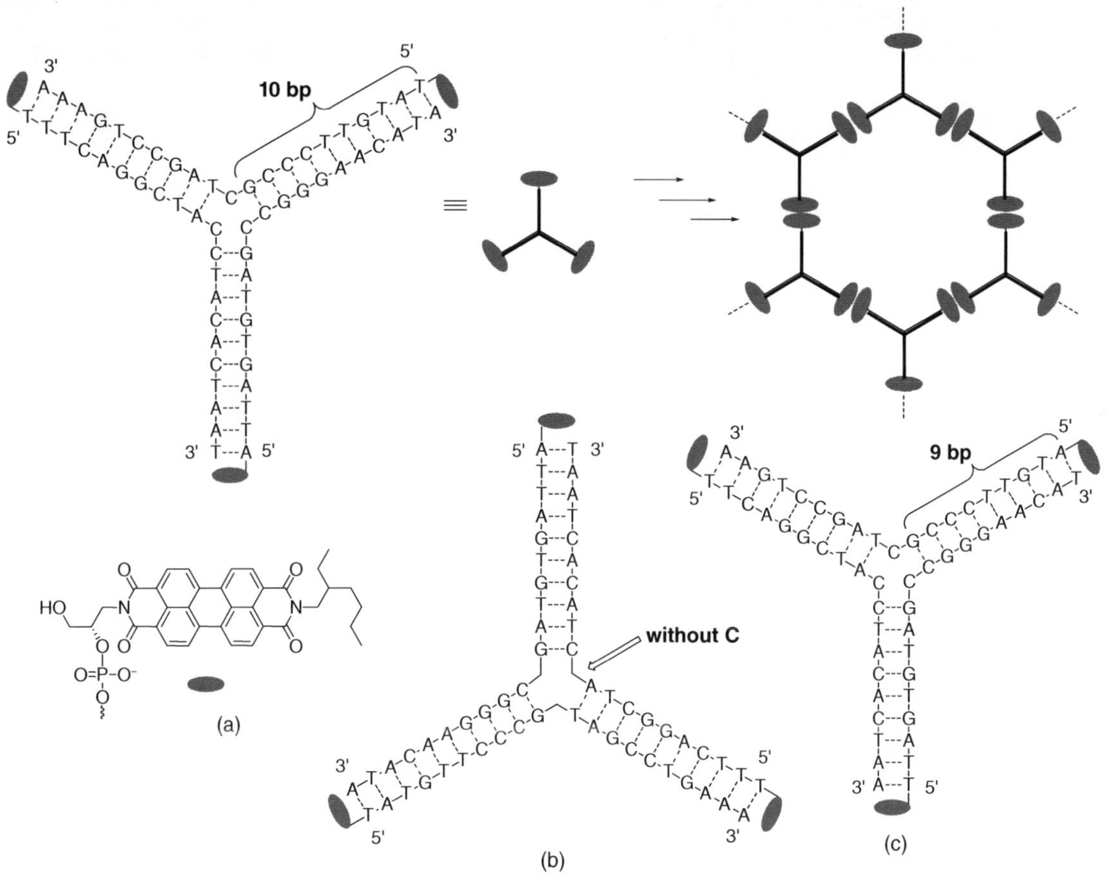

Figure 1.3.11 *Three different DNA 3WJs with PDI caps at the 5′-position of the oligonucleotides that differ in their sequence*

Wagenknecht *et al.* integrated PDI caps at the 5′-position of oligonucleotides that either hybridize to Y-shaped triple strands [56] (Figure 1.3.11a–c) or X-shaped quadruple strands (unpublished work) (Figure 1.3.12). 3WJs, as shown in Figure 1.3.11a, contain cytidine at the branching point as spacers to prevent steric hindrance as well as ten base pairs per arm of the Y-shaped hybrid. Assuming a helical twist of 36° per base pair in B-DNA, each arm consists of one complete turn of the double helix (360°). If they are dissolved together in aqueous buffer solution at concentrations higher than 5 μM, they aggregate spontaneously within minutes to form an insoluble red solid material [43a]. Shortening the DNA sequences by deletion of cytidine at the branching point (Figure 1.3.11b) leads to a stabilization of the Y-construct but decreases the critical concentration. In comparison, the removal of one base pair per arm (Figure 1.3.11c) induces a less twisted double helix, but even though the critical concentration decreases, it has no influence on the stability. From this result it can be assumed that the solubility of PDI-modified DNA assemblies (3WJs) is not independent of the helical twist or packing between the "DNA-arms" but of the hydrophilic DNA strand length. Furthermore, unpaired cytidines at the branching point in 3WJs are not necessary and their removal leads to a more stable construct.

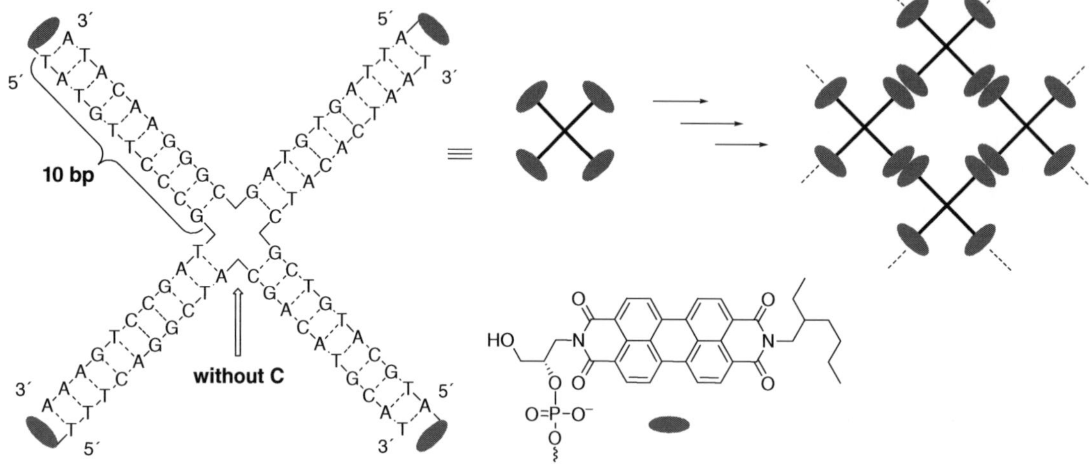

Figure 1.3.12 DNA 4WJs with PDI caps at the 5′-position of the oligonucleotides

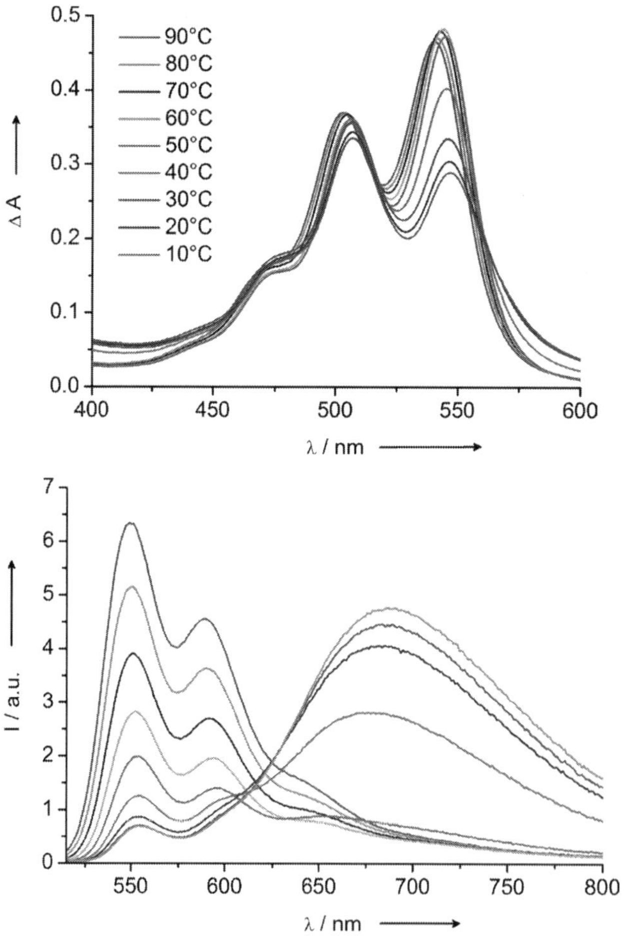

Figure 1.3.13 Representative temperature-dependent UV–Vis absorption spectra (left) and fluorescence spectra (right) for by the PDI-modified 3WJ of Figure 1.3.11 (See color figure in color plate section)

Additionally, PDI-mediated 4WJs have been designed based on four oligonucleotides containing two separate regions of ten bases in length, but which do not incorporate a spacer in the middle quadrangular DNA construct (Figure 1.3.12).

In all DNA constructs shown here, the PDI caps serve as "molecular glue" in the spontaneous assembly of larger DNA ensembles through the aggregation of the PDI chromophores (Figures 1.3.11 and 1.3.12). This yields a hypsochromic shift in the absorption and a bathochromic shift in the fluorescence as characteristic optical readouts (Figure 1.3.13). By thermal dehybridization of the DNA scaffold, the aggregation of the PDI aggregates can be destroyed and reassembled by reannealing of the DNA junctions. The formation of the DNA ensembles occurs without the application of enzymatic ligation and without the use of overhanging DNA as sticky ends [43a].

1.3.4 Conclusions

Metal–ion complexes and π–π interacting perylene diimides represent two important types of modifications that are able to introduce not only additional structural motifs to the purely hydrogen-bonded base pairing in DNA architectures but also have the potential to improve the electron transfer abilities of DNA. Our new C-nucleoside, which is structurally based on the hydroxyquinoline ligand, is able to form stable pairs in DNA both in the absence and in the presence of metal ions. Fluorescence studies with a donor–DNA–acceptor system indicate that photoinduced charge transfer processes across these metal-ion mediated base pairs in DNA occur more efficiently than over natural base pairs. The complexity of potential charge transfer pathways could possibly be increased by using the terpyridine 2'-deoxyuridine, since this DNA base modification would allow the charge to move from one DNA double strand to the other. On the other hand, with perylene diimides as covalently attached organic chromophores inside and outside DNA, new DNA architectures can be produced that are not purely formed by the hydrogen-bond assembly but also include π–π interactions. It is expected that the conductivity can also be improved compared with the natural DNA, due to the presence of the perylene diimide chromophores as artificial chargeable islands, between which electron transport occurs via a hopping process. Overall, this work further supports the idea of utilizing artificial charge carriers for the application of DNA-based architectures for molecular electronics.

References

[1] (a) H. Yang, K. L. Metera, and H. F. Sleiman, *Coord. Chem. Rev.*, 2010, **254**, 2403–2415; (b) T. J. Brandy, A. Brewer, J. R. Burns, G. Marth, T. N. Nguyen, and E. Stulz, *Chem. Soc. Rev.*, 2011, **40**, 138–148.

[2] P. Sa-Ardyen, A. V. Vologodskii, and N. C. Seeman, *Biophys. J.*, 2003, **84**, 3829–3837.

[3] H. T. Maune, S.-P. Han, R. D. Barish, M. Bockrath, W. A. Goddard III, P. W. K. Rothemund, and E. Winfree, *Nature Nanotechnol.*, 2009, **5**, 61–66.

[4] P. W. K. Rothemund, *Nature*, 2006, **440**, 297–302.

[5] D. Han, S. Pal, J. Nangreave, Z. Deng, Y. Liu, and H. Yan, *Science*, 2011, **332**, 342–346.

[6] A. Somoza, Angew. *Chem., Int. Ed.*, 2009, **48**, 9406–9408.

[7] Y. Ke, L. L. Ong, W. M. Shih, and P. Yin, *Science*, 2012, **338**, 1177–1183.

[8] E. S. Andersen, M. Dong, M. M. Nielsen, K. Jahn, R. Subramani, W. Mamdouh, M. M. Golas, B. Sander, H. Stark, C. L. P. Oliveira, J. S. Pedersen, V. Birkedal, F. Besenbacher, and K. V. Gothelf, J. Kjems, *Nature*, 2009, **459**, 73–76.

[9] (a) C. Prunkl, S. Berndl, C. Wanninger-Weiß, J. Barbaric, and H.-A. Wagenknecht, *Phys. Chem. Chem. Phys.*, 2010, 2, 32–43; (b) I. Vaya, T. Gustavsson, T. Douki, Y. Berlin, and D. Markovitsi, *J. Am. Chem. Soc.*, 2012, **134**, 11366–11368.

[10] X. Guo, A. A. Gorodetsky, J. Hone, J. K. Barton and C. Nuckolls, *Nature Nanotechnol.*, 2008, **3**, 163–167.

[11] W. Schmucker and H.-A. Wagenknecht, *Synlett*, 2012, **23**, 2435–2448.

[12] S. Katz, *Biochim. Biophys. Acta*, 1963, **68**, 240–253.

[13] P. Aich, S. L. Labiuk, L. W. Tari, L. J. T. Delbaere, W. J. Roesler, K. J. Falk, R. P. Steer, and J. S. Lee, *J. Mol. Biol.*, 1999, **294**, 477–485.

[14] (a) K. Tanaka and M. Shionoya, *J. Org. Chem.* 1999, **64**, 5002–5003; (b) M. Tasaka, K. Tanaka, and M. Shiro, M. Shionoya, *Supramol. Chem.*, 2001, **13**, 671–675.

[15] T. Ehrenschwender, A. Barth, H. Puchta, and H.-A. Wagenknecht, *Org. Biomol. Chem.*, 2012, **10**, 46–48.

[16] D. Mitra, N. D. Cesare, and H. F. Sleiman, *Angew. Chem., Int. Ed.*, 2004, **43**, 5804–5608.

[17] E. Meggers, P. L. Holland, W. B. Tolman, F. E. Romesberg, and P. G. Schultz, *J. Am. Chem. Soc.*, 2000, **122**, 10714–10715.

[18] K. Tanaka, A. Tengeiji, T. Kato, N. Toyama, and M. Shionoya, *Science*, 2003, **299**, 1212–1213.

[19] K. Tanaka, G. H. Clever, Y. Takezawa, Y. Yamada, C. Kaul, M. Shionoya, and T. Carell, *Nature Nanotechnol.*, 2006, **1**, 190–194.

[20] T. Ehrenschwender, W. Schmucker, C. Wellner, T. Augenstein, P. Carl, J. Harmer, F. Breher, and H.-A. Wagenknecht, *Chem. - Eur. J.*, 2013, **19**, 12547–12552.

[21] L. Zhang and E. Meggers, *J. Am. Chem. Soc.*, 2004, **127**, 74–75.

[22] S. H. Lee, J. H. Kim, and C. B. Park, *Chem. - Eur. J.*, 2013, **19**, 4392–4406.

[23] H. Yang, C. K. McLaughlin, F. A. Aldaye, G. D. Hamblin, A. Z. Rys, I. Rouiller, and H. F. Sleiman, *Nature Chem.*, 2009, **1**, 390–396.

[24] H. Yang and H. F. Sleiman, *Angew. Chem., Int. Ed.*, 2008, **27**, 2443–2446.

[25] (a) A. Wild, A. Winter, F. Schlutter, and U. S. Schubert, *Chem. Soc. Rev.*, 2011, **40**, 1459–1511; (b) E. C. Constable, *Chem. Soc. Rev.*, 2007, **36**, 246–253; (c) S. D. Cummings, *Coord. Chem. Rev.*, 2009, **253**, 449–478.

[26] K. M. Stewart and L. W. McLaughlin, *Chem. Commun.*, 2003, 2934–2935.

[27] S. Ghosh, I. Pignot-Paintrand, P. Dumy, and E. Defrancq, *Org. Biomol. Chem.*, 2009, **7**, 2729–2737.

[28] J. S. Choi, C. W. Kang, K. Jung, J. W. Yang, Y.-G. Kim, and H. Han, *J. Am. Chem. Soc.*, 2004, **126**, 8606–8607.

[29] M. Göritz and R. Krämer, *J. Am. Chem. Soc.*, 2005, **127**, 18016–18017.

[30] (a) J. R. Burns, J. Zekonyte, G. Siligardi, R. Hussain, and E. Stulz, *Molecules*, 2011, **16**, 4912–4922; (b) M. Kalek, A. S. Madsen, and J. Wengel, *J. Am. Chem. Soc.*, 2007, **129**, 9392–9400.

[31] L. Kalachova, R. Pohl, and M. Hocek, *Synthesis*, 2009, 105–112.

[32] R. H. E. Hudson and A. Ghorbani-Choghamarani, *Org. Biomol. Chem.*. 2007, **5**, 1845–1848.

[33] C. Brotschi and C. J. Leumann, *Angew. Chem., Int. Ed.*, 2003, **42**, 1655–1658.

[34] S. Hainke and O. Seitz, *Angew. Chem., Int. Ed.*, 2009, **48**, 8250–8253.

[35] K. H. W. Herbst, *Industrial Pigments: Production, Properties, Applications*, 3rd edn, Wiley-VCH Verlag GmbH, Weinheim, 2004.

[36] (a) S. B. C. Huang and S. R. Marder, *J. Org. Chem.*, 2011, **76**, 2386–2407; (b) H. Langhals, *Heterocycles*, 2005, **88**, 1309–1343.

[37] (a) A. F. X. Zhan, S. Barlow, T. J. Marks, M. A. Ratner, M. R. Wasielewski, and W. R. Marder, *Adv. Mater.*, 2011, **23**, 268–284; (b) M. S. F. Würthner, *Chem. Commun.*, 2011, **47**, 5109–5115; (c) M. A. Abdalla, J. Bayer, J. O. Rädler, and K. Müllen, *Angew. Chem.*, 2004, **116**, 3967–3970; (d) D. Baumstark and H.-A. Wagenknecht, *Chem. - Eur. J.*, 2008, **14**, 6640–6645; (e) S. Bevers, T. P. O´DEa, and L. W. McLaughlin, *J. Am. Chem. Soc.*, 1998, **120**, 11004–11005; (f) S. Bevers, S. Schutte, and L. W. McLaughlin, *J. Am. Chem. Soc.*, 2000, **122**, 5905–5915; (g) N. Bouquin, V. L. Malinovskii, and R. Häner, *Chem. Commun.*, 2008, 1974–1976; (h) N. Rahe, C. Rinn, and T. Carell, *Chem. Commun.*, 2003, 2120–2121; (i) C. Wagner and H.-A. Wagenknecht, *Org. Lett.*, 2006, **8**, 4191–4194; (j) W. Wang, W. Wan, H.-H. Zhou, S. Niu, and A. D. Q. Liu, *J. Am. Chem. Soc.*, 2003, **125**, 5248–5249; (k) Y. Zheng, H. Long, G. C. Schatz, and F. D. Lewis, *Chem. Commun.*, 2005, 4795–4797; (l) Y. Zheng, H. Long, G. C. Schatz, and F. D. Lewis, *Chem. Commun.*, 2006, 3830–3832; (m) T. A. Zeidan, R. Carmieli, R. F. Kelley, T. M. Wilson, F. D. Lewis, and M. R. Wasielewski, *J. Am. Chem. Soc.*, 2008, **130**, 13945–13955; (n) M. Hariharan, Y. Zheng, H. Long, T. A. Zeidan, G. C. Schatz, J. Vura-Weis, M. R. Wasielewski, X. Zuo, D. M. Tiede, and F. D. Lewis, *J. Am. Chem. Soc.*, 2009, **131**, 5920–5929; (o) B. A. Jones, A. Facchetti, M. R. Wasielewski, and T. J. Marks, *J. Am. Chem. Soc.*, 2007, **129**, 15259–15278; (p) F. Würthner, *Chem. Commun.*, 2004, 1564–1579; (q) D. Baumstark and H.-A. Wagenknecht, *Angew. Chem., Int. Ed.*, 2008, **47**, 2652–2654; (r) T. M. Wilson, T. A. Zeidan, M. Hariharan, F. D. Lewis, and M. R. Wasielewski, *Angew. Chem., Int. Ed.*, 2010, **49**, 2385–2388.

[38] (a) C. T. F. Würthner, S. Diele, and C. Tschierske, *Chem.- Eur. J.*, 2001, **7**, 2245–2253; (b) A. D. Q. Li, W. W. Wang, and L.-Q. Wang, *Chem.- Eur. J.*, 2003, **9**, 4594–4601; (c) J. Q. T. J. Tang, K. Mullen, and S. E. Webber, *Langmuir*, 2006, **22**, 7610–7616; (d) Y. C. L. Zang and J. S. Moore, *Acc. Chem. Res.*, 2008, **41**, 1596–1608; (e) E. S. J. Baram, N. Ben-Shitrit, A. Ustinov, H. Weissman, I. Pinkas, S. G. Wolf, and B. Rybtchinski, *J. Am. Chem. Soc.*, 2008, **130**, 14966–14967; (f) S. R. X. Zhang, M. M. Salfont-Sempere, and F. Würthner, *Nature Chem.*, 2009, **1**, 623–629; (g) T. F. A. D. Greef, M. M. J. Smulders, M. Wolffs, A. P. H. J. Schenning, R. P. Sijbesma, and E. W. Meijer, *Chem. Rev.*, 2009, **109**, 5687–5754; (h) P. P. Neelakandan, Z. Pan, M. Hariharan, Y. Zheng, H. Weissman, B. Rybtchinski, and F. D. Lewis, *J. Am. Chem. Soc.*, 2010, **132**, 15808–15813.

[39] X. Z. D. Görl and F. Würthner Angew. *Chem., Int. Ed.*, 2012, **51**, 2–23.

[40] A. V. Ustinov, V. V. Dubnyakova, and V. A. Korshun, *Tetrahedron Lett.*, 2008, **64**, 1467–1473.

[41] T. A. Zeidan, M. Hariharan, K. Siegmund, and F. D. Lewis, *Photochem. Photobiol. Sci.*, 2010, **9**, 916–922.

[42] R. Carmieli, T. A. Zeidan, R. F. Kelley, Q. Mi, F. D. Lewis, and M. R. Wasielewski, *Phys. Chem. A*, 2009, **113**, 4691–4700.

[43] (a) F. Menacher, V. Stepanenko, F. Würthner, and H.-A. Wagenknecht, *Chem. - Eur. J.*, 2011, **17**, 6683–6688; (b) M. Probst, D. Wenger, S. M. Biner, and R. Häner, *Org. Biomol. Chem.*, 2012, 755–759.

[44] D. Görl, X. Zhang, and F. Würthner, *Angew. Chem., Int. Ed.*, 2012, **51**, 6328–6348.

[45] H. Langhals, *Heterocycles*, 1995, **40**, 477–500.

[46] (a) G. Seybold, G. Wagenblast, *Dyes Pigm.*, 1989, **11**, 303–317; (b) G. Seybold and A. Stange, in *Ger. Pat, Vol. DE 35 45 004*,1987.

[47] (a) M. A. Abdalla, J. Bayer, J. O. Rädler, and K. Müllen, *Angew. Chem., Int. Ed.*, 2004, **43**, 3967–3970; (b) S. Bevers, S. Schutte, and L. W. McLaughlin, *J. Am. Chem. Soc.*, 2000, **122**, 5905–5915.

[48] F. Würthner, Z. Chen, V. Dehm, and V. Stepanenko, *Chem. Commun.*, 2006, 1188–1190.

[49] S. Tyagi and F. R. Kramer, *Nature Biotechnol.*, 1996, **14**, 303–308.

[50] F. Menacher and H.-A. Wagenknecht, *Photochem. Photobiol. Sci.*, 2011, **10**, 1275–1278.

[51] (a) S. M. Langenegger and R. Häner, *Bioor. Med. Chem. Lett.*, 2006, **16**, 5062–5065; (b) S. M. Langenegger and R. Häner, *Chem. Commun.*, 2004, 2792–2793; (c) V. L. Malinovskii, F. Samain, and R. Häner, *Angew. Chem., Int. Ed.*, 2007, **46**, 4464–4467; (d) V. L. Malinovskii, D. Wenger, and R. Häner, *Chem. Soc. Rev.*, 2010, **39**, 410–422.

[52] (a) R. Häner, S. M. Biner, S. M. Langenegger, T. Meng, and V. L. Malinovskii, *Angew. Chem., Int. Ed.*, 2010, **49**, 1227–1230; (b) H. Bittermann, D. Siegemund, V. L. Malinovskii, and R. Häner, *J. Am. Chem. Soc.*, 2008, **130**, 15285–15287; (c) S.-N. Uno, C. Dohno, H. Bittermann, V. L. Malinovskii, R. Häner, and K. Nakatani, *Angew. Chem., Int. Ed.*, 2009, **48**, 7362–7365.

[53] F. Würthner, V. Stepanenko, Z. Chen, C. R. Saha-Möller, N. Kocher, and D. Stalke, *J. Org. Chem.*, 2004, **69**, 7933–7939.

[54] F. Menacher and H.-A. Wagenknecht, *Eur. J. Org. Chem.*, 2011, 4564–4570.

[55] (a) T. Carell, C. Behrens, and J. Gierlich, *Org. Biomol. Chem.* 2003, **1**, 2221–2228; (b) R. Varghese and H.-A. Wagenknecht, *Chem. Commun.*, 2009, 2615–2624; (c) J. Wengel, *Org. Biomol. Chem.*, 2004, **2**, 277–280.

[56] C. Stubinitzky, A. Bijeljanin, L. Antusch, D. Ebeling, H. Hölscher, and H.-A. Wagenknecht, *Chem. - Eur. J.*, 2014, **20**, 12009–12014.

1.4
DNA with Metal-Mediated Base Pairs

Indranil Sinha and Jens Müller

Institute for Inorganic and Analytical Chemistry, University of Münster, Münster, Germany

1.4.1 Introduction

DNA carries the genetic information in all organisms. More than 60 years ago, Watson and Crick first proposed the double helical structure of DNA. Since then, its use as a sequence-specific self-assembling building block has become a promising route for the construction of bio-inspired nanoarchitecture. The backbone of the DNA helix comprises negatively charged phosphate moieties, rendering the nucleic acid extremely water soluble. The interior of the duplex is made up of a stack of aromatic nucleobases, held together by hydrogen bonds. As a result of modern technology that allows routine synthesis of small oligonucleotides via solid-phase automated synthesis, modifications can easily be introduced into the nucleosides or the backbone. Longer oligonucleotide strands are accessible through ligation or by applying the polymerase chain reaction.

The development of artificial nucleobases has increased nucleic acid functionality and is expected to lead to an expansion of the genetic alphabet [1]. Moreover, the unnatural bases represent attractive building blocks for the self-assembly of nanostructured objects (see also Part V of this book). The canonical DNA duplex has two major stabilising parameters, namely, hydrogen bonds between complementary bases and π-stacking interactions between neighbouring bases. Complementary nucleobases may also pair via coordinate bonds to a metal ion located in the centre of the duplex if ligand-based nucleosides are present. The replacement of natural nucleosides by suitable ligands hence leads to the formation of metal-mediated base pairs (Figure 1.4.1) [2]. Eventually, a site-specific functionalisation of these biomolecules with transition metal ions is achieved, enabling a significant number of possible applications [3]. To date, a considerable number of metal-mediated base pairs have been reported not only for DNA but also for other nucleic acids and nucleic acid derivatives such as RNA, PNA and GNA [4–6].

DNA in Supramolecular Chemistry and Nanotechnology, First Edition.
Edited by Eugen Stulz and Guido H. Clever.

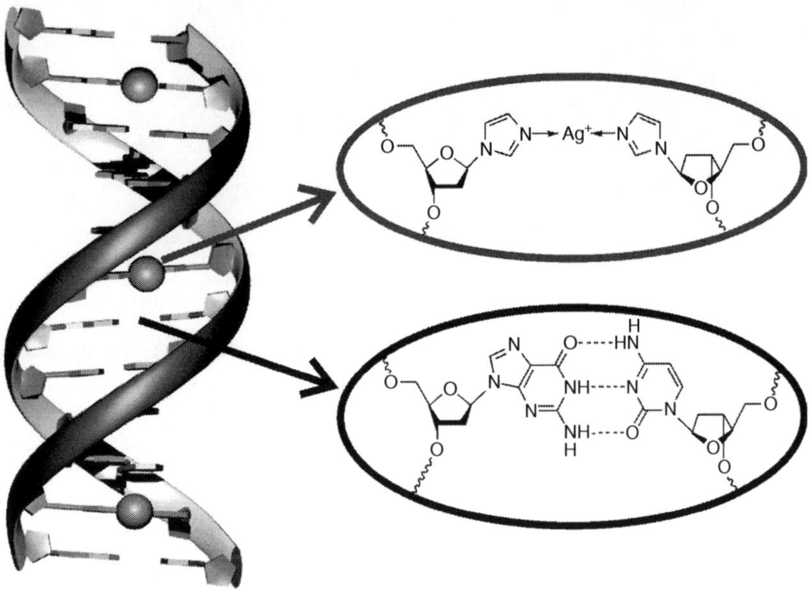

Figure 1.4.1 Schematic representation of a B-DNA duplex with canonical base pairs and artificial metal-mediated base pairs. Adapted from [2]. Used with permission. Copyright © 2008 WILEY-VCH Verlag GmbH & Co. KGaA, Weinheim

1.4.2 Metal-mediated base pairs with natural nucleobases

1.4.2.1 Pyrimidines

The first metal-mediated base pairs were proposed in the 1960s, when the interaction of Hg(II) with polynucleotides was studied [7, 8]. It took several years until the proposed formation of **T**–Hg–**T** base pairs (Scheme 1.4.1a) was corroborated by single crystal X-ray diffraction data and ^{1}H NMR (nuclear magnetic resonance) spectroscopy [9, 10]. Since then, the **T**–Hg–**T** base pairs have become the most intensely investigated and frequently applied metal-mediated base pairs [11]. They are particularly useful in oligonucleotide-based sensors for the toxic Hg(II) ions [3]. It was confirmed by various NMR spectroscopic methods that in **T**–Hg–**T** base pairs the Hg(II) ion is indeed bridging two deprotonated thymine residues on complementary positions [12, 13]. Hg(II) specifically binds to a **T**:**T** mispair with an association constant in the order of 10^5 M^{-1} [14]. As a result of the formation of **T**–Hg–**T** base pairs, thymine residues can even be incorporated into oligonucleotides by DNA polymerases [15]. Hence, it is conceivable that metal-mediated base pairs are directly involved in the mutagenic activity of Hg(II) ions.

In RNA, the thymine is replaced by uracil, hence the formation of **U**–Hg–**U** base pairs has also been investigated. It was confirmed by NMR spectroscopy that these base pairs can be incorporated into RNA double helices [4]. However, probably because of the different conformation of RNA duplexes compared with DNA duplexes, **U**–Hg–**U** base pairs seem to be slightly less stable than **T**–Hg–**T** base pairs [4].

Cytosine preferentially binds to Ag(I) ions under chloride-free buffer conditions. Hence, if two complementary strands contain two cytosine moieties in opposing positions, **C**–Ag–**C** base pairs can be formed. Different orientations are possible depending on the conformation of the double helix. In B-DNA with canonical Watson–Crick base pairs, a *cisoid* orientation of the nucleobases in **C**–Ag–**C** is geometrically enforced (Scheme 1.4.1b) [16]. In parallel-stranded DNA, a *transoid* orientation with one additional hydrogen bond is

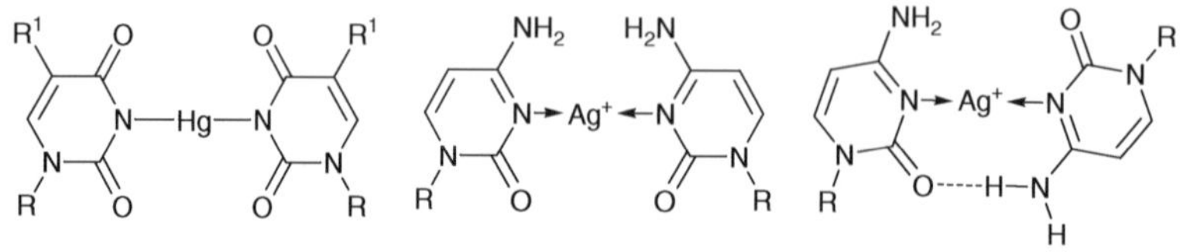

Scheme 1.4.1 *Metal-mediated base pairs formed from entirely natural nucleobases. (a)* ***T*****–*Hg*–*****T*** *($R^1 = CH_3$) [12],* ***U*****–*Hg*–*****U*** *($R^1 = H$) [4]; (b) cisoid* **C–*Ag*–C** *[16]; and (c) transoid* **C–*Ag*–C** *[18]*

feasible (Scheme 1.4.1c) [17]. Computational studies showed that the **C**–Ag–**C** base pair with a *transoid* orientation is more stable (by about 7 kcal mol^{-1}) than that with a *cisoid* orientation [18]. Nonetheless, the Watson–Crick context imposed in a regular B-DNA duplex is expected to compensate for the slightly lower stability of the *cisoid* orientation. The association constant for the specific binding of Ag(I) to a **C**:**C** mispair amounts to about 10^5 M^{-1} [19], hence is comparable to that of a **T**–Hg–**T** base pair.

1.4.2.2 Purines

Natural purine nucleobases have not been investigated in terms of their applicability in metal-mediated base pairs, which is probably a result of the multitude of possible monodentate metal-binding sites. However, if one considers the interstrand cross-links studied in the context of DNA metalation with *trans*-$[PtCl_2(NH_3)_2]$ as metal-mediated base pairs, then a number of purine-containing metal-mediated base pairs are also known [20, 21]. Moreover, soaking experiments performed during a crystallographic study of RNA duplexes led to the (unintentional) observation of a **G**–Au–**C** base pair [22]. A Hoogsteen-type **G**–Cu–**C** base pair has also been suggested to exist in a left-handed DNA duplex [23].

1.4.3 Metal-mediated base pairs with artificial nucleobases

1.4.3.1 Individual metal-mediated base pairs

The first example of a metal-mediated base pair with an artificial nucleoside involved *o*-phenylenediamine as the base surrogate. In the presence of $Pd(NO_3)_2$, a Pd(II)-mediated base pair could be detected in solution [24]. The first incorporation of a metal-mediated base pair into a nucleic acid duplex was achieved by combining the tridentate pyridine-2,6-dicarboxylate (**Dic**) nucleoside and the monodentate pyridine nucleoside (**³Py**) opposite to each other in two complementary strands (Scheme 1.4.2a) [25]. The resulting **Dic:³Py** mispair selectively binds Cu(II) by providing a [3+1] coordination environment. The Cu(II) ion is coordinated in a square planar fashion by the two artificial nucleobases. The coordination sphere is completed by two additional oxygen donor atoms above and below the coordination plane, provided by neighbouring natural nucleosides [26]. This arrangement leads to a Z-type conformation for the DNA duplex (see Chart 1.4.1a for the experimentally determined duplex structure), indicating that the presence of metal-mediated base pairs can significantly influence the overall nucleic acid conformation. Based on the **Dic**–Cu–**³Py** pair, other closely related derivatives (**Dam:³Py**, **Dit:³Py**, Scheme 1.4.2a,b) have also been investigated [27], leading to a family of ligands with an O,N,O-, N,N,N- or S,N,S-donor system. In particular the **Dit** nucleobase was found to strongly stabilise Ag(I)-containing base pairs, such as **Dit**–Ag–**³Py** and **Dit**–Ag–**Dit** [28]. In a conceptually similar [3+1] system, dipicolylamine (**Dipic**, Scheme 1.4.2c) was applied to generate Ag(I)-mediated base pairs, with nucleosides based on imidazole (**Imi**), 1,2,4-triazole (**Tri**) or tetrazole (**Tet**) acting

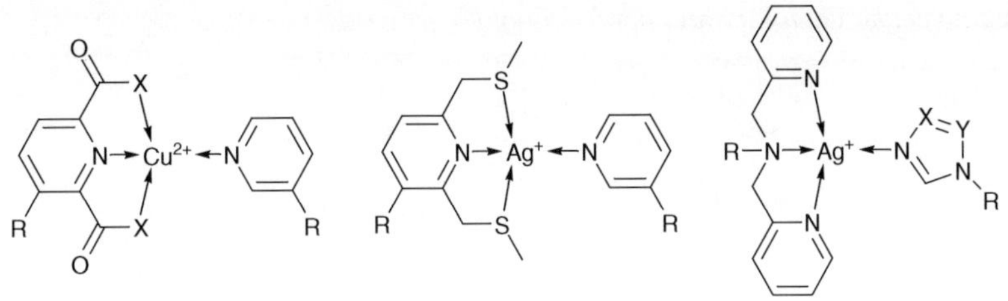

(a) X = O (**Dic**); X = NH_2 (**Dam**)

(b)

(c) X = Y = CH (**Imi**); X = CH, Y = N (**Tri**); X = Y = N (**Tet**)

Scheme 1.4.2 *Selected metal-mediated base pairs with a [3+1] coordination geometry (R = nucleic acid backbone). (a)* ***Dic****–Cu–*^{3}***Py*** *and* ***Dam****–Cu–*^{3}***Py*** *[25, 27]; (b)* ***Dit****–Ag–*^{3}***Py*** *[28]; and (c)* ***Dipic****–Ag–****Imi****,* ***Dipic****–Ag–****Tri*** *and* ***Dipic****–Ag–****Tet*** *[29]*

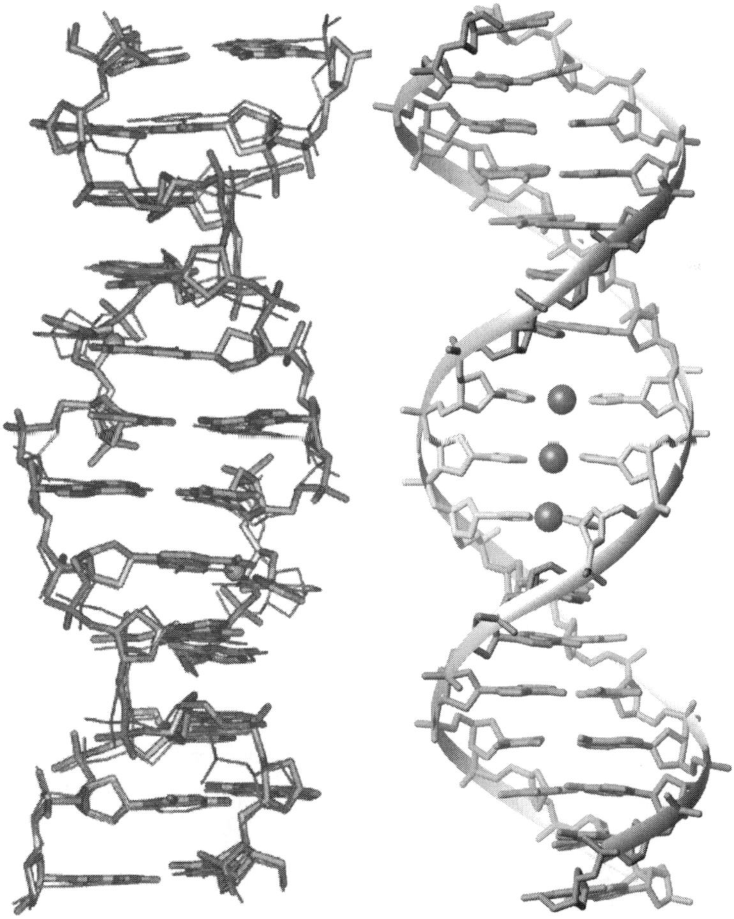

(a) (b)

Chart 1.4.1 *Experimental structures of two DNA double helices with metal-mediated base pairs. (a) Solid-state structure of Z-type DNA duplex with two* ***Dic****–Cu–*^{3}***Py*** *base pairs. Superimposed in blue is the structure of ideal Z-DNA. Reprinted with permission from [26]. Copyright 2001, American Chemical Society. (b) Solution structure of B-type DNA duplex with three consecutive* ***Imi****–Ag–****Imi*** *base pairs. Reproduced with permission from [54]. Copyright © 2010, Rights Managed by Nature Publishing (See colour figure in colour plate section)*

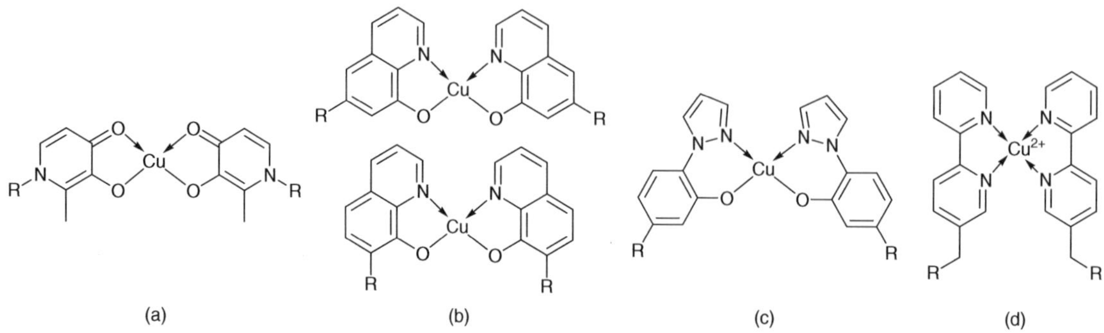

Scheme 1.4.3 *Selected metal-mediated base pairs with a [2+2] coordination environment (R = nucleic acid backbone). (a)* ***H*****–*****Cu*****–*****H*** *[30]; (b)* ^{6}***HQ*****–*****Cu*****–**^{6}***HQ*** *(top) [32] and* ^{7}***HQ*****–*****Cu*****–**^{7}***HQ*** *(bottom) [33]; (c)* ***PzP*****–*****Cu*****–*****PzP*** *[34]; and (d)* ^{5Me}***Bipy*****–*****Cu*****–**^{5Me}***Bipy*** *[35]*

as the complementary monodentate ligand [29]. The thermal stability of these base pairs is controlled by the identity of the monodentate nucleoside, with the most basic nucleobase (**Imi**) giving rise to the highest stabilisation.

Another prominent example of a metal-mediated base pair involves the use of the ligand 3-hydroxy-2-methyl-4-pyridone (**H**, Scheme 1.4.3a). An **H**:**H** mispair provides a [2+2] coordination environment for transition metal ions. It is significantly stabilised in the presence of Cu(II), as evidenced by an increase in the melting temperature of about 13 °C [30]. Analogously, mercaptopyridone and hydroxypyridinethione were found to efficiently incorporate soft transition metal ions such as Pd(II) and Pt(II), albeit with the monomeric nucleosides only and not yet in an oligonucleotide context [31]. Other metal-mediated base pairs with a [2+2] coordination environment have also been reported, for example, based on the ligands 8-hydroxyquinoline (^{6}**HQ** or ^{7}**HQ**, Scheme 1.4.3b) [32, 33], 2-pyrazolylphenol (**PzP**, Scheme 1.4.3c) [34] and 5-methyl-2,2′-bipyridine (^{5Me}**Bipy**, Scheme 1.4.3d) [35]. It is interesting to note that the ^{6}**HQ**:^{6}**HQ** mispair in particular is stabilised considerably by complexing Cu(II), leading to the formation of ^{6}**HQ**–Cu–^{6}**HQ**. The thermal stabilisation is even higher when using the GNA analogue of ^{6}**HQ** [32]. Recent experiments have shown that DNA duplexes with ^{7}**HQ**–Cu–^{7}**HQ** base pairs transfer electric charges more efficiently than unmodified DNA with canonical base pairs [33], thereby presenting interesting applications. The use of a 2,2′-bipyridyl moiety as an artificial nucleobase shows the importance of steric considerations: on using ^{5Me}**Bipy**, the ligand is attached to the sugar by means of a methylene linker, giving rise to a stable ^{5Me}**Bipy**–Cu–^{5Me}**Bipy** base pair [35]. Without the methylene linker, that is, with a direct attachment of the bipyridyl moiety to the sugar (**Bipy**), no stable metal-mediated base pairs are formed [36]. Apparently, ^{5Me}**Bipy** is more flexible than **Bipy** and hence is able to adopt a conformation capable of binding metal ions.

The natural nucleobases can also be modified in a way that increases their affinity towards transition metal ions. Accordingly, various derivatives of purine and pyrimidine nucleosides have been reported with pendant ligands. Scheme 1.4.4 gives an overview of the resulting artificial nucleosides. The nucleobase ^{6py}**Pur** with a pyridyl moiety in the 6-position of the purine ring selectively forms ^{6py}**Pur**–Ni–^{6py}**Pur** base pairs (Scheme 1.4.4a) [37]. Analogously, the 4-substituted pyrimidone ^{4py}**Pyr** (Scheme 1.4.4b) forms stable homo base pairs in the presence of Ni(II) [38]. It is interesting to note that despite a more or less identical steric requirement compared with the **Bipy** nucleobase mentioned earlier, stable metal-mediated base pairs are formed with ^{4py}**Pyr**. This indicates that not only steric but also the electronic properties of the nucleosides are important in determining whether metal-mediated base pairs are formed. Upon attaching two pyridyl moieties or one bipyridyl entity to a purine or pyrimidine, tridentate artificial nucleobases can be generated. These can be applied to create metal-mediated base pairs of the [3+1] type. Variation of the monodentate ligand

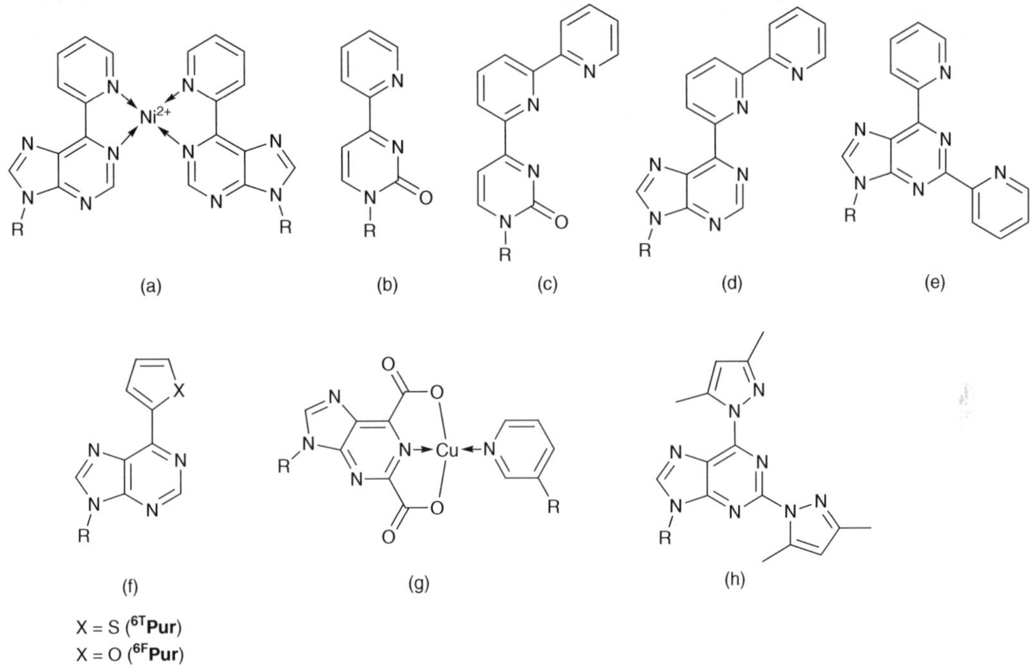

Scheme 1.4.4 *Artificial nucleobases and base pairs derived from purine or pyrimidine (R = nucleic acid backbone). (a)* ^{6py}***Pur***–*Ni*–^{6py}***Pur*** *[37]; (b)* ^{4py}***Pyr*** *[38]; (c)* ^{4bipy}***Pyr*** *[39]; (d)* ^{6bipy}***Pur*** *[40]; (e)* ^{2,6py}***Pur*** *[41]; (f)* ^{6T}***Pur*** *and* ^{6F}***Pur*** *[41]; (g)* ^{2,6dic}***Pur***–*Cu*–^{3}***Py*** *[42]; and (h)* ^{2,6pz}***Pur*** *[43]*

opposite to ^{4bipy}**Pyr** (Scheme 1.4.4c), ^{6bipy}**Pur** (Scheme 1.4.4d) or ^{2,6py}**Pur** (Scheme 1.4.4e) has been achieved by changing the glycosylation site of the pyridine moiety (forming ^{3}**Py** or ^{4}**Py**, respectively). Ag(I)-mediated base pairs were observed when ^{4}**Py** was complemented by ^{4bipy}**Pyr** or ^{6bipy}**Pur** [39, 40]. In these cases, geometrical considerations rule out the possibility of forming stable base pairs with ^{3}**Py**. The latter nucleoside is capable of engaging in metal-mediated base pairing with ^{2,6py}**Pur** [40].

These examples nicely demonstrate the importance of a steric fit of the complementary artificial nucleobases. In addition to these metal-mediated base pairs with only nitrogen donor atoms, two purine derivatives with a pendant thienyl (^{6T}**Pur**) or furyl (^{6F}**Pur**) moiety (Scheme 1.4.4f) have been reported. However, no conclusive evidence has been provided yet that the resulting nucleosides support the formation of stable metal-mediated base pairs [41]. The ^{2,6dic}**Pur**–Cu–^{3}**Py** base pair with a purine moiety bearing two carboxylate substituents (Scheme 1.4.4g) has been shown to be processable by DNA polymerases [42]. Hence, metal-mediated base pairs comprising purely artificial nucleosides are also compatible with DNA polymerases. Other substituents on the purine moiety, such as dimethylpyrazole, have also been used successfully in the generation of purine-derived ligands for metal-mediated base pairs (^{2,6pz}**Pur**, Scheme 1.4.4h) [43]. However, the artificial nucleobase ^{2,6pz}**Pur** is not very selective with respect to its complementary nucleobase, as it can pair with cytosine, uracil, adenine and thymine, depending on the position of the base pair within the oligonucleotide sequence.

An Ag(I)-mediated base pair with the unusual [2+1] coordination environment has been reported with the bidentate artificial nucleobase 4-[(1*H*-1,2,4-triazol-1-yl)methyl]-1*H*-1,2,3-triazole **TriTri**, comprising two triazole moieties connected via a methylene bridge (Scheme 1.4.5a) [44]. Hence, tricoordinate metal ions can also be efficiently incorporated into metal-mediated base pairs.

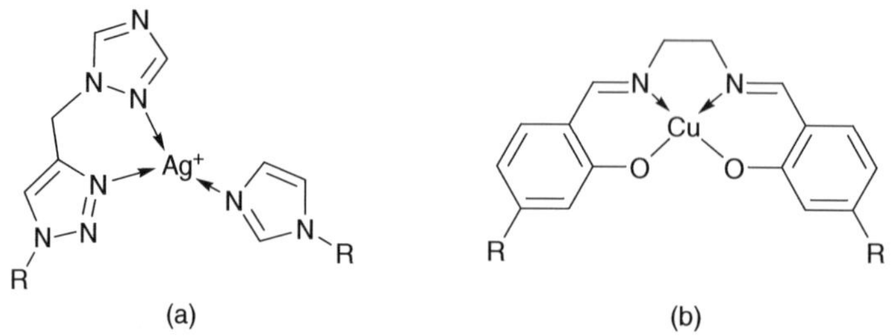

Scheme 1.4.5 *Metal-mediated base pairs with unusual coordination patterns (R = nucleic acid backbone). (a) [2+1] coordination environment brought about by **TriTri** and **Imi** [44]; and (b) Cu(II)-mediated **Salen** base pair with an additional covalent linkage between the complementary DNA strands [45]*

A somewhat different concept of metal-mediated base pairing has been developed based on the well-known salen ligand. The corresponding metal-mediated base pair contains a cross-linking ethylenediamine moiety, resulting in the formation of highly stable **Salen** base pairs [45]. The formation of the metal-mediated salen base pair in DNA is a two-step process. Firstly, the Schiff base is formed from ethylenediamine and two salicylaldehyde nucleosides on the complementary DNA strands. Secondly, the metal ion is added, to be coordinated by the chelating ligand. The most stable metal-mediated base pairs are formed with Cu(II), but other cations such as Mn(III), Fe(II) and Ni(II) can also be bound [46]. The importance of the additional covalent linkage via the ethylene bridge can be estimated from a comparison of various melting temperatures: in the absence of ethylenediamine, the addition of Cu(II) stabilises the mispair of two salicylaldehyde nucleosides by 15.0 °C. In the presence of ethylenediamine *and* Cu(II), that is, upon formation of the metal-mediated **Salen** base pair, a stabilisation of 42.5 °C is observed [45]. This represents the highest duplex stabilisation ever observed for a metal-mediated base pair. Interestingly, this base pair is fully orthogonal to the canonical base pairs, and it can also be replicated and amplified by a polymerase [47].

1.4.3.2 Stacks of metal-mediated base pairs

The incorporation of a series of consecutive metal-mediated base pairs into a nucleic acid duplex is of particular interest, as an interaction of the neighbouring transition metal ions is expected to lead to interesting new physical properties of the nucleic acid. Accordingly, various arrays of metal-mediated base pairs have been reported to date. One of the most intensely investigated base pairs in this context is **H**–Cu–**H** (Scheme 1.4.3a). In the course of the very first study on arrayed metal-mediated base pairs, several oligonucleotide sequences were designed, and the number of **H** moieties in the centre of the duplex was varied from 1 to 5 [48]. In all five duplexes, all possible **H**–Cu–**H** pairs were formed. The unpaired *d* electrons on the Cu(II) ions were found to couple ferromagnetically with one another in the stacked motifs. Hence, the Cu(II)-modified DNA may be considered as a self-assembling molecular magnet. A computational study nicely explains the ferromagnetic coupling by showing the presence of Cu–O bonding interactions between two neighbouring **H**–Cu–**H** base pairs [49]. DNA duplexes with **H**–Cu–**H** base pairs have also been investigated in the context of charge transfer through DNA: single-molecule measurements indicated that the molecular conductance through DNA with three contiguous **H**–Cu–**H** pairs tends to be larger than that through DNA with only one **H**–Cu–**H** pair [50]. Consequently, metal-mediated base pairs are promising routes for the improvement of the intrinsic charge transfer properties of DNA.

The **Salen** base pairs have also been used extensively in arrays of metal-mediated base pairs. For the first time, an entire helical turn of metal-mediated base pairs was created by introducing ten contiguous Cu(II)-mediated **Salen** base pairs (Scheme 1.4.5b) into a DNA duplex [51]. Interestingly, an antiferromagnetic

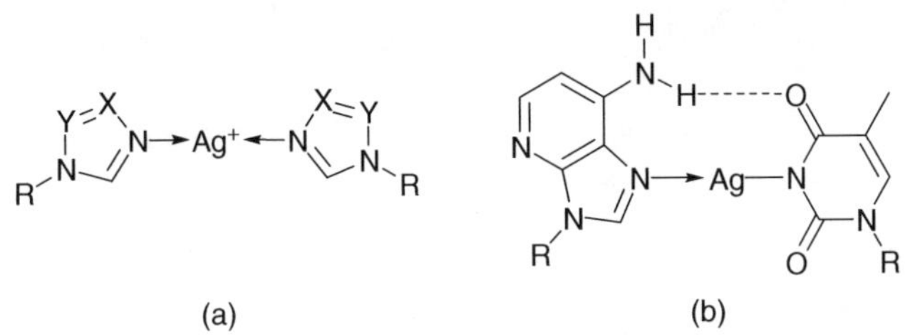

X = Y = CH (**Imi**)
X = CH, Y = N (**Tri**)
X = Y = N (**Tet**)

Scheme 1.4.6 *Selected metal-mediated base pairs introduced into DNA with contiguous metal-mediated base pairs (R = nucleic acid backbone). (a) Ag(I)-mediated azole base pairs (e.g.* ***Imi****–Ag–****Imi****) [53, 54]; and (b)* ^{1D}***A****–Ag–****T*** *[58]*

coupling of unpaired *d* electrons of the Cu(II) ions was observed for these duplexes [52], contrasting the observation for the duplexes involving **H**–Cu–**H** base pairs. Again, the antiferromagnetic coupling was confirmed in a computational study [49], which showed no strong coupling between neighbouring Cu(II) ions.

The first structural insight into B-DNA with contiguous metal-mediated base pairs was achieved with **Imi**–Ag–**Imi** base pairs. **Imi** nucleosides belong to the family of azole nucleosides, capable of binding linearly coordinating metal ions in a linear fashion (Scheme 1.4.6a) [53]. An NMR-based solution structure of a DNA duplex with three neighbouring **Imi**–Ag–**Imi** base pairs proved for the first time that metal-mediated base pairing is compatible with B-type DNA (Chart 1.4.1b) [54]. In this duplex, the Ag(I) ions are arrayed along the helical axis almost like pearls on a string (Figure 1.4.2a). The binding of the Ag(I) ions to the **Imi** nucleoside was unambiguously derived via the direct observation of a $^{1}J(^{15}N,^{107/109}Ag)$ coupling of 86 Hz (Figure 1.4.2b). A re-refined structure including QM/MM-derived constraints for the intermetallic distances shows an Ag···Ag distance of 3.45(2) Å, indicating the possibility of weak argentophilic interactions [55]. In this context, a thermodynamic study showed that neighbouring **Imi**–Ag–**Imi** base pairs are formed cooperatively, with the binding constant for the binding of the second Ag(I) ion being more than 20-fold larger than that for the first Ag(I) ion [56]. The stacking of metal-mediated azole base pairs is not limited to **Imi**–Ag–**Imi** base pairs only. For example, three contiguous **Tri**–Ag–**Tri** have also been arrayed in the centre of a DNA duplex [57].

The longest stack of contiguous metal-mediated base pairs has been established applying 1-deazaadenine ^{1D}**A** (Scheme 1.4.6b). The ^{1D}**A**–Ag–**T** base pair, also involving a deprotonated thymine residue, is mediated not only by coordinate bonds but also by an additional hydrogen bond [58]. A duplex with a continuous stack of 19 metal-mediated base pairs was created with the ^{1D}**A**–Ag–**T** base pair. The reversible self-assembly of this duplex is possible because of the moderate stability of the ^{1D}**A**–Ag–**T** base pair. One such pair contributes approximately 2 °C to the melting temperature of the duplex, similar to the contribution of the natural **A**:**T** pair.

Combinations of different metal-mediated base pairs within the same DNA duplex are possible, too. By exploiting the preferential binding of two metal ions to two different nucleosides, heterogeneous arrays of the two metals were obtained by either combining **T**–Hg–**T** and Cu(II)-mediated **Salen** base pairs, or by combining ^{3}**Py**–Hg–^{3}**Py** and **H**–Cu–**H** base pairs [59].

1.4.3.3 Doubly metalated base pairs

The metalation of DNA is not just restricted to one metal ion per base pair. Accordingly, a few doubly metal-mediated base pairs have been established. One prominent example involves 1,3-dideazaadenine (^{1,3D}**A**) and thymine (Scheme 1.4.7a). This base pair has also been applied in the sole example of duplexes

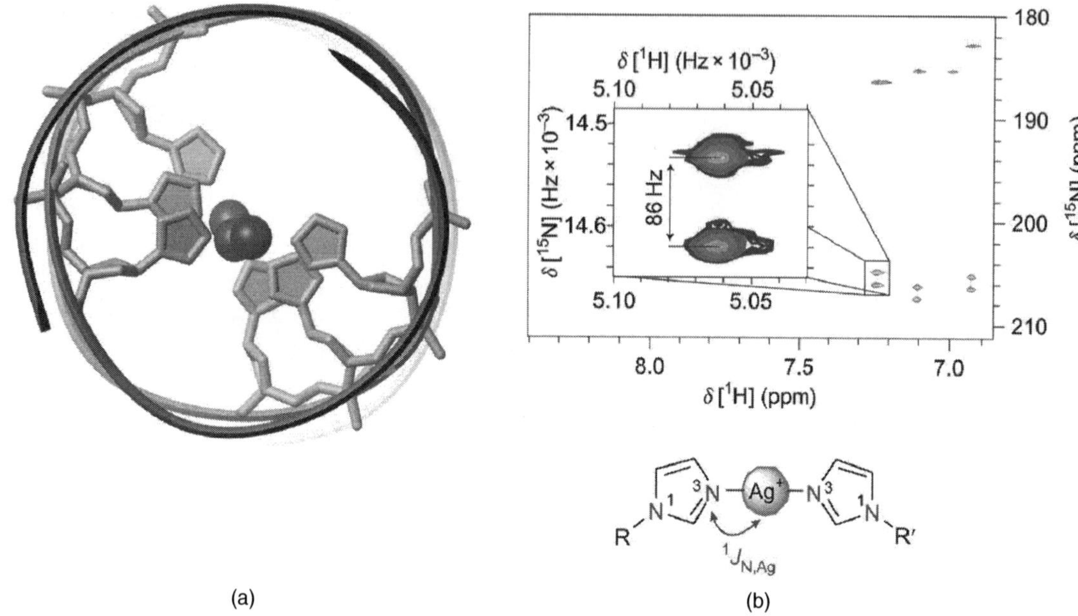

Figure 1.4.2 *(a) View along the helical axis of a DNA duplex with three consecutive **Imi**–Ag–**Imi** base pairs, showing the alignment of the metal ions along the helical axis; and (b) section of the [^{1}H,^{15}N] HSQC spectrum illustrating the $^{1}J(^{15}N,^{107/109}Ag)$ coupling of 86 Hz. Reproduced with permission from [54]. Copyright © 2010, Rights Managed by Nature Publishing*

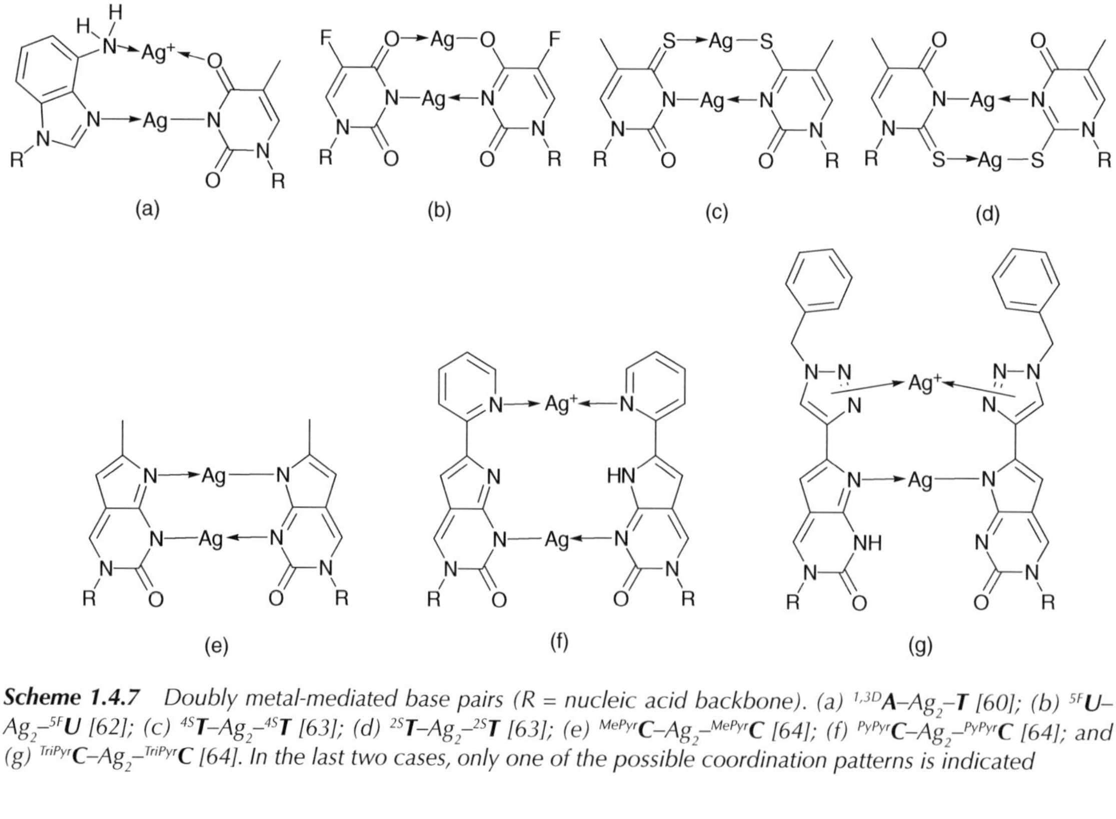

Scheme 1.4.7 *Doubly metal-mediated base pairs (R = nucleic acid backbone). (a) ^{1,3D}**A**–Ag$_2$–**T** [60]; (b) ^{5F}**U**–Ag$_2$–^{5F}**U** [62]; (c) ^{4S}**T**–Ag$_2$–^{4S}**T** [63]; (d) ^{2S}**T**–Ag$_2$–^{2S}**T** [63]; (e) ^{MePyr}**C**–Ag$_2$–^{MePyr}**C** [64]; (f) ^{PyPyr}**C**–Ag$_2$–^{PyPyr}**C** [64]; and (g) ^{TriPyr}**C**–Ag$_2$–^{TriPyr}**C** [64]. In the last two cases, only one of the possible coordination patterns is indicated*

comprising metal-mediated base pairs only [60]. DFT calculations revealed that the Ag···Ag distance within the base pair amounts to 2.88 Å and that the resulting argentophilic interaction contributes about 16 kcal mol^{-1} to the overall stability of the doubly metal-mediated base pair [60]. In an independent study, the double metalation was also confirmed for the individual model nucleobases [61].

Derivatives of natural nucleobases are also applicable for the formation of doubly metal-mediated base pairs. Examples include 5-fluorouracil (^{5F}**U**, Scheme 1.4.7b) [62], 4-thiothymine (^{4S}**T**, Scheme 1.4.7c) [63], 2-thiothymine (^{2S}**T**, Scheme 1.4.7d) [63], 6-methylpyrrolocytosine (^{MePyr}**C**, Scheme 1.4.7e) [64], 6-(2-pyridyl)pyrrolocytosine (^{PyPyr}**C**, Scheme1.4.7f) [64] and 6-(1-benzyl-1*H*-1,2,3-triazol-5-yl)pyrrolocytosine (^{triPyr}**C**, Scheme 1.4.7g) [64]. All their homo base pairs can bind two Ag(I) ions. It is interesting to note that the thermal stabilisation of duplexes comprising pyrrolocytosine derivatives depends on the nature of the substituent. Within otherwise identical sequences, the melting temperature increases by 18.5, 21.5 and 36.0 °C upon the formation of ^{MePyr}**C**–Ag_2–^{MePyr}**C**, ^{PyPyr}**C**–Ag_2–^{PyPyr}**C** and ^{TriPyr}**C**–Ag_2–^{TriPyr}**C**, respectively, compared with the metal-free mispairs [64]. It could be speculated that the greater flexibility of the benzyltriazolyl moiety is in part responsible for this tremendous stabilisation. Another possible explanation is a stabilisation via hydrophobic interactions of the pendant phenyl rings.

1.4.4 Outlook

Since the beginning of the century, research on metal-mediated base pairs has come a long way. Within the last decade or so, the number of established metal-mediated base pairs has risen from around a handful to more than 30. At the same time, the number of nucleic acids applied in metal-mediated base pairing has increased significantly: instead of investigating DNA only, other nucleic acids and nucleic acid derivatives such as RNA, PNA and GNA are regularly utilised. This development has been accompanied by an increasing knowledge of the underlying principles of the formation of metal-mediated base pairs and of the possibilities for fine-tuning their stability. Several examples exist in which DNA with metal-mediated base pairs is successfully processed by polymerases. It is the combination of metal-based functionality and self-assembling capabilities that makes nucleic acids with metal-mediated base pairs so popular. Nowadays, metal-mediated base pairs, particularly those involving natural nucleosides, are used in numerous sensing applications. Other areas of application, such as self-assembling molecular magnets or nucleic acids with an improved charge transfer capability, are emerging, and many more applications are feasible. For instance, the implementation of metal-mediated base pairs into DNA nanostructures or DNA origami is expected to lead to a fruitful cross-fertilisation of these research areas. Evidently, metal-mediated base pairs are heading towards a promising future!

References

[1] Y. J. Seo, S. Matsuda, F. E. Romesberg, Transcription of an expanded genetic alphabet, *J. Am. Chem. Soc.* 2009, **131**, 5046–5047.

[2] J. Müller, Metal-ion-mediated base pairs in nucleic acids, *Eur. J. Inorg. Chem.* 2008, 3749–3763.

[3] P. Scharf, J. Müller, Nucleic acids with metal-mediated base pairs and their applications, *ChemPlusChem* 2013, **78**, 20–34.

[4] S. Johannsen, S. Paulus, N. Düpre, J. Müller, R. K. O. Sigel, Using *in vitro* transcription to construct scaffolds for one-dimensional arrays of mercuric ions, *J. Inorg. Biochem.* 2008, **102**, 1141–1151.

[5] W. He, R. M. Franzini, C. Achim, Metal-containing nucleic acid structures based on synergetic hydrogen and coordination bonding, *Prog. Inorg. Chem.* 2007, **55**, 545–611.

[6] M. K. Schlegel, L. Zhang, N. Pagano, E. Meggers, Metal-mediated base pairing within the simplified nucleic acid GNA, *Org. Biomol. Chem.* 2009, **7**, 476–482.

[7] T. Yamane, N. Davidson, On the complexing of desoxyribonucleic acid (DNA) by mercuric ion, *J. Am. Chem. Soc.* 1961, **83**, 2599–2607.

[8] S. Katz, Mechanism of the reaction of polynucleotides and Hg^{II}, *Nature* 1962, **194**, 569.

[9] L. D. Kosturko, C. Folzer, R. F. Stewart, The crystal and molecular structure of a 2:1 complex of 1-methylthymine-mercury(II), *Biochemistry* 1974, **13**, 3949–3952.

[10] Z. Kuklenyik, L. G. Marzilli, Mercury(II) site-selective binding to a DNA hairpin. Relationship of sequence-dependent intra- and interstrand cross-linking to the hairpin-duplex conformational transition, *Inorg. Chem.* 1996, **35**, 5654–5662.

[11] D. A. Megger, N. Megger, J. Müller, Metal-mediated base pairs in nucleic acids with purine- and pyrimidine-derived nucleosides, *Met. Ions Life Sci.* 2012, **10**, 295–317.

[12] Y. Miyake, H. Togashi, M. Tashiro, H. Yamaguchi, S. Oda, M. Kudo, Y. Tanaka, Y. Kondo, R. Sawa, T. Fujimoto, T. Machinami, A. Ono, $Mercury^{II}$-mediated formation of thymine-Hg^{II}–thymine base pairs in DNA duplexes, *J. Am. Chem. Soc.* 2006, **128**, 2172–2173.

[13] Y. Tanaka, S. Oda, H. Yamaguchi, Y. Kondo, C. Kojima, A. Ono, ^{15}N-^{15}N *J*-coupling across Hg^{II}: Direct observation of Hg^{II}-mediated T-T base pairs in a DNA duplex, *J. Am. Chem. Soc.* 2007, **129**, 244–245.

[14] H. Torigoe, A. Ono, T. Kozasa, Hg^{II} ion specifically binds with T:T mismatched base pair in duplex DNA, *Chem. Eur. J.* 2010, **16**, 13218–13225.

[15] H. Urata, E. Yamaguchi, T. Funai, Y. Matsumura, S.-i. Wada, Incorporation of thymine nucleotides by DNA polymerases through T–Hg^{II}–T base pairing, *Angew. Chem., Int. Ed.* 2010, **49**, 6516–6519.

[16] A. Ono, S. Cao, H. Togashi, M. Tashiro, T. Fujimoto, T. Machinami, S. Oda, Y. Miyake, I. Okamoto, Y. Tanaka, Specific interactions between silver(I) ions and cytosine–cytosine pairs in DNA duplexes, *Chem. Commun.* 2008, 4825–4827.

[17] D. A. Megger, J. Müller, Silver(I)-mediated cytosine self-pairing is preferred over Hoogsteen-type base pairs with the artificial nucleobase 1,3-dideaza-6-nitropurine, *Nucleosides Nucleotides Nucleic Acids* 2010, **29**, 27–38.

[18] D. A. Megger, C. Fonseca Guerra, F. M. Bickelhaupt, J. Müller, Silver(I)-mediated Hoogsteen-type base pairs, *J. Inorg. Biochem.* 2011, **105**, 1398–1404.

[19] H. Torigoe, I. Okamoto, T. Dairaku, Y. Tanaka, A. Ono, T. Kozasa, Thermodynamic and structural properties of the specific binding between Ag^+ ion and C:C mismatched base pair in duplex DNA to form C-Ag-C metal-mediated base pair, *Biochimie* 2012, **94**, 2431–2440.

[20] F. Paquet, M. Boudvillain, G. Lancelot, M. Leng, NMR solution structure of a DNA dodecamer containing a trans-platin interstrand GN7-CN3 cross-link, *Nucleic Acids Res.* 1999, **27**, 4261–4268.

[21] J. Müller, M. Drumm, M. Boudvillain, M. Leng, E. Sletten, B. Lippert, Parallel-stranded DNA with Hoogsteen base pairing stabilized by a *trans*-$[Pt(NH_3)_2]^{2+}$ cross-link: characerization and conversion into a homodimer and a triplex, *J. Biol. Inorg. Chem.* 2000, **5**, 603–611.

[22] E. Ennifar, P. Walter, P. Dumas, A crystallographic study of the binding of 13 metal ions to two related RNA duplexes, *Nucleic Acids Res.* 2003, **31**, 2671–2682.

[23] M. G. Santangelo, P. M. Antoni, B. Spingler, G. Jeschke, Can copper(II) mediate Hoogsteen base-pairing in a left-handed DNA duplex? A pulse EPR study, *ChemPhysChem* 2010, **11**, 599–606.

[24] K. Tanaka, M. Shionoya, Synthesis of a novel nucleoside for alternative DNA base pairing through metal complexation, *J. Org. Chem.* 1999, **64**, 5002–5003.

[25] E. Meggers, P. L. Holland, W. B. Tolman, F. E. Romesberg, P. G. Schultz, A novel copper-mediated DNA base pair, *J. Am. Chem. Soc.* 2000, **122**, 10714–10715.

[26] S. Atwell, E. Meggers, G. Spraggon, P. G. Schultz, Structure of a copper-mediated base pair in DNA, *J. Am. Chem. Soc.* 2001, **123**, 12364–12367.

[27] N. Zimmermann, E. Meggers, P. G. Schultz, A second-generation copper(II)-mediated metallo-DNA-base pair, *Bioorg. Chem.* 2004, **32**, 13–25.

[28] N. Zimmermann, E. Meggers, P. G. Schultz, A novel silver(I)-mediated DNA base pair, *J. Am. Chem. Soc.* 2002, **124**, 13684–13685.

[29] K. Seubert, C. Fonseca Guerra, F. M. Bickelhaupt, J. Müller, Chimeric GNA/DNA metal-mediated base pairs, *Chem. Commun.* 2011, **47**, 11041–11043.

[30] K. Tanaka, A. Tengeiji, T. Kato, N. Toyama, M. Shiro, M. Shionoya, Efficient incorporation of a copper hydroxypyridone base pair in DNA, *J. Am. Chem. Soc.* 2002, **124**, 12494–12498.

[31] Y. Takezawa, K. Tanaka, M. Yori, S. Tashiro, M. Shiro, M. Shionoya, Soft Metal-mediated base pairing with novel synthetic nucleosides possessing an *O,S*-donor ligand, *J. Org. Chem.* 2008, **73**, 6092–6098.

[32] L. Zhang, E. Meggers, An extremely stable and orthogonal DNA base pair with a simplified three-carbon backbone, *J. Am. Chem. Soc.* 2005, **127**, 74–75.

[33] T. Ehrenschwender, W. Schmucker, C. Wellner, T. Augenstein, P. Carl, J. Harmer, F. Breher, H.-A. Wagenknecht, Development of a metal-ion-mediated base pair for electron transfer in DNA, *Chem. Eur. J.* 2013, **19**, 12547–12552.

[34] M. Su, M. Tomás-Gamasa, S. Serdjukow, P. Mayer, T. Carell, Synthesis and properties of a Cu(II) complexing pyrazole ligandoside in DNA, *Chem. Commun.* 2014, **50**, 409–411.

[35] H. Weizman, Y. Tor, 2,2′-Bipyridine ligandoside: A novel building block for modifying DNA with intra-duplex metal complexes, *J. Am. Chem. Soc.* 2001, **123**, 3375–3376.

[36] C. Brotschi, C. J. Leumann, Transition metal ligands as novel DNA-base substitutes, *Nucleosides Nucleotides Nucleic Acids* 2003, **22**, 1195–1197.

[37] C. Switzer, S. Sinha, P. H. Kim, B. D. Heuberger, A purine-like nickel(II) base pair for DNA, *Angew. Chem., Int. Ed.* 2005, **44**, 1529–1532.

[38] C. Switzer, D. Shin, A pyrimidine-like nickel(II) DNA base pair, *Chem. Commun.* 2005, 1342–1344.

[39] D. Shin, C. Switzer, A metallo base-pair incorporating a terpyridyl-like motif: bipyridyl-pyrimidinone·Ag(I)·4-pyridine, *Chem. Commun.* 2007, 4401–4403.

[40] B. D. Heuberger, D. Shin, C. Switzer, Two Watson-Crick-like metallo base-pairs, *Org. Lett.* 2008, **10**, 1091–1094.

[41] I. Sinha, J. Kösters, A. Hepp, J. Müller, 6-Substituted purines containing thienyl or furyl substituents as artificial nucleobases for metal-mediated base pairing, *Dalton Trans.* 2013, **42**, 16080–16089.

[42] E.-K. Kim, C. Switzer, Polymerase recognition of a Watson–Crick-like metal-mediated base pair: Purine-2,6-dicarboxylate·copper(II)·pyridine, *ChemBioChem* 2013, **14**, 2403–2407.

[43] S. Taherpour, H. Lönnberg, T. Lönnberg, 2,6-Bis(functionalized) purines as metal-ion-binding surrogate nucleobases that enhance hybridization with unmodified 2'-*O*-methyl oligoribonucleotides, *Org. Biomol. Chem.* 2013, **11**, 991–1000.

[44] T. Richters, J. Müller, A metal-mediated base pair with a [2+1] coordination environment, *Eur. J. Inorg. Chem.* 2014, 437–441.

[45] G. H. Clever, K. Polborn, T. Carell, A highly DNA-duplex-stabilizing metal-salen base pair, *Angew. Chem., Int. Ed.* 2005, **44**, 7204–7208.

[46] G. H. Clever, Y. Söltl, H. Burks, W. Spahl, T. Carell, Metal–salen-base-pair complexes inside DNA: Complexation overrides sequence information, *Chem. Eur. J.* 2006, **12**, 8708–8718.

[47] C. Kaul, M. Müller, M. Wagner, S. Schneider, T. Carell, Reversible bond formation enables the replication and amplification of a crosslinking salen complex as an orthogonal base pair, *Nat. Chem.* 2011, **3**, 794–800.

[48] K. Tanaka, A. Tengeiji, T. Kato, N. Toyama, M. Shionoya, A discrete self-assembled metal array in artificial DNA, *Science* 2003, **299**, 1212–1213.

[49] S. S. Mallajosyula, S. K. Pati, Conformational tuning of magnetic interactions in metal–DNA complexes, *Angew. Chem., Int. Ed.* 2009, **48**, 4977–4981.

[50] S. Liu, G. H. Clever, Y. Takezawa, M. Kaneko, K. Tanaka, X. Guo, M. Shionoya, Direct conductance measurement of individual metallo-DNA duplexes within single-molecule break junctions, *Angew. Chem., Int. Ed.* 2011, **50**, 8886–8890.

[51] G. H. Clever, T. Carell, Controlled stacking of 10 transition-metal ions inside a DNA duplex, *Angew. Chem., Int. Ed.* 2007, **46**, 250–253.

[52] G. H. Clever, S. J. Reitmeier, T. Carell, O. Schiemann, Antiferromagnetic coupling of stacked Cu^{II}–salen complexes in DNA, *Angew. Chem., Int. Ed.* 2010, **49**, 4927–4929.

[53] J. Müller, D. Böhme, P. Lax, M. Morell Cerdà, M. Roitzsch, Metal ion coordination to azole nucleosides, *Chem. Eur. J.* 2005, **11**, 6246–6253.

[54] S. Johannsen, N. Megger, D. Böhme, R. K. O. Sigel, J. Müller, Solution structure of a DNA double helix with consecutive metal-mediated base pairs, *Nat. Chem.* 2010, **2**, 229–234.

[55] S. Kumbhar, S. Johannsen, R. K. O. Sigel, M. P. Waller, J. Müller, A QM/MM refinement of an experimental DNA structure with metal-mediated base pairs, *J. Inorg. Biochem.* 2013, **127**, 203–210.

[56] K. Petrovec, B. J. Ravoo, J. Müller, Cooperative formation of silver(I)-mediated base pairs, *Chem. Commun.* 2012, **48**, 11844–11846.

[57] D. Böhme, N. Düpre, D. A. Megger, J. Müller, Conformational change induced by metal-ion-binding to DNA containing the artificial 1,2,4-triazole nucleoside, *Inorg. Chem.* 2007, **46**, 10114–10119.

[58] F.-A. Polonius, J. Müller, An artificial base pair, mediated by hydrogen bonding and metal-ion binding, *Angew. Chem., Int. Ed.* 2007, **46**, 5602–5604.

[59] K. Tanaka, G. H. Clever, Y. Takezawa, Y. Yamada, C. Kaul, M. Shionoya, T. Carell, Programmable self-assembly of metal ions inside artificial DNA duplexes, *Nat. Nanotechnol.* 2006, **1**, 190–194.

[60] D. A. Megger, C. Fonseca Guerra, J. Hoffmann, B. Brutschy, F. M. Bickelhaupt, J. Müller, Contiguous metal-mediated base pairs comprising two Ag^I ions, *Chem. Eur. J.* 2011, **17**, 6533–6544.

[61] Y. Nosenko, F. Menges, C. Riehn, G. Nieder-Schatteburg, Investigation by two-color IR dissociation spectroscopy of Hoogsteen-type binding in a metalated nucleobase pair mimic, *Phys. Chem. Chem. Phys.* 2013, **15**, 8171–8178.

[62] I. Okamoto, K. Iwamoto, Y. Watanabe, Y. Miyake, A. Ono, Metal-ion selectivity of chemically modified uracil pairs in DNA duplexes, *Angew. Chem., Int. Ed.* 2009, **48**, 1648–1651.

[63] I. Okamoto, T. Ono, R. Sameshima, A. Ono, Metal ion-binding properties of DNA duplexes containing thiopyrimidine base pairs, *Chem. Commun.* 2012, **48**, 4347–4349.

[64] H. Mei, I. Röhl, F. Seela, Ag^+-Mediated DNA base pairing: Extraordinarily stable pyrrolo-dC–pyrrolo-dC pairs binding two silver ions, *J. Org. Chem.* 2013, **78**, 9457–9463.

1.5

Metal-Aided Construction of Unusual DNA Structural Motifs

Yusuke Takezawa, Jean-Louis Duprey, and Mitsuhiko Shionoya

Graduate School of Science, Department of Chemistry, The University of Tokyo, Tokyo, Japan

1.5.1 Introduction

Hydrogen bonding plays a key role in the folding and the self-assembly of biomolecules. One of the prime examples is the Watson–Crick base pairing inside DNA double helices, between adenine (A) and thymine (T), and guanine (G) and cytosine (C). In nature, this complementary hydrogen bonding allows sequence-specific hybridization of oligonucleotides as well as the storage, replication, and transfer of genetic information. The sequence-dependent self-assembly of DNA oligomers affords not only naturally occurring structures but also artificial supramolecular architectures including junctions, polyhedra, and DNA origami structures. A large variety of DNA nanoarchitectures have been constructed so far via self-assembly processes, which can be precisely programmed based on the DNA sequence.

One of the other molecular interactions that is widely exploited when one synthesizes supermolecules is metal coordination bonding. Its geometrical diversity enables the elaborate design of various three-dimensional structures, in which metal complexes serve as "nodes" of the resulting supramolecular architectures. The kinetic reversibility of metal coordination allows for dynamic structural conversion, which is one of the essential requirements for developing non-biological molecular machines. These properties can be controlled to some extent by selecting appropriate metal species and organic ligand moieties, depending on what is known of the coordination chemistry.

Introduction of metal complexes into DNA structures is an effective approach for constructing functional DNA supermolecules. It has thus become an active area of research in recent years [1, 2]. Metal ligand moieties can be readily introduced at both ends as well as in the middle of oligonucleotides through standard

DNA in Supramolecular Chemistry and Nanotechnology, First Edition.
Edited by Eugen Stulz and Guido H. Clever.

solid-phase DNA synthesis or post-synthetic modification. When ligand moieties are located at the terminus of DNA duplexes, metal complexation results in the connection of DNA duplexes to afford larger supramolecular structures, in a similar fashion to typical supramolecular metal complexes. An alternative approach is the introduction of metal complexes *inside* DNA structural motifs such as duplexes, triple helices, quadruplexes, and branched structures. Additional metal coordination bonding enhances the thermal stability of the original DNA structures, induces unusual motifs, and thus increases the conformational diversity of DNA supramolecular structures. In this chapter, we focus on the latter studies, namely metal-aided construction of unusual DNA structural motifs, as promising components for DNA-based nanoarchitectures and molecular machines.

1.5.2 DNA duplexes containing metal-mediated base pairs

Research on the metal-assisted construction of DNA secondary structures began with the incorporation of metal complexes into duplexes. The most common approach is to replace canonical hydrogen-bonded base pairs with artificial metal-mediated base pairs (Figure 1.5.1) [3–5]. A metal-mediated base pair consists of two ligand-type nucleosides, which are located at opposite sites in the DNA duplexes, and an appropriate metal ion, which binds and bridges the nucleobase moieties. Since metal coordination bonds are generally stronger than hydrogen bonds, incorporation of metal-mediated base pairs leads, in most cases, to significant stabilization of the DNA duplex. Over 20 ligand-type artificial nucleosides that form metal-mediated base pairs within DNA duplexes have been developed so far (see also Chapter 1.4). Monodentate to tridentate planar ligands possessing O, N, and/or S donor atoms were chosen as artificial nucleobase moieties and square-planar- and linear-coordinating metal ions such as Mn(II/III), Ni(II), Cu(II), Pd(II), Ag(I), Pt(II), and Hg(II) were employed as the bridging metal ions in order to achieve efficient base pairing. It should be noted that

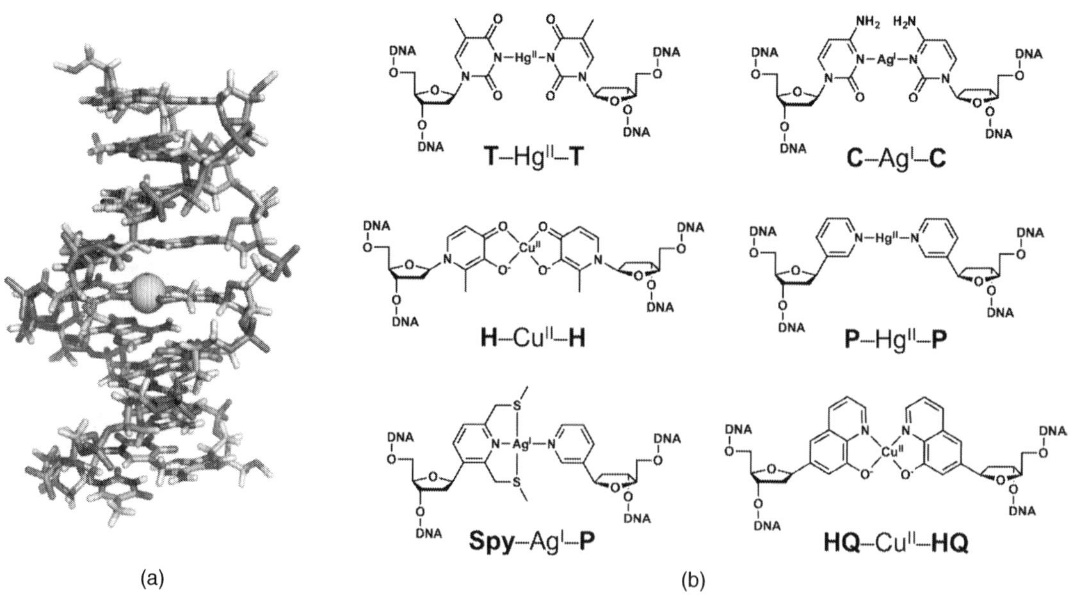

Figure 1.5.1 Artificial DNA duplexes containing a metal-mediated base pair. (a) Structure of metallo-DNA duplex; and (b) typical examples of metal-mediated base pairs

metal-mediated base pairing of natural nucleosides, namely Hg(II)-mediated thymine (**T**–Hg(II)–**T**) base pairs [6–9] and Ag(I)-mediated cytosine (**C**–Ag(I)–**C**) [10], has been widely explored [11].

Successful examples of metallo-base pairs include a Cu(II)-mediated hydroxypyridone base pair (**H**–Cu(II)–**H**) [12]. The bidentate 3-hydroxy-4-pyridone (hydroxypyridone), introduced as the ligand-type nucleobase, forms a square-planar 2:1 complex with a Cu(II) ion, resulting in metal-mediated base pairing. Its flat structure fits well within the hydrophobic base-stacked environment inside a DNA duplex and incorporation of the **H**–Cu(II)–**H** base pair enhances the thermal stability of duplexes, as confirmed by thermal denaturation experiments. The melting temperature (T_m), the temperature at the midpoint of the transition from duplex to single strand, of a 15-bp DNA duplex containing the hydroxypyridone nucleosides (**H**), d(5′-CACATTA**H**TGTTGTA-3′)·d(3′-GTGTAAT**H**ACAACAT-5′), was 37 °C in the absence of metal ions, while that of a natural fully-matched 15-bp duplex d(5′-CACATTA**A**TGTTGTA-3′)·d(3′-GTGTAAT**T**ACAACAT-5′) was 44 °C. This result indicated that the metal-free **H–H** pair behaved like a mismatched pair, which decreases the stability of the duplex. In contrast, the melting temperature of the artificial DNA duplex was increased to 50 °C (ΔT_m = 13 °C) upon addition of one equivalent of Cu(II) ions, whereas that of the natural duplex hardly changed after the metal addition. The Cu(II)-dependent stabilization resulted from the formation of an **H**–Cu(II)–**H** base pair, which crosslinked two DNA strands. It is noteworthy that the duplex containing an **H**–Cu(II)–**H** base pair is more stable than the natural duplex. The degree of the stabilization varies with the exact nature of the metal-mediated base pairs, in other words, the combination of ligand moieties and a metal ion. For example, an Ag(I)-mediated unsymmetrical base pair, **Spy**–Ag(I)–**P**, exhibited a smaller stabilization (ΔT_m = 12 °C) [13], while a Cu(II)-mediated hydroquinone base pair (**HQ**–Cu(II)–**HQ**) showed a larger stabilization (ΔT_m = 29 °C) [14] when incorporated into the same 15-bp DNA duplex. The largest stabilization was achieved with a salen-type Cu(II)-mediated base pair (**S**–Cu(II)(en)–**S**), in which ethylenediamine covalently bridges the nucleobase moieties (ΔT_m = 43 °C) [15]. Multiple incorporation of metal-mediated base pairs is also possible and, in some cases, allows for more effective stabilization of DNA duplexes [13, 16, 17].

Metal-triggered construction of artificial DNA duplexes has been demonstrated using such metal-mediated base pairing systems. Short ligand-bearing oligonucleotides d(5′-G$\mathbf{H}_n$C-3′) (n = 1–5), consisting of one-to-five consecutive **H** nucleosides and flanking G and C, were found to form double helical structures upon addition of Cu(II) ions (Figure 1.5.2) [18]. The oligonucleotides did not form duplexes in the absence of metal ions due to the lack of sufficient natural hydrogen-bonded base pairs. The addition of Cu(II) induced the self-assembly of two strands to afford metallo-DNA duplexes, nCu(II)·d(5′-G$\mathbf{H}_n$C-3′)$_2$ (n = 1–5), via quantitative formation of **H**–Cu(II)–**H** base pairs. The UV spectra of the DNA strands showed a gradual increase in the absorbance at 307 nm upon Cu(II) addition, which is ascribed to the formation of an **H**–Cu(II)–**H** base pair. The UV-based titration studies also confirmed the stoichiometric metal complexation, and electrospray ionization-time-of-flight (ESI-TOF) mass spectrometry provided evidence for the intermolecular assembly of two strands, that is, the formation of the duplex structures. Circular dichroism (CD) spectra indicated periodic stacking of the **H**–Cu(II)–**H** base pairs inside right-handed double helices. Based on EPR spectral simulation, the distance between two adjacent Cu(II) ions was estimated to be 3.7 ± 0.1 Å, suggesting that the Cu(II)-induced metallo-DNA duplexes have a helical pitch similar to natural B-DNA duplexes.

Metal-mediated base pairs have attracted increased attention in recent years for their potential applications, particularly in the field of nanomaterial science. The consecutive incorporation of metallo-base pairs into DNA duplexes is a promising approach to constructing one-dimensional discrete metal arrays, which are expected to exhibit unique physical properties. For example, the stacks of **H**–Cu(II)–**H** base pairs inside the nCu(II)·d(5′-G$\mathbf{H}_n$C-3′)$_2$ duplexes (n = 1–5) exhibited the highest spin states (S = n/2), arising from ferromagnetic coupling of unpaired d electrons of Cu(II) ions (Figure 1.5.2b) [18]. In contrast, stacks of Cu(II)-mediated salen-type base pairs predominantly showed antiferromagnetic coupling [19].

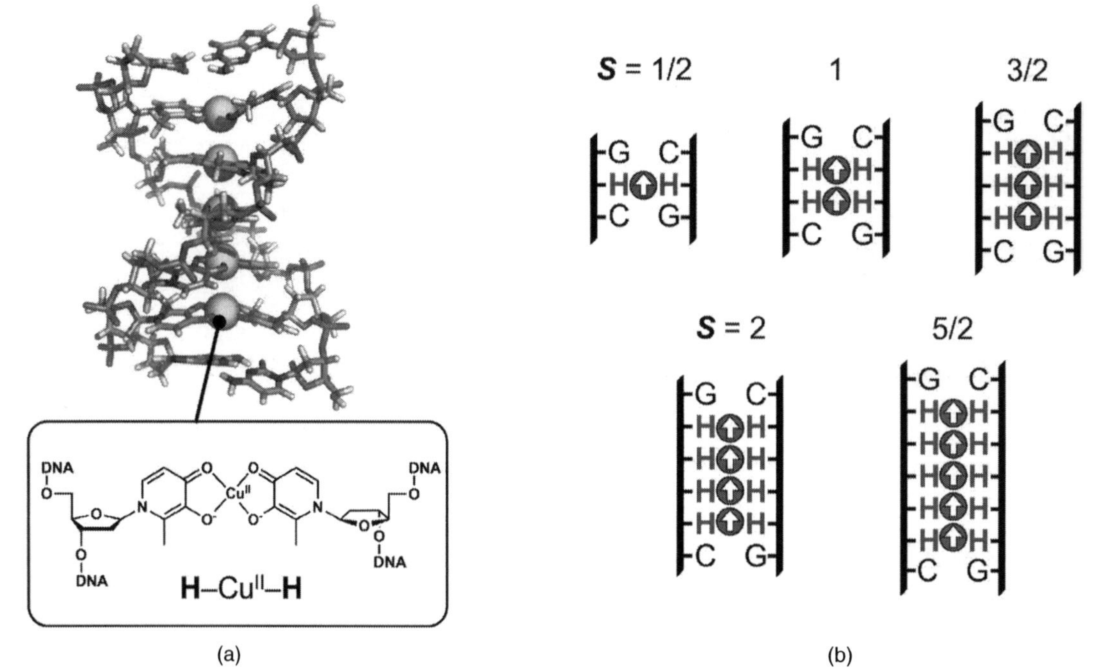

Figure 1.5.2 Metal-aided formation of DNA duplexes. (a) DNA duplex formation through consecutive Cu(II)-mediated hydroxypyridone base pairs (***H***–Cu(II)–***H***); and (b) discrete assembly of Cu(II) ions inside a duplex. ***S*** stands for the spin state as determined by EPR analysis

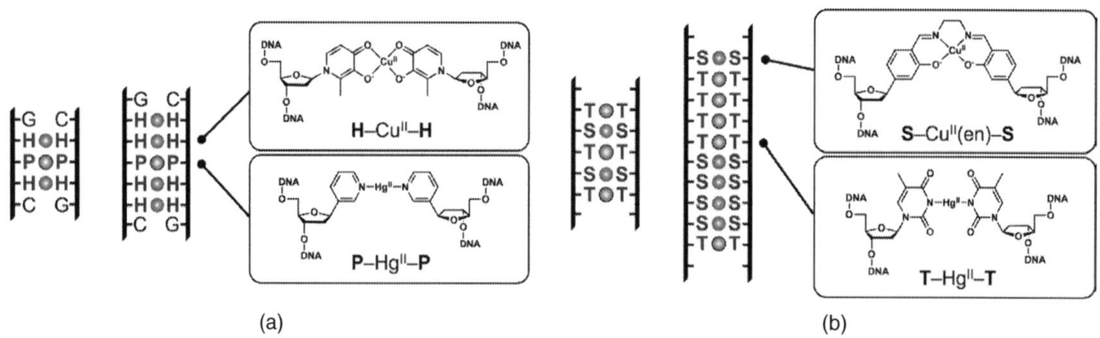

Figure 1.5.3 Heterogeneous metal arrays constructed in artificial DNA duplexes. (a) Metal assembly via formation of ***H***–Cu(II)–***H*** and ***P***–Hg(II)–***P*** base pairs; and (b) metal assembly via formation of ***S***–Cu(II)(en)–***S*** and ***T***–Hg(II)–***T*** base pairs

Heterogeneous metal arrays were constructed by incorporation of both Cu(II)-mediated and Hg(II)-mediated base pairs inside DNA duplexes (Figure 1.5.3) [20]. The artificial oligonucleotide d(5′-G**HPH**C-3′), containing two hydroxypyridone-bearing nucleosides (**H**) and a pyridine-bearing one (**P**), formed a double helix in a self-assembled fashion through the formation of two different types of metal-mediated base pairs, **H**–Cu(II)–**H** and **P**–Hg(II)–**P**. As a result, a heterogeneous metal array consisting of Cu(II)–Hg(II)–Cu(II) was constructed inside the duplex. In a similar manner, longer metal arrays of Cu(II) and Hg(II) ions were

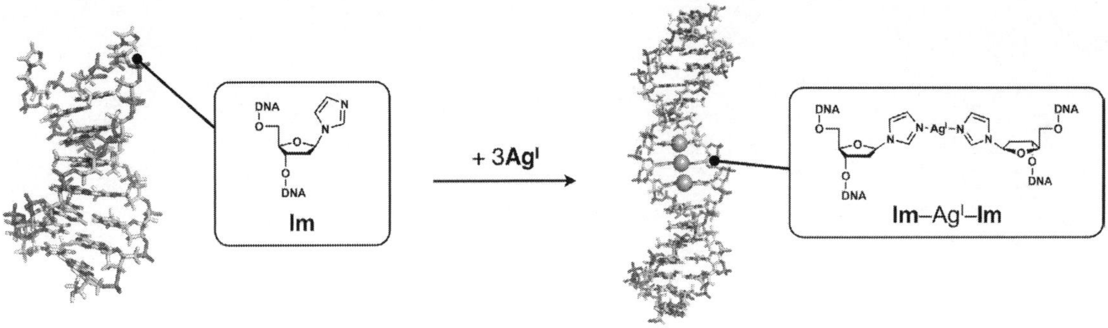

Figure 1.5.4 *Structural conversion from a hairpin to a duplex through the formation of Ag(I)-mediated imidazole base pairs (**Im**–Ag(I)–**Im**). DNA structures are drawn from PDB 2K68 and 2KE8 (NMR structures)*

constructed by using a combination of Cu(II)-mediated salen-type base pairs and thymine-based **T**–Hg(II)–**T** base pairs.

The electrical conductivity of DNA duplexes containing **H**–Cu(II)–**H** base pairs was recently investigated by using a carbon nanotube device in which a metallo-DNA duplex was covalently immobilized within a nanotube gap [21]. The current measurements revealed that the DNA duplex containing **H**–Cu(II)–**H** base pairs showed higher conductivity than ligand-containing but metal-free duplexes. It is important to note here that alternating treatments of the devices with Cu(II) ions and a chelating agent (EDTA) allowed on–off switching of the electrical conductance.

Some of the metal-mediated base pairs were used to induce conformational changes to DNA structures. A structural change between an intermolecular double helix and an intramolecular hairpin structure was demonstrated by using **T**–Hg(II)–**T** base pairs [7] as well as Ag(I)-mediated triazole [22] or imidazole [23] base pairs (Figure 1.5.4). Such metal-triggered structural conversion can be applied for constructing stimuli-responsive functional DNAs, such as metal sensors, logic gates, allosteric deoxyribozymes, and DNA-based molecular machines.

1.5.3 Metal-aided formation of triple-stranded structures

Natural DNA strands are known to fold not only into duplexes but also into triple-stranded structures, depending on their base sequences (Figure 1.5.5). For example, a homopyrimidine oligonucleotide binds in the major groove of a homopurine–homopyrimidine duplex through Hoogsteen hydrogen bonding to give a triple helix in which T–A–T and C^+–G–C base triplets are stacked (where C^+ denotes an N3-protonated cytosine base). The idea of metal-mediated base pairing overviewed earlier can be applied to such triple-stranded structures, leading to an increase in the thermal stability and to metal-triggered induction of the structure.

The hydrogen-bonded Hoogsteen base pairing between the third strand and the central strand can be replaced by metal-mediated base pairing. For instance, an Ag(I) ion has replaced the N3 proton of the C^+ base in the C^+–G–C base triplet, which results in the formation of a metal-mediated base triplet, C–Ag(I)–G–C, in the DNA triplex (Figure 1.5.5c). The formation of natural DNA triplexes containing a C^+–G–C base triplet generally requires slightly acidic conditions due to the requirement for protonation of the N3 of cytosine ($pK_a = 4.5$). However, Ag(I)-mediated base triplet (C–Ag(I)–G–C) formation was observed even at higher pH and resulted in the stabilization of the DNA triplex [24]. The UV melting curve of a DNA triplex d(5′-T_5**C**T_5-3′)·d(3′-A_5**G**A_5-5′)·d(3′-T_5**C**T_5-5′), containing a C–G–C in the middle, showed a two-step transition

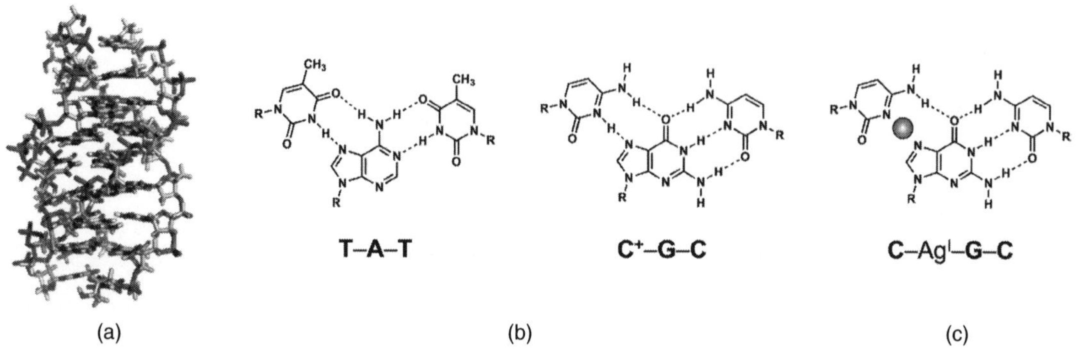

Figure 1.5.5 DNA triple-stranded structure. (a) Natural DNA triple-stranded structure; (b) natural DNA triplets found in the triplexes; and (c) Ag(I)-mediated base triplet, **C**–Ag(I)–**G**–**C**

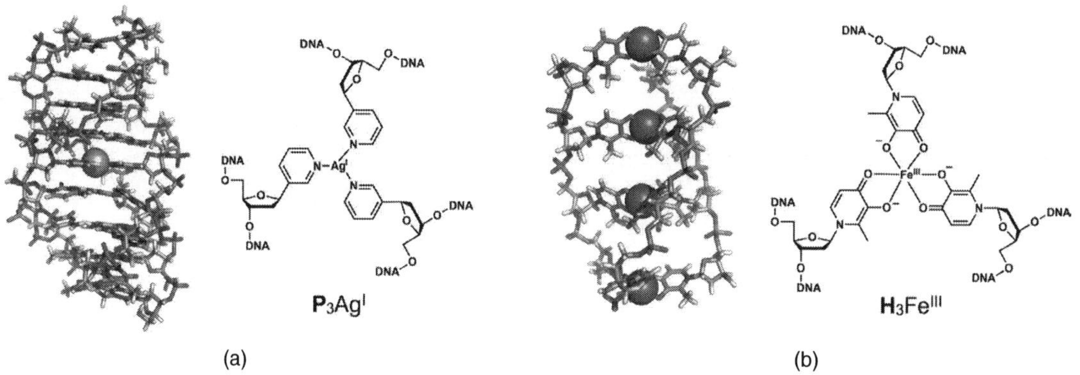

Figure 1.5.6 Metal-aided formation of DNA triple-stranded structures. (a) Metallo-DNA triplex containing an Ag(I)-mediated base triplet $\mathbf{P}_3$Ag(I); and (b) metallo-DNA triplex containing Fe(III)-mediated base triplet $\mathbf{H}_3$Fe(III)

in the absence of metal ions at pH 8.5, which can be attributed to the dissociation of the third strand (with lower T_m) and the dehybridization of the remaining duplex (with higher T_m). In the presence of Ag(I) ions, the first transition step seemed to almost disappear, indicating that the third strand dissociated at higher temperature and the triplex was substantially stabilized. The Ag(I)-mediated base triplet formation was further evidenced by titration of Ag(I) ions, ESI mass spectrometry, and other control experiments.

Another example of metallo-DNA triplexes relates to artificial ligand-bearing nucleosides that form metal-mediated base triplets inside DNA triple-stranded structures (Figure 1.5.6a). Pyridine-bearing nucleosides (**P**) were incorporated into oligo-T and oligo-A strands to prepare a DNA triplex d(5′-T_{10}**P**T_{10}-3′)·d(3′-A_{10}**P**A_{10}-5′)·d(3′-T_{10}**P**T_{10}-5′) [25]. The melting curve showed a two-step transition, as is the case for natural DNA triplexes, and the melting temperature (T_m) corresponding to the third-strand dissociation was determined to be 15.5 °C in the absence of transition metal ions. When Ag(I) ions were added, both transitions (triplex–duplex and duplex–single strand) were observed to occur at higher temperatures. The T_m value of the triplex–duplex transition, that is, the third-strand dissociation, was estimated to be 17.5 °C (ΔT_m = +2.0 °C), whereas that of natural triple-stranded DNA d(5′-T_{21}-3′)·d(3′-A_{21}-5′)·d(3′-T_{21}-5′) was decreased from 28.2 to 25.8 °C (ΔT_m = –2.4 °C). The stabilization of the ligand-modified triplex by Ag(I) addition is accounted for in terms of the formation of a 3:1 complex of the pyridine ligands and an Ag(I) ion, namely a base triplet $\mathbf{P}_3$Ag(I).

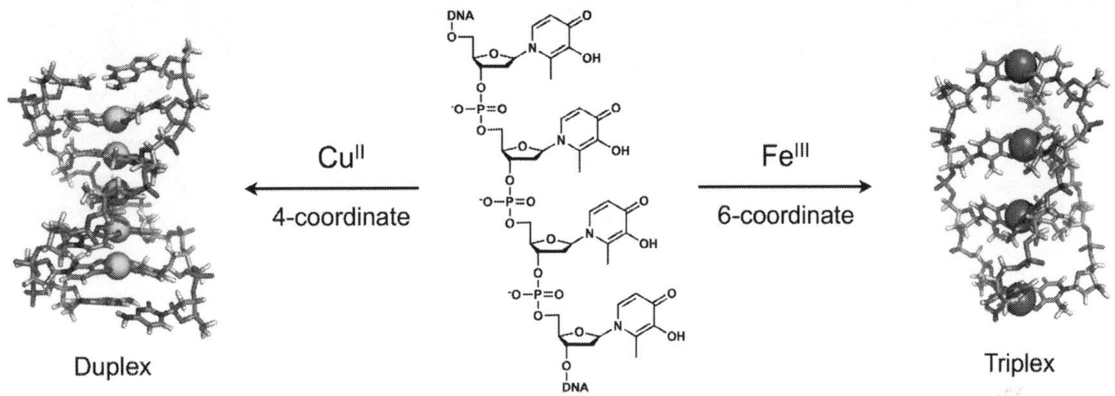

Figure 1.5.7 *Schematic representation of metal-dependent formation of metallo-DNA duplex and triplex*

Totally unnatural DNA triple-stranded structures were constructed with homooligomers of artificial ligand-bearing nucleosides (Figure 1.5.6b). The hydroxypyridone-bearing nucleoside (**H**), which has been previously used for Cu(II)-mediated base pairing, was chosen as a metal ligand for artificial base triplets because bidentate hydroxypyridone and its analogues are known to form 3:1 complexes with some 6-coordinate transition metal ions. Dimeric to tetrameric oligomers of the **H** nucleoside, d(5′-$\mathbf{H}_n$-3′), were synthesized and subjected to complexation with octahedral Fe(III) ions at pH 7.0 [26]. UV-visible spectral measurements showed the emergence of a new absorption band around 457 nm upon addition of Fe(III) ions, which indicated the metal binding to the hydroxypyridone moieties. In the UV-based titration experiments of the tetramer d(5′-**HHHH**-3′), for example, the absorption at 457 nm increased linearly in the range of from 0 to 4.0 ([Fe(III)]/[d(5′-**HHHH**-3′)$_3$]). This result suggested the stoichiometric formation of a triple-stranded structure (Fe(III)$_4$·d(5′-**HHHH**-3′)$_3$) through the formation of an $\mathbf{H}_3$Fe(III) base-triplet. Further evidence for Fe(III)-mediated triple-strand formation came from ESI-TOF mass spectrometry. Similar results were obtained for shorter strands, d(5′-**HH**-3′) and d(5′-**HHH**-3′), confirming the formation of metallo-DNA triplexes, Fe(III)$_2$·d(5′-**HH**-3′)$_3$ and Fe(III)$_3$·d(5′-**HHH**-3′)$_3$, respectively. Although their detailed structures are still unclear, CD spectra showed that a helical configuration was induced upon metal complexation as a consequence of the chiral deoxyribose backbone of the DNA oligomer.

It should be additionally mentioned here that the homooligomer d(5′-$\mathbf{H}_n$-3′) can form duplexes through the **H**–Cu(II)–**H** base pairing when Cu(II) ions are added. Taken together, these results suggest that metallo-DNA duplexes are induced by Cu(II)-mediated base pair formation, and metallo-DNA triplexes are induced by Fe(III)-mediated base triplet formation (Figure 1.5.7). Such metal-dependent self-assembly of the DNA helices has a potential application for developing metal-responsive supramolecular systems, including DNA-based molecular machines whose dynamics are regulated by metal coordination.

1.5.4 Metal-aided formation of four-stranded structures

Four-strand DNA structures are also attractive targets to construct, not only due to their biological importance but also because their well-defined structures are eminently suited for developing DNA-based supermolecules. The most thoroughly studied structures are the G-quadruplex motifs, which consist of four G-rich DNA strands. A tetrad of guanine (**G**) bases forms a cyclic planar tetramer that is stabilized via interstrand hydrogen bonding between the Hoogsteen and Watson–Crick faces. The stability of G-quadruplexes is highly

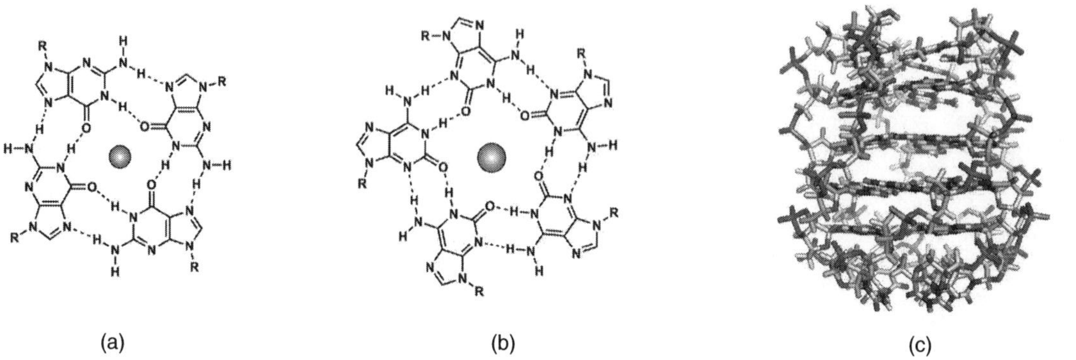

Figure 1.5.8 Metal-aided formation of DNA quadruplexes and pentaplexes. (a) K(I)-aided formation of G-quartet structure that forms a DNA quadruplex; (b) Cs(I)-aided formation of isoguanine-based quintet (**iG**-quintet) structure that forms a DNA pentaplex; and (c) structure of the **iG**-based pentaplex. Drawn from PDB 2LUJ (NMR structure)

dependent upon the concentration of cations, such as Na(I) and K(I) ions. This can be explained by the coordination of these ions to eight O6 atoms from two adjacent G-quartet planes (Figure 1.5.8a).

An analogous coordination mode whereby an alkaline metal ion coordinates to a noncanonical nucleobase has allowed the assembly of five-stranded DNA helical structures [27, 28]. Isoguanine (**iG**), in which the positions of the amino group and the carbonyl group in the guanine base are switched, was employed instead of **G** for this purpose. Five **iG** bases can form a planar pentamer through hydrogen bonding between the Watson–Crick face and the minor-groove face, and thus metal-mediated formation of oligo-**iG** pentaplex was expected (Figure 1.5.8b). A mixture of d(5′-T_4(**iG**)$_4$T-3′) and d(5′-T_8(**iG**)$_4$T-3′) incubated in the presence of K(I) ions gave five distinct bands under native gel electrophoresis conditions. Each band is attributed to one of the five possible four-stranded structures d(5′-T_4(**iG**)$_4$T-3′)$_n$·d(5′-T_8(**iG**)$_4$T-3′)$_{4-n}$ (n = 0–4), thus indicating the formation of quadruplex structures similar to a **G**-quadruplex. In contrast, when incubated in the presence of Cs(I) ions, six bands were clearly observed on the gel. This result is evidence of the selective formation of a five-stranded structure mediated by Cs(I) ions. Recently, the solution structure of the **iG**-pentaplex was determined by means of NMR spectroscopy (Figure 1.2.6c). The formation of a fivefold symmetric (C_5) helical structure, possibly in a parallel orientation with the **iG** pentad stacks, was suggested. The Cs(I) ions are considered to be located between the pentad layers and coordinate to the surrounding O4 atoms of the **iG**-quintets. It is particularly interesting to note that smaller K(I) ions (whose ionic radius is 1.33 Å) mediate quadruplex formation, whereas larger Cs(I) ions (1.67 Å) induce a pentaplex motif, which possesses a larger central cavity.

Incorporation of 4:1 ligand–metal complexes into G-quadruplexes was recently demonstrated (Figure 1.5.9) [29]. Interstrand metal complexation, that is, metal-mediated "tetrad" formation, resulted in the stabilization of a quadruplex, as is the case with metal-mediated base pairs within DNA duplexes. The melting temperature of the parallel quadruplex d(5′-**P**GGGG-3′)$_4$, in which each 5′-terminal of the strand was tethered with a pyridine ligand (**P**), was increased by 15 °C upon addition of Ni(II) ions, and by 20 °C upon Cu(II) addition. The stabilization was attributed to metal-mediated crosslinking via the formation of a tetrakis pyridine complex, which was confirmed by a UV-based titration of Cu(II) ions as well as EPR spectroscopy. Reversible stabilization and destabilization were further demonstrated by sequential addition of Cu(II) ions and a chelating reagent (EDTA), which removes the metals to recover the metal-free structure. In addition, Cu(II)-induced self-assembly of the quadruplex and subsequent EDTA-triggered denaturation was demonstrated by using a shorter strand d(5′-**P**GGG-3′) with a low melting temperature.

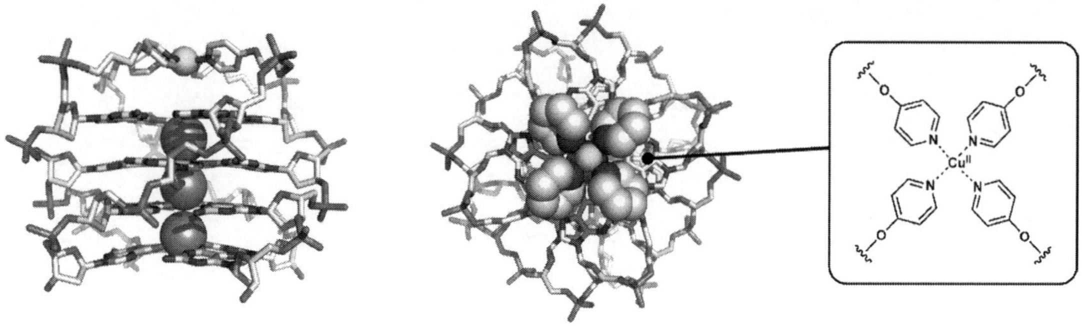

(a) (b)

Figure 1.5.9 Metal-induced stabilization of DNA G-quadruplexes. (a) Side view; and (b) top view

A four-stranded DNA structure composed of C-rich oligonucleotides, known as an i-motif, has also gained much interest recently in the field of DNA nanotechnology. The i-motif quadruplex involves two parallel duplexes, in which hemi-protonated C–H$^+$–C base pairs are formed through three hydrogen bonds. The two duplexes are intercalated into each other in an opposite direction to form an interpenetrated quadruplex. The i-motif is normally formed under acidic conditions due to the proton-mediated C–H$^+$–C base pairing. The strategy of replacing the H$^+$ that is involved in hydrogen bonding with an Ag(I) ion can be employed to achieve metal-mediated i-motif formation, in a similar way to DNA triplexes containing a metal-mediated base triplet, C–Ag(I)–G–C. This approach, where the Ag(I)-mediated formation of i-motif structures at neutral pH was observed, has been recently reported [30].

1.5.5 Metal-aided formation of DNA junction structures

Another class of DNA structures that no doubt contribute to the field of supramolecular chemistry and DNA nanotechnology is that of branched junction motifs. A DNA junction consists of more than two strands and has three or more double helical DNAs emanating from its branching point. DNA junction structures are found in natural systems and are known to play important roles in genetic processes, such as three-way replication junctions and four-way Holliday junctions in genetic recombination. DNA junction motifs are essential building blocks of various DNA architectures because the assembly of linear DNA duplexes yields only one-dimensional structures, which lack structural diversity. Since Seeman published the first paper on the construction of immobile four-way junctions [31], DNA branching structures have been employed for the construction of two-dimensional DNA lattices, three-dimensional nanostructures such as DNA polyhedra, and more intricate architectures fabricated by the now-common DNA origami method.

Metal-aided construction of DNA junction structures is of great interest to DNA chemists [32, 33] because it potentially allows induction and stabilization of not only junction motifs themselves but also whole DNA nanoarchitectures in response to metal coordination. Metal-dependent stabilization of DNA three-way junction (3WJ) structures was recently achieved in an analogous fashion to the metallo-DNA helices described earlier (Figure 1.5.10) [34]. The principle of molecular design of the metallo-DNA 3WJ is the use of 3:1 ligand–metal complexation at the branching point. The three strands are crosslinked in a similar manner to the 2:1 complexation, that is, metal-mediated base pairing, which occurs inside duplexes. A bidentate bipyridine ligand (bpy) was attached at the 2′-position of a nucleoside rather than the nucleobase moiety to provide a novel ligand-bearing nucleoside $\mathbf{U}_{\mathrm{bpy}}$. This design is based on previously reported crystal structures of

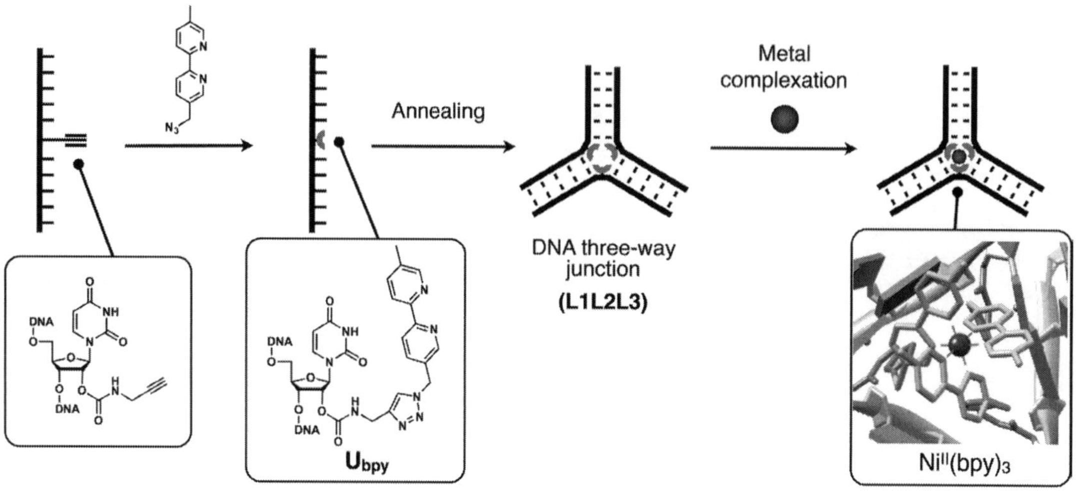

(a)

L1/S1 5'-GAA GGA ACG **X**AC ACT CGC AG-3'
L2/S2 5'-GTT CCA CGC **X**AC GTT CCT TC-3'
L3/S3 5'-CTG CGA GTG **X**AG CGT GGA AC-3'

L1–L3: X = U$_{bpy}$, S1–S3: X = T

(b)

Figure 1.5.10 *(a) Schematic representation of the construction of a metallo-DNA three-way junction; and (b) the sequences of the DNA oligomers used for the junction formation*

unmodified 3WJs [35], which direct their 2′-hydrogen atoms into the junction core. Thus the 2′-modified nucleoside is thought to be suitable for metal complex formation at the center of a junction. Three DNA strands possessing a **U**$_{bpy}$ nucleoside at the middle position (**L1**, **L2**, and **L3**) were prepared through post-synthetic DNA modification using a Huisgen cycloaddition. Subsequent hybridization of the three strands afforded the 3WJ structure (**L1L2L3**), having three bpy ligands preorganized at the branching point.

Metal-mediated crosslinking of the modified 3WJ structure was demonstrated in the presence of various transition metal ions. For example, addition of equimolar amounts of Ni(II) ions remarkably enhanced the thermal stability of the bpy-modified 3WJ, **L1L2L3**, while no stabilization was observed for an unmodified 3WJ (**S1S2S3**) in which **U**$_{bpy}$ nucleotides were replaced by deoxythymidine. The melting temperature (T_m) of the Ni(II)-coordinated 3WJ was estimated to be 9 °C higher than the metal-free **L1L2L3**, and 13 °C higher than the natural 3WJ **S1S2S3**. The stabilization was attributed to metal-mediated crosslinking among the three DNA strands through the formation of an $Ni(bpy)_3^{2+}$ complex at the branching point, as further evidenced by titration experiments and UV spectral measurements. The fact that 3WJ structures with only one or two bpy ligands (**S1S2L3** or **S1L2L3**) showed little stabilization upon Ni(II) addition also reinforced the idea of interstrand crosslinking by 3:1 ligand–metal complexation. Additionally, CD spectroscopy indicated the predominant formation of the Λ-isomer of the $Ni(bpy)_3^{2+}$ complex, albeit in low diastereomeric excess, as well as its resistance to racemization. This could arise from the chiral DNA environment at the core of the 3WJ.

Some of the other divalent transition metal ions also allowed for the stabilization of the bpy-modified junction. The melting temperatures in the presence of equimolar metal ions decreased in the order Ni(II) > Fe(II) > Co(II) (ΔT_m = +8.9, +5.0, and +3.3 °C, respectively, compared with the metal-free 3WJ). This progression agrees with the order of the β_3 values of the corresponding trisbipyridine complexes $[M(bpy)_3]^{2+}$ in aqueous solution (log β_3 = 20.2, 17.2, and 13.2, respectively). This result suggests the possibility that the thermal stability of the metallo-3WJ motifs can be regulated based on the conventional coordination thermodynamics.

A similar molecular design strategy could be further applied to other DNA branched structures, such as four-way junctions. Such a metallo-DNA junction motif will have future applications as a vertex or a node of 2D and 3D DNA architectures, and can in principle be embedded into DNA origami with ease. The stability and rigidity of metallo-DNA junctions can be altered by reversible metal coordination, namely by adding or removing metal ions from the structures. Therefore, the metallo-DNA junction is a potential component of DNA nanomachines and DNA nanodevices whose dynamics and functions can be regulated in response to metal coordination.

1.5.6 Summary and outlook

Construction of DNA supramolecular architectures has been conventionally based on the sequence-specific hybridization of oligonucleotides, which can be programmed using theoretical predictions of the thermodynamic stability of DNA duplexes. The introduction of metal complexes into DNA structures improves the programmability of DNA supermolecules because of the wide variety of geometrical, thermodynamic, and kinetic characteristics of metal coordination that can be incorporated into the self-assembly process. Ligand-bearing DNA strands, precursors of metallo-DNAs, can be readily prepared on a standard DNA synthesizer or via post-synthetic modification. Annealing of the artificial DNA oligomers and appropriate metal ions leads to the self-assembly of metal-containing DNA supramolecules, in which hydrogen bonding between natural nucleobases as well as metal coordination of the additional ligand moieties are formed in a concerted manner. Such an approach was initially presented as a replacement of natural nucleobases by metal ligands that provide metal-mediated artificial base pairs within DNA duplexes. As overviewed in this chapter, various types of metallo-DNA triplexes, quadruplexes and also junction structures were constructed using the same design strategy.

The advantages of using metallo-DNA motifs are summarized as follows: (i) the stability of DNA structures is enhanced by the additional metal coordination bonding; (ii) the hybridization or self-assembly of DNA strands can be modulated based on the metal coordination; and (iii) the conversion of DNA structural motifs can be induced by addition and removal of metal ions. These properties are in theory tunable by selecting appropriate combinations of ligand moieties and metal ions, as has already been demonstrated in the molecular design of supramolecular metal complexes. Using such design ideas, DNA self-assembly could be designed based upon the different thermodynamics of hydrogen bonding and metal coordination, so as to build up more complicated nanoarchitectures. Fine-tuning of the duplex stability would enable metal complexation-triggered strand exchange reactions as well as conformational changes. Therefore, the metallo-DNA structural motifs presented here will be applied to the development of metal-driven or stimuli-responsive DNA molecular machines and devices, which should be among the next and most challenging targets in supramolecular chemistry and nanotechnology. The collaboration of DNA nanotechnology and metal coordination supramolecular chemistry, both of which are rapidly evolving fields, will open a new chapter in the construction of DNA-based nanoarchitectures.

References

[1] H. Yang, K. L. Metera, and H. F. Sleiman, DNA modified with metal complexes: Applications in the construction of higher order metal-DNA nanostructures, *Coord. Chem. Rev.* 2010, **254**, 2403–2415.

[2] C. K. McLaughlin, G. D. Hamblin, and H. F. Sleiman, Supramolecular DNA assembly, *Chem. Soc. Rev.* 2011, **40**, 5647–5656.

[3] Y. Takezawa and M. Shionoya, Metal-mediated DNA base pairing: Alternatives to hydrogen-bonded Watson-Crick base pairs, *Acc. Chem. Res.* 2012, **45**, 2066–2076.

[4] G. H. Clever and M. Shionoya, Metal-base pairing in DNA, *Coord. Chem. Rev.* 2010, **254**, 2391–2402.

[5] P. Scharf and J. Müller, Nucleic acids with metal-mediated base pairs and their applications, *ChemPlusChem* 2010, **78**, 20–34.

[6] S. Katz, The reversible reaction of Hg(II) and double-stranded polynucleotides. A step-function theory and its significance, *Biochim. Biophys. Acta* 1963, **68**, 240–253.

[7] Z. Kuklenyik and L. G. Marzilli, Mercury(II) site-selective binding to a DNA hairpin. relationship of sequence-dependent intra- and interstrand cross-linking to the hairpin-duplex conformational transition, *Inorg. Chem.* 1996, **35**, 5654–5662.

[8] Y. Miyake, H. Togashi, M. Tashiro, H. Yamaguchi, S. Oda, M. Kudo, Y. Tanaka, Y. Kondo, R. Sawa, T. Fujimoto, T. Machinami, and A. Ono, $Mercury^{II}$-mediated formation of thymine-Hg^{II}-thymine base pairs in DNA duplexes, *J. Am. Chem. Soc.* 2006, **128**, 2172–2173.

[9] Y. Tanaka, S. Oda, H. Yamaguchi, Y. Kondo, C. Kojima, and A. Ono, ^{15}N-^{15}N J-coupling across Hg(II): direct observation of Hg(II)-mediated T-T base pairs in a DNA duplex, *J. Am. Chem. Soc.* 2007, **129**, 244–245.

[10] A. Ono, S. Cao, H. Togashi, M. Tashiro, T. Fujimoto, T. Machinami, S. Oda, Y. Miyake, I. Okamoto, and Y. Tanaka, Specific interactions between silver(I) ions and cytosine-cytosine pairs in DNA duplexes, *Chem. Commun.* 2008, 4825–4827.

[11] A. Ono, H. Torigoe, Y. Tanaka, and I. Okamoto, Binding of metal ions by pyrimidine base pairs in DNA duplexes, *Chem. Soc. Rev.* 2011, **40**, 5855–5866.

[12] K. Tanaka, A. Tengeiji, T. Kato, N. Toyama, M. Shiro, and M. Shionoya, Efficient incorporation of a copper hydroxypyridone base pair in DNA, *J. Am. Chem. Soc.* 2002, **124**, 12494–12498.

[13] N. Zimmermann, E. Meggers, and P. G. Schultz, A novel silver(I)-mediated DNA base pair, *J. Am. Chem. Soc.* 2002, **124**, 13684–13685.

[14] L. Zhang and E. Meggers, An extremely stable and orthogonal DNA base pair with a simplified three-carbon backbone, *J. Am. Chem. Soc.* 2005, **127**, 74–75.

[15] G. H. Clever, K. Polborn, and T. Carell, A highly DNA-duplex-stabilizing metal-salen base pair, *Angew. Chem., Int. Ed.* 2005, **44**, 7204–7208.

[16] N. Zimmermann, E. Meggers, and P. G. Schultz, A second-generation copper(II)-mediated metallo-DNA-base pair, *Bioorg. Chem.* 2004, **32**, 13–25.

[17] C. Switzer, S. Sinha, P. H. Kim, and B. D. Heuberger, A purine-like nickel(II) base pair for DNA, *Angew. Chem., Int. Ed.* 2005, **44**, 1529–1532.

[18] K. Tanaka, A. Tengeiji, T. Kato, N. Toyama, and M. Shionoya, A discrete self-assembled metal array in artificial DNA, *Science* 2003, **299**, 1212–1213.

[19] G. H. Clever, S. J. Reitmeier, T. Carell, and O. Schiemann, Antiferromagnetic coupling of stacked Cu(II)-salen complexes in DNA, *Angew. Chem., Int. Ed.* 2010, **49**, 4927–4929.

[20] K. Tanaka, G. H. Clever, Y. Takezawa, Y. Yamada, C. Kaul, M. Shionoya, and T. Carell, Programmable self-assembly of metal ions inside artificial DNA duplexes, *Nat. Nanotechnol.* 2006, **1**, 190–194.

[21] S. Liu, G. H. Clever, Y. Takezawa, M. Kaneko, K. Tanaka, X. Guo, and M. Shionoya, Direct conductance measurement of individual metallo-DNA duplexes within single-molecule break junctions, *Angew. Chem., Int. Ed.* 2011, **50**, 8886–8890.

[22] D. Bohme, N. Dupre, D. A. Megger, and J. Müller, Conformational change induced by metal-ion-binding to DNA containing the artificial 1,2,4-triazole nucleoside, *Inorg. Chem.* 2007, **46**, 10114–10119.

[23] S. Johannsen, N. Megger, D. Bohme, R. K. Sigel, and J. Müller, Solution structure of a DNA double helix with consecutive metal-mediated base pairs, *Nat. Chem.* 2010, **2**, 229–234.

[24] T. Ihara, T. Ishii, N. Araki, A. W. Wilson, and A. Jyo, Silver ion unusually stabilizes the structure of a parallel-motif DNA triplex, *J. Am. Chem. Soc.* 2009, **131**, 3826–3827.

[25] K. Tanaka, Y. Yamada, and M. Shionoya, Formation of silver(I)-mediated DNA duplex and triplex through an alternative base pair of pyridine nucleobases, *J. Am. Chem. Soc.* 2002, **124**, 8802–8803.

[26] Y. Takezawa, W. Maeda, K. Tanaka, and M. Shionoya, Discrete self-assembly of iron(III) ions inside triple-stranded artificial DNA, *Angew. Chem., Int. Ed.* 2009, **48**, 1081–1084.

[27] J. C. Chaput and C. Switzer, A DNA pentaplex incorporating nucleobase quintets, *Proc. Natl. Acad. Sci. U.S.A.* 1999, **96**, 10614–10619.

[28] M. Kang, B. Heuberger, J. C. Chaput, C. Switzer, and J. Feigon, Solution structure of a parallel-stranded oligoisoguanine DNA pentaplex formed by $d(T(iG)_4T)$ in the presence of Cs^+ ions, *Angew. Chem., Int. Ed.* 2012, **51**, 7952–7955.

[29] D. M. Engelhard, R. Pievo, and G. H. Clever, Reversible stabilization of transition-metal-binding DNA G-quadruplexes, *Angew. Chem., Int. Ed.* 2013, **52**, 12843–12847.

[30] H. A. Day, C. Huguin, and Z. A. Waller, Silver cations fold i-motif at neutral pH, *Chem. Commun.* 2013, 7696–7698.

[31] N. C. Seeman, Nucleic-acid junctions and lattices, *J. Theor. Biol.* 1982, **99**, 237–247.

[32] H. Yang and H. F. Sleiman, Templated synthesis of highly stable, electroactive, and dynamic metal-DNA branched junctions, *Angew. Chem., Int. Ed.* 2008, **47**, 2443–2446.

[33] A. Oleksi, A. G. Blanco, R. Boer, I. Uson, J. Aymami, A. Rodger, M. J. Hannon, and M. Coll, Molecular recognition of a three-way DNA junction by a metallosupramolecular helicate, *Angew. Chem., Int. Ed.* 2006, **45**, 1227–1231.

[34] J. L. Duprey, Y. Takezawa, and M. Shionoya, Metal-locked DNA three-way junction, *Angew. Chem., Int. Ed.* 2013, **52**, 1212–1216.

[35] K. C. Woods, S. S. Martin, V. C. Chu, and E. P. Baldwin, Quasi-equivalence in site-specific recombinase structure and function: Crystal structure and activity of trimeric Cre recombinase bound to a three-way Lox DNA junction, *J. Mol. Biol.* 2001, **313**, 49–69.

Part II

DNA Wires and Electron Transport Through DNA

2.1

Gating Electrical Transport Through DNA

Kazushige Yamana

Department of Materials Science and Chemistry, University of Hyogo, Himeji, Japan

2.1.1 Introduction

DNA is an attractive material as a one-dimensional conductor and can be applied for use in molecular wires [1–3], since the frontier orbitals of the bases, the highest occupied molecular orbital (HOMO), and the lowest unoccupied molecular orbital (LUMO) are π-electrons residing perpendicular to the molecular plane. Taking advantage of the properties of DNA and integrating DNA with current semiconductor technology could lead to the development of nanoscale electronic circuits, switches, and memory devices.

Barton and coworkers began to study the charge transport properties of DNA through the photoinduced charge migration between donors and accepters appended onto DNA. They obtained the experimental evidence for distance-independent charge transfer, which indicated that the charge migration in DNA could occur over a significant distance [4]. This result attracted attention from many researchers and triggered active experimental and theoretical research in the area of charge transfer through DNA [5–18]. It is now accepted that both positive (hole) and negative (electron) charges injected into DNA can move through DNA over significantly long distances.

While a large number of charge-transfer experiments support the view that DNA acts as a conducting material, direct conductivity measurements of DNA have yielded conflicting results. The electrical measurements of DNA, where DNA chains were coupled to voltage-biased leads, have shown different types of conduction ranging from insulating [19, 20] via semiconducting [21–24] to quasi-metallic [24–28] and superconducting [29] behaviors.

In order to consider DNA-based electronics, scientific and technical challenges need to be addressed to enable the electrical transport in DNA to be studied through the development of a suitable structure for a DNA-electrode contact and to sustain a stable DNA double helix. More importantly, the external modulation of electrical transport in DNA is a crucial objective for an active electronic device, such as a field-effect transistor. The aim of this chapter is to give an overview of the challenges involved in realizing DNA-based electronic devices.

DNA in Supramolecular Chemistry and Nanotechnology, First Edition.
Edited by Eugen Stulz and Guido H. Clever.

2.1.2 DNA structure

Naturally occurring DNA with a random sequence can adopt several conformations, such as A-type, B-type, and others (such as Z-type), depending on the DNA sequences and the environmental conditions, such as counterions, relative humidity, and ionic strength or solvent polarity in the solution [30]. In contrast, RNAs are found only in two related conformations, A and A′, both of which belong to A-type double helical structures. As shown in Figure 2.1.1, a right-handed DNA double helix takes a B-type conformation under relatively high humidity and an A-type conformation under low humidity. The conformation of the A-DNA structure closely resembles that of A-RNA, but the axial rise per residue in A-DNA, 2.56 Å, is shorter than that of A-RNA, 2.86 Å [30]. In a B-type double helix, stacking is largely limited to interactions between bases in the same polynucleotide chain. In an A-type double helix, both intra- and interstrand stacking occur and the stacking involves bases belonging to two different chains. Sequence-dependent deviations scarcely occur in the A-DNA conformation. In contrast, they are very obvious in B-DNA Structural features of DNA should have a large effect on the electrical properties of DNA.

2.1.3 Direct electrical measurements of DNA

This section focuses on the direct electrical measurements of DNA in a dry state and therefore the DNA conformation is of the A-type. De Pablo *et al.* suggested that λ-DNA with a random sequence is an insulator [19]. As shown in Figure 2.1.2, they carried out the electric conduction measurement of λ-DNA bundles that are placed onto a mica substrate. The sample was dried before measurements were taken. One gold electrode

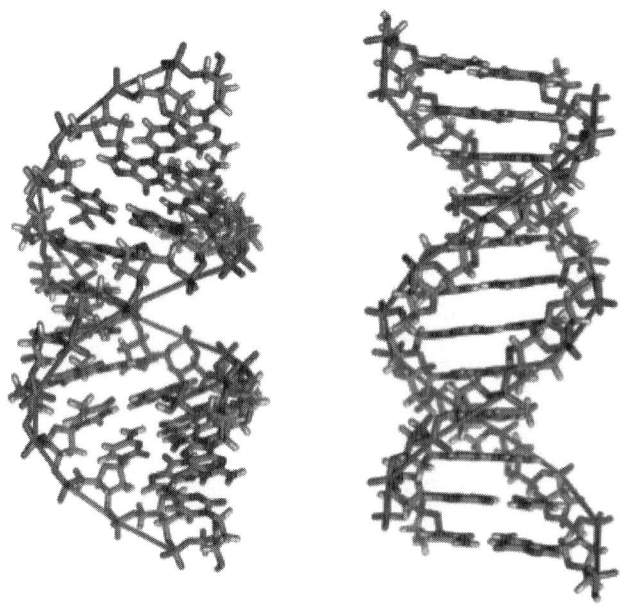

Figure 2.1.1 *Structures of A-type (left) and B-type DNA (right). The structure of DNA depends on its environment. In aqueous environments, including the majority of DNA in a cell, B-type DNA is the most common structure. The A-type DNA structure is dominates in dehydrated samples and is similar to the double-stranded RNA and DNA/RNA hybrids.*

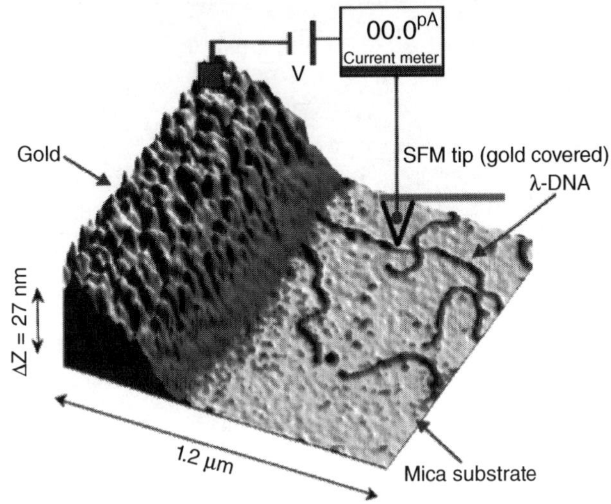

Figure 2.1.2 *Three-dimensional SFM image of the channel border, showing two DNA molecules in contact with the left gold electrode. The image size is 1.2 μm × 1.2 μm. A scheme of the electrical circuit used to measure the DNA resistivity is shown. Reproduced with permission from [19], from American Physical Society*

was produced by evaporation on the end of the DNA and a gold-coated tip was used to measure the current flow on the DNA between two electrodes. A lower resistivity limit of 10^6 Ω cm for a DNA molecule was obtained. However, Kasumov *et al.* argued that the unusual structure of DNA on the mica surface is responsible for such a low resistivity [28]. Storm *et al.* reported the absence of electronic transport through λ-DNA, even when the were DNAs bridged via the thiol-metal contact between Au and Pt nanogaps (40–500 nm) [20].

In other direct electrical measurements, it was found that DNA acts as a semiconducting material. Porath *et al.* measured the conductivity of a $(dG)_{30}$–$(dC)_{30}$ DNA duplex [21]. An electrostatic trapping technique was used to position single DNA molecules between an 8 nm gap in the metal electrodes, where the estimated length of the DNA used is about 10 nm in solution. The sample was dried with a flow of nitrogen before the electrical measurement. The *I–V* curves obtained from this experiment are shown in Figure 2.1.3. As is clear from this figure, the DNA oligomer does not conduct charge for biases below about 1 V at room temperature, which shows that poly(dG)–poly(dC) DNA behaves like a semiconductor with a large bandgap. It was argued that DNA molecular orbital bands are responsible for the mediation of electrical transport on DNA. The conductive AFM measurement of 26-mer DNA resulted in the non-linear S-shaped *I–V* characteristics [24].

Okahata and coworkers showed that the conductivity of aligned DNA film from salmon testes is comparable to that of conventional conductive polymers, such as polyacetylene and polyphenylene, with no doping [25]. Fink and Schonenberger performed a direct measurement of the current flow through DNA using a modified low-energy electron point-source microscope and indicated that DNA acts as a conducting molecular wire [26]. They used a gold-coated carbon grid as the sample holder. The DNA sample structure consists of an array of 2-μm holes and 600-nm long λ-DNA to span these holes. The measurements were carried out at room temperature under vacuum. Linear current voltage (*I–V*) curves in the range of from –20 to +20 mV were obtained as indicated in Figure 2.1.4. By connecting thiol-modified ends of λ-DNA between the 8 μm gap Au electrodes [27], it was shown (Figure 2.1.5) that linear current–voltage (*I–V*) characteristics are obtained on the intact DNA. In a similar electrode system, nonlinear S-shaped *I–V* curves with a conductivity gap were obtained for the DNA possessing nicks.

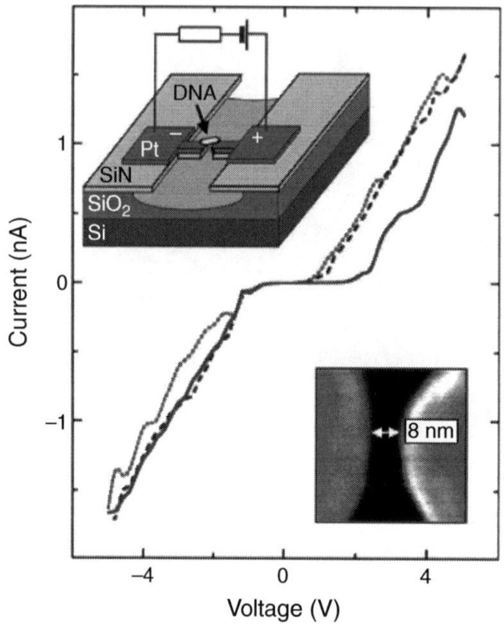

Figure 2.1.3 *Current–voltage curves measured at room temperature on a DNA molecule trapped between two metal nanoelectrodes. The DNA molecule is 30 base pairs, double-stranded poly(dG)–poly(dC), and the nanoelectrodes are separated by 8 nm. The upper inset shows a schematic of our sample layout. The lower inset is a scanning-electron-microscope image of the two metal electrodes (light area) and the 8-nm gap between them (dark area) [21]. Reprinted by permission from Macmillan Publishers Ltd, Copyright 2000. (See color figure in color plate section)*

In addition to insulating and conducting behaviors, Kasumov *et al.* reported proximity-induced superconductivity in DNA (Figure 2.1.6) [29]. Using sputtering techniques, rhenium/carbon electrodes were deposited on a freshly cleaved mica surface. A solution of 16 μm long λ-DNA was then allowed to flow onto the surface, which introduced about 100–200 DNA molecules, bridging the gap between the electrodes. The resistance of the DNA bridges measured in a dry state at room temperature was found to be a few hundred ohms per DNA molecule, indicating that DNA is a conductive material. Resistance measurements at low temperatures of λ-DNA between superconducting electrodes showed that proximity-induced superconductivity could be realized in DNA.

2.1.4 Gate modulation of current flow in DNA

Yoo *et al.* measured the conductance of poly(dG)–poly(dC) (1.7–2.9 μm) and poly(dA)–poly(dT) (0.5–1.5 μm) [22]. DNA was electrostatically trapped between Au/Ti electrodes with a 20 nm gap fabricated onto SiO_2 and dried with nitrogen. The *I–V* curves observed for poly(dA)–poly(dT) DNA were linear at room temperature. At the low temperatures, below 161 K, the *I–V* curves for poly(dA)–poly(dT) became nonlinear with a large band gap. Poly(dG)–poly(dC) DNA displays similar temperature-dependent conducting behavior to that of poly(dA)–poly(dT), but with a much lower resistance of 1.3 MΩ at room temperature, compared with 100 MΩ for poly(dA)–poly(dT). Yoo *et al.* also investigated the effect of back gate voltage on the *I–V* characteristics.

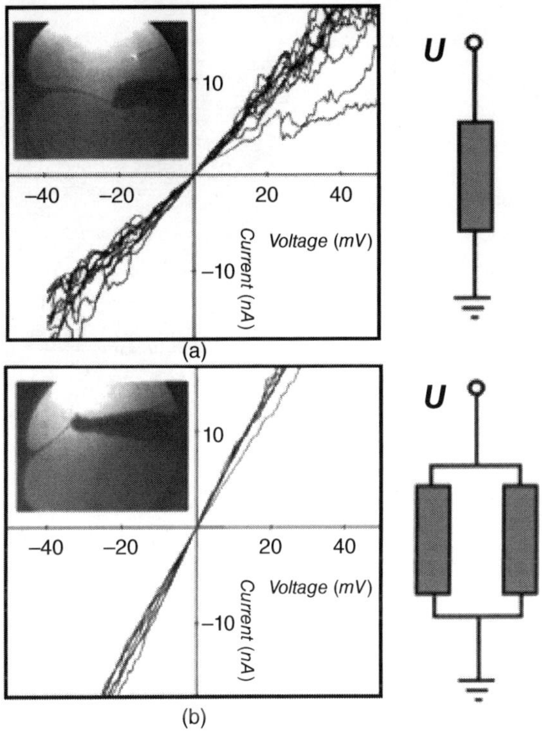

Figure 2.1.4 *I-V characteristics of DNA ropes. Panel a: I–V curve taken for a 600 nm long DNA rope. In the range of ± 20 mV, the curves are linear to give a resistance of about 2.5 MΩ. Panel b: I–V curve when the manipulation-tip is attached to both DNA ropes. The measured resistance drops to 1.4 MΩ [26]. Reprinted by permission from Macmillan Publishers Ltd, Copyright 1999*

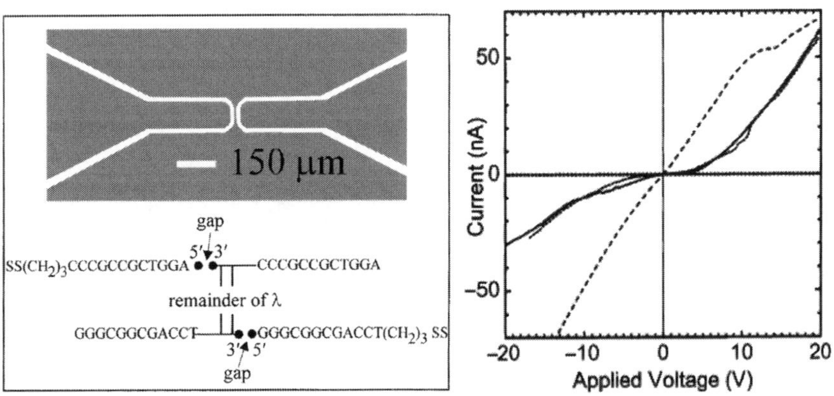

Figure 2.1.5 *Current–voltage characteristics measured at room temperature on disulfide-labeled λ-DNA molecules. Dashes, repaired DNA; dots, nicked DNA, swept from negative to positive potential; solid line, nicked DNA, swept from positive potential to negative potential; filled circles, electrodes without DNA (zero line) [27]*

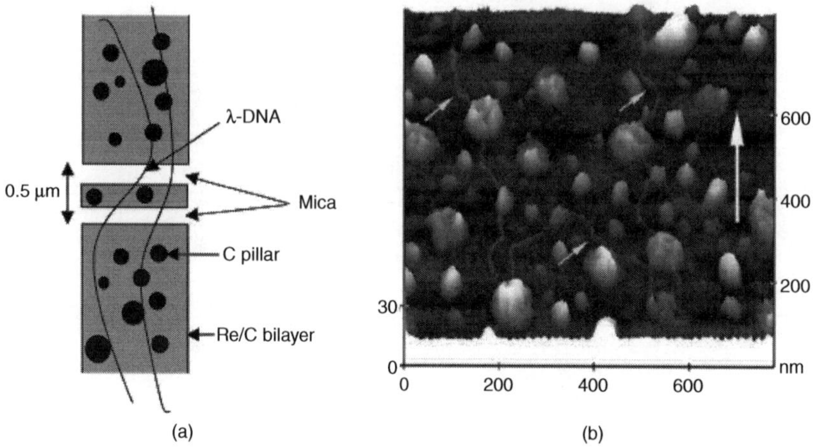

Figure 2.1.6 *(a) Schematic drawing of the measured sample, with DNA molecules combed between Re/C electrodes on a mica substrate. (b) Atomic force microscopy image showing DNA molecules combed on the Re–C bilayer. The large vertical arrow indicates the direction of the solution flow. The small arrows point toward the combed molecules. Note the forest structure of the carbon film [29]*

As shown in Figure 2.1.7, poly(dA)–-poly(dT) DNA has a larger conductance upon application of a positive gate voltage, while poly(dG)–poly(dC) DNA has enhanced conductance under negative gate voltage conditions. Thus, in these experiments, poly(dA)–poly(dT) DNA acts as an n-type semiconducting material, whereas poly(dG)–poly(dC) DNA displays p-type semiconducting behavior.

Watanabe *et al.* measured the electric properties of a single molecule DNA device with a triple-probe atomic force microscope (T-AFM) [23]. For fabrication of the device, a solution of salmon sperm DNA was deposited on an SiO_2/Si substrate, rinsed with water, and then dried in a flow of nitrogen gas. As shown in Figure 2.1.8, carbon nanotube (CNT) electrodes as the source (S), drain (D), and gate (G) were used to connect the single DNA molecule. The electrical measurements were carried out at room temperature. They observed that an increase in the gate bias voltage resulted in enhancement of the source–drain current in the DNA device, where tDNA with a random sequence was used for the device.

2.1.5 DNA transistors

The basic unit of an integrated circuit, such as a computer processor or a memory chip, is the transistor, which operates as a switch between a conducting and a non-conducting state. Until recently, the number of transistors that could be fabricated on an ultra-large scale integrated circuit (ULSI) doubled approximately every 2 years, resulting in dramatic increases in ULSI performance. However, it is known that the fabrication process involved in the miniaturization of transistors faces physical, technological, and economical limitations. To overcome these limitations, new materials other than silicon are in high demanded in order to integrate more materials into a given space. It has been shown that carbon nanotubes and graphene are promising materials for future commercial applications in molecular electronics [31–34]. DNA is another promising material because it is straightforward to use in syntheses and modifications also because of its self-assembling properties to form pre-determined 1D–multi-dimensional structures that are useful for the fabrication into nanospaces.

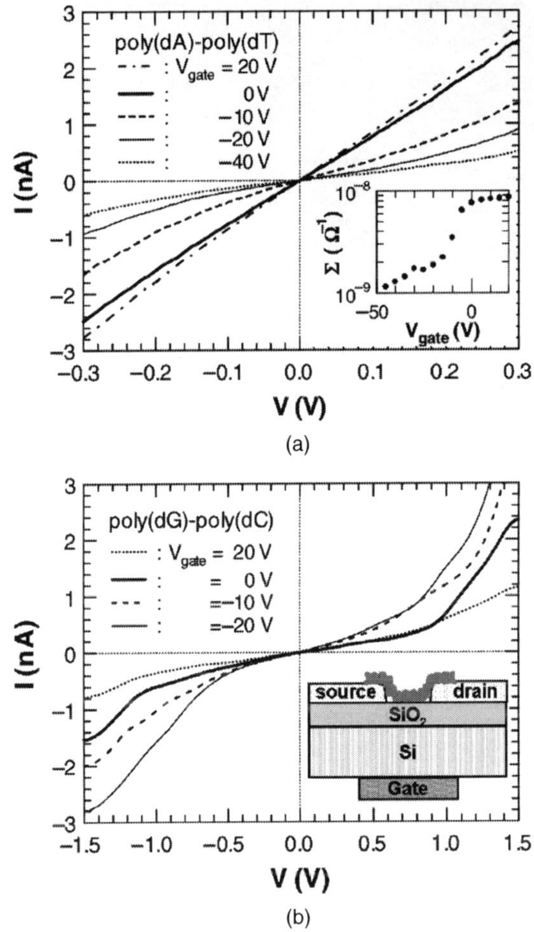

Figure 2.1.7 *I–V characteristics for poly(dA)–poly(dT) (a) and poly(dG)–poly(dC) (b) at various gate voltages. In the inset (a), the conductance at V = 0 is plotted as a function of gate voltage for poly(dA)–poly(dT). The inset of (b) is the schematic diagram of electrode arrangement for gate dependent transport experiments [22]. Reproduced with permission from [22] from American Physical Society*

Yamana, Matsuo and coworkers have been interested in the development of an active electronic device, such as a field-effect transistor (FET), consisting of DNA as a molecular semiconducting wire [35, 36]. Advantage has been taken of recent silicon technology as a source for three-terminal (source, drain, and gate electrodes) devices. Fabrication of DNA-FET devices began with a 4-μm thick, n-type silicon-on-insulator with a resistivity of 40–60 Ω cm. The SiO_2 insulator thickness is 500 nm. The thickness of the top Si was reduced from 4.0 μm to 150 nm by the chemical dry etching. The source and drain electrodes with nanogaps (channel lengths = 120–200 nm) were then formed using a lithography process. The width of the gap, which is the channel width, is 100 μm. Aluminium was sputtered onto the specific region of the remaining n-Si surface to ensure electrical contact to the source and drain electrodes. The optical and SEM images of the silicon-based three terminal electrodes and the DNA device configuration are depicted in Figure 2.1.9.

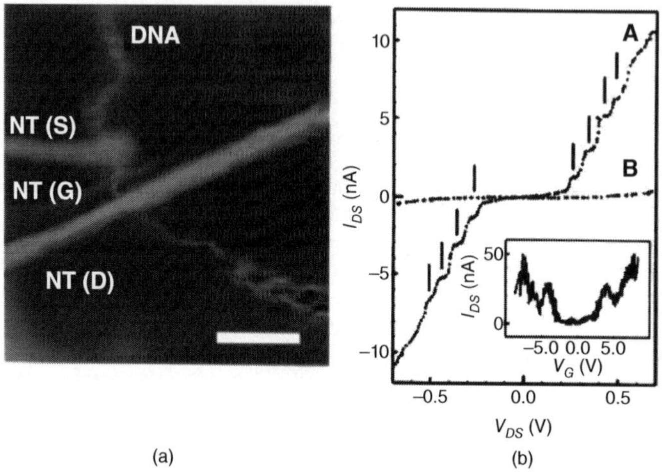

Figure 2.1.8 (a) AFM image (scale bar, 10 nm) of single DNA molecule crossed with a SWCNT. CNT(S), CNT(D), and CNT(G) indicate the source, drain, and gate CNT electrodes, respectively. (b) I_{DS}–V_{DS} curve (curve A) at V_G = 2.0 V. I_{DS}–V_{DS} curve without DNA at V_G = 5 V (curve B) is shown. Inset shows the I_{DS}–V_G curve at V_{DS} = 0.5 V. Reproduced with permission from [23] from American Physical Society

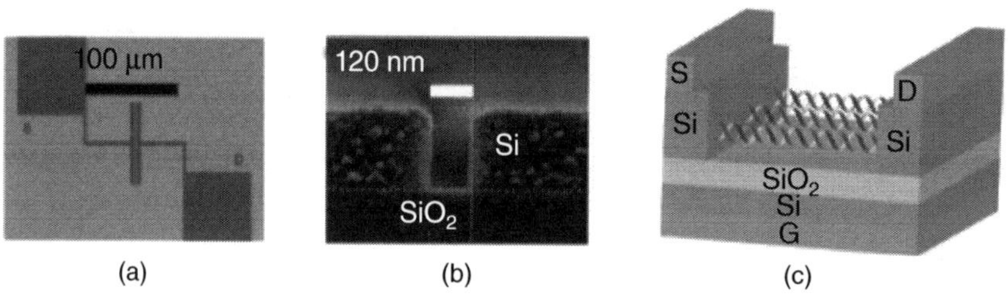

Figure 2.1.9 (a) Optical microscope image of the top view of silicon based three-terminal electrodes. (b) SEM image of the side view of the electrodes. (c) Schematic representation of DNA devices. S, D, and G indicate source, drain, and gate electrodes, respectively. Reproduced with permission from [35]. The Royal Society of Chemistry

For the DNA bridging between the Si nanogap electrodes, both the electrodes and the DNA were chemically modified (Figure 2.1.10). An epoxy function was attached through hydrosilylation of allyl glycidyl ether on the Si surface of the electrodes. Disulfide-terminated double strand (ds) DNA (400 base pairs: 205 A–T bp and 195 G–C bp) was prepared by a PCR technique using 5′-disulfide-modified DNA primers (5′-$HO(CH_2)_3SS(CH_2)_3$-TGCACCGCCAGATATTCC and 5′-$HO(CH_2)_3SS(CH_2)_3$-ATCAACACG GTTCAGCAACA) with λ-DNA as a template. The conversion from disulfide- to thiol-terminated dsDNA (HS-dsDNA–SH) was carried out by treatment with dithiothreitol (DTT) prior to use. For comparison, dsDNA with an SH group at one terminus (HS-dsDNA) was prepared. The thiol-terminated dsDNA, dissolved in an aqueous buffer (pH 8), was deposited onto the epoxy-functionalized electrodes at room temperature to yield the DNA-bridged structures. The DNA devices were quickly washed with pure water, dried, and then characterized at room temperature under atmospheric conditions using a Keithley semiconductor analyzer.

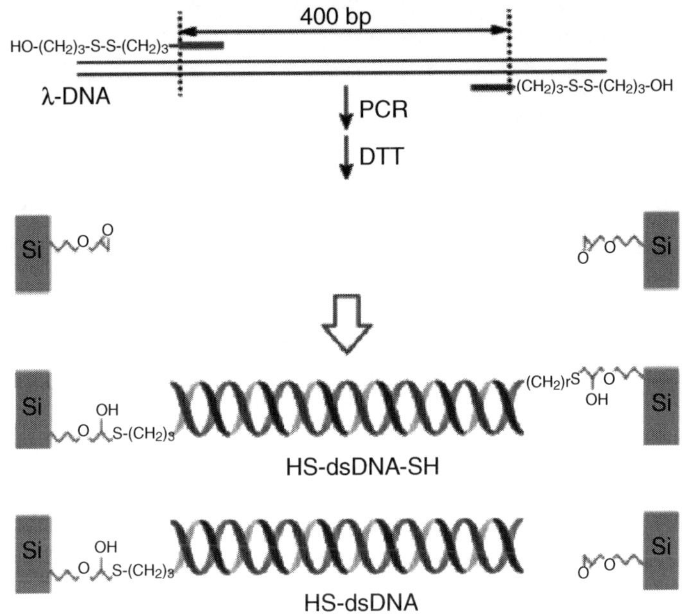

Figure 2.1.10 *DNA connection between silicon nanogap. The Si surfaces were modified by hydrosilylation using allylglycidyl ether. Thiol (SH)-terminated 400 bp long dsDNA (HS-dsDNA-SH or HS-dsDNA) was prepared by using λ-DNA as the template and disulfide modified PCR primers [35]. Reproduced with permission from [35]. The Royal Society of Chemistry*

The contact effects of DNA-electrodes on the electrical transport in the DNA devices were examined first. Figure 2.1.11 indicates the drain current–voltage, I_d–V_d, characteristics for the dsDNA devices at the gate voltage, $V_g = 0$. The devices composed of HS-dsDNA-SH in the 120 nm channel display S-shaped *I*–*V* curves with a voltage gap at the low applied bias. The curves were symmetrical. A current of 5×10^{-7} A at 0.8 V was obtained. No drop in the current was observed for repeated measurements. The DNA devices were not degraded at room temperature over several months. In contrast, in the HS-dsDNA-SH devices with a channel length of 150 nm, a current of 5×10^{-8} A at 0.8 V was obtained. The 180 nm devices of HS-dsDNA-SH showed very low currents against the applied drain voltages (5×10^{-12} A at 0.8 V). We tested the long length ($L \geq 190$ nm) devices in a similar way and obtained very low currents (3×10^{-12} A at 0.8 V). The devices produced by the one HS-terminal dsDNA (HS-dsDNA) with lengths of 120 and 150 nm gave very low currents of 3×10^{-11} and 6×10^{-12} A, respectively, at 0.8 V.

The length of the DNA samples in the dry state is considered to be close to 120 nm (A-form) rather than to 135 nm (B-form). Therefore, the results of *I*–*V* measurements imply that the chemical bonding between both terminals of the dsDNA and the electrodes are necessary for the high current flows (500 nA at 0.8 V) through the DNA molecules. No contributions of residual salts to the high currents are evident because the single HS-modified dsDNA and the long gap devices (gap length ≥ 180 nm) for the DNA bridges between the electrodes exhibit 1/10000 lower current.

To clarify the impact of double helix stability of DNA in the devices, the following experiments were carried out. The electrodes (120 nm channel length) were first attached to short single strand (ss) DNA (15 bases) through the epoxy-thiol bond formation. The dsDNA (400 base pairs) possessing two overhang sequences (15 bases) was then hybridized with the ssDNAs. In this procedure, the formation of the electrode–dsDNA–electrode structures relied on the duplex formation between the short DNA sequences (15 base

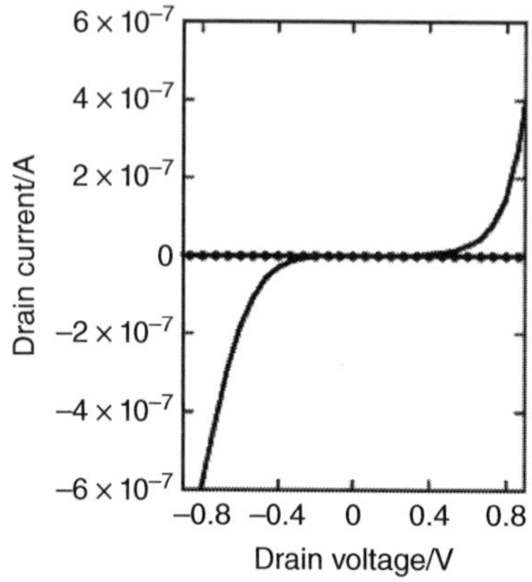

Figure 2.1.11 *I_d–V_d curves measured on dsDNA molecules (HS-dsDNA-SH or HS-dsDNA) embedded within the nanogap source–drain electrodes. Bold line: HS-dsDNA-SH in 120 nm gap. Fine line: HS- dsDNA-SH in 180 nm gap. Scattered line: HS-dsDNA in 120 nm gap. All the measurements were carried out at room temperature in air (50–60% humidity) [35]. Reproduced with permission from [35]. The Royal Society of Chemistry*

pairs). However, with this DNA connection procedure there were some problems obtaining reproducible current–voltage curves with such DNA devices. This is because the 15-mer dsDNA is not stable and may thus be dehybridized by pure water in the final washing step. These results indicate that stable double helical DNA structures are important for obtaining DNA devices showing reliable *I–V* output characteristics.

We then investigated the electronic field effect on the current flows in the DNA devices. Figure 2.1.12a and b shows the representative I_d–V_d characteristics for the device of HS-dsDNA–SH (400 base pairs) in the 120 nm channel measured at different V_g values. The plot of drain current, I_d, versus gate voltage, V_g, for a drain voltage, V_d, of 0.7 V is shown in Figure 2.1.12c. The asymmetric dependence of I_d on V_g was observed: the drain current, I_d, increases as V_g becomes increasingly negative and it decreases at positive V_g. These output characteristics indicate that the DNA in this device acts as a p-type semiconducting channel. The current ratio for between V_g = +5 V and V_g = –5 V is larger than 1×10⁴. The threshold voltage, V_{th}, is 0.6 V.

The observed gate voltage dependences indicate that the negative gate voltages lead to the hole accumulation in this DNA nanowire transistor. The DNA device is "on" at $V_g = 0$, which indicates that the Fermi energy of the electrode is close to the molecular energy bands of DNA. The molecular band conduction model has been used to explain the semiconducting *I–V* characteristics for dry dsDNA ($(dG)_{30}$–$(dC)_{30}$ DNA duplex) [19]. In the present case, the holes are injected by tunneling through the contact barrier to the DNA. If indeed the band-bending length were to be comparable to the DNA length, a positive gate voltage would generate an energy barrier of an appreciable fraction of εV_g in the center of the DNA. The observed threshold voltage, V_{th}–0.6 V, required to suppress hole conduction would be correlated with the thermal energy to overcome this barrier. Based on the external electrostatic effects, our DNA nanowire transistor performs similarly to a carbon nanotube field-effect transistor (CNT-FET) [32, 36].

An attempt was made to obtain a rough estimate of the hole mobility of the DNA nanowire transistor. Assuming that *N* parallel DNA molecules are embedded between the gap, the hole mobility can be given by Equation 2.1.1:

$$\mu_h = \left(\frac{L^2}{NC_g \bullet V_d} \right) \frac{dI_d}{dV_g} \tag{2.1.1}$$

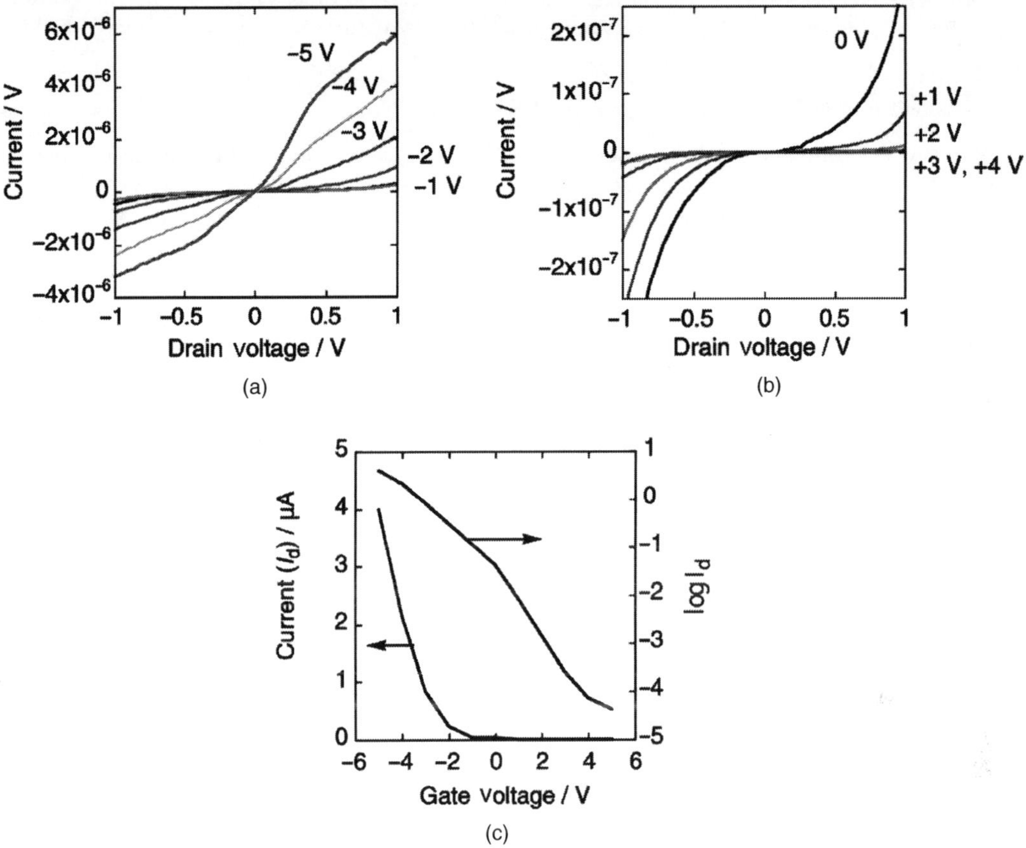

Figure 2.1.12 *(a) The representative I_d–V_d characteristics for the DNA device (HS-dsDNA-SH) in a 120 nm gap electrodes measured at the negative V_g values. (b) The I_d–V_d characteristics for the same device measured at the positive V_g values. (c) The plot of drain current, I_d, and log(I_d) versus gate voltage, V_g, for a drain voltage, V_d, of 0.7 V. All the measurements were carried out at room temperature in air (50–60% humidity) [35]. Reproduced with permission from [35]. The Royal Society of Chemistry (See color figure in color plate section)*

From the linear region of the I_d–V_g dependence in Figure 2.1.12, the transconductance value, $dI_d/dV_g = 4.2\times10^{-6}$ A V^{-1}, was obtained. In the extreme case when a single DNA molecule is bridged in the gap, a wire parallel to the wall can be used to calculate the gate capacitance per unit length using Equation 2.1.2 [37, 38]:

$$\frac{C_g}{L} = \frac{2\pi\varepsilon\varepsilon_0}{\cosh^{-1}\left(\frac{r+h}{r}\right)} \tag{2.1.2}$$

where L and r are the source–drain distance and the DNA radius, respectively. Using $\varepsilon = 2.5$ (average of silicon oxide (3.9) and air (1)), $L = 120$ nm, $h = 650$ nm, and $r = 1$ nm, C_g is calculated to be ~2.3×10^{-18} F. If we assume that the number of DNA nanowires is 1000–10 000 molecules, the calculated total capacitance, NC_g, is 2.3–0.23 × 10^{-15} F. Therefore, the hole mobility is 3.7 × 10^{-2}–10^{-1} cm^2 V s^{-1}. This value is similar to those reported for organic semiconductors (10^{-3}–10^{-1} $cm^2 V^{-1} s^{-1}$). It is interesting to note that Majima and coworkers reported the electron mobility through consecutive thymidines to be 2.0×10^{-3} $cm^2 V^{-1} s^{-1}$ [13].

2.1.6 Summary and outlook

Several experiments on the charge transport through DNA have been conducted to answer the question of whether DNA could be used as a conducting molecular wire. The solution-phase studies on DNA-mediated charge transport have revealed that both anionic and cationic charges injected into DNA can migrate over distances of several nanometers. Through the direct conductance measurements of DNA, it is shown that double stranded (ds) DNA acts as a conducting or semiconducting material when the contacting structures between the DNA and bias-leads fulfill the conditions that charge injection occurs from the electrodes into DNA molecular bands.

The external modulation of the DNA conductance is crucial with regards to DNA-based electronic devices, such as field-effect transistors. The combination of current semiconductor technology and DNA chemistry is one of numerous possible approaches to the development of nanoscale electronic switches and memory devices.

We have found that large gate modulation of electrical transport in DNA is possible in the DNA devices where dsDNAs are fabricated into three-terminal silicon-based electrodes. However, in the silicon three-terminal devices, DNA with a random sequence acts as a p-type semiconducting material. It is not known why the silicon DNA devices show only p-type output. There also remain several important questions as to how structure, sequence, base composition, and length of DNA affect the electrical transport in dry DNA. It is therefore important to study the charge transport on DNA in the A-type conformation. Through answering these questions, we will endeavor to achieve the next generation of DNA-based electronics.

References

[1] M. Yamaguchi and T. Kawai, DNA electronics, *Physica E*, **33**, 1–12 (2006).

[2] H.-A. Wagenknecht (Ed.), *Charge Transfer in DNA: From Theory to Applications*, Wiley-VCH Verlag, Weinheim, 2006.

[3] G. Schuster (Ed.), *Long-Range Charge Transfer in DNA, Topics in Current Chemistry*, Springer-Verlag, Berlin, 2004.

[4] C. J. Murphy, M. R. Arkin, Y. Jenkins, N. D. Ghatlia, S. H. Bossmann, N. Turro, and J. K. Barton, Long-range photoinduced electron transfer through a DNA helix, *Science,* **262**, 1025–1029 (1993).

[5] F. D. Lewis, T. Wu, Y. Zhang, R. L. Letsinger, S. R. Greenfield, and M. R. Wasielewski, Distance-dependent electron transfer in DNA hairpins, *Science*, **277**, 673–676 (1997).

[6] R. N. Barnett, C. L. Cleveland, A. Joy, U. Landman, and G. B. Schuster, Charge migration in DNA: Ion-gated transport, *Science*, **294**, 567–571 (2001).

[7] B. Giese, J. Amaudrut, A.-K. Köhler, M. Spormann, and S. Wessely, Direct observation of hole transfer through DNA by hopping between adenine bases and by tunneling, *Nature*, **412**, 318–320 (2001).

[8] B. Giese, B. Carl, T. Carl, T. Carell, C. Behrens, U. Hennecke, O. Schiemann, and E. Feresin, Excess electron transport through DNA: A single electron repairs more than one UV-induced lesion, *Angew. Chem., Int. Ed.*, **116**, 1884–1887 (2004).

[9] C. Behrens, L. T. Burgdorf, A. Schwogler, and T. Carell, Weak distance dependence of excess electron transfer in DNA, *Angew. Chem., Int. Ed.,* **41**, 1763–1766 (2002).

[10] B. Elias, F. Shao, and J. K. Barton, Charge migration along the DNA duplex: Hole versus electron transport, *J. Am. Chem. Soc.*, **130**, 1152–1153 (2008).

[11] J. V. Weis, M. R. Wasielewski, A. K. Thazhathveetil, and F. D. Lewis, Efficient charge transport in DNA diblock oligomers, *J. Am. Chem. Soc.*, **131**, 9722–9727 (2009).

[12] K. Kawai, H. Kodera, Y. Osakada, and T. Majima, Sequence-independent and rapid long-range charge transfer through DNA, *Nature Chem.*, **1**, 156–159 (2009).

[13] M. J. Park, M. Fujitsuka, K. Kawai, and Y. Majima, Direct measurement of the dynamics of excess electron transfer through consecutive thymine sequence in DNA, *J. Am. Chem. Soc.*, **133**, 15320–15323 (2011).

[14] D.-C. U. Wenge, J. Wengel, and H.-A. Wagenknecht, Photoinduced reductive electron transfer in LNA:DNA hybrids: A compromise between conformation and base stacking, *Angew. Chem., Int. Ed.*, **51**, 10026–10029 (2012).

[15] N. Renaud, Y. A. Berlin, F. D. Lewis, and M. Ratner, Between superexchange and hopping: An intermediate charge-transfer mechanism in poly(A)-poly(T) DNA hairpins, *A. J. Am. Chem. Soc.*, **135**, 3953–3963 (2013).

[16] M. Ratner, Photochemistry: electronic motion in DNA, *Nature*, **397**, 480–481 (1999).

[17] H.-A. Wagenknecht, Reductive electron transfer and transport of excess electrons in DNA, *Angew. Chem., Int. Ed.*, **42**, 2454–2460 (2003).

[18] J. Genereux and J. K. Barton, Mechanisms for DNA charge transport, *Chem. Rev.*, **110**, 1642–1662 (2010).

[19] P. J. De Pablo, F. M. Herrero, J. Colchero, J. G. Herrero, P. Hererro, A. M. Baro, P. Ordejon, J. M. Soler, and E. Artacho, Absence of dc-conductivity in λ-DNA, *Phys. Rev. Lett.*, **85**, 4992–4995 (2000).

[20] A. J. Storm, J. van Noort, S. De Vries and C. Dekker, Insulating behavior for DNA molecules between nanoelectrodes at the 100 nm length scale, *Appl. Phys. Lett.*, **79**, 3881–3883 (2001).

[21] D. Porath, A. Beryadin, S. De Veries, and C. Dekker, Direct measurement of electrical transport through DNA molecules, *Nature*, **403**, 635–638 (2000).

[22] K.-H. Yoo, D. H. Ha, J.-O. Lee, J. W. Park, J. Kim, J. J. Kim, H.-Y. Lee, T. Kawai, and H. Y. Choi, Electrical conduction through poly(dA)-poly(dT) and poly(dG)-poly(dC) DNA molecules, *Phys. Rev. Lett.*, **87**, 198102–198106 (2001).

[23] H. Watanabe, C. Manabe, T. Shigematsu, K. Shimotani, and M. Shimizu, Single molecule DNA device measured with triple-probe atomic force microscope, *Appl. Phys. Lett.*, **79**, 2462–2464 (2001).

[24] H. Cohen, C. Nogues, R. Naaman, and D. Porath, Direct measurement of electrical transport through single DNA molecules of complex sequence, *Proc. Natl. Acad. Sci. U.S.A.*, **102**, 11589–11593 (2005).

[25] Y. Okahata, T. Kobayashi, K. Tanaka, and M. Shimomura, Anisotropic electric conductivity in an aligned DNA cast film, *J. Am. Chem. Soc.*, **120**, 6165–6166 (1998).

[26] H. W. Fink and C. Schonenberger, Electrical conduction through DNA molecules, *Nature*, **398**, 407–410 (1999).

[27] B. Hartzell, B. McCord, D. Asare, H. Chen, J. J. Heremans, and V. Soghomonian, Comparative current–voltage characteristics of nicked and repaired λ-DNA, *Appl. Phys. Lett.*, **82**, 4800–4803 (2003).

[28] A. Yu. Kasumov, D. V. Klinov, P.-E. Roche, S. Gueron, and H. Bouchiat, Thickness and low-temperature conductivity of DNA molecules, *Appl. Phys. Lett.*, **84**, 1007–1010 (2004).

[29] A. Yu. Kasumov, M. Kociak, S. Gueron, B. Reulet, V. T. Volkov, D. Klinov, and H. Bouchiat, Proximity-induced superconductivity in DNA, *Science*, **291**, 280–282 (2001).

[30] W. Saenger, *Principles of Nucleic Acid Structure*, Springer-Verlag, New York, 1983.

[31] M. F. L. De Volder, S. H. Tawfick, R. H. Baughman, and A. J. Hart, Carbon nanotubes: Present and future commercial applications, *Science*, **339**, 535–539 (2013).

[32] K. Kim, J.-Y. Choi, T. Kim, S.-H. Cho, and H.-J. Chung, A role for graphene in silicon-based semiconductor devices, *Nature*, **479**, 338–344 (2011).

[33] R. M. Westervelt, Graphene nanoelectronics, *Science*, **320**, 324–325 (2008).

[34] S. J. Tans, A. R. M. Verschueren, and C. Dekker, Room-temperature transistor based on a single carbon nanotube, *Nature*, **393**, 49 –52 (1998).

[35] S. Takagi, T. Takada, N. Matsuo, S. Yokoyama, M. Nakamura, and K. Yamana· Gating electrical transport through DNA molecules that bridge between silicon nanogaps, *Nanoscale*, **4**, 1975–1977 (2012).

[36] N. Matsuo, S. Takagi, K. Yamana, A. Heya, T. Takada, and S. Yokoyama, Electrical property of DNA field-effect transistor: Charge retention property, *J.J. Appl. Phys.*, **51**, 04DD13 (2012).

[37] Y. Cui, Z. Zhong, D. Wang, W. U. Wang, and C. M. Lieber, High performance silicon nanowire field effect transistors, *Nano Lett.*, **3**, 149–152 (2003).

[38] R. Martel, T. Schmidt, H. R. Shea, T. Hertel, and P. Avourisa, Single- and multi-wall carbon nanotube field-effect transistors, *Appl. Phys. Lett.*, **73**, 2447–2449 (1998).

2.2

Electrical Conductance of DNA Oligomers — A Review of Experimental Results

A. Erbe

Helmholtz-Zentrum Dresden-Rossendorf, Institute of Ion Beam Physics and Materials Research, Division Scaling Phenomena, Dresden, Germany

2.2.1 Introduction

The question whether single-stranded or double-stranded DNA can conduct electric currents along a molecule has been raised by theoretical predictions that the overlap between the π -orbitals of neighbouring bases can lead to efficient charge transport along the DNA [1]. The DNA must exist in a specific conformation for this overlap to occur, therefore most measurements have been performed using ds-DNA in the so-called B-conformation, in which the strands form a double helix. Some other measurements have investigated the conductance of specially folded single-stranded DNA, in which a conjugation of the orbitals along a certain direction of the molecule can be expected to lead to large conductance values [2, 3]. Throughout this chapter, we will deal with ds-DNA unless otherwise noted. Conductance measurements in the direction perpendicular to the DNA strand have been used for detection of DNA [4] and show characteristics depending on the base pairs at which the measurement is taken

Early measurements of the conductance have shown that the electrical characterisation of a single molecule, which is rather long for techniques requiring nanometer precision, is not straightforward [5]. It took a long period of time to develop several measurement setups to arrive at the first conclusive statements on the conductance of these structures. One major challenge on the way to developing reliable contacts was the need for ultimate control of the environmental conditions, which depends strongly on the chosen measurement technique. In addition, different techniques are needed for contacting molecular structures of different

DNA in Supramolecular Chemistry and Nanotechnology, First Edition.
Edited by Eugen Stulz and Guido H. Clever.

geometries. Thus, the comparison of oligomers with a small number of base pairs on the nanometer scale with oligomers composed of hundreds of base pairs on the micrometer scale is almost impossible to achieve.

We will therefore split the chapter into different parts, which will mainly compare measurements performed using measurement techniques operating under similar conditions. One major difference depends on whether the DNA is placed in direct contact with the underlying substrate or suspended above the substrate. It has been shown that the interaction between the oligomer and the substrate can lead to a substantial deformation of the secondary molecular structure and thus exerts a major influence on the transport properties [6].

2.2.2 DNA structures

We have already mentioned in the introduction that the prediction for electrical conductance through DNA molecules was mainly based on the coupling of π -orbitals, which occurs in the double helix conformation. This tertiary structure of the DNA is typically formed when all bases of two complementary DNA sequences (primary structure) are coupled to their matching base (secondary structure) to form a double stranded (ds) DNA. Therefore, all measurements characterising the conductance of DNA aim at conditions in which the double helix can be formed. This implies that mostly ds-DNA are observed. In addition, the molecules should be longer than 10 base pairs [7], because shorter molecules do not complete a single turn of the double helix and occur in a different conformation. The persistence length of ds-DNA has been measured to be on the order of 30 nm [8], setting an upper limit to techniques contacting single molecules.

For reliable measurements of the conductance through a metal–DNA–metal junction, the molecules need to be anchored to the metal electrodes. In a large number of experiments, organic molecules were connected to Au electrodes using thiol–gold bonds [9] and provided stable electrical properties [10, 11]. This thiol group is still the most frequently used anchoring group, although in more recent years other anchoring groups such as amino groups have been investigated [12]. Therefore, the vast majority of conductance measurements through DNA molecules have been performed on modified DNA, which include a thiol anchoring group at the 3′- or 5′-end [13] of the molecule. Usually the thiol group is coupled via short alkane linkers to the DNA base in order to ensure good mechanical properties [14]; more recently, however, molecules with direct incorporation of the thiol anchors in the thymidine base have demonstrated good mechanical coupling to Au surfaces [15]. Single strands with the anchoring group attached to one end can be coupled with complementary strands to form ds-DNA with anchoring groups at both ends.

Measurements on ss-DNA with arbitrary base sequence usually show insulating behaviour of the structures [16]. For specific primary structures, however, the base sequence induces a geometrical structure of the single strand, which favours coupling between the bases. Special examples are given by guanine-rich oligomers, which form a four-stranded structure leading to a good coupling between the guanine bases. These structures are called G-quadruplexes [17].

2.2.3 Scanning probe microscopy

Scanning probe microscopy (SPM) techniques provide accurate information on the position and structure of a single small organic molecule on a flat surface [18]. Highest spatial resolution can be achieved in ultra-high vacuum (UHV) conditions, only [19], which do not allow a reliable formation of the double stranded tertiary structure. Therefore, it can be questioned whether the DNA structure can be imaged correctly under UHV conditions. In addition, the formation of electrical contacts to single small organic molecules has been successfully achieved in only very few special cases, in which direct current transport through the molecular species could be exploited [20]. To date, no comparable measurements have been achieved for DNA molecules. The use of scanning probe techniques for the electrical characterisation of DNA molecules, however, has been successfully demonstrated using either scanning tunnelling microscopy (STM) or conductive atomic force microscopy (cAFM) based break junction techniques. Therefore, we will give a brief introduction to the two techniques here and then report on the major findings that can be achieved by using these techniques for the characterisation of DNA.

2.2.3.1 STM break junction

The STM break junction technique is based on the use of a conventional STM setup in a liquid environment. The metallic tip (usually the top of the tip is made of Au, while the remainder of the tip is covered with an insulator in order to prevent undesired leakage currents) is repeatedly moved into the substrate and withdrawn, resulting in the repeated formation and breaking of a metal–metal contact. Such contacts are well known from the technique of mechanically controllable break junctions (which will be discussed in more detail later) and allow the formation of a tunnelling contact with an adjustable gap. When a Au single atom contact is formed in a liquid environment and in the presence of dissolved molecules, those molecules can be contacted between the Au atoms and the substrate repeatedly [12]. Using this method, several conductance properties of organic molecules have been investigated, for example the reliability of the anchoring groups [21, 22], the influence of the geometry of the molecules on the hybridisation of current carrying orbitals [23, 24], and modifications of the conductance by side groups which are attached to the main backbone of the molecules on the conductance [25]. More recently, even the heat dissipation in molecular junctions could be determined using a modified STM setup [26].

In this setup, short oligomers (up to 15 base pairs) can be contacted, when the ends of the DNA are modified to contain linkers to the Au electrodes. This has been demonstrated by the group of Tao [27], who were able to measure conductance through individual DNA molecules. Predictions based on the ionisation potential of the bases lead to the expectation that neighbouring G bases can couple well amongst each other, while A bases reduce this coupling [28]. These predictions were experimentally confirmed by comparing pure GC double strands with strands into which AT base pairs had been inserted. In those configurations the AT bases act like tunnelling bases between the coupled parts of the molecule, which are given by the GC pairs.

The repeated formation of contacts between the conducting tip and the substrates in the presence of the DNA double-strands leads to a series of conductance–distance curves, which can be evaluated statistically. The number of occurrences for the various conductance values are plotted in a histogram for the different molecules. For each molecule, this procedure yields a series of conductance peaks, as shown exemplarily in Figure 2.2.1. The peak exhibiting the lowest conductance is associated with the conductance of a single DNA double-strand connected between the tip and the substrate. From this evaluation, the conductance can be plotted as a function of the molecular length. It turns out that the conductance of pure GC sequences decreases linearly with increasing length of the molecule. This behaviour is expected for a conjugated system, while exponential decrease of the conductance with increasing length is expected for non-conjugated systems. The conductance decays exponentially with the number of AT pairs, which are inserted into the double strands at the centre. This confirms the initial assumption that the AT pairs act like tunnelling barriers in a fully conjugated system of GC pairs.

2.2.3.2 Conductive AFM

Atomic force microscopy (AFM) allows the imaging of single molecules on a flat surface [29] with highest resolution. Therefore, the idea to use a conducting AFM tip and contact individual molecules in this setup promises to deliver reliable data on the conductance of the contacted molecules. A reliable connection between the molecules and the tip, however, relies on the mechanical coupling between the metal atoms and

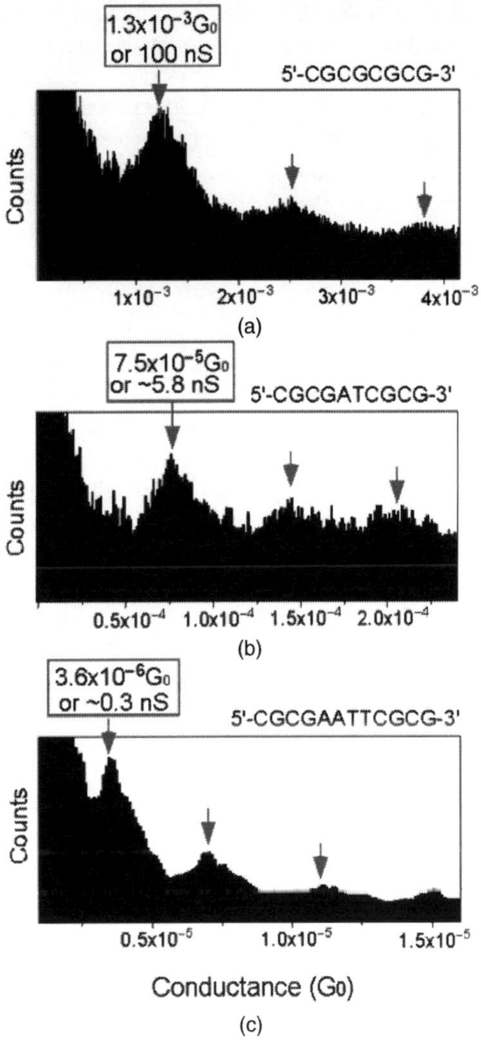

Figure 2.2.1 *Conductance histograms of three different ds-DNA sequences containing different amounts of AT pairs. The molecules are anchored to the substrate and the tip via thiol groups. Each of the measurements shows a series of conductance maxima, the lowest of which is associated to the conductance of a single molecule. Reprinted with permission from [27]. Copyright (2004) American Chemical Society*

the anchoring group of the molecules. Thus, molecules equipped with strongly binding anchoring groups, as for example the thiol groups, stick to the tip during the imaging of the surface, as well, and make reliable imaging impossible. The group of Lindsay found an elegant way out of this difficulty and demonstrated measurements on single organic molecules [30]. They achieved the contact to the molecule by placing a small number of molecules, which are functionalized with thiol anchoring groups on both ends (dithiols), into a monolayer of molecules with only one anchoring group (monothiols). The monolayer is formed by attaching the thiol groups of the monothiols to the Au substrate. Therefore, only the anchoring groups of the dithiols stick out of the monolayer. Au nanoparticles are subsequently bound to the thiol anchors and can be contacted by the tip of the cAFM. Here, similar to the STM breakjunction method, measurements on a series of contacts were repeatedly performed, yielding a series of $I(V)$ curves that can be sorted according to their current value at 1V. The $I(V)$ curves appear to be multiples of a basic $I(V)$ curve, which can be associated with a measurement achieved on a single molecule. This interpretation is further supported by tests taken at varying contacting strengths and various sizes of Au particles, which do not change the observed characterstics.

Based on this technique, DNA double strands were characterised, which were inserted into a carpet of single-stranded DNA molecules that forced the ds-DNA to stand up such that it could be contacted in a vertical setup. The ss-DNa was equipped with only one thiol end-group contacting the substrate, therefore the Au nanoparticles, which were used to achieve the contact to the ds-DNA molecules, were situated on top of the ds-DNA, only. Comparison of various $I(V)$ curves measured at different positions of the substrate confirmed this statement [31]. When the $I(V)$ curves were measured in direct contact between the nanoparticles and the tip, so-called s-shaped $I(V)$ curves were regularly measured, which could not be found for direct contacts to the Au substrate, or for contacts to the top of a layer of ss-DNA. The exact shape and the current amplitude at a given voltage in these curves depended on the details of the contact formation. The conclusion drawn from these experiments was that ds-DNA conducts current, but it does so at a rather low conductance. The molecular states carrying the current are energetically far away from the Fermi energy of the contacts or poorly coupled to the contacts. A detailed understanding of the transport along the DNA can only be achieved if the influence of the dissipative environment on the processes leading to charge motion are taken into account, as well [32, 33].

The results were further supported by comparing the current transport through various DNA monolayers, consisting of ss-DNA and ds-DNA, which were equipped with anchoring groups on one side only and on both sides, respectively [34]. It turns out that efficient current transport is possible only along ds-DNA that is solidly coupled to metallic electrodes via a thiol–Au bond. The internal electronic structure of ds-DNA as well as of G4 quadruplex DNA (see more detailed discussion of the electronic properties of these molecules later) were determined using scanning tunnel spectroscopy [35–37].

2.2.4 Lithographically defined junctions

2.2.4.1 Mechanically controllable break junctions

The main disadvantage of scanning probe based techniques is the instability over long timescales, which is directly connected to the flexibility of the approach and the possibility to amass a large amount of data in rather short periods of time. Electrical contacts to single molecules, which are fabricated using lithography, are much more stable during long measurement times, but one cannot change the contacted molecule without producing a new junction. Thus, a technique which can combine the flexibility of the scanning probe approach with the long-time stability of the lithographically produced junctions may be able to elucidate some details of the electrical properties of single molecules. Mechanically controllable break junctions [38] have shown this ability on the scale of single atoms. The nature of current transport through single atoms of various metals could thus be identified [39]. This concept was successfully employed for the characterisation of single organic molecules [10, 11, 40] and subsequently used for some impressive demonstrations of molecular electronic effects [41–43]. Further development of the technique allowed electrical characterisation of single molecules in a liquid environment [44], a necessary prerequisite for the successful characterisation of DNA molecules in a controlled environment.

Naturally, the length of molecules measured in an MCBJ geometry is subjected to some restrictions. The lower boundary of the length is on the scale of single atoms, the simplest molecule measured so far is a hydrogen molecule [45]. The upper limit is given by the mechanical stability of the molecules and the resistance of the molecule. The persistence length of ds-DNA is of the order of 30 nm, thus the length of measured molecules has to be considerably shorter. As discussed earlier, the resistance of the molecules increases exponentially with the length of the molecules for tunnelling processes and linearly for hopping processes, which can be assumed for the case of conductance through DNA. Initial measurements on ds-DNA in MCBJ geometries have indicated that the resistance of a molecule with a length of 10 nm is higher than 1 GΩ [14]. Resistance values that exceed this threshold will not be detectable in a setup containing buffer solution, which is necessary to stabilise the double helix (B-)conformation of the DNA. This sets an upper limit to the length of the molecules at around 10 nm. Thus, most of the conductance studies used DNA oligomers consisting of approximately 30 base pairs, thus ensuring both good mechanical properties of the molecules and a measurable contribution to the conductance. Stable contacts can be repeatedly formed using MCBJ setups and show conductance values around 1×10^{-4} to 2×10^{-4} G_0 [46].

The poorly conducting molecules measured in the initial experiments were coupled to Au electrodes via thiol anchoring groups, which were connected to the backbone of the DNA using long alkane chains [14, 47]. It has been proven experimentally that alkane molecules act as tunnelling barriers, because the bonds between the carbon atoms are pure σ -bonds [48]. Thus, the low conductance can in part be caused by the limited transport through the anchoring molecules. In order to test this assumption and to develop DNA contacts with larger conductance, new anchoring groups were developed, which couple the thiol group directly to the thymidine base of the DNA. The formation of monolayers from DNA modified by this anchoring was shown [15] prior to conductance measurements on these molecules. The conductance of the molecules was characterised in a buffer, under dried conditions, and in vacuum. In all of these measurements conductance values of around $1\times10^{-3} G_0$ were found, which were not present when the junctions were measured without DNA [49]. This indicates that the conductance of DNA with shorter linker groups is indeed increased compared with alkane linked DNA. $I(V)$ curves measured in junctions with the modified DNA show stable molecular contacts to single DNA molecules, which develop from tunnelling to a molecular contact while the junction is closed (see Figure 2.2.2). These contacts show pure tunnelling through the environments for large distances of the electrodes. Once a molecule is captured at one of the two electrodes, and comes in close proximity to the second, the $I(V)$ curves show an asymmetric behaviour. This asymmetry vanishes when the molecule is situated in-between the two electrodes and the resulting $I(V)$ curve can be well described by assuming current transport through a single molecular transport channel [44].

The mechanical variability of the MCBJ setup allows for stretching the molecular contacts during the measurements. While this leads to a loss of electrical contact for the ds-DNA [49], the result is rather different when G4 quadruplex structures are measured [3]. In these measurements the G4 quadruplex was equipped with thiol linkers on both ends of the sequence. As described earlier, the anchoring groups were incorporated into a modified thymidine base in order to provide reliable mechanical and good electrical coupling to the metallic electrodes. The DNA–metal contact can be stretched over a distance of more than 2 nm without changing the resistance. Instead, a long resistance plateau develops as the distance between the electrodes is increased. Thus, the quadruplexes can be seen as molecular wires with variable length. This property may prove useful once electronic circuits are assembled using the self-organisation of DNA molecules and some variability in the design needs to be accounted for.

2.2.4.2 Direct contacts

Since the secondary double helix structure forms only beyond a molecular length of 5 nm, it is possible to fabricate contacts to the molecules by direct electron beam lithography (EBL). On the other hand, such structures tend to leave the contacted molecule in close contact with the underlying substrate. Therefore, apart from initial studies and studies of metallized DNA [50], no further studies have used fixed metallic electrodes fabricated using EBL.

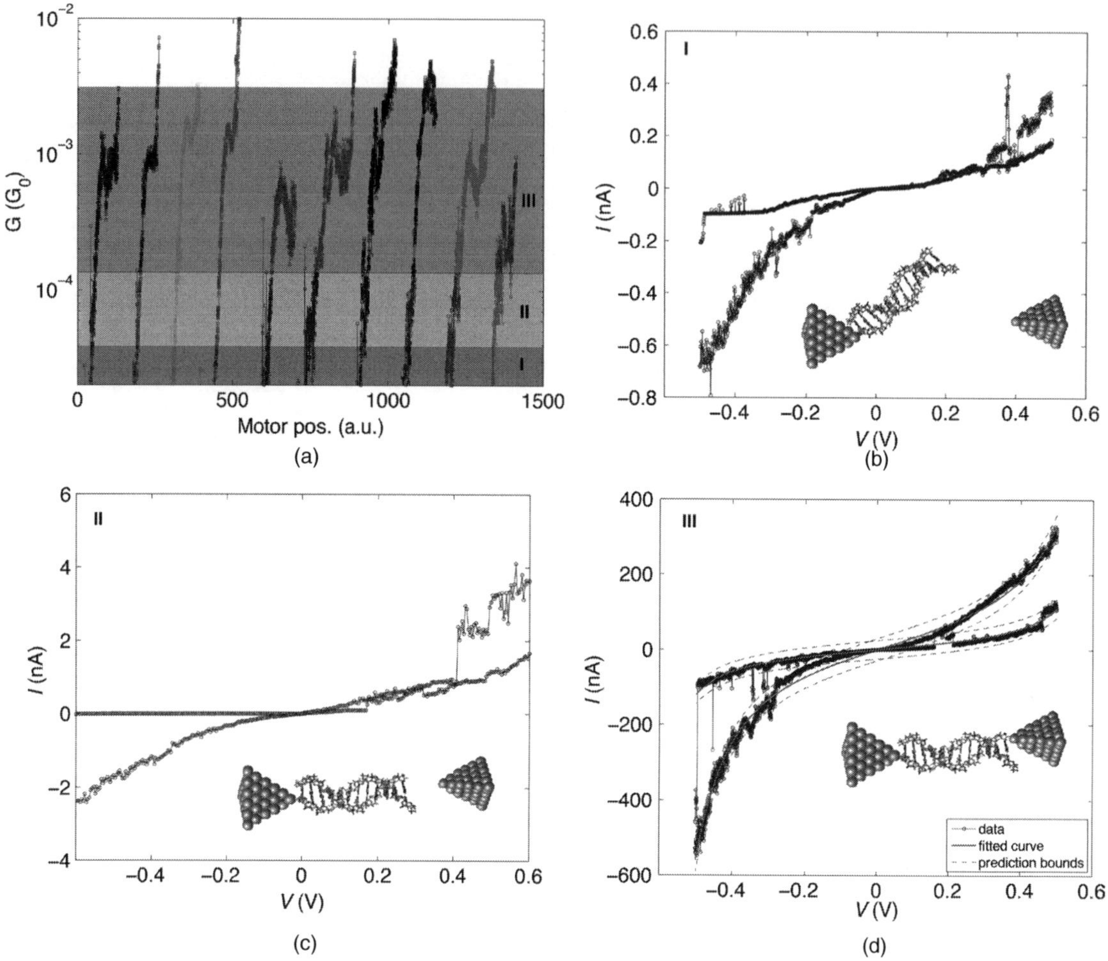

Figure 2.2.2 *Opening curves (a) and I(V) curves taken from junctions to modified ds-DNA molecules. It is shown how the contact develops from a pure tunnel junction (b) to a junction with an asymmetrically coupled molecule (c) and finally to a junction with stable bonds of the molecule to both electrodes (d). Fits to a single level model of the I(V) curves in (d) are shown in red. Reprinted with permission from [49]. Copyright Wiley-VCH (2013). (See colour figure in colour plate section)*

2.2.4.3 Carbon nanotube contacts

An outstanding demonstration of DNA conductance measurements has been achieved by connecting the DNA molecules in-between two ends of a carbon nanotube (CNT), which was cleaved by oxidative processing into two pieces immediately before the assembly of the DNA [51]. The DNA was coupled to the open ends of the nanotube by amine groups that are connected to the double strand at the 3′- or the 5′-end via short alkane linkers. The conductance value of the semiconducting nanotubes lies around 40 nS, the value of the metallic nanotubes around 200 nS. The conductance of the whole junction drops to zero once the carbon nanotubes are cut. After reconnecting the nanotubes with ds-DNA, the conductance rises again to a conductance value which is of the order of 10 nS, that is, below the initial value recorded for the CNT. Similar conductance values can be achieved when an ss-DNA is first linked to both ends of the nanotube and then

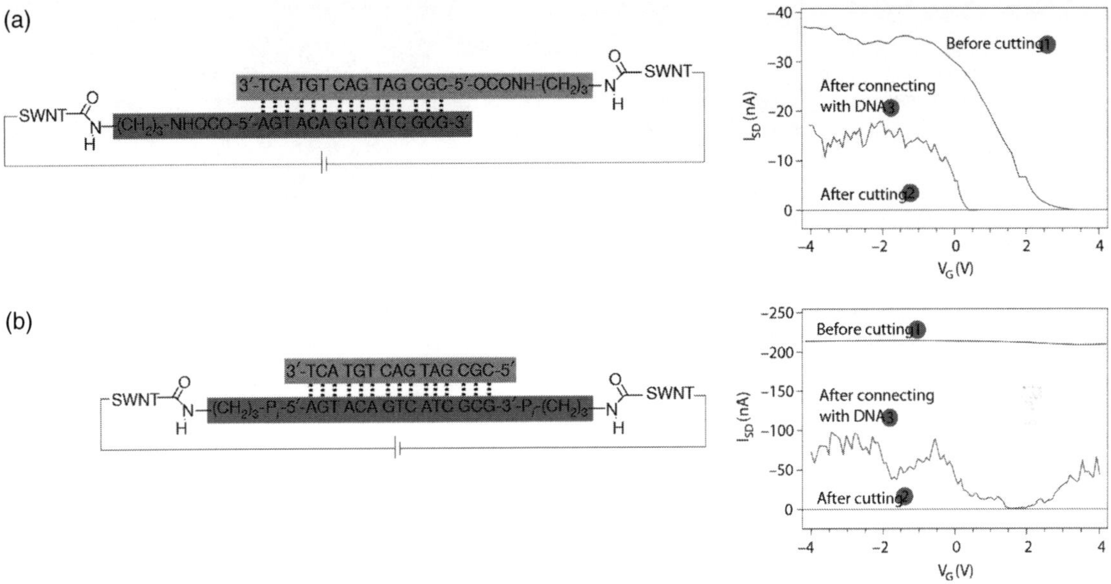

Figure 2.2.3 Sketches of ds-DNA coupled to a semiconducting CNT (a) and an ss-DNA coupled to a metallic CNT and subsequently paired with a complementary strand (b) are shown on the left. On the right the corresponding I(V) curves are shown before and after cutting of the CNT and after reconnecting the CNT with the respective DNA molecules. Reprinted by permission from Macmillan Publisher Ltd. Nature Nanotechnology [51], copyright (2008)

paired with a complementary strand (a comparison of the two cases is shown in Figure 2.2.3. Using that approach, the influence of base pair mismatches on the conductance could be identified, as a strong decrease of the conductance compared with well-matched DNA was obtained in mismatched DNA.

2.2.5 Conclusions

The great progress made in contacting techniques during the past decade has led to advances in understanding charge transport in DNA molecules. The picture of the processes involved in electrical current transport is much more coherent than it was ten years ago. Mechanically and electrically stable contacts can nowadays be produced on the scale of molecular objects. This allows systematic comparisons between various molecular structures. As a result it can be said that ss-DNA most likely shows insulating behaviour even for rather short lengths of the molecules under 5 nm. The conductance changes when ds-DNA in the double-helix conformation is attached to metal electrodes via thiol anchoring groups. Resistance values found in such junctions range from 100 MΩ to 1 GΩ for DNA consisting of the order of 30 base pairs and depend on the chosen base sequence. This resistance is not larger than the resistance found in long conjugated wires [52] and can in part be explained by assuming hopping conductance in a dissipative environment, as given by the surrounding buffer solution. In addition, the contact resistance between metal electrodes and the molecular structure causes a major part of the resistance. Some experiments show clearly that the base sequence (primary structure), the formation of the double strands (secondary structure) and the conformation of the molecule (tertiary structure) influence the conductance of DNA molecules. This shows that in general a wide range of electrical functionalities can be generated by self-assembled generation of DNA networks.

References

[1] D.D. Eley and D.I. Spivey. Semiconductivity of organic substances. Part 9. – Nucleic acid in the dry state. *Trans. Faraday Soc.*, **58**, 411–415 (1962).

[2] A.B. Kotlyar, N. Borovok, T. Molotsky, H. Cohen, E. Shapir, and D. Porath. Long, monomolecular guanine-based nanowires. *Adv. Mater.*, **17**, 1901–1904 (2005).

[3] S.-P. Liu, S.H. Weisbrod, Z. Tang, A. Marx, E. Scheer, and A. Erbe. Direct measurement of electrical transport through G-quadruplex DNA with mechanically controllable break junction electrodes. *Angew. Chem. Int. Ed.*, **49**, 3313–3316 (2010).

[4] R.M.M. Smeets, S.W. Kowalczyk, A.R. Hall, N.H. Dekker, and C. Dekker. Translocation of RecA-coated double-stranded DNA through solid-state nanopores. *Nano Lett*, **9**, 3089–3095 (2009).

[5] R.G. Endres, D.L. Cox, and R.R.P. Singh. Colloquium: The quest for high-conductance DNA. *Rev. Mod. Phys.*, **76**, 195–214 (2004).

[6] A.Y. Kasumov, D.V. Klinov, and P.E. Roche. Thickness and low-temperature conductivity of DNA molecules. *Appl. Phys. Lett.*, **84**, 1007–1009 (2004).

[7] H.R. Drew, R.M. Wing, T. Takano, C. Broka, S. Tanaka, K. Itakura, and R.E. Dickerson. Structure of a B-DNA dodecamer: Conformation and dynamics. *Proc. Natl. Acad. Sci. U.S.A.*, **78**, 2179–2183 (1982).

[8] S. Brinkers, H.R.C. Dietrich, F.H. de Groote, I.T. Young, and B. Rieger. The persistence length of double stranded DNA determined using dark field tethered particle motion. *J. Chem. Phys.*, **130**, 215105 (2009).

[9] J.M. Tour, L.R. Jones, D.L. Pearson, J.J.S. Lamba, T.P. Burgin, G.M. Whitesides, D.L. Allara, A.N. Parikh, and S. Atre. Self-assembled monolayers and multilayers of conjugated thiols, α,ω -dithiols, and thioacetyl-containing adsorbates. Understanding attachments between potential molecular wires and gold surfaces. *J. Am. Chem. Soc.*, **117**, 9529–9534 (1995).

[10] M.A. Reed, C. Zhou, C.J. Muller, T.P. Burgin, and J.M. Tour. Conductance of a molecular junction. *Science*, **278**, 252–254 (1997).

[11] C. Kergueris, J.P. Bourgoin, S. Palacin, D. Esteve, C. Urbina, M. Magoga, and C. Joachim. Electron transport through a metal-molecule-metal junction. *Phys. Rev. B*, **59**, 12505–12513 (1999).

[12] L. Venkataraman, J.E. Klare, I.W. Tam, C. Nuckolls, M.S. Hybertsen, and M.L. Steigerwald. Single-molecule circuits with well-defined molecular conductance. *Nano Lett.*, **6** 458–462 (2006).

[13] Y. Zhang, R.H. Austin, J. Kraeft, E.C. Cox, and N.P. Ong. Insulating behavior of lambda-DNA on the micron scale. *Phys. Rev. Lett.*, **89**, 198102 (2002).

[14] N. Kang, A. Erbe, and E. Scheer. Electrical characterization of DNA in mechanically controlled break-junctions. *New J. Phys.*, **10**, 023030 (2008).

[15] B. Bornemann, S.-P. Liu, A. Erbe, E. Scheer, and A. Marx. Thiolated nucleotides for immobilisation of DNA oligomers on gold surfaces. *Chem. phys. chem.*, **9**, 1241–1244 (2008).

[16] D. Porath, G. Cuniberti, and R. Di Felice. Charge transport in DNA-based devices. *Top Curr. Chem.*, **237**, 183–227 (2004).

[17] C.W, Greider and E.H. Blackburn. Identification of a specific telomere treminal transferase-activity in tetrahymena extracts. *Cell*, **43**, 405–413 (1985).

[18] L. Grill, K.H. Rieder, F. Moresco, G. Rapenne, S. Stojkovic, X. Bouju, and C. Joachim. Rolling a single molecular wheel at the atomic scale. *Nat. Nanotechnol.*, **2**, 95–98 (2007).

[19] J. Repp, G. Meyer, S.M. Stojkovic, A. Gourdon, and C. Joachim. Molecules on Insulating Films: Scanning-Tunneling Microscopy Imaging of Individual Molecular Orbitals. *Phys. Rev. Lett.*, **94**, 026803 (2005).

[20] R. Temirov, A. Lassise, F.B. Anders, and F.S. Tautz. Kondo effect by controlled cleavage of a single-molecule contact. *Nanotechnology*, **19**, 065401 (2008).

[21] S.Y. Quek, L. Venkataraman, H.J. Choi, S.G. Louie, M.S. Hybertsen, and J.B. Neaton. Amine-gold linked single-molecule circuits: Experiment and theory. *Nano Lett.*, **7**, 3477–3482 (2007).

[22] X.L. Li, J. He, J. Hihath, B.Q. Xu, S.M. Lindsay, and N.J. Tao. Conductance of single alkanedithiols: Conduction mechanism and effect of molecule-electrode contacts. *J. Am. Chem. Soc.*, **128**, 2135–2141 (2006).

[23] L. Venkataraman, J.E. Klare, C. Nuckolls, M.S. Hybertsen, and M.L. Steigerwald. Dependence of single-molecule junction conductance on molecular conformation. *Nature*, **442**, 904–907 (2006).

[24] A. Mishchenko, D. Vonlanthen, V. Meded, M. Buerkle, C. Li, I.V. Pobelov, A. Bagrets, J.K. Viljas, F. Pauly, F. Evers, M. Mayor, and T. Wandlowski. Influence of conformation on conductance of biphenyl-dithiol single-molecule contacts. *Nano Lett.*, **10**, 156–163 (2010).

[25] W. Hong, H. Li, S.-X. Liu, Y. Fu, J. Li, V. Kaliginedi, S. Decurtins, and T. Wandlowski. Trimethylsilyl-terminated oligo(phenylene ethynylene)s: An approach to single-molecule junctions with covalent Au-C sigma-bonds. *J. Am. Chem. Soc.*, **134**, 19425–19431 (2012).

[26] W. Lee, K. Kim, W. Jeong, L.-A. Zotti, F. Pauly, J.-C. Cuevas, and P. Reddy. Heat dissipation in atomic-scale junctions. *Nature*, **498**, 209–212 (2013).

[27] B. Xu, P. Zhang, X. Li, and N. Tao. Direct conductance measurement of single DNA molecules in aqueous solution. *Nano Lett.*, **4**, 1105–1108 (2004).

[28] M. Bixon, B. Giese, S. Wessely, T. Langenbacher, M.E. Michel-Beyerle, and J. Jortner. Long-range charge hopping in DNA. *Proc. Natl. Acad. Sci. U.S.A.*, **96**, 11713–11716 (1999).

[29] D.G. de Oteyza, P. Gorman, Y.-C. Chen, S. Wickenburg, A. Riss, D.J. Mowbray, G. Etkin, Z. Pedramrazi, H.-Z. Tsai, and A. Rubio. Direct imaging of covalent bond structure in single-molecule chemical reactions. *Science*, **340**, 1434–1437 (2013).

[30] X.D. Cui, A. Primak, X. Zarate, J. Tomfohr, O.F. Sankey, A.L. Moore, T.A. Moore, D. Gust, G. Harris, and S.M. Lindsay. Reproducible measurement of single-molecule conductivity. *Science*, **294**, 571–574 (2001).

[31] H. Cohen, R. Naaman, and D. Porath. Direct measurement of electrical transport through single DNA molecules of complex sequence. *Proc. Natl. Acad. Sci. U.S.A.*, **102**, 11589–11593 (2005).

[32] R. Gutierrez, R.A. Caetano, B.P. Woiczikowski, T. Kubar, M. Elstner, and G. Cuniberti. Charge transport through biomolecular wires in a solvent: Bridging molecular dynamics and model Hamiltonian approaches. *Phys. Rev. Lett.*, **102**, 208102 (2009).

[33] R. Gutierrez, S. Mandal, and G. Cuniberti. Quantum transport through a DNA wire in a Dissipative Environment. *Nano Lett.*, **5**, 1093–1097 (2005).

[34] H. Cohen, C. Nogues, D. Ullien, S. Daube, R. Naaman, and D. Porath. Electrical characterization of self-assembled single- and double-stranded DNA monolayers using conductive AFM. *Faraday Discuss.*, **131**, 367–376 (2007).

[35] D. Porath. Scanning tunneling microscopy: A DNA sequence scanned. *Nat. Nanotechnol.*, **4**, 476–477 (2009).

[36] D.A. Ryndyk, E. Shapir, D. Porath, A. Calzolari, R. Di Felice, and G. Cuniberti. Scanning Tunneling Spectroscopy of Single DNA Molecules. *ACS Nano*, **3**, 1651–1656 (2009).

[37] D. Rotem, G. Eidelshtein, A. Kotlyar, and D. Porath. Novel Materials for Molecular Electronics-Synthesis and Characterization of Long G4-DNA, in *Guanine Quartets: Structure and Application* The Royal Society of Chemistry (2013).

[38] P. Joyez, M.H. Devoret, D. Esteve, and C. Urbina. Adjustable nanofabricated atomic size contacts. *Rev. Sci. Instrum.*, **67**, 108–111 (1996).

[39] E. Scheer, N. Agrait, J.C. Cuevas, A.L. Yeyati, B. Ludoph, A. Martin-Rodero, G.R. Bollinger, J.M. van Ruitenbeek, and C. Urbina. The signature of chemical valence in the electrical conduction through a single-atom contact. *Nature*, **394**, 154–157 (1998).

[40] J. Reichert, R. Ochs, D. Beckmann, H.B. Weber, M. Mayor, and H. von Lohneysen. Driving current through single organic molecules. *Phys. Rev. Lett.*, **88**, 176804 (2002).

[41] D. Dulic, S.J. van der Molen, T. Kudernac, H.T. Jonkman, J.J.D. de Jong, T.N. Bowden, J. van Esch, B.L. Feringa, and B.J. van Wees. One-way optoelectronic switching of photochromic molecules on gold. *Phys. Rev. Lett.*, **91**, 207402 (2003).

[42] M. Elbing, R. Ochs, M. Koentopp, M. Fischer, C. von Hänisch, F. Weigend, F. Evers F, H.B. Weber, and M. Mayor, A single-molecule diode. *PNAS*, **102**, 8815–8820 (2005).

[43] M.L. Perrin, C.J.O. Verzijl, C.A. Martin, A.J. Shaikh, R. Eelkema, J.H. van Esch, J.M. van Ruitenbeek, J.M. Thijssen, H.S.J. van der Zant, and D. Dulic. Large tunable image-charge effects in single-molecule junctions. *Nat. Nanotechnol.*, **8**, 282–287 (2013).

[44] L.A. Zotti, T. Kirchner, J.-C. Cuevas, F. Pauly, T. Huhn, E. Scheer, and A. Erbe. Revealing the role of anchoring groups in the electrical conduction through single-molecule junctions. *Small*, **6**, 1529–1535 (2010).

[45] R.H.M. Smit, C. Untiedt, N.D. Lang, M.C. van Hemert, and J.M. van Ruitenbeek. Measurement of the conductance of a hydrogen molecule. *Nature*, **419**, 906–909 (2002).

[46] D. Dulic, S. Tuukkanen, C.L. Chung, A. Isambert, P. Lavie, and A. Filoramo. Direct conductance measurements of short single DNA molecules in dry conditions. *Nanotechnology*, **20** (2009).

[47] N. Kang, A. Erbe, and E. Scheer. Observation of negative differential resistance in DNA molecular junctions. *Appl. Phys. Lett.*, **96**, 023701 (2010).

[48] W.Y. Wang and M.A. Reed. Mechanism of electron conduction in self-assembled alkanethiol monolayer devices. *Phys. Rev. B*, **68** 035416 (2003).

[49] S.P. Liu, J. Artois, D. Schmid, M. Wieser, B. Bornemann, S. Weisbrod, A. Marx, E. Scheer, and A. Erbe. Electronic transport through short dsDNA measured with mechanically controlled break junctions: New thiol-gold binding protocol improves conductance. *Phys. Status Solidi. B*, **250** 2342–2348 (2013).

[50] E. Braun, Y. Eichen, U. Sivan, and G. Ben-Yoseph, DNA-templated assembly and electrode attachment of a conducting silver wire. *Nature.* **391**, 775–778 (1998).

[51] X. Guo, A.A. Gorodetsky, J. Hone, J.K. Barton, and C. Nuckolls. Conductivity of a single DNA duplex bridging a carbon nanotube gap. *Nat. Nanotechnol.*, **3**, 163–167 (2008).

[52] S. Choi, B. Kim, and C.D, Frisbie, Electrical resistance of long conjugated molecular wires. *Science*. **320**, 14821486 (2008).

2.3

DNA Sensors Using DNA Charge Transport Chemistry

Jacqueline K. Barton, Ariel L. Furst, and Michael A. Grodick

Division of Chemistry and Chemical Engineering, California Institute of Technology, Pasadena, CA, USA

2.3.1 Introduction

From the first proposal of the structure of DNA [1, 2], debates about what properties DNA may hold beyond the simple transfer of genetic information have arisen. The structure of the stacked DNA bases within the double helix led many to predict that this macromolecular assembly could conduct charge. The stacked base pairs closely resemble the structure of graphene sheets, as both contain aromatic heterocycles stacked at 3.4 Å (Figure 2.3.1) [3]. However, the notion that the DNA helix can conduct charge was long met with skepticism. Today, as a result of extensive experimentation, DNA charge transport (DNA CT) is well-established chemistry, though the full mechanistic understanding still requires development [4].

Experiments with DNA CT first involved the observation of long-range, excited-state charge transport through a DNA duplex between well-stacked donors and acceptors [5–7]. In an early experiment, electron transfer between covalently tethered metallointercalators was observed over a distance of 40 Å through DNA (Figure 2.3.2) [7]. A 15-base DNA duplex was labeled at one terminus with $[Ru(phen)_2dppz]^{2+}$ (dppz = dipyrido[3,2-*a*:2′,3′-*c*]phenazine), with the excited state acting as an electron donor, and $[Rh(phi)_2phen]^{3+}$ (phi = 9,10-phenanthrenequinone diimine) at the opposite terminus acting as an electron acceptor. In the absence of the electron acceptor, the ruthenium complex tethered to DNA luminesces. However, upon incorporation of the rhodium complex, the luminescence is completely quenched. In the years since this experiment, the ability of DNA to conduct charge through its π-stacked bases has been studied extensively

DNA in Supramolecular Chemistry and Nanotechnology, First Edition.
Edited by Eugen Stulz and Guido H. Clever.

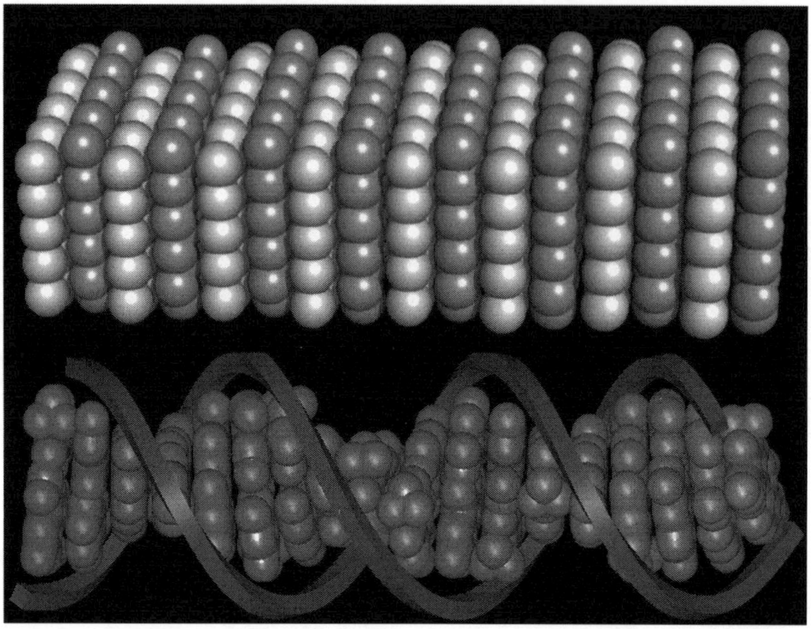

Figure 2.3.1 Schematic illustrations of the structures of graphene (top) and DNA (bottom). The planar sheets of graphene are shown in grey, as are the aromatic DNA bases; for the DNA, the sugar-phosphate backbone has been schematized as a dark ribbon. In both cases, the layers are stacked at a distance of 3.4 Å, enabling orbital overlap, and therefore the flow of charge

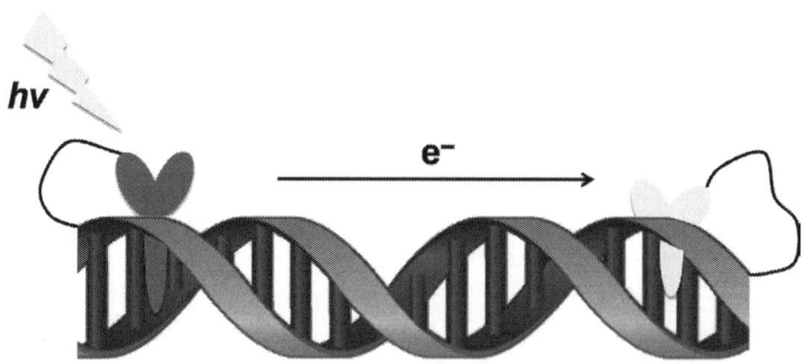

Figure 2.3.2 A DNA modified with two metallointercalators to test photoinduced DNA CT. Electron transfer over 40 Å was shown through DNA between covalently tethered metallointercalators, $[Ru(phen)_2dppz]^{2+}$ (left) as an electron donor and $[Rh(phi)_2phen]^{3+}$ (right) at the opposite terminus as an electron acceptor [7]

using a variety of platforms [7–12]. Moreover, as the power of this chemistry became apparent, experiments focused on biological applications of DNA CT chemistry came to the forefront [13].

The remarkable utility of this chemistry became evident as DNA CT was shown to be extremely efficient over long molecular distances on fast time scales, yet exquisitely sensitive to perturbations in base–base stacking [14]. Single base mismatches and other damaged products have been shown to significantly attenuate CT both in ground- and excited-state experiments [15–17]. DNA CT has also been directly measured in single molecule experiments in the ground state [18]. Using an oxygen plasma, molecular size gaps can be inserted into carbon nanotubes and individual DNA molecules functionalized with terminal amines covalently attached within the gaps using amide chemistry. These robust devices can then be used to measure the current flow in the nanotube containing the covalently attached molecule of interest versus the current in the original nanotube (Figure 2.3.3). In these devices, duplex DNA was attached either by functionalizing the 5′-end of both strands of the DNA with alkyl amines or both 3′- and 5′-ends of only one of the strands of the duplex, with the complementary strand non-covalently associated. Using this device, we found that the resistance generated in the gap with a DNA duplex inserted was quite similar to that expected for a stacked graphite insert (~1 MΩ resistance for a ~6 nm gap).

Even more interesting was how this assembly could be used to test the effect of a mismatch on DNA CT. With one strand covalently attached to the device through both the 5′- and 3′-ends, various complements, with or without a mismatch, could be interchanged into the duplex and the resultant current tested (Figure 2.3.3). Several different complements could be cycled in this robust device. The presence of a mismatch was found to yield a 300-fold attenuation in current relative to the current found for the well-matched complement. Moreover, all current was lost upon DNA cleavage with a blunt-end restriction enzyme, illustrating that the conformation of the DNA duplex in the gap was intact and recognizable by the DNA-binding protein.

While early experiments focused on the distance dependence of DNA CT using largely spectroscopic experiments involving excited-state transport, ground-state measurements repeatedly illustrated the high sensitivity of DNA CT to intervening perturbations in stacking [12]. While the shallow distance dependence of DNA CT was remarkable, so too was the exquisite sensitivity of DNA CT to perturbations in base stacking. This led to significant applications of DNA CT chemistry in bio-sensing. DNA CT has allowed the sensitive detection of single base mismatches irrespective of sequence context, the monitoring of binding of DNA transcription factors, and even following, electrochemically, the reactions of various enzymes on DNA.

However, these sensing studies also led us to question how Nature may monitor similar activities. How, for example, does the cell insure the integrity of its genome? Are these issues ones that benefit from DNA CT chemistry? Here we describe various applications of DNA CT in sensing, illustrating how this chemistry may be used for the construction of sensitive devices for biochemical applications. In addition, we describe our first models for how this sensing technology might also be applied within the cell, how Nature may take advantage of DNA CT chemistry.

2.3.2 DNA-functionalized electrochemical sensors

The most common technique employed for electrochemical detection based on DNA CT involves the immobilization of duplexed DNA on a gold surface at one terminus and modified with a redox-active probe at the distal terminus (Figure 2.3.4). A range of redox-active probes have been employed, and not surprisingly, the most sensitive reporters of the integrity of the DNA duplex are those that are themselves well stacked and hence well coupled into the DNA π-stack.

In one electrochemical experiment, the DNA duplex was shown to carry out ground state charge transport over 100 base pairs, 34 nm [19]. This experiment involved a particularly long molecular wire, but the extraordinary sensitivity of DNA to small perturbations was additionally shown through the significant effect of a single mismatched base incorporated into the DNA duplex. The 100-mer was terminally modified with a covalent Nile blue redox probe and assembled on the gold electrode, and the incorporation of a single mismatched base pair resulted in a significant attenuation of signal, 0.8 ± 0.1 nC for the cathodic peak containing a single base mismatch, as compared with 1.7 ± 0.1 nC for that of the well-matched duplex [19]. Interestingly, the signal attenuation observed through 100 base pairs for the single base mismatch was equal to that observed for the same mismatch incorporated into a 17-mer. Remarkably, while the effect of the mismatch is substantial and independent of duplex length and sequence context, no perturbation in current is

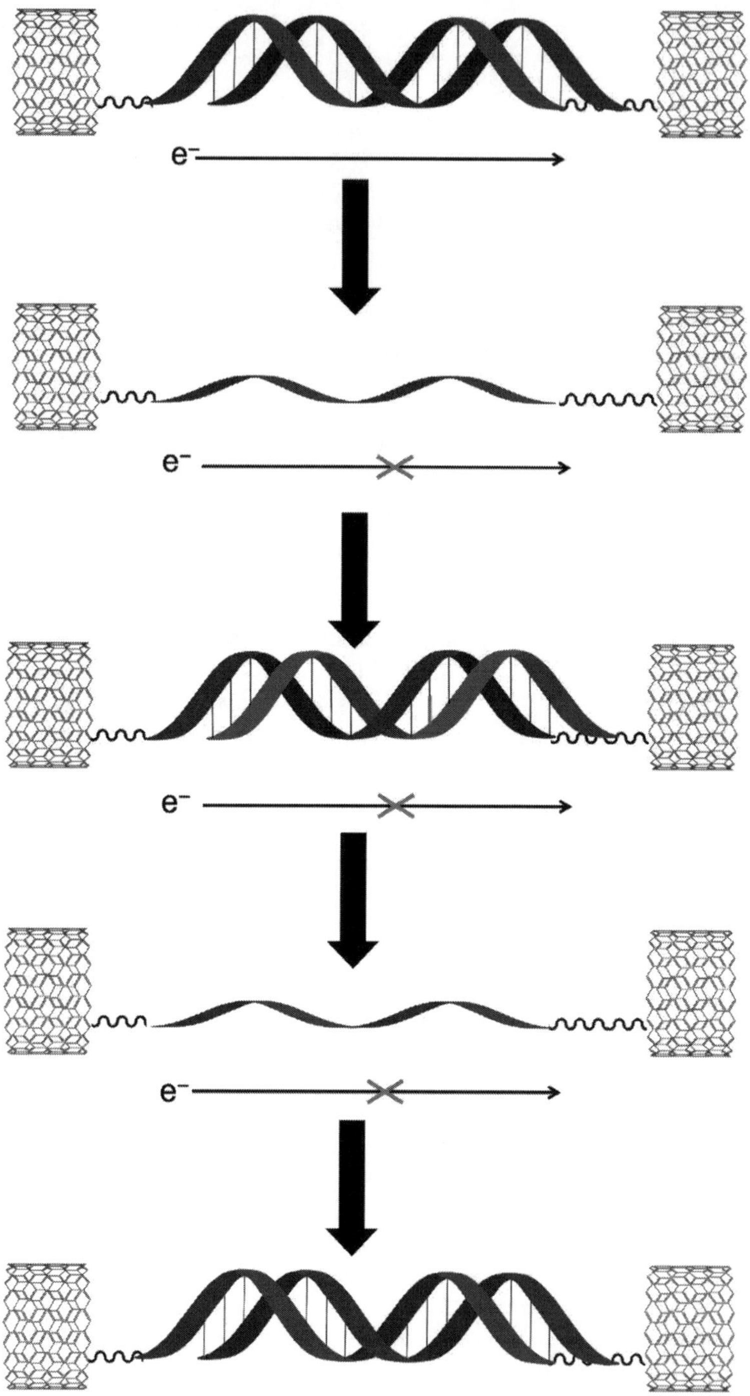

Figure 2.3.3 Illustration of a single molecule experiment with DNA tethered to carbon nanotubes to test ground state DNA CT [18]. Well-matched DNA is covalently attached to carbon nanotubes through the termini of the DNA functionalized with amines. The DNA can then be denatured in the device so that only one strand remains covalently attached. A complementary strand that contains a single base mismatch can then be floated in and annealed to the covalently tethered strand. For the well matched duplex DNA (top), significant current is obtained, but this current is attenuated in the presence of a mismatch (middle with mismatch to the right); addition of the well matched complement (bottom) restores full current flow This cycle of unannealing and reannealing alternative complements can be repeated and the conductivity reproducibly measured

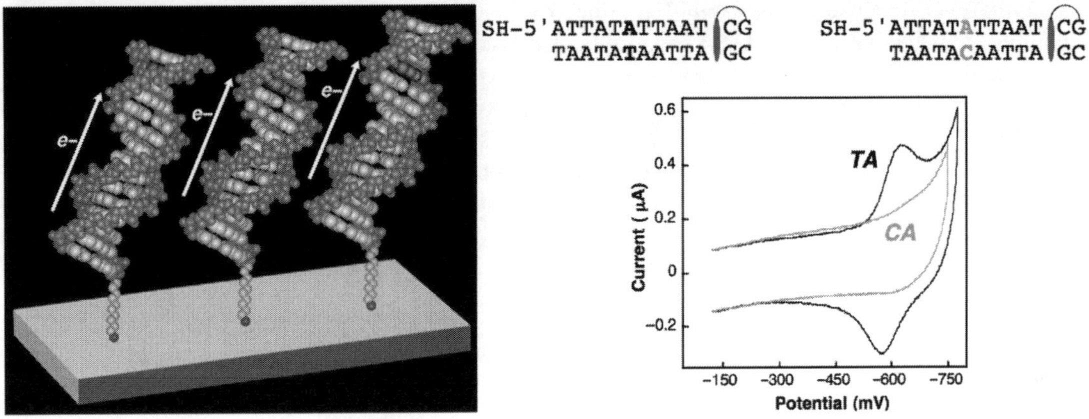

Figure 2.3.4 *Electrochemical signal from well matched DNA and DNA containing a single-base mismatch using DNA-modified electrodes. DNA assembled on a gold electrode (left) containing a covalent redox reporter daunomycin was electrochemically monitored by cyclic voltammetry (right) [15]. The well-matched DNA produces a large, reversible signal. Upon incorporation of a single-base mismatch, the electrochemical signal is significantly attenuated (See color figure in color plate section)*

observed with a nick in the DNA backbone. The 100-mer used in this experiment was actually constructed from stacking together several smaller pieces containing sticky ends. What is essential for DNA CT is effective base stacking; CT is through the base-pair stack, not the sugar-phosphate backbone.

2.3.2.1 Redox probes for ground state DNA-mediated charge transport detection

Effective detection by DNA CT is dependent on the interaction between a redox probe and the base stack, whether the molecule is covalently tethered to DNA or free in solution. There are multiple modes of non-covalent interaction between small molecules and DNA, including groove binding, electrostatic association and intercalation. For effective DNA CT, the redox probes must be well coupled into the base-pair stack. Intercalation, where the probe is itself π-stacked in the duplex is a particularly sensitive coupling mode for electrochemical applications. A series of redox-active probes that interact with DNA through either intercalation or groove binding were tested for signal attenuation upon incorporation of a mismatch. The compounds capable of intercalation, including $[Ir(bpy)(phen)(phi)]^{3+}$, daunomycin, and methylene blue, lead to a differential in the electrochemical current between well-matched duplexes and duplexes containing a single-base mismatch. Ruthenium hexammine, however, which is only capable of hydrogen binding in the groove, shows no difference in current between the well-matched and mismatched DNA; here CT is not *through* the base stack. Moreover, the intercalative complexes that are less likely to groove bind, $[Ir(bpy)(phen)(phi)]^{3+}$ and daunomycin, have a significantly greater signal differential for mismatch discrimination (CA:TA signal ratio of 0.3) than methylene blue (CA:TA signal ratio of 0.5). This mismatch specificity for the intercalative DNA-binding mode speaks to the sensitivity of the π-stack to perturbations and the importance of charge transport directly through the π-stack as a method of specific detection.

In order to detect low-abundance species, it can become necessary to amplify the electrochemical signal obtained directly from a DNA-interacting reporter. One method of signal amplification is electrocatalysis [16]. Electrocatalysis of methylene blue signals can be achieved by adding ferricyanide to the solution. Methylene blue is reduced to leucomethylene blue via DNA-mediated reduction. Freely diffusing ferricyanide, acting as an electron sink, can then oxidize leucomethylene blue back to methylene blue. This facilitates the faster turnover of methylene blue, amplifying the DNA-mediated reaction. The ferricyanide is prevented from directly interacting with the DNA because of electrostatic repulsion. Electrocatalysis coupled to DNA CT has additionally been accomplished with hemoglobin acting as an electron sink [20].

Although freely diffusing DNA intercalators have been extensively used for DNA CT-based detection, covalently-tethered redox reporters offer the same signal specificity without the need for such extensive surface passivation. The two most common covalent redox reporters are Nile blue and methylene blue. Nile blue can be covalently tethered to DNA such that the linker contains an alkene moiety, enabling direct electronic conjugation to the π-stack. However, as previously mentioned, the best signal differential between unperturbed DNA and DNA containing a mismatch or a lesion is achieved with intercalative redox probes. Thus, recently, a method of directly tethering methylene blue to the DNA through either a flexible C_6 or C_2 linker, which enables this probe to intercalate into the DNA base stack, has been developed [21].

2.3.2.2 Different platforms for DNA electrochemistry

Sensitive detection of biomarkers is necessary for fundamental biological studies as well as for the development of effective diagnostic tools. As DNA can be used to specifically capture DNA, RNA, and proteins, nucleic acid sensors provide a flexible platform that can be easily manipulated to detect a variety of targets. Moreover, their linear structure is amenable to multiplexed formats.

Many modern DNA sensors involve modifying capture or target nucleic acids with fluorophores and observing changes in fluorescence upon a hybridization event [22]. Because these platforms rely solely on hybridization, probe sequences can be varied in an array that can contain hundreds of thousands of individual DNA sequences in a single square centimeter [23, 24]. While these platforms provide a significant amount of information, such as gene expression levels, and cannot currently be matched in information content with multiplexed electrochemical chips, the fluorescence assays lack the sensitivity and specificity required to directly detect biomarkers at low concentrations for both fundamental biological studies and diagnostic applications. Furthermore, these assays are expensive, require sophisticated instrumentation, and thus are not well suited for point of care diagnostics.

Electrochemically-based DNA platforms are very well suited for diagnostic applications, from the research lab to clinic, as they are generally very simple, sensitive, and do not require the complex labeling of targets [25]. DNA CT offers a powerful means to interrogate and report on the integrity and conformation of the base stack. Traditionally, DNA-modified electrodes are formed from thiolated DNA self-assembled onto a gold surface (Figure 2.3.4). The gold is then passivated with mercaptohexanol to minimize direct interactions between redox-active moieties and the gold electrode surface. DNA CT sensors are based on the flow of electrons from the surface of electrodes through the DNA base stack to redox-active reporters. Importantly, the flow of electrons through DNA is inhibited by anything that perturbs the base stack, including a single base mismatch, as described earlier, a DNA lesion, or the binding of a protein that disrupts the DNA base stack upon binding, as in proteins that kink DNA or flip bases out of the helix [26]. This sensitivity enables the detection of many classes of biomolecules, from single-stranded DNA, to RNA, and proteins [27].

Electrochemical devices utilizing DNA CT have evolved over time. The first DNA CT-based detection platforms contained only a single electrode. This was advantageous, as devices were simple and could be constructed using commercially available materials. However, single-electrode platforms were limited because multiple experimental parameters could not be directly compared, making subtle differences between samples difficult to discern. More complex electrochemical systems have been developed to address this issue. A multiplexed platform allowing for the simultaneous analysis of different experimental conditions on the same chip has been developed. The current multiplexed silicon chip is fabricated with 16 individually addressable gold electrodes divided into four isolated quadrants [28]. This platform has enabled the incorporation of significantly more complex experiments due to the ability to run multiple experimental conditions in parallel.

We have also recently developed a two-electrode platform for sensing, which enables spatial addressing of DNA on an electrode through the patterning of multiple sequences of DNA onto a single electrode surface [29]. This two-electrode setup involves a large substrate electrode onto which DNA can be specifically patterned in an array via site-specific activation of a click catalyst by a secondary electrode. Detection is subsequently performed through scans across the array with a microelectrode to detect DNA-mediated electrochemistry. Through a simple method of fabrication, this platform allows uniformity in arrays as well as highly sensitive, localized detection.

2.3.2.3 Detection of single base mutations and DNA lesions

DNA-modified electrodes thus provide a powerful technique to monitor mismatches in DNA and thus genomic mutations. Electrochemical signal attenuation has been shown with every possible base mismatch, irrespective of sequence context [16, 30]. Chronocoulometry, with signal amplification through electrocatalysis of ferricyanide by methylene blue, was used to observe all mismatched base pairs [16]. Purine–purine, pyrimidine–purine, and pyrimidine–pyrimidine mismatches all lead to significantly attenuated signals by electrocatalysis. This chemistry can also be applied in the detection of DNA lesions. The majority of DNA lesions have only a small thermodynamic and structural impact on the DNA helix, making them especially difficult to detect with many platforms. However, similarly to single-base mismatches, lesions disrupt the long-range π-stacking of the bases, making them detectable using DNA CT. With chronocoulometry, many common DNA lesions have been shown to significantly attenuate charge accumulation [17]. These include a hydroxylation product of thymine (5,6-hydroxy thymine), an abasic site, an adenine oxidation product (8-oxo-adenine), and a cytosine deamination product (deoxy-uracil). All lesions tested lead to signal attenuations in the order of those observed for mismatches. It is interesting to consider how this lesion detection may be applied biologically.

2.3.3 Detection of DNA-binding proteins

DNA CT platforms are also advantageous for the detection of proteins that interact with DNA. Here we describe a variety of proteins that bind to DNA in different ways but which can all be detected sensitively using DNA electrochemistry.

2.3.3.1 Detection of transcriptional regulators

Transcription factors are vital components of cellular genetic regulation. Transcriptional activators and repressors control the recruitment of RNA polymerase to commence RNA transcription. Many of these proteins mainly interact with DNA simply by bending the helix at the binding site. Because their binding is completely reversible and they do not permanently alter the DNA in any way, transcription factors can be difficult to detect with many DNA-based platforms. However, these proteins can be an important component of pathogenesis, as many of them influence regulation of tumor suppressor genes or oncogenes, making mutations in these proteins potentially extremely deleterious in the cell [31]. As the major mode of interaction between some of these proteins and DNA is helical bending, which distorts the π-stacking of the bases, DNA CT-based detection can be advantageous for their detection.

The transcriptional activator TATA-binding protein (TBP) has been easily detected on DNA-modified electrodes, given the large perturbation in DNA stacking associated with binding to TBP. TBP binds to a TATA sequence in DNA and kinks the helix 80° at that location, leading to a significant DNA-mediated signal attenuation [32]. Figure 2.3.5 shows the result [26]. A DNA-modified electrode containing a covalently

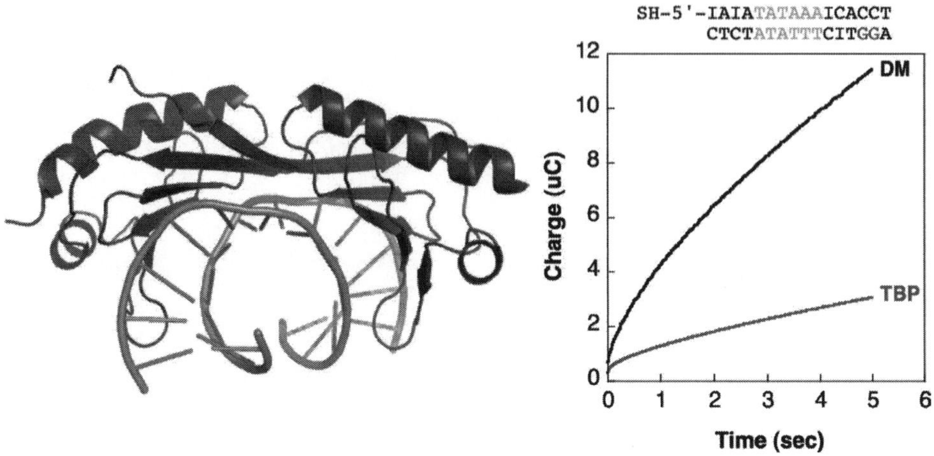

Figure 2.3.5 *Electrochemistry of DNA with TATA-binding protein (TBP). Left: Illustration of the crystal structure of TBP (upper) bound to DNA (lower) [32]; a significant kink in the DNA helix is observable. Right: Chronocoulometry of the DNA-modified electrode without protein (DM) and with TBP bound [26]. As can be seen, the total charge accumulation in the presence of TBP is significantly smaller than in its absence*

bound redox probe shows a large accumulation in charge by chronocoulometry. In the presence of TBP, however, which binds to the specific 5′-TATA-3′ site and kinks the DNA, the charge accumulation is significantly attenuated. Protein binding acts essentially as a switch, turning off DNA CT. In contrast, some proteins that regulate DNA expression bind without perturbing the DNA helix. Helix–turn–helix proteins are one example. These proteins have no significant effect on DNA CT.

2.3.3.2 Methyltransferase and methylation detection

Methyltransferases are proteins responsible for methylation of the genome and are gaining wide interest given their importance in regulation of gene expression. It has recently been shown that aberrant levels of methyltransferase protein are often early indicators of cancer [33–35]. Methyltransferases generally require the flipping of a base out of the π-stack in order to accomplish methylation; after the methyl group has been successfully added to the nucleotide, the base is returned to the DNA stack [36]. While the methylated DNA base product does not hinder DNA CT, CT is significantly diminished when the base is flipped out of the π-stack and a protein residue that is not aromatic is inserted into the base stack, seemingly holding the place while methylation occurs. This attenuation was shown through the detection of a methyltransferase on a DNA-modified electrode surface [26]. Interestingly, if a mutant methylase is used in which an aromatic residue is inserted in the stack, no attenuation of CT is evident.

Methyltransferases have additionally been detected using the carbon nanotube devices, and a reduced affinity of the protein for the DNA after methylation was found [37]. When a single DNA helix containing the binding site for *Sss*I, a bacterial methyltransferase, was exposed to *Sss*I without the necessary cofactor, SAM, a small attenuation in the current through the device was observed. However, upon the addition of the cofactor, the current dropped significantly. This decrease in current was attributed to the de-stacking of the bases, as the protein flips a base out of the π-stack in order for methylation to occur. Restoration of current occurs as the protein is washed from the DNA. Interestingly, given that this corresponds to a single molecule measurement, when the protein was again added to the duplex with its cofactor, no significant current attenuation was observed since the DNA was already methylated. This result indicates that the affinity of the protein for methylated DNA is significantly lower than for its unmethylated counterpart.

Ideally, however, protein detection is performed with a "signal-on" system, as many nonspecific events can cause signal attenuation. Such an assay has been developed, in which DNA that has been successfully methylated maintains its electrochemical signal, while DNA that remains unmethylated is cut by a restriction enzyme [38]. This assay has specifically been used for the detection of the human methyltransferase DNMT1 and bacterial methyltransferase *Sss*I. DNMT1 is the methyltransferase responsible for both the establishment and maintenance of cytosine methylation patterns in the human genome. Aberrant methylation patterns caused by underexpression or overexpression of methyltransferases has been linked to the proliferation of cancers. A bacterial methyltransferase, *Sss*I, an SAM-dependent protein with a preference for unmethylated DNA, was first used, as *Sss*I has significantly higher activity than DNMT1. At 20 nM protein concentration, almost full signal protection is achieved upon addition of the restriction enzyme *BstU*I, which has a preference for unmethylated DNA. When surfaces were treated with either the SAM cofactor alone or the *Sss*I protein alone, large signal decreases were observed upon treatment with *BstU*I. The surfaces were then treated with the restriction enzyme *Rsa*I, which cuts both unmethylated and hemi-methylated DNA. All quadrants had significant signal decreases upon treatment, establishing that the DNA was hemi-methylated in the presence of *Sss*I and SAM (Figure 2.3.6).

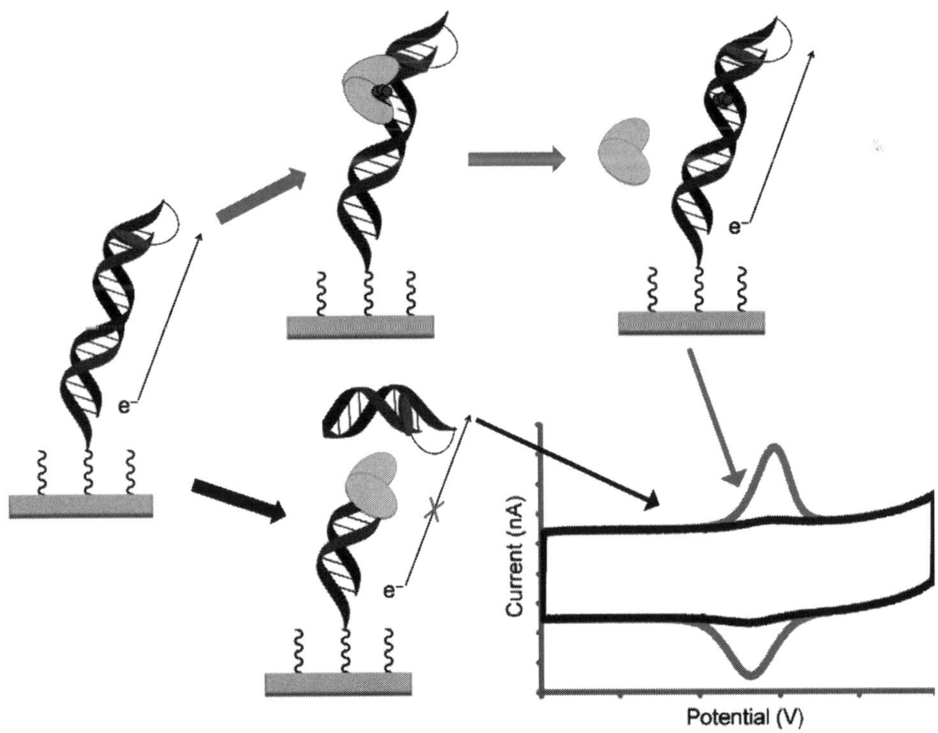

Figure 2.3.6 *Electrochemical assay for methyltransferase activity. DNA duplexes that either contain a binding site for the methyltransferase protein (upper arrows) or DNA that does not contain the binding site (lower arrows) are assembled on electrodes. Methyltransferase protein and then S-adenosyl methionine (SAM) cofactor are added to the surface, and the protein is allowed to methylate the DNA. If the DNA is methylated, restriction enzymes selective for unmethylated DNA will not cut the DNA, maintaining an "on" signal (peaks on cyclic voltammograms). If the DNA remains unmethylated, upon addition of the restriction enzyme, the DNA is cleaved (bottom route), and the electrochemical signal turns off (black cyclic voltammogram)*

This assay was then tested with the human protein, DNMT1, which only methylates hemi-methylated DNA and is associated with genomic methylation maintenance [38]. In a similar manner to *Sss*I, DNMT1 was allowed to methylate DNA on surfaces containing either unmethylated or hemi-methylated DNA. *BssH*II, a restriction enzyme that cuts unmethylated or hemi-methylated DNA but not fully methylated DNA, was then added. Hemi-methylated surfaces that had DNMT1 and SAM added, at protein concentrations as little as 10 nM, saw protection from restriction enzyme cutting. Because methylation patterns and the proteins responsible for this are linked to the development of cancers, the further development of platforms such as this one could be very useful in developing early cancer diagnostic tools.

2.3.3.3 Photolyase activity and detection

Photolyases repair UV damage to DNA, notably pyrimidine–pyrimidine dimers (thymine or cytosine dimers) [39]. Thymine dimers (T<>T) are some of the most common lesions caused by UV damage and greatly distort the structure of DNA. They can result in mutagenesis and are linked to the development of melanomas. Photolyases repair such lesions in bacteria and fungi using visible light and a flavin cofactor. Because thymine dimers disrupt the π-stack, not surprisingly, they attenuate DNA CT. Upon photolyase repair, however, CT is restored through the repaired DNA. This result was shown on DNA-modified electrodes with DNA containing a pre-formed T<>T [40]. Rather than needing to incorporate a redox probe, a signal is observed from the flavin cofactor of the protein bound to the DNA. However, this signal is diminished when a T<>T is incorporated. As the protein–DNA complex is irradiated over time, facilitating DNA repair, the electrochemical signal increases and then levels off, consistent with the repair time for photolyase. Repair of photolyase was confirmed by HPLC (high performance liquid chromatography). Thus, DNA repair by photolyase can be monitored electrochemically.

2.3.3.4 ATP-dependent XPD activity on surfaces

Using similar strategies, we have now been able to monitor the enzymatic activities of several DNA-binding enzymes electrically. Restriction enzymes, where the signal from a covalently bound redox probe decreases once the DNA has been cut and floats away, are a good example [19]. However, perhaps more interesting has been our ability to monitor the activity of a repair enzyme containing a 4Fe-4S cluster. The helicase XPD, from *S. acidocaldarius* (SaXPD), is an ATP-dependent helicase with a 4Fe-4S cluster that is part of the nucleotide excision repair pathway [41]. Its DNA-bound potential was found to be ~80 mV versus NHE [42] similar to the potentials we had seen earlier for MutY and Endonuclease III (EndoIII), two base excision repair proteins from *E. coli* [43]. In electrochemistry experiments on a surface containing the substrate for the helicase (a DNA duplex with a nine base 3′- or 5′-overhang), we could monitor an increase in signal upon addition of ATP to the DNA-bound XPD [42]. In these experiments, there is also no need for a covalent redox probe; as earlier for the photolyase, here the 4Fe-4S cluster of the protein serves that purpose. When the XPD protein binds to the DNA-modified electrode, its redox signal is apparent, and the presence of an intervening abasic site attenuates the signal, establishing that the signal is DNA-mediated. Nevertheless, remarkably, when ATP is added, the signal increases. If, instead, ATP-γS, a poorly hydrolyzed analog, is added, no increase is evident. In fact, we directly correlated the biochemically measured helicase activity of several XPD mutants with their electrochemical activities. The results thus pointed directly to the notion that this protein was electrochemically signaling its activity, with its 4Fe-4S cluster becoming better coupled electrically to the DNA during the course of enzymatic reaction. We have also recently seen this electrochemical monitoring of activity with another repair helicase, DinG, from *E. coli* [44]. These results suggest the possibility that signaling may be something that Nature takes advantage of within the cell.

2.3.4 DNA CT within the cell

DNA CT can occur over long molecular distances, is sensitive to and can be modulated by DNA-binding proteins, and importantly, reports electrically on the integrity of DNA. This chemistry can therefore be powerfully applied in sensing. However, these observations also beg the following question: Is DNA CT utilized within the cell?

2.3.4.1 DNA CT can occur within biologically relevant environments

We began asking whether DNA CT might occur within the cell using our DNA-binding photooxidant, $[Rh(phi)_2(bpy)]^{3+}$. The complex binds to DNA by intercalation, and upon photoactivation, the rhodium complex promotes damage at the 5′-G of guanine doublets; DNA damage at these sites of low oxidation potential have become a signature for one-electron oxidation by DNA CT [45]. We then asked whether $[Rh(phi)_2(bpy)]^{3+}$ causes long-range oxidative damage to genomic DNA within nuclei isolated from HeLa cells via DNA CT [46]. We found that the oxidative damage occurred primarily at these guanine doublets and triplet sites. Moreover, the damage occurred at sites where proteins were constitutively bound, requiring damage caused by DNA CT. Furthermore, using similar techniques, it was shown that DNA CT can occur within mitochondria and at sites also requiring long range CT for efficient reaction [47]. In fact, even when DNA is wound around histones in a nucleosome core particle, as it would be within eukaryotic cells, DNA-mediated oxidative damage caused by long-range photooxidation is observed [48].

However, our small molecule photooxidants are not naturally occurring within the cell. Are there proteins that can serve as redox donors and acceptors, facilitating long-range oxidations and reductions across the genome?

A redox-active cofactor that has increasingly been associated with DNA-processing enzymes is the 4Fe-4S cluster. The first DNA-processing enzymes shown to contain 4Fe-4S clusters were EndoIII and MutY, two base excision repair enzymes in *E. coli*. Since that discovery, however, 4Fe-4S clusters have been found in numerous DNA-processing enzymes including helicases, DNA primases, RNA polymerases, and even DNA polymerases from across the phylogeny [49–52].

When this cofactor was found in EndoIII, its role was initially thought to be solely structural, as the reduction potential for EndoIII was found to be outside the physiological regime [52, 53]. However, those studies were carried out in the absence of DNA. What is more relevant is the *DNA-bound* potential. We therefore measured the DNA-bound potential using DNA-modified electrodes [43]. We found that on gold electrodes modified with duplex DNA, the DNA-bound potential was ~80 mV versus NHE, just what one would want for a physiological switch. MutY and another base excision repair protein showed similar DNA-bound potentials. Direct electrochemistry of EndoIII on highly-oriented pyrolytic graphite surfaces moreover allowed a direct comparison of potentials in the absence and presence of DNA [54]. These results were consistent with a shift negative of ~200 mV upon binding to DNA; in the absence of DNA, there is a redox couple at ~350 mV versus NHE, corresponding to the $[4Fe\text{-}4S]^{3+/2+}$ couple, which is largely irreversible with oxidation. EPR experiments reinforced that the 80 mV redox couple, observed in the presence of DNA, corresponds to the $[4Fe\text{-}4S]^{3+/2+}$ couple of EndoIII. Thus the shift in potential upon DNA binding activates the protein toward oxidation, bringing the $[4Fe\text{-}4S]^{3+/2+}$ couple into a biologically relevant range. The shift in the oxidation potential requires a corresponding increase in the binding affinity of the protein for DNA in the oxidized form versus the reduced form of at least three orders of magnitude, likely caused by the interaction between the positively charged cluster and the polyanionic backbone of DNA.

We have proposed a model to describe how DNA CT may be used by these proteins as a first step in the search for lesions within the genome [13]. These repair proteins are in very low copy number within the cell and have quite low specificity for their target substrates versus unmodified DNA. Yet on the time scale of the

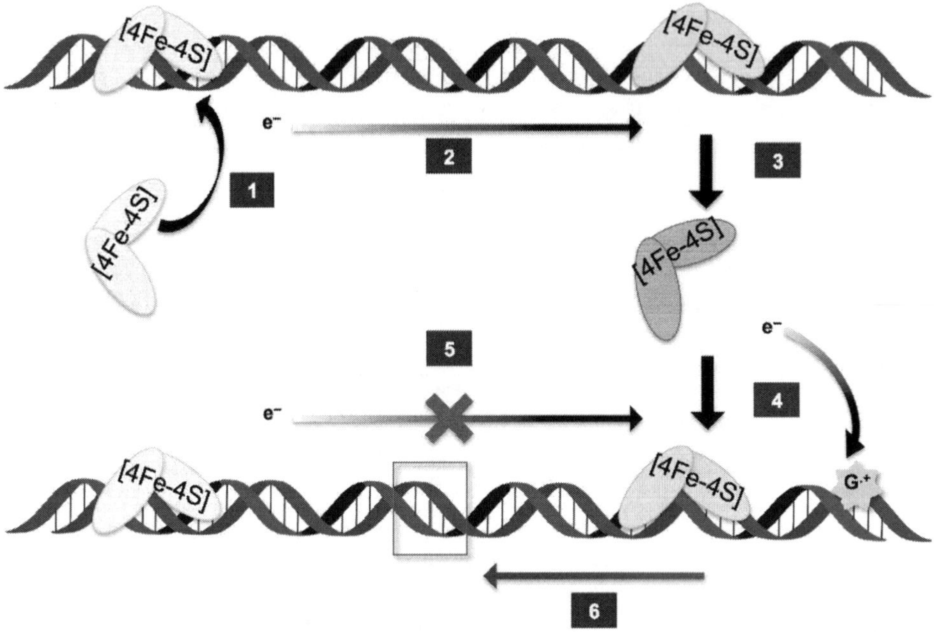

Figure 2.3.7 *Model for redistribution of repair proteins to sites of damage. (1) A protein would be expected first to contain a 4Fe-4S cluster in the 2+ oxidation state. (2) After binding to DNA, this protein, now activated towards oxidation, could be oxidized via DNA CT by a distally bound protein containing a [4Fe-4S]$^{3+}$ cluster. (3) The distally bound protein would then be reduced and its dissociation from DNA would be promoted. (4) Upon binding to DNA, the protein would be activated towards oxidation and could be oxidized by guanine radicals or endogenous reactive oxygen species. (5) If another protein were to bind nearby, a DNA lesion (red) would prevent charge transport through the lesion. (6) Since the distally bound protein would not be reduced, it would stay bound to the DNA and diffuse along the DNA to the lesion in need of repair (See color figure in color plate section)*

E. coli, the proteins must scan the genome, find their cognate sites, and excise the base lesion. DNA CT could aid in the search process as a first scan of the genome, making the search process far more efficient.

Figure 2.3.7 illustrates the model we have proposed. The model posits that freely diffusing proteins contain a [4Fe-4S]$^{2+}$ cluster, but upon DNA binding, clusters are activated towards oxidation. After binding to DNA, a protein could be oxidized by guanine radicals (as we have seen spectroscopically) [55] or endogenous oxidants. If a second protein binds DNA within CT distance of the first protein (1), it could transfer an electron (2) to the oxidized, distally bound protein, reducing it and thereby promoting its dissociation from DNA due to the decreased binding affinity of the reduced protein (3). This electron transfer event would only occur if the intervening DNA is undamaged, and hence really represents a scan of the intervening DNA for damage. The first protein would effectively be signaling that the region of the genome between the proteins is free of damage. This process would continue until proteins bind in the vicinity of a DNA CT-attenuating lesion (4), in which case the electron transfer event would be blocked (5) and the oxidized protein would remain bound to the DNA, proceeding towards the DNA lesion to activate its repair (6). DNA CT thereby offers a potential mechanism for the first step in the detection of DNA damage in the genome by proteins with 4Fe-4S clusters, facilitating a more efficient search and repair process.

To test this model, single-molecule atomic force microscopy (AFM) has been used. In this experiment, the redistribution of the protein EndoIII to DNA strands containing damage that attenuates DNA CT was directly observed [56]. In one case, long and short strands, 3.8 and 1.9 kilobase pairs, respectively, of fully well matched DNA were mixed and incubated with EndoIII. The solutions were then deposited onto mica surfaces and visualized by AFM. As expected, the number of proteins bound per base pair of DNA, the binding density, is equal for both the long and short strands. When the experiment is repeated with long strands that contain a single C:A mismatch, however, the binding density on the long strands of DNA is larger than on the short strands, indicating that the proteins redistribute to the long strands of DNA that contain a mismatch. When the experiment is conducted with a mutant of EndoIII that is poorly coupled to the DNA (i.e., defective in carrying out DNA CT) yet can bind DNA and carry out its enzymatic reaction, this mutant Y82A[56] is unable to redistribute to the mismatched strand. In fact, by testing a variety of EndoIII mutants, we were able to determine a correlation between how effective the protein was in finding the mismatched strand by AFM and its signal on a DNA-modified electrode, reflecting its ability to carry out DNA CT [56, 57]. Interestingly, when the proteins SaXPD and *E. coli* EndoIII were assayed through this AFM redistribution experiment, signaling between the proteins drove redistribution to DNA strands containing a mismatch, despite the fact that they are from separate repair pathways and vastly different organisms [58]. The key was their electronic coupling to the DNA and being at the same DNA-bound potential.

This process has recently been shown to have an influence *in vivo* in *E. coli* between DNA repair proteins from multiple pathways [44]. The iron–sulfur cluster of DinG, a helicase that repairs R-loops, has been shown also to have a DNA-bound redox potential of ~80 mV versus NHE and thus be redox-active at biologically relevant potentials. DinG and EndoIII were then shown to cooperate at long-range via DNA CT in the AFM redistribution assay. Additionally, genetics experiments revealed that this signaling occurs *in cellulo* and is necessary for cellular viability under conditions of stress. If the gene encoding EndoIII is silenced in a strain of *E. coli* where DinG repair is essential, a significant growth defect is observed, which is recovered by complementation with native EndoIII but not with the Y82A mutant.

2.3.5 Conclusions

The ability of DNA to conduct charge is fascinating and powerful chemistry. Many electrochemical detection platforms have been developed to probe enzymatic activity and the fidelity of DNA. DNA CT can occur over long molecular distances, at least up to 34 nm, has a shallow distance dependence, and is exquisitely sensitive to perturbations to the DNA helix. The variety of biological elements that have been shown to affect DNA CT, from the detection of lesions and mismatches to transcription factors and proteins containing iron–sulfur clusters, speaks to the potential importance of DNA CT in a broad spectrum of biological processes. Electrochemical, biochemical, and genetics studies have now shown that DNA CT can occur within biologically relevant environments and may be important chemistry for long-range signaling within the cell. Hence DNA sensing is indeed being applied effectively, both in devices to monitor the integrity of DNA and perhaps even in the cell to monitor the integrity of DNA within our own genomes.

Acknowledgements

We are grateful to all of our coworkers and collaborators who over many years have made elucidation of this chemistry a fascinating journey. We also thank the NIH and the Moore foundation for their funding.

References

[1] J. D. Watson and F. H. Crick, Molecular structure of nucleic acids; a structure for deoxyribose nucleic acid, *Nature* **171**, 737–738 (1953).
[2] S. Neidle, *DNA Structure and Recognition*, Oxford University Press, Oxford, (1994).

[3] D. D. Eley and D. I. Spivey, Semiconductivity of organic substances. Part 9. Nucleic acid in the dry state, *Trans. Faraday Soc.* **58**, 411–415 (1962).
[4] J. G. Genereux and J. K. Barton, Mechanisms for DNA charge transport, *Chem. Rev.* **110**, 1642–1662 (2010).
[5] M. A. O'Neill, H.-C. Becker, C. Wan, J. K. Barton, and A. H. Zewail, Ultrafast dynamics in DNA-mediated electron transfer, base gating and the role of temperature *Angew. Chem., Int. Ed.* **42**, 5896–5900 (2003).
[6] C. Wan, T. Fiebig, S. O. Kelley, C. R. Treadway, J. K. Barton, and A. H. Zewail, Femtosecond dynamics of DNA-mediated electron transfer, *Proc. Natl. Acad. Sci. U.S.A.* **96**, 6014–6019 (1999).
[7] C. J. Murphy, M. R. Arkin, Y. Jenkins, N. D. Ghatlia, S. H. Bossmann, N. J. Turro, and J. K. Barton, Long-range photoinduced electron transfer through a DNA helix, *Science* **262**, 1025–1029 (1993).
[8] A. A. Voityuk, N. Rösch, M. Bixon, and J. Jortner, Electronic coupling for charge transfer and transport in DNA, *J. Phys. Chem. B.* **104**, 9740–9745 (2000).
[9] R. E. Holmlin, P. J. Dandliker, and J. K. Barton, Charge transfer through the DNA base stack, *Angew. Chem., Int. Ed.* **36**, 2714–2730 (1997).
[10] D. B. Hall, R. E. Holmlin, and J. K. Barton, Oxidative damage through long range electron transfer, *Nature* **382**, 731–735 (1996).
[11] F. D. Lewis, L. Xiayoyang, S. E. Miller, and M. R. Wasielewski, Electronic interactions between π-stacked DNA base pairs and diphenylacetylene-4,4′4,4nylacetyleneehrhairpin DNA, *J. Am. Chem. Soc.* **121**, 9746–9747 (1999).
[12] N. B. Muren, E. D. Olmon, and J. K. Barton, Solution, surface and single molecule platforms for the study of DNA-mediated charge transport, *Phys. Chem. Chem. Phys.* **14**, 13754–13771 (2012).
[13] P. A. Sontz, N. B. Muren, and J. K. Barton, DNA charge transport for sensing and signaling, *Acc. Chem. Res.* 1792–1800 (2012).
[14] S. O. Kelley, R. E. Holmlin, E. D. A. Stemp, and J. K. Barton, Photoinduced electron transfer in ethidium-modified DNA duplexes: Dependence on distance and base stacking, *J. Am. Chem. Soc.* **119**, 9861–9870 (1997).
[15] S. O. Kelley, N. M. Jackson, M. G. Hill, and J. K. Barton, Long-range electron transfer through DNA films, *Angew. Chem., Int. Ed.* **38**, 941–945 (1999).
[16] E. M. Boon, D. M. Ceres, T. G. Drummond, M. G. Hill, and J. K. Barton, Mutation detection by electrocatalysis at DNA-modified electrodes, *Nature Biotech.* **18**, 1096–1100 (2000).
[17] A. K. Boal and J. K. Barton, Electrochemical detection of lesions in DNA, *Bioconj. Chem.* **16**, 312–321 (2005).
[18] X. Guo, A. A. Gorodetsky, J. Hone, J. K. Barton, and C. Nuckolls, Conductivity of a Single DNA duplex bridging a carbon nanotube gap, *Nat. Nanotech.* **3**, 163–167 (2008).
[19] J. D. Slinker, N. M. Muren, S. E. Renfrew, and J. K. Barton, DNA charge transport over 34 nm, *Nat. Chem.* **3**, 230–235 (2011).
[20] C. G. Pheeney, L. F. Guerra, and J. K. Barton, DNA sensing by electrocatalysis with hemoglobin, *Proc. Natl. Acad. Sci. U.S.A.* **109**, 11528–11533 (2012).
[21] C. G. Pheeney and J. K. Barton, DNA electrochemistry with tethered methylene blue, *Langmuir* **28**, 7063–7070 (2012).
[22] M. J. Heller, DNA microarray technology: devices, systems, and applications, *Annu. Rev. Biomed. Eng.* **4**, 129–153 (2002).
[23] M. Schena, D. Shalon, R. W. Davis, and P. O. Brown, Quantitative monitoring of gene expression patterns with a complementary DNA microarray, *Science* **270**, 467–470 (1995).
[24] M. A. Behr, M. A. Wilson, W. P. Gill, H. Salamon, G. K. Schoolnik, S. Rane, and P. M. Small, Comparative genomics of BCG vaccines by whole-genome DNA microarray, *Science* **284**, 1520–1523 (1999).
[25] T. G. Drummond, M. G. Hill, and J. K. Barton, Electrochemical DNA sensors, *Nat. Biotechnol.* **21**, 1192–1199 (2003).
[26] E. M. Boon, J. E Salas, and J. K. Barton, An electrical probe of protein–DNA interactions on DNA-modified surfaces, *Nat. Biotechnol.* **20**, 282–286 (2002).
[27] A. A. Gorodetsky, M. C. Buzzeo, and J. K. Barton, DNA-mediated electrochemistry, *Bioconj. Chem.* **19**, 2285–2296 (2008).
[28] J. D. Slinker, N. B. Muren, A. A. Gorodetsky, and J. K. Barton, Multiplexed DNA-modified electrodes, *J. Am. Chem. Soc.* **132**, 2769–2774 (2010).

[29] A. L. Furst, S. Landfield, M. G. Hill, and J. K. Barton, Electrochemical patterning and detection of DNA arrays on a two-electrode platform, *J. Am. Chem. Soc.* **51**, 19099–19102 (2013).

[30] S. O. Kelley, E. M. Boon, J. K. Barton, N. M. Jackson, and M. G. Hill, Single-base mismatch detection based on charge transduction through DNA, *Nucl. Acids. Res.* **27**, 4830–4837 (1999).

[31] J. E. Darnell Jr., Transcription factors as targets for cancer therapy, *Nat. Rev. Cancer* **2**, 740–749 (2002).

[32] Z. S. Juo, T. K. Chiu, P. M. Leiberman, I. Baikalov, A. J. Berk, and R. E. Dickerson, How proteins recognize the TATA box, *J. Mol. Biol.* **261**, 239–254 (1996).

[33] K. D. Robertson, DNA methylation, methyltransferases, and cancer, *Oncogene* **20**, 3139–3155 (2001).

[34] K. Ghoshal and S. Bai, DNA methyltransferases as targets for cancer therapy, *Drugs Today (Barc.)* **43**, 395–422 (2007).

[35] B. Brueckner, D. Kuck, and F. Lyko, DNA methyltransferase inhibitors for cancer therapy, *Cancer J.* **13**, 17–22 (2007).

[36] S. S. Smith, B. E. Kaplan, L. C. Sowers, and E. M. Newman, Mechanism of human methyl-directed DNA methyltransferase and the fidelity of cytosine methylation, *Proc. Natl. Acad. Sci. U.S.A.* **89**, 4744–4748 (1992).

[37] H. Wang, N. B. Muren, D. Ordinario, A. A. Gorodetsky, J. K. Barton, and C. Nuckolls, Transducing methyltransferase activity into electrical signals in a carbon nanotube–DNA device, *Chem. Sci.* **3**, 62–65 (2012).

[38] N. B. Muren and J. K. Barton, Electrochemical assay for the signal-on detection of human DNA methyltransferase activity, *J. Am. Chem. Soc.* **135**, 16632–16640 (2013).

[39] A. Sancar, Structure and function of photolyase and *in vivo* enzymology: 50th Anniversary, *J. Biol. Chem.* **283**, 32153–32157 (2008).

[40] M. C. DeRosa, A. Sancar, and J. K. Barton, Electrochemically monitoring DNA repair by photolyase, *Proc. Natl. Acad. Sci. U.S.A.* **102**, 10788–10792 (2005).

[41] L. Fan, J. O. Fuss, Q. J. Cheng, A. S. Arvai, M. Hammel, V. A. Roberts, P. K. Cooper, and J. A. Tainer, XPD helicase structures and activities: insights into cancer and aging phenotypes from XPD mutations, *Cell* **133**, 789–800 (2008).

[42] T. P. Mui, J. O. Fuss, J. P. Ishida, J. A. Tainer, and J. K. Barton, ATP-stimulated, DNA-mediated redox signaling by XPD, a DNA repair and transcription helicase. *J. Am. Chem. Soc.* **133**, 16378–16381 (2011).

[43] A. K. Boal, E. Yavin, O. A. Lukianova, V. L. O'Shea, S. S. David, and J. K. Barton, DNA-bound redox activity of DNA repair glycosylases containing [4Fe-4S] clusters, *Biochemistry* **44**, 8397–8407 (2005).

[44] M. A. Grodick, H. M. Segal, T. J. Zwang, and J. K. Barton, DNA-mediated signaling by proteins with 4Fe–4S clusters is necessary for genomic integrity, *J. Am. Chem. Soc.* **136**, 6470–6478 (2014).

[45] S. Delaney and J. K. Barton, Long-range DNA charge transport, *J. Org. Chem.* **68**, 6475–6483 (2003).

[46] M. E. Núñez, G. P. Holmquist, and J. K. Barton, Evidence for DNA charge transport in the nucleus, *Biochemistry* **40**, 12465–12471 (2001).

[47] E. J. Merino and J. K. Barton, DNA oxidation by charge transport in the mitochondria, *Biochemistry* **47**, 1511–1517 (2008).

[48] M. E. Núñez, K. E. Noyes, and J. K. Barton, Oxidative charge transport through DNA in nucleosome core particles, *Chem. Biol.* **9**, 403–415 (2002).

[49] Y. Wu, A. N. Suhasini, and R. M. Brosh, Jr., Welcome to the family of FancJ-like helicases to the block of genome stability maintenance proteins, *Cell. Mol. Life Sci.* **66**, 1209–1222 (2009).

[50] B. E. Weiner, H. Huang, B. M. Dattilo, M. J. Nilges, E. Fanning, and W. J. Chazin, An iron-sulfur cluster in the C-terminal domain of the p58 subunit of human DNA primase, *J. Biol. Chem.* **46**, 33444–33451 (2007).

[51] D. J. A. Netz, C. M. Stith, M. Stümpfig, G. Köpf, D. Vogel, H. M. Genau, J. L. Stodola, R. Lill, P. M. J. Burgers, and A. J. Pierik, Eukaryotic DNA polymerases require an iron-sulfur cluster for the formation of active complexes, *Nature Chem. Biol.* **8**, 125–132 (2011).

[52] R. P. Cunningham, H. Asahara, and J. S. Bank, Endonuclease III is an iron-sulfur protein, *Biochemistry* **28**, 4450–4455 (1989).

[53] W. Fu, S. O'Handley, R. P. Cunningham, and M. K. Johnson, The role of the iron-sulfur cluster in *Escherichia coli* endonuclease III. A resonance Raman study, *J. Biol. Chem.* **267**, 16135–16137 (1992).

[54] A. A. Gorodetsky, A. K. Boal, and J. K. Barton, Direct electrochemistry of Endonuclease III in the presence and absence of DNA, *J. Am. Chem. Soc.* **128**, 12082–12083 (2006).

[55] E. Yavin, A. K. Boal, E. D. A. Stemp, E. M. Boon, A. L. Livingston, V. L. O'Shea, S. S. David, J. K. Barton, Protein_DNA charge transport: Redox activation of a DNA repair protein by guanine radical, *Proc. Natl. Acad. Sci. U.S.A.* **102**, 3546–3551 (2005).

[56] A. K. Boal, J. C. Genereux, P. A. Sontz, J. A. Gralnick, D. K. Newman, and J. K. Barton, Redox signaling between DNA repair proteins for efficient lesion detection, *Proc. Natl. Acad. Sci. U.S.A.* **106**, 15237–15242 (2009).

[57] C. A. Romano, P. A. Sontz, and J. K. Barton, Mutants of the base excision repair glycosylase, endonuclease III: DNA charge transport as a first step in lesion detection, *Biochemistry* **50**, 6133–6145 (2011).

[58] P. A. Sontz, T. P. Mui, J. O. Fuss, J. A. Tainer, and J. K. Barton, DNA charge transport as a first step in coordinating the detection of lesions by repair proteins, *Proc. Natl. Acad. Sci. U.S.A.* **109**, 1856–1861 (2012).

2.4

Charge Transfer in Non-B DNA with a Tetraplex Structure

Jungkweon Choi and Tetsuro Majima

The Institute of Scientific and Industrial Research (SANKEN), Osaka University, Ibaraki, Osaka, Japan

2.4.1 Introduction

In the human genome, repeating nucleotide sequences occupy more than 50% of the total genomic DNA [1]. For example, a telomere, which protects the ends of the chromosome from deterioration, consists of repetitive nucleotide sequences. Under certain conditions *in vitro*, however, these repetitive DNA sequences can form unique structures other than the double helix of B-DNA with Watson–Crick base pairing [2–8]. These unusual secondary structures, so-called non-B DNA, have been reported to affect the gene metabolism process and also to participate in several biologically important processes. According to previous studies, the non-B DNA structure-forming sequences can induce genetic instability and consequently may cause human diseases [4, 5, 9, 10]. Significantly, non-B DNAs induce not only genetic expansions and deletions, but also DNA strand breaks and rearrangements. Therefore, in the pharmaceutical and medical fields, the molecular mechanism of non-B DNAs has been studied extensively. Important studies on the non-B DNA structure-induced genetic instability have been summarized excellently in recent reviews [9, 11–13]. Among the non-B DNAs, it is G-quadruplex and i-motif DNAs, which have a tetraplex structure formed from guanine (G)- and cytosine (C)-rich sequences, respectively, that have attracted particular attention in the biomedical field, because these G- and C-rich sequences are frequently observed in the promoter region of oncogene and human telomeric DNA.

DNA in Supramolecular Chemistry and Nanotechnology, First Edition.
Edited by Eugen Stulz and Guido H. Clever.

On the other hand, it is currently recognized that G-quadruplex and i-motif DNAs with a unique tetraplex structure are promising materials for the development of DNA-based molecular electronic devices, because they do not produce any toxic byproducts during the working cycle of a molecular electronic device using their reversible conformational change. [14, 15]. Both theoretical and experimental studies have shown the possibility that G-quadruplex and i-motif DNAs can act as a hole captor and an electron carrier in molecular electronic nano-devices, respectively [16–22]. In particular, because the long G-rich sequences can form long and stable G-quadruple-wires with lengths of up to 300 nm in the presence or absence of stabilizing ions (K^+ or Na^+) [23, 24], the development of G-quadruplex-based nanoscale-molecular electronic devices has attracted considerable attention from many research groups. Significantly, G-quadruplex DNA attached to a gold surface showed a larger polarizability than duplex DNA, indicating that G-quadruplex DNA may have a higher conductance than duplex DNA [25]. Liu *et al.* demonstrated that the G-quadruplex covalently wired between two gold electrodes can produce a relatively high conductance, and that a structural change between an extended structure (single-stranded DNA, ssDNA) and a compact structure (G-quadruplex) could possibly be utilized as a reversible electronic switch [26]. In addition, i-motif DNA is also considered to be a useful nanotechnology material, including use as a nano-electronic device, because of the opportunity for reversible pH-induced conformational change [14, 15]. In this regard, understanding the charge transfer (CT) processes occurring within G-quadruplexes or in i-motif DNA itself is a prerequisite for the application of DNA in nano-biotechnology. In this chapter we will demonstrate the mechanism and dynamics of the CT processes occurring in G-quadruplexes and i-motif DNA.

2.4.2 CT in dsDNA (B-DNA)

Although the electrical conductivity of DNA with Watson–Crick base pairing is still being debated, DNA-mediated CT has been extensively investigated through many theoretical and experimental approaches and has been explained extremely well in several recent reviews [27–30]. Thus, the CT process in DNA will be briefly described here.

Essentially, the CT process in DNA is a very important phenomenon with respect to the oxidation and reduction of DNA, which are closely related to the damage of DNA and to the repair of damaged DNA, respectively. In general, the rate constant (k_{CT}) for the CT process in DNA is expressed by an exponential dependence of the donor (D)–acceptor (A) distance (Δr) as follows [31–33]:

$$k_{CT} = k_0 \exp(-\beta \Delta r) \tag{2.4.1}$$

where β is the distance dependence parameter. On the other hand, the k_{CT} for the CT process in long sequence DNA is described by a power law [27, 34–38]:

$$k_{CT} = k_0 (\Delta r)^{-\eta} \tag{2.4.2}$$

Although, as shown in Equation 2.4.2, k_{CT} is significantly affected by the D–A distance (Δr), the photophysical properties of the photosensitizer that is used, as well as the DNA sequences and structures, are other important factors for the CT dynamics in DNA.

In the DNA-mediated CT process, holes and excess electrons act as positive and negative charge carriers, respectively. DNA-mediated hole transfer, which is strongly dependent on the nucleotide inserted between the G–C base pairs, has been shown to take place over a large distance, longer than 200 Å, through a hopping mechanism [29, 39–42]. In addition, delocalization of the charge over the stacked G bases along the DNA stem has been reported [43, 44]. Although the hole transfer in DNA is strongly dependent on the characteristics of the photosensitizer used as well as the DNA sequences and structures, the long-distance hole transfers have been seen to occur with a rate constant in the range of 10^3–10^9 s^{-1} [41, 45–49]. In addition, the hole hopping rates in A-to-A and G-to-G have been estimated to be 1.2×10^9 and 4.3×10^9 s^{-1}, respectively [50].

On the other hand, in DNA, electron transfer is not as well established compared with hole transfer. Although Tainaka *et al.* have recently shown that the excess electron can migrate over 34 Å through the base pairs [51], it is generally accepted that in DNA the distance for an excess electron transfer is less than that of a hole. These recent studies also revealed that the hopping rate of the excess electron between consecutive thymines (T) is faster than the hole-trapping rate between adenines (A) and guanidines (G) [52]. Meanwhile, the single-step electron transfer (via a superexchange mechanism) between a *D* in the singlet excited state and an *A* proceeds efficiently, and depends on the *D–A* distance in the nicked-dumbbell DNA sequence [52, 53]. The DNA-mediated electron transfer by a superexchange mechanism takes place predominantly in DNA sequences with one to three base pairs between the *D* and *A*, while electron transfer by the hopping mechanism takes place in DNA sequence with a long *D–A* distance.

2.4.3 CT in non-B DNA with a tetraplex structure

2.4.3.1 G-quadruplex DNA

G-quadruplex DNA consists of π–π stacking of planar G-tetrads, cyclically bound to each other through eight hydrogen bonds via the Hoogsteen base pairs (see Figure 2.4.1a). Its structure depends greatly on the nucleotide sequences, the orientation of the strands, the *syn*/*anti* glycosidic conformation of guanines, the loop connectivities,

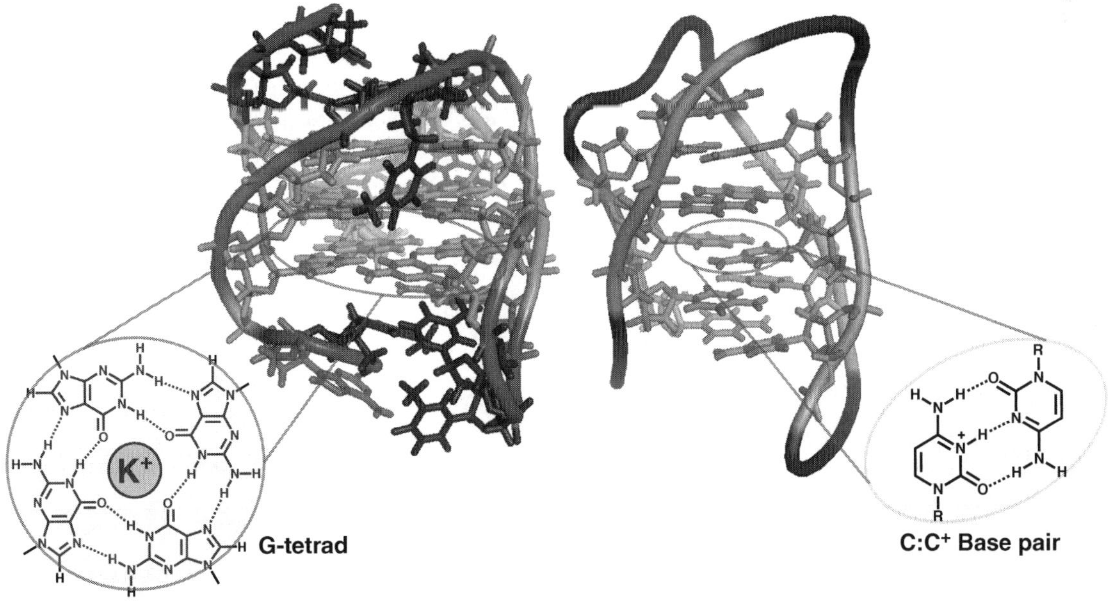

Figure 2.4.1 Molecular structures. (a) G-quadruplex (PDB id: 2KZD) and G-tetrads composed of four G bases (green) and (b) i-motif (PDB id: 1EL2) and hemiprotonated C:C base pair (orange) (See color figure in color plate section)

and environmental factors such as cations, molecular crowding, and dehydration, and thus shows a high degree of structural polymorphism [54–59].

Theoretical studies revealed that the stacking of two or more G bases induces a lower ionization potential (*IP*) compared with that of a single isolated G, and consequently can induce a favorable hole trapping [60–62]. From considerations, it was expected that a G-quadruplex with π–π stacking of planar G-tetrads could act as an excellent hole captor. Interestingly, Delaney and Barton noted that in a duplex-antiparallel G-quadruplex composite, oxidizing radicals are predominantly trapped in the external G-tetrads of the G-quadruplex rather than the stacking site of two G bases within the duplex DNA, demonstrating the efficient hole trapping ability of the G-quadruplex [21]. Ndlebe and Schuster also revealed that the radical cation (hole) is about 50% more likely to be trapped at G bases within G-tetrads than at a –GG– site in duplex DNA, indicating the migration of the radical cation into the G base within the G-tetrads through a duplex DNA region [18]. They also showed that G bases at the top and bottom of the stack of G-tetrads are slightly more reactive than at the three interior G-tetrads, which is consistent with results reported by Delaney and Barton [21]. These results imply that the hole can migrate between the G-tetrad stacking of the individual G-quadruplex.

On the other hand, Sen's research group showed that in duplex–G-quadruplex–duplex (D–Q–D) composites, the hole conduction from the anthraquinone (AQ)–proximal duplex to the AQ–distal duplex via the intervening G-quadruplex was determined to be approximately half (~50%) as efficient as charge flow through an equivalently sized, purely Watson–Crick base-paired duplex [19]. However, they showed that the oxidizing damage of G-bases within G-tetrads is significantly lower than –GG– sites in duplex DNA. This result is appreciably different from those reported by Delancy and Barton [21], and by Ndlebe and Schuster [18]. They interpreted that the disagreement is due to the difference in the G-quadruplex structure, such as parallel and antiparallel structures, and the low reactivity with water for the G-quadruplex sandwiched between two duplex DNAs. Indeed, using a combined quantum mechanics and molecular dynamics (MD) approach, Lech *et al.* revealed that the electron and hole transfer rates in G-quadruplex DNA are largely affected by the stacking geometries of the G-tetrads. They demonstrated that hole hopping within single G-tetrads is slower by at least two orders of magnitude than that between stacked guanines, indicating that hole transfer within individual planar G-tetrads should not affect the hole mobility in the G-quadruplex [22]. In addition, they found that among geometries of the stacked G-tetrads within the G-quadruplex, the stacks of the same polarity (parallel-stranded configuration) are less conductive than those of opposite polarity (non-parallel-stranded configuration), and that hole transfer at the interface of stacked G-quadruplex DNAs is relatively faster than that within G-tetrad stacking of the individual G-quadruplex (see Figure 2.4.2). Furthermore, Sen's research group showed that the G-quadruplex (or pinched duplex), which is formed from a duplex DNA with consecutive G–G mismatches through the binding of K^+ or Sr^{2+}, shows a 40-times higher conductivity than duplex DNA, although duplex DNA has several consecutive G–G mismatches to disrupt the charge migration [20]. The theoretical and experimental results mentioned earlier imply that the G-quadruplex as well as the long and stable G-quadruple-wires can serve as a good electron and hole carrier in nano-electronic device.

However, most studies on the CT process within a G-quadruplex have been made using polyacrylamide gel electrophoresis (PAGE) and product analyses [16, 18–21]. Although these analyses have provided much important information on the mechanism of the CT process within the G-quadruplex, the kinetics of the CT process occurring within the G-quadruplex is not fully understood. Hence, we recently investigated the hole transfer and trapping in a riboflavin (Rf)-labeled G-quadruplex using femtosecond (fs) laser flash photolysis and pulse radiolysis [63]. The Rf-labeled oligomer, Rf-5′-GGTTGGTGTGGTTGG-3′ (Rf–G-q), folds into a monomolecular antiparallel G-quadruplex in the presence of K^+ ions (see Figure 2.4.3a). Upon formation of the G-quadruplex, a decrease in the fluorescence intensity of the Rf was observed, suggesting that Rf interacts efficiently with the G-bases in the G-quadruplex. The fluorescence quenching of the Rf accompanied by the formation of the G-quadruplex probably results from the formation of the charge transfer state due to the hole transfer from Rf in the singlet excited state (^{1}Rf*) to the G-bases. Consequently, the CT process leads to the

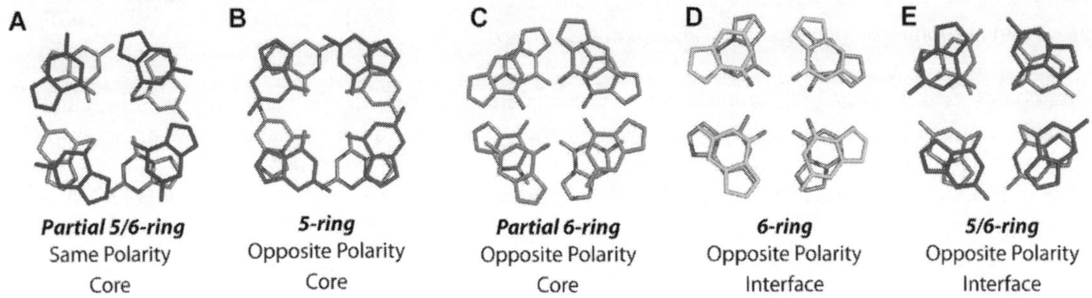

	A	B	C	D	E
k_{HT}/ns^{-1}	26.0	69.4	55.3	163.6	1912.2
k_{rel}	1	2.7	2.1	6.3	73.5

Figure 2.4.2 *G-tetrad stacking modes, and absolute and relative hole transfer rates (k_{HT} and k_{rel}, respectively) between adjacent tetrads in a G-quadruplex. Reprinted with permission from [22]. Copyright, 2013 American Chemical Society*

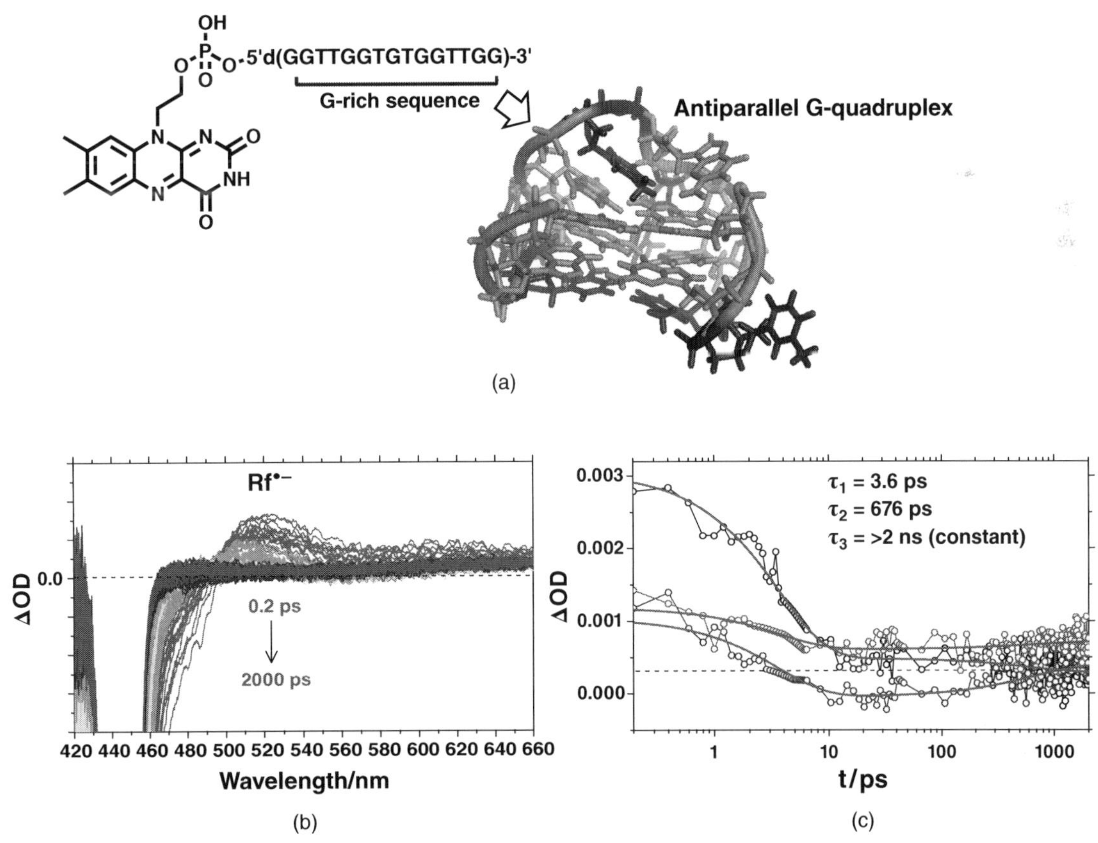

Figure 2.4.3 *(a) Chemical structure of riboflavin-labeled oligonucleotide (Rf–G-q) and G-quadruplex (PDB id: 1C35). (b) Transient absorption spectra of G-quadruplex Rf–G-q after 450 nm laser excitation in 10 mM Tris-HCl buffer (pH 7.4) at room temperature. (c) Decay profiles of G-quadruplex Rf–G-q monitored at 510 (upper trace), 570 (lower trace) and 650 nm (middle trace). Theoretical fits obtained from the global fitting analysis are shown in red. Reprinted with permission from [63]. Copyright Wiley-VCH (2013)*

formation of a riboflavin radical anion and a G radical cation. Indeed, the G-quadruplex Rf–G-q exhibits a distinctly different transient absorption spectra compared with those for free Rf and single-stranded Rf–G-q (see Figure 2.4.3b and reference [63]). At the early delay times, the G-quadruplex Rf–G-q shows a positive signal over a wide wavelength range, as shown in Figure 2.4.3. The absorption maxima around 520 nm is attributed to the riboflavin radical anion generated by the hole injection from ^{1}Rf* to the G-bases. As depicted in Figure 2.4.3c, the decay profiles of the G-quadruplex Rf–G-q were expressed by a tri-exponential function with relaxation times of 3.6 ± 0.2 ps, 676 ± 205 ps, and >2 ns (constant), suggesting that the CT in the G-quadruplex takes place within a few picoseconds. The efficient charge separation in the G-quadruplex is probably due to the excellent hole trapping ability of planar G-tetrads.

To further elucidate the hole trapping ability of a planar G-tetrad of a G-quadruplex, we used the human telomere sequence (5′-TAGGG-(TTAGGG)$_3$-TT-3′) and measured the transient absorption spectra of the G-quadruplex during pulse radiolysis in 20 mM potassium phosphate buffer, containing 20 mM $K_2S_2O_8$ and 0.1 M *tert*-butyl alcohol at many delay times (Δt = 50–1000 ns) [63]. As shown in Figure 2.4.4a, the absorption spectra observed at Δt = 50 ns is characterized by a broad signal with an absorption maxima of ~460 nm. Because this signal is coincident with the spectrum of the G radical cation ($G^{\bullet+}$) reported by Kobayashi *et al.* [64, 65], the absorption spectra observed at Δt = 50 ns is assigned to the $G^{\bullet+}$. That is, the hydrated electron (e_{aq}^-) generated by pulse radiolysis quickly reacts with a peroxydisulfate ($S_2O_8^{2-}$) to produce a sulfate radical anion ($SO_4^{\bullet-}$). The G is then oxidized to $G^{\bullet+}$ by the sulfate radical anion [64–66] (see Equations 2.4.3 and 2.4.4):

$$e_{aq}^- + S_2O_8^{2-} \rightarrow SO_4^{2-} + SO_4^{\bullet-} \tag{2.4.3}$$

$$SO_4^{\bullet-} + G \rightarrow SO_4^{2-} + G^{\bullet+} \tag{2.4.4}$$

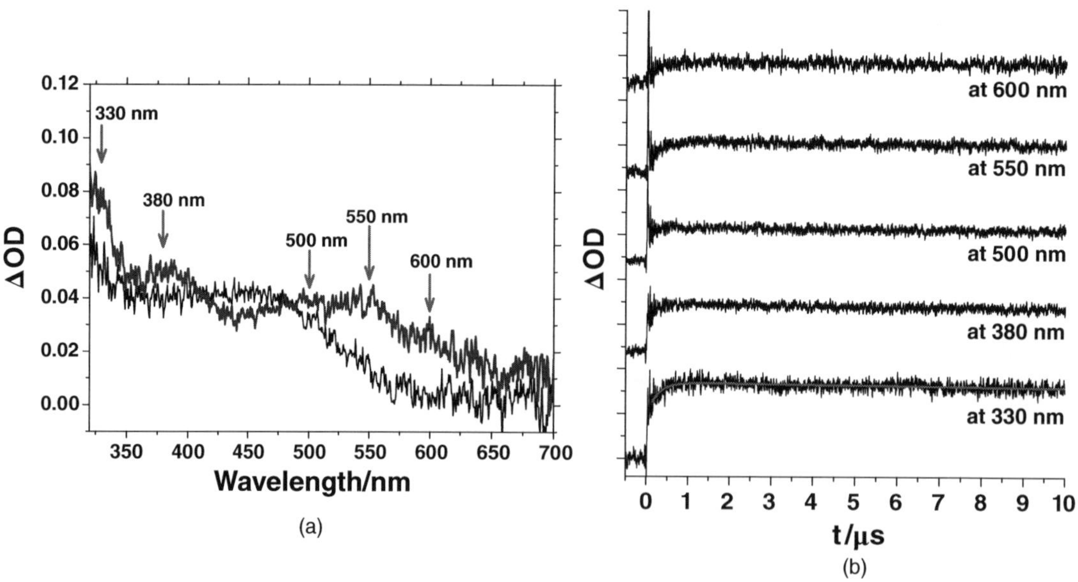

Figure 2.4.4 (a) Transient absorption spectra of G-quadruplex formed from a 5′-TAGGG- (TTAGGG)$_3$TT-3′ sequence at 50 ns (lower) and 1 μs (upper) delay times during pulse radiolysis in 20 mM potassium phosphate buffer (pH 7.4). (b) Decay profiles of G-quadruplex formed from a 5′-TAGGG(TTAGGG)$_3$TT-3′ sequence in 10 mM potassium phosphate buffer (pH 7.4) monitored at 330, 380, 500, 550, and 600 nm, respectively. Theoretical fits obtained from the global fitting analysis are shown along the lowest trace. Reprinted with permission from [63]. Copyright Wiley-VCH (2013)

With one-electron oxidation, the formation of $G^{\bullet+}$ in the G-quadruplex implies that the radical cation (hole) is efficiently trapped in the G-tetrads. As mentioned earlier, the π–π stacking of planar G-tetrads can induce a lower oxidation potential compared with the stacking of two or more G bases within duplex DNA. It is clear that the spectrum of the $G^{\bullet+}$ observed from the G-quadruplex is somewhat red shifted compared with that for two or three consecutive G bases, indicating that the positive charge is delocalized along the more extended π-orbitals of the G-tetrad. The delocalization of the positive charge consequently leads to the favorable hole trapping within the planar G-tetrad because of the lower oxidation potential. Additionally, the spectrum corresponding to the neutral radical of G [$G^{\bullet+}$(-H)], which is the deprotonated species of $G^{\bullet+}$, is observed at the longer delay times. The rate constant for the formation of $G^{\bullet+}$(-H) is determined to be $4.0 \pm 0.2 \times 10^6$ s^{-1}, which is slightly less than values reported in the literatures [65, 67]. This means that the deprotonation of $G^{\bullet+}$ is more difficult in the G-quadruplex than in the duplex DNA structure.

We will now consider why the CT process within the G-quadruplex is more efficient compared with the duplex DNA. Because the G-quadruplex formed with metal ions is more thermally stable than the duplex DNA, it has been suggested that the efficient CT process within the G-quadruplex is attributed to an increased structural stability and a higher number of overlapping π-orbitals. Additionally, the presence of metal ions in G-quadruplexes can contribute to the efficient CT process in the G-quadruplex. However, using a combination of electronic structure calculations, MD simulations, and the formulation of an effective tight-binding model Hamiltonian, Woiczikowski *et al.* showed that the presence of central metal ions in the G-quadruplex has little effect on the hole transfer in the G-quadruplex [17]. Consequently, they suggested that the improved electrical conduction of the G-quadruplex is not due to its structural stability and rigidity, but to a larger number of charge-transfer active conformations, implying that the higher stability of the G-quadruplex than for the duplex DNA may help to maintain a conducting conformation. Indeed, their theoretical calculations showed that the G-quadruplex has a large number of CT active conformations that can induce large interstrand couplings between four strands in the G-quadruplex, although the intrastrand coupling within the G-quadruplex is smaller than in the duplex DNA (see Table II in reference [17]). Considering previous studies and our results, we thus conclude that the efficient CT within the G-quadruplex is mainly due to π–π stacking of planar G-tetrads and to a large number of CT active conformations that induce the interstrand couplings between four strands in the G-quadruplex. However, the contribution of metal ions to the CT within the G-quadruplex is still not clear and requires more study.

2.4.3.2 i-motif DNA

The i-motif structure, which is formed from C-rich strands at slightly acidic pH or even at a neutral pH, consists of two parallel-stranded C:C^+ hemiprotonated base-paired duplexes that are intercalated in an antiparallel manner (see Figure 2.4.1b). Its structure is significantly affected by the number of cytosine bases [68], the loop length [69], environmental conditions [70, 71], and material attached or interacting with the DNA strands [72–74].

As with the G-quadruplexes, i-motif DNA is a promising material for nanotechnology, including as nanoelectronic devices or switches, because of its reversible pH-induced conformational changes between an extended structure (ssDNA) and a compact structure (i-motif) [14]. Nevertheless, there have been few reports of the CT dynamics in i-motif DNA because its structure is more stable under slightly acidic conditions than under neutral conditions. Recently, Keane *et al.* investigated the long-lived excited-state of i-motif DNA using picosecond time-resolved IR spectroscopy [75]. They suggested that the long-lived transient species observed at 1545 cm^{-1} may result from the formation of a CT state ($C^{\bullet-}$····$C^{\bullet}$) between the C and C^+ moieties. The formation of a CT state is explained in terms of the stacking geometry of the i-motif, positioning the electron-rich amino group near the electron-deficient pyrimidine ring of the nucleobase below. Meanwhile, we studied the electron-transfer dynamics of i-motif DNA conjugated with pyrene (Py) and AQ using the

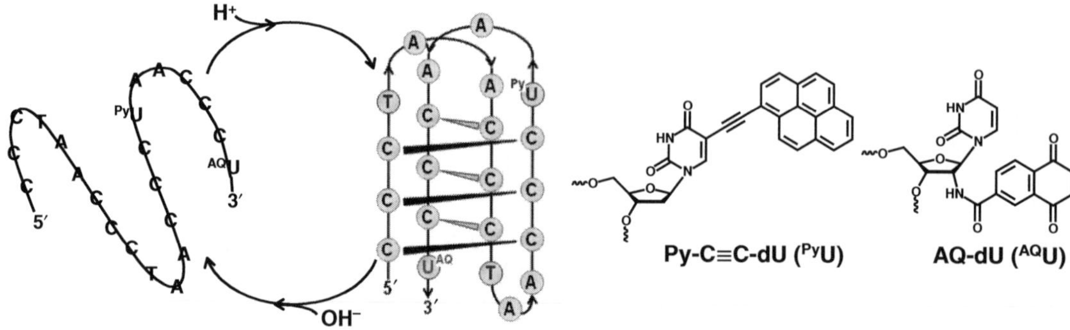

	Structure	Sequences
Py-i-motif	Tetraplex	5′-CCCTAA CCCTAA CCC-^{Py} UAA CCC- T-3′
Py-i-motif-AQ	Tetraplex	5′-CCCTAA CCCTAA CCC-^{Py} UAA CCC-^{AQ} U-3′
Py-dsDNA-AQ	Duplex	5′-CCCTAA CCCTAA CCC-^{Py} UAA CCC-^{AQ} U-3′ 3′-GGGATTGGGATT GGG-ATT GGG-5′

Figure 2.4.5 Schematic illustration of the reversible pH-induced conformational change of i-motif DNA (Py–i-motif–AQ) and sequences. Py and AQ are used as D and A, respectively. Reprinted with permission from [76]. Copyright Wiley-VCH (2013)

fluorescence up-conversion and transient absorption spectroscopic methods [76] (Figure 2.4.5). The formation of an i-motif from the synthesized oligonucleotide (5′-(CCCTAA)$_2$CCC-^{Py}UAA CCC-^{AQ}U-3′, Py–i-motif–AQ) was confirmed by its characteristic CD spectrum, which is consistent with that reported in a previous study [77]. On the other hand, the steady-state fluorescence spectrum of Py–i-motif–AQ showed a slightly structured emission band at ≤425 nm and a broader emission band above 453 nm, which may be attributed to the singlet excited state of Py (^{1}Py*) and the charge separated (CS) state, respectively (see Figure 2.4.6). To confirm the existence of two emissive states observed from Py–i-motif–AQ, the fluorescence up-conversion technique was used to measure the emission lifetime of Py–i-motif–AQ in 100 mM sodium phosphate buffer (pH 5.2). As shown in Figure 2.4.6b, two emission decay profiles of Py–i-motif–AQ monitored at 460 and 480 nm were fitted by a double-exponential function with two relaxation times of 0.9 ± 0.2 ps (1.11×10^{12} s^{-1}) and 9.4 ± 2.4 ps. From the quantitative analysis,the fast and slow decay component is originated from ^{1}Py* and the CS state.

To further elucidate the electron transfer occurring in i-motif DNA, the fs-transient absorption spectra for Py–i-motif–AQ are measured in 100 mM sodium phosphate buffers (pH 5.2). Py–i-motif–AQ showed only positive signals at ~510, 600, and in regions higher than 670 nm, without any negative signal, as depicted in Figure 2.4.7a. According to the results reported by previous studies [78–81], strong absorption peaks at ~510 and 600 nm are attributed to the contact ion pair (CIP) state (Py$^{•+}$–dU$^{•-}$) formed by the electron transfer from ^{1}Py* to dU, while the positive absorption signal at wavelengths in the region above 675 nm is assigned to the formation and decay of ^{1}(Py–C≡C–dU)*. Interestingly, upon labeling of AQ to the Py–i-motif, the negative signal observed at <450 nm for Py–i-motif was converted into a positive signal in Py–i-motif–AQ. This spectral change suggests that AQ is involved in the electron transfer process that occurs in i-motif DNA. According to the results reported by previous studies [78, 79, 82], the positive signal around 450 nm observed for Py–i-motif–AQ may be assigned to AQ$^{•-}$ or Py$^{•+}$. However, because the positive signal around 450 nm was only observed for Py–i-motif–AQ and not the Py–i-motif, the positive signal in the region below 450 nm observed for Py–i-motif–AQ can be readily assigned to AQ$^{•-}$ formed by the CT within the i-motif DNA.

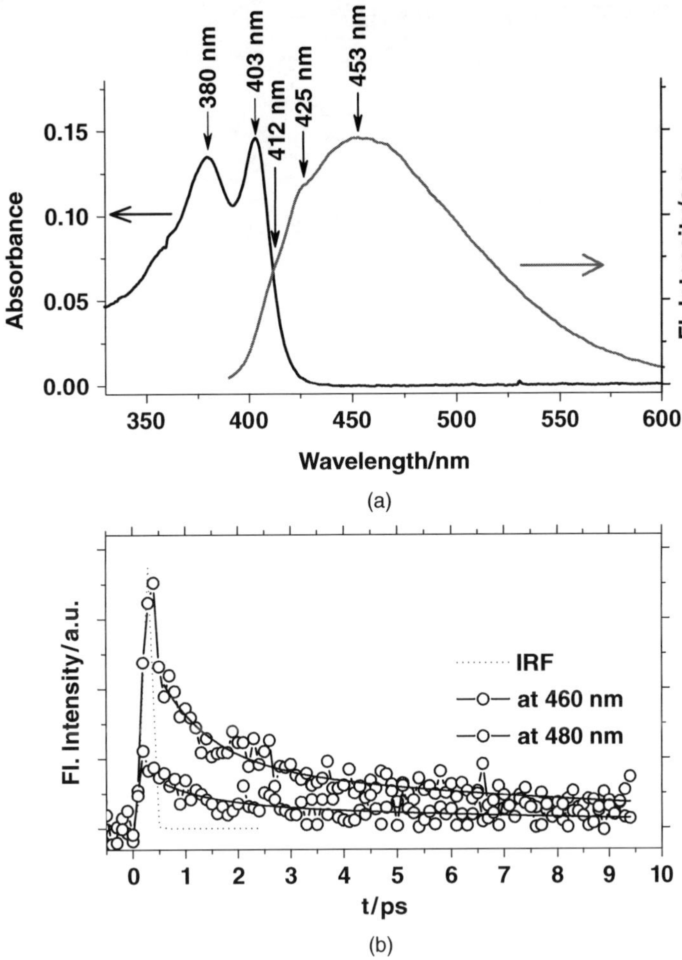

Figure 2.4.6 *(a) Absorption and emission spectra of Py–i-motif–AQ measured in 100 mM sodium phosphate buffer (pH 5.2) (λ_{Ex} = 380 nm). (b) Fluorescence decay profiles of Py–i-motif–AQ monitored at 460 (black, upper) and 480 nm (blue, lower) with λ_{Ex} = 400 nm in 100 mM sodium phosphate buffer (pH 5.2) measured by a fluorescence up-conversion technique. Theoretical fitting curves are shown by the solid (red) line. Reprinted with permission from [76]. Copyright Wiley-VCH (2013)*

In Py–i-motif–AQ, the CS state ($Py^{\bullet+}\cdots AQ^{\bullet-}$) can be formed through two processes, as depicted in Figure 2.4.8a: direct electron transfer from $^1Py^*$ to AQ (superexchange mechanism) and electron transfer from $dU^{\bullet-}$ in the CIP state to AQ via a hopping mechanism. It is known that in pyrene-1-yl-2′-deoxyuridine (Py–dU) substituted DNA, most conformers undergo intramolecular electron transfer from $^1Py^*$ to dU to form $Py^{\bullet+}$–$dU^{\bullet-}$ within a timescale of a few picoseconds [78]. In addition, previous studies pointed out that only $dT^{\bullet-}$ (or $dU^{\bullet-}$) but not $dC^{\bullet-}$ can participate as an intermediate charge carrier for the excess electron transfer in DNA, and the protonated $dC^{\bullet-}$ can interrupt the excess electron transfer in DNA [80, 83]. Accordingly, the CS state should be formed by the direct electron transfer from $^1Py^*$ to AQ but not by the electron transfer from $dU^{\bullet-}$ in the CIP state to AQ. Thus, the rate constant for the superexchange process (k_s) in Py–i-motif–AQ was directly determined from the rising profile of the absorption band at 440 nm,

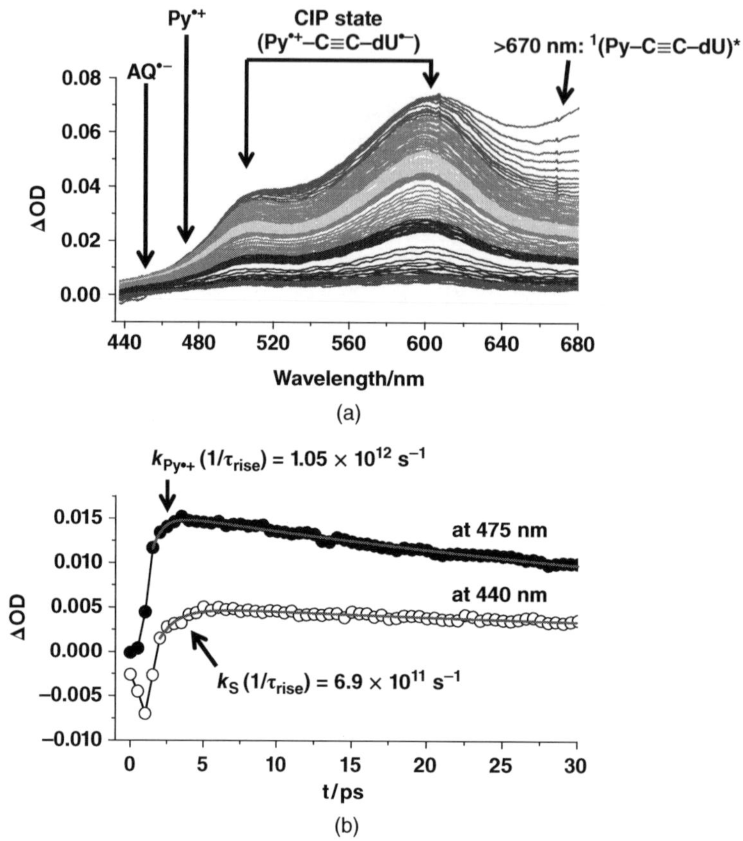

Figure 2.4.7 (a) Transient absorption spectra observed after the 400 nm laser excitation during fs-laser flash photolysis of Py–i-motif–AQ in 100 mM sodium phosphate buffer (pH 5.2) in the time range of from 0.5 to 2030 ps at room temperature, respectively. (b) Decay profiles of Py–i-motif–AQ monitored at 440 (○) and 475 nm (●). Theoretical fitting curves are shown in red. Reprinted with permission from [76]. Copyright Wiley-VCH (2013) (See color figure in color plate section)

corresponding to $AQ^{\bullet-}$ as being $k_s = 6.9 \pm 1.0 \times 10^{11}\ s^{-1}$ (Figure 2.4.7b), which is significantly larger than those observed from duplex DNA including hairpin DNA [84, 85]: $(1.8–3.5) \times 10^{11}\ s^{-1}$ for the nicked-dumbbell DNA sequences with one to four base pairs between *D* and *A* [84], and $1.7 \times 10^{8}–3.7 \times 10^{10}\ s^{-1}$ for the bridge-mediated electron transfer in hairpin DNA sequences [85]. This indicates that the direct electron transfer from $^{1}Py^{*}$ to AQ within i-motif DNA occurs faster than in a duplex DNA, including hairpin DNA.

It is worth noting that Py–dsDNA–AQ at pH 7.0 did not show a positive transient absorption signal in the wavelength range lower than 450 nm, indicating that the electron transfer from $^{1}Py^{*}$ to AQ in a duplex DNA does not take place. Comparing two structures of i-motif and duplex DNA, i-motif DNA has less consecutive base pairs embedded between *D* and *A* than the duplex DNA: five consecutive base pairs (two A–T and three G–C base pairs) for a duplex DNA; and three consecutive $C:C^{+}$ base pairs for i-motif DNA (Figure 2.4.8b). Furthermore, the stacking distance between the $C:C^{+}$ base pairs in i-motif DNA is 3.1 Å [86], while the distance between two bases in dsDNA is 3.4 Å. Therefore, the structural change from dsDNA (or ssDNA) to i-motif could induce a decrease in the *D–A* distance. On this basis, we conclude that the fast electron transfer observed in

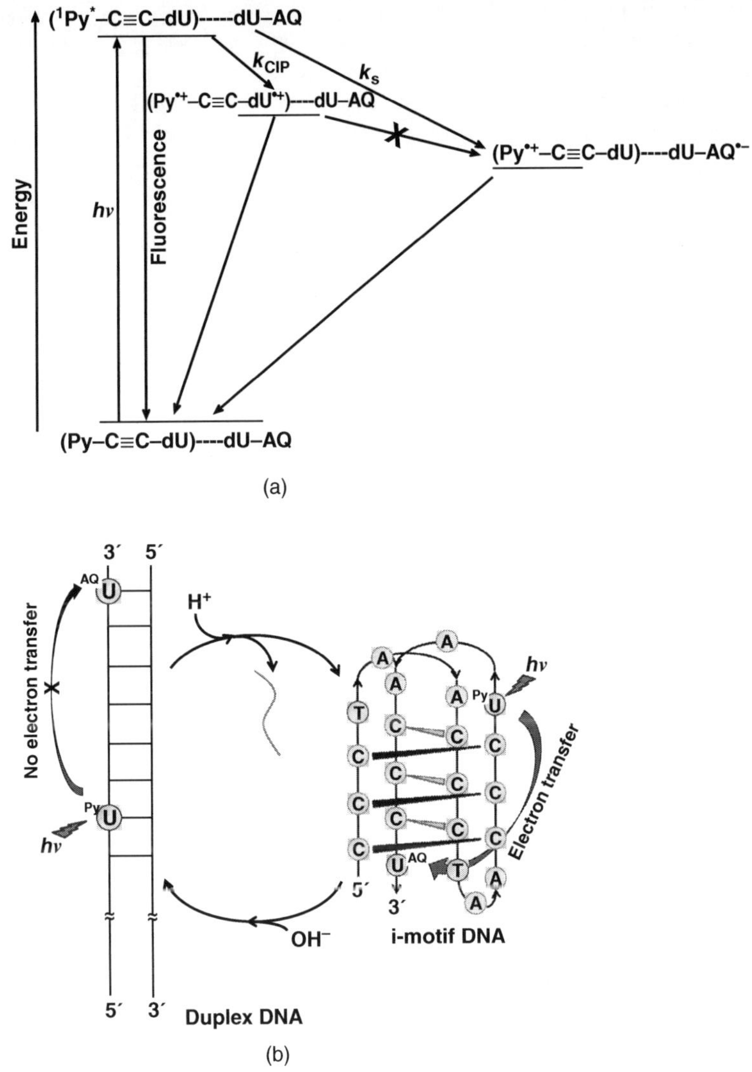

Figure 2.4.8 *(a) Schematic energy diagram for the photo-induced electron transfer in Py–i-motif–AQ DNA. (b) The electron transfer process for the structural change between the dsDNA and i-motif structure. Reprinted with permission from [76]. Copyright Wiley-VCH (2013)*

i-motif DNA ($k_s = 6.9 \times 10^{11}\ s^{-1}$) resulted not from the electron transfer from $dU^{\bullet-}$ in the CIP state but from the direct electron transfer from $^1Py^*$ to AQ due to the hemiprotonated $C{:}C^+$ base pairs and the relatively compact structure of the i-motif.

On the other hand, the generation rate of $Py^{\bullet+}$ ($k_{Py\bullet+}$) in Py–i-motif–AQ was estimated to be $k_{Py\bullet+} = 1.05 \pm 0.28 \times 10^{12}\ s^{-1}$ from the rising profile of the absorption band at 475 nm corresponding to $Py^{\bullet+}$ (Figure 2.4.7b). The value of $1.05 \times 10^{12}\ s^{-1}$ is larger than that measured for Py–C≡C–dU substituted DNAs ($3.3–5.0 \times 10^{11}\ s^{-1}$) [78]. In case of Py-i-motif-AQ DNA, the $Py^{\bullet+}$ in i-motif DNA can be generated by the electron injection yielding the CIP state and electron transfer from $^1Py^*$ to AQ with rate constants of k_{CIP} and k_s, respectively (see Figure 2.4.8a). Therefore, $k_{Py\bullet+}$ is the sum of k_{CIP} and k_s, $k_{Py\bullet+} = k_{CIP} + k_s$, resulting in the large $k_{Py\bullet+}$. Significantly, the determined $k_{Py\bullet+}$ of 1.05×10^{12} s^{-1} is consistent with the ^{1}Py* lifetime of Py–i-motif–AQ obtained by the fluorescence up-conversion technique. Considering that the $k_{CIP} = 4.0 \pm 0.4 \times 10^{11}$ s^{-1} measured for Py-i-motif DNA, which shows only the electron injection process yielding the CIP state, a $k_{Py\bullet+}$ of 1.09×10^{12} s^{-1} was calculated for Py-i-motif-AQ. The calculated value (1.09×10^{12} s^{-1}) is consistent with the experimentally determined $k_{Py\bullet+} = 1.05 \times 10^{12}$ s^{-1}. This agreement supports that the $k_{Py\bullet+}$ measured in this study is very reasonable for the generation rate of $Py^{\bullet+}$ in Py–i-motif–AQ and that no excess electron transfer occurs from $dU^{\bullet-}$ in the CIP state to AQ, but that direct electron transfer occurs efficiently from ^{1}Py* to AQ in i-motif DNA.

2.4.4 Conclusions

We have summarized the CT dynamics occurring within a G-quadruplex and i-motif DNA, that have a tetraplex structure and which can act as good electron carriers or hole captors in DNA-based electronic devices. The theoretical and experimental studies have shown that the efficient CT within the G-quadruplex is probably due to π–π stacking of planar G-tetrads and a large number of CT active conformations to induce the interstrand couplings between four strands in the G-quadruplex. On the other hand, the electron transfer in i-motif DNA proceeds more efficiently than in duplex DNA, suggesting that the i-motif can act as a good electron carrier in nano-electronic devices, due to the hemiprotonated $C{:}C^+$ base pairs as well as to its compact structure. Although the CT dynamics with G-quadruplex and i-motif DNA have become clear to some extent, based on our experimental results, much more research needs to be done to understand the various aspects, compared with those in duplex DNA. Thus, there is a growing need for a time-resolved spectroscopic study of the CT within G-quadruplex and i-motif DNA. We believe that the CT dynamics with G-quadruplex and i-motif DNA will become clear in the near future, and will certainly contribute to the development of molecular electronic devices.

Acknowledgments

The authors would like to thank their collaborators, especially Professor Mamoru Fujitsuka, Mr Atsushi Tanaka, and members of the Research Laboratory for Quantum Beam Science of ISIR, Osaka University for running the linear accelerator. This work has been supported in part by the Innovative Project for Advanced Instruments, Renovation Center of Instruments for Science Education and Technology, Osaka University, and a Grant-in-Aid for Scientific Research (Projects 24550188, 25220806 and others) from the Ministry of Education, Culture, Sports, Sciences and Technology (MEXT) of the Japanese Government.

References

[1] E. S. Lander, *et al.*, Initial sequencing and analysis of the human genome, *Nature*, **409**, 860–921 (2001).

[2] R. D. Wells, R. Dere, M. L. Hebert, M. Napierala, and L. S. Son, Advances in mechanisms of genetic instability related to hereditary neurological diseases, *Nucleic Acids Res.*, **33**, 3785–3798 (2005).

[3] G. Wang and K. M. Vasquez, Non-B DNA structure-induced genetic instability, *Mutat. Res.*, **598**, 103–119 (2006).

[4] A. Bacolla and R. D. Wells, Non-B DNA conformations as determinants of mutagenesis and human disease, *Mol. Carcinog.*, **48**, 273–285 (2009).

[5] J. Zhao, A. Bacolla, G. Wang, and K. M. Vasquez, Non-B DNA structure-induced genetic instability and evolution, *Cell Mol. Life Sci.*, **67**, 43–62 (2010).

[6] J. Choi and T. Majima, Conformational changes of non-B DNA, *Chem. Soc. Rev.*, **40**, 5893–5909 (2011).

[7] S. Kim, J. Choi, and T. Majima, Self-assembly of polydeoxyadenylic acid studied at the single-molecule level, *J. Phys. Chem. B*, **115**, 15399–15405 (2011).

[8] G. W. Collie and G. N. Parkinson, The application of DNA and RNA G-quadruplexes to therapeutic medicines, *Chem. Soc. Rev.*, **40**, 5867–5892 (2011).

[9] G. Wang and K. M. Vasquez, Non-B DNA structure-induced genetic instability, *Mutat Res.*, **598**, 103–119 (2006).

[10] Y. Wu and R. M. Brosh, Jr., G-quadruplex nucleic acids and human disease, *FEBS J.*, **277**, 3470–3488 (2010).

[11] A. Bacolla and R. D. Wells, Non-B DNA conformations as determinants of mutagenesis and human disease, *Mol. Carcinog.*, **48**, 273–285 (2009).

[12] J. Zhao, A. Bacolla, G. Wang, and K. M. Vasquez, Non-B DNA structure-induced genetic instability and evolution, *Cell Mol. Life Sci.*, **67**, 43–62 (2010).

[13] Y. Wu and R.M. Brosh, Jr., G-quadruplex nucleic acids and human disease, *FEBS J.*, **277**, 3470–3488 (2010).

[14] J. Choi and T. Majima, Reversible conformational switching of i-motif DNA studied by fluorescence spectroscopy, *Photochem. Photobiol.*, **89**, 513–522 (2013).

[15] G. Pelossof, R. Tel-Vered, S. Shimron, and I. Willner, Controlling interfacial electron transfer and electrocatalysis by pH- or ion-switchable DNA monolayer-modified electrodes, *Chem. Sci.*, **4**, 1137–1144 (2013).

[16] M. de Champdore, L. De Napoli, D. Montesarchio, G. Piccialli, C. Caminal, Q. G. Mulazzani, M. L. Navacchia, and C. Chatgilialoglu, Excess electron transfer in G-quadruplex, *Chem. Commun.*, 1756–1757 (2004).

[17] P. B. Woiczikowski, T. Kubar, R. Gutierrez, G. Cuniberti, and M. Elstner, Structural stability versus conformational sampling in biomolecular systems: why is the charge transfer efficiency in G4-DNA better than in double-stranded DNA? *J. Chem. Phys.*, **133**, 035103 (2010).

[18] T. Ndlebe and G. B. Schuster, Long-distance radical cation transport in DNA: Horizontal charge hopping in a dimeric quadruplex, *Org. Biomol. Chem.*, **4**, 4015–4021 (2006).

[19] Y. C. Huang, A. K. Cheng, H. Z. Yu, and D. Sen, Charge conduction properties of a parallel-stranded DNA G-quadruplex: implications for chromosomal oxidative damage, *Biochemistry*, **48**, 6794–6804 (2009).

[20] Y. C. Huang and D. Sen, A contractile electronic switch made of DNA, *J. Am. Chem. Soc.*, **132**, 2663–2671 (2010).

[21] S. Delaney and J. K. Barton, Charge transport in DNA duplex/quadruplex conjugates, *Biochemistry*, **42**, 14159–14165 (2003).

[22] C. J. Lech, A. T. Phan, M. E. Michel-Beyerle, and A. A. Voityuk, Electron-hole transfer in G-quadruplexes with different tetrad stacking geometries: a combined QM and MD study, *J. Phys. Chem. B*, **117**, 9851–9856 (2013).

[23] A. B. Kotlyar, N. Borovok, T. Molotsky, H. Cohen, E. Shapir, and D. Porath, Long, monomolecular guanine-based nanowires, *Adv. Mater.*, **17**, 1901–1905 (2005).

[24] N. Borovok, T. Molotsky, J. Ghabboun, D. Porath, and A. Kotlyar, Efficient procedure of preparation and properties of long uniform G4-DNA nanowires, *Anal. Biochem.*, **374**, 71–78 (2008).

[25] H. Cohen, T. Sapir, N. Borovok, T. Molotsky, R. Di Felice, A. B. Kotlyar, and D. Porath, Polarizability of G4-DNA observed by electrostatic force microscopy measurements, *Nano. Lett.*, **7**, 981–986 (2007).

[26] S. P. Liu, S. H. Weisbrod, Z. Tang, A. Marx, E. Scheer, and A. Erbe, Direct measurement of electrical transport through G-quadruplex DNA with mechanically controllable break junction electrodes, *Angew. Chem., Int. Ed. Engl.*, **49**, 3313–3316 (2010).

[27] F. D. Lewis, R. L. Letsinger, and M. R. Wasielewski, Dynamics of photoinduced charge transfer and hole transport in synthetic DNA hairpins, *Acc. Chem. Res.*, **34**, 159–170 (2001).

[28] M. Fujitsuka and T. Majima, Hole and excess electron transfer dynamics in DNA, *Phys. Chem. Chem. Phys.*, **14**, 11234–11244 (2012).

[29] K. Kawai and T. Majima, Hole transfer kinetics of DNA, *Acc. Chem. Res.*, **46**, 2616–2625 (2013).

[30] H. A. Wagenknecht, Electron transfer processes in DNA: mechanisms, biological relevance and applications in DNA analytics, *Nat. Prod. Rep.*, **23**, 973–1006 (2006).

[31] Y. A. Berlin, A. L. Burin, and M. A. Ratner, Charge hopping in DNA, *J. Am. Chem. Soc.*, **123**, 260–268 (2001).

[32] F. D. Lewis, H. Zhu, P. Daublain, T. Fiebig, M. Raytchev, Q. Wang, and V. Shafirovich, Crossover from superexchange to hopping as the mechanism for photoinduced charge transfer in DNA hairpin conjugates, *J. Am. Chem. Soc.*, **128**, 791–800 (2006).

[33] S. Priyadarshy, S. M. Risser, and D. N. Beratan, DNA is not a molecular wire: Protein-like electron-transfer predicted for an extended pi-electron system, *J. Phys. Chem.*, **100**, 17678–17682 (1996).

[34] J. Jortner, M. Bixon, T. Langenbacher, and M. E. Michel-Beyerle, Charge transfer and transport in DNA, *Proc. Natl. Acad. Sci. U.S.A.*, **95**, 12759–12765 (1998).

[35] Y. A. Berlin and M. A. Ratner, Intra-molecular electron transfer and electric conductance via sequential hopping: Unified theoretical description, *Radiat. Phys. Chem.*, **74**, 124–131 (2005).

[36] K. Kawai, T. Takada, S. Tojo, T. Majima, Kinetics of weak distance-dependent hole transfer in DNA by adenine-hopping mechanism, *J. Am. Chem. Soc.*, **125**, 6842–6843 (2003).

[37] B. Giese, S. Wessely, M. Spormann, U. Lindemann, E. Meggers, and M. E. Michel-Beyerle, On the mechanism of long-range electron transfer through DNA, *Angew. Chem., Int. Ed. Engl.*, **38**, 996–998 (1999).

[38] N. Renaud, Y. A. Berlin, F. D. Lewis, and M. A. Ratner, Between superexchange and hopping: an intermediate charge-transfer mechanism in poly(A)-poly(T) DNA hairpins, *J. Am. Chem. Soc.*, **135**, 3953–3963 (2013).

[39] B. Giese, Long-distance electron transfer through DNA, *Annu. Rev. Biochem.*, **71**, 51–70 (2002).

[40] G. B. Schuster, Long-range charge transfer in DNA: transient structural distortions control the distance dependence, *Acc. Chem. Res.*, **33**, 253–260 (2000).

[41] T. Takada, K. Kawai, M. Fujitsuka, and T. Majima, Direct observation of hole transfer through double-helical DNA over 100 Å, *Proc. Natl. Acad. Sci. U.S.A.*, **101**, 14002–14006 (2004).

[42] K. Kawai, H. Kodera, and T. Majima, Long-range charge transfer through DNA by replacing adenine with diaminopurine, *J. Am. Chem. Soc.*, **132**, 627–630 (2010).

[43] E. M. Conwell, S. M. Bloch, P. M. McLaughlin, and D. M. Basko, Duplex polarons in DNA, *J. Am. Chem. Soc.*, **129**, 9175–9181 (2007).

[44] T. A. Zeidan, R. Carmieli, R. F. Kelley, T. M. Wilson, F. D. Lewis, and M.R. Wasielewski, Charge-transfer and spin dynamics in DNA hairpin conjugates with perylenediimide as a base-pair surrogate, *J. Am. Chem. Soc.*, **130**, 13945–13955 (2008).

[45] T. Takada, K. Kawai, M. Fujitsuka, and T. Majima, Contributions of the distance-dependent reorganization energy and proton-transfer to the hole-transfer process in DNA, *Chem. - Eur. J.*, **11**, 3835–3842 (2005).

[46] Y. Osakada, K. Kawai, M. Fujitsuka, and T. Majima, Charge transfer through DNA nanoscaled assembly programmable with DNA building blocks, *Proc. Natl. Acad. Sci. U.S.A.*, **103**, 18072–18076 (2006).

[47] Y. Osakada, K. Kawai, M. Fujitsuka, and T. Majima, Kinetics of charge transfer in DNA containing a mismatch, *Nucleic Acids Res.*, **36**, 5562–5570 (2008).

[48] K. Kawai, M. Hayashi, and T. Majima, HOMO energy gap dependence of hole-transfer kinetics in DNA, *J. Am. Chem. Soc.*, **134**, 4806–4811 (2012).

[49] K. Kawai, H. Kodera, Y. Osakada, and T. Majima, Sequence-independent and rapid long-range charge transfer through DNA, *Nat. Chem.*, **1**, 156–159 (2009).

[50] S. M. Conron, A. K. Thazhathveetil, M. R. Wasielewski, A. L. Burin, and F. D. Lewis, Direct measurement of the dynamics of hole hopping in extended DNA G-tracts. An unbiased random walk, *J. Am. Chem. Soc.*, **132**, 14388–14390 (2010).

[51] K. Tainaka, M. Fujitsuka, T. Takada, and K. Kawai, T. Majima, Sequence dependence of excess electron transfer in DNA, *J. Phys. Chem. B*, **114**, 14657–14663 (2010).

[52] M. J. Park, M. Fujitsuka, K. Kawai, and T. Majima, Direct measurement of the dynamics of excess electron transfer through consecutive thymine sequence in DNA, *J. Am. Chem. Soc.*, **133**, 15320–15323 (2011).

[53] M. J. Park, M. Fujitsuka, H. Nishitera, K. Kawai, and T. Majima, Excess electron transfer dynamics in DNA hairpins conjugated with *N,N*-dimethylaminopyrene as a photosensitizing electron donor, *Chem. Commun.*, **48**, 11008–11010 (2012).

[54] J. L. Huppert, structure, location and interactions of G-quadruplexes, *FEBS J.*, **277**, 3452–3458 (2010).

[55] A. T. Phan, V. Kuryavyi, and D. J. Patel, DNA architecture: from G to Z, *Curr. Opin. Struct. Biol.*, **16**, 288–298 (2006).

[56] V. Viglasky, K. Tluckova, and L. Bauer, The first derivative of a function of circular dichroism spectra: biophysical study of human telomeric G-quadruplex, *Eur. Biophys. J.*, **40**, 29–37 (2011).

[57] J. B. Chaires, Human telomeric G-quadruplex: thermodynamic and kinetic studies of telomeric quadruplex stability, *FEBS J.*, **277**, 1098–1106 (2010).

[58] N. Smargiasso, F. Rosu, W. Hsia, P. Colson, E. S. Baker, M. T. Bowers, E. De Pauw, and V. Gabelica, G-quadruplex DNA assemblies: loop length, cation identity, and multimer formation, *J. Am. Chem. Soc.*, **130**, 10208–10216 (2008).

[59] B. Heddi, A. T. Phan, Structure of human telomeric DNA in crowded solution, *J. Am. Chem. Soc.*, **133**, 9824–9833 (2011).

[60] E. Cauet, Unique hole-trapping property of the human telomere sequence, *J. Biomol. Struct. Dyn.*, **29**, 557–561 (2011).

[61] A. A. Voityuk, J. Jortner, M. Bixon, and N. Rosch, Energetics of hole transfer in DNA, *Chem. Phys. Lett.*, **324**, 430–434 (2000).

[62] Y. Yoshioka, Y. Kitagawa, Y. Takano, K. Yamaguchi, T. Nakamura, and I. Saito, Experimental and theoretical studies on the selectivity of GGG triplets toward one-electron oxidation in B-form DNA, *J. Am. Chem. Soc.*, **121**, 8712–8719 (1999).

[63] J. Choi, J. Park, A. Tanaka, M. J. Park, Y. J. Jang, M. Fujitsuka, S. K. Kim, and T. Majima, Hole trapping of G-quartets in a G-quadruplex, *Angew. Chem., Int. Ed.*, **52**, 1134–1138 (2013).

[64] K. Kobayashi and S. Tagawa, Direct observation of guanine radical cation deprotonation in duplex DNA using pulse radiolysis, *J. Am. Chem. Soc.*, **125**, 10213–10218 (2003).

[65] K. Kobayashi, R. Yamagami, and S. Tagawa, Effect of base sequence and deprotonation of guanine cation radical in DNA, *J. Phys. Chem. B*, **112**, 10752–10757 (2008).

[66] L. P. Candeias and S. Steenken, Structure and acid-base properties of one-electron-oxidized deoxyguanosine, guanosine, and 1-methylguanosine, *J. Am. Chem. Soc.*, **111**, 1094–1099 (1989).

[67] E. D. A. Stemp, M. R. Arkin, and J. K. Barton, Oxidation of guanine in DNA by $Ru(phen)_2(dppz)^{3+}$ using the flash-quench technique, *J. Am. Chem. Soc.*, **119**, 2921–2925 (1997).

[68] J.-L. Mergny, L. Lacroix, X. Han, J.-L. Leroy, and C. Helene, Intramolecular folding of pyrimidine oligodeoxynucleotides into an i-DNA motif, *J. Am. Chem. Soc.*, **117**, 8887–8898 (1995).

[69] S. Kendrick, Y. Akiyama, S. M. Hecht, and L. H. Hurley, The i-motif in the bcl-2 P1 promoter forms an unexpectedly stable structure with a unique 8:5:7 loop folding pattern, *J. Am. Chem. Soc.*, **131**, 17667–17676 (2009).

[70] J. Zhou, C. Wei, G. Jia, X. Wang, Z. Feng, and C. Li, Formation of i-motif structure at neutral and slightly alkaline pH, *Mol. Biosyst.*, **6**, 580–586 (2010).

[71] D. Miyoshi, H. Karimata, and N. Sugimoto, Hydration regulates thermodynamics of G-quadruplex formation under molecular crowding conditions, *J. Am. Chem. Soc.*, **128**, 7957–7963 (2006).

[72] S. R. Shin, K. S. Jin, C. K. Lee *et al.*, Fullerene attachment enhances performance of a DNA nanomachine, *Adv. Mater.*, **21**, 1907–1910 (2009).

[73] X. Li, Y. Peng, J. Ren, and X. Qu, Carboxyl-modified single-walled carbon nanotubes selectively induce human telomeric i-motif formation, *Proc. Natl. Acad. Sci. U.S.A.*, **103**, 19658–19663 (2006).

[74] O. Y. Fedoroff, A. Rangan, V. V. Chemeris, and L. H. Hurley, Cationic porphyrins promote the formation of i-motif DNA and bind peripherally by a nonintercalative mechanism, *Biochemistry*, **39**, 15083–15090 (2000).

[75] P. M. Keane, *et al.*, Long-lived excited states in i-motif DNA studied by picosecond time-resolved IR spectroscopy, *Chem. Commun.*, **50**, 2990–2992 (2014).

[76] J. Choi, A. Tanaka, D. W. Cho, M. Fujitsuka, and T. Majima, Efficient electron transfer in i-Motif DNA with a tetraplex structure, *Angew. Chem., Int. Ed.*, **52**, 12937–12941 (2013).

[77] X. Li, Y. Peng, J. Ren, and X. Qu, Carboxyl-modified single-walled carbon nanotubes selectively induce human telomeric i-motif formation, *Proc. Natl. Acad. Sci. U.S.A.*, **103**, 19658–19663 (2006).

[78] P. Kaden, E. Mayer-Enthart, A. Trifonov, T. Fiebig, and H. A. Wagenknecht, Real-time spectroscopic and chemical probing of reductive electron transfer in DNA, *Angew. Chem., Int. Ed.*, **44**, 1636–1639 (2005).

[79] N. Amann, E. Pandurski, T. Fiebig, and H. A. Wagenknecht, Electron injection into DNA: synthesis and spectroscopic properties of pyrenyl-modified oligonucleotides, *Chem. Eur. J*, **8**, 4877–4883 (2002).

[80] M. Raytchev, E. Mayer, N. Amann, H. A. Wagenknecht, and T. Fiebig, Ultrafast proton-coupled electron-transfer dynamics in pyrene-modified pyrimidine nucleosides: model studies towards an understanding of reductive electron transport in DNA, *ChemPhysChem*, **5**, 706–712 (2004).

[81] A. Trifonov, M. Raytchev, I. Buchvarov, M. Rist, J. Barbaric, and H. A. Wagenknecht, and T. Fiebig, Ultrafast energy transfer and structural dynamics in DNA, *J. Phys. Chem. B*, **109**, 19490–19495 (2005).

[82] F. D. Lewis, A. K. Thazhathveetil, T. A. Zeidan, J. Vura-Weis, and M. R. Wasielewski, Dynamics of ultrafast singlet and triplet charge transfer in anthraquinone-DNA conjugates, *J. Am. Chem. Soc.*, **132**, 444–445 (2010).

[83] N. Amann, E. Pandurski, T. Fiebig, and H. A. Wagenknecht, A model nucleoside for electron injection into DNA: 5-pyrenyl-2′-deoxyribose, *Angew. Chem., Int. Ed.*, **41**, 2978–2980 (2002).

[84] M. J. Park, M. Fujitsuka, K. Kawai, and T. Majima, Excess-electron injection and transfer in terthiophene-modified DNA: terthiophene as a photosensitizing electron donor for thymine, cytosine, and adenine, *Chem. - Eur. J*, **18**, 2056–2062 (2012).

[85] F. D. Lewis, R. S. Kalgutkar, Y. S. Wu, X. Y. Liu, J. Q. Liu, R. T. Hayes, S. E. Miller, and M. R. Wasielewski, Driving force dependence of electron transfer dynamics in synthetic DNA hairpins, *J. Am. Chem. Soc.*, **122**, 12346–12351 (2000).

[86] L. Chen, L. Cai, X. Zhang, and A. Rich, Crystal structure of a four-stranded intercalated DNA: d(C4), *Biochemistry*, **33**, 13540–13546 (1994).

Part III

Oligonucleotides in Sensing and Diagnostic Applications

3.1

Development of Electrochemical Sensors for DNA Analysis

Hanna Radecka and Jerzy Radecki

Department of Biosensors, Institute of Animal Reproduction and Food Research of Polish Academy of Sciences, Olsztyn, Poland

3.1.1 Introduction

According to the IUPAC definition, a biosensor is a self-contained integrated device that is capable of providing specific quantitative or semi-quantitative analytical information using a biological recognition element, which is in direct spatial contact with a transducer element [1]. Generally the biosensors consist of two main elements:

- the first is a molecular recognition layer that enables the selective recognition of a particular analyte or a group of analytes;
- the second is a signal transducer that converts an energetic signal resulting from an intermolecular recognition process into another form of energy that is readable by the readout device.

The main parameters describing the quality of biosensors are selectivity, sensitivity (low limit of detection), reproducibility and time of response.

Electrochemical biosensors belong to a subclass of biosensors that contain the electrochemical transductor. In the last decade we have observed an impressive increase in the number of papers concerning the development of such types of sensors for the determination of DNA. These sensors are based on a self-assembly layer of DNA strands as a recognition element at the surface of a transducer and are known as genosensors

DNA in Supramolecular Chemistry and Nanotechnology, First Edition.
Edited by Eugen Stulz and Guido H. Clever.

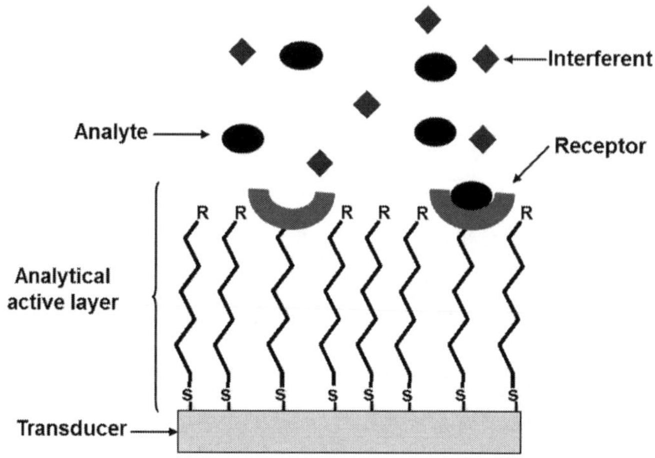

Figure 3.1.1 General scheme of a biosensor

(Figure 3.1.1). Because of the high sensitivity, user-friendliness, the relatively low price of the determination and possibilities for miniaturization, the electrochemical genosensors are very attractive from a medical diagnosis, environmental monitoring and food quality control point of view.

The immobilization of the DNA probe at the surface of the electrode plays a crucial role in the analytical parameters of the future genosensor.

The probes are usually short oligonucleotides (24–25 mer) that are able to hybridize with a specific and unique region of the target nucleotide sequence. The control of the surface chemistry connected with the immobilization processes is essential for the sensitivity and selectivity of the sensor. In spite of the many papers published on this subject, until now the relationship between the surface environment of the biosensors and their analytical quality has not been fully established.

Several useful procedures for probe immobilization on the electrode surface have been described in the scientific literature. The most popular rely on a biotin–avidin interaction [2], self-assembly of a thiol functionalized probe on a gold electrode surface [3], or carbodiimide covalent binding to an active surface [4]. The electrochemical genosensors introduced in this short chapter can be classified into two groups, based on the sensing strategy:

- genosensors in which direct reduction and oxidation of DNA bases as a recognition event is used;
- genosensors in which the sensing strategy is based on complementary base paring between the sensor's nucleic acid sequence and the analyte of interest.

We are aware that this is a severe limitation in relation to the very impressive developments in the electrochemical genosensors areas of research. However, in our opinion these two strategies of sensing are the basis of the new directions for electrochemical genosensors development.

3.1.2 Genosensors based on direct electrocactivity of nucleic bases

The fundament for the first strategy of electrochemical DNA sensing was given by Paleček in a paper concerning the electroactivity of deoxyribonucleic acid [5]. The first electrochemical DNA sensor was developed by Millian and Mikkelson in 1993 [6]. It consists of a carbon electrode and exogenous, redox active hybridization indicators: tris(2,2′-bipyridyl)cobalt(III) perchlorate, tris(1,10-phenanthroline)cobalt(III) perchlorate ($Co(bpy)_3^{3+}$ and $Co(phen)_3^{3+}$. These complexes are reversibly electroactive ($1e^-$) and preconcentrate at the electrode surface through association with double-stranded DNA. Using this system these workers were able to distinguish hybridized double-stranded (dsDNA) from single-stranded (ss)-probe DNA. Five years later, Paleček published a paper on the application of adsorption stripping voltammetry for the determination of nanograms of DNA [7]. By means of adsorptive transfer stripping voltammetry (AdTSV), these workers applied a two-step procedure for the analysis. The first step relied on the adsorption of DNA at the electrode surface. In the next step, the electrode was washed and the adsorbed layer from the surface of the electrode was transferred into the medium not containing the DNA. Voltammetric analysis was performed in this medium. According to this procedure, nanogram amounts of DNA could be determined.

Since this first publication, over the next decades many papers concerning the detection of DNA by direct oxidation or reduction of nucleic acids using various electrodes have been published [8–15]. In the case of this type of sensors, workers focused their attention on improving the efficiency of the redox reactions of nucleic acids.

Raoof published the results of a comparative study of electrochemical detection techniques using carbon nanotube paste (CNTPE) and carbon paste electrodes (CPE) as methods for indicator-free DNA sensors [16]. The tested sensors relied on the immobilization of a 20-mer single stranded oligonucleotide probe for detection of target DNA as a model. The hybridization event was observed by differential pulse voltammetry (DPV) and electrochemical impedance spectroscopy (EIS) techniques. The results showed that the application of EIS for the detection of hybridization on the CNTPE surface was impossible and on the CPE surface it was not beneficial. However, the DPV method demonstrated better potential for detection of the hybridization event on the surface of the electrodes. The results that they obtained demonstrated that CNTPE had some advantages over CPE. Utilization of carbon nanotubes as particles incorporated into a carbon paste electrode improved the electrochemical signal of this biosensing procedure. The enhanced current values can be attributed to the high local density of the electronic states in the CNTs. The proposed biosensor selectively responded to the complementary target and eliminated external indicator accumulation [16].

The DNA sensing procedures based on direct redox reaction of nucleic bases are fairly sensitive and selective, but their applicability is rather limited. The main drawback is the background current at the relatively high potentials required for direct oxidation of DNA. In the case of reduction, the serious limitation is the necessity to use a mercury electrode.

3.1.3 Genosensors based on electrochemical mediators

One of the methods is based on direct electroactivity of nucleic bases, which takes into account the limitations of genosensors and improves their application, relies on of the same electrochemical mediators that facilitate electron transfer between the electro-active base and the electrode surface. This method provides sensitivity at the attomol level in terms of concentration of the target DNA [17]. The most commonly used electron mediators are ferrocene, $Fe[(CN)]_6^{3-/4-}$, $Ru(bpy)_3^{3+/2+}$, $Os(bpy)_3^{3+/2+}$ and methylene blue.

Elliott and coworkers described an electrochemical approach to detect target DNA molecules from solutions based on the catalytic oxidation of $Co(DTB)_3^{2+}$, where DTB is 4,4′-di-t-butyl-2,2′-bipyridine [18]. In this case the probe/target duplex DNA bound to the ITO electrode surface selectively recruits the redox catalysts that turns on the redox chemistry of the reporter molecules, and the catalytic current results in a signal amplification. They were able to detect a 20-mer ssDNA target oligonucleotide at picomolar concentrations [18].

Recently, Xionga described a very interesting biosensor that consists of a graphite electrode covered with adsorbed DNA. Voltammetric responses stemmed from the redox reaction of anthraquinone monosulfonate co-adsorbed on the same electrode. It was demonstrated that there was an inverse linear relationship between the surface concentrations of the two species. This approach made the indirect measurement of the quantity of adsorbed DNA at the electrode through the voltammetric signal of the co-adsorbed anthraquinone monosulfonate possible. This system was developed through the use of multi-walled carbon nanotube screen-printed electrodes and pyrolytic graphite electrodes. The described sensor allowed the determination of DNA in solution with a limit of detection of 8.8 μM (equivalent to 5.9 μg mL^{-1}) [19]. This report showed that the application of redox mediators allowed improvements to be made to the sensitivity of DNA sensors based on the electroactivity of the nucleic acids.

3.1.4 Genosensors based on free diffusional redox markers

One interesting approach for voltammetric signal amplification was demonstrated by Aoki and Umezawa [20]. In the sensor they proposed, the mechanism of analytical signal generation was connected with a binding event between the electro-inactive analyte and the receptor monolayer located at the electrode surface. This process controlled the redox active marker's access to the electrode surface through electrostatic interactions or by creation of steric hindrance at the surface of the electrode. Finally, the electron transfer from the marker to the surface of the electrode was affected by the binding of the analytes to the receptor. The heterogeneous rate constant for electron transfer from the marker to the electrode surface was higher for the case of electrostatic attraction or lower due to the electrostatic repulsion between the analyte–receptor complex formed at the surface of the electrode and the diffusional redox marker. Therefore, the redox current increased or decreased accordingly. The electrochemical sensors based on this mechanism are called ion-channel mimetic sensors (Figure 3.1.2).

There are numerous examples in the scientific literature of biosensors based on this mechanism. In the case of DNA sensors as the analytically active element (probe), an appropriate sequence of ssDNA immobilized on the surface of an electrode can serve this purpose. $[Fe(CN)_6]^{3-}/[Fe(CN)_6]^{4-}$ or $[Ru(NH_3)_6]^{3+}$ are the most commonly used redox active marker ions [21, 22].

Aoki and Umezawa described a very sensitive sensor for detection of oligonucleotides using peptide nucleic acid (PNA) modified electrodes. A self-assembled monolayer composed of a PNA probe and 8-amino-1-octanethiol was applied. The detection limit of this sensor recorded for a complementary oligonucleotide sequence was 5.1×10^{-10} M, with a relative standard deviation of 1.5% [23].

Liu and coworkers introduced a genosensor that works according to the ion-channel mechanism, and is based on a gold electrode modified with multi-walled carbon nanotubes covered with polypyrrole nanowires and gold nanoparticles. A thiol terminated DNA aptamer was immobilized on this hybrid material. The recognition process of target DNA was observed by the DPV technique using cobalt(III) perchlorate as a redox marker. The detection limit was 0.43 pM [24].

Recently, in our laboratory we developed genosensors based on a gold electrode intended for detection of the specific DNA sequence of the Avian Influenza Virus H5N1 using NH_2-ssDNA (NH_2-NC3) or HS-ssDNA (SH-NC3) probes for modification of gold electrodes [25a, 25b]. These sensors were applied for the detection of 20-mer and 180-bp (PCR products) oligonucleotide complementary sequences. In the case of the first system (amino-DNA), the immobilization was performed via amide formation with the surface bound thioacid. In the second system (thiol-DNA), the probe was immobilized directly onto the gold surface via S–Au bond formation. In both cases we applied an $[Fe(CN)_6]^{3-/4-}$ redox couple as the marker ion. The sensor based on the SH-NC3 probe displayed a detection limit in the10 pM range [25b], whereas in the case of the NH_2-NC3 the detection limit was in the fM range [25a]. These results show that electrodes modified with longer spacer molecules give a higher hybridization signal. This is mainly due to easier accessibility of the target ssDNA to the probe DNA, but the cost of this approach is lower selectivity. In the case of the shorter spacer,

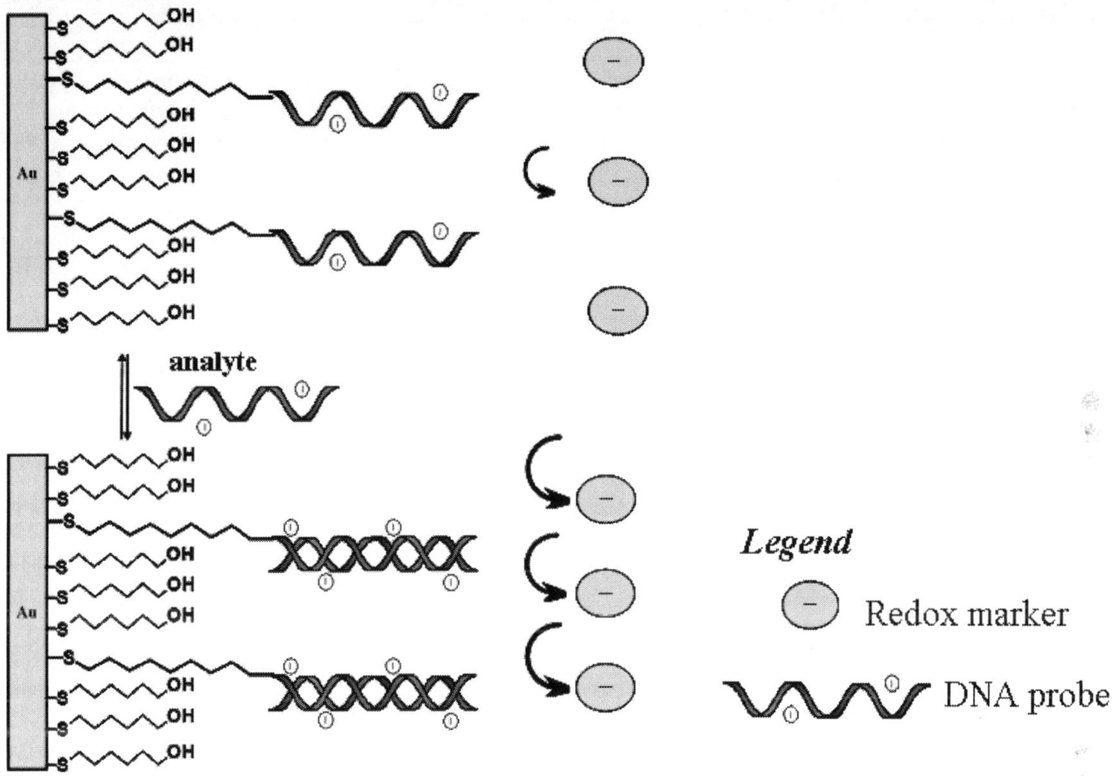

Figure 3.1.2 Working principle of the ion-channel mimetic genosensor

the sensor was able to distinguish between the PCR products with mismatches in the complementary parts, whereas the electrode modified with longer spacer molecules was not able to recognise such mismatched sequences.

Choi and coworkers fabricated biosensors based on glassy carbon electrodes modified via avidin–biotin binding of a biotinylated DNA probe [26]. The sensing efficiency of this genosensor was tested by cyclic voltammetry using various concentrations of target DNA in the presence of 1.0 mM $Fe(CN)_6^{3-/4-}$. The sensor was characterized by a wide dynamic detection range of from 1.0×10^{-13} to 1.0×10^{-10} M and a low detection limit equal to 8.5×10^{-14} M. These workers demonstrated that the proposed sensor was able to distinguish between a fully complementary strand of DNA and a two-base mismatched DNA. The redox peak current of the E-DNA biosensor stimulated with a two-base mismatched DNA sequence was larger than that obtained from the hybridization with the complementary target DNA sequence. These results demonstrate that a DNA biosensor displays a very high selectivity for DNA hybridization detection [26].

The detection of some mutations or damaged DNA bases is very important from the point of view of an early diagnosis of genetic diseases. Application of small DNA-intercalating or groove binding compounds opens the possibility of distinguishing between single stranded probes from the double stranded hybrid probes located at the surface of the electrode. According to Barton's definition, the intercalators are small organic molecules or metal complexes that unwind DNA in order to π-stack between two base pairs [27].

The application of intercalators as markers for hybridization is based on the fact that the DNA base pairs combine to form a continuous π-stacked conduit for charge transfer (CT). The disruption of a single base pair, as in a mismatch, is able to interrupt the CT pathway, which is suitable for the determination of DNA sequence mutations.

The system using the intercalator as a marker for the hybridization event requires a special procedure for the measurement. In the first step the modified electrodes are immersed in the indicator solution containing the intercalator compound before and after hybridization. Because of the high affinity of intercalators to the double-strand hybrid DNA, the redox active marker intercalates into dsDNA, which enhances the electrochemical response.

One of the earliest reports describing the application of this scheme for the detection of DNA hybridization was given by Hashimoto and coworkers [28]. They demonstrated that the oxidation signal of Hoechst dye increased as a result of its intercalation into dsDNA located at the surface of the electrode [28].

One of the most commonly used intercalators that has been successfully applied as a redox marker for DNA duplex formation in gene sensors is daunomycin. The detection limit of such genosensors is in the region of 1 μg mL^{-1} of the target sequence [29, 30].

One way to improve the sensitivity of biosensors intended for the determination of base mismatches in the DNA duplexes is to make use of bis-intercalators, which are a group of intercalators consisting of two intercalating rings (aromatic) covalently linked via chains of variable length.

Electrochemical studies of the synthetic threading intercalator ferrocenyl naphthalene diimide [31] and the naturally occurring bis-intercalator ECHI (antibiotics and antitumour agent) have been undertaken by many research groups [32, 33]. Sensors based on a ferrocenyl-modified naphthalene diimide are capable of discriminating between single-strand and hairpin DNA at the gold electrode surface [34].

Recently, Lorenzo and coworkers developed a genosensor based on a bifunctional ruthenium complex that was able to detect the position of a single-mismatch in a DNA duplex [35]. The probe consisted of a pentaamine ruthenium[3-(2-phenentren-9-yl-vinyl)pyridine] complex (RuL), which was generated *in situ*. The Ru centre provided the redox probe and the L supplied a fluorescent tag. These workers demonstrated that the presence of an aromatic group in the ligand, allowing for intercalation of the complex, made it capable of binding to double-stranded DNA stronger than to single-stranded DNA. The Ru centres were responsible for electrochemical signal generation after the hybridization event. The ligand-based fluorescence allowed for characterization of the complex formation. This type of sensor was sensitive not only for hybridization with the fully complementary target sequence with a detection limit of 92 ± 0.4 pM, but also for the detection of single mismatches in the DNA sequence [35].

In recent years, many studies describing electrochemical sensors designed for single base mismatch determination based on intercalators as markers have been published [36–43]. The strength of this type of genosensors is the lack of the necessity to label the target or DNA probe. Generally, they are very sensitive and selective, particularly those that are based on the ion-channel mechanism or that have intercalators as redox markers.

The weakness of these two types of sensors is the necessity to add marker molecules to the sample solution. Because of this, in many cases, in spite of very good parameters such as sensitivity, selectivity, reproducibility and short time of response, their applicability in practice is very limited.

The main disadvantages of using free diffusional indicators are that they tend to be toxic (e.g. daunomycin) and there are other charge-transfer pathways that are not DNA mediated. Also, complete knowledge of the electrochemistry of these mediators and their interactions with DNA molecules is still not available [44].

To overcome the problems associated with non-covalently bonded mediators, while keeping the advantages of the long-range DNA charge-transport properties, Mousavi and coworkers [45] described a genosensor for the detection of single-base mismatch based on the 2,5-dihydroxybenzoic acid (DHBA) as a new redox-active reporter, conjugated to the end of the DNA signalling probe. To demonstrate the advantages of this approach over those with non-covalent mediators, they compared their results with the responses obtained by MB, which is a well known and standard redox reporter in DNA biosensors. Comparison of the results showed that, for classical intercalative mediators, only the mediators that are intercalated on top of the DNA double helix (i.e. after the mismatched base pairs) could discriminate between matched and mismatched targets. In the case of the DHBA system, the redox signal showed that the position of the mismatches, either near the electrode surface (N-type) or further away from it (F-type), does not affect the electrochemical signal of DHBA.

The probe, covalently bonded to the end of amino-labelled DNA, is sufficiently sensitive to any perturbation in the electronic coupling of the DNA base stack, and allows for the detection of mismatches, including the thermodynamically stable G–A mismatch, without any catalytic method being required [45].

3.1.5 Genosensors incorporating DNA probes modified with redox active molecules – 'signal-off' and 'signal-on' working modes

In 2003, Fan, Plaxco and Heeger described a new class of electrochemical DNA sensors (E-DNA) [46]. They used single stem–loop DNA structures immobilized at the surface of a gold electrode via an S–Au bond. The unbonded terminus was decorated with the redox active label ferrocene. A scheme of this structure is illustrated in Figure 3.1.3.

As can be seen from Figure 3.1.3, before hybridization the stem–loop structure holds the ferrocene close to the electrode surface. After hybridization the stem– loop was transferred to a linear duplex structure and, as a consequence, the distance between the electrode surface and the redox marker increased, which lead to a decrease in the faradic current signal.

The selectivity and sensitivity of the E-DNA sensors arises from a combination of a conformational change upon hybridization, together with the redox labels being active at potentials far from those of the most electroactive biomolecules typical for clinical and environmental samples, thus being resistant to interfering contaminants. The E-DNA sensors can detect picomoles of ssDNA. Generally, the genosensors based on the described mechanism belong to a family with a very wide 'signal-off' mode.

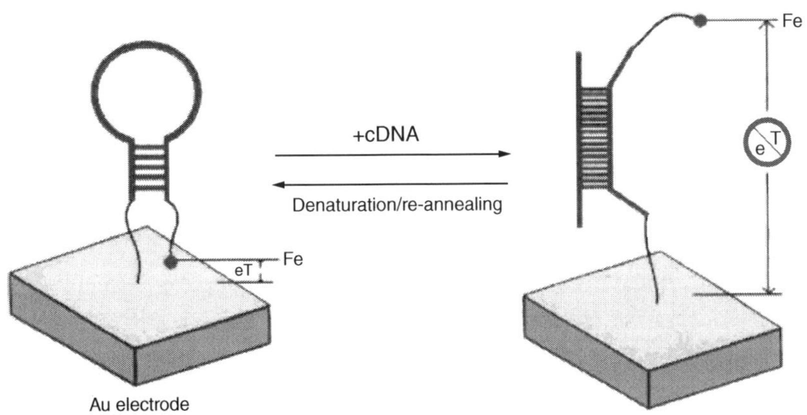

Figure 3.1.3 *A stem–loop oligonucleotide terminal thiol and a ferrocene group is immobilized at a gold electrode through self-assembly. In the absence of a target, the stem–loop structure holds the ferrocene tag in close proximity to the electrode surface, thus ensuring rapid electron transfer and efficient redox reaction of the ferrocene label. Upon hybridization with the target sequence, a large change in redox current is observed, presumably because the ferrocene label is separated from the electrode surface Reprinted with permission from [46]. Copyright 2003 National Academy of Sciences, U.S.A*

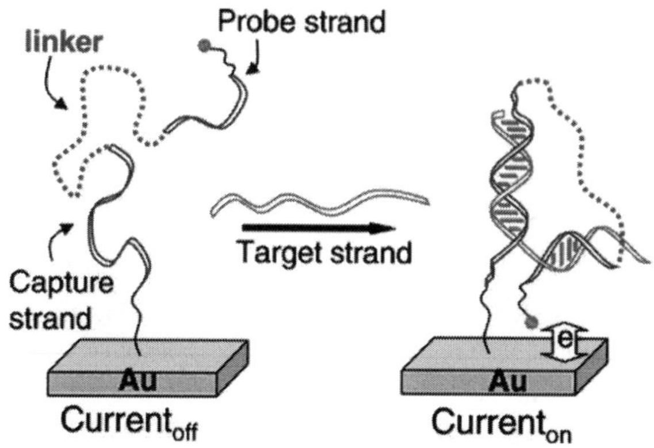

Figure 3.1.4 Electrochemical detection of target nucleic acid sequences using a DNA wrap assay as opposed to a conventional sandwich assay Reprinted with permission from [50]. Copyright 2004, American Chemical Society (See colour figure in colour plate section)

Another example of a 'signal-off' E-DNA consists of a gold electrode modified with linear DNA probes decorated with redox active markers. These sensors appear to be more sensitive relative to stem–loop based ones. Linear probe E-DNA sensors exhibit an 85% signal reduction at a given target concentration, whereas the equivalent stem–loop sensor exhibits only 71% signal suppression at this target concentration. The explanation of this phenomenon might be that the accessibility of the target DNA towards linear probes is easier in comparison with a stem–loop system.

The linear probe E-DNA sensor is reusable, sequence specific and selective. It is possible to use it directly in complex samples, such as blood serum [46]. A very important parameter that is critical to the sensitivity of E-DNA sensors is the density of the DNA probes at the surface of the electrode. The optimal density is in the range of from 10^{10} to 10^{12} molecules cm^{-2}. It was observed that with an increase in the probe density, the signal generated according to the 'signal-off' mechanism also increases. The crowding between neighbouring probe–target duplexes minimizes the possibility of collision of the redox moiety with the electrode surface. As a consequence, this leads to suppression of the electron transfer efficiency [47, 48].

It is worth noting that the detection limit of E-DNA working according to the 'signal-off' mechanism is around 10 pM. Recently, we were able to observe the dynamic development of E-genosensors based on the 'signal-off' mechanism [47, 48, 49].

The limitation of 'signal-off' sensors is that the recognition processes are signalled by the loss of the initial current. Therefore, it is not possible to suppress the original current by more than 100%. Thus, this is a strong limitation to the sensitivity of these types of sensors.

The first genosensor working according to the 'signal-on' mechanism was described by Limmoos and coworkers (Figure 3.1.4) [50]. This group described a sensor consisting of two strands of ssDNA. The capture and probe strands are linked together via a flexible linker (polyethylene glycol). The capture strand contained a 3′-terminal thiol for immobilization on the gold surface. The probe strand was decorated with ferrocene at the 5′-terminus. Hybridization with target DNA that was complementary to both of the immobilized sequences (probe and capture strand) moved the labelled end of the probe into the proximity of the electrode, increasing the electron transfer rate. This approach resulted in a detection limit of 200 pM and a current signal increase of 600%.

The next example of a 'signal-on' E- DNA sensor was described by Xiao *et al.* [51]. Their approach is based on a target-induced strand-displacement mechanism. The sensor consists of two parts. The first component is a single-stranded probe covalently attached to the surface of a gold electrode via 5′-thiol

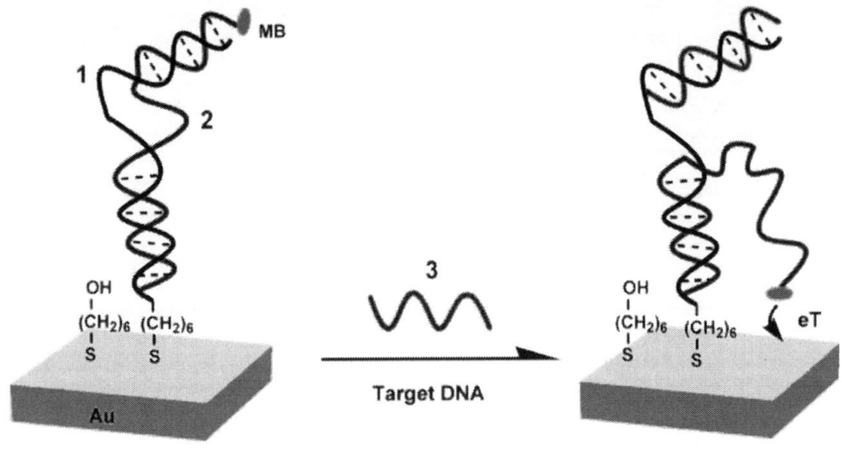

Figure 3.1.5 *Schematic of the signal-on E-DNA sensor, which is based on a conformational change in methylene blue (MB) modified duplex DNA that occurs after target-induced strand displacement. In the absence of a target, the two double-stranded regions formed between the capture and signaling probes sequester the MB from the electrode surface, producing a relatively small MB redox current. When the sensor is challenged with a complementary target, the observed MB redox current increases significantly, presumably because the flexible, single-stranded element liberated in the signaling probe increase the efficiency with which the MB can transfer electrons to the electrode surface. Reprinted with permission from [51]. Copyright 2006 National Academy of Sciences, U.S.A*

modification (Figure 3.1.5). The second component is a 'signalling probe' with a methylene blue redox moiety at the 5′-terminus. The probe consists of two domains: a 15-base segment (5′-terminus) for stable linking to the electrode surface, and a 7-base segment (3′-terminus), which forms the capture probe for the DNA sequence to be analysed. Hybridizing the signalling probe, which is complementary to both segments, in the absence of a target DNA keeps the methylene blue further away from the electrode surface and, as a consequence, the observed initial current is relatively low.

In the presence of a 15-base complementary target, the 7-hybridized bases at the 5′-terminus of the signalling probe were displaced. As a consequence, the flexible methylene blue decorated end of the signalling probe allowed the redox marker to collide with the electrode surface, which resulted in the increase of the transfer of electrons. As a result, after target binding the faradic current increased significantly. The detection limit of this E-DNA sensor was 400 fM.

Both types of E-DNA sensors, 'signal-on' and 'signal-off', generate an analytical signal after stimulation with target DNA without the addition of exogenous reagents. This property is very important from a medical diagnosis point of view.

These types of sensors have their own strengths, but also their limitations. They are readily reusable, sequence specific and selective. They can be applied for measurements even in blood serum [52]. The weakness of the 'signal-off' sensors is suppression of the signal generated after stimulation with the target. Maximal suppression can only be 100% of the original current. 'Signal-on' architecture incorporating redox decorated DNA sequences have the potential to greatly improve the sensitivity, because the stimulation of these type of sensors with the target does not have a limit to the amount the signal can increase, that is, >100%.

The major weak point of these types of E-DNA sensors is their rather complicated and not very stable architectures. This leads to poor reusability. The majority of the 'signal-on' systems contain signal generating strands non-covalent attached to the surface. However, this approach does not work well in complex samples.

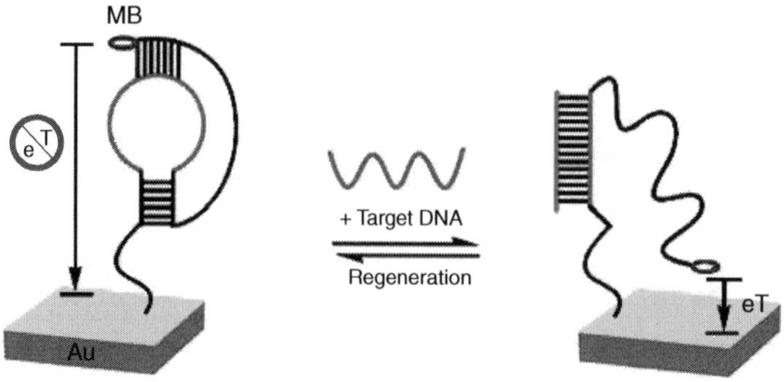

Figure 3.1.6 Schematic of the 'signal-on' E-DNA sensor, which is based on the target-induced resolution of an electrode-bound, methylene blue-modified, DNA pseudoknot. Reprinted with permission from [53]. Copyright 2007, American Chemical Society

Sensors based on the pseudoknot architecture, which work according to the 'signal-on' mechanism, are a good example of significant improvements to the analytical parameters [53, 54] (Figure 3.1.6).

This system consists of a strand, which is covalently attached to the surface of an electrode, and forms a pseudoknot structure consisting of two stem–loops.

The first stem–loop, which is covalently bound to the gold electrode via the 5′-terminus, forms part of the second loop. The first stem–loop was decorated with methylene blue redox marker at the 3′-terminus. In the absence of a target the pseudoknot structure keeps the redox reporter further away from the electrode surface, reducing the faradic current. Hybridization with the target DNA liberates a single-stranded element allowing the reporter to come close to the electrode and efficiently transfer the electrons. The described E-sensor was able to detect the target in blood plasma with a detection limit of 2 nM and a working dynamic range of from 2 to 100 nM.

Because signalling was based on a target hybridization-induced change (rather than the absorption of mass or charge at the sensor surface), and as the DNA probe was covalent attached, the architecture of the new sensor is relatively insensitive to the nonspecific binding of contaminants and is stable in complex media. This allows for detection of a perfectly matched target doped in 50% serum (diluted with buffered saline to control the pH and ionic strength). This E-DNA sensor is able to discriminate between 100 nM of a perfectly matched target and 2 μM of a three-base mismatched DNA even in a complex, contaminant-ridden sample such as blood serum [53, 54].

Using an alternative approach, Liu *et al.* developed an enzyme based E-DNA sensor that generated the analytical signal according to the 'signal-on' mechanism [55] (Figure 3.1.7).

This sensor consists of a stem–loop probe dually labelled with biotin and digoxygenin (DIG) at the 3′- and 5′-ends, respectively. The electrochemical signal is generated by using horseradish peroxidaze linked-anti-DIG antibody (anti-DIG-HRP). Initially, in the absence of a target, the immobilized stem–loop is closed and is shielding the DIG from being approached by the bulk anti-DIG-HRP conjugate, because of the steric effect. After hybridization, the loop sequence forms a rigid duplex. As a result, DIG is transferred further away from the surface of the electrode and becomes accessible for the anti-DIG-HRP.

An HRP enzyme brought about by a single hybridization event can efficiently catalyse thousands of reduction reactions of hydrogen peroxide, leading to significantly amplified electrochemical current signals. This group found an experimental detection limit of less than 10 fM. This high sensitivity is a result of high signal amplification by an enzymatic reaction. The weak side of this sensor is a rather complicated structure of the sensing layer and the necessity to add hydrogen peroxide to the sample solution.

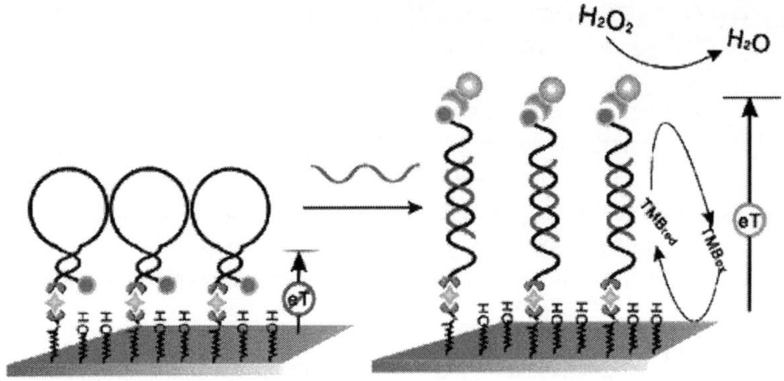

Figure 3.1.7 Scheme for the enzyme-based E-DNA sensor. In the closed conformation, the DIG label is sterically shielded and thus inaccessible to the reporter enzyme. Upon target binding, the disruption of the stem–loop and the formation of the duplex make the DIG label accessible for HRP binding, which catalyses the electrochemical reduction of hydrogen peroxide. Reprinted with permission from [55]. Copyright 2008, American Chemical Society

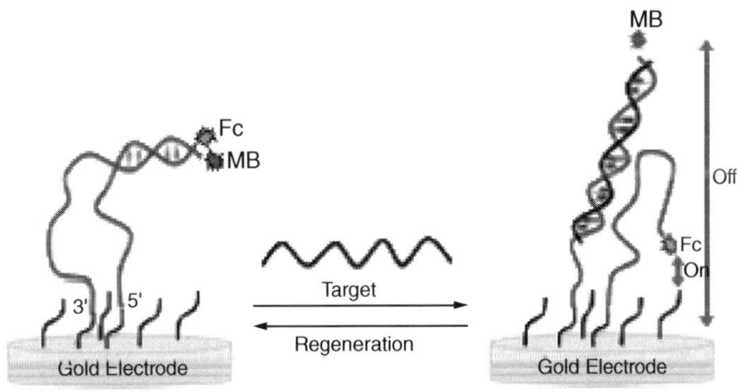

Figure 3.1.8 Schematic representation of the dual-signalling E-DNA sensor. Reprinted with permission from [56]. Copyright 2012, Royal Society of Chemistry

A very interesting E-DNA sensor was recently introduced by Yang and Lai [56] (Figure 3.1.8). In this sensor, a system that works according to both 'signal-off' and 'signal-on' mechanisms is applied simultaneously. This fully covalent dual-signalling sensor can provide more robust target recognition, potentially limiting false positive results that are commonly observed in biosensors.

This sensor consists of a gold electrode simultaneously modified with two strands of DNA, where one is decorated with methylene blue (MB), while the second one is decorated with ferrocene (Fc). The MB-probes posses a 17-base sequence that specifically binds to the target DNA. Four additional thymine bases were added to the 3′-end of the MB-probe to improve its flexibility. The Fc-probe is somewhat complementary to the MB-probe. This allows for the formation of a short DNA duplex at the distal end of the probes. In the presence of a fully complementary target for the MB-probe, these workers observed 71% of signal suppression of the MB current and, simultaneously, a significant increase in the Fc current (267%). This is a consequence of disrupting the 7-base duplex and liberating the Fc-probe, which facilitates the collision of the Fc marker with the surface of the electrode. Formation of the 17-base duplex with the MB-probe sequence made this probe more rigid, resulting in the MB marker being kept further away from the electrode surface.

Consequently, the presence of target DNA caused the reduction of the MB redox current and an increase in the Fc redox current at the same time. The sensitivity and selectivity of this sensor were probed using μM concentrations of target DNA [56].

Recently, in our laboratory we have developed a new type of genosensor based on gold electrode modified oligonucleotide probes that are decorated with 3-iron bis(dicarbollide) [57] (Figure 3.1.9).

This provides an alternative redox active complex that is probably more stable in comparison with those previously reported. Moreover, the novelty of the approach presented was the attachment of the redox label at the 'foot' of the oligonucleotide probe, very close to the electrode surface. This new type of genosensor was intended for the determination of the genetic material of the H5 subtype of Avian Influenza Virus (AIV).

The proposed genosensor displayed good selectivity and sensitivity towards complementary targets, 20-mer ssDNA and PCR products derived from Avian Influenza Virus type H5N1. Osteryoung square-wave voltammetry was applied as a sensing technique, and the changes in the redox activity of Fe(III) centres generated by the hybridization event were used as the analytical signals. The formation of the double helix structure on the electrode surface changes the thickness of the double layer at the interface of the surface of the electrode/analyte solution. As a consequence, the 3-iron bis(dicarbollide) centres, which are attached at the bottom of the oligonucleotide probe, become more 'immersed' in the double helix micro-environment. This further hindered the electron transfer. The detection limits recorded for the 20-mer complementary ssDNA and PCR products having complementary sequences at the 3′-end and in the middle of the oligonucleotide were equal to 0.03 and 0.08 fM, respectively, which are superior to many other systems reported

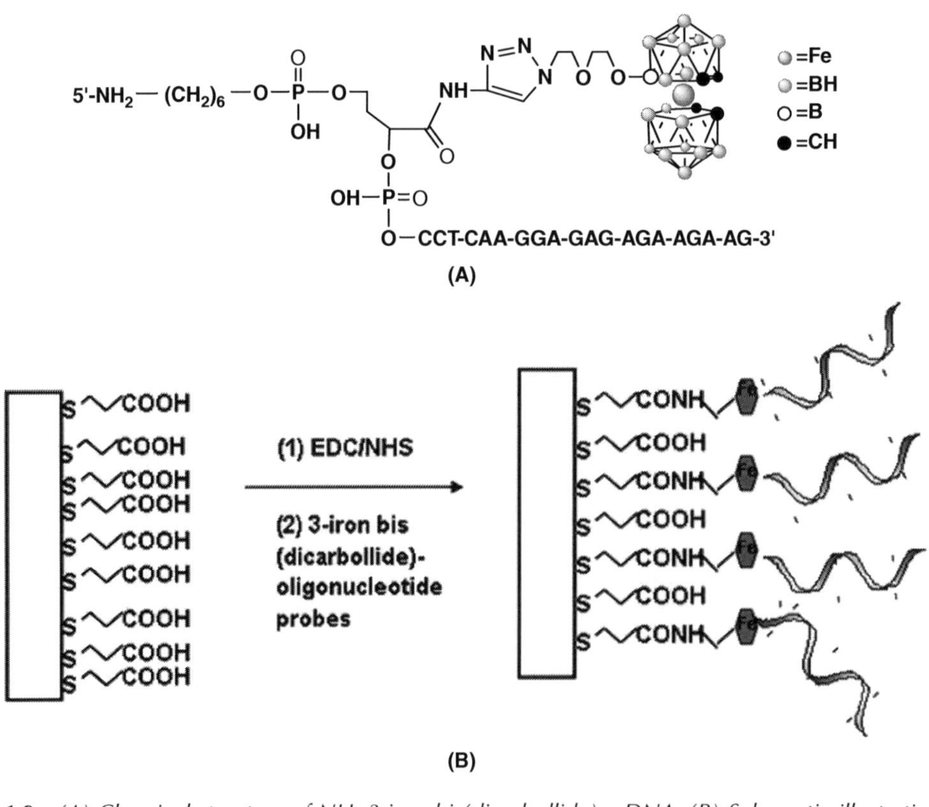

Figure 3.1.9 *(A) Chemical structure of NH_2-3-iron bis(dicarbollide)-ssDNA. (B) Schematic illustration of gold electrode modification with NH_2-3-iron bis(dicarbollide)-ssDNA via EDC/NHS coupling to carboxythiol SAM [57]*

previously. An additional advantage of this genosensor is its ability to differentiate PCR products containing complementary sequences in different positions [57].

3.1.6 Genosensors for simultaneous detection of two different DNA targets

Today, particularly in medical diagnoses, an ideal biosensor is required not only to be miniaturized and cost effective, but also capable of simultaneous detection of multiple analytes.

Büttow and Listat described a system that allowed for a parallel electrochemical analysis of different DNA sequences [58]. For this purpose, the two different DNA probes were immobilized on the surface of one gold electrode. The target DNA modified with methylene blue as a redox marker was present in the sample solution. The principle of the detection of two different ssDNA-targets through this approach is illustrated in Figure 3.1.10.

Generally, the proposed parallel detection method of two sequences of DNA relies on the comparison of voltammetric signal values generated by the electrode stimulated separately with each target and in a second step with their mixture. The detection limits for both targets were 3 nM, corresponding to 3 pM of MB-labelled DNA in 1 mL of buffer solution in an electrochemical cell. This proposition is interesting from a fundamental research point of view, but in practice, its usability is rather limited because of the complicated measurement procedures [58].

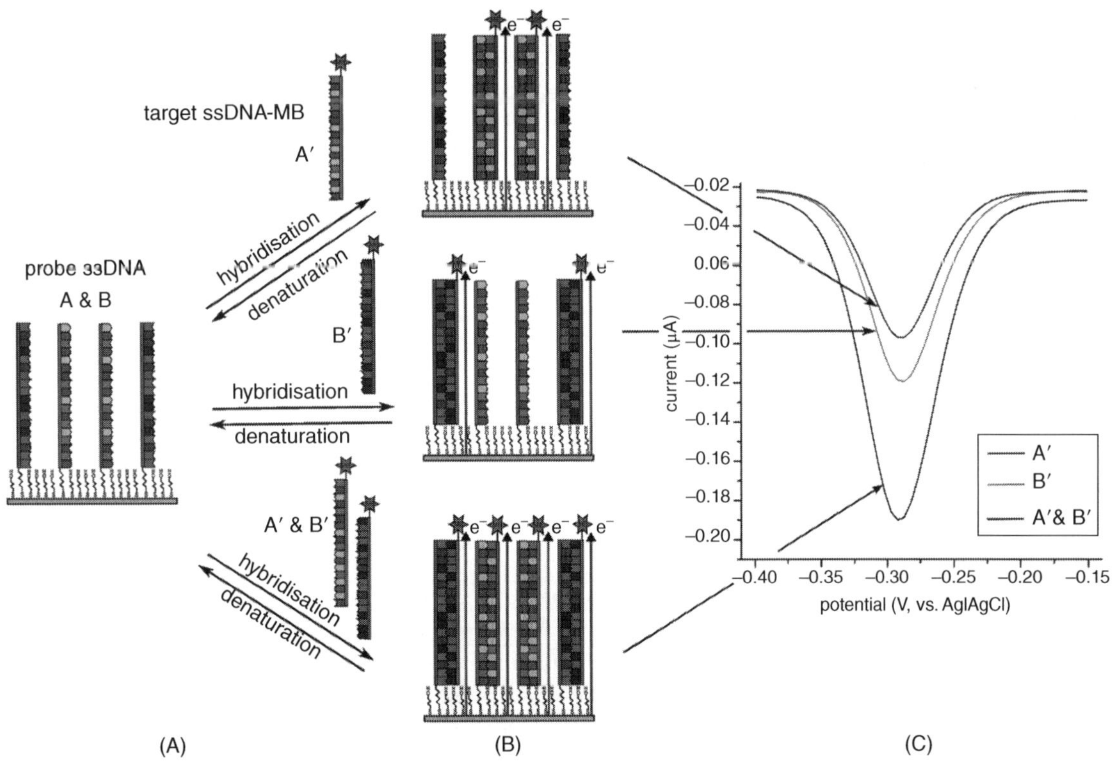

Figure 3.1.10 *Concept of the dual sensor. (A) Sensor after preparation, (B) sensor after hybridization with different sequences, and (C) resulting DPV after hybridization with target DNA (1 mM A′ or B′ and 1 mM A′ and B′ each) [58]*

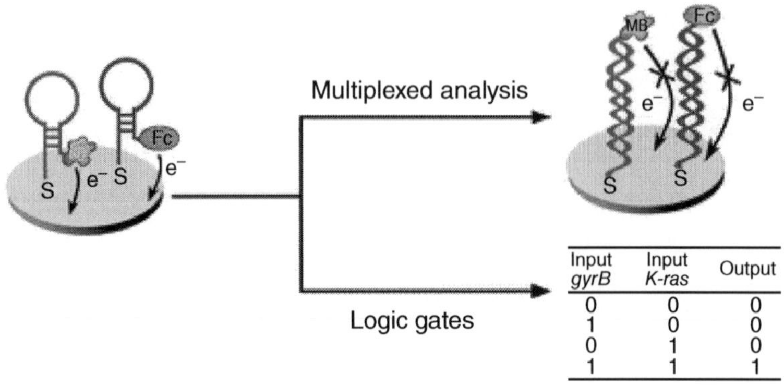

Input gyrB	Input K-ras	Output
0	0	0
1	0	0
0	1	0
1	1	1

Figure 3.1.11 *Schematic representation of the 'AND' logic gate system based on current intensity suppression of MB and Fc at −0.28 and +0.25 V by the* gyrB *and* K-ras *gene inputs: 500 nM* gyrB *gene (1,0), 500 nM* K-ras *gene (0,1), 500 nM* gyrB *and* K-ras *gene (1,1). Reprinted with permission from [59]. Copyright 2011, Royal Society of Chemistry (See colour figure in colour plate section)*

Xiang and Yuan proposed a very interesting approach for the preparation of a reagentless disposable and sensitive multiplexed genosensor for one-spot simultaneous monitoring of two gene biomarkers from the *Salmonella typhimurium* pathogen (*gyrB* gene) and the colorectal tumour (*K-ras* gene). This genosensor consists of two stem–loops of DNA probes, one labelled with methylene blue and the other labelled with ferrocene, immobilized at the surface of a disposable screen printed carbon electrode (Figure 3.1.11). The signal generated by this sensor after stimulation with DNA targets was according to the 'signal-off' mechanism [59].

Hybridization of the surface immobilized probes with the target analyte sequences causes the unfolding of the stem–loop structure. Consequently, the redox markers are forced away from the electrode surface, which leads to the faradic current suppressions. These workers determined the sensitivity of this sensor towards each sequence of target DNA separately. The reported defection limits were 8.3 and 10.0 nM for the *gyrB* gene and *K-ras* gene, respectively. In the case of dual-target gene biomarker detection, the current intensities exhibited inverse dependence upon the target concentration. The MB current (*gyrB* gene) and Fc current (*K-ras* gene) decreased as the concentrations of the target sequences increased from 25 to 100 nM. The sensor selectivity was impressive and was able to distinguish target sequences in the presence of a 50-fold excess of random DNA sequences [59].

Recently, in our laboratory we have developed a novel dual E-DNA sensor, which generates the analytical signal according to the 'signal-off' and 'signal-on' scheme for the simultaneous detection of two different sequences of DNA derived from Avian Influenza Virus (AIV) type H5N1 by means of one electrode [60]. Two sequences of ssDNA that are characteristic of hemagglutinin labelled with ferrocene (ss-DNA–Fc) and for neuraminidase labelled with methylene blue (ss-DNA–MB) were immobilized covalently together on the surface of a gold electrode (Figure 3.1.12).

In order to test if the system under study demonstrates any sensitivity towards complementary ssDNA, the responses of each electrode modified with a single gene sequence stimulated with a complementary 20-mer ssDNA target over a range of concentrations of from 10^{-8} to 10^{-7} M DNA were evaluated, for DNA–Fc as well as for DNA–MB modified electrodes separately. In the case of the electrode modified with DNA–Fc, a peak at 395 mV after stimulation with a 10^{-8} M concentration of target DNA was observed. This peak increased with increasing target DNA concentrations. This indicates the 'signal-on' scheme of analytical signal generation. In the case of the electrode modified with ssDNA–MB, stimulation with increasing concentrations of the complementary DNA target caused a decrease in the current value. This indicated that this electrode generated a signal according to the 'signal-off' mode.

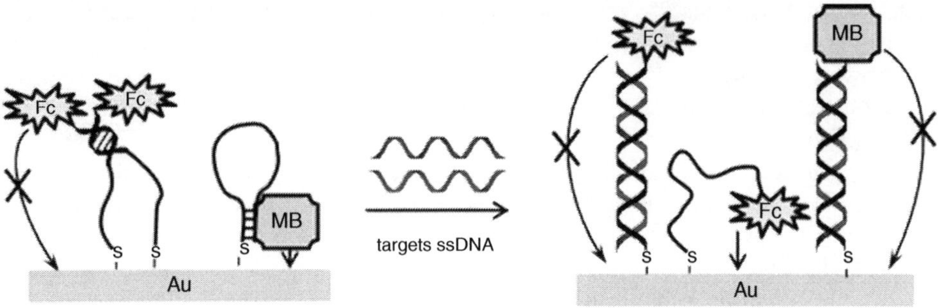

Figure 3.1.12 Schematic representation of the working principle of 'signal-on' and 'signal-off' modes of genosensors consisting of two different ssDNA probes decorated with ferrocene (Fc) and methylene blue (MB). Reprinted with permission from [60]. Copyright © 2013, American Chemical Society

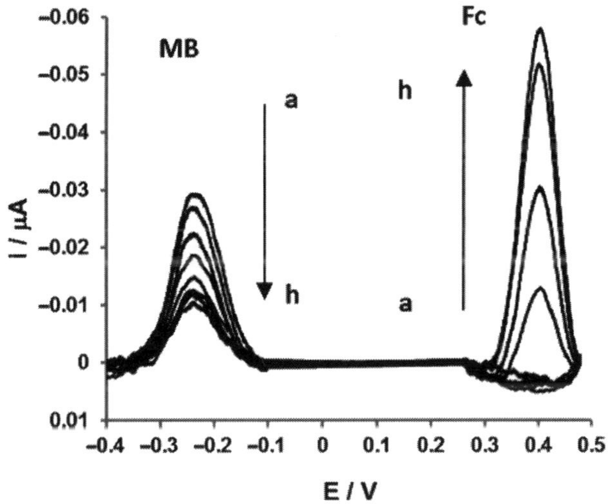

Figure 3.1.13 Representative Osteryoung square-wave voltammograms recorded with electrodes modified with an SH-ssDNA–Fc probe and an SH-ss-DNA–MB probe (a), and after hybridization with 8 nM (b), 10 nM (c), 20 nM (d), 40 nM (e), 60 nM (f), 80 nM (g) and 100 nM (h) oligonucleotide sequences complementary to SH-ssDNA–Fc and SH-ssDNA–MB. Reprinted with permission from [60]. Copyright © 2013, American Chemical Society

An explanation of this phenomenon could be as follow. In the case of the sequence characteristic for the hemagglutinin of the influenza H5N1 virus (ss- DNA–Fc) there was a possibility for formation of dimers between the immobilized DNA strands. The energy of dimer formation is –12.87 kcal mol^{-1}. As a consequence of this, the redox probe is kept a relatively large distance away from the electrode surface. After hybridization with the complementary strand the dimer is disassembled. One of the released strands forms a duplex with the target DNA. As a consequence, the distance between the redox probe and electrode surface increases, which causes a decrease in the current. The second component of the dimer recovers its natural flexibility and the distance between the redox probe decreases, which results in a current increase. Our experimental data showed that after stimulation of the electrode with target DNA at a concentration of $\geq 10^{-8}$ M, the second effect was dominant and the electrode generated an analytical signal according to the 'signal-on' scheme. In the case of the sequence characteristic of the neuraminidase of the influenza H5N1 virus (ssDNA–MB), the formation of the previously mentioned dimers was less probable, due to two reasons. The energy of dimer formation was only –7.47 kcal mol^{-1}, and, additionally, immobilization of ssDNA–MB via the 3′-end creates a steric obstacle to the dimerization. Thus, the response of this electrode was generated according to the 'signal-off' scheme.

In the next step, an electrode modified with both probes was stimulated with a mixture of 20-mer ssDNA containing two components: one complementary to ssDND–Fc, and the other complementary to ssDNA–MB. The range of concentrations of each target ssDNA was from 10^{-10} to 10^{-6} M. Representative voltammograms are presented in Figure 3.1.13.

The changes to the height and position of the peaks showed the same tendency as in the case of stimulation of the electrodes separately by the ssDNA target. Detection limits determined by a graphic method were 4.0×10^{-8} and 2.0×10^{-8} M, for the simultaneous analysis of both sequences and for a single one, respectively.

These values, in particular the detection limit for the parallel determination of two sequences, are very promising from a diagnostic point of view.

3.1.7 Conclusions

By describing selected genosensors, we have been trying to present a logical trend in their development. There are currently numerous published scientific reports that show the extremely high level of research interest in the interdisciplinary fundamental studies on detection methods for hybridization events as well as on gene mismatch detection.

On the other hand, the number of genosensors that find applications in real sample analysis is rather limited. This indicates the strong need for scientific efforts directed at the development of sensitive and selective genosensors, which could be mass-produced at a reasonable price. Medical diagnoses, food control and environmental monitoring are waiting for such analytical devices.

Acknowledgements

This work was supported by Project WND-POIG .01.01.02-00-007.

References

[1] D. R. Thevenot, K. Toth, R. A. Durst, and G. S. Wilson, Electrochemical biosensors: Recommended definitions and classification, *Pure Appl. Chem.*, **7**, 2333–2348 (1999).

[2] R. C. Ebersole , J. A. Miller, J. R. Moran, and M. D. Ward, Spontaneously formed functionally active avidin monolayers on metal surfaces: A strategy for immobilizing biological reagents and design of piezoelectric biosensors, *J. Am. Chem. Soc.*, **112**(8), 3239–3241 (1990).

[3] R. Levicky , T. M. Herne , M. J. Tarlov, and S. K. Satija, Using self-assembly to control the structure of DNA monolayers on gold: A neutron reflectivity study, *J. Am. Chem. Soc.*, **120**(38), 9787–9792 (1998).

[4] K. M. Millan, A. J. Spurmanis, and S. R. Mikkelsen, Covalent immobilization of DNA onto glassy carbon electrodes, *Electroanalysis*, **4**, 929–932 (1992).

[5] E. Palecek, Oscilographic polarography of highly polymerized deoxyribonucleic acid, *Nature*, **188**, 656–657 (1960).

[6] K. M. Millian and S. R Mikkelson, Sequence-selective biosensor for DNA-based on electroactive hybridization indicators, *Anal. Chem.*, **65**, 2317–2323 (1993).

[7] E. Palecek, Adsorptive transfer stripping voltametry: Determination of nanogram quantities of DNA immobilized at the electrode surface, *Anal. Biochem.*, **170**, 421–431 (1988).

[8] C. Hinnen, A. Rousseau, R. Parsons, and J. A. Reynaud, Comparison between behaviour of native and denaturated DNA at mercury and gold electrodes by capicity measurments and cyclic voltametry, *J. Electroanal. Chem.*, **125**, 193–203 (1981).

[9] E. E. Ferapontova and E. Dominguez, Direct electrochemical oxidation of DNA on polycrystalline gold solution, *Electroanalysis*, **15**, 629–634 (2003).

[10] E. Sharifi, A. Salimi, and E. Shams, DNA/nickel oxide nanoparticles/osmium(III)-complex modified electrode toward selective oxidation of L-cysteine and simultaneous detection of L-cysteine and homocysteine, *Bioelectrochemistry*, **86**, 9–21(2012).

[11] J. Wang, Electrochemical nucleic acid biosensors, *Anal. Chim. Acta*, **469**, 63–71 (2002).

[12] P. De-los-Santos-Álvarez, M. J. Lobo-Castañón, A. J. Miranda-Ordieres, and P. Tuñón Blanco, Electrochemistry of nucleic acids at solid electrodes and its applications, *Electroanalysis*, **16**, 1193–1204 (2004).

[13] E. Palecek and F. Jelen, Electrochemistry of nucleic acids and development of DNA sensors, *Crit. Rev. Anal. Chem.*, **32**, 261–270 (2002).

[14] H.-S. Wang, H.-X. Ju, and H.-Y. Chen, Voltammetric behavior and detection of DNA at electrochemically pretreated glassy carbon electrode, *Electroanalysis*, **13**, 1103–1109 (2000).

[15] X. Zhang, Y. Qu, G. Piao, J. Zhao, and K. Jiao, Reduced working electrode based on fullerene C60 nanotubes@ DNA: Characterization and application, *Mater. Sci. Eng., B* **175**, 159–163 (2010).

[16] J.-B. Raoof, M. S. Hejazi, R. Ojani, and E. H. Asl, A comparative study of carbon nanotube paste electrode for development of indicator-free DNA sensors using DPV and EIS: Human interleukin-2 oligonucleotide as a model, *Int. J. Electrochem. Sci.*, **4**, 1436–1451(2009).

[17] P. M. Armistead and H. H. Thorp, Electrochemical detection of gene expression in tumor samples: overexpression of Rak nuclear tyrosine kinase. *Bioconjug. Chem.*, **13**, 172–176 (2002).

[18] D. Xue, C. M. Elliott, P. Gong, D. W. Grainger, C. A. Bignozzi, and S. Caramori, Indirect electrochemical sensing of DNA hybridization based on the catalytic oxidation of cobalt(II), *J. Am. Chem. Soc.*, **129**(7), 1854–1855 (2007).

[19] L. Xiong, C. Batchelor-McAuley, L. M. Gonçalves, J. A. Rodrigues, and R. G. Compton, The indirect electrochemical detection and quantification of DNA through its co-adsorption with anthraquinone monosulphonate on graphitic and multi-walled carbon nanotube screen printed electrodes, *Biosens. Bioelectron.*, **26**, 4198–4203 (2011).

[20] H. Aoki and Y. Umezawa, Ion channel sensors based on artificial receptors, *Anal. Chem.*, **76**(17), 320A–326A (2004).

[21] H. Aoki and H. Tao, Gene sensors based on peptide nucleic acid (PNA) probes: Relationship between sensor sensitivity and probe/target duplex stability, *Analyst*, **130**, 1478–1482 (2005).

[22] H. Aoki and Y. Umezawa, High sensitive ion-channel sensors for detection of oligonucleotides using PNA modified gold electrodes, *Electroanalysis*, **14**(19-20), 1405–1410 (2002).

[23] H. Aoki and H. Tao, Signal enhancement for gene detection based on a redox reaction of $[Fe(CN)_6]^{-4}$ mediated by ferrocene at the terminal of a peptide nucleic acid as a probe with hybridization-amenable conformational flexibility, *Anal. Sci.*, **24**(7), 929–933 (2008).

[24] X. Liu, Z. Cheng, H. Fan, S. Ai, and R. Han, Electrochemical detection of avian influenza virus H5N1 gene sequence using a DNA aptamer immobilized onto a hybrid nanomaterial-modified electrode, *Electrochim. Acta*, **56**, 6266– 6270 (2011).

[25] (a)K. Malecka, I. Grabowska, J. Radecki, A. Stachyra, A. Góra-Sochacka, A. Sirko, and H. Radecka, Voltammetric detection of specific DNA sequence of Avian influenza virus H5N1 using HS-ssDNA probe deposited onto gold electrode, *Electroanalysis*, **24**(2), 439–446 (2012);(b)K. Malecka, I. Grabowska, J. Radecki, A. Stachyra, A. Góra-Sochacka, A. Sirko, and H. Radecka, Electrochemical detection of Avian influenza virus genotype using amino-ssDNA probe modified gold electrodes, *Electroanalysis*, **25**(8), 1871–1878 (2013).

[26] D.-J. Chung, K.-Ch. Kim, and S-H Choi, Electrochemical DNA biosensor based on avidin–biotin conjugation for influenza virus (type A) detection, *Appl. Surf. Sci.*, **257**, 9390–9396 (2011).

[27] B. M. Zeglis, V. C. Pierre, and J. K. Barton, Metallo-intercalators and metallo-insertors, *Chem. Commun.*, 4565–4579 (2007).

[28] K. Hashimoto, K. Ito, and Y. Ishimori, Sequence-specific gene detection with a gold electrode modified with DNA probes and an electrochemically active dye, *Anal. Chem.*, **66**(21), 3830–3833 (1994).

[29] M. G. Blackburn and M. J. Gait, *Nucleic Acids in Chemistry and Biology*, IRL Press, New York, 1990.

[30] (a)M. Mascini, I. Palchetti, and G. Marraza, DNA electrochemical biosensors, *Fresenius' J. Anal. Chem.*, **369**, 15–22 (2001);(b)S. Palanti, G. Marrazza, and M. Mascini, Electrochemical DNA probe, *Anal. Lett.*, **29**, 2309–2331(1996).

[31] H. Miyahara, K. Yamashita, M. Kanai, K. Uchida, M. Takagi, H. Kondo, and S. Takenaka, Electrochemical analysis of single nucleotide polymorphisms of p53 gene, *Talanta*, **56**, 829–835 (2002).

[32] S. Hasoň, J. Dvořák, F. Jelen, and V. Vetterl, Interaction of DNA with echinomycin at the mercury electrode surface as detected by impedance and chronopotentiometric measurements, *Talanta*, **56**, 905–913 (2002).

[33] F. Jelen , A. Erdem, and E. Palecek, Cyclic voltammetry of echinomycin and its interaction with double-stranded and single-stranded DNA adsorbed at the electrode, *Bioelectrochemistry*, **55**(1-2), 165–167 (2002).

[34] M. Nakayama, T. Ihara, K. Nakano, and M. Maeda, DNA sensors using a ferrocene-oligonucleotide conjugate, *Talanta*, **56**, 857–866 (2002).

[35] T. García , M. Revenga-Parra , H. D. Abruña, F. Pariente, and E. Lorenzo, Single-mismatch position-sensitive detection of DNA based on a bifunctional ruthenium complex, *Anal. Chem.*, **80**(1), 77–84 (2008).

[36] L. E. Ahangar and M. A. Mehrgardi, 3,4-Diaminobenzoic acid (DABA) as a redox label for electrochemical detection of single base mismatches, *Electrochim. Acta*, **56**, 10264–10269 (2011).

[37] L. E. Ahangar and M. A. Mehrgardi, Nanoparticle-functionalized nucleic acids: A strategy for amplified electrochemical detection of some single-base mismatches, *Electrochim. Acta*, **56**, 2725–2729 (2011).

[38] M. A. Mehrgardi and L. E. Ahangar, Silver nanoparticles as redox reporters for the amplified electrochemical detection of the single base mismatches, *Biosens. Bioelectron.*, **26**, 4308–4313 (2011).

[39] J. C. Genereux and J. K. Barton, Mechanisms for DNA charge transport, *Chem. Rev.*, **110**, 1642–1662 (2010).

[40] E. Paleček and M. Bartošík, Electrochemistry of nucleic acids, *Chem. Rev.*, **112**, 3427–3481 (2012).

[41] N. Hüsken, M. Gębala, A. Battistel, F. La Mantia, W. Schuhmann, and N. Metzler-Nolte, Impact of single basepair mismatches on electron-transfer processes at Fc-PNA·DNA modified gold surfaces, *ChemPhysChem*, **13**(1), 131–139 (2012).

[42] E. Spain, R. Kojima, R. B. Kaner, G. G. Wallace, J. O'Grady, K. Lacey, T. Barry, T. E. Keyes, and R. J. Forster, High sensitivity DNA detection using gold nanoparticle functionalised polyaniline nanofibres, *Biosens. Bioelectron.*, **26**(5), 2613 –2618 (2011).

[43] E. E. Ferapontova, Electrochemical indicators for DNA electroanalysis, *Curr. Anal. Chem.*, **7**, 51–62 (2011).

[44] N. B. Muren, E. D. Olmon, and J. K. Barton, Solution, surface, and single molecule platforms for the study of DNA-mediated charge transport, *Phys. Chem. Chem. Phys.*, **14**, 13754–13771(2012).

[45] N. Moradi, M. F. Mousavi, M. A. Mehrgardi, and A. Noori, Preparation of a new electrochemical biosensor for single base mismatch detection in DNA, *Anal. Methods*, **5**, 6531–6538 (2013).

[46] C. Fan, K. W. Plaxco and A. J. Heeger, Electrochemical interrogation of conformational changes as a reagentless method for the sequence-specific detection of DNA, *PNAS*, 9134–9137 (2003).

[47] C. E. Immoos, S. J. Lee, and M. W. Grinstaff, Conformationaly gated electrochemical gene detection, *Chem. BioChem.*, **5**, 1100–1103 (2004).

[48] F. Ricci, R. Y. Lai, and K. W. Plaxco, Linear, redox modified DNA probes as electrochemical DNA sensors, *Chem. Commun.*, 3768–3770 (2007).

[49] F. Ricci and K.W. Plaxco, E-DNA sensors for convenient, label free electrochemical detection of hybridization, *Microchim. Acta*, **163**, 149–155 (2008).

[50] C. E. Immoos, S. J. Lee, and M. W. Grinstaff, DNA-PEG-DNA triblock macromolecules for reagentless DNA detection, *J. Am. Chem. Soc.*, **126**, 10814–10815 (2004).

[51] Y. Xiao, A. A. Lubin, B. R. Baker, K. W. Plaxco, and A. J. Heeger, Single-step electronic detection of femtomolar DNA by target-induced strand displacement in an electrode-bound duplex, *PNAS*, **103**(45) 16677–16680 (2006).

[52] A. A. Lubin, R. Y. Lai, B. R. Baker, A. J. Heeger, and K. W. Plaxco, Sequence-specific, electronic detection of oligonucleotides in blood, soil, and foodstuffs with the reagentless, reusable E-DNA sensor, *Anal. Chem.*, **78**(16), 5671–5677 (2006).

[53] Y. Xiao, X. Qu, K.W. Plaxco, and A. J. Heeger, Label-free electrochemical detection of DNA in blood serum via target-induced resolution of an electrode-bound DNA pseudoknot, *J. Am. Chem. Soc.*, **129**, 11896–11897 (2007).

[54] K. J. Cash, A. J. Heeger, K. W. Plaxco, and Y. Xiao, Optimalization of reusable DNA pseudoknot-based electrochemical sensor for sequence-specific DNA dedection in blood serum, *Anal. Chem.*, **81**, 656–661 (2009).

[55] G. Liu, Y. Wan, V. Gau, J. Zhang, L. Wang, S. Song, and C. Fan, An enzyme-based E-DNA sensor for sequence-specific detection of femtomolar DNA targets, *J. Am. Chem. Soc.*, **130**(21), 6820–6825 (2008).

[56] W. Yang and R. Y. Lai, A dual-signalling electrochemical DNA sensor based on target hybridization-induced change in DNA probe flexibility, *Chem. Commun.*, **48**, 8703–8705 (2012).

[57] I. Grabowska, A. Stachyra, A. Góra-Sochacka, A. Sirko, A. B. Olejniczak, Z. J. Leśnikowski, J. Radecki, and H. Radecka, DNA probe modified with 3-iron bis(dicarbollide) for electrochemical determination of DNA sequence of Avian Influenza Virus H5N1, *Biosens. Bioelectron.*, **51**, 170–176 (2014).

[58] S. Büttow and F. Lisdat, Parallel detection of different DNA sequences on one gold electrode, *Electroanalysis*, **22**(9), 931–937 (2010).

[59] Y. Xiang, X. Qian, Y. Chen, Y. Zhang, Y. Chai, and R. Yuan, A reagentless and disposable electronic genosensor: from multiplexed analysis to molecular logic gates, *Chem. Commun.*, **47**, 2080–2082 (2011).

[60] I. Grabowska, K. Malecka, A. Starycha, A. Góra-Sochacka, A. Sirko, W. Zagórski Ostoja, H. Radecka, and J. Radecki, Single electrode genosensor for simultaneous determination of sequences encoding hemagglutinin and neuraminidase of Avian Influenza Virus Type H5N1, *Anal. Chem.*, **85**, 10167–10173 (2013).

3.2

Oligonucleotide Based Artificial Ribonucleases (OBANs)

Alice Ghidini[a], Merita Murtola[a,b], and Roger Strömberg[a]

[a] *Karolinska Institute, Department of Biosciences and Nutrition, Novum, Sweden*
[b] *Department of Chemistry, University of Turku, Turku, Finland*

3.2.1 Introduction

Artificial ribonucleases can be visualized as molecular scissors cleaving ribonucleic acids, typically at the internucleoside phosphodiester linkages. Substantial research has gone into the development of small catalytic entities that cleave RNA and, although of high value for understanding the mechanisms and for further development of artificial nucleases, these small catalytic units typically do not discriminate between different RNA sequences and are only, at best, selective with respect to neighboring heterocyclic bases. Several reviews have covered such catalytic groups and other catalytic systems [1–7]. DNA cutters have also been reviewed recently [8] as well as the new development of zinc finger based nucleases [9]. None of these will be covered here, since this overview mainly considers key developments, as well as an account of our own work on artificial ribonucleases that recognize sufficiently long specific sequences of a target RNA which could therefore develop not only into tools for molecular biology but also into therapeutics for hereditary diseases.

Synthetic nucleic acids have been, and still are, crucial for the development of life sciences research. Modified oligonucleotides (ONs) and/or conjugates are not only used for research but are also used for diagnostics and have been developed as drugs for the treatment of patients with diseases of genetic origin. Most oligonucleotide therapies that have been successful, such as siRNA [10–12] (short interfering RNA) approaches and antisense technologies [13–17], including splice-switching [18, 19], target specific RNA sequences. Oligonucleotide therapy has the potential for treatment of a broad range of diseases, such as Alzheimer's disease [20, 21] , Huntington's disease [22, 23], cancers [24, 25], and muscular dystrophies [26, 27]. After a

DNA in Supramolecular Chemistry and Nanotechnology, First Edition.
Edited by Eugen Stulz and Guido H. Clever.

relatively slow start in the first couple of decades of research, the interest in oligonucleotide therapy is now booming, largely due to there now being several different possible approaches and, with many new regulatory RNAs being discovered, miRNAs are also being explored as targets [28]. Much current research is focused on developing efficient cellular delivery of ONs, but other issues are stability, higher specificity towards the target, and of course overall efficiency, which includes the efficient use of the ON.

Efficiency in regulation of gene expression should be more readily achieved if turnover of the target RNA is obtained. This can occur if native enzymes (e.g., RNAse H for antisense [13–17] and RISC complex for siRNA [10–12]) can recognize the relevant oligonucleotide complex. However, most modifications of oligonucleotides are not recognized by RNase H and hence gapmers are needed to take advantage of the cellular enzyme [13–17]. If the antisense oligonucleotide could carry a group that causes cleavage of the target RNA upon hybridization, this problem would be avoided and independence from native enzymes would be achieved. Thus, use of an oligonucleotide based artificial nuclease [1, 3–6] (OBAN) [29] could then lead to recognition and cleavage of RNA sequences responsible for genetic or viral diseases. Such artificial enzymes could allow for complete modifications that produce stable oligonucleotides that are not degraded by the host enzymes. Recognition of the target substrate could be reliably achieved through Watson–Crick base pairing and the concept should work in theory. In reality, the difficulty lies in developing a sufficiently efficient artificial enzyme.

3.2.2 Early development of OBANs

For RNA the cleavage can be achieved most effectively through intramolecular transesterification of the phosphodiester linkages with the 2′-hydroxyl function. One obvious task is to develop catalytic groups to be able to achieve efficient cleavage. However, if the catalyst is too active, cleavage of non-hybridized non-complementary RNA could become a problem. For this reason, the proximity factor is crucial, that is, a high intramolecularity between the cleavage agent and the target RNA is desirable. For a given system this is largely determined by the nature of the linker and its position of attachment. Furthermore, the choice of target sequence may also be crucial in this respect. The nature of the substrate is also critical for overall reactivity, as it is well known that cleavage of RNA has a high dependence on the sequence [30] and is considerably more reactive in the single-stranded form (also as bulges or loops) than as a duplex [31, 32]. Thus, the vulnerability of the substrate is a key issue.

Many of the initial so-called artificial nucleases, as pioneered by the groups of Komiyama [33, 34], Bashkin [35, 36], Vlassov [37, 38], and Häner [39], had molecular scissors (both as non-metal polyamines or imidazole derivatives, and as metal ion chelate based catalysts) at the termini, to cleave the RNA outside the sequence that is complementary to the carrier oligonucleotide, or if the "artificial nuclease" was used in excess. These systems do not display turnover of substrate and should thus really be called cleavers and not catalysts or artificial enzymes.

3.2.3 Metal ion based artificial nucleases

Magda *et al.* developed europium (EuTx) [40] and dysprosium thexaphyrin (DyTx) [41] based cleavers that are reasonably active. However, the most important contribution was that a DyTx derivative attached to an internally positioned glyceryl linker (Figure 3.2.1) in the carrier oligonucleotide gave turnover of the substrate RNA [42], but when attached to end of the oligonucleotide there was no turnover, presumably because the cleaved substrate was not released.

Double-stranded RNA is considerably less susceptible to cleavage than single-stranded RNA [43, 44]. This was clearly identified by the Ciba-Geigy group of researchers as being extremely important for the development of artificial nucleases and that creation of bulges in the target RNA should promote cleavage [45].

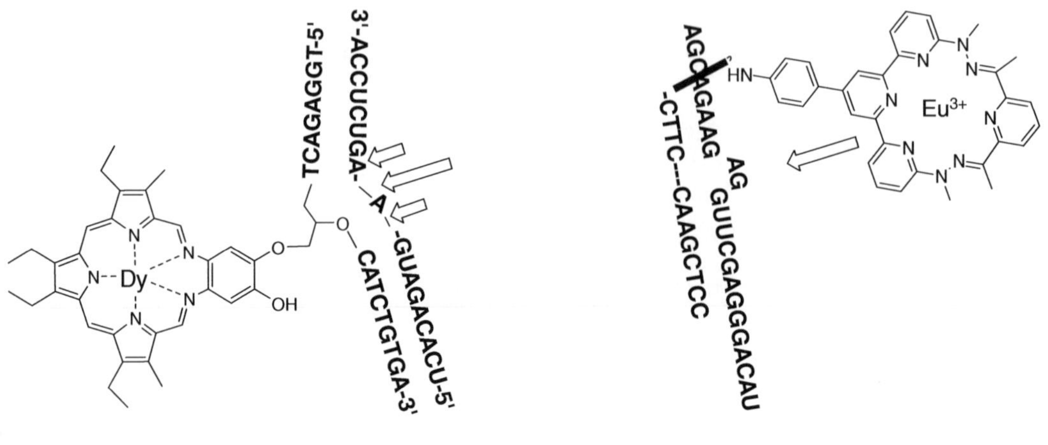

Figure 3.2.1 *Lanthanide metal ion complex and oligonucleotide based artificial nucleases by Magda* et al. *and Häner's group, which early on displayed turnover of the RNA substrate*

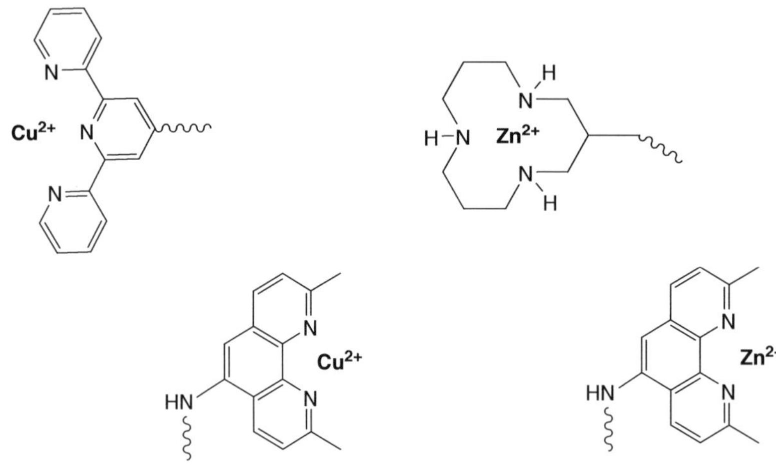

Figure 3.2.2 *Examples of copper and zinc ion chelates of terpyridine, 1,5,9-triazacyclodo-decane and 5-amino-2,9-dimethyl-1,10-phenanthroline used as the cleaving moiety in oligonucleotide based artificial nuclease constructs*

That there is no energy penalty from disruption of the base paring, which would be inevitable when the intramolecular reaction leading to a five membered cyclic phosphate is formed, should of course be beneficial. It was also suggested that prearrangement of the 2-hydroxyl into a more favorable position for in-line attack on the phosphate is important [45], and Häners group also showed that oligonucleotide and lanthanide chelate based cleavers (Figure 3.2.1) indeed gave enhanced cleavage of RNA at bulged-out sites [46]. With these first pioneering systems, turnover was not realized, but switching from a deoxyribonucleotide backbone to one with 2′-*O*-methoxyethoxyribonucleosides gave turnover of the RNA substrate (39 times in 64 h) [47].

Komiyama is one of the true pioneers, with many contributions to the field of artificial RNA cutters [5, 33, 34]. Among the most interesting work is activation of the RNA strand by introduction of acridine groups into the recognizing strand, which then makes the RNA target more susceptible to cleavage by lanthanide ions in the vicinity of the acridine moieties, and hence gives high selectivity [48, 49]. This "pin-point" activation further substantiates that not only is the cleaver important, but also the structure of the substrate RNA.

Most of the work on the development of artificial nucleases has relied on metal ion based cleaver groups and there are of course a number of phosphate cleaving enzymes utilizing metal ions in the active site, some of which have two co-acting metal ions. There is very limited work on oligonucleotides conjugated with two metal chelates [50], but most notable is the work with two Cu(II) terpyridine [51] and that with two Zn^{2+} 1,5,9-triaza-cyclododecane [52] complexes (Figure 3.2.2). In the former case, there was some rate enhancement compared with a single Cu(II) terpyridine system [53], but turnover of the RNA substrate was shown. In the latter case the rate was an order of magnitude higher than with the system having a single azacrown inclusion [54]. Development of artificial nuclease with two metal ions does have some pitfalls, and there are examples of when the rate goes down [55] or is completely quenched [56] upon introduction of a second metal ion or chelate.

3.2.4 Non-metal ion based systems

In mimicking RNA cleaving enzymes, one can of course also consider use of non-metal catalytic groups, such as imidazole moieties. Even from the early developments of oligonucleotide based artificial RNA cleavers, there were reports on the use of non-metal catalysts [33, 34, 37, 38]. Multivalent metal ions benefit from readily finding the negatively charged target phosphodiesters through electrostatic attraction. However, for future *in vivo* applications the advantage of using non-metal ion based systems is that, apart from restrictions due to metal-ion toxicity, this does not depend on finding the metal ion in the living cell or on the complex being stable enough that it does not dissociate. Incorporation of bisimidazole moieties [37, 38, 57] can be considered as RNase A mimics. In particular, Vlassov's group have been persistent in pursuing 5′-linked bisimidazole constructs (Figure 3.2.3), and site-selective cleavage (a major and a minor site) in a fragile sequence within yeast $tRNA^{Phe}$ has been achieved [38, 58]. It was noted that the catalytic imidazole groups were flexible and with no stable contacts to the cleavage site, suggesting that such systems can be much improved by more exact positioning of the imidazole moieties [54]. Oligonucleotides conjugated to oligoarginine–leucine peptide conjugates ($G(RL)_4$) have also been shown to cleave RNA and to act as true catalysts [59]. However, no specificity dependent on the oligonucleotide guided sequence recognition was found. Instead, the oligonucleotide seems to act as a scaffold, which enhances activity. However, cleavage occurs at multiple sites, mainly adjacent to guanosine moieties, and in this way resembling RNase T1 [60].

Another interesting development of non-metal ion cleavers, trisbenzimidazoles (Figure 3.2.3) [61] has been pursued by Göbel's group. Trisbenzimidazoles have also been conjugated to DNA oligonucleotides that were found to bind and cleave complementary RNA sequences [62]. Turnover was not realized but substrate specificity due to oligonucleotide hybridization was achieved.

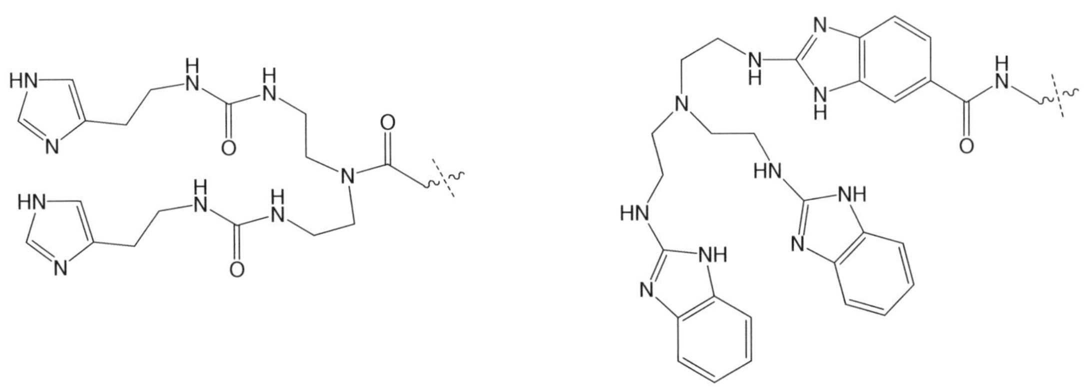

Figure 3.2.3 *Examples of non-metal ion cleavers of oligonucleotide based artificial nucleases, for example, by Vlassov and Göbel, respectively*

3.2.5 Creating bulges in the RNA substrate

When beginning the development of oligonucleotide based artificial nucleases at the Karolinska Institute, we selected neocuproine (2,9-dimetylphenanthroline) as a chelating agent for a catalytic metal ion. This had been used before as a Cu(II) complex in oligodeoxyribonucleotide constructs, where cleavage of target RNA, but no turnover, was accomplished since the oligonucleotide was used in excess [63]. The study did show that the Cu^{2+}–neocuproine conjugate was more active than the corresponding phenanthroline and terpyridine constructs and correlated with the same trend that was found for the parent complexes in the cleavage of a dinucleotide [64]. We realized that in order to achieve turnover it would be advisable to cleave the target in a central part of the sequence so that shorter more readily dissociated fragments would be obtained. In addition, it was desirable to cleave single-stranded rather than duplexed RNA, since this is more susceptible to cleavage. It then seemed clear that we should position the moiety with the cleaving ability centrally while the target RNA should be single stranded in the vicinity of the cleaver.

The oligonucleotide based artificial nuclease sequence should then be such that it is complementary to the RNA but is not in a central portion of the target. Systems without a bulge and those with a single nucleotide bulge and with a Zn^{2+}–neocuproine attached to the 2′-position of a central nucleoside were unimpressive and the complex had a low melting point [65]. The starting structure for the first real constructs of the "bulge concept" was instead based on an NMR structure of a bulge–loop of a group I ribozyme (Figure 3.2.4) [66].

A number of different 2′-*O*-methyloligoribonucleotide constructs with Zn^{2+}–neocuproine attached from different positions of the oligonucleotide were made, and these were investigated with respect to cleavage of target RNA sequences, which, upon binding, formed either a full duplex or a 1–5 nucleotide (nt) bulge [29]. The target RNA was only cleaved at phosphodiesters in the portion of the RNA that formed a bulge, and with some selectivity where the most efficient system gave about 70% cleavage at one site. The rate of cleavage was also clearly dependent on the position on the oligonucleotide the metal ion complex was attached; the importance of this positioning was substantiated further by additional studies [67]. The highest rate of cleavage was obtained with the metal chelate protruding from the 5-position of a centrally placed uridine of a GU wobble bulge closing base pair and the optimal bulge sizes were the 3 and 4 nt bulges (Figure 3.2.5). It was also shown that OBAN systems based on this concept can give cleavage of a target RNA with real catalysis and turnover. A Michaelis–Menten type of kinetics and dependence on the metal ion was also shown [29].

Since the activity depends upon the target and the OBAN sequence, it could be advantageous from the start to actually develop the OBAN with the aim being a potential therapeutic target sequence, before the catalyst is optimized and positioned for efficiency. The M-BCR/ABL mRNA transcript from the Philadelphia chromosome (Ph), t(9;22) has for a long time been associated with human cancer and is the cytogenetic hallmark

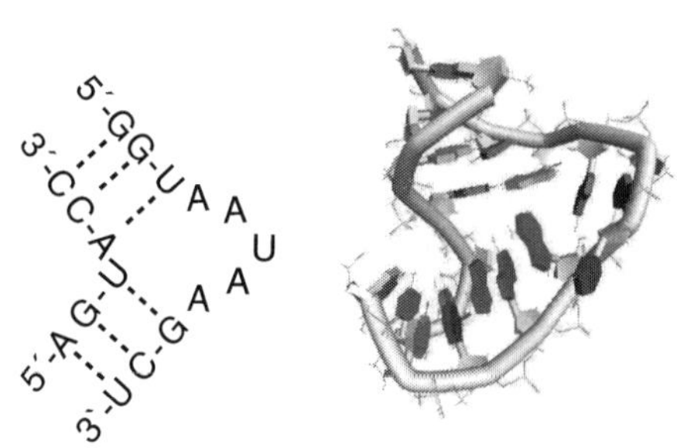

Figure 3.2.4 The five nucleotide (AAUAA) RNA bulge from the NMR structure [66] of the bulge–loop in the Group I ribozyme domain, which was the starting point for the concept of OBAN–RNA complexes where the OBAN binds in such a way that a bulge is formed in the RNA strand

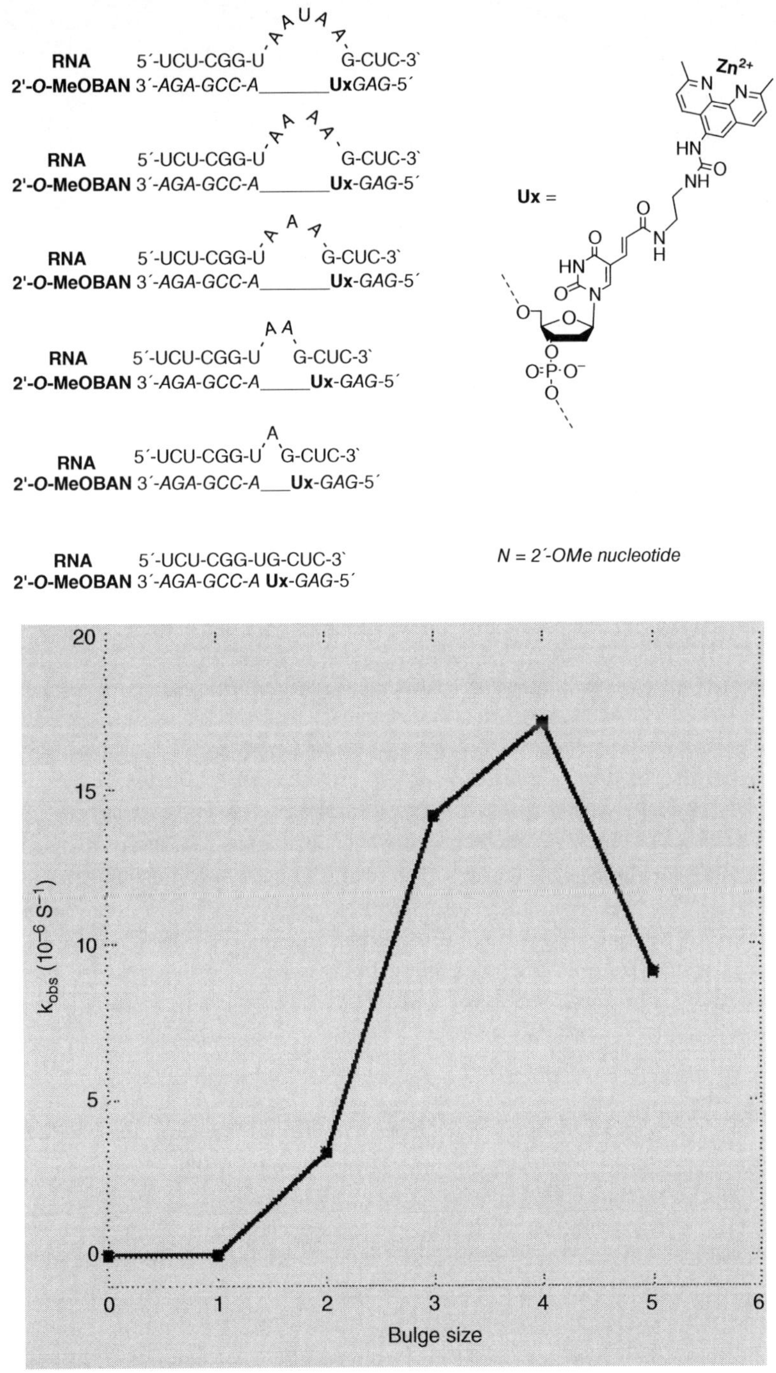

Figure 3.2.5 Bulge size dependence of rate of cleavage for a 2′-O-MeOBAN [29]

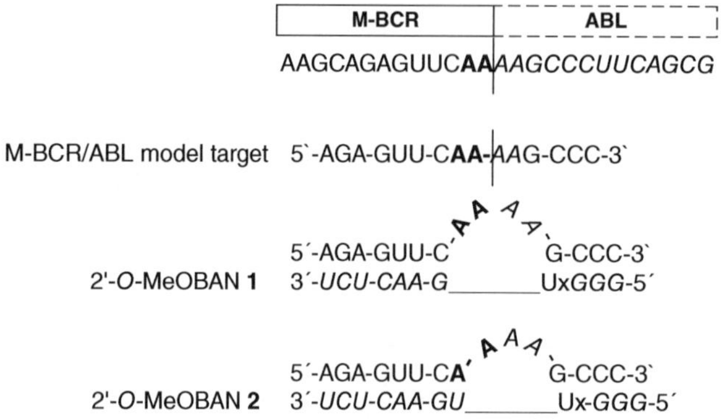

Rate constants and half-lives for cleavage of the M-BCR/ABL mRNA model by 2'-*O*-MeOBANs at a 1:1 ratio (4 μM of each)

	k_{obs} (10^{-6} s^{-1})	$t_{1/2}$ (h)
2'-*O*-MeOBAN **1**	13.8	14
2'-*O*-MeOBAN **2**	22.6	8.5

Figure 3.2.6 *2′-O-MeOBANs that cleave a leukemia related RNA target with turnover [70]*

of Chronic Myeloid Leukemia (CML) [68]. It also has been demonstrated that exogenously delivered chemically synthesized dsRNA (siRNA) directed at the M-BCR/ABL fusion site, is effective in killing leukemic cells [69]. 2′-*O*-MeOBANs directed at cleavage of the M-BCR/ABL RNA have been constructed and these were shown to work with a similar efficiency (Figure 3.2.6) to the earlier developed systems described previously [70]. The parts bulging out were the same but the Watson–Crick base paired stems flanking the bulge where different, showing that the duplex stems can be varied at will. This suggests that, provided a position in the target is chosen so that three or four unpaired nucleotides in the target RNA are positioned at the cleaver, any target RNA can be selected and that OBANs can be tailor-made.

3.2.6 PNAzymes and creation of artificial RNA restriction enzymes

Peptide nucleic acids (PNA) [71] are characterized by an uncharged *N*-(2-aminoethyl)-glycine backbone, which is stable in biological fluids, able to hybridize in a sequence-specific manner to RNA targets, and, largely because of its neutral backbone, give more stable complexes with RNA than DNA or 2′-*O*-MeRNA [72, 73]. PNA is also readily conjugated through peptide-type chemistry, but has been used sparingly as a carrier for RNA cleaving agents. Examples include conjugation with oligoamines [74, 75], neocuproine [76, 77], neamine [78] or a peptide [79] where the cleaver is attached to the *N*-terminal of PNA. In all these examples the PNA conjugate was used in equimolar or excess amounts over the RNA and gave modest cleavage rates with no turnover of the target RNA. With the objective of forming centrally placed bulges in the RNA target, PNA based artificial nucleases that give catalytic cleavage of RNA with turnover of substrate should be obtained. Conjugation of 5-amino-2,9-dimethyl-1,10-phenanthroline (neocuproine) to PNA was made possible by a solid-phase strategy involving a centrally positioned diaminopropionic acid moiety [80]. To have a clear comparison with 2′-*O*-MeOBANs, PNA conjugates targeting the M-BCR/ABL RNA model sequence were constructed [81] and these displayed an RNA cleavage activity (Figure 3.2.7) and selectivity similar to 2′-*O*-MeOBAN systems [29, 70]. These Zn^{2+} based PNAzymes were the first truly catalytic artificial nucleases based on PNA, as excess RNA was also shown to be cleaved with turnover of the substrate [80].

Further development of these stable PNAzymes was obtained by changing the neocuproine bound metal ion from Zn^{2+} to Cu^{2+} [82]. This lead to a remarkable improvement in the cleavage specificity. Single-site cleavage within the RNA bulge was obtained (Figure 3.2.8). In addition, the rate of cleavage was an order of magnitude higher than for the zinc ion based system [82].

Further elaborations of the system by varying the bases in the bulge lead to even higher rates and half-lives that are now in the minute scale (about 30 min), with most bulge sequences being cleaved at a considerably

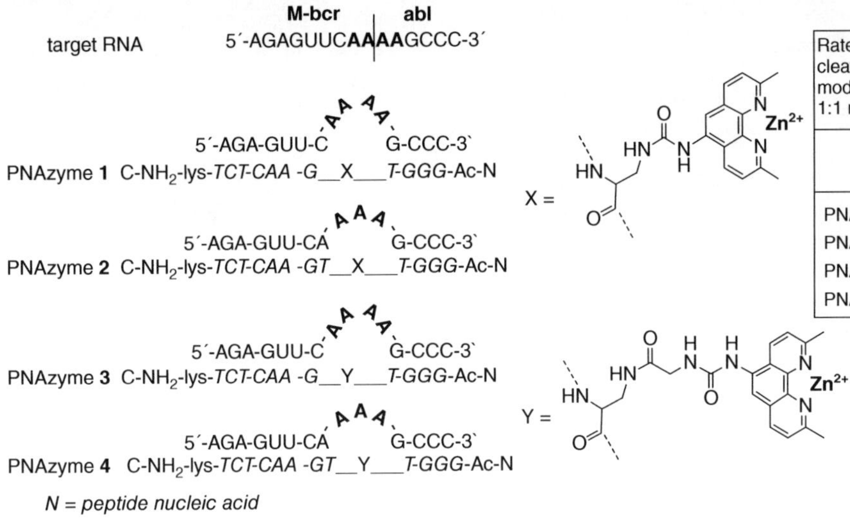

Rate constants and half-lives for cleavage of the M-BCR/ABL RNA model by PNAzymes at a 1:1 ratio (4 μM of each)		
	k_{obs} (10^{-6} s^{-1})	$t_{1/2}$ (h)
PNAzyme **1**	17.6	11
PNAzyme **2**	9.2	21
PNAzyme **3**	16.0	12
PNAzyme **4**	13.1	15

Figure 3.2.7 *The first reported "PNAzymes", that is, peptide nucleic acid based artificial nucleases carrying out cleavage with turnover of RNA substrate [80]*

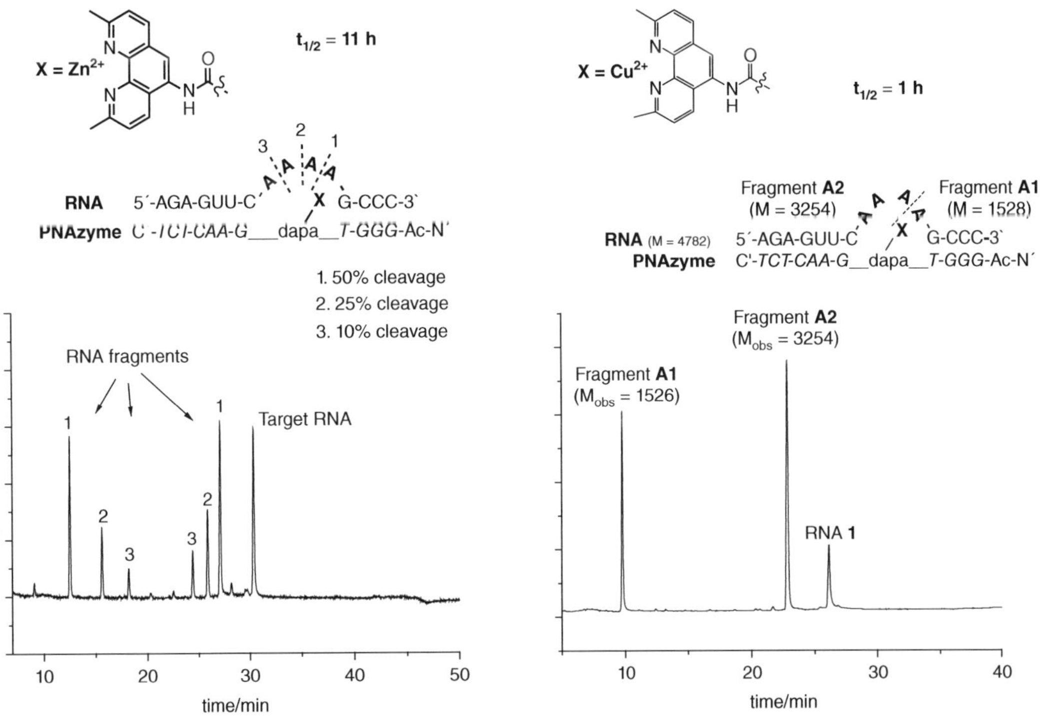

Figure 3.2.8 *Left panel: HPLC analysis of a Zn^{2+}–PNAzyme promoted cleavage of the M-BCR/ABL RNA model, after 22 h at 37 °C (hal-flife = 11 h). Right panel:. HPLC analysis showing selective cleavage of the M-BCR/ABL RNA model with the Cu^{2+}–PNAzyme after 3 h at 37 °C. The cleavage sites are indicated by the dashed lines*

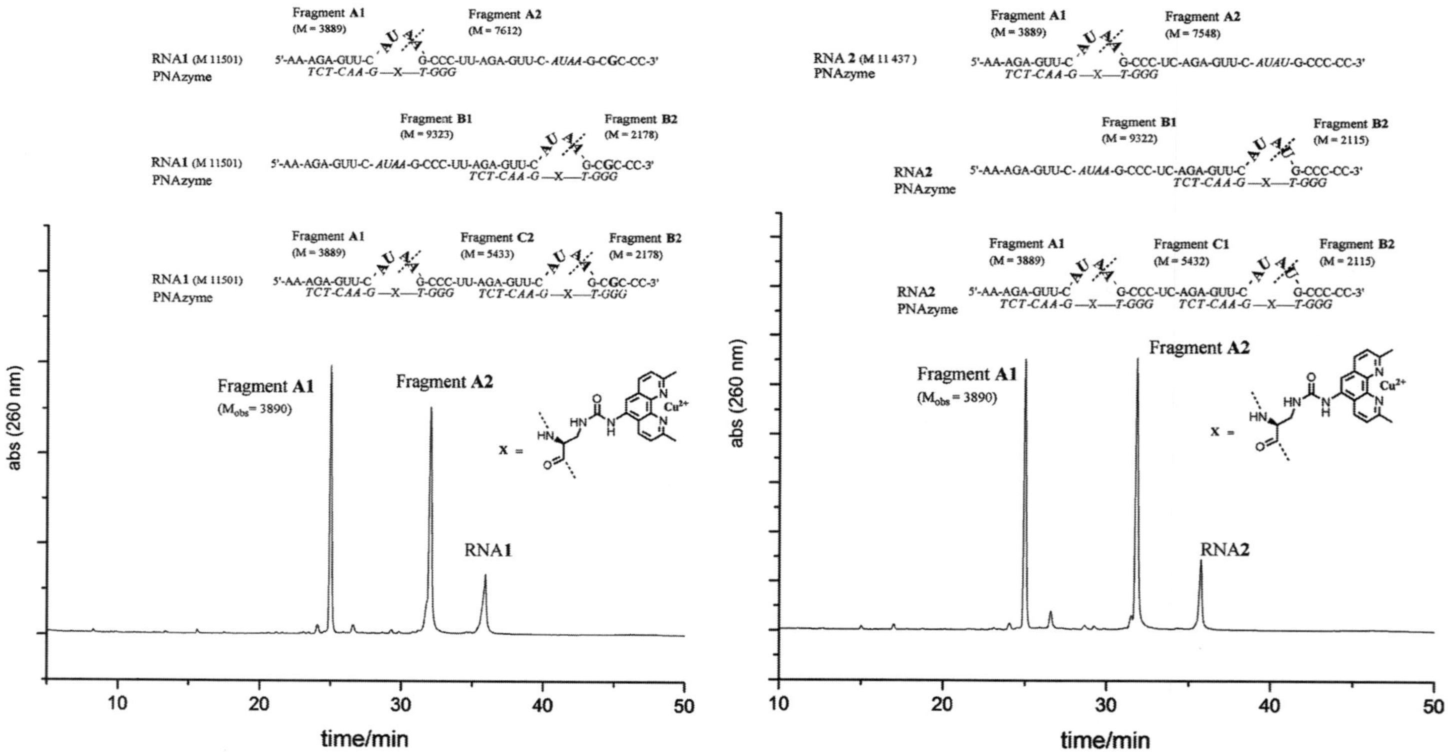

Figure 3.2.9 *Left: HPLC analysis of PNAzyme cleavage of an RNA carrying two sites with the potential of forming identical bulges but with a mismatch in one of the stems. Right: HPLC analysis of PNAzyme cleavage of an RNA carrying two sites with the potential of full Watson–Crick pairing but with different bulges*

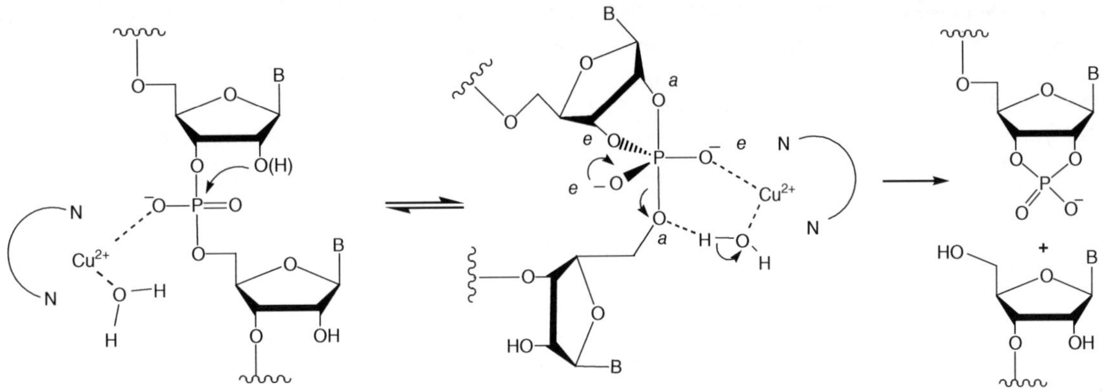

Figure 3.2.10 *Plausible mechanism for Cu^{2+}–PNAzyme catalyzed cleavage of RNA phosphodiester bonds (*a = *apical ligand and* e = *equatorial ligand)*

lower rate. This gives a further kinetic resolution of the sequence selectivity in the bulge in addition to that governed by the hybridization of the duplex stems. These PNAzymes display real catalytic behavior and turnover of RNA substrate, even when used with 100-times excess of the substrate versus the artificial enzyme [82]. A possible explanation for the higher selectivity of the PNAzyme with Cu^{2+} in comparison with the previous Zn^{2+} system could be distortions in the geometry that do not allow the copper ions to reach the other phosphate diester bond, and possible interaction of the copper ion with the nucleobases.

It was also clear that some structural features of the complex were important for efficient cleavage, for example, the closing GT wobble base pair (inherited from the NMR structure of the ribozyme bulge–loop) was crucial, because if replaced with a standard Watson–Crick pair it resulted in a considerable penalty in the rate of RNA cleavage. The sensitivity to the bulge sequence was high, and mismatches in the Watson–Crick paired stems gave large decreases in RNA cleavage rates [82]. When targetign longer RNA targets with two potential cleavage sites where one had a mismatch in one stem, cleavage occurred only at the site without the mismatch (Figure 3.2.9).

With another RNA substrate with two possible sites that both had full Watson–Crick complementarity in the stems but different bulge sequences, cleavage occurred only at the site that was found to be more readily cleaved in the study of bulge sequence dependence (Figure 3.2.9). Thus, the Cu^{2+}–PNAzymes that have been developed gave a high degree of discrimination for the target RNA sequence [82]. The cleavage mechanism has not been studied in detail and thus is unknown. However, a reasonable hypothesis is that water bound to the acidic copper ion acts as an acid catalyst to enhance departure of the leaving group (Figure 3.2.10), in a similar fashion to what has been suggested for both the pH independent buffer catalyzed cleavage [83] and divalent metal aquo ion catalyzed cleavage [84] of nucleoside alkylphosphates. Together with the excellent turnover, these PNAzymes can be considered to be highly specific artificial RNA restriction endonucleases that can be tailor-made to fit any desired sequence and can become useful tools in biotechnology and biomedicine.

3.2.7 Conclusions

Oligonucleotide based artificial nuclease systems have been developed over the last three decades. Progress has been slow and most systems have been quite inefficient and often just act somewhat as selective cleavers and not as enzymes. However, some systems have displayed real catalytic behavior and turnover, and the recently developed PNAzymes give single-site cleavage and excellent mismatch rejection, which make them the first RNA restriction enzymes. When bound to the substrate, RNA cleavage half-lives are reduced to the minute scale. The sequence dependence in both bulge and base-paired stems means that these are tailor-made RNases and thus should now be usable tools in molecular biology. Hence, it may seem as if these OBANs/PNAzymes are not that far from reaching a stage when they can be used for gene silencing. However, it is probable that for this, a higher rate of RNA cleavage is needed. It is a difficult task to enhance the rate further and it is possible that it will be necessary to position active groups more accurately and perhaps combine several effectors of catalysis. This would be made more likely by obtaining detailed NMR or crystal structures of the OBAN/PNAzyme–RNA complexes that could serve as starting points for further refinement. The currently most efficient and specific systems that utilize the Cu^{2+} ion may not be a toxicity problem if the required concentrations are low enough. However, the complex constant for metal ion binding probably needs to be higher than in the neocuproine systems, in order to keep the metal from diffusing away. Non-metal ion based systems, or those based on tight metal complexes, do not suffer from this, but have so far been substantially less effective and cleave with lower selectivity. The development of OBANs/PNAzymes is, nevertheless, getting closer to being attempted for suppression of gene expression and hopefully studies in this direction will appear in the not too distant future.

References

[1] R. Häner and J. Hall, The sequence-specific cleavage of RNA by artificial chemical nucleases, *AntisenseNucl. Acid Drug Dev.*, **7**, 423–430 (1997).
[2] M. Oivanen, S. Kuusela, and H. Lönnberg, Kinetics and mechanisms for the cleavage and isomerization of the phosphodiester bonds of RNA by Brønsted acids and bases, *Chem. Rev.*, **98**, 961–990 (1998).
[3] B. N. Trawick, A. T. Daniher, and J. K. Bashkin, Inorganic mimics of ribonucleases and ribozymes: From random cleavage to sequence-specific chemistry to catalytic antisense drugs, *Chem. Rev.*, **98**, 939–960 (1998).
[4] T. Niittymäki and H. Lönnberg, Artificial ribonucleases, *Org. Biomol. Chem.*, **4**, 15–25 (2006).
[5] A. Kuzuya and M. Komiyama, Site-selective artificial ribonucleases, *Curr. Org. Chem.*, **11**, 1450–1459 (2007).
[6] J. R. Morrow, Speed limits for artificial ribonucleases, *Commun. Inorg. Chem.*, **29**, 169–188 (2008).
[7] F. Mancin, P. Scrimin, and P. Tecilla, Progress in artificial metallonucleases, *Chem. Commun.*, 5545–5559 (2012).
[8] Y. Aiba, J. Sumaoka, and M. Komiyama, Artificial DNA cutters for DNA manipulation and genome engineering, *Chem. Soc. Rev.*, **40**, 5657–5668 (2011).
[9] C. L. Ramirez and J. K. Joung, Engineered zinc finger nucleases for targeted genome editing. *Top. Curr. Genet.*, **23**, 121–145 (2013).
[10] Y. Dorsett and T. Tuschl, siRNAs: applications in functional genomics and potential as therapeutics, *Nature Rev. Drug Disc.*, **3**, 318–329 (2004).
[11] M. Manoharan, RNA interference and chemically modified small interfering RNAs, *Curr. Opin. Chem. Biol.*, **8**, 570–579 (2004).
[12] S. Bakhtiyari, K. Haghani, G. Basati, and M. H. Karimfar, siRNA therapeutics in the treatment of diseases, *Ther. Delivery*, **4**, 45–57 (2013).
[13] S. T. Crooke, Progress in antisense technology, *Annu. Rev. Med.*, **55**, 61–95 (2004).
[14] A. Matsuda, Oligonucleotide therapeutics, *CSJ Curr. Rev.*, **6**, 166–172 (2011).
[15] R. Z. Yu, J. S. Grundy, and R. S. Geary, Clinical pharmacokinetics of second generation antisense oligonucleotides, *Expert Opin. Drug Metab. Toxicol.*, **9**, 169–182 (2013).
[16] V. K. Sharma, R. Kumar, P. Rungta, V. S. Parmar, and A. K. Prasad, Modified oligonucleotides: strides towards antisense drugs, *Trends Carbohydr. Res.*, **5**, 1–7 (2013).
[17] P. Hair, F. Cameron, and K. McKeage, Mipomersen sodium: first global approval, *Drugs*, **73**, 487–93 (2013).
[18] S. M. Hammond and M. J. A. Wood, Genetic therapies for RNA mis-splicing diseases, *Trends Genet.*, **27**, 196–205 (2011).
[19] I. Eperon, Antisense therapeutics. New ways to nudge splicing, *Nature Chem. Biol.*, **8**, 507–508 (2012).

[20] S. L. DeVos and T. M. Miller, Antisense oligonucleotides: Treating neurodegeneration at the level of RNA, *Neurotherapeutics*, **10**, 486–497 (2013).

[21] S. A. Farr, M. A. Erickson, W. A. Banks, M. L. Niehoff, and J. E. Morley, Central and peripheral administration of antisense oligonucleotide targeting amyloid-β protein precursor improves learning and memory and reduces neuroinflammatory cytokines in Tg2576 (AβPPswe) mice, *J. Alzheimer's Dis.*, **40**(4), 1005–1016 (2014).

[22] X.-H. Lu and X. W. Yang, "Huntingtin Holiday": Progress toward an antisense therapy for Huntington's disease, *Neuron*, **74**, 964–966 (2012).

[23] M. Sass and N. Aronin, RNA- and DNA-based therapies for Huntington's disease, *Neurobiol. Huntington's Dis.*, 225–253 (2011).

[24] J. Wan, J. A. Bauman, M. A. Graziewicz, P. Sazani, and R. Kole, Oligonucleotide therapeutics in cancer, *RNA Cancer*, 213–233 (2013).

[25] M. Ruden and N. Puri, Novel anticancer therapeutics targeting telomerase, *Cancer Treat. Rev.*, **39**, 444–456 (2013).

[26] T. Koo and M. J. Wood, Clinical trials using antisense oligonucleotides in duchenne muscular dystrophy, *Hum. Gene Ther.*, **24**, 479–488 (2013).

[27] O. Danos, Antisense oligonucleotides for therapeutic interventions in neuromuscular diseases, *Hum. Gene Ther.*, **24**, 470–471 (2013).

[28] J. Li, S. Tan, R. Kooger, C. Zhang, and Y. Zhang, MicroRNAs as novel biological targets for detection and regulation, *Chem. Soc. Rev.*, **43**, 506–517 (2014).

[29] H. Åström, N. H. Williams, and R. Strömberg, Oligonucleotide based artificial nuclease (OBAN) systems. Bulge size dependence and positioning of catalytic group in cleavage of RNA–bulges, *Org. Biomol. Chem.*, **1**(9), 1461–1465 (2003).

[30] U. Kaukinen, T. Venäläinen, H. Lönnberg, and M. Peräkylä, The base sequence dependent flexibility of linear single-stranded oligoribonucleotides correlates with the reactivity of the phosphodiester bond, *Org. Biomol. Chem.*, **1**, 2439–2447 (2003).

[31] D. Huesken, G. Goodall, M. J. J. Blommers, W. Jahnke, J. Hall, R. Häner, and H. E. Moser, Creating RNA bulges: Cleavage of RNA in RNA/DNA duplexes by metal ion catalysis, *Biochemistry*, **35**(51), 16591–16600 (1996).

[32] U. Kaukinen, I. Bielecki, S. Mikkola, R. W. Adamiak, and H. Lönnberg, The cleavage of phosphodiester bonds within small RNA bulges in the presence and absence of metal ion catalysts, *J. Chem. Soc., Perkin Trans.* **2**, (7), 1024–1031 (2001).

[33] K. Yoshinari, K. Yamazaki, and M. Komiyama, Oligoamines as simple and efficient catalysts for RNA hydrolysis, *J. Am. Chem. Soc.*, **113**, 5899–901 (1991).

[34] M. Komiyama and T. Inokawa, Selective hydrolysis of tRNA by ethylenediamine bound to a DNA oligomer, *J. Biochem.*, **116**, 719–20 (1994).

[35] A. S. Modak, J. K. Gard, M. C. Merriman, K. A. Winkeler, J. K. Bashkin, and M. K. Stern, Toward chemical ribonucleases. 2. Synthesis and characterization of nucleoside-bipyridine conjugates. Hydrolytic cleavage of RNA by their copper(II) complexes, *J. Am. Chem. Soc.*, **113**, 283–291 (1991).

[36] W. C. Putnam and J. K. Bashkin, De novo synthesis of artificial ribonucleases with benign metal catalysts, *Chem. Comm.*, 767–768 (2000).

[37] D. A. Konevetz, I. E. Beck, N. G. Beloglazova, I. V. Sulimenkov, V. N. Sil'nikov, M. A. Zenkova, G. V. Shishkin, and V. V. Vlassov, Artificial ribonucleases: synthesis and RNA cleaving properties of cationic conjugates bearing imidazole residues, *Tetrahedron*, **55**, 503–512 (1999).

[38] V. V. Vlassov, T. Abramova, T. Godovikova, R. Giege, and V. Silnikov, Sequence specific clevage of yeast tRNAPhe with oligonucleotides conjugated to a diimidazole construct, *Antisense Nucl. Acid Drug Dev.*, **7**, 39–42 (1997).

[39] J. Hall, D. Huesken, U. Pieles, H. E. Moser, and R. Häner, Efficient sequence–specific cleavage of RNA using novel europium complexes conjugated to oligonucleotides, *Chem. Biol.*, **1**, 185–190 (1994).

[40] D. Magda, R. A. Miller, J. L. Sessler, and B. L. Iverson, Site-specific hydrolysis of RNA by europium(III) texaphyrin conjugated to a synthetic oligodeoxyribonucleotide, *J. Am. Chem. Soc.*, **116**, 7439–7440 (1994).

[41] D. Magda, S. Crofts, A. Lin, D. Miles, M. Wright, and J. L. Sessler, Synthesis and kinetic properties of ribozyme analogs prepared using phosphoramidite derivatives of dysprosium(III) texaphyrin, *J. Am. Chem. Soc.*, **119**, 2293–2294 (1997).

[42] D. Magda, M. Wright, S. Crofts, A. Lin, and J. L. Sessler, Metal complex conjugate of antisense DNA which displays ribozyme-like activity, *J. Am. Chem. Soc.*, **119**, 6947–6948 (1997).

[43] K. A. Kolasa, J. R. Morrow, and A. P. Sharma, Trivalent lanthanide ions do not cleave RNA in DNA-RNA hybrids, *Inorg. Chem.*, **32**, 3983–3984 (1993).

[44] S. Mikkola, U. Kaukinen, and H. Lönnberg, The effect of secondary structure on cleavage of the phosphodiester bonds of RNA, *Cell Biochem. Biophys.*, **34**, 95–118 (2001).

[45] D. Huesken, G. Goodall, M. J. J. Blommers, W. Jahnke, J. Hall, R. Haener, and H. E. Moser, Creating RNA bulges: Cleavage of RNA in RNA/DNA duplexes by metal ion catalysis, *Biochemistry*, **35**, 16591–16600 (1996).

[46] J. Hall, D. Huesken, and R. Häner, Towards artificial ribonucleases: the sequence-specific cleavage of RNA in a duplex, *Nucl. Acids Res.*, **24**, 3522–3526 (1996).

[47] R. Häner, J. Hall, A. Pfuezer, and D. Huesken, Development of artificial ribonucleases, *Pure Appl. Chem.*, **70**, 111–116 (1998).

[48] A. Kuzuya, K. Machida, R. Mizoguchi, and M. Komiyama, Conjugation of various acridines to DNA for site-selective RNA scission by lanthanide ion, *Bioconjugate Chem.*, **13**, 365–369 (2002).

[49] A. Kuzuya, R. Mizoguchi, F. Morisawa, K. Machida, and M. Komiyama, Metal ion-induced site-selective RNA hydrolysis by use of acridine-bearing oligonucleotide as cofactor, *J. Am. Chem. Soc.*, **124**, 6887–6894 (2002).

[50] S. Matsuda, A. Ishikubo, A. Kuzuya, M. Yashiro, and M. Komiyama, Conjugates of a dinuclear zinc(II) complex and DNA oligomers as novel sequence-selective artificial ribonucleases, *Angew. Chem., Int. Ed.*, **37**, 3284–3286 (1998).

[51] S. Sakamoto, T. Tamura, T. Furukawa, Y. Komatsu, E. Ohtsuka, M. Kitamura, and H. Inoue, Highly efficient catalytic RNA cleavage by the cooperative action of two Cu(II) complexes embodied within an antisense oligonucleotide, *Nucl. Acids Res.*, **31**, 1416–1425 (2003).

[52] T. Niittymäki, P. Virta, K. Ketomäki, and H. Lönnberg, Di(aza-crown) conjugates of 2′-O-methyl oligoribonucleotides as sequence-selective artificial ribonucleases, *Bioconjugate Chem.*, **18**(5), 1583–1592 (2007).

[53] L. A. Jenkins Autry and J. K. Bashkin, Transesterification of RNA by Cu(II) terpyridine. *Inorg. Chim. Acta*, **263**, 49–52 (1997).

[54] T. Niittymäki and H. Lönnberg, Sequence-selective cleavage of oligoribonucleotides by 3d transition metal complexes of 1,5,9-triazacyclododecane-functionalized 2′-O-methyl oligoribonucleotides, *Bioconjugate Chem.*, **15**(6), 1275–1280 (2004).

[55] M. Kalek, P. Benediktson, B. Vester, and J. Wengel, Identification of efficient and sequence specific bimolecular artificial ribonucleases by a combinatorial approach, *Chem. Commun.*, 762–764 (2008).

[56] D. A. Ossipov and R. Strömberg, Studies in oligonucleotide-based artificial nuclease systems. Intramolecular copper(II) complexe formation in an oligonucleotide bis-phenanthroline conjugate, *Nucleosides, Nucleotides, Nucleic Acids*, **24**, 901–905 (2005).

[57] M. A. Reynolds, T. A. Beck, P. B. Say, D. A. Schwartz, B. P. Dwyer, W. J. Daily, M. M. Vaghefi, M. D. Metzler, R. E. Klem, and L. J. Arnold Jr., Antisense oligonucelotides containing an internal, non-nucleotide-based linker promotes site-specific cleavage of RNA, *Nucl. Acids. Res.*, **24**, 760–765 (1996).

[58] N. G. Beloglazova, M. M. Fabani, M. A. Zenkova, E. V. Bichenkova, N. N. Polushin, V. V. Sil'nikov, K. T. Douglas, V. V. Vlassov, Sequence-specific artificial ribonucleases. I. Bis-imidazole-containing oligonucleotide conjugates prepared using precursor-based strategy, *Nucl. Acids Res.*, **32**, 3887–3897 (2004).

[59] N. L. Mironova, D. V. Pyshnyi, E. M. Ivanova, M. A. Zenkova, H. J. Gross, and V. V. Vlassov, Covalently attached oligodeoxyribonucleotides induce RNase activity of a short peptide and modulate its base specificity, *Nucl. Acids Res.*, **32**, 1928–1936 (2004).

[60] N. L. Mironova, D. V. Pyshnyi, D. V. Shtadler, A. A. Fedorova, V. V. Vlassov, and M. A. Zenkova, RNase T1 mimicking artificial ribonuclease, *Nucl. Acids Res.*, **35**, 2356–2367 (2007).

[61] U. Scheffer, A. Strick, V. Ludwig, S. Peter, E. Kalden, and M. W. Gobel, Metal-free catalysts for the hydrolysis of RNA derived from guanidines, 2-aminopyridines, and 2-aminobenzimidazoles, *J. Am. Chem. Soc.*, **127**, 2211–2217 (2005).

[62] C. Gnaccarini, S. Peter, U. Scheffer, S. Vonhoff, S. Klussmann, and M. W. Gobel, Site-specific cleavage of RNA by a metal-free artificial nuclease attached to antisense oligonucleotides, *J. Am. Chem. Soc.*, **128**, 8063–8067 (2006).

[63] W. C. Putnam, A. T. Daniher, B. N. Trawick, and J. K. Bashkin, Efficient new ribozyme mimics: direct mapping of molecular design principles from small molecules to macromolecular, biomimetic catalysts, *Nucl. Acids Res.*, **29**, 2199–2204 (2001).
[64] B. Linkletter and J. Chin, Rapid hydrolysis of RNA with a Cu^{II} complex, *Angew. Chem., Int. Ed.*, **34**, 472–474 (1995).
[65] H. Åström and R. Strömberg, A method for synthesis of an artificial ribonuclease, *Nucleosides, Nucleotides Nucleic Acids*, **20**(4-7), 1385–1388 (2001).
[66] K. J. Luebke, S. M. Landry, and I. Tinoco, Jr., Solution conformation of a five-nucleotide RNA bulge loop from a Group I intron, *Biochemistry*, **36**(33), 10246–10255 (1997).
[67] H. Åström and R. Strömberg, Synthesis of new OBAN's and further studies on positioning of the catalytic group, *Org. Biomol. Chem.*, **2**, 1901–1907 (2004).
[68] J. D. Rowley, A new consistent chromosomal abnormality in chronic myelogenous leukaemia identified by quinacrine fluorescence and giemsa staining, *Nature*, **243**, 290 (1973).
[69] M. Wilda, U. Fuchs, W. Wossmann, and A. Borkhardt, Killing of leukemic cells with a BCR/ABL fusion gene by RNA interference (RNAi), *Oncogene*, **21**, 5716 (2002).
[70] M. Murtola and R. Strömberg, 2′-O-Methyloligoribonucleotide based artificial nucleases (2′-O-MeOBANs) cleaving a model of the leukemia related M-BCR/ABL m-RNA, *ARKIVOC*, 84–94 (2009).
[71] P. E. Nielsen, M. Egholm, R. H. Berg, and O. Buchardt, Sequence-selective recognition of DNA by strand displacement with a thymine-substituted polyamide, *Science*, **254**, 1497–1500 (1991).
[72] K. E. Lundin, L. Good, R. Strömberg, A. Gräslund, and C. I. E. Smith, Biological activity and biotechnological aspects of peptide nucleic acid, *Adv. Genet.*, **56**, 1–51 (2006).
[73] P. E. Nielsen and M. Egholm, *An Introduction to PNA*, Horizon Bioscience, pp. 1–36 (2004).
[74] J. C. Verheijen, B. A. L. M. Deiman, E. Yeheskiely, G. A. van der Marel, and J. H. van Boom, Efficient hydrolysis of RNA by a PNA-diethylenetriamine adduct, *Angew. Chem., Int. Ed.*, **39**(2), 369–372 (2000).
[75] L. Petersen, M. C. de Koning, P. van Kuik-Romeijn, J. Weterings, J. C. Pol, G. Platenburg, M. Overhand, G. A. van der Marel, and J. H. van Boom, Synthesis and *in vitro* evaluation of PNA-peptide-DETA conjugates as potential cell penetrating artificial ribonucleases, *Bioconj. Chem.*, **15**, 576–582 (2004).
[76] A. Whitney, G. Gavory, and S. Balasubramanian, Site-specific cleavage of human teleomerase RNA using PNA-neocuproine-Zn(II) derivatives, *Chem. Commun.*, 36–37 (2003).
[77] M. Murtola, D. Ossipov, J. Sandbrink, and R. Stroemberg, RNA Cleavage by 2,9-diamino-1,10-phenanthroline PNA conjugates, *Nucleosides, Nucleotides, Nucleic Acids*, **26**, 1479–1483 (2007).
[78] E. Riguet, S. Tripathi, B. Chaubey, J. Desire, V. N. Pandey, and J.-L. Decout, A peptide nucleic acid-neamine conjugate that targets and cleaves HIV-1 TAR RNA inhibits viral replication, *J. Med. Chem.*, **47**, 4806–4809 (2004).
[79] M. Gaglione, G. Milano, A. Chambery, L. Moggio, A. Romanelli, and A. Messere, PNA-based artificial nucleases as antisense and anti-miRNA oligonucleotide agents, *Mol. BioSyst.*, **7**, 2490–2499 (2011).
[80] J. Sandbrink, M. Murtola, and R. Strömberg, Solid-support conjugation of amino acids and a phenanthroline derivative to a central position in peptide nucleic acids, *Nucleosides, Nucleotides, Nucleic Acids*, **26**, 1485–1489 (2007).
[81] M. Murtola and R. Strömberg, PNA based artificial nucleases displaying catalysis with turnover in the cleavage of a leukemia related RNA model, *Org. Biomol. Chem.*, **6**(20), 3837–3842 (2008).
[82] M. Murtola, M. Wenska, and R. Strömberg, PNAzymes that are artificial RNA restriction enzymes, *J. Am. Chem. Soc.*, **132**, 8984–8990 (2010).
[83] M. Kosonen, E. Yousefi-Salakdeh, R. Strömberg, and H. Lönnberg, pH- and buffer-independent cleavage and mutual isomerization of uridine 2′- and 3′-alkylphosphodiesters: Implications for the buffer catalyzed cleavage of RNA, *J. Chem. Soc. Perkin Trans* **2**, 1589–1596 (1998).
[84] S. Mikkola, E. Stenman, K. Nurmi, E. Yousefi-Salakdeh, R. Strömberg, and H. Lönnberg, The mechanism of the metal ion promoted cleavage of RNA phosphodiester bonds involves a general acid catalysis by the metal aquo ion on the departure of the leaving group, *J. Chem. Soc. Perkin Trans* **2**, 1619–1625 (1999).

3.3

Exploring Nucleic Acid Conformations by Employment of Porphyrin Non-covalent and Covalent Probes and Chiroptical Analysis

Alessandro D'Urso[a], Ana G. Petrovic[b,c], Maria Elena Fragalà[a], Manuel A. Tamargo[b], George A. Ellestad[b], Roberto Purrello[a], and Nina Berova[b]

[a] *Department of Science, University of Catania, Catania, Sicily, Italy*
[b] *Department of Chemistry, Columbia University, Havemeyer Hall, New York, NY, USA*
[c] *Department of Life Sciences, New York Institute of Technology (NYIT), New York, NY, USA*

3.3.1 Introduction

DNA mainly forms double helical structures, with oligonucleotide strands arranged in antiparallel orientation. Depending on the sequence of the bases and the environmental conditions, DNA can adopt various duplex conformations: from the canonical right-handed B-form, to the compact A-form, and the unusual left-handed Z-form (Figure 3.3.1). Additionally, Watson–Crick hydrogen bonds can form alternative base pairs (Hoogsteen, reverse Hoogsteen, etc.) that allow DNA to exist in various secondary structures (triplex, quadruplex, i-motif, …) (Figure 3.3.1).

Some of these structures might be responsible for the regulation of genes [1]. It is well known that, depending on the sequences and biological roles, RNA can adopt single or double helical structures, as well as several conformations [2]. Identification of the different structures and specific sequences of the nucleic acids, using functional probes, might allow one to know the mechanism of some diseases and the most suitable treatments. Moreover, this versatility of DNA can be exploited to design addressable DNA logic gates, which are at the center of significant research efforts to construct DNA computations [3].

DNA in Supramolecular Chemistry and Nanotechnology, First Edition.
Edited by Eugen Stulz and Guido H. Clever.

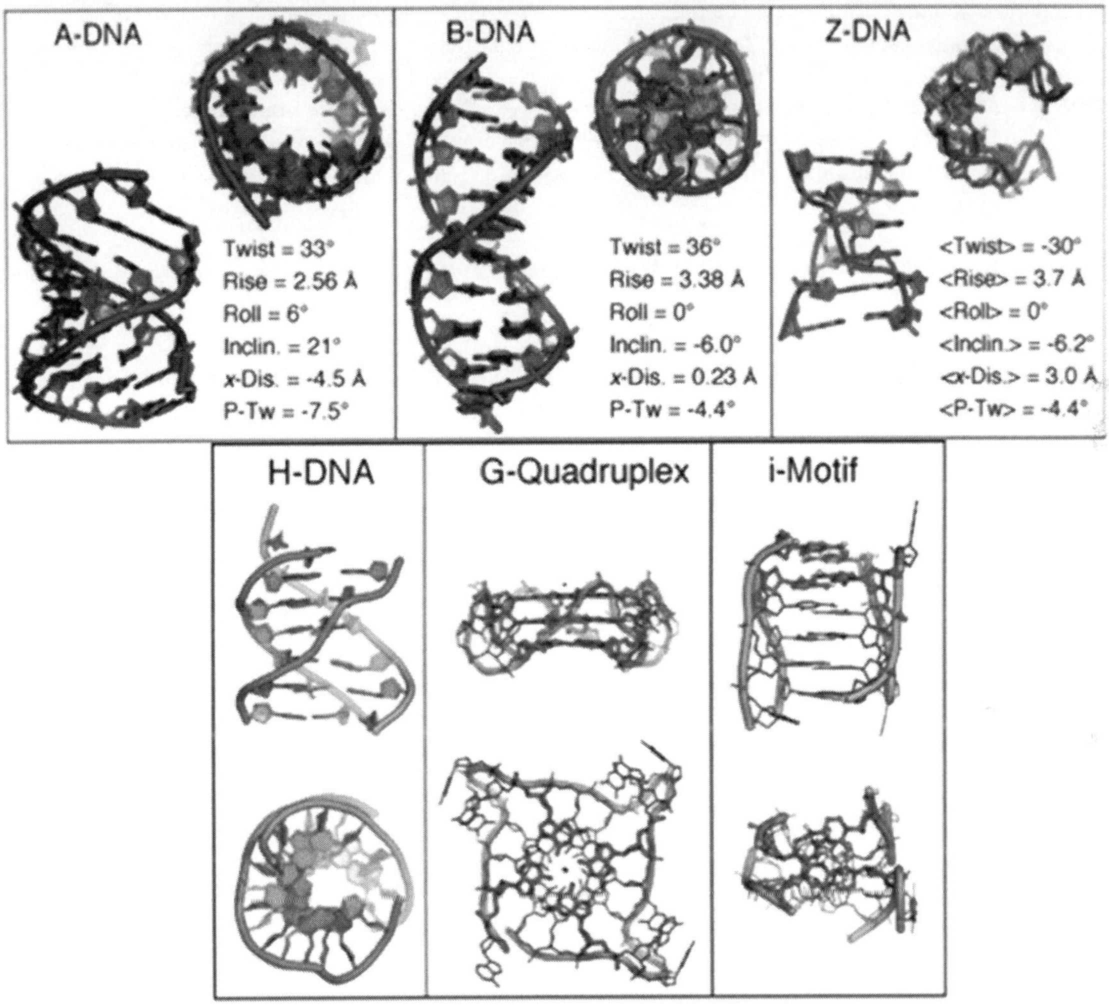

Figure 3.3.1 *Different conformations (A, B, and Z) and several secondary structures (triplex, tetraplex, and i-motif) adopted for DNA*

The past two decades have seen a number of studies in which the free base or metallated porphyrins have been covalently and non-covalently incorporated into a DNA scaffold, as a consequence of their unique photochemical and photophysical properties. The type and the location of a covalent linker, degree of porphyrin hydrophobicity, along with environmental factors such as ionic strength and temperature, and porphyrin to DNA ratio, all play vital roles in modulating porphyrin–DNA interactions. Thus, the objectives of such studies have been to examine the effect of the porphyrins on DNA for their use as spectroscopic sensors in conformational studies of DNA, and on their stability and electronic properties using circular dichroism (CD), UV-melting, UV–Vis, and fluorescence spectroscopy [4]. CD measurements are, in general, more specific because interactions of these macrocycles with polynucleotides (which are conformationally chiral) induce a dichroic band (ICD) in the absorption region of the achiral ligands. The shape and intensity of ICD is a diagnostic of the type of interactions. These findings are important for the development of porphyrin–DNA constructs, which potentially have photochemical and physicochemical properties that would be useful in nanotechnology.

Examples presented in this section include an initial part exploring the non-covalent interaction of porphyrins with single- double- (both B- and Z-conformations) and tetra-helixes of DNA, then a second part where the DNA is 5′- and 3′-end-labelled with a porphyrin attached via a phosphate or an amide linker, or it is attached directly to a nucleobase or to the phosphate groups. We have selected examples from each type of these arrays to provide the reader with an overview of this rapidly expanding area of investigation with the application of electronic CD as a diagnostic tool for monitoring the conformational properties of these DNA constructs.

3.3.2 Non-covalent interaction of porphyrin–DNA complexes

3.3.2.1 Interaction with single-stranded DNA

In spite of the large number of studies on the interaction of porphyrinoids with duplex DNA, only a few works have appeared in the literature reporting the interaction with single-stranded nucleic acids. Single-stranded segments of DNA play important biological roles as they are involved in both replication and transcription processes. Single-stranded homopolymer nucleic acids are not just synthetic model systems, they are also found in nature as stretches of polyadenilic acid (polyA), of about 200 nucleotides long. These are linked covalently to the 3′-end of polydisperse nuclear RNA, messenger RNA of animal cells, and numerous animal viruses [5, 6]. In eukaryotes, polyadenylation is part of the process that produces mature messenger RNA (mRNA) for translation; the function of the 3′-tail is probably protective, in fact mRNAs without a 3′-tail are rapidly degraded. The polycytidine (C)(n) repeats (polyC) have been found within the D-loop region of the mitochondrial DNA [7]. Somatic mutations at a mitochondrial non-coding polycytidine (C)(n) repeat have been associated with tumor progression [8]. Thus, studies on the binding of porphyrinoids with single-stranded nucleic acids have been used to identify specific probes for different sequences and conformations. The work of Pasternack and coworkers with poly(dA), poly(A), and poly(C) with several metallo- and non-metallo-porphyrins led to promising probes for these single-stranded homopolymers. CuTMPyP4 in the presence of poly(dA) and poly(A) shows a large conservative CD feature in the Soret region that is dependent on the porphyrin/poly(dA) ratio and the ionic strength, which is not observed for the DNA duplex, suggesting that CuTMPyP4 is a reporter molecule for single-stranded DNA (Figure 3.3.2). A large conservative CD

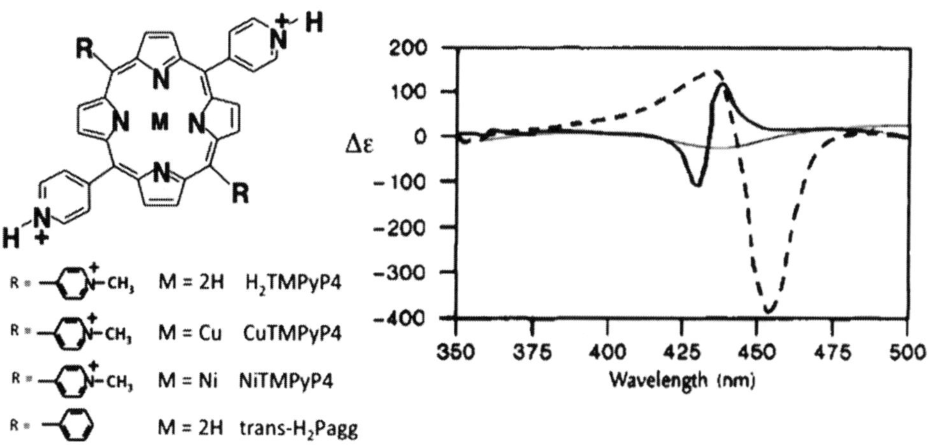

Figure 3.3.2 *Molecular structures of cationic porphyrins: induced CD spectra of CuTMPyP4 3.44 μM (continuous black line), NiTMPyP4 3.44 μM (grey line), and trans-H_2Pagg 3.44 μM (dashed line) in the presence of 86.0 mM poly(dA) (pH 7.0,* m *= 0.20 M, 25 °C). Modified from reference [13]*

signal is characteristic of porphyrin–porphyrin self-stacking on a chiral matrix [9]. This binding mode of CuTMPyP4 with single-stranded DNA has also been confirmed by resonance light scattering (RLS) data, which show an intense band in the porphyrin absorption region [10]. In the presence of poly(C), CuTMPyP4 produces a small negative CD spectrum, similar to that observed for double-stranded DNA, which is consistent with a model involving monodispersed CuTMPyP4 oriented parallel to adjacent bases, as in a pseudo-intercalation event [11]. CD and RLS techniques have been used to identify H_2TMPyP4 as an effective probe to distinguish the A-form of polynucleotides, poly(A), from the B-form, poly(dA); in fact, this porphyrin forms extended aggregates on single-stranded poly(A), which produces a pseudo-intercalation in the presence of the poly(dA). No spectroscopic changes are observed for H_2TMPyP4 in the presence of poly(C). t-H_2Pagg aggregates extensively in the presence of duplex DNA [12], as well as with both poly(dA) and poly(A), leading to enormous CD features (Figure 3.3.2) [13]. On increasing the salt concentration in the latter system, the induced circular dichroism spectrum in the Soret region is reversed, thus indicating a conformational change of the porphyrin assembly [10].

As to the other porphyrins that have been studied (AuTMPyP4, NiTMPyP4, PtTMPyP4, and ZnTMPyP4) an intense hypochromic effect of the Soret absorption band is observed in the presence of the poly(dA), most probably due to porphyrin–adenine interactions arising from pseudointercalative events [11]. Only for NiTMPyP4 and PtTMPyP4 are very small negative ICD signals observed, consistent with the porphyrin chromophore being oriented parallel to the adjacent adenine bases (Figure 3.3.2). It is possible that even five-coordinate ZnTMPyP4 is able to form such a complex using the distal side of the porphyrin plane, suggesting that the steric restrictions of a duplex are partially relaxed for a single-stranded polymer. Accordingly, a negative ICD signal in the porphyrin absorption region is observed for ZnTMPyP4 in the presence of poly(A). Such a feature is not observed in the presence of other single or double helices, thus suggesting ZnTMPyP4 as a specific probe for single-stranded poly(A) stretches.

The spectral differences observed between poly(A) and poly(C) systems can be rationalized by invoking the greater extension of the purine π-system leading to more effective stacking interactions. However, such an explanation is not valid to explain the differences observed between porphyrin interactions with poly(A) and poly(dA) (e.g., the greater bathochromic shift and the presence of negative CD signals for H_2TMPyP4 and ZnTMPyP4-poly(A) complexes are not observed for the analogous poly(dA) system). These unusual aspects are likely a manifestation of the structure and properties of the polyribopurine such as, for example, the smaller axial rise per residue of the poly(A) helix (A-family, pitch 3.0 Å) [14] versus the poly(dA) helix (B-family, pitch 3.3 Å) [15, 16]. The smaller pitch of the poly(A) helix could permit closer and stronger contacts between adjacent π-systems [17].

The concept that the steric restrictions of a duplex are partially relaxed for a single-stranded polymer is used by Slama-Schwork and Lehn [18] to explain the selective binding of the porphyrin derivative used with single-stranded DNA. They synthesized a macrotetracyclic cryptand molecule containing two diaza[19]crown-6 ethers linked by a biphenyl bridge and attached to the porphyrin macrocycle in the β-position, to design a probe with high affinity for single-stranded DNA. Thermal denaturation experiments showed that this porphyrin derivative binds efficiently to denaturized DNA and does not stabilize the double-helical structure. As porphyrin derivatives, porphyrinoids are also employed as probes for single stranded nucleic acids. Very recently, Purrello and coworkers characterized the binding of tri-*N*-methylpyridyl corrole (TMPyC) and the Ge(III) derivative (GeTMPyC) with single-stranded homopolymers poly(A) and poly(C) by using a multi-technique approach (CD, UV, RLS, and fluorescence measurements) [19]. It is worth noting that this approach allows H aggregates of TMPyC in the presence of poly(A) and J aggregates in the presence of poly(C) to be distinguished, as well as the identification of the formation of GeTMPyC dimers in the presence of single-stranded poly(A) and pseudo-intercalation with single-stranded poly(C). GeTMPyC is, in fact, penta-coordinated due to the presence of a weakly bound axial-OH group [20, 21]. The facile formation of μ-oxo corrole dimers [22, 23] supports the proposed dimerization in the presence of poly(A).

3.3.2.2 Double helix conformations B- and Z-DNA

The non-covalent interaction between porphyrins and double-stranded DNA has been widely investigated since 1945, with pioneering work performed by Gupta and coworkers [24]. This led to a large amount of work appearing in the literature on porphyrin–DNA non-covalent complexes. A crude generalization from all these studies established that porphyrins bind to DNA in at least three ways [25]. One binding mode is the intercalation of porphyrin into the base pairs, characterized by large hypochromicity, a red shift of the Soret band, and a negatively induced CD in the Soret region. The second mode is external groove binding, characterized by minor hypochromicity and Soret band shift, and a very small or even no induced CD. The third mode is outside stacking of the porphyrins along the DNA surface to produce a split CD spectrum (Figure 3.3.3).

It is clear that the porphyrin–DNA binding modes depend on both the structural properties of the porphyrinoid derivatives and the DNA sequences and conformations. The experimental conditions, that is, the molar ratio of porphyrin to DNA and the ionic strength also influence the binding mode. As a detailed report on porphyrinoids–duplex DNA non-covalent complexes has recently been published by Purrello and coworkers [26], non-covalent interactions of porphyrins with different conformations of DNA will be explored in this chapter.

We decided to mention only one example of an interaction of porphyrins with a double helix with respect to the aggregation process on DNA. This aspect excites interested investigators due to the photochemical and photophysical properties of the aggregate chromophores. For extended, electronically interacting chromophore arrays, a remarkable enhancement of light scattering is observed within these absorption envelopes. These so-called enhanced RLS signals are useful not only for identifying such assemblies, but for characterizing them as well [27]. In fact, this technique was used by Pasternack and coworkers to control the extent of porphyrin aggregation on a nucleic acid scaffold [28]. Liu and Chen examined the formation of porphyrin hetero-aggregates by adding anionic H_2TPPS4 to a DNA–cationic H_2TMPyP4 complex [29]. Through UV, CD, and RLS measurements they suggested different model structures for the binding modes of H_2TMPyP4–DNA complexes upon addition of H_2TPPS4, depending on the porphyrin to DNA ratio (r). When $r < 1$, the ICD spectrum of the ternary complex was similar to that of H_2TMPyP4 intercalated into DNA. For $r = 1$, the induced

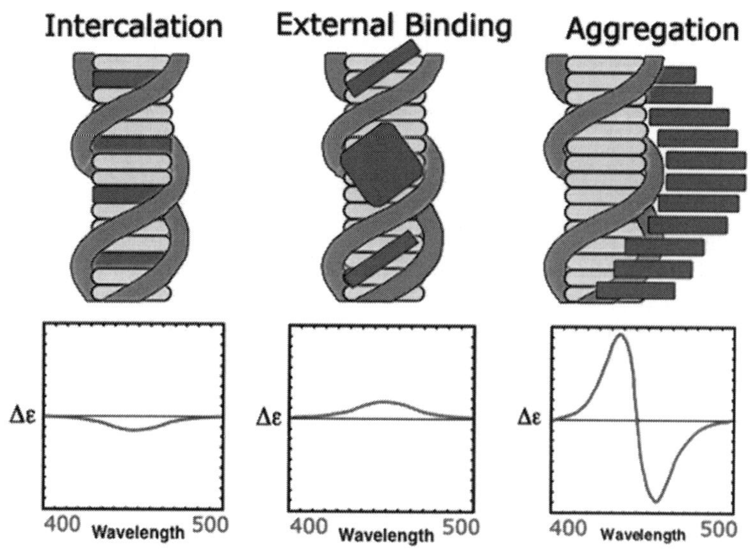

Figure 3.3.3 Schematization and ICD features of different porphyrins (square)/duplex-DNA binding modes

CD spectrum showed a reversed biphasic signal due to the complex of $H_2TMPyP4$ and H_2TPPS4 stacking along the DNA surface. At a higher *r* value (>1), there was an induced CD signal at 482 nm attributed to a lateral shifted arrangement of the heteroaggregate of H_2TPPS4 and $H_2TMPyP4$ on the DNA matrix where $H_2TMPyP4$ acted as a spacer to mediate the growth of the heteroaggregates (Figure 3.3.4).

Double-helix DNA can adopt several conformations; the most unusual is the left-handed Z-form. Since the Z-form of DNA was first detected in 1972 [30], much work has been done to gain a scientific understanding of its biological role [31–35]. *In vitro* studies of Z-DNA have provided information regarding its structure and properties [14, 36, 37]. It is a left-handed higher-energy conformation of double-stranded DNA [38, 39]. The helix contains 12 base pairs per turn; the dG residues were in the *syn* conformation, whereas the dC residues are in the *anti* conformation; the position of the minor groove was reversed, and there was no major groove, which is occupied by DNA bases; it was named Z-DNA because the deoxyribose-phosphate backbone follows a zigzag course instead of the regular one found in the B-structure.

It is known that, *in vitro*, the transition from B- to Z-DNA in alternating pyrimidine–purine sequences can be induced by molar or millimolar concentrations of cationic species (e.g., Na^+, Ni^{2+}, $Co[NH_3]_6^{3+}$) or

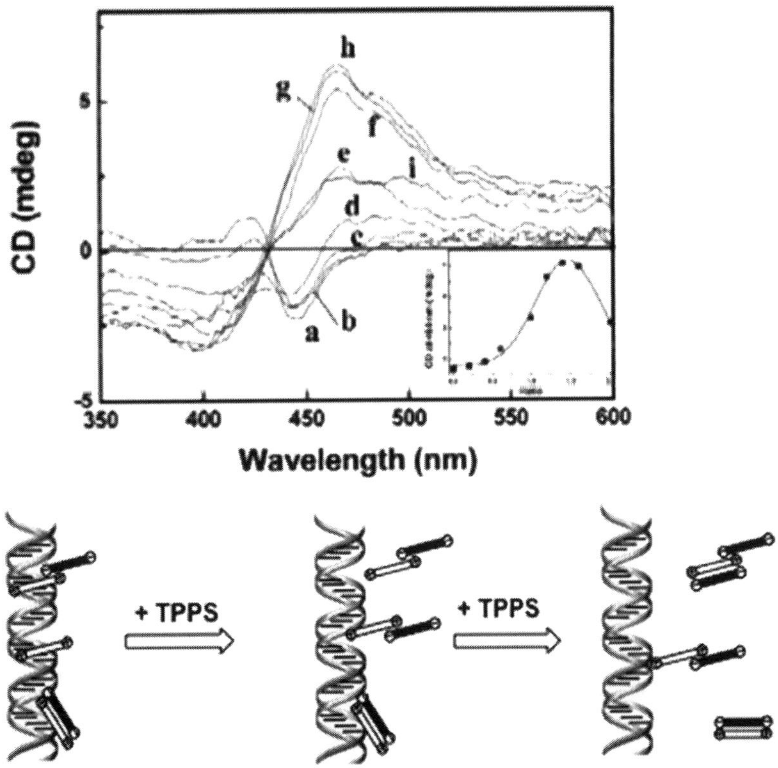

Figure 3.3.4 CD spectra of H_2TPPS4–$H_2TMPyP4$ ternary complexes at the various concentrations of H_2TPPS4: (a) 0, (b) 1, (c) 2, (d) 3, (e) 5, (f) 6, (g) 7, (h) 8, and (i) 10 μM. The inset shows the relationship between CD intensity at 464 nm and the ratio of H_2TPPS4 and $H_2TMPyP4$. For the model structure of the binding modes of the $H_2TMPyP4$–DNA complexes, $r < 1$, H_2TPPS4 mainly complexed with the intercalated $H_2TMPyP4$, and the intercalated $H_2TMPyP4$ was "pulled out" gradually; $1.4 > r \geq 1$, the intercalated $H_2TMPyP4$ was pulled out completely, and binary aggregates of H_2TPPS4 and $H_2TMPyP4$ stacked along the DNA surface; $r > 1.4$, some binary aggregates of H_2TPPS4 and $H_2TMPyP4$ departed from the DNA surface. Modified from reference [29]

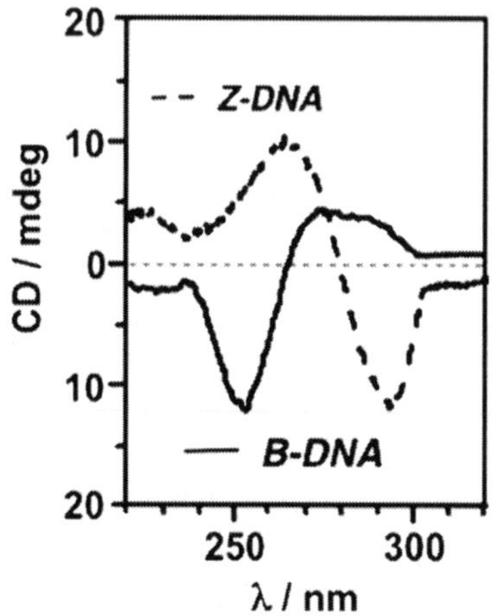

Figure 3.3.5 CD spectra of DNA in B- (black line) and Z- (dashed line) conformations

micromolar concentrations of polycationic amines ($spermine^{4+}$) [40, 41]. This conformational transition can be studied experimentally using CD. As shown in Figure 3.3.5, B-DNA exhibits a positive dichroic band at 280 nm and a negative band at 250 nm, while Z-DNA has a negative band at 290 nm and a positive band at 260 nm [42–45].

Experimental results indicate that Z-DNA might play a role in transcription, [46–48] and recently it has been demonstrated that some proteins selectively interact with left-handed DNA and stabilize it [35]. In particular, proteins with charged side chains in close proximity to the DNA can produce, during the binding event, the necessary environment to account for the conformational changes. Intracellular cationic concentrations do not approach the experimental levels to induce this transition in unmodified DNA, yet aluminium-induced Z-DNA has been associated with the pathology of neurological disorders and autoimmune diseases [49, 50]. However, the potential biological role of Z-DNA has not yet been fully clarified [51–59], due to its challenging detection related to unfavorable B/Z DNA ratios and spectroscopic interferences (in the UV region) of proteins and other biological materials. Therefore, the identification of chiroptical probes able to discriminate between B- and Z-DNA, is a subject of great interest.

Binding of porphyrins with the Z-form of DNA was initially studied by Pasternack *et al.* [60] who demonstrated that H_2TMPyP4 and Cu(II) derivatives are able to interact with Z-poly(dG–dC) via intercalation. Moreover, H_2TMPyP4 and CuTMPyP4 induce the Z-form to transition to the B-form of poly(dG–dC). The transition rate constant is generally not influenced by the concentration of porphyrin and poly(dG–dC) over a considerable range, except when the ratio of the base pairs to porphyrin ($1/r_0$) is less than about six. However, the reaction is extremely fast, as evidenced by a color change in the Soret maxima that occurs within the time required to hand-mix the reagents. The reaction of H_2TMPyP4 and CuTMPyP4 with duplex (dG–dC) follows a simple first-order kinetics at low $1/r_0$ values ($K \approx 0.6$ s^{-1}). After interaction with porphyrin, the B-form and Z- form regions coexist within the duplexes. The remaining fraction of the Z-forms is dependent on the ratio of the concentration of porphyrin to base pairs. H_2TMPyP4 and CuTMPyP4 are able to convert around 10–20 pairs of Z-form DNA into the B-form. The interaction with the Z-form begins with a porphyrin unassisted distortion, characteristic of the polymer, related to the Z-form preparation procedure that leads to a fraction of the polymer remaining in a B-form. These regions represent preferential binding sites for porphyrins.

At a $1/r_0 > 6$ a dramatic increase in conversion rate only occurs for H_2TMPyP4 and CuTMPyP4. Pasternack suggested that, under these conditions, virtually all of the porphyrins intercalate into B-regions, thus leading to slow conversion of Z into B. By reducing the number of base pairs per porphyrin (to below 6),

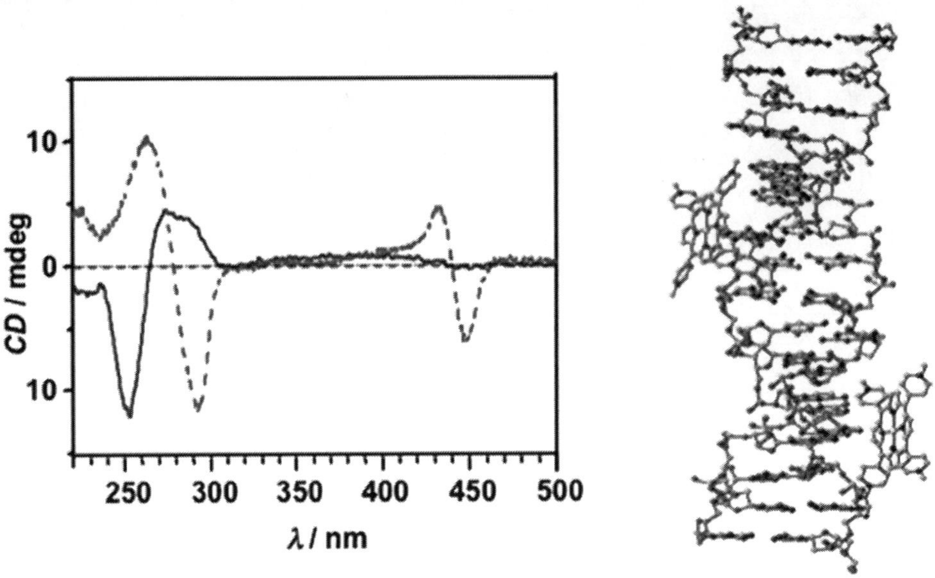

Figure 3.3.6 Left: CD spectra of ZnTMPyP4 (4 μM) in the presence of poly(dG–dC) in the B-conformation (solid curve) and in the Z-form induced by 9 μM of spermine (dashed curve). Right: Schematic model proposed for ZnTMPyP4/Z–DNA complex. Modified from reference [62]

42-mer: GGTTATCXCX$(CG)_{12}$ATAAACC Z/B=2 (28:14)

58-me: C$(TA)_3$AATTATGGCXCX$(CG)_{12}$CCATAATT$(TA)_3$G Z/B=1.23 (32:26)

108-mer: GGTTTATAATTATAA$(AT)_3$AATTTATGG$(CG)_{23}$CCATAATT$(AT)_3$TTATAATTATAA-ACCC-3' Z/B=0,86 (60:46)

Figure 3.3.7 BZB sequences with different B/Z ratios. X indicates 8-bromoguanines substitutes

the B-region becomes saturated and the porphyrin begins to populate a distorted Z-portion, thus bringing a rapid conversion to the B-form.

However, since cationic porphyrin intercalation promotes the Z- to B-transition [60, 61], investigation of the interaction of these systems with Z-DNA is not trivial. However, it has been demonstrated that non-intercalative porphyrins H_2TMPyP2 and MnTMPyP4 are not able to convert Z-poly(dG-dC) into the B-form.

Beginning with these considerations, Purrello and collaborators, using CD, UV adsorption, fluorescence, and RLS data [62], have shown that ZnTMPyP4 (non-intercalative porphyrins due to the presence of the axial ligand) is an excellent chiroptical probe for detecting Z-DNA in alternating cytosine–guanine polynucleotides, poly(dG–dC). The ZnTMPyP4 in the presence of Z-DNA shows a bisignate ICD signal in the Soret region (Figure 3.3.6) and a red shift (about 7 nm) accompanied by a hypochromic effect of the Soret band, a blue shift and enhancement of the fluorescence intensity, and an increase in the RLS intensity. All this spectroscopic evidence suggests that the interaction mechanism is mainly driven by the exposure of N7 guanine atoms, which are available for coordination with the central metal of ZnTMPyP4 through substitution of the axial coordinated water molecule. Such coordination can be reversed by the presence of Ni^{2+} ions able to coordinate the N7 and thus to release the porphyrin (Figure 3.3.6).

The recognition capability of ZnTMPyP4, has been proven under more competitive conditions, using short Z-DNA tracts and designing several BZB sequences, where the GC portion (which can be easily converted into the Z-form) is embedded in B-DNA sequences with different B/Z ratios [63] (Figure 3.3.7). To favor the formation of the Z-form, 8-bromoguanines (X in the sequence) are used [64–66].

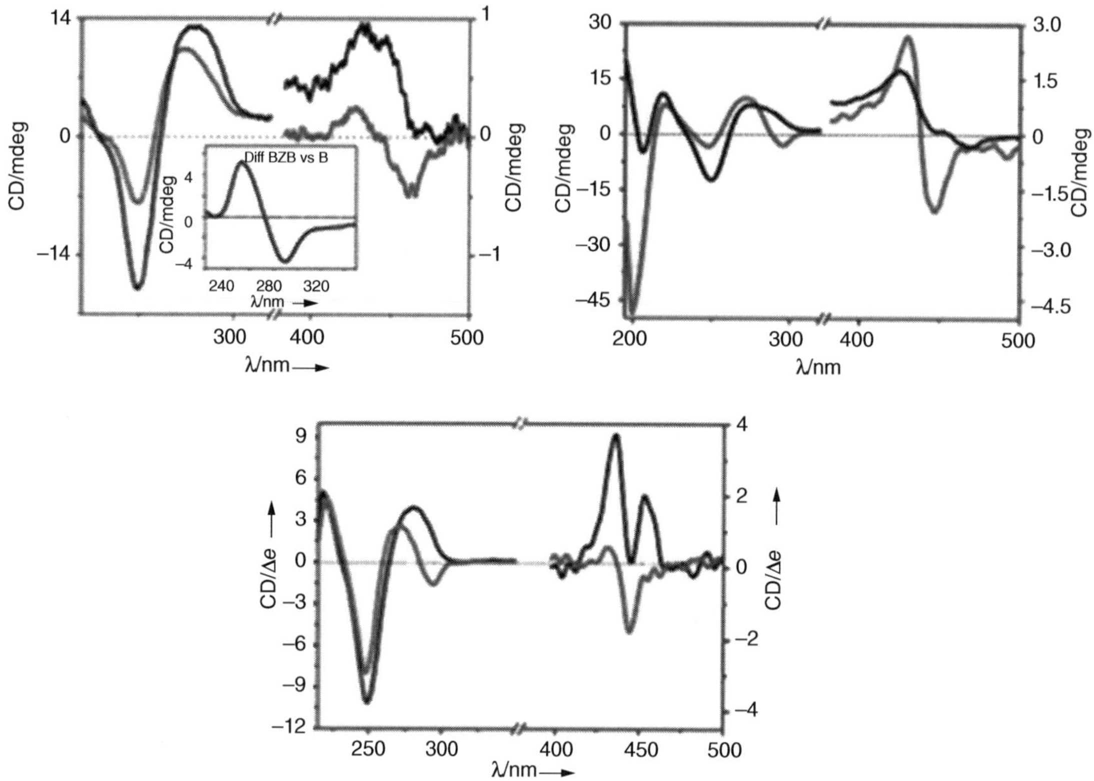

Figure 3.3.8 *CD spectra of ZnTMPyP4 in the presence of different DNA sequences from left to right: 42-mer, 58-mer, 108-mer (shown in Figure 3.3.7), in B-conformation (black curves) and in Z-conformation (grey curves). Modified from reference [63]*

In the presence of all the studied sequences, ZnTMPyP4 shows a bisignate ICD signal, when the GC portion of the BZB sequences is induced in the Z-form, or a positive feature, when all sequences are in the B-conformation. This different behavior confirms the capability of ZnTMPyP4 to detect the Z-form in the presence of B. Similar results have been obtained in the presence of DNA sequences with lower Z/B ratios, where the absence of bromoguanines makes the induction of Z-conformation more difficult. However, although high NaCl concentrations are required, ZnTMPyP4 continues to interact with the Z-portion through binding to the N7 of the guanines (electrostatics are very weak) (Figure 3.3.8).

Another chiroptical probe to detect the left-handed Z-DNA (at micromolar concentrations) is the anionic NiTPPS4 (Figure 3.3.9). A strong bisignated ICD signal in the Soret region is associated with the Z-DNA recognition, tested both with oligonucleotides (number of base pairs ~50) [67] and polynucleotides (number of base pairs ~1000) [68]. NiTPPS4 can detect the Z-form of 48-mer CG ODN at concentration <100 nM, while it is completely insensitive to CG and AT B-DNA, thus confirming its selectivity and sensitivity. It is able to detect Z-DNA located at the end of a B-DNA tract (BZ) or embedded in B-DNA (BZB) [69]. Balaz and coworkers, using the DNA sequence shown in Figure 3.3.7, demonstrated that NiTPPS4 detects Z-DNA fragments in short non-condensed oligonucleotides with high sensitivity and selectivity. The shape of the ICD signal provides information about the stage of Z-DNA condensation, thus allowing for a clear distinction

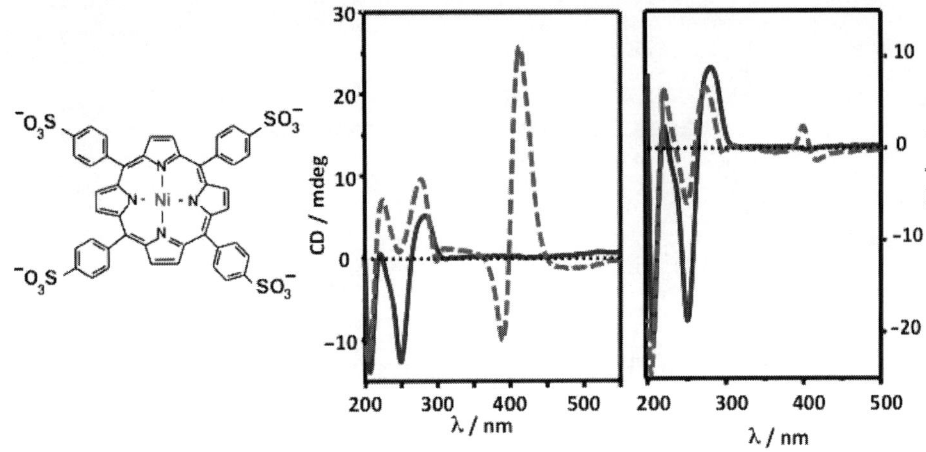

Figure 3.3.9 *Molecular structure of NiTPPS4 and CD spectra in the presence of an 48-mer DNA sequence in B-conformation (solid curves) and in Z-conformation induced with 50 mM of $NiCl_2$ (dashed curve left panel) or with 0.1 mM of $NiCl_2$ + 6 μM of spermine (dashed curve right panel). Modified from reference [70]*

between condensed Z-DNA (positive bisegnate CD couplet) [70] and non-condensed Z-DNA (negative bisegnate CD couplet) (Figure 3.3.9).

In this case, the interaction between anionic porphyrin and negatively charged DNA is promoted by the charge-shielding effect of spermine that stabilizes porphyrin–DNA complex formed by axial coordination of the nitrogen N7 of guanine with nickel. Moreover, by triggering spermine protonation, it is possible to modulate the interaction between NiTPPS4 and Z-DNA as well as the DNA helicity (Figure 3.3.10) [68]. This behavior makes the complex switchable by pH, allowing for a supramolecular reversible information storage system and an AND logic gate.

3.3.2.3 G-quadruplex

The formation of four-stranded DNA structures by guanine rich DNA oligomers has received considerable interest in the recent literature.

G-quartets (GQ) are highly polymorphic (Figure 3.3.11) and can be classified in terms of stoichiometry (uni, bi, and tetramolecular) and orientation (parallel, antiparallel, and mixed) [14]. The type of GQ structure depends on the DNA composition and length as well as on the conditions under which it has been prepared (i.e., nature of cations ($Li^+ < Na^+ < K^+$), the presence of ligands, and the annealing conditions These structures play an important biological role. GQ conformations are, in fact, adopted by telomeres (the end portion of chromosomal DNA) and protect the DNA from degradation. Through stabilization of the telomeric G-quadruplex structure, for example by capping it with a porphyrin, it is possible to inhibit the activity of telomerase, an enzyme responsible for the elongation of telomeres. This enzyme is over-expressed in cancer cells, and, in fact, it has been associated with the longevity of most cancer cells. Therefore, telomerase inhibition, by stabilization of the GQ portion, can help to restrict the growth of cancer cells [71–73].

There is a structure-based quest for telomerase inhibitors, mainly based on planar or extended-planar molecules.

Porphyrin and metallo-porphyrins were among the first ligands studied for GQ binding and stabilization capability. The metal-free tetracation H_2TMPyP4 possesses molecular dimensions that resemble G-tetrads and act as an effective inhibitor of human telomerase by an *in vitro* assay.

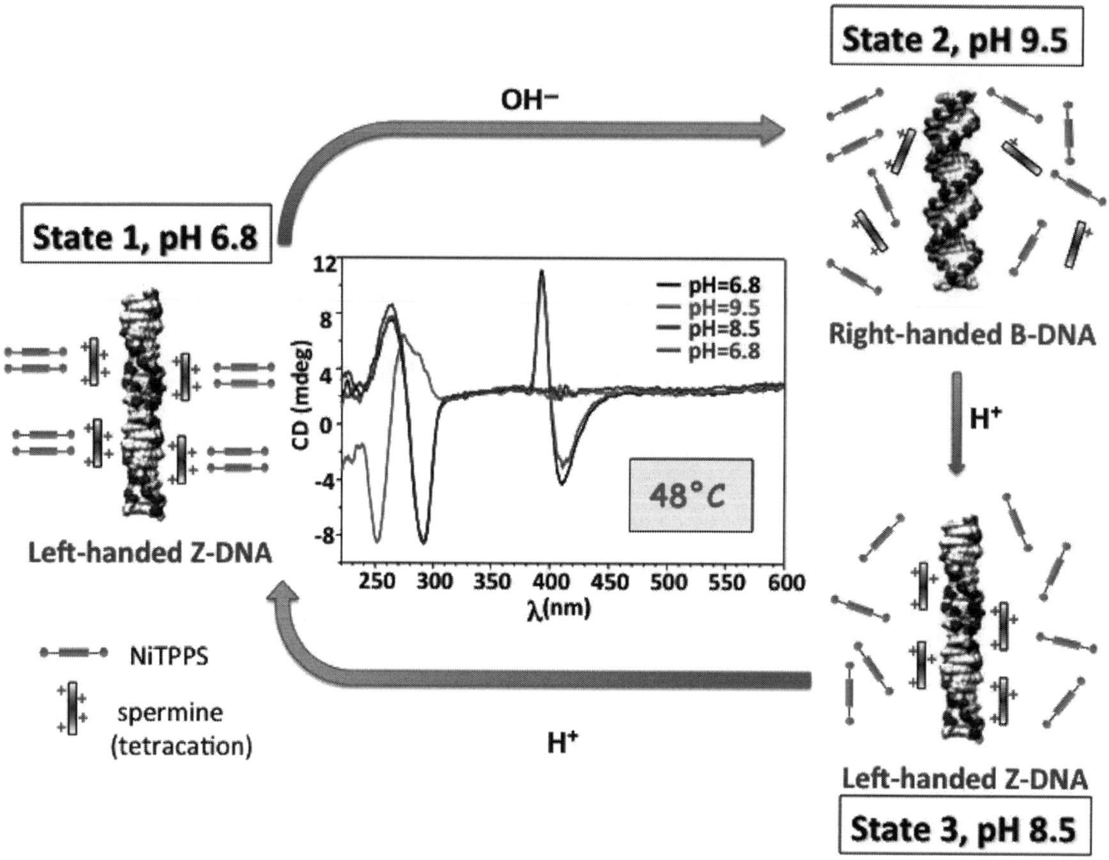

Figure 3.3.10 *CD spectra and schematic representation of poly(dG–dC)–spermine–NiTPPS4 complex, changing the pH of the solution*

The binding mode of porphyrin–GQ depends on several factors (i.e., type of GQ, DNA sequence, and the arrangement of strand orientations). Among them, the groove sizes of different antiparallel G-quadruplexs as well as the structures of the loops and adjacent non-G-tetrads might also influence the binding specificity.

Three modes of binding have been proposed: (i) binding to the top and the bottom of the quadruplex (capping) [74], (ii) external binding on strands of G-quadruplexes [75], or (iii) intercalation between guanine tetrads within a quadruplex [76]. All of these binding modes have been shown to increase the stability of the GQ structure. The capping ability of porphyrins at the end of the GQ results in an enhanced stability of the GQ.

Sherdy, using UV–Vis titration, showed that the Soret absorption band of H_2TMPyP4 in excess of T4G4 (which forms simple parallel stranded quadruplexes with K^+) and the self-complementary 12-mer $(CG)_2$ATAT$(CG)_2$, is red shifted (10–12 nm) and shows a significant hypochromic effect (35–40%) [77]. In particular, the Soret red shift of the H_2TMPyP4 is intermediate between what is observed for intercalation (≥15 nm) and that for outside binding (≤8 nm), whilst the hypochromic is typical for an intercalation mechanism. These spectroscopic features and energy transfer studies confirm that H_2TMPyP4 interacts with T4G4 via intercalation at low concentrations (0.25–0.4 mM DNA and 3–5 μM H_2TMPyP4), while at higher concentrations a more complex mixed binding mechanism occurs.

Intercalation of H_2TMPyP4 has also been observed with distinct DNA GQ (G2, G3 and G4, Figure 3.3.12) [78].

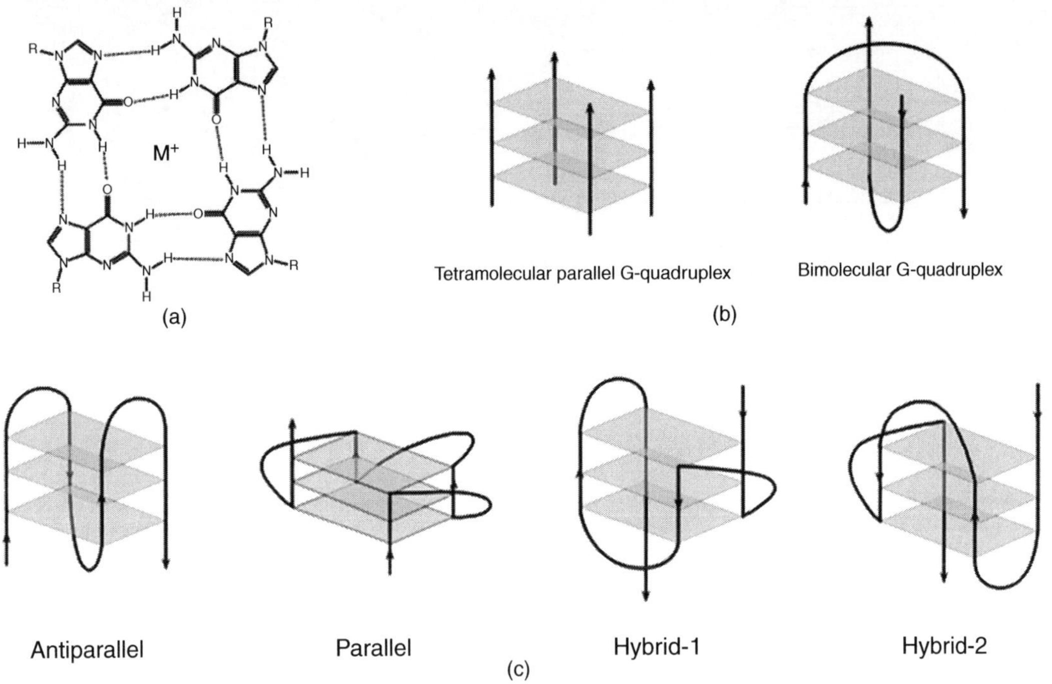

Figure 3.3.11 *Different types of GQs structures*

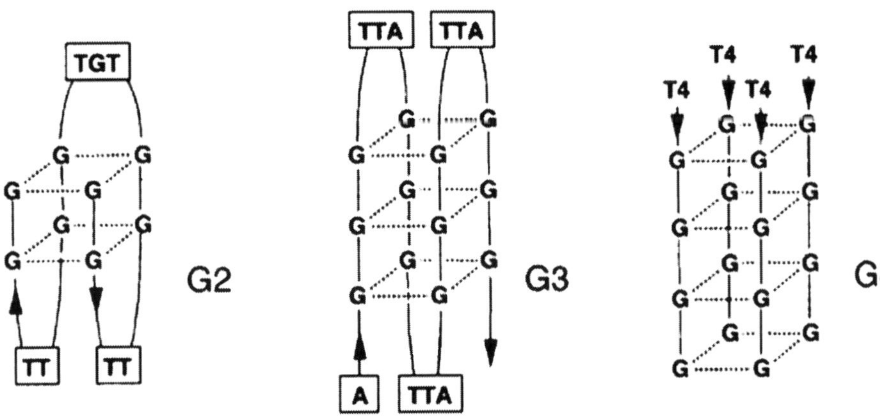

Figure 3.3.12 *Scheme of the sequences studied in reference [78]. Modified from reference [78]*

The assembly of G rich DNA sequences to form a tetraplex is favored under high [K^+] conditions. Binding ITC data in K-BES buffer show an affinity order G2 > G4 >> G3, with distinct binding stoichiometries of 1:1, 2:1, and 3:1 H_2TMPyP4 to GQ for G2, G3, and G4, respectively. In Na-BPES buffer the interactions with G3 and G4 are characterized by a high-affinity binding event requiring low [H_2TMPyP4]/[DNA] ratios, followed by a weaker secondary process that requires higher ligand concentrations to achieve saturation. The affinity of the parallel stranded G4 tetraplex toward the porphyrin is greatly enhanced in the presence of Na^+ ions, such that the first binding process dominates the overall profile interaction.

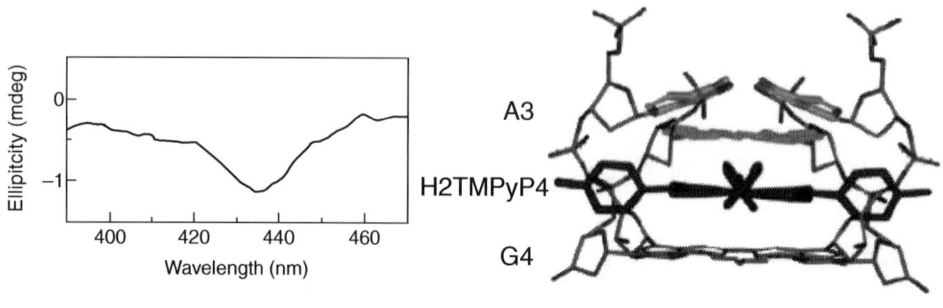

Figure 3.3.13 *CD spectra of H_2TMPyP4, 6 μM, in the presence of d(TTAGCG). Models of intercalation of H_2TMPyP4 into the A3pG4 step proposed from the molecular mechanics calculation. Modified from reference [79]*

Intercalative binding of ligands at GpG sites, allowing for a high ligand to DNA loading, rather than competitive modes involving pasting or stacking to either end of the tetraplex, is supported by molecular modeling of the possible complexes. Simultaneous occupation of adjacent intercalation sites in the DNA requires that exclusion effects on neighboring sites are not relevant for the binding of the ligand. The proposed model is based on asymmetric positioning of successive porphyrins without either significant inter-ligand contact or disruption of the global tetraplex integrity.

However, the interactions are strongly influenced by the solution conditions. Using Na^+ conditions, differential sites for stepwise binding become apparent, where an initial 1:1 high affinity event dominates the overall interaction and masks the subsequent uptake of further ligand. The porphyrin–tetraplex binding behaviors are strongly influenced by K^+/Na^+ buffer conditions.

Nevertheless, intercalative binding has been reported by Yamamoto and coworkers [79], who studied the interaction of H_2TMPyP4 with all-parallel type G-quadruplex d(TTAGGG), formed by using potassium buffer as reported previously [80] (a single repeat sequence of the human telomere, exhibiting a melting temperature (T_m) of >50 °C) by NMR, UV-Vis absorption, and CD. Intercalation of H_2TMPyP4 into the A3pG4 step of the G-quadruplex DNA, [d(TTAGGG)]4, with a K_a value of 6.2×10^6 M^{-1} is supported by a red shift (18 nm) and 66% hypochromicity of the Soret band in the UV–Vis region as well as by a negative induced CD band (Figure 3.3.13).

Nuclear Overhauser Effect (NOE) studies confirm that the binding of H_2TMPyP4 to [d(TTAGGG)]4 is significantly stabilized by the stacking interaction between the π-system of the porphyrin moiety and the GQ plane formed by GQ guanine bases, as well as by the electrostatic attraction between pyridinium ions and phosphate ions of the DNA backbone. A model study suggested that the π–π stacking between the porphyrin moiety of H_2TMPyP4 and A3 adenine bases also contributes to complex formation (Figure 3.3.13) [79]. The relatively flexible 5′-terminal TTA region of d(TTAGGG)4 could be suitable for the accommodation of H_2TMPyP4 at the A3pG4, with the porphyrin sandwiched between G4 and A3 bases. Since d(TTAGGG)4 forms a dimer in solution through intermolecular end-to-end stacking of the G-quartets formed by the G6 guanine bases, the interface between GQ formed by G6 could be a potential binding site for H_2TMPyP4, leading to a 1:2 stoichiometry for the complex formation.

A1 crystal structure of the bimolecular human telomeric quadruplex of the d(TAGGGTTAGGG) sequences with H_2TMPyP4 has been provided by Neidle and Balasubramanian [81]. This crystal structure indicates that H_2TMPyP4 does not intercalate, but one molecule is stacked onto an AT base pair, thus forming a TT propeller loop. The second H_2TMPyP4 molecule is stacked externally onto thymine bases at the edges of both TTA and TT loops. There is no direct ligand contact with any G-tetrads. The structure presented shows that the stable parallel quadruplex topology is preserved on ligand binding. H_2TMPyP4 forced a major change in one loop, from a trinucleotide TTA loop to a dinucleotide TT one. The loop topology thus results in providing a more flexible and complex interface for the H_2TMPyP4 binding.

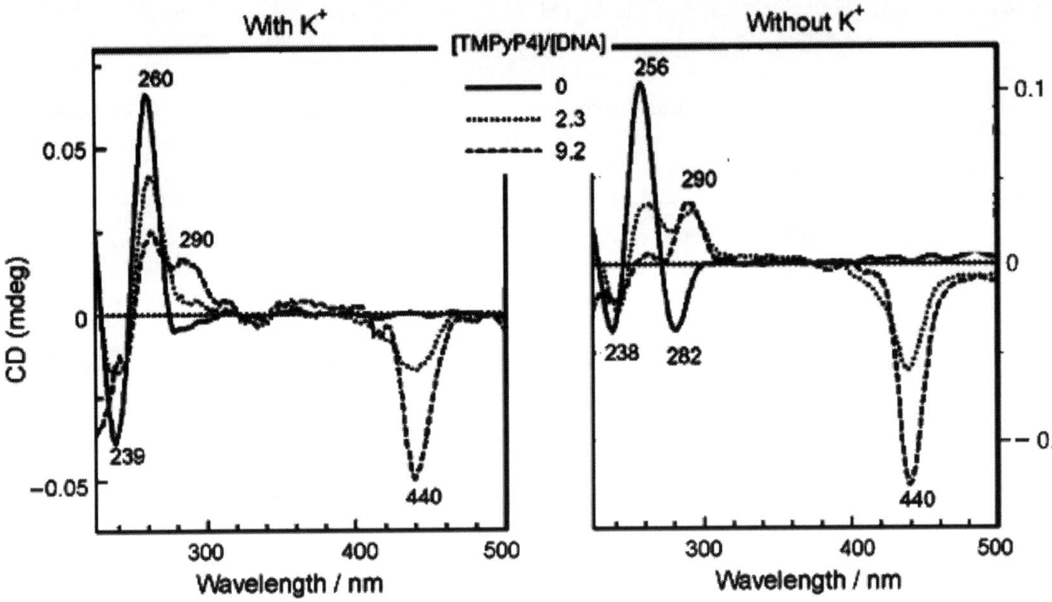

Figure 3.3.14 *CD spectra of d(TTAGGG)2 at a concentration of 8.8 × 10^{-7} M in 10 mM Tris–HCl buffer (pH = 7.2) titrated with H_2TMPyP4 (0–1.67 × 10–^{5} M) in the presence (left panel) or absence of K^+ (right panel). Modified from reference [82]*

The influence of the loop region has also been studied by Zhang and coworkers [82]. CD measurements indicate the transformation of the GQ structures of the single-strand oligonucleotide d(TTAGGG)2 (12-mer) from the parallel to a mixed parallel/antiparallel, in the presence of K^+, with the binding of H_2TMPyP4 [82]. Titration with K^+ shows a turning point at ratio (H_2TMPyP4/DNA) $R \approx 5$, thus suggesting the number of porphyrins required to induce the GQ structural conversion. In the absence of K^+, for $0 < R < 5$ the interaction between H_2TMPyP4 and d(TTAGGG)2 induces the formation of an antiparallel GQ structure. $R = 5$ is taken as the saturated binding number of H_2TMPyP4 molecules to induce such a transition and stabilize it (Figure 3.3.14).

Similar experiments have been performed using shorter (6-mer 5′-TTAGGG-3′) and longer (24-mer 5’-TAGGG(TTAGGG)$_3$T-3′) DNA strands. While a 6-mer remains in a random form upon addition of porphyrin, the 24-mer sequence transforms into an antiparallel GQ. The 24-mer sequence has the potential to form intramolecular antiparallel or parallel GQ DNA, the 12-mer sequence has the ability to form hairpin antiparallel GQ or the parallel GQ with two strands, while 6-mer sequence can only form intermolecular parallel GQ. For the case of 12- and 24-mer DNAs, both the antiparallel GQ structures have end loops for the stacking of H_2TMPyP4, which will stabilize GQ further by the aid of the stacked porphyrins. On the contrary, parallel GQ merely has external loops and cannot provide a suitable space for accommodation of H_2TMPyP4.

It is thus reasonable to speculate that the end-loops play a crucial role in stabilizing GQ induced by H_2TMPyP4.

Shieh *et al.* presented a novel strategy of using GQ as a drug carrier to target cancer cells for photodynamic therapy (PDT) [83]. They used an AS1411 aptamer as the drug carrier to deliver H_2TMPyP4 into the MCF7 breast cancer cells. Because the GQ structure is known to target the overexpressed nucleolin in cancer cells, the effect of the GQ structure as a carrier for the delivery of H_2TMPyP4 into cancer cells by nucleolin-mediated internalization was investigated. The target and delivery mechanism is based on the specific interaction of the aptamer-H_2TMPyP4 complex with nucleolin, a protein overexpressed in breast cancer.

UV measurements show a red shift of the H_2TMPyP4 Soret, whose entity is at an intermediate level between the typical intercalated binding and the outside binding modes and a more evident hypochromicity than that associated with the intercalation binding mode. CD measurements suggest a mixed binding mode (intercalative–outside binding). Energy transfer measurements show that a significant energy transfer occurs from excited G bases to H_2TMPyP4, thus suggesting an intercalative binding mode. Scatchard plot fitting indicated that the binding number of H_2TMPyP4 per AS1411 aptamer was in the 5–7 range, while Job plot fitting suggests that one AS1411 aptamer molecule binds with 6.1 H_2TMPyP4 molecules. From these results it can be assumed that H_2TMPyP4 molecules intercalate into four groups of adjacent stacked GQ of aptamers, and the other two H_2TMPyP4 molecules present end-pasted arrangements.

In vitro studies showed that MCF7 cells with high nucleolin expression have a large aptamers–H_2TMPyP4 complex uptake and serious photodamage in comparison with normal epithelium cells. The GQ of the AS1411 aptamers appears to be an ideal vector for nucleolin targeting and breast cancer specific drug delivery.

Interest then moved to the influence of chemical modification of the porphyrin in order to increase the stabilization of the GQ.

The differences in DNA quadruplex interactions between the positional isomers H_2TMPyP4 and H_2TMPyP2 have been scrutinized by ligase assay, DMS foot printing, and CD spectroscopy [84]. UV thermal melting profiles of human telomere repeats GGTTA, forming antiparallel GQ, show that both H_2TMPyP4 and H_2TMPyP2 interact with and stabilize antiparallel quadruplex DNA, but by structurally distinct modes. In fact H_2TMPyP4 binds by external stacking with the guanine tetrads (intercalation between the guanine tetrads is less probable) with intramolecular quadruplex DNA with a diagonal loop (in K^+ buffer). In Na^+ buffer DNA forms a hairpin-dimer quadruplex and H_2TMPyP4 binds through external stacking to a three-layered hairpin-dimer quadruplex. Conversely, H_2TMPyP2 mainly binds to TTA loops. In H_2TMPyP2 there is a high-energy barrier for the pyridyl rings to become coplanar with the porphyrin core. Thus H_2TMPyP2 cannot easily stack externally to the bottom guanine tetrad layer and fits more easily in the larger niche created by the diagonal loop and the thymine loops at the quadruplex–duplex junction. Ortho-substitution groups at the meso porphyrin ring position are, thus, not favored for telomerase inhibition.

Hurley and coworkers studied the interactions of H_2TMPyP2 (ortho), H_2TMPyP3 (meta), and H_2TMPyP4 (para) with parallel and antiparallel GQ, using gel mobility shift experiments and a helicase assay [85]. Photocleavage methods demonstrated that H_2TMPyP3 ostensibly has the strongest binding affinity with parallel GQ, followed by H_2TMPyP4. H_2TMPyP2 has almost no interaction with these structures. H_2TMPyP4, according to photocleavage, binds with monomolecular antiparallel GQ through external stacking, rather than by GpG intercalation. Porphyrins with very similar structures showed different binding affinities to antiparallel GQ DNA. H_2TMPyP2 does not appreciably interact with either parallel or antiparallel GQ. H_2TMPyP4 is more specific to antiparallel GQ. Therefore, the different position of the *N*-pyridyl group determines the free rotation of the pyridyl groups at the meso positions and the relative dihedral angles of the porphyrin core and pyridyl groups, thus affecting their interactions with the GQ. H_2TMPyP3 fits the pockets between the end G-tetrads and the loops in a parallel GQ better than H_2TMPyP4, whilst H_2TMPyP2 does not fit any site in the GQ-parallel.

Finally, the influence of the central metal was investigated by Yatsunyk and coworkers [86]. H_2TMPyP4 and its Zn(II), Cu(II), and Pt(II) derivatives induce GQ folding in the oligonucleotide d(TAGGG)2 in K^+ buffer. CuTMPyP4 and PtTMPyP4 form square planar complexes, as does H_2TMPyP4, while ZnTMPyP4 adopts a square pyramidal geometry, which should preclude an intercalative binding mode. CD and UV spectroscopies demonstrated that H_2TMPyP4, ZnTMPyP4, and CuTMPyP4 are capable of stabilizing GQ DNA, but only ZnTMPyP4 induces a GQ structure in d(TAGGG)2. The presence of K^+ is important in such processes. d(TAGGG)2 exists in a propeller type parallel bimolecular GQ form under a high K^+ concentration, while a buffer with a low K^+ concentration or any amount of Li^+ or Na^+ are unable to induce GQ structures of d(TAGGG)2. ZnTMPyP4 induces a mixture of parallel–antiparallel GQ structures. CD titration of preformed

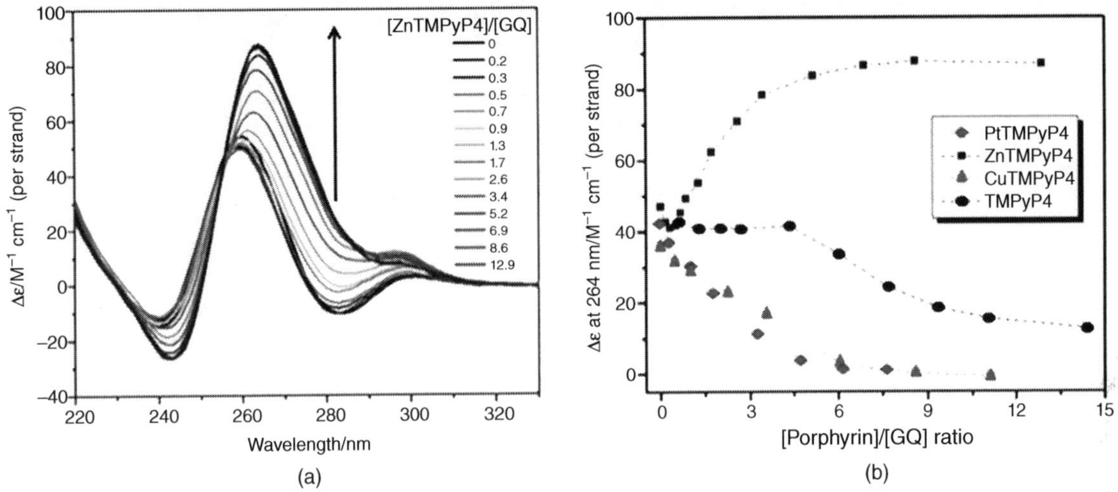

Figure 3.3.15 *(a) Titration of 4.5 μM d(TAGGG)2 in LS KPi with ZnTMPyP4 at 25 °C monitored by CD. (b) Change in molar ellipticity of d(TAGGG)2 at 264 nm upon addition of H_2TMPyP4, ZnTMPyP4, CuTMPyP4, and PtTMPyP4 at 25 °C. Modified from reference [86]*

GQ structures with ZnTMPyP4 led to low hypochromicity and no Soret shift, thus suggesting a minimal rearrangement of the GQ structure imposed by Zn porphyrin (Figure 3.3.15).

The H_2TMPyP4, CuTMPyP4, and PtTMPyP4 addition to d(TAGGG)2 led to a complete disintegration of the secondary DNA structure at the end of the titration. H_2TMPyP4 is able to promote single stranded dTAGGG(TTAGGG)3T into an antiparallel quadruplex conformation or to promote, in a K^+ rich buffer, transition from a hybrid to an antiparallel form. On the contrary, H_2TMPyP4 is not able to promote and stabilize folding of d(TAGGG)2 parallel GQ (Figure 3.3.15).

ZnTMPyP4 is capable of promoting parallel GQ structures of human telomeric DNA. ZnTMPyP4 is not expected to be capable of intercalating between G-tetrads. Therefore, the most likely mode of binding is end-stacking as opposed to groove or loop binding. These modes do not produce large changes in the UV–Vis spectra.

A thermodynamic profile suggests an external binding that does not perturb quartet stacking, specifically an end-stacking mode. The high entropic contribution is probably related to the displacement of water from the surface of GQ upon ligand binding. UV–Vis and ITC titrations suggest two equivalent and independent ZnTMPyP4 binding sites on each end of the quadruplex. ZnTMPyP4 facilitates the folding of d(TAGGG)$_2$ into parallel GQ, driving the equilibrium between the unfolded and folded GQ structure toward the latter under otherwise unfavorable conditions (low amount of K^+). The role of the two axial water molecules can be important.

3.3.3 Porphyrins covalently linked to DNA

3.3.3.1 Porphyrins attached to 5′- and 3′-termini of DNA with phosphates and amides

3.3.3.1.1 5′-Phosphate linkage

Traditionally, CD in the region below 300 nm has been used to identify DNA transitions resulting from ligand binding and environmental factors, such as temperature and ionic strength. In order to gain greater sensitivity it was envisioned that porphyrins attached to the opposite ends of the DNA scaffolding might be useful

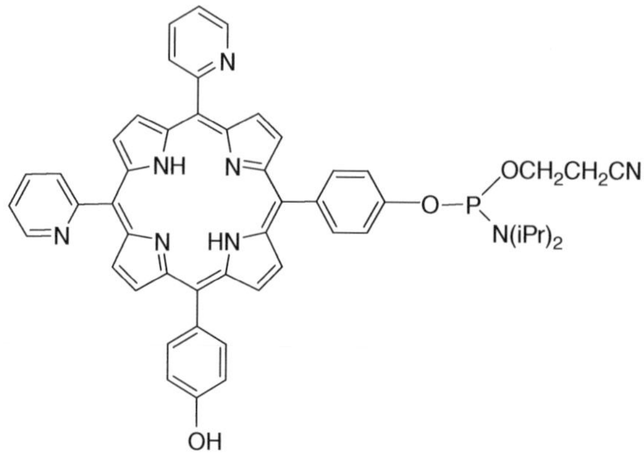

Figure 3.3.16 *Tetraarylporphyrin **P1** attached to the cytosine of the $(GC)_4$ DNA octamer*

reporter groups for exciton-coupled CD to detect geometrical changes in the backbone of a DNA sequence. As such, the DNA polymer serves as a chiral spacer for the bis-porphyrin system. The consequent through-space interaction of the porphyrins across the DNA duplex can give rise to an exciton couplet CD signal in the Soret band region. The Soret region is useful because, unlike that of the DNA below 300 nm, the 420 nm region is usually free of undesired spectral overlaps from chromophoric ligands. Hence, this part of the spectrum should reflect a DNA conformational change upon ligand binding, depending on the length of the sequence and the degree of interporphyrinic twist.

5′-Porphyrin end-labeled DNA sequences with a phosphate linker were prepared using the phosphoramidite **P1** (Figure 3.3.16) and automated solid phase chemistry [87]. Pyridyl groups were incorporated into the porphyrin scaffold to increase water solubility, and it was then demonstrated that 5′-porphyrin end-labeling via phosphates does not induce a conformational change of the B-form duplex.

As shown in Figure 3.3.17, the typical +/–couplet at ~260 nm is observed, similar to the porphyrin-free duplex [4, 87, 88]. In the Soret region a characteristic positive exciton couplet (right-handed twist) can be observed at low temperatures, which converts into a strongly positive, monosignate band upon denaturation of the duplex at 80 °C.

For confirmation of the through-space coupling between the porphyrins, a double-stranded sequence was formed from the non-self-complementary 5′-end-labelled $(CT)_4$ and its complementary porphyrin-free sequence 5-$(GA)_4$, so no porphyrin–porphyrin coupling would be possible. As anticipated no exciton couplet in the Soret portion of the CD spectrum was seen. Only a negligible positive CD band was observed at 0 °C in the DNA region in addition to a strong positive band for a single strand at 80 °C in the 400 nm part of the spectrum.

It is intriguing to note that, generally speaking, phosphate–porphyrin modified sequences with B-DNA or A-DNA conformations display a positive couplet in the Soret region. Conversely, amide–porphyrin modified sequences with B-DNA conformation, which will be discussed in more detail later, display a negative couplet in the Soret region. The sign of the couplet does not seem to depend on the nature of the porphyrin (metallated versus free-base). The following examples provide evidence. A positive couplet in the Soret region is observed (Figure 3.3.18a) in a study of 5′-phosphate–porphyrin end-capped $(dCdG)_n$ (n = 3–6) duplexes, where the DNA region clearly indicates a B-DNA conformation [87]. A positive couplet is also displayed (Figure 3.3.18b) in the Soret region of the 5′-phosphate–porphyrin end-capped $(dG\text{–}dA)_4$ duplex, which, as it is based on the DNA region, displays the spectral character of an A-DNA conformation [89]. Intriguingly, end-capped amide–porphyrin sequences display a negative couplet as

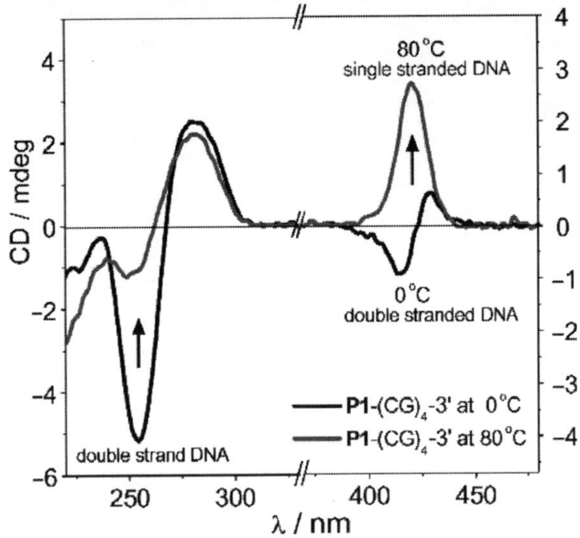

Figure 3.3.17 CD of **P1**–$(GC)_4$ at 0 and 80 °C. Reprinted with permission from [4], Copyright 2007 IUPAC

seen in studies of a 3′-end-capped 3′-dT-5′-$(dG–dC)_3$-dA-3′ duplex (Figure 3.3.18c) [88] as well as of a 5′-end-capped 5′-dT-$(dG–dC)_3$-dA-3′duplex [90].

3.3.3.1.2 Detection of the B- to Z-transition

GC rich sequences are well known to undergo B- to Z-transitions at high ionic strengths. Thus a 5′-end-labelled self-complementary $(CG)_4$ 8-mer was prepared to determine whether the CD analysis of the Soret region of the spectrum would detect the B- to Z-transition in these sequences under high salt conditions [91]. Figure 3.3.19 shows that indeed the Soret region sensitively reflects this transition from a right- to left-handed DNA conformation. A weak signal is observed in the 400–425 nm region with no added salt while the Z-form shows a very strong and positive bisignate signal (right-handed twist) in the presence of a saturated amount of sodium chloride. Thus CD analysis of the Soret region is useful to quantify the amounts of B- and Z-form DNAs during the salt-induced transition.

As the porphyrin–DNA conjugate with **P1**-5′-phosphate-$(CG)_4$-3′ becomes titrated with salt to higher ionic strengths, the UV–Vis spectral response displays a strong hypochromic shift of the Soret and DNA absorption bands (Figure 3.3.20). The hypochromic shift is interpreted as being due to an increased shielding between the porphyrin and the nucleotide π-electron systems in the Z-conformation. Two DNA forms (B- and Z-forms) coexist in equilibrium, as shown by the presence of one isosbestic point in the DNA absorption region (~212 nm). The same equilibrium is reflected by an isosbestic point in the Soret band region (~436 nm).

Even though in principle the exciton CD diagnostic methodology holds promise for following DNA conformational changes, in practice it has so far only been demonstrated to work for the monitoring of salt-induced B→Z conformational transitions. Additionally, it is worth noting that the success of case studies analogous to the one just presented is contingent on the fact that the inter-porphyrin twist and distance are favorable (up to 66 Å) [92–97]. The optimal distance insures that porphyrin chromophores found on the opposite sides of the DNA duplex can serve as real-time reporters by the exciton couplet CD in the Soret band region.

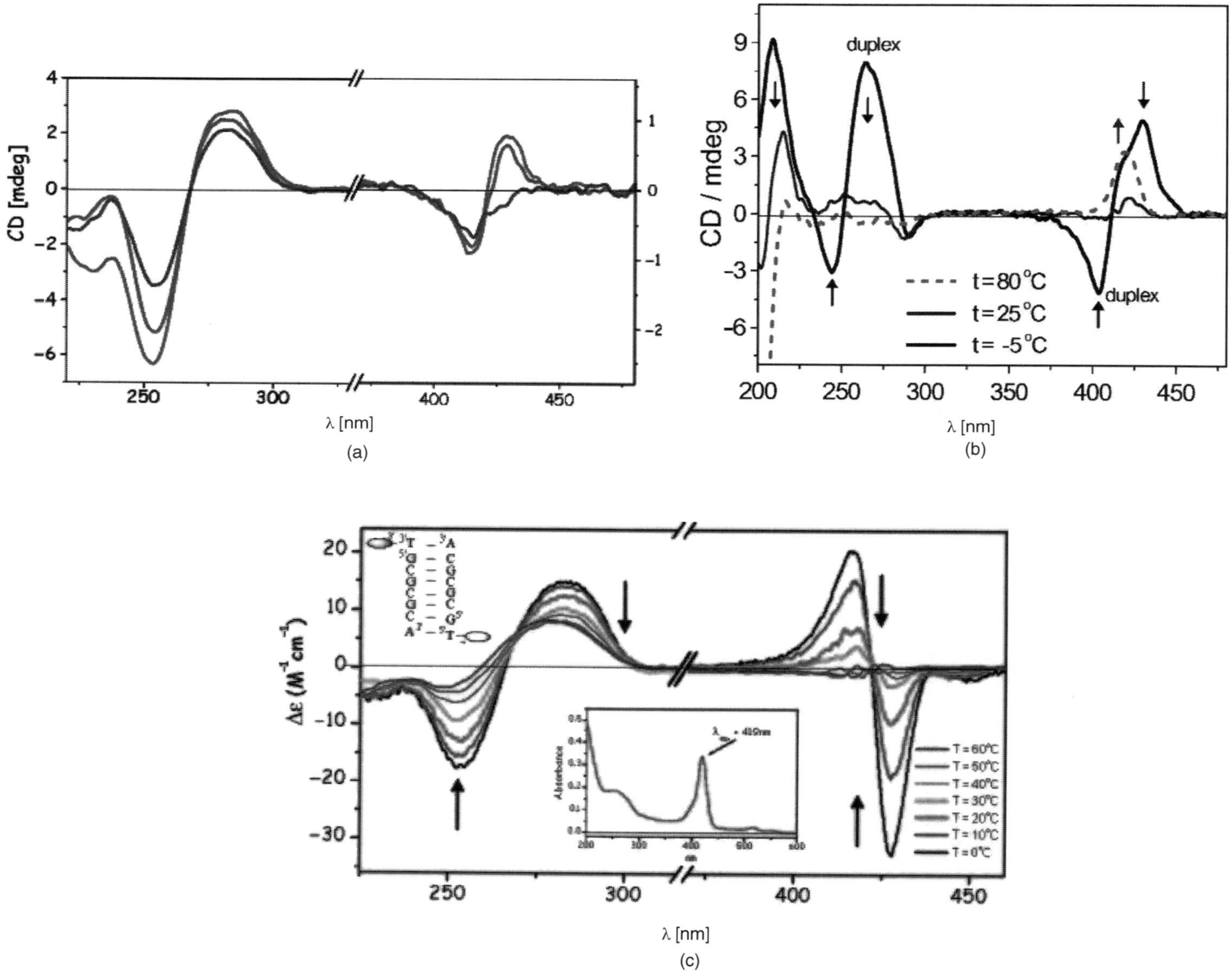

Figure 3.3.18 DNA and Soret region signals for end-termini porphyrin labeled DNA sequences with phosphates (a) and (b), and amide (c) linkages. Reprinted with permission from [4], Copyright 2007 IUPAC, and from [87] and [88] with permission from American Chemical Society

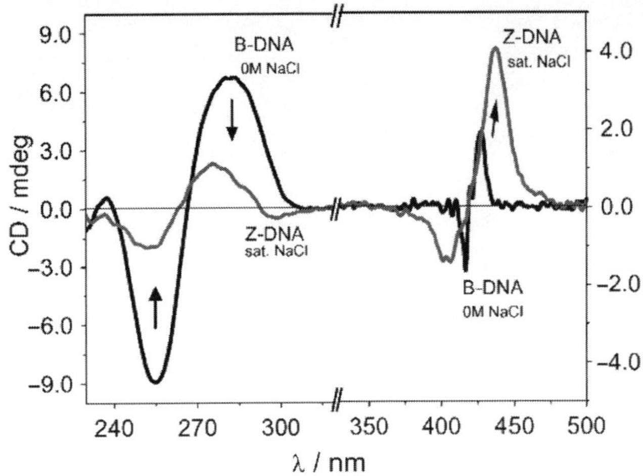

Figure 3.3.19 *CD spectra of 8-mer sequence* ***P1****-5′-phosphate-(CG)*$_4$*-3′ at different salt concentrations [4, 91]. Reprinted with permission from [4], Copyright 2007 IUPAC*

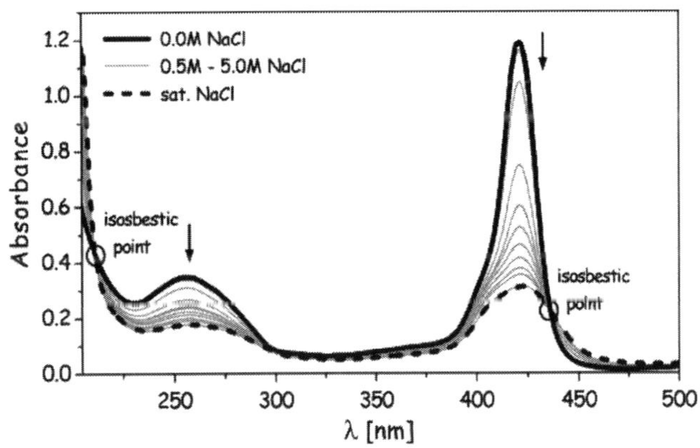

Figure 3.3.20 *Hypochromic shift and isobestic point in both DNA and porphyrin absorption regions of sequence* ***P1****-5′-phosphate-(CG)*$_4$*-3′. Reproduced from [91] with permission from The Royal Society of Chemistry*

3.3.3.1.3 3′-Amide linkage

In order to explore the effect on the Soret region of a 3′-porphyrin linkage, a thymidine phosphoramidite, **P2-3′-T** (Figure 3.3.21), was attached to a solid-supported 5′-GCGCGCA-3′ sequence [98].

Based on the CD, a porphyrin in the 3′-position does not appear to disrupt the conformation of the duplex conformation. The CD of the modified and unmodified sequences are essentially the same in the DNA region below 300 nm. The CD spectrum taken at 0 °C of an annealed 8-mer conjugate **P2-3′-T**-5′-GCGACGCA-3′ showed a negative exciton split signal in this 3′-amide-linked construct, characteristic of a left-handed inter-porphyrin twist (Figure 3.3.22) [4, 98]. In contrast to the porphyrin **P1** attached to the 5′-end of a duplex via a phosphate linker, the single-stranded porphyrin **P2** attached to the 3′-end of the sequence did not show any signal in the Soret region.

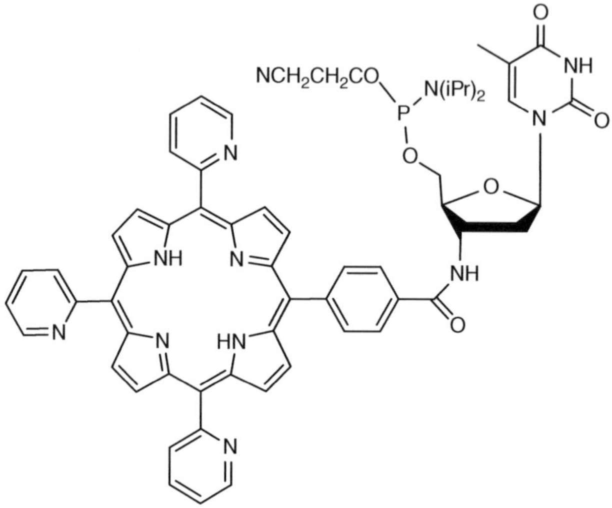

Figure 3.3.21 Tetraarylporphyrin phosphoramidte **P2-3′-T** attached through 5′–5′ coupling on a solid supported 5′-GCGCGCA-3′ 7-mer by conventional 3′–5′ synthesis

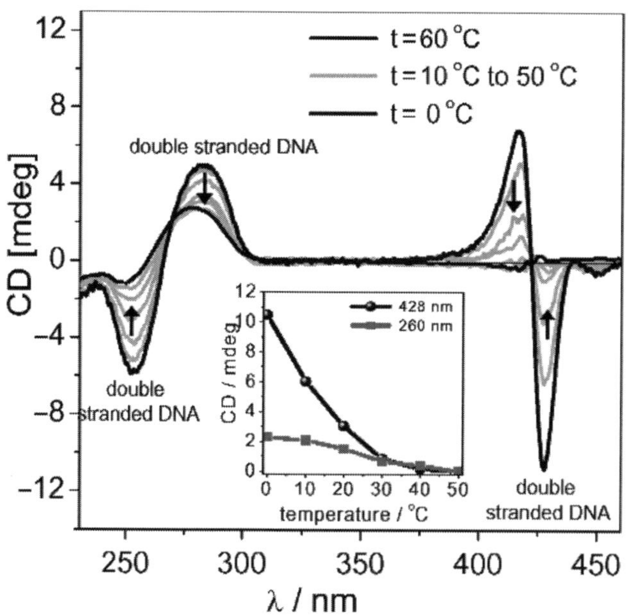

Figure 3.3.22 CD melting spectra of 8-mer sequenced **P2-3′-T**-5′-amide-GCGCGCA-3′. Inset: Difference in sensitivity in the detection of conformational changes at 260 and 428 nm. Reprinted with permission from [4], Copyright 2007 IUPAC

3.3.3.1.4 5′-Amide linkage

For the replacement of the natural 5′-phosphate linker with a more rigid amide linker, conjugate **P2-T**-5′-$(GC)_3$A-3′ was prepared (Figure 3.3.23) [4, 99]. The annealed CD spectra showed a right-handed B-DNA

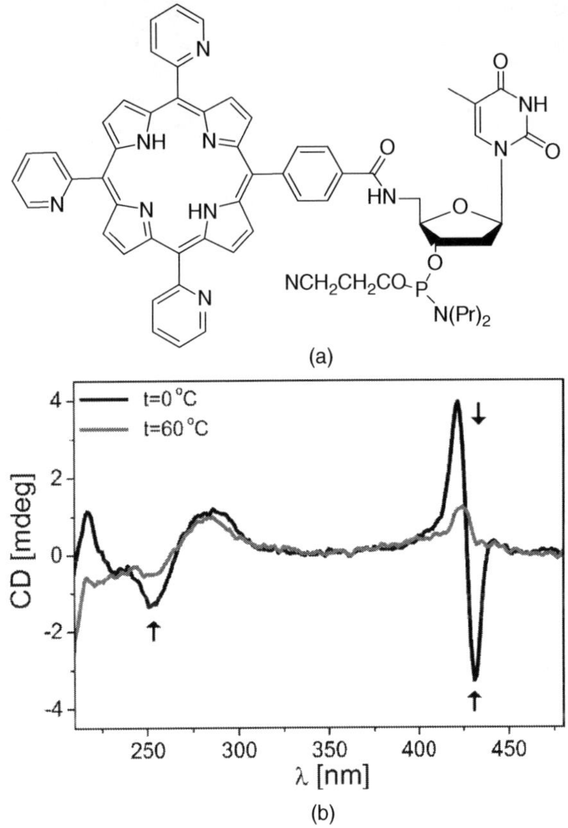

Figure 3.3.23 *(a) Structure of phosphoramidite **P2-5′-T** for the preparation of **P2-5′-T**-5′ (GAC)$_3$A-3′; and (b) CD spectrum of the 8-mer at two different temperatures [4, 99]. Reprinted with permission from [4], Copyright 2007 IUPAC*

signal in the 220–320 nm region and, as in the 3′-amide linker case, a very strong negative couplet in the 400–440 nm region (Figure 3.3.23b). Increasing the temperature from 0 to 60 °C caused the bisignate signal to disappear and at the same time a positive band at 424 nm developed at 60 °C.

Figure 3.3.23b reveals that the sensitivity of the 5′-amide-linked sequence is slightly greater than that for the corresponding 3′-amide-linked sequence.

3.3.3.2 Capping effect

It has been shown [4, 89] that a terminal porphyrin can even stabilize non-self-complementary sequences, for example, OH-$(GA)_4$-OH, as depicted in the left side of Figure 3.3.24. In this case, the CD spectrum of an unmodified sequence shows a curve reflecting an unorganized single-stranded structure. The CD spectrum of the 5′-end-labelled sequence shows a spectrum characteristic of A-like conformational features (Figure 3.3.18b). The bisignate signal in the Soret region at –5 °C indicates long-range porphyrin–porphyrin interactions originating from the antiparallel duplex shown in the figure. The Soret band signal changed markedly upon raising the temperature from –5 to 80 °C as the bisignate signal disappeared at temperatures higher than 30 °C, at which point a positive band began to develop. As anticipated, the CD spectrum of the

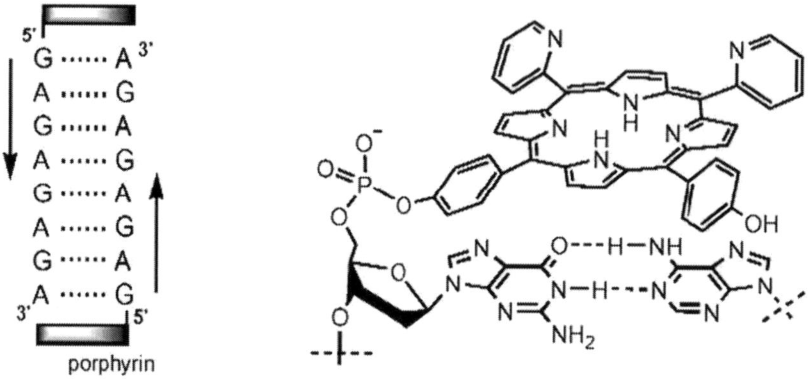

Figure 3.3.24 Depiction of an 8-mer non-self-complementary sequence P1-phosphate-d(GA). Reprinted with permission from [89]. Copyright © 2006 WILEY-VCH Verlag GmbH & Co. KGaA, Weinheim

porphyrin-free sequence was only marginally affected on raising the temperature from –5 to 80 °C. In summary, the results show that the tetrarylporphyrin **P1** can serve as a hydrophobic molecular cap for the stabilization of a non-self-complementary non-Watson–Crick 5′-$(GA)_4$-3′ sequence.

A comparative study by Balaz's group with constructs prepared with an amide linker to the 5′-DNA termini led to more significant hybridization with a non-self-complementary $(dA\text{–}dG)_4$ DNA than a construct with a charged phosphate linker [100]. In this study they observed interesting multisignate CD profiles in the porphyrin region in the presence of 100 mM NaCl, similar to previous studies, which has been assigned to strong electronic interactions between helically stacked porphyrins. The fact that DNA termini are vulnerable to low base-pairing fidelity can be resolved by introducing porphyrins as protective capping agents.

Interestingly, it has been found that free-base porphyrin, Cu–porphyrin, and Zn–porphyrin DNA conjugates gave very different CD responses in the porphyrin region in a low-salt environment. This has been interpreted as resulting from the different porphyrin hydrophobicities, which affect the extent of the interaction with adjacent nucleobases even under low or moderate salt concentrations. The three different CD profiles shown in Figure 3.3.25 imply that these signals do not just originate from degenerate porphyrin–porphyrin through-space interactions, but also in part from some non-degenerate porphyrin-adjacent nucleobase interactions. From the very intense and narrow CD couplet shown in Figure 3.3.25, it was anticipated that a Zn–porphyrin construct in which the Zn is axially liganded to a water molecule will be less involved in closely attached nucleobases.

To provide further insight into the preferred conformation of the porphyrins around the DNA duplexes, Molecular Dynamics (MD) and Monte Carlo (MC) molecular modeling simulations [90] were carried out. These simulations revealed a significant difference in the dynamic behavior of the conjugates depending on the type of porphyrin (presence and nature of the metal at the porphyrin center). The free-base porphyrin conjugate displayed a reduced structural flexibility at one of the DNA termini, with one of the porphyrins showing less mobility while retaining some degree of capping-effect over the terminal A–T base pair to which it is covalently tethered. The capping or at least semi-capping effects induced by free-base porphyrin, as evidenced by MD and MC, can be observed in Figure 3.3.26b and c. In contrast, the Zn-based conjugate showed both tethered porphyrins to be equally mobile (Figure 3.3.26d), uncapping fully from the terminal A–T base pairs and rotating freely in the solvent, only transiently capping the A–T base pairs.

These MD and MC results corroborate the observed CD spectral results. In the case of the Zn-porphyrin conjugate a strong non-degenerate coupling becomes possible because the distance between the porphyrin plane and closest A–T base pair is about 12 Å. However, in the case of the Cu and non-metallated conjugates

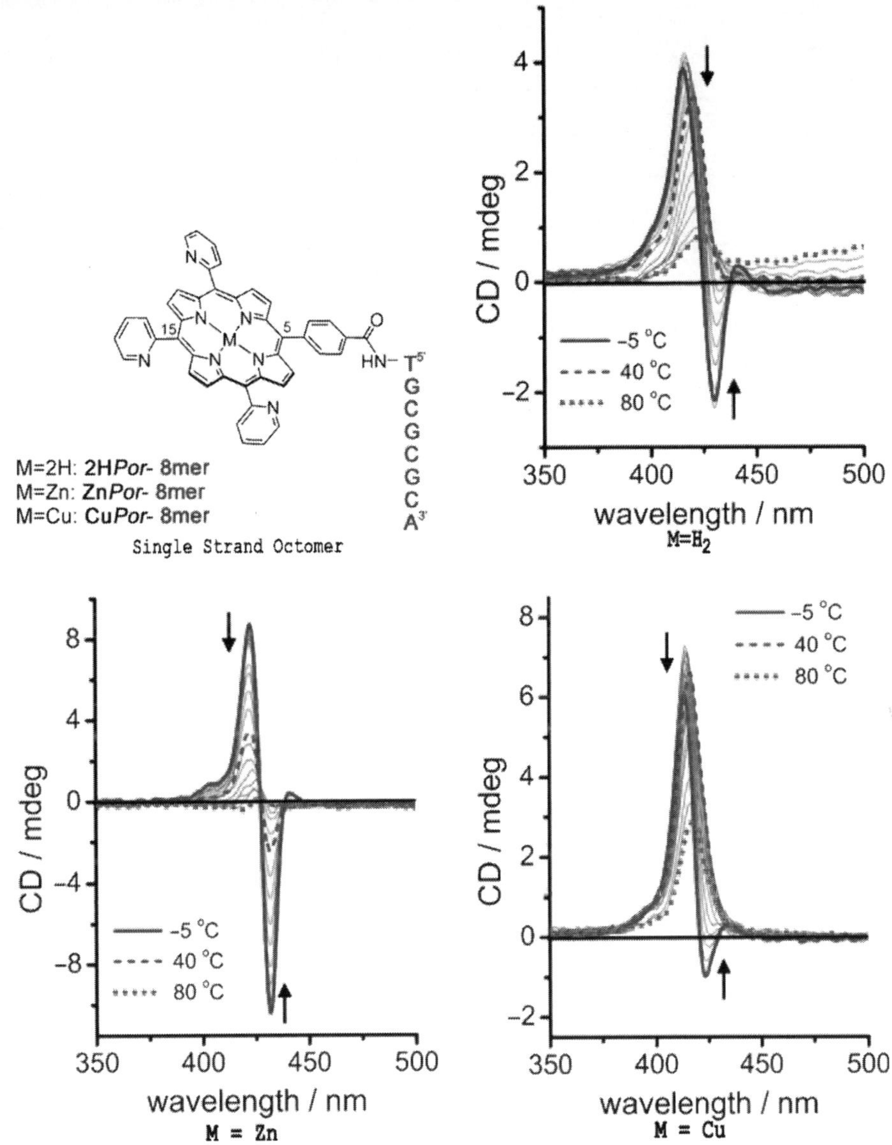

Figure 3.3.25 Top: Sequence of 5′-DNA porphyrin conjugate. Bottom: CD response of free-base porphyrin (left panel); Zn–porphyrin (middle panel); Cu–porphyrin (right panel) attached to a 5′-dT-(dG-dC)$_3$-dA-3′ DNA octamer via a 5′-amide linker. Reprinted with permission from [90]. Copyright © 2009 WILEY-VCH Verlag GmbH & Co. KGaA, Weinheim

the distance between the porphyrins and the closest A–T base pairs becomes relatively close (4–5 Å). Therefore, a second strong contribution due to non-degenerate coupling appears and leads to overall monosignate bands. Under these conditions the B-form conformation of the DNA is preserved based on the CD profile below 300 nm. An increase in the ionic strength from 150 to 400 mM NaCl promotes the formation of porphyrin–DNA non-aggregates with particularly characteristic, although unusual, multisignate CD Cotton

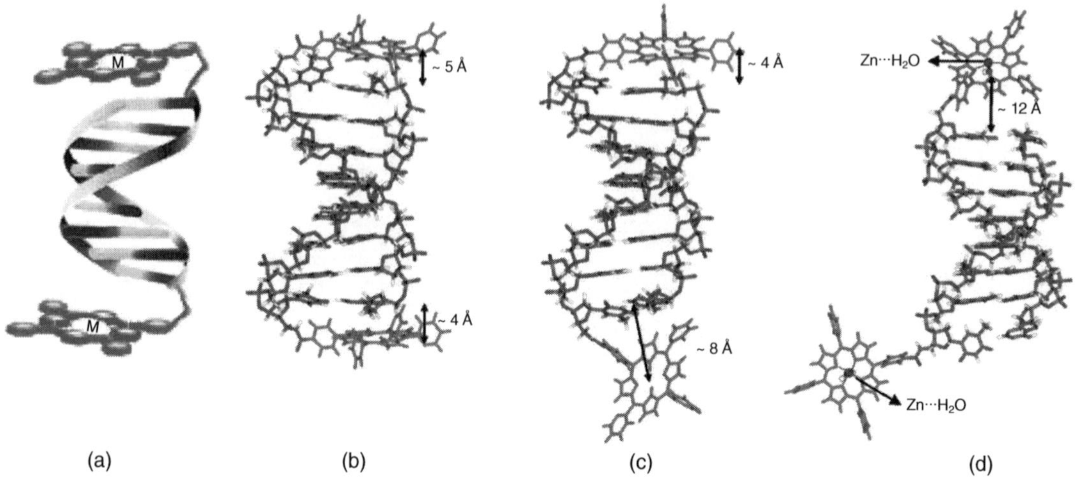

Figure 3.3.26 (a) Schematic representation of 5′-DNA metalloporphyrin conjugates; (b) most energetically stable capped free-base porphyrin 8-mer and (c) its conformation resulting from 2 ns MD run; and (d) lowest energy structure from conformational of Zn–porphyrin 8-mer. Distances between porphyrin and A–T planes are indicated in Å. Reprinted with permission from [90]. Copyright © 2009 WILEY-VCH Verlag GmbH & Co. KGaA, Weinheim (See color figure in color plate section)

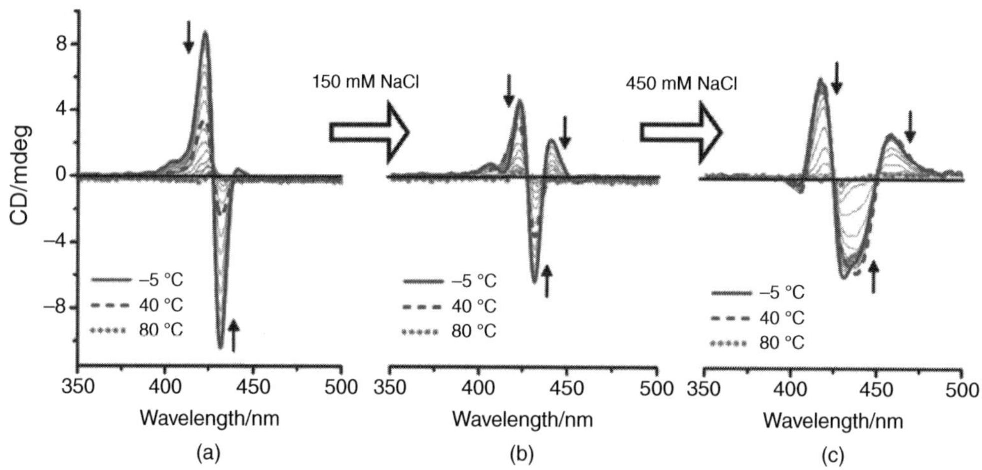

Figure 3.3.27 Variable temperature CD spectra of ZnP-8mer from −5 to 80 °C: (a) in the absence of NaCl; (b) the presence of 150mm NaCl; and (c) 450 mm NaCl. Reprinted with permission from [90]. Copyright © 2009 WILEY-VCH Verlag GmbH & Co. KGaA, Weinheim

effects within the porphyrin Soret region (trisignate or tetrasignate CD curves) (Figure 3.3.27). The origin of the observed multisignate CD profile is the strong electronic interaction between stacked porphyrins due to head-to-tail interduplex aggregation. This emphasizes the sensitivity of CD to structural changes from normal DNA duplexes to large interduplex aggreagates as well as the role of the phorphyrin to serve as an efficient glue to promote this transition.

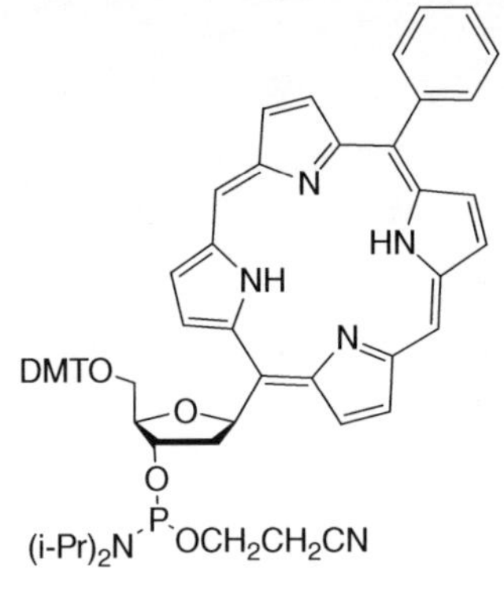

Figure 3.3.28 *Structure of **P** monomer **1***

This duplex aggregation has been supported by steady- and time-resolved fluorescence data while DeVoe calculations [90] provided additional conformation for the origin of the multisignate CD porphyrin bands.

3.3.3.3 Porphyrin C-nucleoside replacement of natural nucleobases

Kool's group studied the replacement of natural DNA nucleobases with different non-polar aromatics and heterocyles and how this might influence the electronic and physicochemical properties of the hybrid duplex [101]. In their work, they linked a free porphryin (**P**) directly to 2-deoxy-D-ribose by a C–C bond for attachment to the natural backbone of the polymer via phosphoramidite chemistry (Figure 3.3.28) [102]. The availability of the phenylporphyrin monomer **1** permitted the synthesis of oligonucleotides where the porphyrin could be placed anywhere within the duplex. In the event, they prepared a number of duplexes where the **P** was paired against **T**, **A**, and an abasic site, **a,** as well as an abasic nucleoside duplex terminated with a porphyrin nucleoside.

Duplexes with the porphyrin placed in the middle of the sequence and paired with **A** and **T** were less stable relative to the natural **A–T** containing a control duplex. The duplex paired with an abasic site, **a,** was more stable (~10 °C) than a duplex containing **a** paired with a natural base. CD spectra of these duplexes showed spectral features similar to those of a B-form control duplex in the DNA region of the spectra (Figure 3.3.29) indicating no obvious effect on the DNA backbone conformation by porphyrin substitution. A strong, negative induced CD signal in the Soret region at 400 nm for duplexes where the porphyrin is at or near the center of the duplex, suggests intercalation of the porphyrin within the major groove of the DNA. Single strands with **P** showed a strong positive band at the same wavelength consistent with no intercalation.

3.3.3.4 Porphyrins embedded in the backbone of DNA

Richert's group were the first to report the synthesis and physicochemical properties of oligodeoxynucleotides with an imbedded porphyrin attached to phosphates in order to determine the extent of the interaction between the porphyrin and neighboring nucleobases [103]. Specifically, they joined the alkylporphyrin **2** to the deoxyribosyl moiety, via the phosphate at the end of the oxypropyl side chains of the porphyrin, as indicated in Figure 3.3.30.

An octadecamer was hybridized with several complementary DNA oligomers and the duplex was characterized extensively by UV–Vis, CD, fluorescence, and NMR spectroscopy. All the duplexes synthesized were

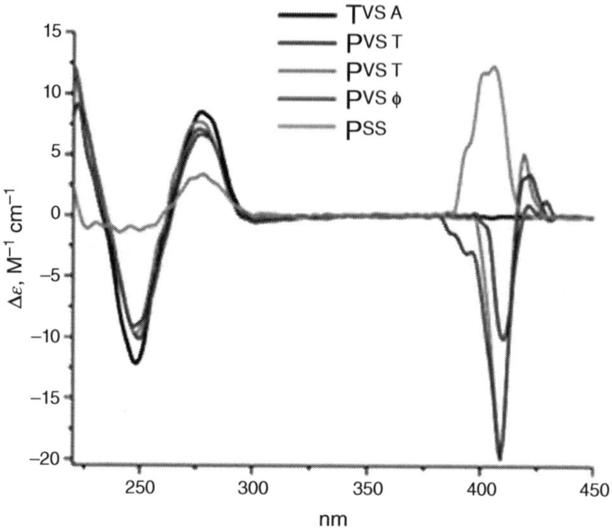

Figure 3.3.29 *Circular dichroism of a control duplex (**T** versus **A**); duplexes **1** and **2** with **P** paired **A** and **T**; duplex **3** with **P** opposite an abasic site, Φ **a**; and single-stranded **4** with **P**. Reprinted with permission from [102]. Copyright 2002, American Chemical Society*

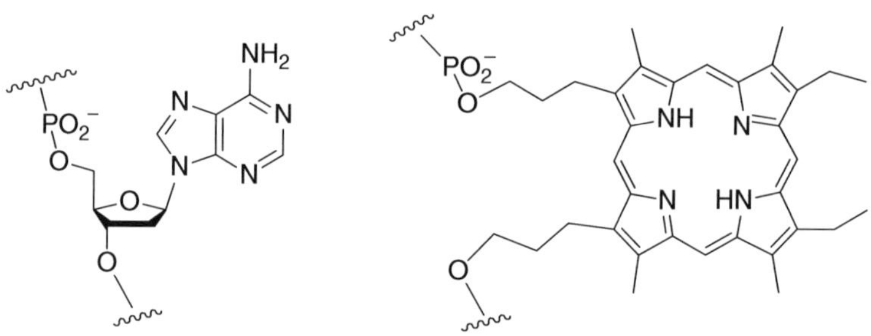

Figure 3.3.30 *Structure of monomer **2***

somewhat less stable than the unmodified control. The CD spectra in the DNA region of the duplexes were similar to that observed with normal B-form DNA, demonstrating no significant Watson–Crick structural distortions. A UV-melting temperature study showed loss of elipticity that is typical of normal DNA undergoing transition between duplex and random coil forms of the polymer. These absorbance, CD, and fluorescence experimental results show that the imbedded porphyrin interacts electronically with adjacent nucleobases with a stronger effect with duplex DNA relative to single-stranded DNA.

3.3.3.5 Diastereochemically pure anionic porphyrin–DNA dimers

The Majima group [104] were the first to prepare diastereochemically pure anionic porphyrin–DNA dimer complexes in which the porphyrin is tethered through a chiral phosphorus atom in the middle of a self-complementary DNA 10-mer, 5′-GCGTAT**P**ACGC-3′, and a non-self-complementary 5′-GCCT**P**AGTCG-3′. Both constructs were prepared with monomer **3**. The R_p-phosphoramidate configuration directs the attached

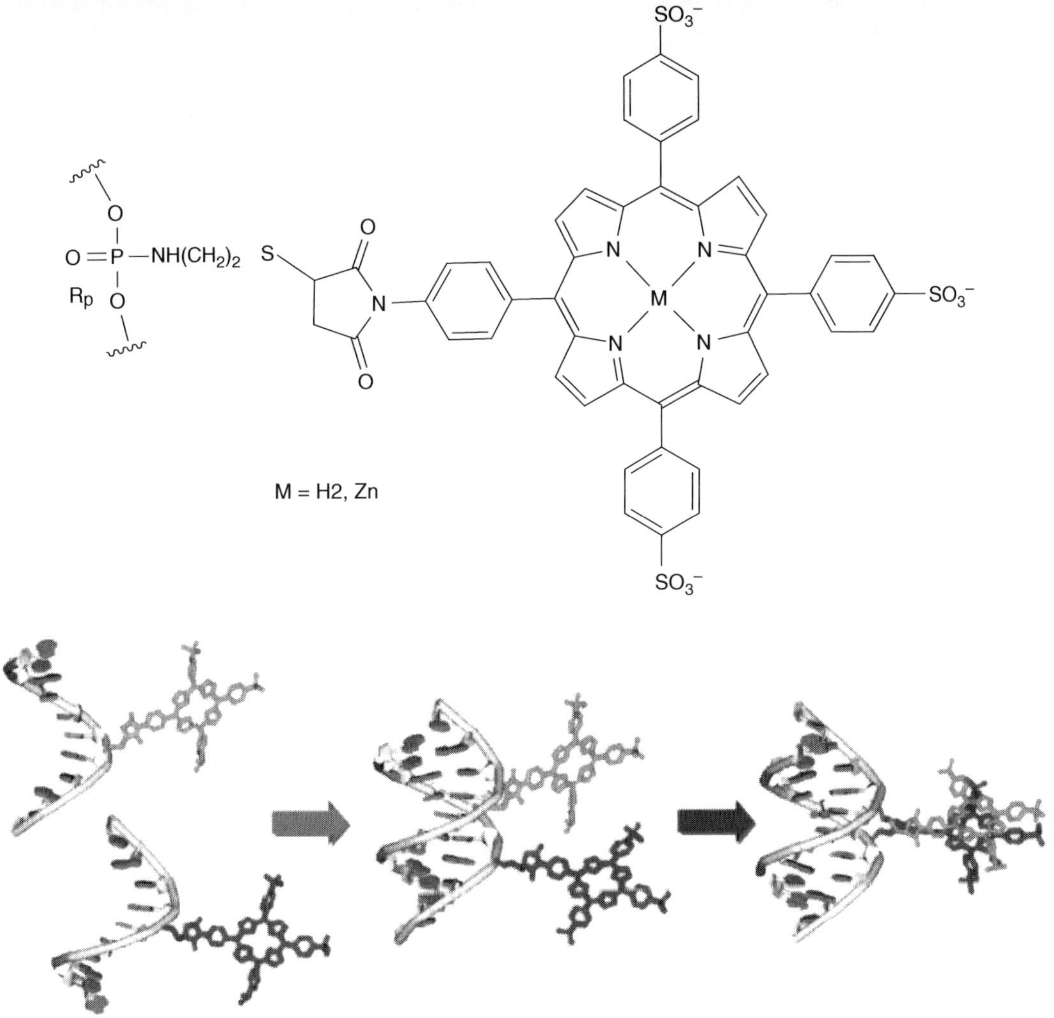

Figure 3.3.31 *Top: Structure of monomer 3; Bottom: Interaction of DNA and porphyrin moieties during self-assembly of DNA and porphyrin moieties by decreasing the temperature. The DNA formation in the initial step is followed by porphyrin H-like aggregate formation in the final step. Reprinted with permission from [104]. Copyright 2008, American Chemical Society*

porphyrin to the minor groove of the DNA helix whereas the S_p-phosphoramidate configuration directs the porphyrin toward the major groove of the duplex (Figure 3.3.31). Thus, having configurationally pure phosphoramidates available, either R_p or S_p, provides the means for the preparation of two configurationally distinct duplexes and to study their electronic and physicochemical properties.

CD measurements indicate that the DNA region is essentially the same as a classical B-form duplex with the usual positive and negative bands at 260 and 280 nm, respectively. The R_p-porphyrin-DNA constructs are stacked in a right-handed, sandwich-type, array as evidenced by the excitonic coupling between the positive and negative bands at 443 and 410 nm, respectively. In addition, this group prepared the S_p constructs that showed excitonic bands at 438 and 410 nm that also indicate a clockwise arrangement of the porphyrins.

The non-superimposable couplets of the two constructs, however, reflect the different configurations of the phosphoramidites from which the constructs were made.

3.3.3.6 Incorporation of rigid and flexible linked porphyrins to DNA nucleobases

Stulz and coworkers have been very active in this field and have synthesized a number of porphyrin–DNA conjugates. They attached tetraphenylporphyins to DNA nucleobases by attaching conformationally rigid acetylene porphyrins **4** as well as the more flexible propargyl amide linked porphyrins **5** to a 5-iodo-2′-deoxyuridine using Sonogashira chemistry [105–108], which is then incorporated into the DNA scaffold with automated phosphoramidite chemistry (Figure 3.3.32).

In a study with rigid acetylene porphyrin linked DNA zipper arrays, for example, **4**, stacking was observed to be reversible in the mixed free-base and zinc-metallated porphyrins, and energy transfer occurred in the annealed but not in the denatured state [108]. In another study [106] they were able to incorporate 11 contiguous tetraphenyl acetylene-linked porphryins **4** (after deprotection of the trityl group), in a row (10 nm scale) into a 21-mer duplex, **a–b,** with a loss of stability of 5–7 °C per porphyrin added (Figure 3.3.33). The CD spectrum in this duplex in the UV region gave evidence for a normal B-form helix with the usual bisignate positive signal pattern at (+276)/(–250) nm. The porphyrin region was characterized by a negative band at 422 nm with a very weak bisignate pattern indicating essentially no excitonic coupling. Force-field minimized structures agree with this explanation with the porphyrins located in the major groove with little or no influence on the DNA helicity. Thus CD is not applicable for a duplex with a large number of contiguously incorporated porphyrins joined to the DNA major groove by a rigid linker.

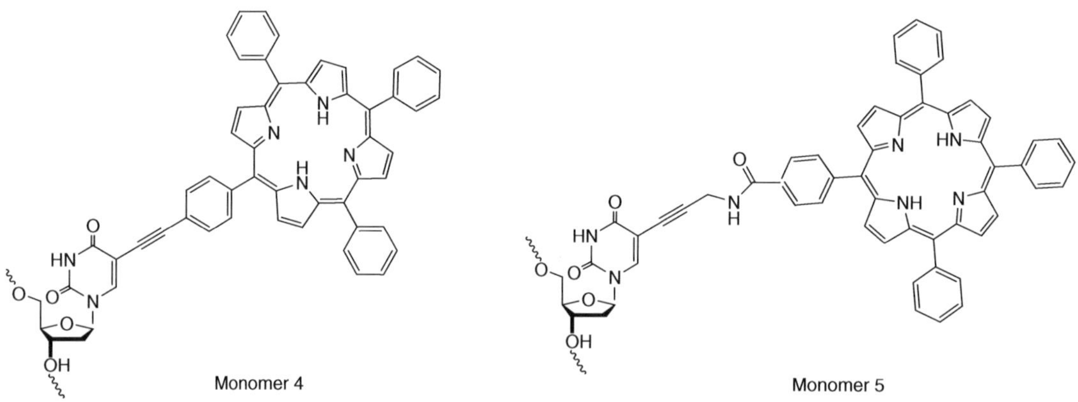

Figure 3.3.32 Structures of monomers **4** and **5**

a, 3'-ATG **CT1 111 111 111 1**GA TGC-5'
b, 5'-TAC CTA AAA AAA AAA ACT ACG-3'
c, 5'-GGC CAT CGT CGC **1A1 A1A 1A1 A1A** TTA T-3'
d, 3'-A GCG **A1A 1A1 A1A 1A1** AAT ACC GTA TGG-5'
e, 5'-GGC CAT CGT CGC **2A2 A2A 2A2 A2A** TTA T-3'
f, 3'-A GCG **A2A 2A2 A2A 2A2** AAT ACC GTA TGG-5'
g, 5'-GGC CAT CGT CGC TAT ATA TAT ATA TTA T-3' TA T-3'
h, 3'-A GCG ATA TAT ATA TAT AAT ACC GTA TGG-5'

Figure 3.3.33 Sequence of constructs **a** to **f** synthesized with monomers **4** and **5** and control constructs **g** and **h**. **1** = monomer **4**; **2** = monomer **5**

The CD of the above single-stranded porphyrin-linked DNA **a** suggests that a significant amount of nucleobase stacking imparts a helical structure in the porphyrin region, as evidenced by the bisignate signal in the UV part of the spectrum. This helicity is induced by the attached porphyrins, as the unmodified construct shows only a negative band at 300 nm, consistent with a random coil structure. Here again an induced negative band at around 420 nm is observed in the CD porphyrin region, supporting the idea that the porphyrins induce a helical structure for the porphyrin part of the construct.

In subsequent studies by the same group [107, 108] the rigid acetylene porphyrin linked DNA duplex **c–d** was compared with the more flexible propargyl amide linked DNA duplex **e–f** by UV-Vis, UV-melting, CD, and fluorescence spectroscopy (Figure 3.3.33). Placement of the rigid acetylene linked porphyrins in alternate complementary strands in zipper-like arrays led to stabilization of the DNA duplex, whereas placement of porphyrins in one strand led to destabilization of the duplex. Greater stability was observed with propargyl amide linked porphyrin DNA conjugates over the more rigid acetylene linked arrays. These duplexes all exhibit essentially the same UV spectroscopic behavior, indicating that the two types of linkers have little or no influence on the electronic properties of the porphyrin. The fluorescence intensity of the porphyrins in the zipper arrays increased on heating, indicating that the porphyrins do interact electronically when stacked in the major groove.

The DNA region of the CD spectra for both types of duplexes, in which the porphyrins are attached via the linkers in zipper-like constructs, indicate that the zipper DNA arrays are in normal B-form-like helical structures, as characterized by the usual bisignate pattern at (+276)/(–252) nm indicting little distortion of the duplexes. The pronounced negative Soret signal at 425 nm is observed, indicating the porphyrins are located in the major groove although models (see later) show they extend somewhat beyond the DNA helix. Upon heating, the signals sharpen significantly, indicative of unstacking of the zippered DNA. CD spectra of the Soret region for several porphyrin-linked DNA duplexes with the two types of linkers are shown in Figure 3.3.34, and the effect of cobalt, copper, and zinc on the Soret region in Figure 3.3.35.

It should be noted that duplex **e–f**, which contains only the conformationally flexible propargyl amide-linked monomer **5**, exhibits normal –/+ exciton couplet behavior in the Soret region, characteristic of linear oscillators where the B_x and B_y transitions behave as simple dipoles and are not discrete. Duplex **c–f** with both a flexible propargyl amide- and rigid acetylene-linked porphyrins also exhibited a normal +/– exciton couplet. In contrast, duplex **c–d,** made up of only conformationally rigid acetylene-linked porphyrins, shows a +/–/+ trisignate signal, characteristic of circular oscillators where both perpendicular B_x and B_y transitions contribute to through-space excitonic coupling.

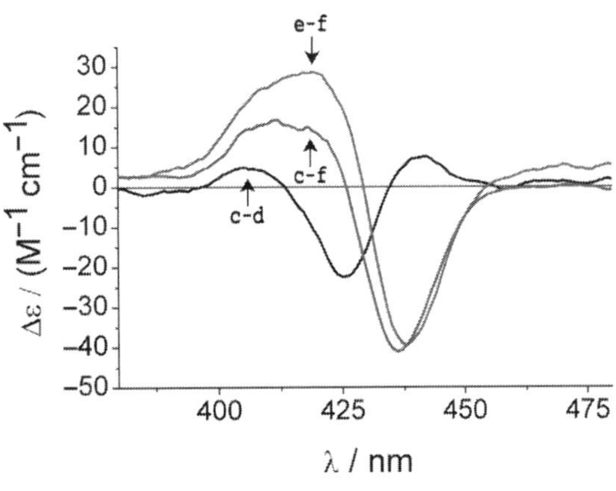

Figure 3.3.34 *Comparison of CD Soret region of duplexes* **c–d**, **e–f**, *and* **c–f**. *Reproduced from [107] with permission from The Royal Society of Chemistry*

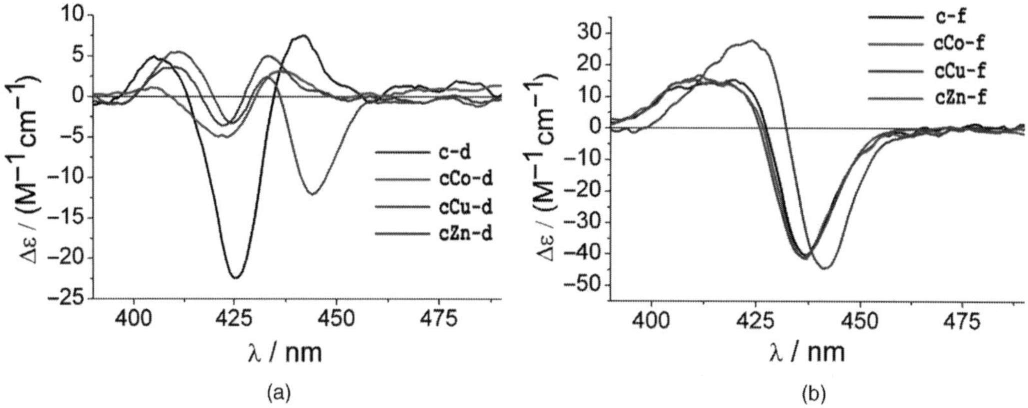

Figure 3.3.35 *Comparison of CD Soret region spectra of metallated-porphyrin DNA duplexes (a)* ***c–d*** *and (b)* ***c–f****. Reproduced from [107] with permission from The Royal Society of Chemistry (See color figure in color plate section)*

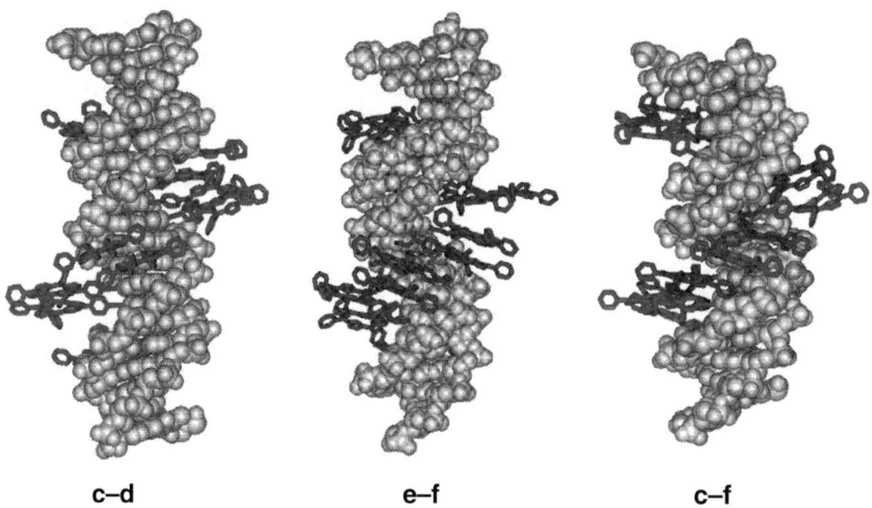

Figure 3.3.36 *Molecular models of porphyrin–DNA zipper systems* ***c–d, e–f, c–f****. Red is monomer* ***4*** *and blue is monomer* ***5****. Reproduced from [107] with permission from The Royal Society of Chemistry (See color figure in color plate section)*

The CD traces in the Soret region of cobalt-, copper-, and zinc-metallated duplexes **c–d** and **e–f** are shown in Figure 3.3.35. It is interesting that the cobalt and copper metallated duplex **c–d**, where both strands contain the rigid acetylene linked porphyrin, retains a +/–/+ trisignate pattern almost unchanged from the free base spectrum (Figure 3.3.35a). In contrast, the zinc metallated **c–d** spectrum shows a more complex –/+/–/+ tetrasignate signal. The zinc complex, being more hydrophilic and less planar with axially-liganded water, is quite different to the more hydrophobic cobalt- and copper-containing complexes. This is a good example of where CD spectroscopy is more diagnostic than UV–Vis for, as mentioned previously, the UV–Vis spectra of the various constructs are almost identical. Upon metallation of the mixed linker **c–f** duplex, with both rigid acetylene and the more flexible propargyl amide linker, the resulting CD spectra are essentially the same compared with the spectrum of the free base duplex **g–h**.

Molecular models of these porphyrin–DNA conjugates that are in agreement with the spectral data have been constructed and are depicted in Figure 3.3.36. The zipper-like arrays stem from the zigzag arrangement of the porphyrins, which are stacked in pairs in duplexes **c–d** and **e–f**. As can be seen in the models, the structures are different for the three combinations. When two different linkers are combined as in duplex **c–f**, the porphyrin rings become almost perpendicular to the helical axis and the pairwise stacking is reduced.

3.3.4 Conclusions

In conclusion, we have summarized stability and spectroscopic findings from some selected examples of recently synthesized porphyrin–DNA non-covalent and conjugates complexes. Porphyrinoids are ideal compounds to be incorporated into DNA due to their peculiar spectroscopic characteristics, fascinating photochemistry, and synthetic versatility. The ability to act as photosensitizers in the presence of oxygen promoted interest of the scientific community to study the non-covalent binding of porphyrins with DNA in order to use them in photodynamic therapy. It has been demonstrated that a photosensitizer needs to be bound to DNA in order to produce photodynamic damage, and the efficiency of this process depends on the type of interaction of the photosensitizer. Thus, the understanding of the mechanisms involved in the binding of porphyrins with DNA can help to find a porphyrin with specific base degradation activity.

One of the spectroscopic techniques mainly used for their characterization is circular dichroism. This technique turned out to be very useful not only for its intrinsic ability in detecting DNA conformations and their changes upon complexation, but (in the case of the achiral porphyrins used in the studies described) also for the possibility of monitoring porphyrins owing to the ICD signal increasing in the porphyrin absorption region (Soret). The appearance of an ICD is unquestionable evidence of porphyrin–DNA interactions. The shape of this ICD signal depends on the binding mode and on the DNA conformation, making porphyrins excellent chiroptical probes.

In addition to medical applications, porphyrinoid–DNA systems can be promoted in technological fields as molecular devices. In fact, it is possible to design reversible information storing systems by using the right combination of modulable non-covalent interactions (electrostatic, hydrophobic, Van der Waals, hydrogen bonds, etc.). The conjugate DNA–porphyrin complexes required the development of new synthetic chemistry in combination with traditional phosphoramidite and solid state automated DNA synthesis, which was successful in providing the targeted sequences in reasonable yields.

A common theme in all these examples is that attaching free base and metallated porphyrins to the DNA scaffolding does not result in significant distortion of the DNA backbone conformation, thus simplifying the structural study of the DNA conjugates and the corresponding inter-duplex nanostructures. Thus, these cases provide a solid foundation for further modulation and tuning of the photochemical and physicochemical properties of these constructs for potential use in molecular electronics.

References

[1] S. Neidle and S. Balasubramanian *Quadruplex Nucleic Acids,* The Royal Society of Chemistry, London (2006).

[2] P. Belmont, J.-F. Constant, and M. Demeunynck, Nucleic acid conformation diversity: from structure to function and regulation, *Chem. Soc. Rev.*, **30**, 70–81 (2001).

[3] Q. H. Liu, L. M. Wang, A. G. Frutos, A. E. Condon, R. M. Corn, and L. M. Smith, DNA computing on surfaces, *Nature*, **403**, 175–179 (2000).

[4] M. Balaz, K. Bitsch-Jensen, A. Mamana, G. A. Ellestad, and N. Berova, Porphyrins as spectroscopic sensors for conformational studies of DNA, *Pure Appl. Chem.*, **79**, 801-809 (2007).

[5] M. Edmonds and M. A. Winter, A poly(A) polymerase from calf thymus characterization of the reaction product and the primer requirement, *Prog. Nucleic Acid Res. Mol. Biol.*, **17**, 149–179 (1976).

[6] P. Ahlquist and P. Kaesberg, Determination of the length distribution of poly(A) at the 3′ terminus of the virion RNAs of EMC virus, poliovirus, rhinovirus, RAV-61 and CPMV and of mouse globin mRNA, *Nucleic Acid Res.*, **7**, 1195–1204 (1979).

[7] W. Habano, T. Sugai, T. Yoshida, and S. Nakamura, Mitochondrial gene mutation, but not large-scale deletion, is a feature of colorectal carcinomas with mitochondrial microsatellite instability, *Int. J. Cancer*, **83**, 625–629 (1999).

[8] S. Schwartz Jr, H. Alazzouzi, and M. Perucho, Mutational dynamics in human tumors confirm the neutral intrinsic instability of the mitochondrial D-loop poly-cytidine repeat, *Genes Chromosomes Cancer*, **45**(8), 770–780 (2006).

[9] V. L. Makarov, A. Poletaev, and M. V. Vol'kenshtein, Circular dichroism of dna complexes with dyes. Part 1 theoretical treatment of chromophore chromophore interaction of ligands, *Mol. Biol.*, **11**(1), 228–236 (1977).

[10] R. F. Pastemack, S. Gurrieri, R. Lauceri, and R. Purrello, Single-stranded nucleic acids as templates for porphyrin assembly formation, *Inorg. Chim. Acta,* **246**, 7–12 (1996).

[11] M. Dourlent and J. F. Hogrel, Competitive cooperative bindings of a small ligand to a linear polymer. II. Investigations on the mechanisms of proflavine binding to poly(A) and DNA, *Biopolymers*, **15**, 29–41 (1976).

[12] E. J. Gibbs, I. Tinoco Jr, M. F. Maestre, P. A. Ellinas, and R. F. Pasternack, Self-assembly of porphyrins on nucleic acid templates, *Biochem. Biophys. Res. Commun.*, **157**, 350–358 (1988).

[13] R. F. Pasternack, R. A. Brigandi, M. J. Abrams, A. P. Williams, and E. J. Gibbs, Interactions of porphyrins and metalloporphyrins with single-stranded poly (dA), *Inorg. Chem.*, **29**, 4483–4486 (1990).

[14] W. Seanger, *Principles of Nucleic Acid Structure*, Springer-Verlag, New York (1988).

[15] C. M. Olsthoorn, L. J. Bostelaar, J. F. M. De Rooij, J. H. Van Boom, and C. Altona, Circular dichroism study of stacking properties of oligodeoxyadenylates and polydeoxyadenylate. A three-state conformational mode, *Eur. J. Biochem.*, **115**, 309–321 (1981).

[16] C. M. Olsthoorn, L. J. Bostelaar, J. H. Van Boom, and C. Altona, Circular dichroism studies of 6-N-methylated adenylyladenosine and adenylyluridine and their parent compounds: Thermodynamics of stacking, *Eur. J. Biochem.*, **106**, 85–95 (1980).

[17] C. Bustamante, S. Guerrieri, R. F. Pasternack, R. Purrello, and E. Rizzarelli, Interaction of water-soluble porphyrins with single- and double-stranded polyribonucleotides, *Biopolymers*, **34**, 1099–1104 (1994).

[18] A. Slama–Schwok, J.-M. Lehn, Interaction of a porphyrin-containing macrotetracyclic receptor molecule with single-stranded and double-stranded polynucleotides. A photophysical study, *Biochemistry*, **29**, 7895–7903 (1990).

[19] A. D'Urso, S. Nardis, G. Pomarico, M. E. Fragalà, R. Paolesse, and R. Purrello, Interaction of tricationic corroles with single/double helix of homopolymeric nucleic acids and DNA, *J. Am. Chem. Soc.*, **135**, 8632–8638 (2013).

[20] R. Paolesse, S. Licoccia, and T. Boschi, Towards the periodic table of metallocorrolates: synthesis and characterization of main group metal complexes of octamethylcorrole, *Inorg. Chim. Acta*, **178**, 9–12 (1990).

[21] L. Simkhovich, A. Mahammed, I. Goldberg, and Z. Gross, Synthesis and characterization of germanium, tin, phosphorus, iron, and rhodium complexes of tris(pentafluorophenyl)corrole, and the utilization of the iron and rhodium corroles as cyclopropanation catalysts, *Chem. Eur. J.*, **7**, 1041–1055 (2001).

[22] M. Mastroianni, W. Zhu, M. Stefanelli, S. Nardis, F. R. Fronczek, K. M. Smith, Z. Ou, K. M. Kadish, and R. Paolesse, β-Nitro derivatives of germanium(IV) corrolates, *Inorg. Chem.*, **47**, 11680–11687 (2008).

[23] S. Nardis, F. Mandoj, R. Paolesse, F. R. Fronczek, K. M. Smith, L. Prodi, M. Montalti, and G. Battistini, Synthesis and functionalization of germanium triphenylcorrolate: The first example of a partially brominated corrole, *Eur. J. Inorg. Chem.*, **16**, 2345–2352 (2007).

[24] R. J. Fiel, E. H. Mark, and N. D. Gupta, Radiation modifying effects of meso-tetraphenyl porphyrins, *Res. Commun. Chem. Pathol. Pharmacol.*, **10**, 65–76 (1975).

[25] (a)R. F. Pasternack and E. J. Gibbs, *Metal–DNA Chemistry*, T. Tullius (Ed.), ACS Symposium Series **402**, American Chemical Society, Washington, DC, pp. 59–73 (1989);(b)B. P. Hudson, J. Sou, D. J. Berger, and D. R. McMillin, Luminescence studies of the intercalation of Cu(TMpyP4) into DNA, *J. Am. Chem. Soc.*, **114**, 8997–9002 (1992).

[26] A. D'Urso, M. E. Fragalà, and R. Purrello, Non-covalent interactions of porphyrinoids with duplex DNA, *Top Heterocycl. Chem.*, DOI: 10.1007/7081_2013_113 (2013).

[27] (a)R. F. Pasternack and P. J. Collings, Resonance light scattering: a new technique for studying chromophore aggregation, *Science*, **269**, 935–939 (1995);(b)P. J. Collings, E. J. Gibbs, T. E. Starr, O. Vafek, C. Yee, L. A. Pomerance, and R. F. Pasternack, Resonance light scattering and its application in determining the size, shape, and aggregation number for supramolecular assemblies of chromophores, *J. Phys. Chem. B*, **103**, 8474–8481 (1999).

[28] L. Monsù Scolaro, A. Romeo, and R. F. Pasternack, Tuning porphyrin/DNA supramolecular assemblies by competitive binding, *J. Am. Chem. Soc.*, **126**, 7178–7179 (2004).

[29] X. Chen and M. Liu, Induced chirality of binary aggregates of oppositely charged water-soluble porphyrins on DNA matrix, *J. Inorg. Biochem.*, **94**, 106–113 (2003).

[30] F. M. Pohl and T. M. Jovin, Salt-induced co-operative conformational change of a synthetic DNA: equilibrium and kinetic studies with poly (dG-dC), *J. Mol. Biol.*, **67**, 375–396 (1972).

[31] J. V. Ditlevson, S. Tornaletti, B. P. Belotserkovskii, V. Teijeiro, G. Wang, K. M. Vasquez, and P. C. Hanawalt, Inhibitory effect of a short Z-DNA forming sequence on transcription elongation by T7 RNA polymerase, *Nucleic Acid Res.*, **36**, 3163–3170 (2008).

[32] P. Dröge, Protein tracking-induced supercoiling of DNA: A tool to regulate DNA transactions *in vivo*? *BioEssays*, **16**, 91–99 (1994).

[33] Y. G. Kim, K. Lowenhaupt, D. B. Oh, K. K. Kim, and A. Rich, Evidence that vaccinia virulence factor E3L binds to Z-DNA *in vivo*: Implications for development of a therapy for poxvirus infection, *Proc. Natl. Acad. Sci. U.S.A.*, **101**, 1514–1518 (2004).

[34] J. A. Kwon and A. Rich, Biological function of the vaccinia virus Z-DNA-binding protein E3L: Gene transactivation and antiapoptotic activity in HeLa cells, *Proc. Natl. Acad. Sci. U.S.A.*, **102**, 12759–12764 (2005).

[35] A. Herbert, K. Lowenhaupt, J. Spitzner, and A. Rich, Chicken double-stranded RNA adenosine deaminase has apparent specificity for Z-DNA, *Proc. Natl. Acad. Sci. U.S.A.*, **92**, 7550–7554 (1995).

[36] E. Dickerson, H. R. Drew, B. N. Conner, R. M. Wing, A. V. Fratini, and M. L. Kopka, The anatomy of A-, B-, and Z-DNA, *Science*, **216**, 475–485 (1982).

[37] P. Belmont, J. F. Constant, and M. Demeunyck, Nucleic acid conformation diversity: from structure to function and regulation, *Chem. Soc. Rev.*, **30**, 70–81 (2001).

[38] A. Rich, A. Nordheim, and A. H.-J. Wang, The chemistry and biology of left-handed Z-DNA, *Annu. Rev. Biochem.*, **53**, 791–846 (1984).

[39] A. Jaworski, W. T. Hsieh, J. A. Blaho, J. E. Larson, and R. D. Wells, Left-handed DNA *in vivo*, *Science*, **238**, 773–777 (1987).

[40] A. Parkinson, M. Hawken, M. Hall, K. J. Sanders, and A. Rodger, Amine induced Z DNA in poly(dG-dC)·poly(dG-dC): Circular dichroism and gel electrophoresis study, *Phys. Chem. Chem. Phys.*, **2**, 5469–5478 (2000)

[41] A. Krzyaniak, P. Salaski, J. Jurczak, and J. Barciszewski, B-Z DNA reversible conformation changes effected by high pressure, *FEBS Lett.*, **279**, 1–4 (1991).

[42] K. E. Van Holde, W. C. Johnson, and P. S. Ho, *Principles of Physical Biochemistry*, 3rd edn, Pearson/Prentice Hall, Upper Saddle River, NJ (1998).

[43] V. I. Ivanov and E. E. Minyat, The transitions between left- and right-handed forms of poly(dG-dC), *Nucleic Acids Res.*, **9**, 4783–4798 (1981).

[44] K. B. Hall and M. F. Maestre, Temperature-dependent reversible transition of poly(dCdG)·poly(DcdG) in ethanolic and methanolic solutions, *Biopolymers*, **23**, 2127–2139 (1984).

[45] M. E. Harder and W. C. Johnson, Stabilization of the Z' form of poly(dGdC):poly(dGdC) in solution by multivalent ions relates to the ZII form in crystals, *Nucleic Acids Res.*, **18**, 2141–2148 (1990).

[46] F. Kouzine and D. Levens, Supercoil-driven DNA structures regulate genetic transactions, *Front Biosci.*, **1**, 4409–4423 (2007).

[47] T. M. Jovin, D. M. Soumpasis, and L. P. McIntosh, The transition between B-DNA and Z-DNA, *Annu. Rev. Phys. Chem.*, **38**, 521–558 (1987).

[48] A. Rich, A. Nordheim, and A. H.-J. Wang, The chemistry and biology of left-handed Z-DNA, *Annu. Rev. Biochem.*, **53**, 791–846 (1984).

[49] K. S. Latha, S. Anitha, K. S. J. Rao, and M. A. Viswamitra, Molecular understanding of aluminum-induced topological changes in (CCG)12 triplet repeats: relevance to neurological disorders, *Biochim. Biophys. Acta*, **1588**, 56–64 (2002).

[50] P. Helden, Potential Z-DNA-forming elements in serum DNA from human systemic lupus erythematosus, *J. Immunol.*, **134**, 177–179 (1985).

[51] A. H. J. Wang, G. J. Quigley, F. J. Kolpak, J. L. Crawford, J. H. van Boom, G. van der Marel, and A. Rich, Molecular structure of a left-handed double helical DNA fragment at atomic resolution, *Nature*, **282**, 680–686 (1979).

[52] A. Rich and S. Zhang, Z-DNA: the long road to biological function, *Nat. Rev. Genet.*, **4**, 566–572 (2003).

[53] D. B. Oh, Y. G. Kim, and A. Rich, Z-DNA-binding proteins can act as potent effectors of gene expression invivo, *Proc. Natl. Acad. Sci. U.S.A.*, **99**, 16666–16671 (2002).

[54] S. Rothenburg, F. Koch-Nolte, A. Rich, and F. Haag, A polymorphic dinucleotide repeat in the rat nucleolin gene forms Z-DNA and inhibits promoter activity, *Proc. Natl. Acad. Sci. U.S.A.*, **98**, 8985–8990 (2001).

[55] B. Wong, S. Chen, J. A. Kwon, and A. Rich, Characterization of Z-DNA as a nucleosome-boundary element in yeast Saccharomyces cerevisiae, *Proc. Natl. Acad. Sci. U.S.A.*, **104**, 2229–2234 (2007).

[56] M. M. Garner and G. Felsenfeld, Effect of Z-DNA on nucleosome placement, *J. Mol. Biol.*, **196**, 581–590 (1987).

[57] H. Liu, N. Mulholland, H. Fu, and K. Zhao, Cooperative activity of BRG1 and Z-DNA formation in chromatin remodeling, *Mol. Cell. Biol.*, **26**, 2550–2559 (2006).

[58] G. Wang and K. M. Vasquez, Non-B DNA structure-induced genetic instability, *Mutat. Res.*, **598**, 103–119 (2006).

[59] G. Wang, L. A. Christensen, and K. M. Vasquez, Z-DNA-forming sequences generate large-scale deletions in mammalian cells, *Proc. Natl. Acad. Sci. U.S.A.*, **103**, 2677–2682 (2006).

[60] R. F. Pasternack, D. Sidney, P. A. Hunt, E. A. Snowden, and E. J. Gibbs, Interactions of water soluble porphyrins with Z-poly(dG.dC), *Nucleic Acids Res.*, **14**, 3927–3943 (1986).

[61] R. E. McKinnie, J. D. Choi, J. W. Bell, E. J. Gibbs, and Robert F. Pasternack, Porphyrin induced Z to B conversion of poly$(dG\text{-}dC)_2$ in ethanol, *J. Inorg. Biochem.*, **32**, 207–224 (1988).

[62] M. Balaz, M. De Napoli, A. E. Holmes, A. Mammana, K. Nakanishi, N. Berova, and R. Purrello, A cationic zinc porphyrin as a chiroptical probe for Z-DNA, *Angew. Chem.*, **117**, 4074–4077 (2005); *Angew. Chem., Int. Ed.*, **44**, 4006–4009 (2005).

[63] A. D'Urso, A. E. Holmes, N. Berova, M. Balaz, and R. Purrello, Z-DNA recognition in B-Z-B sequences by a cationic zinc porphyrin, *Chem. Asian J.*, **6**, 3104–3109 (2011).

[64] A. Moller, A. Nordheim, S. A. Kozlowski, D. J. Patel, and A. Rich, Bromination stabilizes poly(dG-dC) in the Z-DNA form under low-salt conditions, *Biochemistry*, **23**, 54–62 (1984).

[65] A. Nadler and U. Diederichsen, Guanosine analog with respect to Z-DNA stabilization: Nucleotide with combined C8-bromo and C2′-ethynyl modifications, *Eur. J. Org. Chem.*, **9**, 1544–1549 (2008).

[66] T. Kimura, K. Kawai, S. Tojo, and T. Majima, One-electron attachment reaction of B- and Z-DNA modified by 8-bromo-2′-deoxyguanosine, *J. Org. Chem.*, **69**, 1169–1173 (2004).

[67] A. D'Urso, J. K. Choi, M. Shabbir-Hussain, F. N. Ngwa, M. I. Lambousis, R. Purrello, and M. Balaz, Recognition of left-handed Z-DNA of short unmodified oligonucleotides under physiological ionic strength conditions, *Biochem. Biophys. Res. Commun.*, **397**, 329–332 (2010).

[68] A. D'Urso, A. Mammana, M. Balaz, A. E. Holmes, N. Berova, R. Lauceri, and R. Purrello, Interactions of a tetraanionic porphyrin with DNA: from a Z-DNA sensor to a versatile supramolecular device, *J. Am. Chem. Soc.*, **131**, 2046–2047 (2009).

[69] A. E. Holmes, J. K. Choi, J. Francis, A. D'Urso, and M. Balaz, Sulfonated Ni(II)porphyrin improves the detection of Z-DNA in condensed and non-condensed BZB DNA sequences, *J. Inorg. Biochem.*, **110**, 18–20 (2012).

[70] J. K. Choi, G. Sargsyan, M. Shabbir-Hussain, A. E. Holmes, and M. Balaz, Chiroptical detection of condensed nickel(II)-Z-DNA in the presence of the B-DNA via porphyrin exciton coupled circular dichroism, *J. Phys. Chem. B*, **115**, 10182–10188 (2011).

[71] K. A. Olausse, K. Dubrana, J. Domot, J.-P. Spano, and L. Sabatier, J.-C. Soria, Telomeres and telomerase as targets for anticancer drug development, *Crit. Rev. Oncol. Hematol.*, **57**, 191–214 (2006).

[72] M. A. Blasco, Telomeres and human disease: Ageing, cancer and beyond, *Nat. Rev, Genet.*, **6**, 611–622 (2005).

[73] S. Neidle and G. Parkinson, Telomere maintenance as a target for an anticancer drug discovery, *Nat. Rev. Drug Discovery*, **1**, 383–393 (2002).

[74] I. M. Dixon, F. Lopez, A. M. Tejera, J-P. Esteva, M. A. Blasco, G. Pratviel, and B. Maunier, A G-quadruplex ligand with 10000-fold selectivity over duplex DNA, *J. Am. Chem. Soc.*, **129**, 1502–1503 (2007).

[75] H.-J. Zhang, X.-F. Wang, P. Wang, S.-P. Pang, X.-C. Ai, and J.-P. Zhang, Interactions between meso-tetrakis (4-(N-methyl pyridiumyl)) porphyrin TMPyP4 and DNA G-quadruplex of telomeric repeated sequence TTAGGG, *Sci. China Ser. B:Chem.*, **51**, 452–456 (2008).

[76] I. Lubitz, N. Borovok, and A. Kotlyar, Interaction of monomolecular G4-DNA nanowires with TMPyP: evidence for intercalation, *Biochemistry*, **46**, 12925–12929 (2007).

[77] V. A. Narayanan, M. Azam, and R. D. Sheardy, Porphyrin binding to Quadruplexed T4G4, *Biochemistry*, **37**, 2709–2714 (1998).

[78] I. Haq, J. O. Trent, B. Z. Chowdhry, and T. C. Jenkins, Intercalative G-tetraplex stabilization of telomeric DNA by a cationic porphyrin, *J. Am. Chem. Soc.*, **121**, 1768–1779 (1999).

[79] H. Mita, T. Ohyama, Y. Tanaka, and Y. Yamamoto, Formation of a complex of 5,10,15,20-tetrakis(**N**-methylpyridinium-4-yl)-21**H**,23**H**-porphyrin with G-quadruplex DNA, *Biochemistry*, **45**, 6765–6772 (2006).

[80] Y. Wang and D. J. Patel, Guanine residues in d(T2AG3) and d(T2G4) form parallel-stranded potassium cation stabilized G-tetraplexes with anti glycosidic torsion angles in solution, *Biochemistry*, **31**, 8112–8119 (1992).

[81] G. N. Parkinson, R. Ghosh, and S. Neidle, Structural basis for binding of porphyrin to human telomeres, *Biochemistry*, **46**, 2390–2397 (2007).

[82] H. Zhang, X. Xiao, P. Wang, S. Pang, F. Qu, X. Ai, and J. Zhang, Conformational conversion of DNA G-quadruplex induced by a cationic porphyrin, *Spectrochim. Acta Part A*, **74**, 243–247 (2009).

[83] Y.-A. Shieh, S.-J. Yang, M.-F. Wei, and M.-J. Shieh, Aptamer-based tumor-targeted drug delivery for photodynamic therapy, *ACS Nano*, **4**, 1433–1442 (2010).

[84] F. X. Han, R. T. Wheelhouse, and L. H. Hurley, Interactions of TMPyP4 and TMPyP2 with quadruplex DNA. Structural basis for the differential effects on telomerase inhibition, *J. Am. Chem. Soc.*, **121**, 3561–3570 (1999).

[85] H. Han, D. R. Langley, A. Rangan, and L. H. Hurley, Selective interactions of cationic porphyrins with G-quadruplex structures, *J. Am. Chem. Soc.*, **123**, 8902–8913 (2001).

[86] A. J. Bhattacharjee, K. Ahluwalia, S. Taylor, O. Jin, J. M. Nicoludis, R. Buscaglia, J. B. Chaires, D. J. P. Kornfilt, D. G. S. Marquardt, and L. A. Yatsunyk, Induction of G-quadruplex DNA structure by Zn(II)5,10,15,20-tetrakis (N-methyl-4-pyridyl)porphyrin, *Biochimie*, **93**, 1297–1309 (2011).

[87] M. Balaz, J. D. Steinkruger, G. A. Ellestad, and N. Berova, 5′-Porphyrin-oligonucleotide conjugates: neutral porphyrin-DNA interactions, *Org. Lett.*, **7**, 5613–5616 (2005).

[88] M. Balaz, A. E. Holmes, M. Benedetti, P. C. Rodriguez, N. Berova, K. Nakanishi, and G. Proni, Synthesis and circular dichroism of tetraarylporphyrin-oligonucleotide conjugates, *J. Am. Chem. Soc.*, **127**, 4172–4173 (2005).

[89] M. Balaz, B. C. Li, S. Jockusch, G. A. Ellestad, and N. Berova, Tetraarylporphyrin as a selective molecular cap for non-Watson–Crick guanine–adenine base-pair sequences, *Angew. Chem., Int. Ed.*, **45**, 3530–3533 (2006).

[90] A. Mammana, G. Pescitelli, T. Asakawa, S. Jockusch, A. G. Petrovic, R. R. Monaco, R. Purrello, N. J. Turro, K. Nakanishi, G. A. Ellestad, M. Balaz, and N. Berova, Role of environmental factors on the structure and spectroscopic response of 5′-DNA-porphyrin-conjugates due to changes in porphyrin-porphyrin interaction, *Chem. Eur. J.*, **15**, 11853–11866 (2009).

[91] M. Balaz, B. C. Li, J. D. Steinkruger, G A. Ellestad, K. Nakanishi, and N. Berova, Porphyrins conjugated to DNA as CD reporters of the salt-induced B to Z-DNA transition, *Org. Biomol. Chem.*, **4**, 1865–1867 (2006).

[92] S. Oancea, F. Formaggio, S. Campestrini, Q. B. Broxterman, B. Kaptein, and C. Toniolo, Distance dependency of exciton coupled circular dichroism using turn and helical peptide spacers, *Biopolymers*, **72**, 105 (2003).

[93] F. X. Redl, M. Lutz, and J. Daub, Chemistry of porphyrin-appended cellulose strands with a helical structure: spectroscopy, electrochemistry, and *in situ* circular dichroism spectroelectrochemistry, *Chem. Eur. J.*, **7**, 5350–5358 (2001).

[94] J. B. MacMillan and T. F. Molinski, Long-range stereo-relay: Relative and absolute configuration of 1,*n*-glycols from circular dichroism of liposomal porphyrin esters, *J. Am. Chem. Soc.*, **126**, 9944–9945 (2004).

[95] S. Matile, N. Berova, K. Nakanishi, J. Fleischhauer, and R. W. Woody, Structural studies by exciton coupled circular dichroism over a large distance: Porphyrin derivatives of steroids, dimeric steroids, and brevetoxin B, *J. Am. Chem. Soc.*, **118**, 5198–5206 (1996).

[96] S. Matile, N. Berova, and K. Nakanishi, Exciton coupled circular dichroic studies of self-assembled brevetoxin-porphyrin conjugates in lipid bilayers and polar solvents, *Chem. Biol.*, **3**, 379–392 (2004).

[97] K. Tsubaki, I. H. Tanaka, M. Miura, K. Takahash, and T. Kawabata, Long-range exciton-coupled circular dichroism: Application for determination of the absolute configuration of oligonaphthalenes, *Org. Lett.*, **8**, 2587–2590 (2006).

[98] M. Balaz, A. E. Holmes, M. Benedetti, G. Proni, and N. Berova, Porphyrin substituted phosphoramidites: new building blocks for porphyrin–oligonucleotide syntheses, *Bioorg. Med. Chem.*, **13**, 2413–2421 (2005).

[99] A. Mammana, T. Asakawa, K. Bitsch-Jensen, A. Wolfe, S. Chaturantabut, Y. Otani, X. Li, Z. Li, K. Nakanishi, M. Balaz, G. A. Ellestad, and N. Berova, Synthesis and characterization of water-soluble free-base, zinc and copper porphyrin-oligonucleotide conjugates, *Bioorg. Med. Chem.*, **16**, 6544–6551 (2008).

[100] G. Sargsyan and M. Balaz, Porphyrin–DNA conjugates: porphyrin induced adenine–guanine homoduplex stabilization and interduplex assemblies, *Org. Biomol. Chem.*, **10**, 5533–5540 (2012).

[101] E. T. Kool, Replacing the nucleobases in DNA with designer molecules, *Acc. Chem. Res.*, **35**, 936–943 (2002).

[102] H. Morales and E. T. Kool, A porphyrin C-nucleoside incorporated into DNA, *Org. Lett.*, **4**, 4377–4380 (2002).

[103] K. Berlin, R. K. Jain, M. D. Simon, and C. Richert, A porphyrin embedded in DNA, *J. Org. Chem.*, **63**, 1527–1435 (1998).

[104] M. Endo, M. Fujitsuka, and T. M. Majima, Diastereochemically controlled porphyrin dimer formation on a DNA scaffold, *J. Org. Chem.*, **73**, 1106–1112 (2008).

[105] T. N. Nguyen, A. Brewer, and E. Stulz, Duplex stabilization and energy transfer in zipper porphyrin-DNA, *Angew. Chem., Int. Ed.*, **48**, 1974–1977 (2009).

[106] L.-A. Fendt, I. Bouamaied, S. Thöni, and E. Stulz, DNA as supramolecular scaffold for porphyrin arrays on the namometer scale, *J. Am. Chem Soc.*, **129**, 15319–15329 (2007).

[107] A. Brewer, G. Siliggardi, C. Neylon, and E. Stulz, Introducing structural flexibility into porphyrin-DNA zipper arrays, *Org. Biomol. Chem.*, **9**, 777–782 (2011).

[108] T. N. Nguyen, A. Brewer, and E. Stulz, Duplex stabilization and energy transfer in zipper porphyrin-DNA, *Angew. Chem., Int. Ed.*, **48**, 1974–1977 (2009).

3.4

Chemical Reactions Controlled by Nucleic Acids and their Applications for Detection of Nucleic Acids in Live Cells

Andriy Mokhir

Friedrich-Alexander-University Erlangen-Nürnberg, Department of Chemistry and Pharmacy, Germany

3.4.1 Introduction

Nucleic acid controlled chemical reactions of various types are known. They include enantioselective transformations using DNA-based catalysts (described by Roelfes in Chapter 5.7), DNA-induced formation of nanoparticles consisting, for example, of silver nanoclusters (see Chapter 4.4 by Somoza), and other hybridization-triggered or templated reactions. In the last case, the reactants are two moieties (**A** and **B**), which are covalently attached to short oligonucleotides or their analogues (**A**–ON1, ON2–**B**). When sequences of ON1 and ON2 are selected to be complementary to each other and fragments **A** and **B** are attached to opposite termini of these ONs, conjugates **A**–ON1 and ON2–**B** form double helical structures in solution, in which **A** and **B** are positioned in close proximity to each other. Therefore, the reaction between **A** with **B** in an **A**–ON1/ON2–**B** associate is strongly facilitated with respect to the reaction between unconjugated **A** and **B** (Figure 3.4.1). This design has been used for combinatorial synthesis of small molecules, development of new chemical reactions, and testing the compatibility of known chemical reactions with nucleic acids [1]. A related approach has been also applied to search for specific ligands to bind to protein targets [2].

In other design, sequences ON1 and ON2 are selected to be complementary to some (e.g., naturally occurring) nucleic acids. In the presence of this nucleic acid (TEMPLATE) the associate TEMPLATE/(**A**–ON1, ON2–**B**) is formed usually quantitatively even at low concentrations of the reagents, due to the high affinity

DNA in Supramolecular Chemistry and Nanotechnology, First Edition.
Edited by Eugen Stulz and Guido H. Clever.

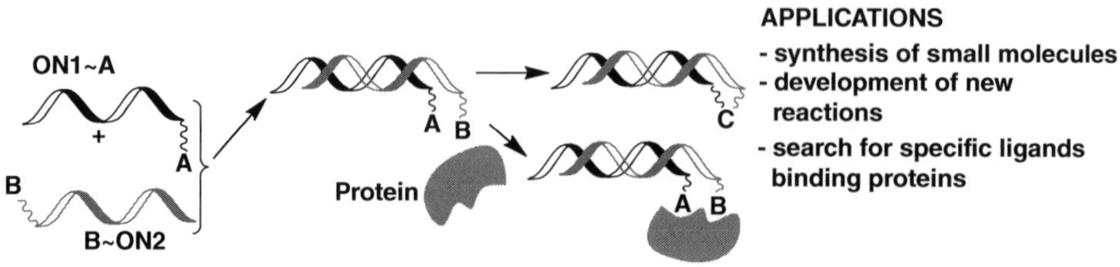

Figure 3.4.1 Hybridization-triggered chemical transformations of reactants **A** and **B** into products and a list of demonstrated applications of these reactions

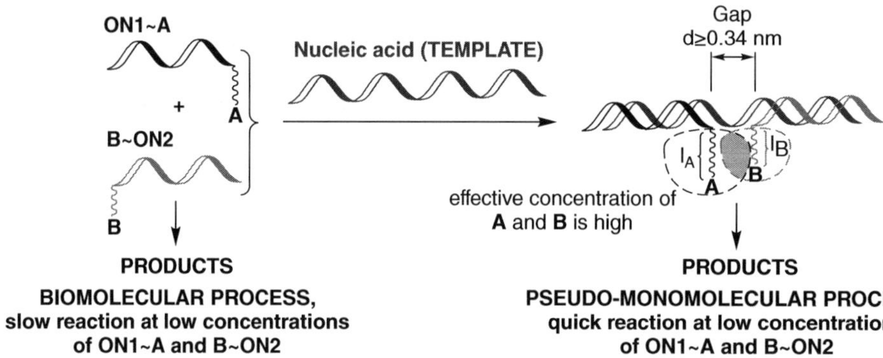

Figure 3.4.2 A nucleic acid (TEMPLATE)-induced reaction between chemical moieties **A** (black colored) and **B** (grey). In the associate TEMPLATE/(**A**–ON1, ON2–**B**) these fragments can diffuse in the areas indicated in the right-hand figure by the corresponding dotted lines (black lines for **A** and grey lines for **B**) providing the linkers between them and oligonucleotides ON1 and ON2 are fully flexible. The reaction occurs in the grey region, where black and grey dotted areas overlap with each other

of the complementary nucleic acids for each other. Here moieties **A** and **B** are also positioned in the vicinity to each other that facilitates an interaction between them (Figure 3.4.2).

The hybridization-induced acceleration can be understood in terms of effective concentrations of reagents **A** and **B**. In particular, since **A** and **B** are linked to the oligonucleotides covalently, they cannot freely diffuse in the TEMPLATE/(**A**–ON1, ON2–**B**) associate. As is shown in Figure 3.4.2, they occupy zones having half spherical shapes with radii l_A and l_B, defined by the linkers between these chemical entities and oligonucleotide termini, providing these linkers are fully flexible. Shapes of these zones change for the linkers, which are not fully flexible and can adopt preferred structures. The reaction occurs in the space where these spheres overlap with each other: labeled in grey in Figure 3.4.2. In the case when the overlapping area can be approximately described as a sphere with a radius of 0.34 nm, its volume is equal to ~2 x 10^{-22} *L*. One molecule in this volume would correspond to a concentration of ~10 mM. Since the effective concentrations of **A** and **B** in the grey area should correlate with the rate of the templated reaction, one can estimate that in this simplified case with the concentration of reagents in solution equal to 1 µM, template-induced acceleration will be 10 000-fold. Based on these considerations, one may rationally approach the problem of increasing efficiency of templated reactions by varying/optimizing:

- length and rigidity of linkers l_A and l_B;
- reactivities of moieties **A** and **B**;

- the number of nucleotide units in the gap area of the template (variation of parameter *d*, Figure 3.4.2);
- for the reaction, occurring catalytically with respect to the template, rates of association and dissociation of ON1 and ON2 with and from the template should also be considered.

This chapter is devoted to the templated reactions, which are applicable for detection of nucleic acids in live cells. Their other applications, including modification of surface of nano-objects and design of functional nanostructures, have recently been described in a series of articles by Gothelf, Mokhir and colleagues [3]. The chemistry and applications of the templated reactions have been extensively reviewed [4].

3.4.2 Intracellular nucleic acid targets

Nucleic acids, including deoxyribonucleic acids (DNAs) and ribonucleic acids (RNAs), are ubiquitous biomolecules playing a number of roles in Nature. For example, a double stranded DNA molecule is the major carrier of genetic information in cells, whereas single stranded, typically folded, RNA strands act as mediators between the DNA and proteins (mRNAs) and as transporters of amino acids to ribosomes (tRNAs). Moreover, RNAs participate in the synthesis of proteins as part of the ribosomes (rRNAs) and act as regulators of gene expression (e.g., micro-RNAs: miRNAs). Functions of some recently discovered RNAs (antisense RNAs, pseudogenes) have not yet been fully identified. All these intracellular nucleic acids can potentially be analyzed by using templated reactions. For example, application of single stranded nucleic acid templates as triggers of chemical transformations is common [4]. In this case, formation of the associate TEMPLATE/(**A**–ON1, ON2–**B**) is governed by Watson–Crick hydrogen bonding between AT and GC base pairs. Some reactions of this type have been developed that are applicable for monitoring single-stranded ribonucleic acids in live cells, which will be described in the next section. Moreover, a few reactions are known in which double-stranded DNA (dsDNA) acts as a template. These include [2+3] cycloaddition between azide- and alkyne-modified polyamide probes [5], Cu^{2+}-catalyzed hydrolysis of activated ester substrates attached to triplex-forming DNA sequences [6], and Staudinger reduction, using triplex-forming probes [7] (Figure 3.4.3). However, their applicability for monitoring intracellular dsDNA has not yet been demonstrated [4]. The reason for this may be the low accessibility of genetic DNA in cells and the absence of sequence specific binders, which can efficiently compete *in vivo* with intracellular dsDNA-binders. However, recent promising sequence-specific, *in vivo*, effects that have been demonstrated for polyamides may fill this gap in the near future [8].

3.4.3 Methods for monitoring ribonucleic acids in live cells

An intracellular pool of RNAs is highly dynamic. These biomolecules are constantly cleaved, or otherwise transformed and newly generated in response to changes in the external stimuli, such as temperature, biochemical signals, and interactions with other cells. Therefore, monitoring RNAs in live cells is of great value and can help to better understand numerous RNA-dependent intra- and extra-cellular biochemical processes. RNAs can be detected in live cells by using:

1. genetically encoded reporters
2. hybridization-responsive oligonucleotide probes and
3. fluorogenic templated reactions.

Although the first two methods are not based on templated reactions, they will be discussed briefly here to outline their advantages and disadvantages and to compare their performance with that of fluorogenic templated reactions.

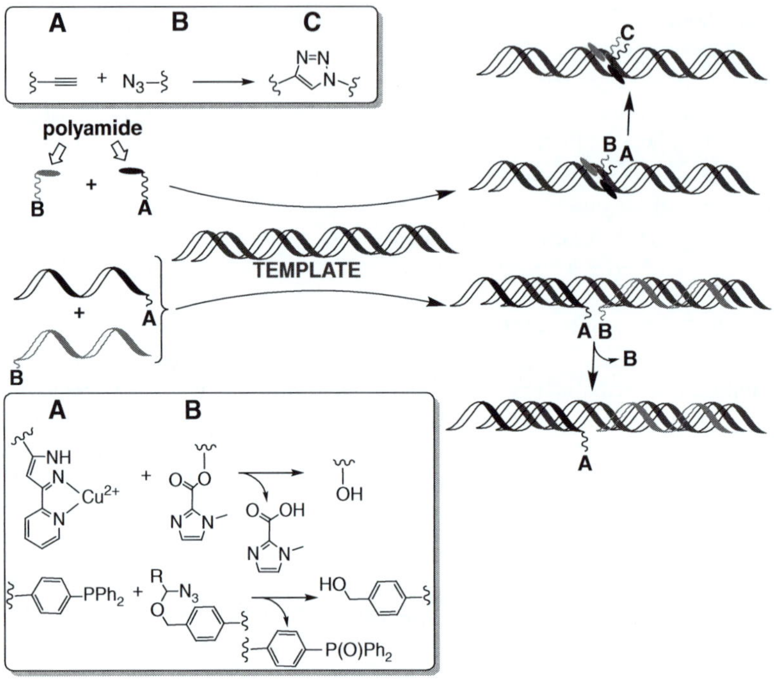

Figure 3.4.3 Outline of known chemical reactions that are induced by double-stranded nucleic acid templates [5–7]

3.4.3.1 Genetically encoded reporters

Genetically encoded reporters are specific, usually folded, RNA sequences, which are attached to the target ribonucleic acid. They are visualized via their particular interactions with fluorescent proteins or small molecules (Figure 3.4.4). For example, 19-nucleotide stem–loop folded sequence of the MS2 translational operator can be linked to an mRNA of interest. Expression of this mRNA is then visualized in live cells by using a fluorescent hybrid protein, for example, YFP-MS2, which specifically binds to MS2-RNA-hairpins. To enable such imaging, the hybrid proteins should be co-expressed with the target [9]. Another related label includes the 18-nucleotide stem–loop RNA sequence U1A, which can be specifically bound by human splicing protein U1A fused to a fluorescent protein. This system has been applied for imaging of bacterial and yeast mRNAs [10]. While fluorescent proteins in hybrids with MS2 and U1A usually exhibit low photostability, it has been demonstrated that it is still high enough for the monitoring of mRNA with single-molecule sensitivity [11].

Alternatively, aptamer RNA sequence 24-2 can be used as a label. This RNA binds 3,5-difluoro-4-hydroxybenzylidene imidazolinone (a dye), with the formation of a green fluorescent complex, termed "Spinach" (Figure 3.4.4B) [12]. The dye is cell-membrane permeable and the quantum yield of its fluorescence is substantially lower in the absence of the specific RNA sequence. Therefore, it can be simply introduced into live cells by its addition to the medium. This RNA–dye construct resembles well established green fluorescent protein (GFP). It has been demonstrated that 5S RNA labeled with the 24-2 sequence and 3,5-difluoro-4-hydroxybenzylidene imidazolinone can be imaged in live cells.

Both of the approaches described here allow convenient imaging of ribonucleic acids in live cells. However, labels attached to endogenous RNAs can potentially change the properties of these biomolecules (stability,

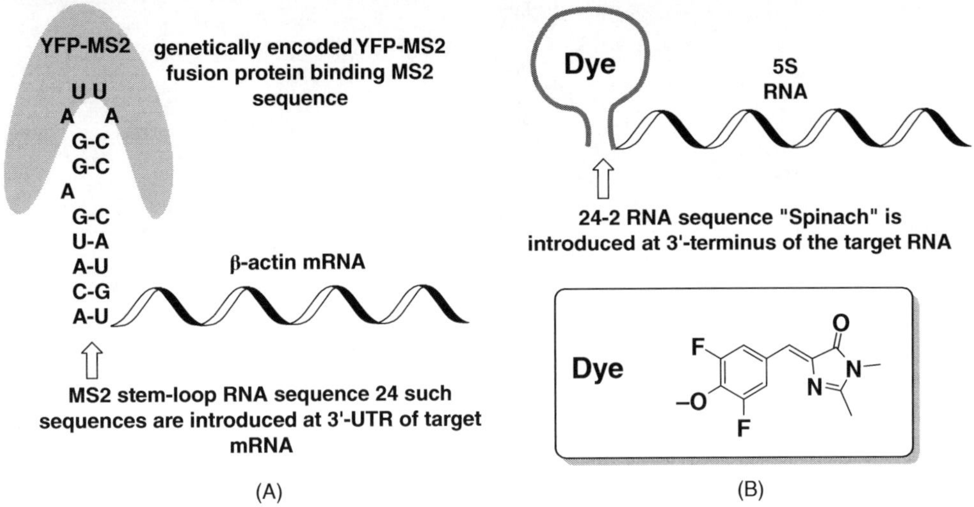

Figure 3.4.4 Summary of selected labels used to monitor intracellular RNAs: (A) MS2 hairpin, which has been applied to label β-actin-mRNA for live cell imaging [9]; (B) aptamer 24-2–dye construct, which has been applied to label 5S RNA in live cells [12]

localization, biological function, etc.), which is an intrinsic disadvantage of all methods that are based on the genetic alteration of nucleic acids.

3.4.3.2 Hybridization-responsive oligonucleotide probes

Hybridization-responsive probes are weakly fluorescent in solution. Upon binding to specific nucleic acids, their fluorescence is substantially enhanced, which can then be used to monitor nucleic acids in the cells. For example, hairpin-folded or stem–loop oligonucleotides, which carry a fluorophore and a quencher at their termini, are known as molecular beacons (MBs) [13]. Their mode of action is outlined in Figure 3.4.5. These reagents can provide for up to 200-fold fluorescence intensity increases upon binding to complementary nucleic acids [14]. The length of the stem and loop regions of the MBs can be easily optimized to achieve high target specificity. For example, even single nucleotide mismatches in the target can be recognized. However, high stability of the hairpin often leads to slow binding kinetics of the beacons to their targets. Moreover, MBs are known to bind to some proteins (e.g., single-strand DNA-binding protein, SSB), which leads to their nucleic acid independence and, therefore, unspecific activation. To solve this problem two MBs can be used, which bind to neighboring sites on the mRNA and whose fluorophores can participate in the fluorescence resonance energy transfer (FRET) between one another. In this case one fluorophore acts as a donor and the other as an acceptor. Unspecific opening of each individual beacon, or of both of them, does not lead to the FRET, therefore, the specific binding can be distinguished from that which is unspecific [15]. MBs are not cell-membrane permeable. They can be introduced into the cells by electroporation, reversible membrane permeabilization, by using streptolysin-O (SLO), and by direct injection. Related probes have recently been developed in which gold nanoparticles act as quenchers and at the same time as carriers of probes through the cell membrane. These nano-objects are termed nano-flares. They permeate the cell membrane without any additional additives and have been found to be applicable for the detection of mRNA in live cells [16].

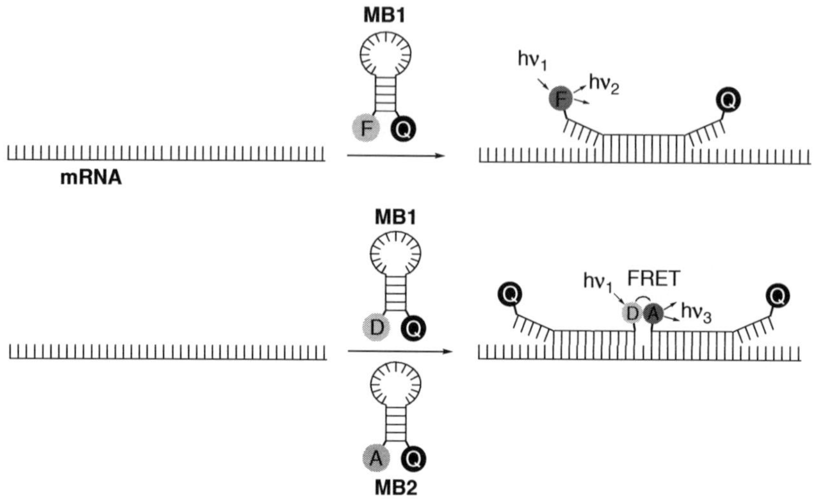

Figure 3.4.5 The principle of detection of nucleic acids by conventional and dual-molecular beacons (MBs)

3.4.3.3 Fluorogenic templated reactions

The mode of action of hybridization-responsive probes relies on their secondary structure. These reagents are usually rather long: stem, 5–7 base pairs (together 10–14 nucleotides are required), and the loop or recognition element, ≥ 16 nucleotides long. To achieve single nucleotide discrimination, the folded conformation of these probes is designed to be reasonably stable, which is achieved in practice by increasing either the overall number of base pairs or the number of GC-base pairs in the stem. In this case, only fully matched target nucleic acids will be able to open up the probe, whereas the mismatched ones will have no effect. However, such highly specific probes bind to the target substantially slower than unfolded single-stranded oligonucleotides and their target affinity is reduced. These are important disadvantages for applications of these reagents in live cells, where they have to bind targets, which themselves are folded (mRNA) or exist in a partially double-stranded form (miRNA). In contrast, in templated reactions two shorter (≥ 6-mers), unfolded reagents are applied [4]. Binding of both reagents to the target occurs quickly and is highly sequence specific. In particular, single-nucleotide specificity is routinely observed. However, for the rare event of the unspecific binding (e.g., one of the reagents did not bind), no product (signal) is generated. Therefore, templated reactions allow for detection of complementary RNA-targets highly accurately and exhibit fewer target-off effects.

3.4.3.3.1 Nucleophilic substitution reactions

The first fluorogenic templated reaction that was suitable for detection of nucleic acids in live cells was described by Kool and coworkers in 2001–2002 (Figures 3.4.6 and 3.4.7) [17]. In this system, a nucleophilic phosphorothioate group attached to the 3′-terminus of one of the oligonucleotide strands reacts via an S_N2-mechanism with either a 5′-dabsyl substituted terminal alcohol or 5′-iodo-5′-deoxy-terminal fragment of another oligonucleotide strand. The substitution takes place only in the presence of the template nucleic acid with the release of the dabsyl group (or iodide) and ligation of two oligonucleotide strands via formation of a new S–C chemical bond (Figure 3.4.6). In the fluorogenic version of this reaction, a fluorescein dye (FAM) is attached to one of the nucleotides in the sequence, which carries the dabsyl fragment. In the resulting intact probe, the fluorophore is quenched by the dabsyl, whereas upon dabsyl cleavage its fluorescence is restored.

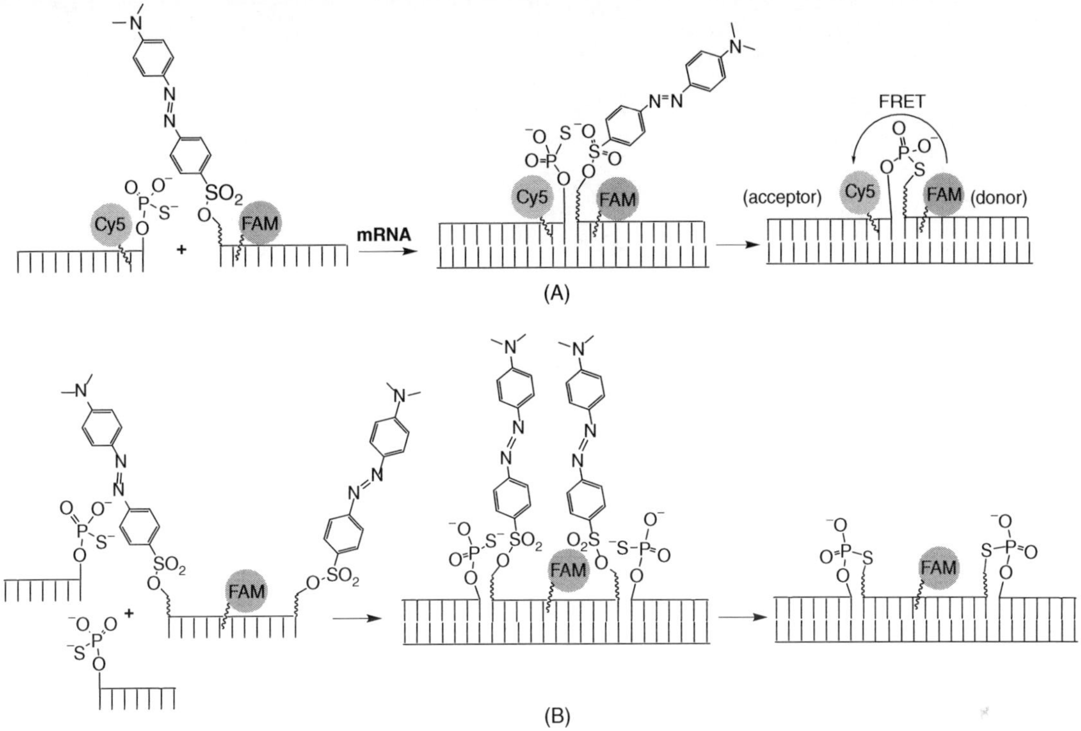

Figure 3.4.6 (A) The first templated chemical reaction to be applied to detect nucleic acids in live cells; (B) the analogous reaction, which exhibits a lower background fluorescence due to the presence of two quencher fragments in the substrate

Using this fluorogenic reaction, Kool and coworkers have successfully detected four selected sites on ribosomal RNA 16S in the fixed *Escherichia coli* K12 strain MG1655. In these first experiments, maximal conversion was reached only 6–8 h after the reaction began. To improve the rate of this transformation and the signal-to-noise ratio, a range of alterations were introduced. For example, the second dye (Cy5) was conjugated to the strand that carries the nucleophile (Figure 3.4.6A). When the resulting substrate reacts with the electrophilic reagent carrying the fluorescein dye (FAM), a ligation product is generated in which the fluorescence resonance energy transfer (FRET) between the FAM dye and the Cy5 dye is observed, providing the fluorescein is excited. In the unreacted starting materials, no FRET occurs. It has been confirmed that this highly specific, nucleic acid dependent FRET signal is suitable for monitoring nucleic acids in mammalian cells [18].

In another study, sandwich probes that carry two dabsyl quenchers were designed (Figure 3.4.6B). Two dabsyl residues quench the fluorescence of FAM more efficiently than one residue does. In the presence of a nucleic acid target and two phosphorothioate-terminated probes, both dabsyl fragments are cleaved off and the fluorescence of the dye is restored [19]. In a related approach, a single quencher was replaced with two such fragments, which were linked together via an optimized linker. The background fluorescence and the signal-to-noise ratio were found to be significantly better in this system (Figure 3.4.7A) [20]. Furthermore, the strength of the nucleophilic reagent was enhanced by substituting phosphorothioate for phosphorodithioate or phosphorotrithioate fragments, which increased the reaction rate both *in vitro* and in bacteria [21] (Figure 3.4.7B).

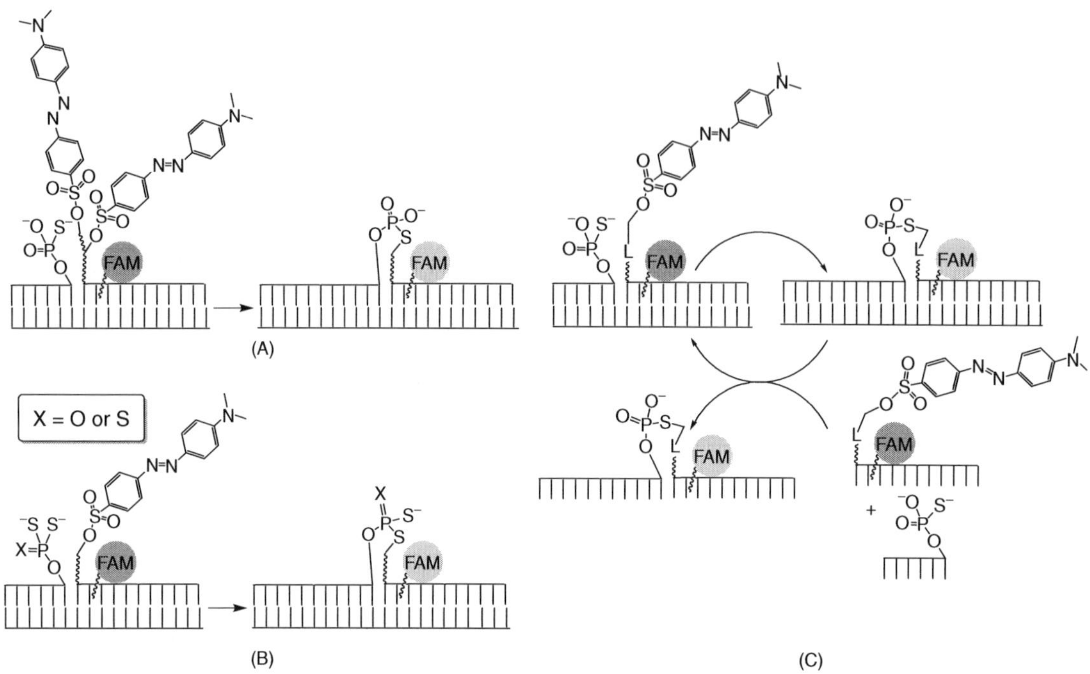

Figure 3.4.7 Improvements of the templated nucleophilic substitution reaction: (A) because of the presence of two quenchers in the substrate, the background fluorescence of the substrate is very low; (B) introduction of nucleophiles stronger than phosphorothioate leads to the reaction rate increasing; (C) introducing the linker between the recognition element (oligonucleotide) and the electrophile does not favor the product inhibition and facilitates the catalytic turnover in the templated reaction

Finally, it has been observed that the problem of product inhibition, which is typical for templated ligation reactions [4], could be at least partially solved by introducing a longer (product-destabilizing) linker between the recognition element and the reactive fragment [22] (Figure 3.4.7C). In the presence of a 10^4-fold probe excess with respect to the template, the reaction delivers 92 turnovers over 24 h, which corresponds to an average rate of 3.83 h^{-1} and 0.92% reagent conversion. Despite being rather slow, the reaction generates a strong fluorescent signal in the presence of specific nucleic acids in live cells, due to the very low intrinsic fluorescence of the starting materials and extremely slow background (template-free) reaction. The probes required for this transformation (termed QUAL-probes) can be easily prepared in every laboratory on a standard DNA/RNA-synthesizer using commercially available phosphoramidites. They are not cell-membrane permeable, but can be brought into mammalian cells by using the reversible membrane permeabilization procedure based on the activity of streptolysin-O, and into bacteria, by conducting the reaction in a buffer containing 0.05% sodium dodecyl sulfate (SDS). According to the reported data the activation of QUAL-probes is applicable to the detection of ribonucleic acids in live bacteria (16S in *E. coli*, *Salmonella enterica*, *Pseudomonas putida*) and mammalian cells (28S rRNA, β-actin mRNA, histone H3 mRNA, JUND mRNA) [18, 23].

3.4.3.3.2 Staudinger reduction

Since QUAL probes react with each other fairly slowly and quenched probes exhibit undesirable spontaneous activation, the application of the reactions based on the nucleophilic displacement via the S_N2-mechanism is limited to the detection of the somewhat abundant nucleic acids in live cells [4, 18]. To address this issue,

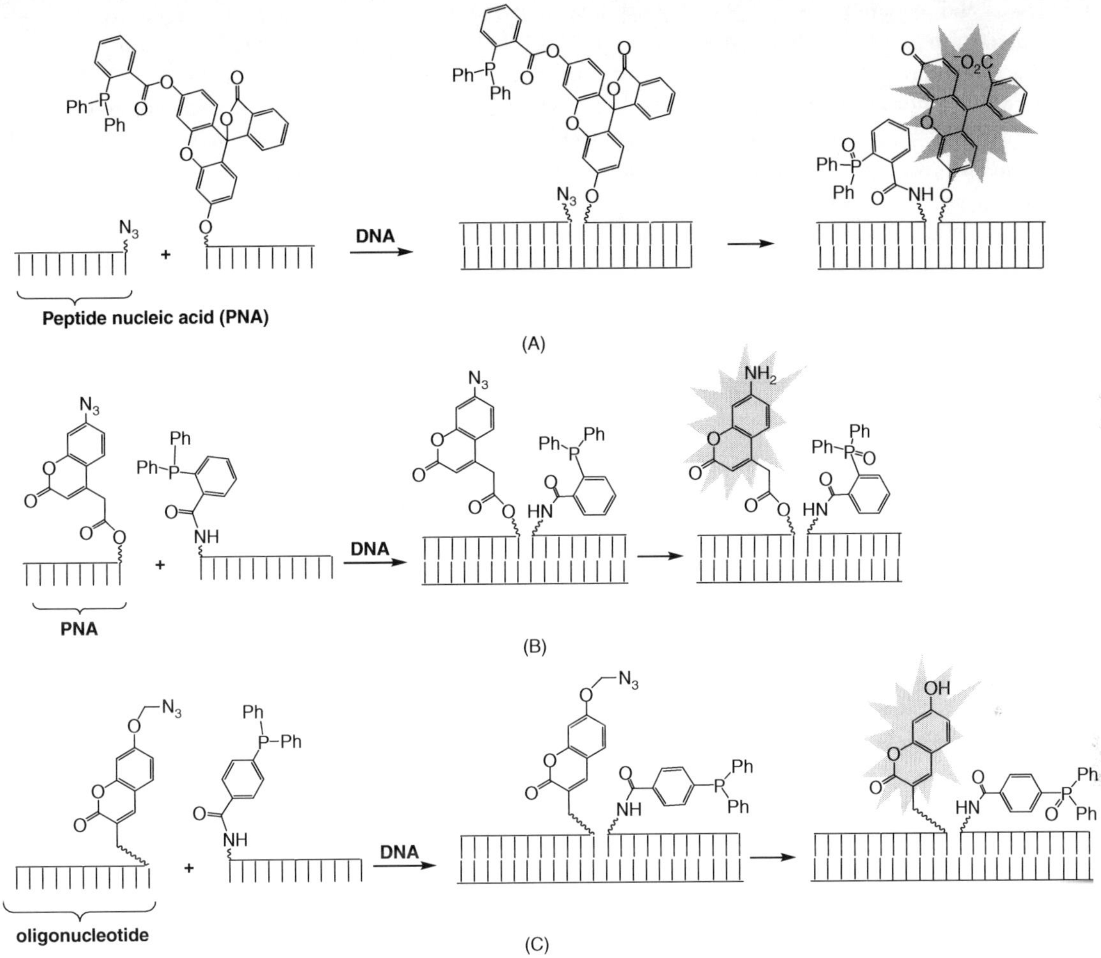

Figure 3.4.8 *(A) The outline of the first templated Staudinger reaction, in which phosphine-based pro-fluorophore was used; (B) and (C) improved templated reductions, in which caging of fluorophores relies on a more stable azido-group*

other reactions are being investigated. One of the most successful alternative chemical transformations, which is compatible with live cells, is the Staudinger reduction. The first application of this type of chemistry, specifically Bertozzi's modification of the Staudinger reduction [24], in a nucleic acid-templated format, has been described by the group working with Taylor, in 2004 (Figure 3.4.8) [25]. In this case the monoalkylated fluorescein ester of 2-(diphenylphosphino)benzoic acid was used as a pro-fluorophore. It was attached to the *N*-terminus of one peptide nucleic acid (PNA) strand. Furthermore, the *C*-terminus of another PNA strand was converted into an alkylazide fragment. Upon binding of these two PNA conjugates to the DNA template, the pro-fluorophore and the azide are positioned in a proximity to each other that facilitates the interaction between them (Figure 3.4.8A). In the first step of this reaction, an aza–ylide intermediate is formed. This then undergoes intramolecular cyclization that results finally in the formation of a stable amide bond. The overall result of these transformations is cleavage of the fluorescein ester bond with formation of the fluorescent dye and transfer and oxidation of the triphenylphosphine from one PNA strand to the other one. This reaction was found to occur in aqueous buffer at pH 7 and to be 188-fold faster than the background reaction.

Although these results are fairly promising, the reaction is not expected to be applicable inside cells, since the fluorescein ester fragment in the pro-fluorophore will decompose in the intracellular milieu, due to the reaction with endogenous nucleophilic species or esterases. Moreover, the conversion observed by Taylor did not exceed 50%, which was attributed to oxidation of the phosphine reagent under the experimental conditions used. Therefore, it was desirable to alter the system to make sure that the fluorescence output is linked to the azide rather than to the phosphine. Three years later, in 2007, the group working with Winssinger introduced such an improvement (Figure 3.4.8B) [26]. In particular, they used a practically non-fluorescent 7-azidocoumarin as a pro-fluorophore, which was conjugated to the alcohol-modified *C*-teminus of a PNA strand via formation of an ester linkage. A triphenylphosphine residue was attached to the *N*-terminus of another PNA strand. Upon binding to the DNA-target, the azide was reduced by the triphenylphosphine with formation of a highly fluorescent, yellow dye. The reaction was found to work with catalytic amounts of the template (~5 turnovers in 30 min, 5% conversion in the presence of 1% of the template), and to be highly sequence specific and quick. However, since in this design the ester linkage between the pro-fluorophore and the PNA was also utilized, the reaction is expected to be unsuitable for intracellular applications.

In 2008, Franzini and Kool designed a related system, in which 7-azidomethoxycoumarin was attached to an oligonucleotide strand via a stable amide linkage (Figure 3.4.8C) [27]. Here, an azidomethoxy group rather than the azido-group was used for caging of the coumarin dye. In this case the reaction was found to be highly sequence specific and to occur catalytically with respect to the template. In particular, 22 turnovers were observed in 30 min in the presence of catalytic amounts of the template (2.5% with respect to the pro-fluorophore) and excess triphenylphosphine-containing oligonucleotide (200%). Under these conditions the substrate conversion was equal to ~27%. In contrast, the templated S_N2-replacement is both slower (>tenfold) and exhibits lower substrate conversion. Higher amplification power of the system based on the azidomethoxy protection compared with that based on the azido-modification of the dye is caused by formation of the intramolecular aza–ylide ligation product in the latter case, which inhibits the reaction. This has been confirmed experimentally in studies on the related system, in which the coumarin dye was replaced with xanthene dyes. In particular, Abe, Ito, and coworkers have studied reductive templated activation of two pro-fluorophores: azide-functionalized rhodamine and azidomethoxy-modified fluorescein (Figure 3.4.9A, B) [28]. They found that only the latter substrate can be converted into the fluorophore in a catalytic reaction, which generates 50 turnovers within 4 h (or 6.25 turnovers in 30 min). In contrast, the former substrate formed an aza–ylide, which slowed the reaction down. Although conversion of azidomethoxy-modified fluorescein is somewhat less efficient than that of the analogous coumarin-based substrate, the xanthene dye is better suited to the cellular applications due to the lesser overlap of its emission spectrum with the cellular autofluorescence. Indeed, the reactions based on xanthene-dyes (Figure 3.4.9A and B) were found to be suitable for detection of nucleic acids in live bacteria (23S rRNA in *E. coli* K12) and mammalian cells (28S rRNA and β-actin mRNA in HL-60 cells). Moreover, the related reaction, in which triphenylphosphine was replaced by a more reactive trialkylphosphine, and membrane-permeable, guanidine-based peptide nucleic acids (GPNAs) were used as recognition elements, was found to be suitable for detection of mRNA of *O*-6-methylguanine-DNA methyltransferase in HEK cells (Figure 3.4.9C) [29]. The latter case is the first example of the application of templated chemical reactions for monitoring nucleic acids in live cells, where no additives, such as SLO or SDS were required to bring the reagents into cells.

Winssinger and coworkers have applied the related reaction to the monitoring of micro-RNA 21 in fixed breast cancer cell lines (MCF-7, BT474, JIMT-1), cervic cancer cell line HeLa, and immortalized human embryonic kidney cell line HEK293T. In this case, mono-azido-fluorescein was substituted for its bis-azido-analogue. The latter pro-fluorophore provides for 120-fold fluorescence enhancement upon its reduction, whereas the former provides for only a 30-fold enhancement under the same conditions

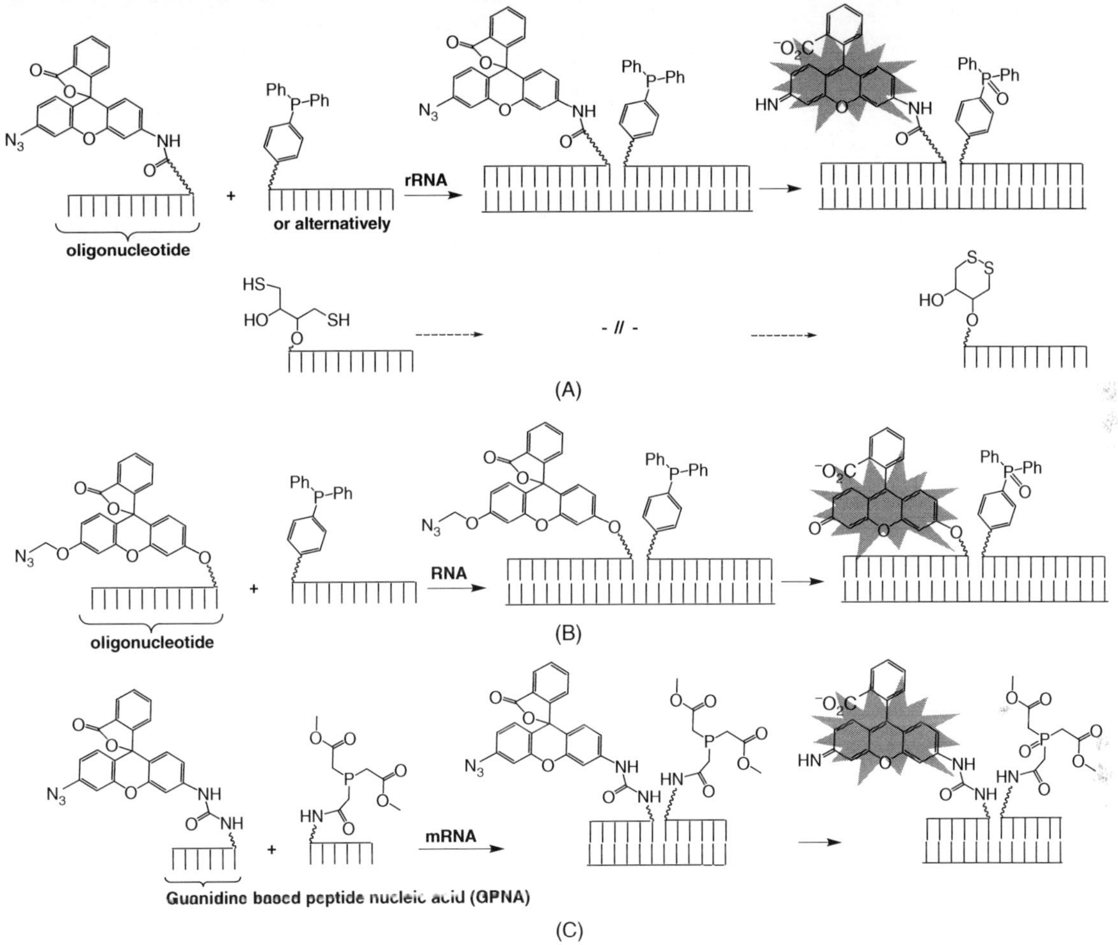

Figure 3.4.9 *Templated Staudinger reduction reactions, in which caged xanthene fluorophores carrying azido- or azidomethoxy-groups are used as substrates*

(Figure 3.4.10A) [30]. Furthermore, Abe, Ito, and colleagues have applied the templated Staudinger reduction to detect intron lariat RNA and pre-mRNA splicing with the formation of the mature mRNA of chicken d-crystallin (CDC) gene *in vitro* (Figure 3.4.10B) [31]. No data on monitoring splicing in live cells are available yet.

Aza–ylide intermediates, which are formed in the templated Staudinger reaction with triphenylphosphine-based reducing agents, are often very stable towards hydrolysis in water at neutral pH. As has already been mentioned, this may preclude catalytic turnover in the templated reactions. The problem can be partially solved by using trialkylphosphines as reducing agents. However, they are not particularly stable under aerobic conditions. Abe and Ito have approached this problem by screening a series of triphenylphosphinecarboxamides as reducing agents [32]. They observed that 2-substituted compounds exhibit the most favorable properties. In particular, these reactants quickly reduce azides with the formation of amines, whereas there is virtually no accumulation of intermediate aza–ylides in the reaction mixture. These results were confirmed in both templated and non-templated reaction formats (Figure 3.4.11A).

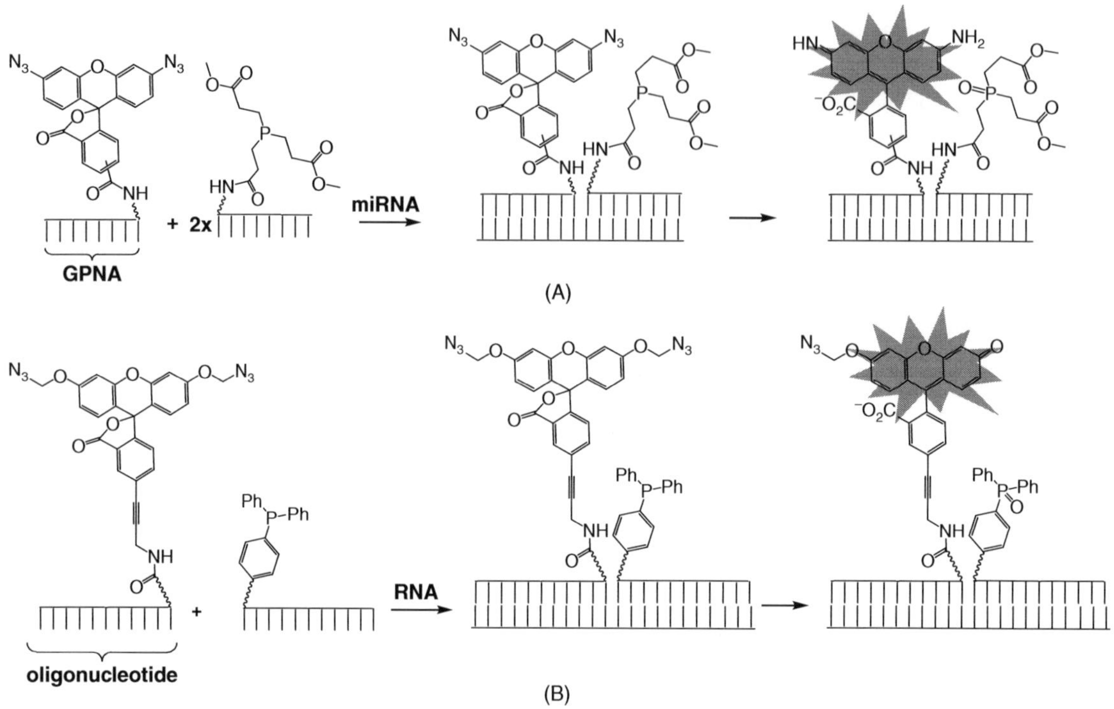

Figure 3.4.10 *Bis-protected ("caged") azido-containing (A) and azidomethoxy-containing (B) fluorophores deliver lower background signals in the templated reaction than their mono-protected analogues*

In the majority of templated reactions, organic fluorophores are used. Since cells autofluoresce, the signal generated in the templated reaction is always superimposed with this background. The autofluorescence fades quickly beyond 500 nm. Therefore, red light emitting organic dyes are particularly well suited for the cellular applications. Another interesting approach for solution of the autofluorescence problems is to use luminophores with long lifetimes. In this case delayed detection can be conducted, which allows the signal to be eliminated from the usually short-lived autofluorescence. Recently, Abe and Ito have managed to combine the templated Staudinger chemistry with a luminophore based on a lanthanide complex (Figure 3.4.11B) [33]. In their reaction a sensitizer for the luminescence of the complex is generated on the template via the cyclization, which is induced by the Bertozzi-type Staudinger reaction. These workers have demonstrated that the reaction is suitable for monitoring 23S rRNA in live *E. coli* cells. As expected, the delayed detection of the emission allowed efficient elimination of the background autofluorescence from cells.

In the reactions discussed so far, each fluorophore was caged individually by either introducing one or two azide moieties or one or two azidomethoxy fragments in place of corresponding $-NH_2$ or –OH groups. Since the synthetic effort that is required for preparation and conjugation of such pro-fluorophores is relatively large, only a few dyes have been tested in the templated Staudinger reaction using this approach, which includes coumarin, fluorescein, and rhodamine. In 2009, Kool and coworkers described a more versatile system, which allowed the use of commercially available dyes for the preparation of substrates for templated reactions. In particular, they prepared an immolative linker, which is cleavable under the reductive conditions used (Figure 3.4.12) [34].

This linker was introduced in between the oligonucleotide sequence and the quencher. The same strand also carried a fluorophore, which was attached to one of the nucleobases. In the intact form the fluorescence

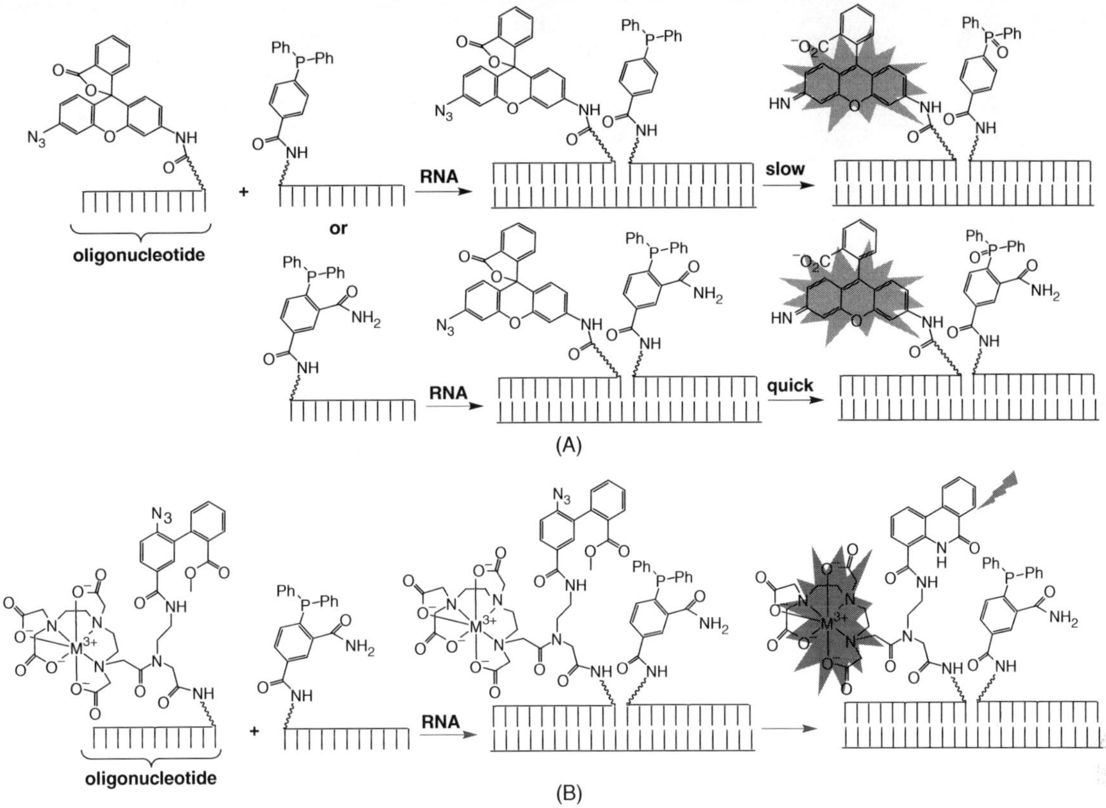

Figure 3.4.11 *(A) 2-Substituted triphenylphosphinecarboxamide is a more efficient reducing agent in the templated Staudinger reaction than its unsubsituted analogue; (B) a templated Staudinger reduction generating a long lived emission signal in the presence of a nucleic acid target*

of the probe is diminished due to the proximity of the fluorophore and the quencher. However, in the presence of a nucleic acid template and an oligonucleotide carrying a phosphine fragment, the linker is cleaved, releasing the quencher and restoring fluorescence of the dye. This approach is highly attractive since a range of different dyes can be introduced within the substrate by using numerous commercially available phosphoramidites, carboxylic acids or succinimides of the dyes. The probes based on the reactivity of the immolative linker are termed Q-STAR. They were demonstrated to be applicable for detection of nucleic acids in live bacteria (16S rRNA in *E. coli* and *S. enterica*). Owing to the modularity of this system it can be easily optimized. For example, Kool and coworkers have improved the background fluorescence of substrates by using double quenchers (Figure 3.4.12B) [35]. In contrast, using double triphenylphosphine residues did not lead to improvement of the reaction efficiency. Furthermore, the immolative linkers were applied to design probes, which release functional molecules in a nucleic acid specific fashion [36].

3.4.3.4 Photochemical reactions

Reactions triggered by light are particularly well suited for cellular applications, since they can be temporally and spatially controlled. In particular, light can be switched on and off at will, its intensity can be easily regulated, and it can be focused on the desired area of the tissue or cell. Moreover, light-controlled and

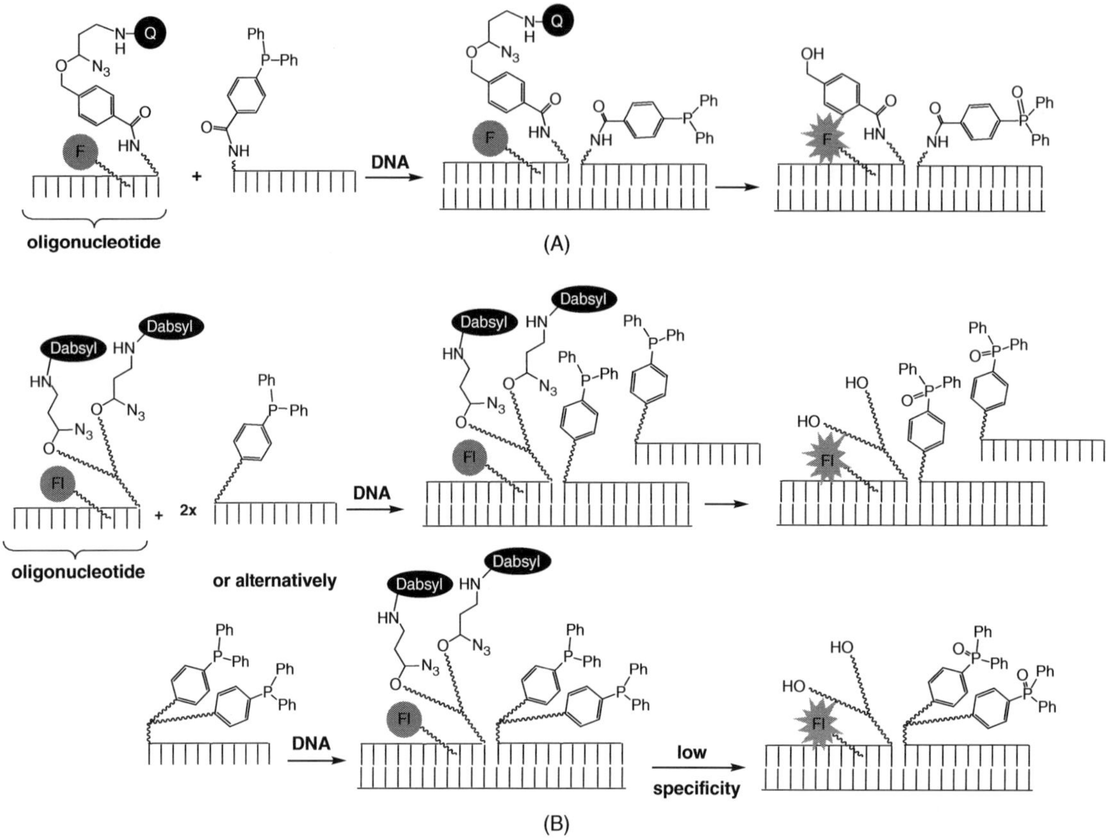

Figure 3.4.12 A templated reduction using modular substrates containing an immolative linker bound to the quencher. (A) A reaction with a single cleavable unit; (B) reactions with a double cleavable unit and a double reducing unit

nucleic acid dependent probes can potentially be applied in super-resolution fluorescence microscopy [37]. Several templated reactions have been reported, which are induced by either UV- or short-wavelength (<450 nm) visible light. Those based on [2+2] cycloaddition between natural or chemically modified pyrimidines are the most common [38]. Other reported transformations of this type include photoligations using coumarin, psoralen, stilbene, and anthracene, and photodeprotection of 2-(2-nitrophenyl)propyloxycarbonyl-modified rhodamine dye as reviewed elsewhere [4]. Neither of these reactions have been applied for monitoring nucleic acids in cells. The possible reason for this could be the significant toxicity of UV and short wavelength visible light to cells.

In 2006 Gothelf and coworkers described a visible light-induced and nucleic acid dependent reaction that can generate singlet oxygen ($^1O_2$) from triplet oxygen ($^3O_2$) [39]. In this system a duplex is formed between a (pyropheophorbide-a)–oligodeoxyribonucleotide conjugate (PS–OD1) and a complementary quencher–oligodeoxyribonucleotide conjugate (Q–ON2). In this form the catalyst PS is not photoactive, since it is inhibited by the closely positioned quencher Q (Figure 3.4.13). However, in the presence of a target nucleic acid, the PS–ON1 conjugate is set free and the catalyst PS is reactivated and can photogenerate $^1O_2$ upon its exposure to visible light. The reaction was monitored by measuring phosphorescence of $^1O_2$ at 1270 nm. This group observed that the reaction in this form is limited to D_2O-based buffers and requires fairly high

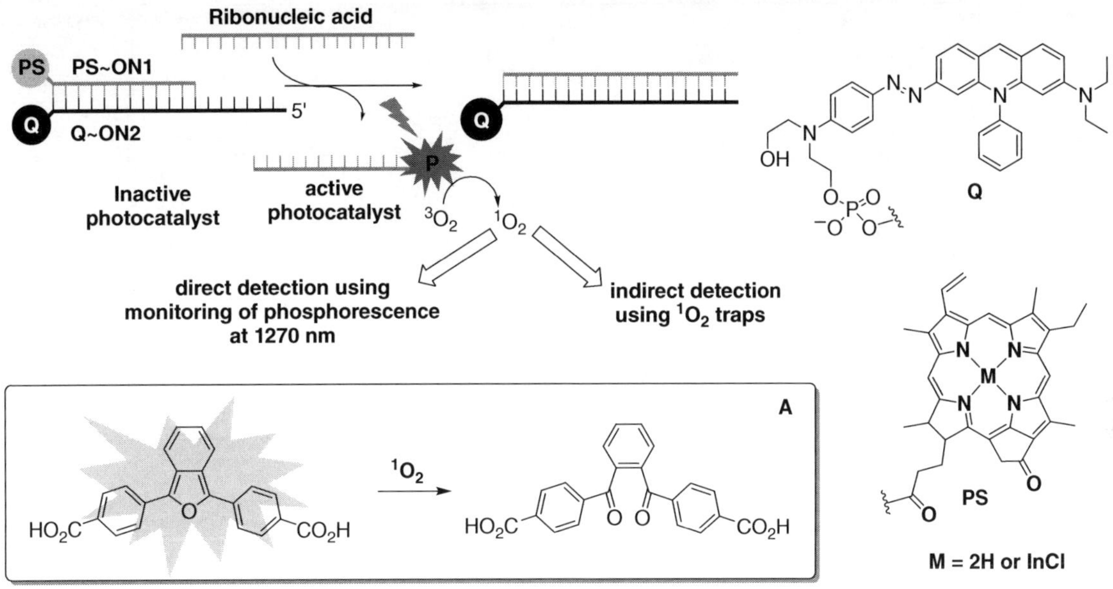

Figure 3.4.13 Nucleic acid-dependent photogeneration of singlet oxygen, which is triggered by red light (λ > 630 nm) that is non-toxic to cells. (A) Structure of the singlet oxygen trap (highlighted grey) and the product formed as the result of its reaction with photogenerated singlet oxygen

concentrations of the reactants, that is, 5 μM. Moreover, owing to the low photostability of pyropheophorbide-a no catalytic turnover was observed. Mokhir and coworkers have improved the reaction by substituting pyropheophorbide-a for its indium(III) complex, which is substantially more photostable (Figure 3.4.13) [40]. Another alteration was the use of water soluble diphenylisobenzofurane-based singlet oxygen traps for monitoring the formation of singlet oxygen. In particular, the traps react with $^1O_2$ quickly, leading to their bleaching, which can be detected *in vitro* by using fluorescence spectroscopy and in cell-flow cytometry (Figure 3.4.13A). The resulting nucleic acid-controlled photocatalyst works well in aqueous buffer at pH 7 *in vitro*. It is activated by non-toxic red light at λ > 630 nm, exhibits a high turnover frequency (3.1 min^{-1}) and number (over 6000), and retains its activity in live cells. For example, this system was applied to detect 28S rRNA in HL-60 cells [40]. However, sensitivity of this reaction is somewhat low: nucleic acids at concentrations below 100 nM cannot be detected. This has been rationalized in the following way: one of the reaction steps including the bleaching of the trap with singlet oxygen is a bimolecular process and is, correspondingly, strongly dependent on the concentrations of the reagents. Since singlet oxygen exhibits a very short lifetime in aqueous buffers (< 3.5 μs), it seems that it is not possible for a sufficient amount to be generated on activation of the photocatalyst in the presence of less than 100 nM of the target nucleic acid. To solve this problem, related reactions have been developed by the same group, which rely exclusively on pseudo-monomolecular reaction steps (Figure 3.4.14) [41]. In these reactions InP–ON1 is used as a photosensitizer and either F-L–ON2 or Q-L–ON2(TMR) is used as the substrate. In the former case, the substrate is photoactivated, which leads to dissociation of the fluorophore from the target. The key process here is cleavage of the 9,10-dialkoxy-substituted anthracene by photogenerated singlet oxygen (Figure 3.4.14A, B) [42]. Sensitivity of this reaction is improved by a factor of 100 with respect to that previously described [40, 41]. For example, it can be used to detect nucleic acids at concentrations down to 1 nM. Since the fluorophore does not stay attached to the target, this reaction cannot be applied for the localization of nucleic acids in cells. To address this issue a

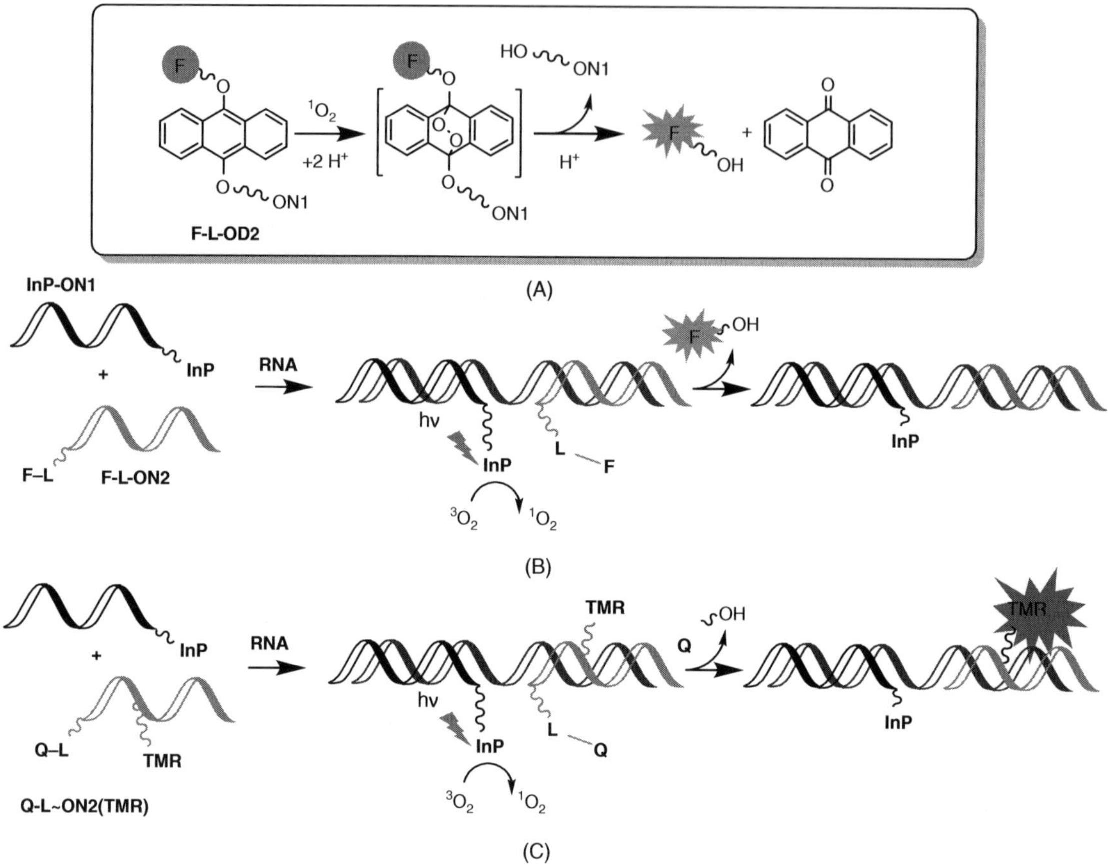

Figure 3.4.14 Photochemical fluorogenic templated reactions induced by non-toxic red light. (A) The mechanism of cleavage of 9,10-dialkoxysubstituted anthracene, which is used as a linker (L) in substrates F-L–ON2 and Q-L–ON2(TMR): F, fluorescein; Q, dabsyl; TMR, N,N,N′N′-tetramethylrhodamine. In one of the templated reactions the activated fluorophore is cleaved from the target (B) and in another one it stays on the target (C)

substrate F-L–ON2 was replaced with Q-L–ON2(TMR) (Figure 3.4.14C). Upon cleavage of the linker in this substrate, the quencher is released and the fluorescence of the TMR dye is restored. The fluorophore that is activated in this way remains on the target, which can potentially be used not only for detection, but also for localization of nucleic acids. Applicability of these templated reactions for monitoring nucleic acids in live cells is currently being investigated.

An alternative photochemical, nucleic acid-dependent reaction, which is triggered by visible light (455 nm) has been developed by the group working with Winssinger. In particular, they have shown that a Ru(II)-complex, containing oligonucleotide RuL–ON1, is an effective catalyst in the templated photocatalytic reduction of an azide fragment bound to another oligonucleotide: S–ON2 (Figure 3.4.15). Ascorbate or NADPH is used as the sacrificial reducing agent in this process. As a result of the reduction of the azide group, an amine is formed, which is spontaneously decomposed, as is outlined in Figure 3.4.15. One of the products generated is the rhodamine dye, whose detection is used to monitor the reaction (Figure 3.4.15) [43]. This chemistry has been demonstrated to be compatible with live cells. In particular, it can be used for the

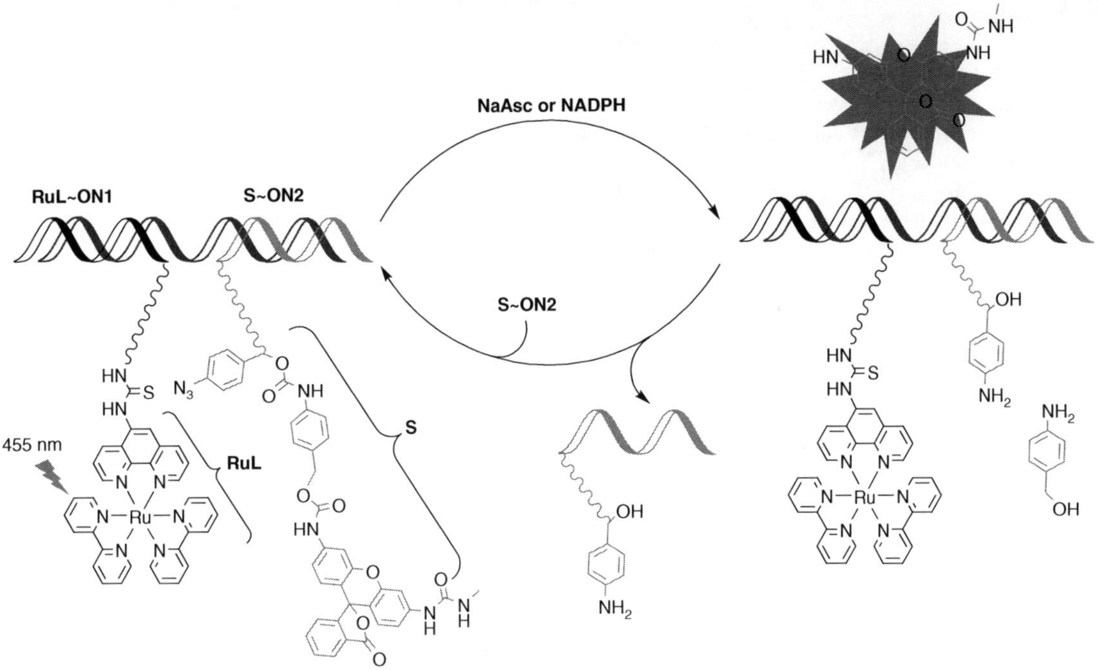

Figure 3.4.15 *Nucleic acid dependent, photochemical reduction of an azide-containing immolative linker coupled to activation of the rhodamine fluorophore. The reaction is catalyzed by the Ru(II)-complex, RuL, and can be controlled by its exposure to 455 nm light*

detection of intracellular micro-RNAs [44]. Interestingly, it was shown that the Ru(II)-complex–ON1 conjugate is capable of high turnover (> 4000) and is non toxic in cell cultures.

3.4.4 Perspectives

The chemical reactions that have been reviewed in this chapter were found to be suitable for the detection of nucleic acids in live cells. They can all be used for the analysis of fairly abundant nucleic acids, such as 28S RNA and β-actin-mRNA. However, less abundant targets are not yet readily detected by these known approaches. Therefore, the chemistry (specificity, signal-to-noise ratio, reaction rate, and other parameters) of the templated reactions have to be further improved to enable monitoring of rare ribonucleic acids, which are often very interesting targets.

Another key problem in the field is poor membrane permeability of the reagents used in templated reactions. Notable exceptions here are GPNAs developed by the group of Professor Winssinger. The conjugates based on these positively charged peptide nucleic acids were confirmed to be membrane permeable. However, all other reagents are typically introduced into the cells by using reversible permeabilization with streptolysin-O, which is fairly toxic to mammalian cells. Permeability of the bacterial membrane can be improved by using sodium dodecyl sulfate. Therefore, finding efficient carriers of these reagents through the cellular membrane and working on new chemical modifications of oligonucleotides or their analogues, which improve their membrane permeability, is highly desirable.

References

[1] (a) Z. J. Gartner and D. R. Liu, The generality of DNA-templated synthesis as a basis for evolving non-natural small molecules, *J. Am. Chem. Soc.*, **123**, 6961–6963 (2001); (b) Z. J. Gartner, M. W. Kanan, and D. R. Liu, Multistep small-molecule synthesis programmed by DNA templatesm, *J. Am. Chem. Soc.*, **124**, 10304–10306 (2002); (c) X. Li, Z. J. Gartner, B. N. Tse, and D. R. Liu, Translation of DNA into synthetic N-acyloxazolidines, *J. Am. Chem. Soc.*, **126**, 5090–5092 (2004); (d) Z. J. Gartner, M. W. Kanan, and D. R. Liu, Expanding the reaction scope of DNA-templated synthesis, *Angew. Chem., Int. Ed.*, **41**(10), 1796–1800 (2002).

[2] L. Mannocci, M. Leimbacher, M. Wichert, J. Scheuermann, and D. Neri, 20 years of DNA-encoded chemical libraries, *Chem. Commun.*, **47**, 12747–12753 (2011).

[3] S. Helmig, A. Rotaru, D. Arian, L. Kovbasyuk, J. Arnbjerg, P. R. Ogilby, J. Kjems, A. Mokhir, F. Besenbacher, and K. V. Gothelf, Single molecule AFM studies of photosensitized singlet oxygen behavior on a DNA origami template, *ACS Nano*, **4**(12), 7475–7480 (2010).

[4] (a) K. Gorska and N. Winssinger, Reactions templated by nucleic acids: more ways to translate oligonucleotide-based instructions into emerging function, *Angew. Chem., Int. Ed.*, **52**, 6820–6843 (2013); (b) A. Shibata, H. Abe, and Y. Ito, Oligonucleotide-templated reactions for sensing nucleic acids, *Molecules*, **17**, 2446–2463 (2012); (c) A. P. Silverman and E. T. Kool, Detecting RNA and DNA with templated chemical reactions, *Chem. Rev.*, **106**, 3775–3789 (2006).

[5] A. T. Poulin-Kerstien, and P. B. Dervan, DNA-templated dimerization of hairpin polyamides, *J. Am. Chem. Soc.*, **125**, 15811–15821 (2003).

[6] I. Boll, E. Jentzsch, R. Krämer, and A. Mokhir, Metal complex catalysis on a double stranded DNA template, *Chem. Commun.*, 3447–3449 (2006).

[7] (a) H. Li, R. M. Franzini, C. Bruner, and E. T. Kool, Templated chemistry for sequence-specific fluorogenic detection of duplex DNA, *ChemBioChem*, **11**, 2132–2137 (2010); (b) S. H. Lee, S. Wang, and E. T. Kool, Templated chemistry for monitoring damage and repair directly in duplex DNA, *Chem. Commun.*, **48**, 8069–8071 (2012).

[8] (a) J. A. Raskatov, N. G. Nickols, A. E. Hargrove, G. K. Marinov, B. Wold, and P. B. Dervan, Gene expression changes in a tumor xenograft by a pyrrole-imidazole polyamide, *Proc. Natl. Acad. Sci. U.S.A.*, **109**(40), 16041–16045 (2012); (b) N. G. Nickols, J. O. Szablowski, A. E. Hargrove, B. C. Li, J. A. Raskatov, and P. B. Dervan, Activity of a Py-Im polyamide targeted to the estrogen response element, *Mol. Cancer Ther.*, **12**(5), 675–684 (2013).

[9] (a) Y. Ben-Ari, Y. Brody, N. Kinor, A. Mor, T. Tsukamoto, D. L. Spector, R. H. Singer, and Y. Shav-Tal, The life of an mRNA in space and time, *J. Cell Sci.*, **123**, 1761–1774 (2010); (b) D. Grünwald and R. H. Singer, *In vivo* imaging of labelled endogeneous β-actin mRNA during nucleocytoplasmic transport, *Nature*, **467**, 604–609 (2010).

[10] A. S. Brodsky and P. A. Silver, Identifying proteins that affect mRNA localization in living cells, *Methods*, **26**(2), 151–155 (2002).

[11] Y. Shav-Tal, X. Darzacq, S. M. Shenoy, D. Fusco, S. M. Janicki, D. L. Spector, and R. H. Singer, Dynamics of single mRNPs in nuclei of living cells, *Science*, **304**, 1797–1800 (2004).

[12] J. S. Paige, K. Y. Wu, and S. R. Jaffrey, RNA mimics of green fluorescent protein, *Science*, **333**, 642–646, (2011).

[13] P. Santangelo, N. Nitin, and G. Bao, Nanostructured probes for RNA detection in living cells, *Ann. Biomed. Eng.*, **34**(1), 39–50 (2006).

[14] S. Tyagi and F. R. Kramer, Molecular beacons: Probes that fluoresce upon hybridization, *Nat. Biotechnol.*, **14**, 303–308 (1996).

[15] (a) P. J. Santangelo, B. Nix, A. Tsourkas, and G. Bao, Dual FRET molecular beacons for mRNA detection in living cells, *Nucleic Acids Res.*, **32**, e57 (2004); (b) A. Tsourkas, M. A. Behlke, Y. Xu, and G. Bao, Spectroscopic features of dual fluorescence/luminescence resonance energy-transfer molecular beacons, *Anal. Chem.*, **75**, 3697–3703 (2003).

[16] (a) D. S. Seferos, D. A. Giljohann, H. D. Hill, A. E. Prigodich, and C. A. Mirkin, Nano-flares: probes for transfection and mRNA detection in living cells, *J. Am. Chem. Soc.*, **129**(50), 15477–15479 (2007); (b) A. E. Prigodich, P. S. Randeria, W. E. Briley, N. J. Kim, W. L. Daniel, D. A. Giljohann, and C. A. Mirkin, Multiplexed nanoflares: mRNA detection in live cells, *Anal. Chem.*, **84**, 2062–2066 (2012).

[17] (a) Y. Xu, N. B. Karalkar, and E. T. Kool, Nonenzymatic autoligation in direct three-color detection of RNA and DNA point mutations, *Nat. Biotechnol.*, **19**, 148–152, (2001); (b) S. Sando and E. T. Kool, Quencher as leaving group: efficient detection of DNA-joining reactions, *J. Am. Chem. Soc.*, **124**(10), 2096–2097 (2002); (c) S. Sando and E. T. Kool, Imaging of RNA in bacteria with self-ligating quenched probes, *J. Am. Chem. Soc.*, **124**(33), 9686–9687 (2002).

[18] H. Abe and E. T. Kool, Flow cytometric detection of specific RNAs in native human cells with quenched autoligating FRET probes, *Proc. Natl. Acad. Sci. U.S.A.*, **103**(2), 263–268 (2006).

[19] D. J. Kleinbaum and E. T. Kool, Sandwich probes: two simultaneous reactions for templated nucleic acid detection, *Chem. Commun.*, **46**, 8154–8156 (2010).

[20] D. J. Kleinbaum, G. P. Miller, and E. T. Kool, Double displacement: an improved bioorthogonal reaction strategy for templated nucleic acid detection, *Bioconj. Chem.*, **21**, 1115–1120 (2010).

[21] G. O. Miller, A. P. Silverman, and E. T. Kool, New, stronger nucleophiles for nucleic acid-templated chemistry: synthesis and application in fluorescence detection of cellular RNA, *Bioorg. Med. Chem. Lett.*, **16**, 56–64 (2008).

[22] H. Abe and E. T. Kool, Destabilizing universal linkers for signal amplification in self-ligating probes for RNA, *J. Am. Chem. Soc.*, **126**, 13980–13986 (2004).

[23] A. P. Silverman and E. T. Kool, Quenched autoligation probes allow discrimination of live bacterial species by single nucleotide differences in rRNA, *Nucleic Acids Res.*, **33**(15), 4978–4986 (2005).

[24] E. Saxon and C. R. Bertozzi, Cell surface engineering by a modified Staudinger reaction, *Science*, **287**, 2007–2010 (2000).

[25] J. Cai, X. Li, X. Yue, and J. S. Taylor, Nucleic acid-triggered fluorescent probe activation by the Staudinger reaction, *J. Am. Chem. Soc.*, **126**, 16324–16325 (2004).

[26] Z. L. Pianowski and N. Winssinger, Fluorescence-based detection of single nucleotide permutation in DNA via catalytically templated reaction, *Chem. Commun.*, 3820–3822 (2007).

[27] R. M. Franzini and, E. T. Kool, 7-Azidomethoxy-coumarins as profluorophores for templated nucleic acid detection, *ChemBioChem*, **9**, 2981–2988 (2008).

[28] (a) H. Abe, J. Wang, K. Furukawa, K. Oki, M. Uda, S. Tsuneda, and Y. Ito, A reduction-triggered fluorescence probe for sensing nucleic acids, *Bioconj. Chem.*, **19**, 1219–1226 (2008); (b) K. Furukawa, H. Abe, K. Hibino, Y. Sako, S. Tsuneda, and Y. Ito, Reduction-triggered fluorescent amplification probe for the detection of endogenous RNAs in living human cells, *Bioconj. Chem.*, **20**, 1026–1036 (2009).

[29] Z. Pianowski, K. Gorska, L. Oswald, C. A. Merten, and N. Winssinger, Imaging of mRNA in live cells using nucleic acid-templated reduction of azidorhodamine probes, *J. Am. Chem. Soc.*, **131**, 6492–6497 (2009).

[30] K. Gorska, I. Keklikoglou, U. Tschulena, and N. Winssinger, Rapid fluorescence imaging of miRNAs in human cells using templated Staudinger reaction, *Chem. Sci.*, **2**, 1969–1975 (2011).

[31] (a) K. Furukawa, H. Abe, Y. Tamura, R. Yoshimoto, M. Yoshida, S. Tsuneda, and Y. Ito, Fluorescence detection of intron lariat RNA with reduction-triggered fluorescent probes, *Angew. Chem., Int. Ed.*, **50**, 12020–12023 (2011); (b) Y. Tamura, K. Furukawa, R. Yoshimoto, Y. Kawai, M. Yoshida, S. Tsuneda, Y. Ito, and H. Abe, Detection of pre-mRNA splicing *in vitro* by an RNA-templated fluorogenic reaction, *Bioorg. Med. Chem. Lett.*, **22**, 7248–7251 (2012).

[32] H. Saneyoshi, T. Ochikubo, T. Mashimo, K. Hatano, Y. Ito, and H. Abe, Triphenylphosphinecarboxamide: an effective reagent for the reduction of azides and its application to nucleic acid detection, *Org. Lett.*, **16**, 30–33 (2014).

[33] H. Saneyoshi, Y. Ito, and H. Abe, Long-lived luminogenic probe for detection of RNA in a crude solution of living bacterial cells, *J. Am. Chem. Soc.*, **135**, 13632–13635 (2013).

[34] (a) R. M. Franzini and E. T. Kool, Efficient nucleic acid detection by templated reductive quencher release, *J. Am. Chem. Soc.*, **131**, 16021–16023 (2009); (b) R. M. Franzini, E. T. Kool, Improved templated fluorogenic probes enhance the analysis of closely related pathogenic bacteria by microscopy and flow cytometry, *Bioconj. Chem.*, **22**, 1869–1877 (2011).

[35] R. M. Franzini and E. T. Kool, Two successive reactions on a DNA template: a strategy for improving background fluorescence and specificity in nucleic acid detection, *Chem. Eur. J.*, **17**, 2168–2175 (2011).

[36] K. Gorska, A. Manicardi, S. Barluenga, and N. Winssinger, DNA-templated release of functional molecules with an azide-reduction-triggered immolative linker, *Chem. Commun.*, 4364–4366 (2011).

[37] (a) S. Dutta, B. Flottmann, M. Heilemann, and A. Mokhir, Hybridization and reaction-based fluorogenic nucleic acid probes, *Chem. Commun.*, 9664–9666 (2012); (b) M. Schikora, S. Dutta, and A. Mokhir, Nucleic acid-specific photoactivation of oligodeoxyribonucleotides labeled with deuterated dihydro-N,N,N′,N′-tetramethylrhodamine using green light, *Histochem. Cell Biol.*, Published online on 05-Feb-2014, DOI: 10.1007/s00418-014-1187-0 (2014).

[38] (a) R. J. Lewis and P. C. Hanawalt, Ligation of oligonucleotides by pyrimidine dimers - A missing "link" in the origin of life? *Nature*, **298**, 393–396 (1982); (b) K. Fujimoto, S. Matsuda, N. Takahashi, and I. Saito, Template-directed photoreversible ligation of deoxyoligonucleotides via 5-vinyldeoxyuridine, *J. Am. Chem. Soc.*, **122**, 5646–5647 (2000); (c) K. Fujimoto, S. Matsuda, N. Ogawa, M. Hayashi, and I. Saito, Templated-directed reversible photocirculization of DNA via 5-vinyldeoxycytidine, *Tetrahedron Lett.*, **41**, 6451–6454 (2000); (d) M. Ogino, Y. Taya, and K. Fujimoto, Highly selective detection of 5-methylcytosine using photochemical ligation, *Chem. Commun.*, 5996–5998 (2008); (e) M. Ogino, Y. Yoshimura, A. Nakazawa, I. Saito, and K. Fujimoto, Templated-directed DNA photoligation via α-5-cyanovinyldeoxyuridine, *Org. Lett.*, **7**(14), 2853–2856 (2005); (f) I. Saito, Y. Miyauchi, Y. Saito, and K. Fujimoto, Template-directed photoreversible ligation of DNA via 7-carboxyvinyl-7-deaza-2′-deoxyadenosine, *Tetrahedron Lett.*, **46**, 97–99 (2005).

[39] E. Cló, J. W. Snyder, N. V. Voigt, P. R. Ogilby, and K. V. Gothelf, DNA-programmed control of photosensitized singlet oxygen production, *J. Am. Chem. Soc.*, **128**, 4200–4201 (2006).

[40] (a) D. Arian, E. Cló, K. V. Gothelf, and A. Mokhir, A nucleic acid dependent chemical photocatalysis in live human cells, *Chem. Eur. J.*, **16**, 288–295 (2010); (b) S. Dutta and A. Mokhir, An autocatalytic chromogenic and fluorogenic photochemical reaction controlled by nucleic acids, *Chem. Commun.*, **47**, 1243–1245 (2011).

[41] (a) S. Dutta, A. Fülöp, and A. Mokhir, Fluorogenic, catalytic, photochemical reaction for amplified detection of nucleic acids, *Bioconj. Chem.*, **24**, 1533–1542 (2013); (b) A. Fülöp, X. Peng, M. M. Greenberg, and A. Mokhir, A nucleic acid-directed, red light-induced chemical reaction, *Chem. Commun.*, **46**, 5659–5661 (2010).

[42] D. Arian, L. Kovbasyuk, and A. Mokhir, 1,9-Dialkoxyanthracene as a $^1O_2$-sensitive linker, *J. Am. Chem. Soc.*, **133**, 3972–3980 (2011).

[43] M. Röthlingshöfer, K. Gorska, and N. Winssinger, Nucleic acid-templated uncaging of fluorophores using Ru-catalyzed photoreduction with visible light, *Org. Lett.*, **14**, 482–485. (2012).

[44] K. K. Sadhu and N. Winssinger, Detection of miRNA in live cells by using templated Ru^{II}-catalyzed unmasking of a fluorophore, *Chem. Eur. J.*, **19**, 8182–8189 (2013).

3.5

The Biotechnological Applications of G-Quartets

David Monchaud

ICMUB (Institut de Chimie Moleculaire Universite de Bourgogne), CNRS UMR6302, Dijon, France

3.5.1 Introduction

Chemists are unparalleled in their ability to strive to construct complex molecular assemblies that mimic biological systems, with the hope of harnessing Nature's insights and synthetic know-how. Nature indeed achieves the amazing *tour de force* of producing a virtually infinite panoply of biopolymers (proteins, nucleic acids, etc.) from a limited number of building blocks (amino acids, nucleobases, etc.), which display not only structural but also functional diversity. Deoxyribonucleic acid (DNA) is certainly the most dazzling example of such know-how: Nature has excelled at using DNA to encode, store, and transmit information, assembling a two billion base-pair biopolymer (in mammalian cells) from only four basic building blocks, namely deoxyguanosine (dG), deoxycytosine (dC), deoxythymidine (dT), and deoxyadenosine (dA). The striking contrast between the functional complexity of DNA and the structural simplicity of its monomers has held a fascination for chemists (notably to supramolecular chemists), who are now on a relentless pursuit to handle and control this unique skill.

3.5.2 Nucleobases and H-bonds

The four aforementioned nucleobases (dG, dC, dT, and dA, Figure 3.5.1) are iconic due to their participation in the so-called Watson–Crick base pairing, in which dG is associated with dC via three hydrogen bonds (or H-bonds), while the dT–dA couple is assembled through the formation of two H-bonds only. The specific

DNA in Supramolecular Chemistry and Nanotechnology, First Edition.
Edited by Eugen Stulz and Guido H. Clever.

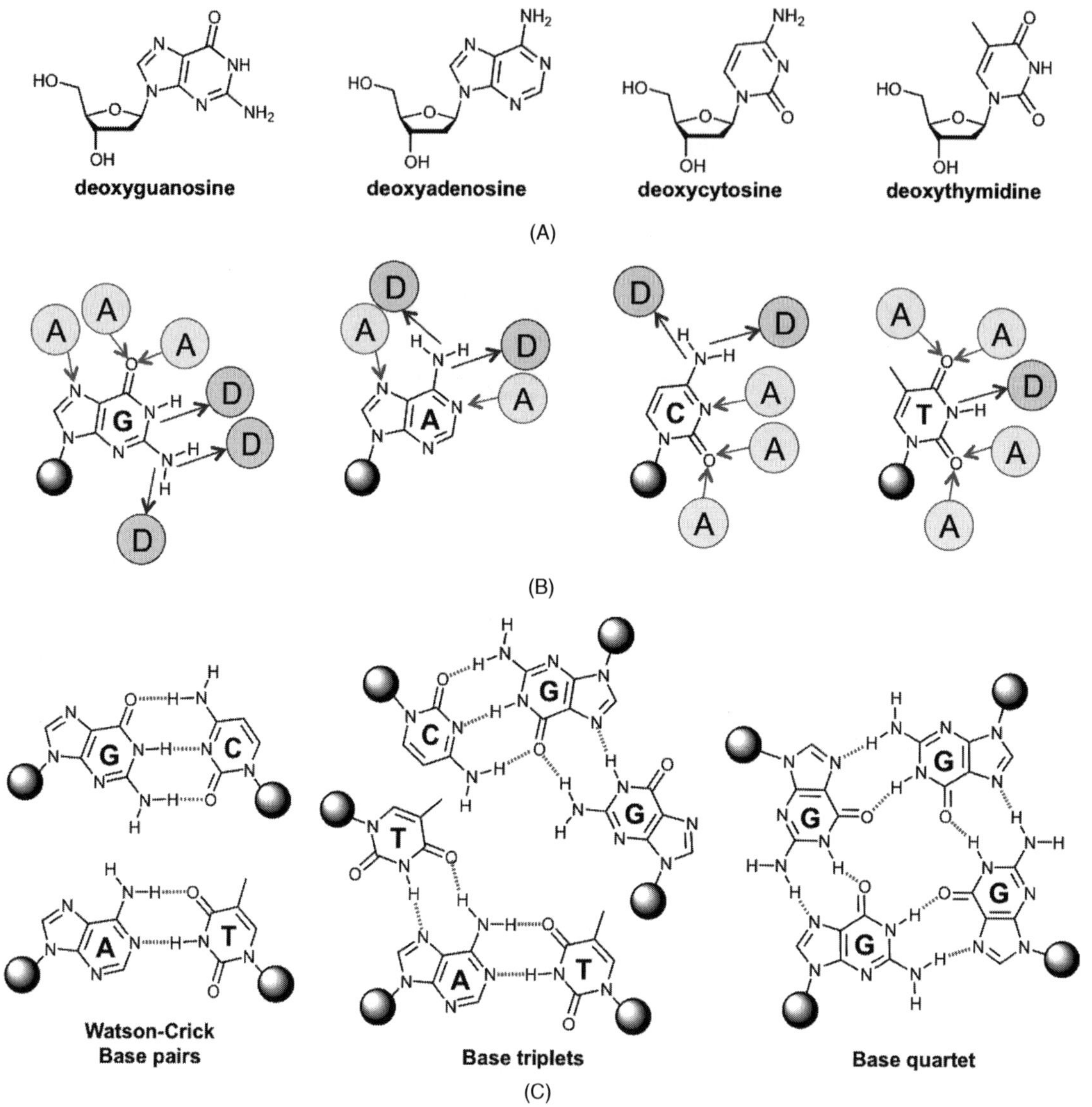

Figure 3.5.1 *Chemical structures of nucleobases (A), details of their H-bond donating and accepting ability (B), and various base pairing possibilities involving two, three or four nucleobases (C)*

dG–dC and dA–dT base pairings were initially surmised correctly by J. D. Watson and F. H. C. Crick in their landmark 1953 article in *Nature* [1], in their quest for the elucidation of the double-helix structure. These specific base associations were thus described as playing a salient role in the formation of the secondary structure of DNA, which is leveraged on the association of two DNA single strands via the formation of interstrand dG–dC and dT–dA base pairs, thereby insuring a high-fidelity, fully programmable association of the two DNA strands.

Interestingly, the Watson–Crick base pairing exploits only a small proportion of the H-bond donor–acceptor potential of nucleobases [2]. As seen in Figure 3.5.1B, the purines display two to three acceptor (A) and donor

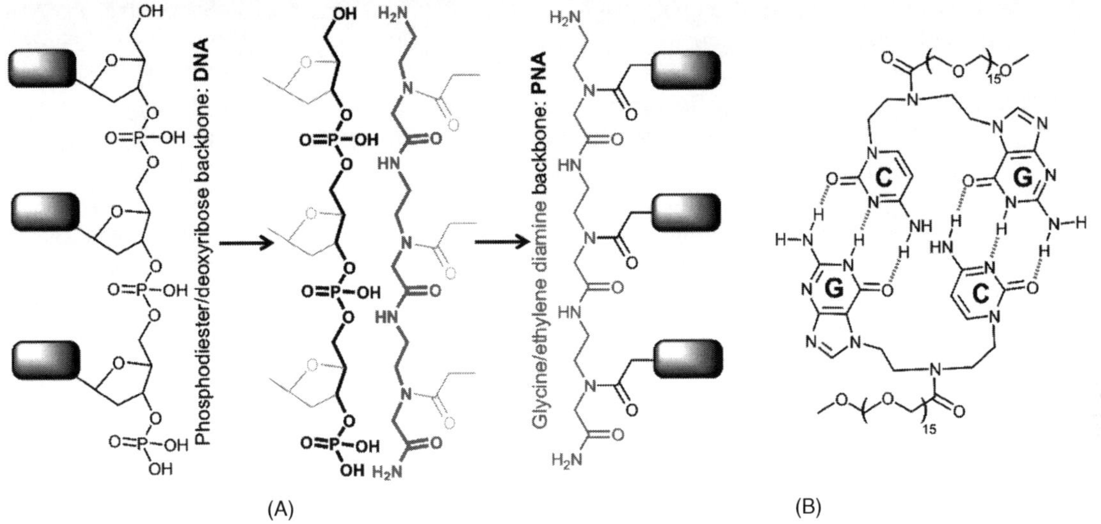

Figure 3.5.2 *Structures of DNA and PNA backbones (A) and the first prototype of a synthetic duplex mimic (B)*

(D) sites, while pyrimidines elicit one to four A–D sites. This versatility is invaluable for making them capable of participating in numerous H-bond-based arrangements, culminating in the formation of 28 distinct base-pairing possibilities involving at least two H-bonds. This H-bond versatility also makes higher-order base association possible, notably with base triplets or quartets (Figure 3.5.1C). The latter results from the self-arrangement of four bases – principally guanines – associated through Hoogsteen H-bonds; G is indeed quite unique in its ability to spontaneously oligomerize to form a cyclic array of four G held together via eight Hoogsteen H-bonds. As will be further discussed subsequently, this macrocyclic supramolecular assembly, indiscriminately termed a G-quartet or G-tetrad, is a fascinating, multitasking nanotool for chemists that have endeavored to harness its quite unique biophysical properties (including tunable stability and cation chelating ability) to develop enticing biotechnological applications.

3.5.3 Duplex-DNA mimics

In an attempt to mimic Nature, chemists have tried to design synthetic DNA; peptidic nucleic acids (or PNA) can be considered as a brilliant demonstration of the chemists' know-how, producing a fully synthetic DNA in which the phosphodiester–deoxyribose backbone is replaced by a neutral peptidic scaffold (Figure 3.5.2A) [3]. However, PNA has not been developed to mimic DNA *per se* but it has been designed to strongly interact with it, either forming PNA–DNA chimeric duplexes or targeting double-stranded DNA through major groove binding, thereby forming an artificial triplex aimed at dampening or ultimately silencing the expression of the targeted sequence. This approach is interesting in that it actually clarifies that stable DNA mimics can only be polymeric: indeed, the energy imparted by the formation of a discrete base pair is weak, with association constants (K_a) of between ~10^2 and 10^4 M^{-1} for dA–dT and dG–dC pairs, respectively. Therefore, the astute strategy implemented by Nature to make DNA a robust biopolymer is to assemble millions of nucleobases in a single double-stranded scaffold, whose stability thus originates from a subtle combination of hydrogen-bonding (base pairing) along with hydrophobic and π-stacking interactions (between the aromatic purine–pyrimidine moieties of consecutives bases).

With the hope of harnessing the great potential of the programmable H-bond capability of nucleobases for the preparation of non-covalent supramolecular assemblies, chemists rapidly decided to gain more insights into the stability and dynamics of the small, readily accessible duplex-DNA models. To this end, early in 1992 Sessler *et al.* reported on the very first, smallest artificial duplex-DNA, comprised of only two G–C base pairs (Figure 3.5.2B) [4]. This initial prototype was fairly unstable (K_a <10 M^{-1}), due to the flexibility of the linker used in between the nucleobases. This approach was pioneering in that it shed a bright light on the trade-off of the chemical approach, between the simplicity of the synthetic access, and the low stability of the resulting artificial DNA models. The stability of the subsequent generations of small synthetic duplex-DNA was improved via the use of more rigid linkers (e.g., an anthracene moiety), which did indeed result in enhanced stabilities at the expense of water solubility [2], thereby limiting the possibilities of bioapplications.

3.5.4 Guanine and G-quartets

In stark contrast to the aforementioned supramolecular assemblies, those relying on base quartets fared far better than their base-pairs counterparts (for authoritative reviews see reference [5]). Indeed, base pair promoted assemblies proved to be somewhat unstable, thereby limiting the scope of their applications; however, when a nucleobase is no longer involved in base pairs (i.e., association with its complementary base via 2–3 Watson–Crick H-bonds) but in base quartets (i.e., surrounded by two homonucleobases via four Hoogsteen H-bonds), the stability of the resulting edifice is high enough to allow for that system to be used as a valuable nanotechnological linchpin (*vide infra*). Interestingly, at present the origins of the elevated stability of G-quartets remain unclear: when a guanine becomes involved in a quartet, the rearrangement of its electron density is a phenomenon referred to as resonance-assisted hydrogen bonding (RAHB) [6]; G-quartet stability (and classical Watson–Crick base pairs alike) is assumed to originate in an electronic redistribution ascribed to interplay between H-bond formation and delocalization of the nucleobase π-electrons (resonance). RAHB has been recently challenged by the “charge separation” theory, which involves not π- but σ-electrons [7], basing the stability of a G-quartet upon a synergy between Hoogsteen H-bond formation and electronic interactions between donor (N and O lone-pairs) and acceptor (N–H antibonding) orbitals in the σ-electron system. According to both theories, the stability of a base association is dependent on the number of H-bonds and on the extent to which π-/σ-electrons are delocalized; since more nucleobases are involved in quartets than in pairs, the stability of the former is exponentially higher than that of the latter.

Even more interesting is the observation that a G-quartet assembly creates a wreath of inwardly pointing guanine oxygen atoms, which are thus structurally prone and electronically eager to interact strongly with cations, predominantly physiologically relevant cations (sodium and potassium). This interaction is not only favored but also requested for the neutralization of the electrostatic repulsion between the neighboring oxygen atoms of the inner cavity of the quartet. As further detailed below, this quite unique property has been exploited not only by Nature to make quadruplex architectures highly stable but also by researchers to develop biotechnological assays enabling, for instance, the detection of quartet assembling cations (such as K^+, used as an *input* signal) using quartet affinic reporters whose emitting state is switched on upon quartet interactions (such as fluorescence or luminescence response, used as an *output* signal).

3.5.5 G-Quartets and G-quadruplexes

Many different avenues of research have recently concurred in highlighting the pivotal role of both naturally occurring and purely synthetic G-quartets in nanotechnological developments. Naturally occurring G-quartets were the first to attract intensified interest, not from a technological but a biological point of view thanks

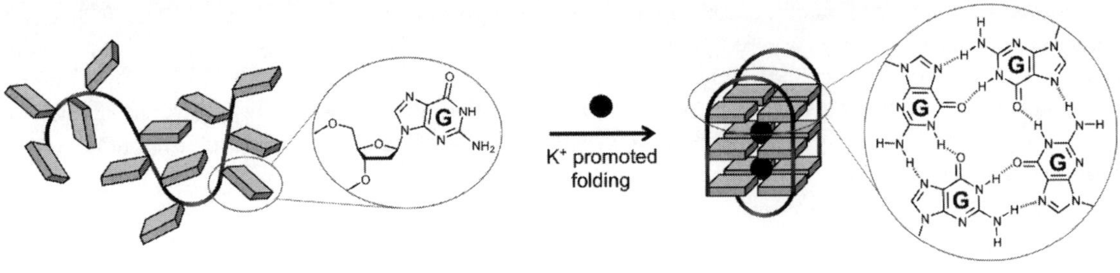

Figure 3.5.3 G-rich quadruplex forming sequence under its unfolded (left) and cation-promoted folded state (right)

to their participation in the formation of both DNA and RNA quadruplexes [8]. Briefly, these higher-order DNA–RNA structures can fold from G-rich DNA and RNA sequences, which appear to be particularly enriched in key regions of both the genome (DNA) and the transcriptome (RNA). The discussion about the biological relevance of DNA–RNA quadruplexes is beyond the scope of the present chapter; readers interested in this fascinating field are invited to refer to some of the recent excellent reviews that have authoritatively covered it [8–16]. The fundamental premises can nevertheless be briefly summarized, by explaining that G-rich DNA and RNA sequences, and more precisely sequences in which G-richness hinges on the presence of blocks of consecutive G units, can give rise to corresponding G-quadruplex structures (Figure 3.5.3) [9]: the folding of the quadruplexes is promoted by physiological cations that are trapped in between the consecutive G-quartet layers (caught by two carbonyl wreaths), which also contribute to the stability of the resulting DNA–RNA edifices along with the self-stacking interactions between the contiguous G-quartets. The quadruplex-forming sequences (QFS) are particularly enriched in key regions of the chromosomes [10], including telomeres [11] and promoter regions of oncogenes [12], as well as in the transcriptome, including telomeric RNA [13] and leader sequences (also called 5′-UTR) of messenger RNA (mRNA) [14]. This has spurred on investigations into the possibility of driving the corresponding cellular events, that is, chromosomal stability and regulation of gene expression at the transcriptional (DNA quadruplexes) or translational level (RNA quadruplexes), respectively, controlling the QFS folded/unfolded states, notably via the quadruplex stabilization with specific small-molecules (termed quadruplex ligands) [15]. For deeper discussions about the biological relevance of DNA–RNA quadruplexes and the therapeutic uses of quadruplex-interacting compounds, readers are invited to refer to references [8–16].

Beyond their biological implications, quadruplexes have also been studied for their biotechnological potential. Indeed, quadruplexes are unique in their ability to firmly bind to small molecules, offering them a structurally singular binding site, that is, their external, readily accessible G-quartets. The external G-quartet–small molecule interaction is a pivotal event in the aim of using quadruplexes as structural switches to control all aforementioned DNA–RNA transactions (i.e., telomere stability and transcription, gene expression, etc.). From this perspective, small molecules must trap and stabilize the QFS folded state, withstanding protein activities such as helicases (that is aim to unwind quadruplexes to insure normal DNA metabolisms) or ribosomes (that assemble on mRNA 5′-UTR to initiate its translation) for instance. In this case, quadruplex ligands must mainly display a high affinity for their DNA–RNA targets. The situation becomes even more interesting if the ligands acquire specific properties once they are in interactions with the quadruplexes, such as an increase in fluorescence or luminescence properties, an enhanced oxidative capability, and so on; in this case, the quadruplex–ligand interactions thus convey easily readable output signals, which can be readily exploited to make quadruplexes valuable molecular devices. DNA–RNA quadruplexes, which have been thoroughly exploited as malleable scaffolds to engineer catalytic centers involved in a wide variety of nanoprocesses, thus represent shining examples of the biotechnological use of naturally occurring G-quartets.

Contrary to synthetic G-quartet-based applications (*vide infra*), the field of quadruplex-based technological developments is regularly covered by excellent reviews; to gain more precise insights into this fascinating field, interested readers are invited to refer to references [17–20] (along with Chapter 2.4 by Choi and Majima and Chapter 5.6 by Willner in this book). The fundamental premises can, however, be briefly summarized by classifying the use of quadruplex-based sensing platforms according to the nature of the output signal, enabling either colorimetric, luminescence–fluorescence or electrochemical detections.

3.5.5.1 Colorimetric detection

This application is certainly the most studied quadruplex-based detection system. The key event of this nature-inspired process relies on the strong binding of an iron(III) porphyrin (namely hemin, the well known cofactor of many hemoproteins) to quadruplexes (via external stacking), an interaction that significantly enhances its oxidative capability [18]. Therefore, the quadruplex–hemin system has been used comprehensively to perform hemoprotein-like reactions, chiefly peroxidase-mimicking reactions, in which a chromogenic substrate is oxidized by a catalytic quadruplex–hemin complex (in the presence of a stoichiometric oxidant, usually H_2O_2), making it a valuable enzyme-like system allowing for readily implemented colorimetric detection (Figure 3.5.4A). From a historical perspective, this pseudo-enzymatic property was discovered in 1998 by Li and Sen [21], during their investigations related to the combinatorial use of short DNA fragments to catalyze the metalation of porphyrins [22]. They not only demonstrated that DNA oligonucleotides were indeed efficient porphyrin metalation catalyzers, but also that the already metalated porphyrins, notably hemin, competitively inhibited the process. From this observation, they wisely mused about further study of the short DNA fragments as hemin aptamers, showing that the hemin-binding properties of the DNA fragments relied on their ability to fold into G-quadruplex structures (thereby offering hemin a privileged binding site). The quadruplex–hemin assemblies were thus evaluated for their hemoprotein-like, so called DNAzyme properties, notably their ability to facilitate the peroxidase-like H_2O_2-promoted oxidation of chromogenic substrates (chiefly ABTS (2,2′-azino-bis(3-ethylbenzothiazoline-6-sulfonic acid)) or TMB (3,3′,5,5′-tetramethylbenzidine)), which is routinely used for ELISA-type diagnosis: the observed catalytic activity provided the proof-of-concept that such a system is capable of performing pseudo-enzymatic reactions. This pioneering work thus ushered in the modern era of what is nowadays called the "DNA enzymology" field, with dozens of new applications reported each year (*vide infra*).

3.5.5.2 Luminescence–fluorescence detection

The logic behind this application relies on the use of quadruplex ligands whose luminescence–fluorescence properties are directly modulated by the quadruplex binding. The situation is especially interesting when the ligand-emitting state is switched on upon quadruplex interaction, making the resulting quadruplex–ligand assembly an invaluable "turn-on" detection method. Several interesting reporter probes have been designed and studied over the past years, including small organic (chiefly cyanine and porphyrin derivatives) or metallo-organic compounds (mainly ruthenium, zinc, and iridium complexes) [19]. Both colorimetric and fluorescence–luminescence methods share many common applications; the choice for a detection modality is mainly driven by the sensitivity required by the assays and/or the instrumentation available for performing the experiments. The numerous reported applications range from the very simple detection of the quadruplex *per se* (e.g., for the detection of cations promoting quadruplex folding, Figure 3.5.4A) [23], to more complex constructs. The QFS could, for instance, be embedded in duplexes (Figure 3.5.4B), one strand of which experiences an enzyme-mediated modification that frees the QFS from its duplex restraint, thereby folding into a quadruplex and triggering the detection signal (e.g., for the detection of exonuclease activity) [24]. Alternatively, the QFS can be "split" into two parts, associated with flanking oligonucleotides that are aimed

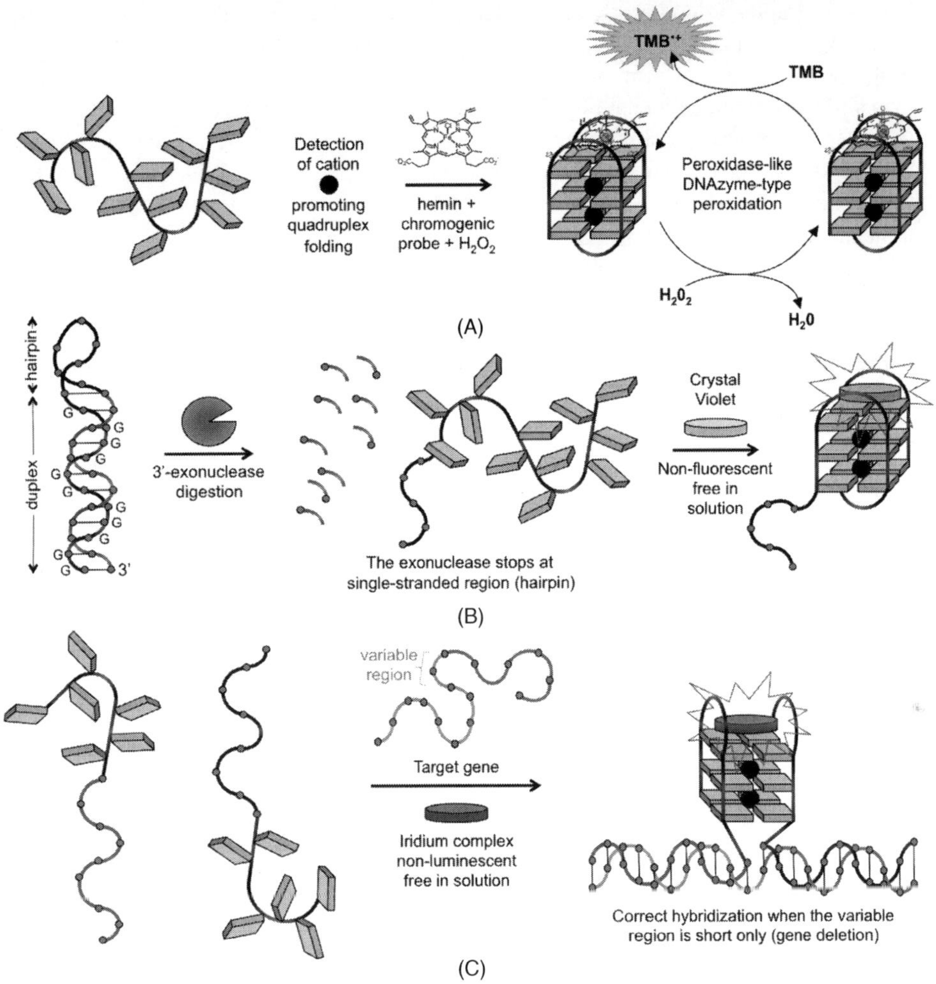

Figure 3.5.4 *Colorimetric (A), fluorescence (B), and luminescence (C) detection systems implemented with QFS, that is, detection of* K^+ *[23], 3′-exonuclease activity [24] and gene deletion [25], respectively (See color figure in color plate section)*

at recognizing a given sequence (Figure 3.5.4C). Upon hybridization with their DNA targets, the two QFS halves come close enough to assemble a quadruplex, which interacts with a reporter probe to produce the appropriate output signal (e.g., for the detection of gene deletion) [25]. Now that the versatility of quadruplex-based biosensors has been firmly established, efforts are being invested in the design of surface-immobilized quadruplex-based systems (mainly via streptavidin to date) [26], to allow for the development of sensing systems suitable for ELISA-type assays for instance (in terms of sensitivity, robustness, recyclability, etc.).

3.5.5.3 Electrochemical detection

The surface immobilization of QFS is even more interesting if the surface itself has specific properties. Notably, QFS can be bound to electrode surfaces, to convey electrochemical stimuli as the output signals [20]. In this particular case, the "on" state originates in the modification of an electrochemical signal that is promoted by the quadruplex fold, directly or indirectly. For instance, the QFS could be covalently modified by a ferrocene (Fc) reporter on one of its extremities, and the contraction that results from the quadruplex formation triggered by an external stimulus (e.g., K^+ cations) restores the electronic communication between the Fc and the gold surface, thereby enabling the direct detection of the molecular motion by monitoring the current changes (Figure 3.5.5A) [27]. "Label free" systems have also been designed, that is, systems based on the use of unmodified (naked) oligonucleotides (as in the aforementioned colorimetric and fluorescence–luminescence methods). These are mostly based on the efficient hemin–quadruplex interactions, either for the detection of the quadruplex fold itself (e.g., for the detection of proteins that promote quadruplex structures, such as thrombin that folds the QFS thrombin-binding aptamers (TBA), making the resulting system an electrochemical aptasensor, Figure 3.5.5B) [28]. Alternatively, the QFS could again be embedded in duplexes, one strand of which experiences an enzyme-mediated modification that frees the QFS from its duplex restraint (Figure 3.5.5C), thereby folding into a quadruplex and triggering hemin-mediated current changes to be used as detection signals (e.g., for the detection of endonuclease activity) [29]. In both instances, the readout signal is conveyed by the catalytic reduction of H_2O_2, the required electrons being provided by the electrode surface. As compared with colorimetric and fluorescence–luminescence methods, electrochemical biosensors have their own trade-offs, relying on specialized instrumentations and skills on the one hand but offering the advantages of being highly sensitive and readily recyclable on the other.

Altogether, the variety, complementarity, reliability, and practical convenience of the three aforementioned detection methods provide a compelling rationale for explaining why dozens of new examples of quadruplex-based sensors are reported each year. These include the detection of monovalent (K^+ or Ag^+) and divalent cations (Cu^{2+}, Hg^{2+} or Pb^{2+}), proteins (such as nucleolin, thrombin, lysozyme or vascular endothelial growth factor), nucleic acid derivatives (DNA–RNA analytes, single-stranded DNA, miRNA, adenosine triphosphate or c-di-GMP (*vide infra*)) or nucleotide modifications (single nucleotide polymorphism or single-base mutations), as well as more complex assays to assess telomerase, methyltransferase, glucose oxidase or cholesterol oxidase activities for instance [17–20].

3.5.6 Quadruplex-DNA mimics

Contrary to the modest successes achieved in the design and synthesis of minimalist synthetic duplex-DNA mimics (*vide supra*), the higher stability of discrete G-quartets (as compared with that of discrete base pairs) endows synthetic quadruplex-DNA mimics with fairly unique biotechnological skills [5]. In a vast majority of the applications developed so far, guanines and non-guanosine derivatives are employed, thereby giving rise to fully synthetic G-quartets (hereafter termed SQ). Guanines are appealing for the versatility of their uses, acting either as molecular staples to assemble supramolecular edifices via the formation of *inter*molecular G-quartets, or as simplified quadruplex-DNA mimics *per se*, when the formation of the G-quartet is *intra*molecular. The biotechnological applications of guanines can thus be categorized with regards to the nature of the guanine entities involved: G-monomers (a single guanine per molecular scaffold) that form intermolecular SQ only, G-dimers (two guanines per scaffold) that also form intermolecular but interconnected SQ, and G-tetramers (four guanines per scaffold) that adopt an intramolecular SQ fold.

3.5.6.1 Intermolecular SQ: G-monomers

The ability of concentrated solutions of guanine to form gels was reported early on in the twentieth century via the pioneering work from Bang (in 1901) [30], who first observed this phenomenon, and Davis and coworkers (in 1962) [31], who provided the molecular basis for the gelation process, demonstrating the self-organization of guanines into G-quartets via X-ray crystallography analysis. This property was further

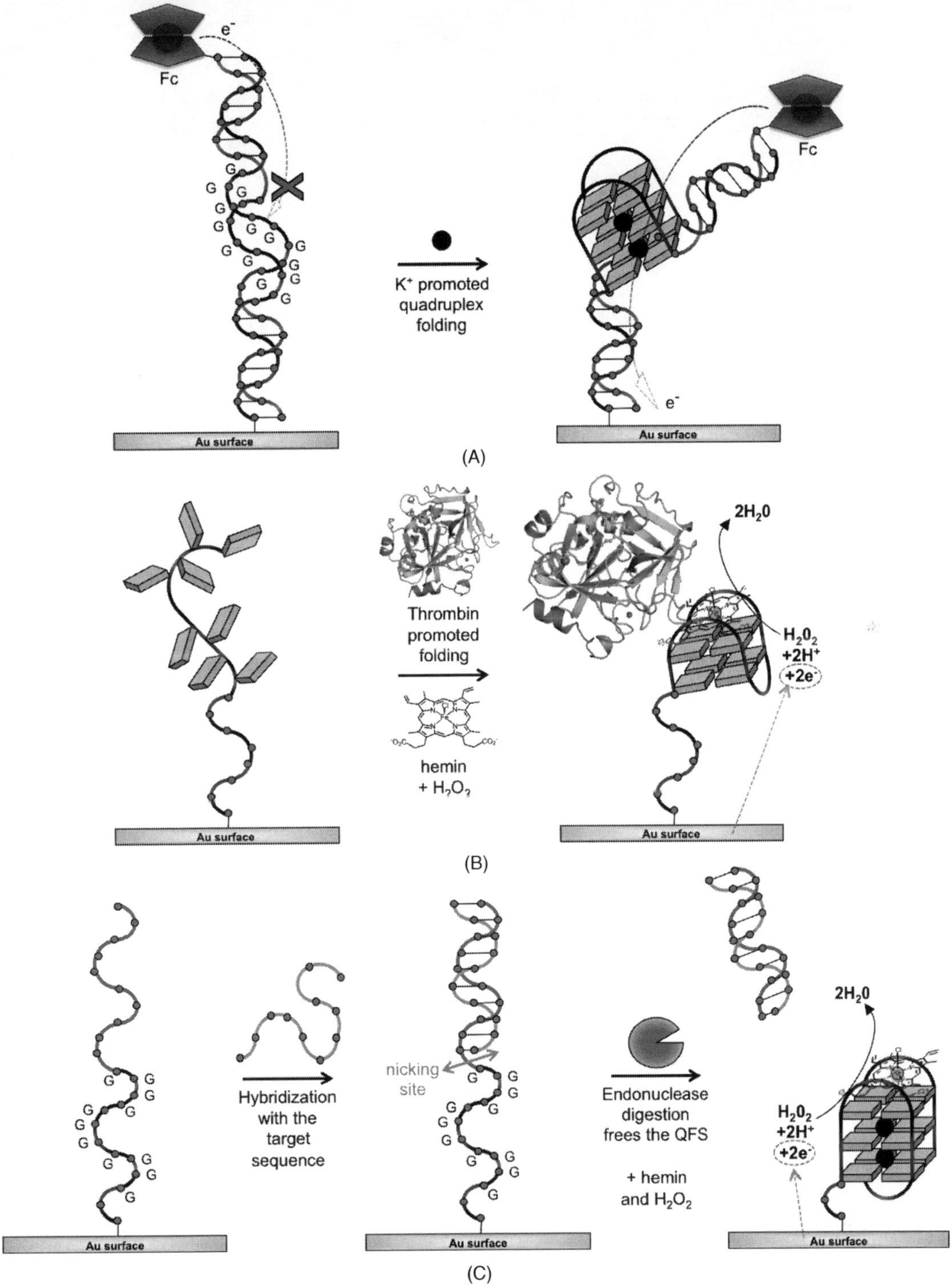

Figure 3.5.5 *Covalent (A) and label free (B,C) electrochemical detection systems implemented with QFS, that is, a contractile nanoswitch [27] and the detection of TBA [28] and endonuclease activity [29], respectively*

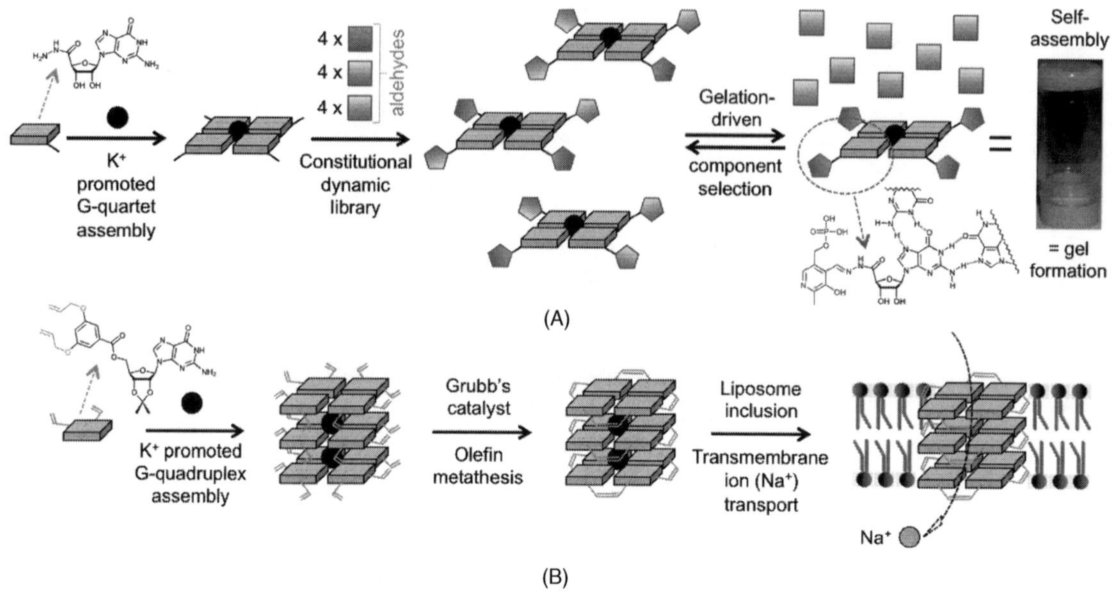

Figure 3.5.6 *Nanotechnological applications of G-monomers, for dynamic polymer (A) [32] and transmembrane ionophore (B) [39] formation*

exploited half a century later, notably by the teams working with Lehn [32] and Barboiu [33], through a series of studies aimed at using G-monomers as building blocks for stable hydrogels (Figure 3.5.6A). Hydrogels, notably pH responsive gels, are of particular interest for biomedical innovations, especially for drug delivery applications. Lehn and coworkers, for instance, demonstrated that guanosine hydrazide derivatives are efficient hydrogelators upon addition of metal cations, the properties of the resulting supramolecular hydrogels being reversibly modulated by external stimuli (chiefly pH and temperature). Even more elegant were the investigations on the chemical reactivity of the hydrazide appendages, which readily and reversibly react with a library of aldehydes to afford decorated G-quartets: interestingly, the gelation process drives the component selection since the system itself selects the aldehyde that provides the most stable gel in a dynamic manner. This gelation-driven component selection harnesses the principles of the constitutional dynamic libraries to trigger self-organization processes at the macromolecular level, thus being a very elegant example of SQ-based supramolecular chemistry applied to materials science.

While it has long been thought that G-quartets could form in water only, independent research from the groups of Davis [34] and Spada [35] pioneered the demonstration, in 1995, that finely structurally tuned (lipophilic) guanines (or closely related analogues) could also self-assemble in organic phases in the presence of alkali cations [36] (a restriction that was actually removed a couple a years later by Sessler and coworkers [37]). This opened a glorious pathway to the use of the resulting SQ in lipophilic media, notably via the use of the subsequent quadruplexes as selective ion extractants, enabling, for instance, the removal of cations from organic phases. Even more interesting were the applications based on lipid bilayer membranes: SQ can self-stack within phospholipid bilayers to form synthetic quadruplexes whose structure (notably their inner ion channel) makes them perfectly suited for transmembrane ion transport [38]. To make such a system even more efficient, Davis and coworkers decided to covalently freeze artificial quadruplexes once assembled, thereby discarding kinetic and thermodynamic stability problems (further discussed below) classically linked to intermolecular associations, and thus making them more robust ionophores [39]. To this end, they used guanosine derivatives with two appended reactive double-bonds and cross-linked all monomers together after they assembled into quadruplexes via ruthenium-catalyzed olefin metathesis technology (Figure 3.5.6B). The resulting covalently captured artificial quadruplexes were found to be highly efficient for enabling Na^+ transport across liposomal membranes. Alternatively, G-monomers have also been used for the construction of many nanoscale objects, including organic nanoparticles [40], nanoparticle assemblies [41], and biomolecular nanowires [42].

3.5.6.2 Intermolecular interconnected SQ: G-dimers

In most of the aforementioned cases, SQ are used to construct artificial quadruplexes and applications are mainly driven by the properties of the quadruplex itself (and not by that of the SQ *per se*). To this end, the use of discrete SQ might appear somewhat unwise since the formation of the resulting synthetic quadruplex requires the self-organization of four separated building blocks (guanines), followed by the self-recognition of at least two of the resulting SQ to form an artificial quadruplex: as both events are non-covalent (i.e., reversible) and intermolecular (notably the first step requires the quadruple association of discrete G), they might thus be somewhat limited or favored, both in kinetic and thermodynamic terms (*vide supra*). Astute strategies have been devised to favor intermolecular assemblies, for example, the aforementioned use of lipophilic G-conjugates whose association is facilitated by hydrophobic forces that make the reacting partners come closer to each other in a highly polar medium. Another elegant way to promote such intermolecular associations is to assemble more than one guanine in a single scaffold: G-dimers have been designed and synthesized, varying in the nature of the molecular link between the G units, these being, for instance, biphenylene [43] or bis-lithocholate [44] linkers, and the corresponding G-dimers being included in the construction of barrel-type or guanosine-sterol artificial ion channels, respectively. Among the most elegant applications, Ghoussoub and Lehn further extended their hydrogel studies (*vide supra*), implementing a process by which they fully seized control of the sol–gel interconversion of a PEG-linked G-dimer-based supramolecular polymer by external stimuli [45]. To this end, they constructed a macroscopic material on the basis of G-dimers whose gelation is triggered by K^+ cations and subsequently disaggregated by the addition of a K^+-affinic cryptand that hijacks the ions from the gel (Figure 3.5.7A). Interestingly, the chelation capability of the selected cryptand [2.2.2] is modulated by its protonation state, sequestering K^+ at neutral and basic pH [$K^+ \subset 2.2.2$] and releasing it at acidic pH [$2H^+ \subset 2.2.2$]. The sol–gel interconversion can thus be fully and easily controlled by sequential acidification–neutralization cycles, making pH-responsive SQ-based hydrogels very promising dynamers [2c].

An exception to the use of G-dimers to form interconnected SQ is the cyclic dimeric guanosine-3′,5′-monophosphate, or c-di-GMP: this naturally occurring G-dimer, in which the two G units are connected through their riboses by two phosphodiester groups, act as signaling agents in bacteria, notably regulating the production of virulence factors [46]. c-di-GMP has not been exploited for its nanotechnological applications *per se*; conversely, Sintim and coworkers benefitted significantly from the wealth of knowledge acquired through the quadruplex-based investigations to implement an easy-to-perform colorimetric assay to detect intrabacterial c-di-GMP concentrations. In contrast to GMP, which tetramerizes (to form G-quartets) at elevated concentrations only, c-di-GMP is quite unique in its ability to tetramerize (to form a G-quadruplex) at low micromolecular concentrations, compatible with its natural occurrence in bacteria ($<10\ \mu M$), notably in the presence of small-molecule intercalators (chiefly proflavine) that non-covalently template the quadruplex assembly, trapping itself in between the two quartet planes (Figure 3.5.7B) [47]. Therefore, c-di-GMP was colorimetrically quantified through its proflavine-induced quadruplex fold, via an H_2O_2-promoted hemin-mediated oxidation of a chromogenic probe (akin to DNAzyme) [48], an unprecedented ternary intercalator–quadruplex–hemin catalytic system that enables the detection of low ($\sim 1\ \mu M$) c-di-GMP concentrations from bacterial cell lysates.

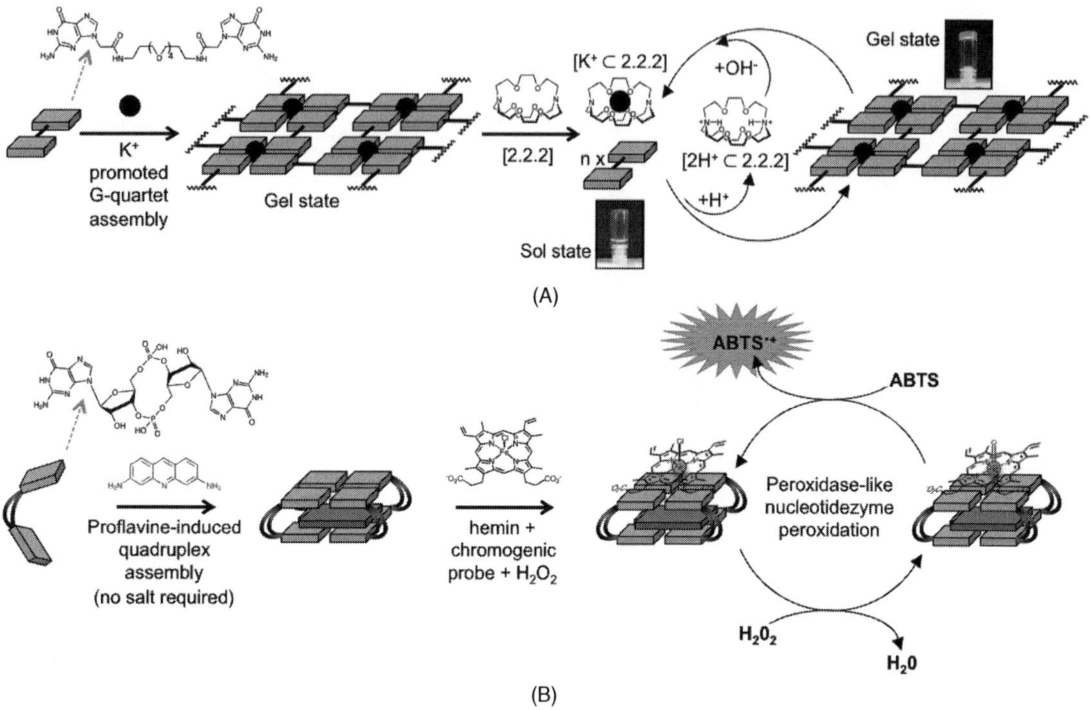

Figure 3.5.7 *Nanotechnological applications of G-dimers for tunable dynamer formation (A) [45] and application of the DNAzyme technological principles for the colorimetric quantification of c-di-GMP (B) [48]*

3.5.6.3 Intramolecular SQ (or iSQ): G-tetramers

The most straightforward way to circumvent the above kinetic and thermodynamic problems is to form G-quartets intramolecularly, assembling four G units in a single scaffold. The first example of a G-tetramer was reported by Davis and coworkers in 2000, in a study related to a 1,3-alternate calix[4]arene-based edifice (Figure 3.5.8A) [49]. However, the alternate nature of the calixarene used as a template to gather the four G oriented two pairs of G orthogonally (in an "up-down-up-down" fashion), promoted the resulting G-tetramers to polymerize to provide an SQ-based tube-like suprastructure. They further studied even more lipophilic calixarene–guanosine assemblies, demonstrating, for instance, that a controlled Na^+-promoted dimerization allowed for the resulting supramolecular ion pair receptor to firmly bind to NaCl and make it soluble in organic solvents. In 2008, Nikan and Sherman reported on a convergent approach, using a cavitand template that allows for the four G to be concomitantly present on the same side of the platform (Figure 3.5.8B), thereby enabling the formation of a *stricto sensu* intramolecular SQ (iSQ), which was termed TASQ (for template-assembled synthetic G-quartet) [50]. This lipophilic TASQ was unique in its ability to fold into its iSQ state in a cation-free manner, while also acting as a very efficient ion receptor. It is worth noting that the remarkable stability of Sherman's quadruplex DNA mimic (TASQ), as compared with that of the Sessler's duplex DNA mimic (*vide supra*), provides not only compelling evidence but also a clear demonstration of the higher stability of base quartets over the base pairs.

An important step on the road towards wider biotechnological applications was taken with the synthesis of water-soluble TASQ. The water solubility is mostly imparted by the nature of the template that is used to assemble the G-tetramer, this being a DOTA-like polyazamacrocycle (DOTASQ and ^{PNA}DOTASQ) [51],

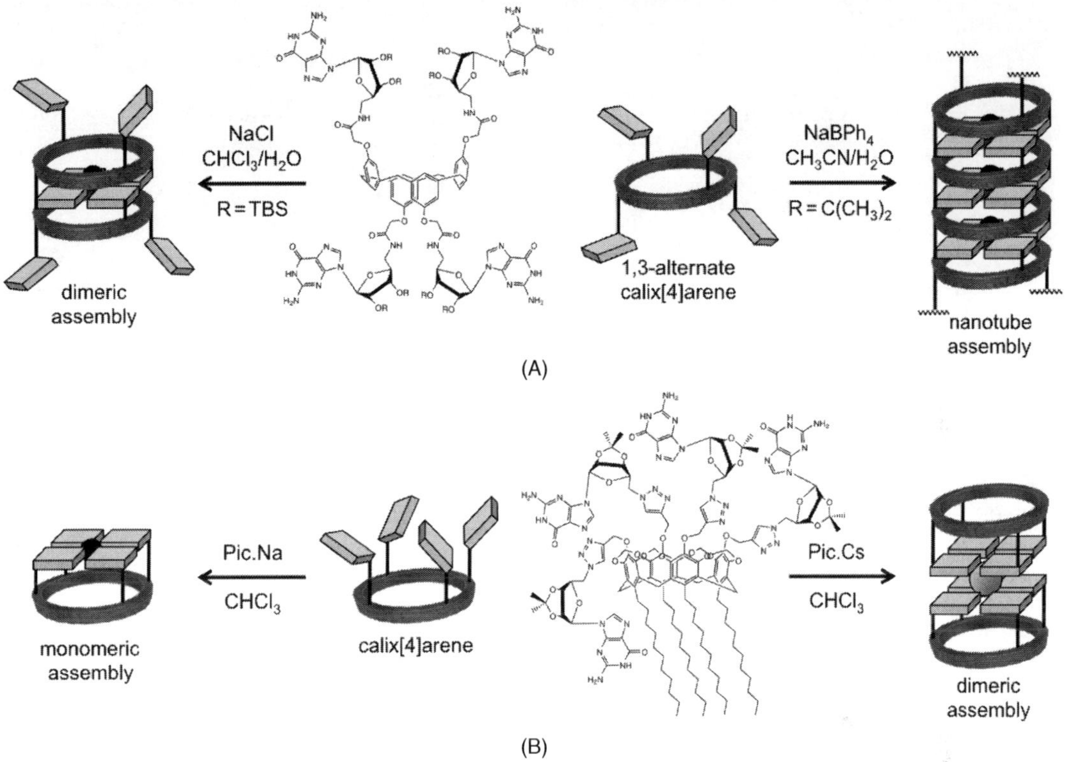

Figure 3.5.8 *Nanotechnological applications of G-tetramers, for ion pair receptor (A) [49] and selective ion extraction (B) [50] applications*

a porphyrin ring (PorphySQ and ^{PNA}PorphySQ) [52], a macrocyclodecapeptide (RAFT-G4) [53], and more recently a series of water-soluble calixarenes [54]. This water-solubility has opened an avenue into biologically directed applications: DOTASQ has, for instance, been used as a G-quadruplex ligand (*vide supra*), which strongly deviates from the classical approach in which designed compounds directly thrive from the erstwhile used duplex-DNA intercalators pool. Using an SQ as a G-quadruplex ligand was directly inspired by the astute strategy implemented by Nature itself to make quadruplex architectures reasonably stable, that is, the self-recognition and self-assembly of G-quartets, promoted by the presence of biologically relevant cations. The relationship between the number of constitutive G-quartet layers and the quadruplex stability being straightforward – the more, the better – quadruplex stabilization was successfully attempted with synthetic G-quartets, notably DOTASQ, which makes it the very first prototype of bioinspired quadruplex ligands [51], interacting with its DNA target according to a "*like likes like*" process driven by the nature-inspired assembly of two G-quartets, one synthetic (TASQ) and the other native (quadruplex).

Another leap was taken with the second-generation prototype, ^{PNA}DOTASQ, which differs structurally from DOTASQ in that its chemical construction involves guanine moieties surrounded by primary amine side-chains, thereby making ^{PNA}DOTASQ cationic in solution (protonated at physiological pH). The wreath of cationic charges around its SQ means ^{PNA}DOTASQ is able to fit snugly into the quadruplex binding site, which is a native G-quartet surrounded by negative charges (from the phosphodiester groups). Its quadruplex

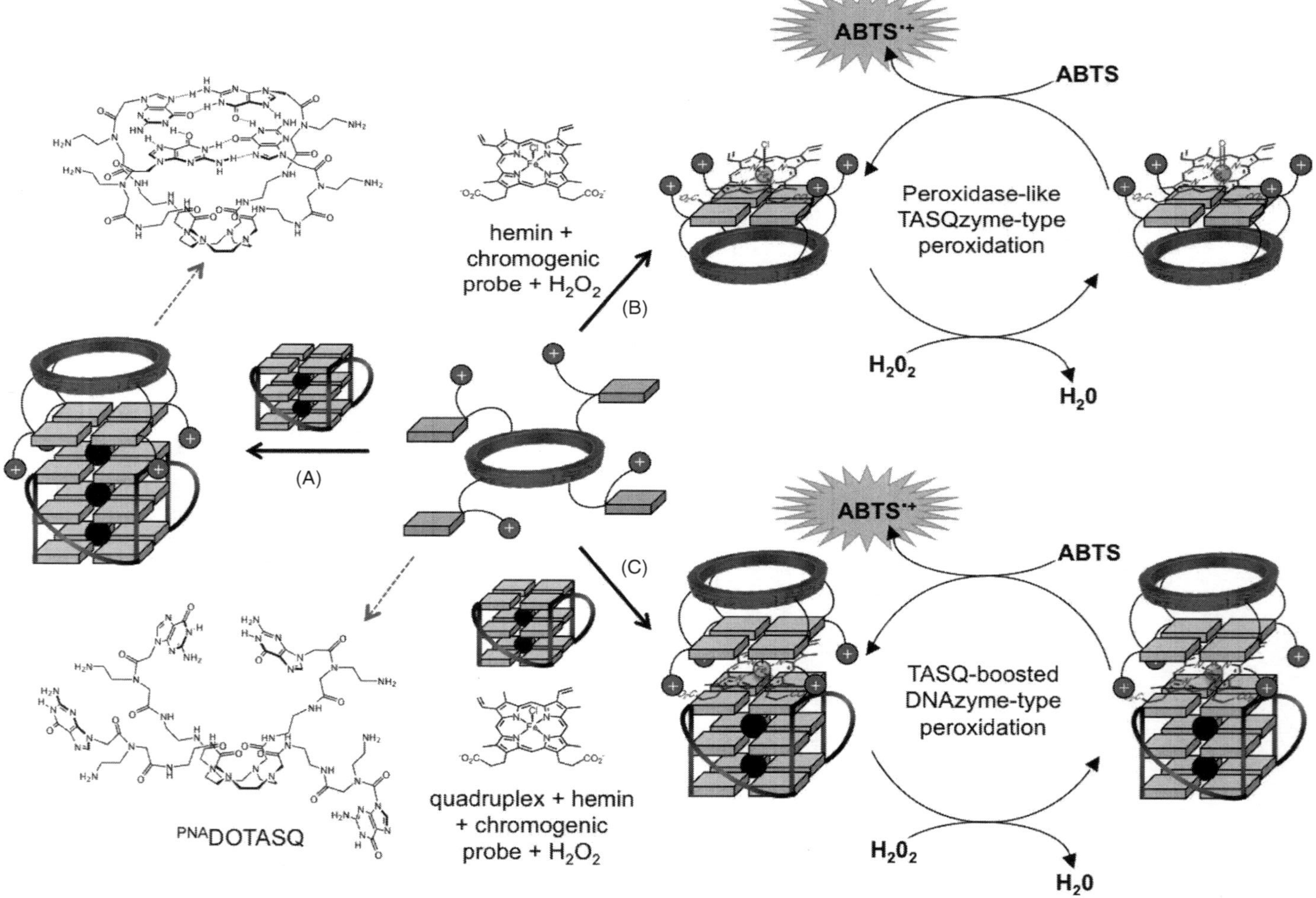

Figure 3.5.9 *Nanotechnological applications of water-soluble G-tetramers, for bioinspired quadruplex interaction (A) [51], as pre-catalyst for peroxidase-like TASQzyme reactions (B) [58] and as a boosting agent for DNAzyme-type catalytic oxidations (C) [59]*

interaction is thus promoted not only by the self-recognition of quartets ("*like likes like*") but also by a well-defined electrostatic interaction network (Figure 3.5.9A). Even more interesting is the observation that amine appendages actively participate in the quadruplex–ligand interaction, leveraging a fairly unique quadruplex-promoted iSQ folding that triggers the overall quadruplex-affinity of the ligand, making ^{PNA}DOTASQ the very first prototype of a smart G-quadruplex ligand [51]. More recently, a third-generation TASQ named PyroTASQ was reported as the very first prototype of "twice-as-smart" quadruplex ligand, thanks to a unique design that makes it both a smart ligand (i.e., a structurally labile ligand that adopts its high-affinity conformation upon contact with quadruplexes only) and a smart probe (i.e., the aforementioned conformational switch triggers its fluorescent properties, making it a quadruplex-selective fluorescent light-up probe) [55].

However, only neutral (DOTASQ and PorphySQ) and cationic TASQ (^{PNA}DOTASQ, ^{PNA}PorphySQ and PyroTASQ) interact with the negatively charged DNA; Sherman and coworkers' cavitand-base TASQ is water-solubilized by the presence of four phosphodiester groups [54], making it an accurate quadruplex-DNA mimic while precluding its interaction with quadruplexes. This TASQ was thus used to evaluate the isolated quartet affinity of quadruplex ligands (being mostly cationic), thereby providing a unique means of delineating the contributions of quartets, loops, and groove interactions to their overall quadruplex affinity.

Reminiscent of what has been described here with quadruplexes, the small molecule–SQ interactions are even more interesting if ligands acquire specific properties once in interactions with the SQ. The unique properties of hemin, which have been highlighted and thoroughly exploited with quadruplexes, make it a perfectly suited candidate for G-tetramer-based biotechnological developments. Indeed, studies aimed at uncovering the mechanism of quadruplex–hemin catalytic systems have highlighted that the hemin binding to the accessible G-quartet of a quadruplex is one of the main key steps of the DNAzyme-type catalysis [56]. Inspired by this observation, quadruplexes have been efficiently substituted by isolated synthetic G-quartets (TASQ), thereby giving rise to a novel catalytic system known as TASQzyme (Figure 3.5.9B). Both DOTASQ and RAFT-G4 [57] have been used as pre-catalysts for peroxidase-mimicking oxidations, upon interaction with hemin and H_2O_2, albeit with a decreased catalytic efficiency (~25-fold) as compared with quadruplex-DNA. Given that hemin is negatively charged at physiological pH, the cationic nature of ^{PNA}DOTASQ make it a better pre-catalyst than DOTASQ and RAFT-G4, with a better overall efficiency that is still, however, lower than that of the quadruplexes (~11-fold) [58]. Even if preliminary, these results are nevertheless invaluable since: (i) they represent an important technological simplification of the DNAzyme process, given that oxidations can be performed on larger scales, without pre-catalyst preparation (no DNA folding steps) and cofactor restrictions (hemin can be replaced by FeTPPS for instance, which is known to display better water solubility and stability); and (ii) they also confirm definitively that the hemin–quartet interaction is the most important step of peroxidase-like oxidations, since acceptable catalytic efficiencies are obtained with pre-catalysts that are actually structurally far simpler than quadruplexes, as they are devoid of both loops and grooves, thereby providing hemin with a less hydrophobic binding site.

Finally, given that the hydrophobicity of the hemin binding site is also a determining factor for the DNAzyme efficiency, iSQ were further exploited for their dual skills, that is, their ability to interact with the quadruplex on the one hand (quadruplex ligand) and hemin on the other (TASQzyme precatalyst). The logic behind further use of the iSQ versatility, implementing catalytic oxidation in the concomitant presence of both TASQ and quadruplexes, was to sandwich hemin between two quartet layers, one natural (quadruplex) the other synthetic (TASQ) to offer hemin a binding pocket with artificially enhanced hydrophobicity (Figure 3.5.9C) [59]. The efficiency of this unprecedented ternary TASQ–hemin–quadruplex catalytic system, being approximately threefold better as compared with the hemin–quadruplex system, is important not only since it provides important insights into the DNAzyme mechanism (highlighting the impact of the hydrophobic nature of the hemin binding site) and a brand new way to enhance the catalytic proficiency of quadruplex-based catalytic reactions (creation of an artificial inter-quartet hydrophobic active site), but also because it represents another, fully innovative example of biotechnological use of SQ (as a "*booster*" of quadruplex–hemin DNAzyme systems).

3.5.7 Conclusions

Alongside its fundamental role as a repository of genetic information, in recent years DNA has become a central player for nanotechnological innovations. The pioneering demonstration that DNA is not merely holding the genetic information but could also be enzymatically proficient was reported exactly 20 years ago by Breaker and Joyce [60], when they demonstrated that short DNA fragments were effectual in promoting the catalytic cleavage of the RNA sequences they are associated with. This peculiar capability of DNA, termed DNAzyme activity, set off a burst of investigations aimed at de-mystifying DNA and making it a routine molecular tool to build multitasking nanodevices, an impetus that was further bolstered by the ever easier commercial access to synthetic oligonucleotides. The knowledge of the nucleic acid community combined with the techniques of chemistry, molecular biology, surface, and biophysical sciences have unambiguously yielded real biotechnological dividends, as vividly demonstrated by the versatility of the selected examples reported in the present chapter. However, some sour notes still remain, such as scalability, chemical diversity, recyclability, and cost issues, which must be addressed for the unquestionable adoption of DNA–G-quartets based technologies in the industrial arena. The emerging field of synthetic G-quartets, discussed thoroughly herein, is providing not only preliminary but also invaluable solutions that help address these issues; notably, the iSQ (or TASQ) bear the lion's share of these promises, since they concomitantly display the unique properties of G-based supramolecular assemblies while circumventing the entropic costs associated with intermolecular associations. The wealth of information that has already been acquired definitively provides a strong message for the exciting nucleotide-based technological area to be launched from a solid scientific basis; however, even if these findings are significant and promising, they admittedly represent only the very first milestones on the road to an overall grasp of the true potential of synthetic G-quartets as bionanotechnological tools. Let us wager that many future studies will shortly prove us right.

So, at present, it is still unclear whether (and to what extent) chemists have succeeded in harnessing Nature's synthetic know-how. It is, however, absolutely clear that Nature was, is, and will remain the most precious and inexhaustible source of inspiration, not only for chemists but also for all researchers in the many, strongly interlinked fields of science that collectively constitute the world of nanotechnology.

References

[1] (a) J. D. Watson and F. H. C. Crick, *Nature,* **171**, 737 (1953); (b) M. H. F. Wilkins, A. R. Stokes, and H. R. Wilson, *Nature,* **171**, 738 (1953); (c) R. E. Franklin and R. G. Gosling, *Nature,* **171**, 740 (1953).

[2] S. Sivakova and S. J. Rowan, *Chem. Soc. Rev.,* **34**, 9 (2005); (b) J. L. Sessler, C. M. Lawrence, and J. Jayawickramarajah, *Chem. Soc. Rev.*, **36**, 314 (2007); (c) M. Fathalla, C. M. Lawrence, N. Zhang, J. L. Sessler, and J. Jayawickramarajah, *Chem. Soc. Rev.*, **38**, 1608 (2009).

[3] (a) P. E. Nielsen, M. Egholm, R. H. Berg, and O. Buchardt, *Science*, **254**, 1497 (1991); (b) B. Hyrup and P. E. Nielsen, *Bioorg. Med. Chem.*, **4**, 5 (1996).

[4] J. L. Sessler, D. J. Magda, and H. Furuta, *J. Org. Chem.*, **57**, 818 (1992).

[5] (a) J. T. Davis, *Angew. Chem., Int. Ed.*, **43**, 668 (2004); (b) J. T. Davis and G. P. Spada, *Chem. Soc. Rev.*, **36**, 296 (2007); (c) S. Lena, S. Masiero, S. Pierraccini, and G. P. Spada, *Chem. Eur. J.*, **15**, 7792 (2009).

[6] G. Gilli, F. Bellucci, V. Ferretti, and V. Bertolasi, *J. Am. Chem. Soc.*, **111**, 1023 (1989).

[7] C. F. Guerra, H. Zijlstra, G. Paragi, and F. M. Bickelhaupt, *Chem. Eur. J.*, **17**, 12612 (2011).

[8] G. W. Collie and G. N. Parkinson, *Chem. Soc. Rev.*, **40**, 5867 (2011).

[9] S. Burge, G. N. Parkinson, P. Hazel, A. K. Todd, and S. Neidle, *Nucleic Acids Res.*, **34**, 5402 (2006).

[10] N. Maizels and L. T. Gray, *PLOS Genet.*, **9**, e1003468 (2013).

[11] (a) A. De Cian, L. Lacroix, C. Douarre, N. Temime-Smaali, C. Trentesaux, J.-F. Riou, and J.-L. Mergny, *Biochimie*, **90**, 131 (2008); (b) Y. Xu, *Chem. Soc. Rev.*, **40**, 2719 (2011).

[12] S. Balasubramanian, L. H. Hurley, and S. Neidle, *Nature Rev. Drug Discov.*, **10**, 261 (2011).

[13] (a) C. M. Azzalin, P. Reichenbach, L. Khoriauli, E. Giulotto, and J. Lingner, *Science*, **318**, 798 (2007); (b) S. Schoeftner and M. A. Blasco, *Nature Cell Biol.*, **10**, 228 (2008); (c) B. Luke, J. Lingner, *EMBO J.*, **28**, 2503 (2009); (d) S. Schoeftner and M. A. Blasco, *EMBO J.*, **28**, 2323 (2009).
[14] A. Bugaut and S. Balasubramanian, *Nucleic Acids Res.*, **40**, 4727 (2012).
[15] (a) D. Monchaud and M.-P. Teulade-Fichou, *Org. Biomol. Chem.*, **6**, 627 (2008); (b) S. N. Georgiades, N. H. Abd Karim, K. Suntharalingam, and R. Vilar, *Angew. Chem., Int. Ed.*, **49**, 4020 (2010); (c) T. Vy Thi Le, S. Han, J. Chae, and H.-J. Park, *Curr. Pharm. Des.*, **18**, 1948 (2012).
[16] (a) S. Neidle, *FEBS J.*, **277**, 1118 (2010); (b)S. Balasubramanian and S. Neidle, *Curr. Opin. Chem. Biol.*, **13**, 345 (2009).
[17] (a) I. Willner, B. Shlyahovsky, M. Zayats, and B. Willner, *Chem. Soc. Rev.*, **37**, 1153 (2008); (b) S. K. Silverman, in *Wiley Encyclopedia of Chemical Biology*, John Wiley & Sons (2008); (c) J. Liu, Z. Cao, and Y. Lu, *Chem. Rev.*, **109**, 1948 (2009); (d) Y. Krishnan and F. C. Simmel, *Angew. Chem., Int. Ed.*, **50**, 3124 (2011); (e) L. Lv, Z. Guo, J. Wang, and E. Wang, *Curr. Pharm. Des.*, **18**, 2076 (2012); (f) D.-M. Kong, *Methods*, **64**, 199 (2013); (g) B. T. Roembke, S. Nakayama, and H. O. Sintim, *Methods*, **64**, 185 (2013).
[18] D. Sen and L. C. H. Poon, *Crit. Rev. Biochem. Mol. Biol.*, **46**, 478 (2011).
[19] D.-L. Ma, H.-Z. He, K.-H. Leung, H.-J. Zhong, D. S.-H. Chan, and C.-H. Leung, *Chem. Soc. Rev.*, **42**, 3427 (2013).
[20] (a) D. Li, S. Song, and C. Fan, *Acc. Chem. Res.*, **43**, 631 (2010); (b) Y. Tang, B. Ge, D. Sen, and H.-Z. Yu, *Chem. Soc. Rev.*, **43**, 518 (2014).
[21] (a)Y. Li and D. Sen, *Chem. Biol.*, **5**, 1 (1998); (b) P. Travascio, Y. Li, and D. Sen, *Chem. Biol.*, **5**, 505 (1998).
[22] Y. Li and D. Sen, *Nat. Struct. Mol. Biol.*, **3**, 743 (1996).
[23] X. Yang, T. Li, B. Li, and E. Wang, *Analyst*, **135**, 71 (2010).
[24] C.-H. Leung, D. S.-H. Chan, B. Y.-W. Man, C.-J. Wang, W. Lam, Y.-C. Cheng, W.-F. Fong, W.-L. W. Hsiao, and D.-L. Ma, *Anal. Chem.*, **83**, 463 (2011).
[25] H.-Z. He, D. S.-H. Chan, C.-H. Leung, and D.-L. Ma, *Chem. Commun.*, **48**, 9462 (2012).
[26] (a) H. Mei, T. Bing, C. Qi, N. Zhang, X. Liu, T. Chang, J. Yan, and D. Shangguan, *Chem. Commun.*, **49**, 164 (2013); (b) Y. He, X. Wang, Y. Zhang, F. Gao, Y. Li, H. Chen, and L. Wang, *Talanta*, **116**, 816 (2013); (c) L. Stefan, T. Lavergnes, N. Spinelli, E. Defrancq, and D. Monchaud, *Nanoscale*, **6**, 2693 (2014).
[27] B. Ge, Y. Chuan Huang, D. Sen, and H.-Z. Yu, *Angew. Chem., Int. Ed.*, **49**, 9965 (2010).
[28] B. Shen, Q. Wang, D. Zhu, J. Luo, G. Cheng, P. He, and Y. Fang, *Electroanalysis*, **22**, 2985 (2010).
[29] J. Chen, J. Zhang, Y. Guo, J. Li, F. Fu, H.-H. Yang, and G. Chen, *Chem. Commun.*, 8004 (2011).
[30] I. Bang, *Z. Physiol. Chem.*, **32**, 201 (1901).
[31] M. Gellert, M. N. Lipsett, and D. R. Davis, *Proc. Natl. Acad. Sci. U.S.A.*, **48**, 2013 (1962).
[32] (a)N. Sreenivasachary and J.-M. Lehn, *Proc. Natl. Acad. Sci. U.S.A.*, **102**, 5938 (2005); (b) E. Buhler, N. Sreenivasachary, S.-J. Candau, and J.-M. Lehn, *J. Am. Chem. Soc.*, **129**, 10058 (2007).
[33] C. Arnal-Herault, A. Pasc, M. Michau, D. Cot, E. Petit, and M. Barboiu, *Angew. Chem., Int. Ed.*, **46**, 8409 (2007).
[34] J. T. Davis, S. Tirumala, J. Jenssen, E. Radler, and D. Fabris, *J. Org. Chem.*, **60**, 4167 (1995).
[35] G. Gottarelli, S. Masiero, and G. P. Spada, *J. Chem. Soc., Chem. Commun.*, 2555 (1995).
[36] (a) A. L. Marlow, E. Mezzina, G. P. Spada, S. Masiero, J. T. Davis, and G. Gottarelli, *J. Org. Chem.*, **64**, 5116 (1999); (b) S. L. Forman, J. C. Fettinger, S. Pieraccini, G. Gottarelli, and J. T. Davis, *J. Am. Chem. Soc.*, **122**, 4060 (2000); (c) E. Mezzina, P. Mariani, R. Itri, S. Masiero, S. Pieraccini, G.-P. Spada, F. Spinozzi, J. T. Davis, and G. Gottarelli, *Chem. Eur. J.*, **7**, 388 (2001); (d) M. S. Kaucher, Y.-F. Lam, S. Pieraccini, G. Gottarelli, and J. T. Davis, *Chem. Eur. J.*, **11**, 164 (2005).
[37] J. L. Sessler, M. Sathiosatham, K. Doerr, V. Lynch, and K. A. Abboud, *Angew. Chem., Int. Ed.*, **39**, 1300 (2000).
[38] S. Bhosale, A. L. Sisson, N. Sakai, and S. Matile, *Org. Biomol. Chem.*, **4**, 3031 (2006).
[39] M. S. Kaucher, W. A. Harrell, and J. T. Davis, *J. Am. Chem. Soc.*, **128**, 38 (2006).
[40] D. Gonzalez-Rodriguez, P. G. A. Janssen, R. Martin-Rapun, I. D. Cat, S. D. Feyter, A. P. H. J. Schenning, and E. W. Meijer, *J. Am. Chem. Soc.*, **132**, 4710 (2009).
[41] Z. Li and C. A. Mirkin, *J. Am. Chem. Soc.*, **127**, 11568 (2005).
[42] (a) A. Calzolari, R. D. Felice, E. Molinari, and A. Garbesi, *Appl. Phys. Lett.*, **80**, 3331 (2002); (b) A. Wong, R. Ida, L. Spindler, and G. Wu, *J. Am. Chem. Soc.*, **127**, 6990 (2005).
[43] L. Chen, N. Sakai, S. T. Moshiri, and S. Matile, *Tetrahedron Lett.*, **39**, 3627 (1998).

[44] (a) L. Ma, W. A. Harrell, and J. T. Davis, *Org. Lett.*, **11**, 1599 (2009); (b) L. Ma, M. Melegari, M. Colombini, and J. T. Davis, *J. Am. Chem. Soc.*, **130**, 2938 (2008).
[45] A. Ghoussoub, and J.-M. Lehn, *Chem. Commun.*, 5763 (2005).
[46] D. Kalia, G. Merey, S. Nakayama, Y. Zheng, J. Zhou, Y. Luo, M. Guo, B. T. Roembke, and H. O. Sintim, *Chem. Soc. Rev.*, **42**, 305 (2013).
[47] (a) S. Nakayama, I. Kelsey, J. Wang, K. Roelofs, B. Stefane, Y. Luo, V. T. Lee, and H. O. Sintim, *J. Am. Chem. Soc.*, **133**, 4856 (2011); (b) S. Nakayama, I. Kelsey, J. Wang, and H. O. Sintim, *Chem. Commun.*, **47**, 4766 (2011).
[48] S. Nakayama, K. Roelofs, V. T. Lee, and H. O. Sintim, *Mol. Biosyst.*, **8**, 726 (2012).
[49] (a) V. Sidorov, F. W. Kotch, M. El-Kouedi, and J. T. Davis, *Chem. Commun.*, 2369 (2000); (b) F. K. Kotch, V. Sidorov, Y.-F. Lam, K. J. Kayser, H. Li, M. S. Kaucher, and J. T. Davis, *J. Am. Chem. Soc.,* **125**, 15140 (2003); (c) A. Wong, F. W. Kotch, I. C. M. Kwan, J. T. Davis, and G. Wu, *Chem. Commun.*, 2154 (2009).
[50] (a)M. Nikan, J. C. Sherman, *Angew. Chem., Int. Ed.*, **47**, 4900 (2008); (b) M. Nikan and J. C. Sherman, *J. Org. Chem.*, **74**, 5211 (2009); (c) M. Nikan, B. O. Patrick, and J. C. Sherman, *ChemBioChem*, **13**, 1413 (2012).
[51] (a) L. Stefan, A. Guedin, S. Amrane, N. Smith, F. Denat, J.-L. Mergny, and D. Monchaud, *Chem. Commun.*, **47**, 4992 (2011); (b) R. Haudecoeur, L. Stefan, F. Denat, and D. Monchaud, *J. Am. Chem. Soc.*, **135**, 550 (2013).
[52] (a) H.-J. Xu, L. Stefan, R. Haudecoeur, S. Vuong, P. Richard, F. Denat, J.-M. Barbe, C. P. Gros, and D. Monchaud, *Org. Biomol. Chem.*, **10**, 5212 (2012); (b) A. Laguerre, N. Desbois, L. Stefan, P. Richard, C. P. Gros, and D. Monchaud, *ChemMedChem*, **9**, 2035 (2014).
[53] P. Murat, B. Gennaro, J. Garcia, N. Spinelli, P. Dumy, and E. Defrancq, *Chem. Eur. J.*, **17**, 5791 (2011).
[54] G. A. L. Bare, B. Liu, and J. C. Sherman, *J. Am. Chem. Soc.*, **135**, 11985 (2013).
[55] A. Laguerre, L. Stefan, M. Larrouy, D. Genest, J. Novotna, M. Pirrotta, and D. Monchaud, *J. Am. Chem. Soc.* **136**, 12406 (2014).
[56] L. Stefan, F. Denat, and D. Monchaud *Nucleic Acids Res.*, **40**, 8759 (2012).
[57] (a) L. Stefan, H.-J. Xu, C. P. Gros, F. Denat, and D. Monchaud *Chem. Eur. J.*, **17**, 10857 (2011); (b) L. Stefan, D. Duret, N. Spinelli, E. Defrancq, and D. Monchaud, *Chem. Commun.*, 1500 (2013).
[58] R. Haudecoeur, L. Stefan, and D. Monchaud, *Chem. Eur. J.*, **19**, 12739 (2013).
[59] L. Stefan, F. Denat, and D. Monchaud, *J. Am. Chem. Soc.*, **133**, 20405 (2011).
[60] R. R. Breaker and G. F. Joyce, *Chem. Biol.*, **1**, 223 (1994).

Part IV

Conjugation of DNA with Biomolecules and Nanoparticles

4.1

Nucleic Acid Controlled Reactions on Large Nucleic Acid Templates

Anika Kern and Oliver Seitz

Department of Organic and Bioinorganic Chemistry, Humboldt-University of Berlin, Berlin, Germany

4.1.1 Introduction

Nucleic acids are evolutionary optimized recognition modules. In the key biological processes (replication, transcription, and translation) nucleic acid recognition is used to arrange mutually reactive functional groups in the close proximity that is required to drive bond-forming reactions. This has inspired many scientists to use nucleic acids as templates that are able to program chemical reactions. In an ever-growing number of investigations it has been shown that nucleic acid templated reactions are useful tools in nucleic acid diagnosis and reaction/drug discovery. Furthermore, applications in materials chemistry are being explored, wherein nucleic acid templates instruct the size-controlled formation of macromolecular materials.

In recent years, significant efforts have been invested into the development of chemistries that enable large, nanosized nucleic acids to act as templates. These efforts have been fueled by the desire to apply nucleic acid templated reactions to the detection of biologically occurring DNA and RNA molecules in challenging environments, such as cell lysates, polymerase chain reaction or within living cells. However, the application of large nucleic acid templated chemistry is by no means restricted to the field of nucleic acid diagnosis and it is tempting to speculate about biogenic instructors of macromolecular functions.

Nucleic acid templated chemical reactions can be broadly divided into three groups (Figure 4.1.1): (i) templated ligation reactions, (ii) templated transfer reactions, and (iii) templated interconversion reactions. What the three principle reaction formats have in common is that the increased effective concentration of the reactive probes enables a chemical reaction to proceed under dilution conditions, which disfavor bimolecular reactions in the absence of the template. In the following we will discuss reactions that have been applied to large templates.

DNA in Supramolecular Chemistry and Nanotechnology, First Edition.
Edited by Eugen Stulz and Guido H. Clever.

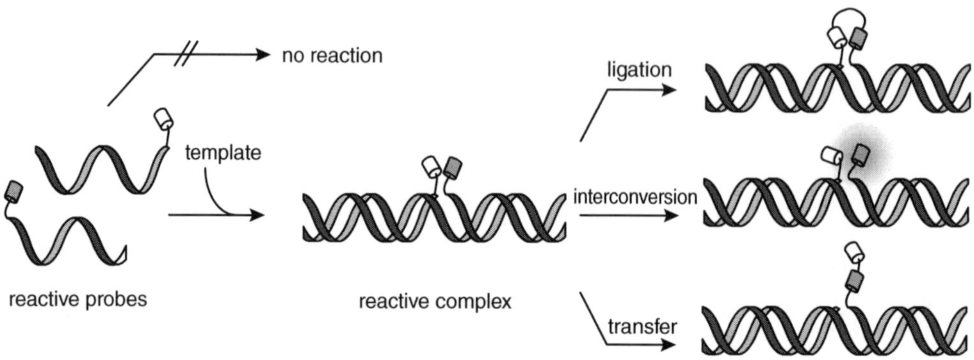

Figure 4.1.1 *Overview over template-controlled chemical reactions; reactive probes with functional moieties brought into close proximity through hybridization to the nucleic acid template, hence enabling a chemical reaction, e.g., ligation reaction, transfer reaction, and interconversion of the reactive groups. In the absence of a template no chemical reaction takes place*

For a comprehensive overview of reactions used in DNA/RNA-templated reactions, the reader is guided to recent reviews [1–5].

4.1.2 Nucleic acid controlled chemical reactions

In templated *ligation reactions* two reactive molecules, whether they consist of DNA, RNA or their synthetic analogues, are connected to form a product of extended length. The pioneering work of the groups working with Naylor and Gilham [6], Shabarova [7], and Orgel [8] focused on the non-enzymatic formation of phosphodiesters. Letsinger and coworkers introduced a non-native structure based on nucleophilic substitution of two DNA strands, one terminated with 3′-phosphorothioate and the second bearing a 5′-bromoacetamide group (Figure 4.1.2A) [9]. In this work, one of the key features of nucleic acid templated reactions, ligation fidelity, was highlighted. It was shown that a single nucleotide mismatch can reduce the efficacy of the templated reaction by one order of magnitude. Current chemistries enable 10^3-fold rate differences (*vide infra*). Several reports describe further development of ligation via halide-sulfur S_N2-displacement [10–17]. A notable ligation method was developed by the groups working with Kool. They introduced the term *autoligation*, which describes reactions that proceed in the absence of added reagents. A 3′-phosphorthioate served as a nucleophile in the reaction with a 5′-iodinated oligonucleotide. The use of 5′-sulfonates allowed the design of a fluorogenic ligation reaction, in which a fluorescence quencher (Y = dabsyl) is released upon templated S_N2 displacement (Figure 4.1.3, see also Figure 4.1.2A) [17]. The so-called QUAL probes (quenched autoligation probes) were applied, among other applications, in the detection of rRNA in bacterial cells (see Section 4.13 of this chapter) [18].

Lynn and coworkers introduced reductive amination to nucleic acid templated reactions (Figure 4.1.2B). In this reaction, a 5′-aldehyde is attacked by a 3′-amino group on the second strand. The imine was formed in 65% yield and is irreversibly trapped as amine upon addition of $NaCNBH_3$ [23, 29–31]. These seminal contributions also focused the attention on templated reactions under turnover conditions, which has become a key issue in contemporary research [5, 32]. Turnover refers to reactions in which a single template instructs the formation of many product molecules. The resulting amplification effect is useful when the availability of templates is limited. DNA-templated reductive amination allows the sequence specific oligomerization of formyl-DNA and its analogues to detect long DNA templates [31].

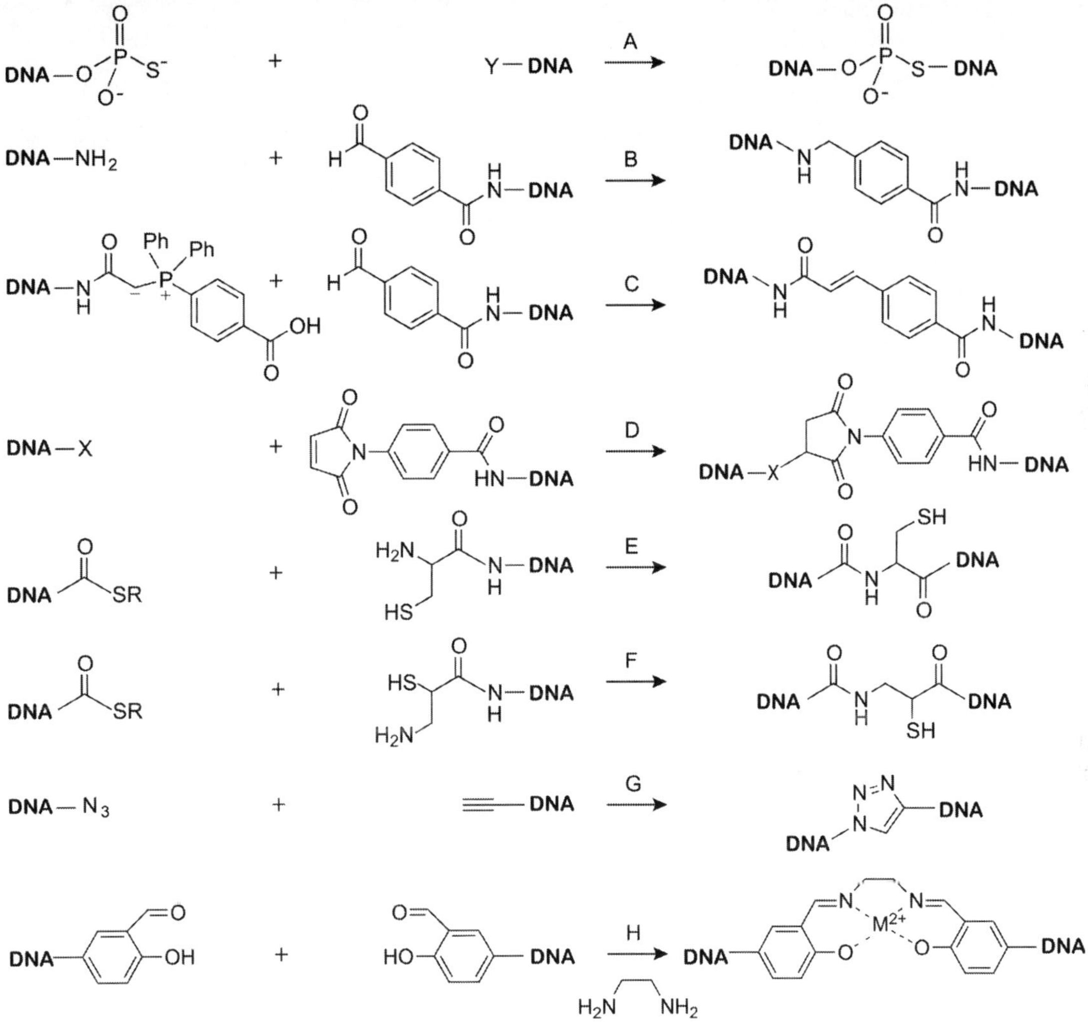

Figure 4.1.2 Nucleic acid templated ligation reaction. A, Autoligation with phosphorothiate-Y (Y = Br, I) [10, 16, 17, 19–21]; B, reductive amination [22, 23]; C, Wittig olefination [22, 24]; D, maleimide-based ligation with different nucleophiles X (X = SH, NH_2) [21]; E and F, native chemical ligation [25, 26]; G, click reaction [27]; and H, metal–salen formation [28]

Liu and coworkers extended the repertoire of DNA-templated chemistry [33]. For example, they introduced DNA-templated Wittig olefination (Figure 4.1.2C) [24], conjugate additions via maleimides (Figure 4.1.2D), and acyl transfer reaction via DNA-linked active esters (*vide infra*). The power of DNA-directed chemistry was demonstrated in impressive multistep template-controlled syntheses of small-molecules (see also Figure 4.1.16) [22, 34–37].

The work of Seitz *et al.* was inspired by the native chemical ligation of peptides introduced by Kent and coworkers [38]. This chemistry was applied to peptide nucleic acid (PNA) based oligomers, which have higher affinity for complementary nucleic acids than DNA or RNA based oligonucleotides. One PNA strand

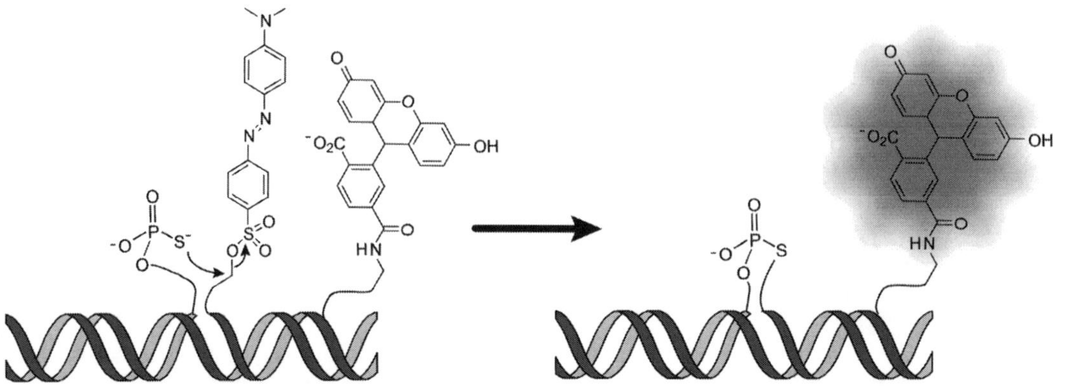

Figure 4.1.3 Schematic illustration of QUAL (quenched autoligation) probes developed by Kool et al. A fluorophore (fluorescein) is turned off by a quencher (dabsyl), which is released when the phosphorothioate-bearing strand attacks the quencher to switch the fluorophore on

was equipped with a *C*-terminal thioester group. The other strand contained an *N*-terminal cysteine (Figure 4.1.2E). The two reactive PNA conjugates are ligated through a thiol exchange reaction to yield a thioester intermediate, which rapidly rearranges to yield an amide-linked ligation product [39]. The reaction is fast (75% yield after 5 min) and proceeds with high sequence specificity [39, 40]. The replacement of cysteine by isocysteine (Figure 4.1.2F) increased the turnover numbers at high probe excess without being detrimental to the specificity and rates [25, 26, 39, 41]. Native chemical PNA ligation was performed on large DNA templates formed during PCR (see also Figure 4.1.12) [42].

Brown and coworkers used copper-promoted click chemistry to ligate two oligonucleotide strands in a template-directed synthesis (Figure 4.1.2G) [43]. The reactions involved a 3′-azido and a 5′-alkinyl modified oligonucleotides and provided triazole-linked ligation products [44]. The reaction was used to covalently crosslink complementary DNA strands [45], to cyclize double stranded DNA [46], and to form catenane-like structures [47]. For further information the reader is guided to recent reviews [44, 48–52].

Furthermore, Saito and coworkers introduced a reversible photoligation reaction, a reagent-free reaction that is triggered by UV light (366 and 302 nm) that does not promote significant DNA damage [53]. A 5′-vinyl modified pyrimidine DNA strand was ligated with a 3′-thymidine bearing DNA strand to form the *cis–syn* [2+2] cyclophotoadduct (93% yield in 1 h) after irradiation at 366 nm. This photoligation reaction was demonstrated to be reversible after irradiation at 302 nm. The scope of applications included reactions on triplex DNA [54], catenane-like plasmid DNA structures [55], and branched oligonucleotides [56] (Figure 4.1.4).

Czlapinski and Sheppard first reported nucleic acid templated reactions forming metal–salen bonds (Figure 4.1.2H). Two DNA strands were equipped with salicylaldehyde groups at both the 3′- and the 5′-ends. Addition of ethylenediamine and metal (e.g., $Mn(OAc)_2$ or $Ni(OAc)_2$) lead to assembly of the metal–salen complex at 65% after 1 h [28]. Later Gothelf *et al.* used this principle for the sequence-instructed assembly of nanomaterials [57].

In *templated transfer reactions*, functional groups are designed to enable a relocation of a group of atoms. The number of nucleobases within a given reactive group remains unchanged. Therefore, the reaction products have similar template affinity to the probes before reaction. Templated transfer reactions can be performed under conditions of dynamic strand exchange. This helps to increase the turnover in the template. The transfer of thioester-linked reporter groups from a donating PNA probe to an isocysteinyl-PNA acceptor has been described by Grossmann and Seitz (Figure 4.1.5A) [58–60]. The transfer of a quencher enabled the real-time

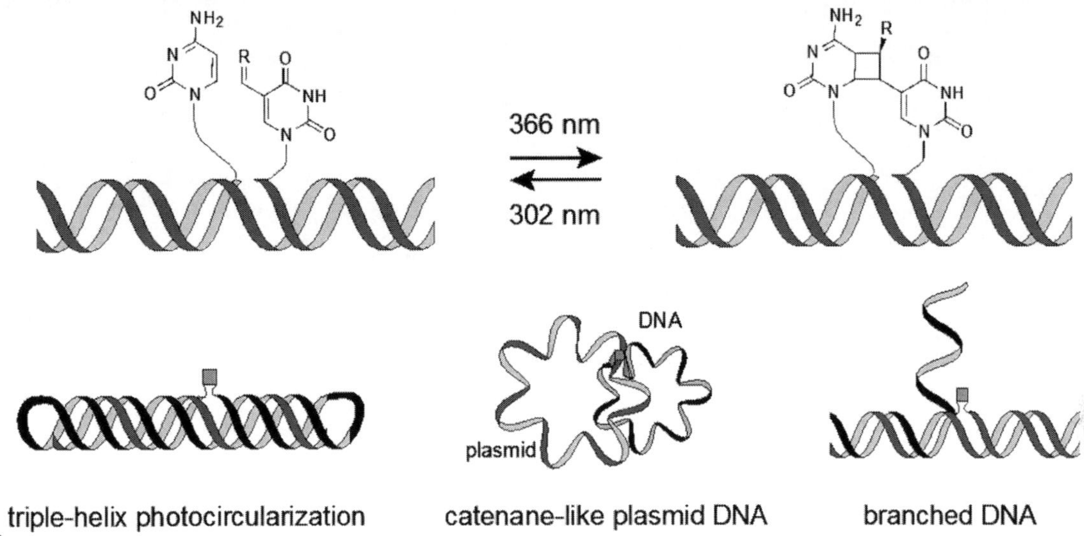

Figure 4.1.4 *Reversible photoligation reaction of 3′-thymidine DNA and 5′-vinyl-modified pyrimidine DNA strand and its scope of application. R = H (31% yield within 1 h) or CO_2H (93% yield within 1 h)*

analysis of the decrease in emission of the acceptor-linked fluorophore and an increase in the emission of the donor-linked fluorophore. The method was applied to the transfer of pyrene and biotine residues [61]. Recently, the templated acyl transfer onto cysteine has been adapted to DNA-based reaction systems allowing the transfer of a fluorophore from donating DNA conjugate to acceptor DNA conjugate [62] or to a quantum dot [63].

The group working with Turberfield and Stulz studied the scope of DNA-templated acyl transfer reactions and assessed various nucleophiles (e.g., amine, hydrazine, hydrazide, aminooxy, benzylamine, and hydrazine-nicotinate) as well as linker lengths under different reaction conditions (pH, temperature, and solvent mixtures) [64]. They found that the acyl transfer reaction works best with α-effect nucleophiles at neutral pH.

Liu and coworkers extended the scope of nucleic acid templated reactions and explored reagent-free reactions (*vide supra*) and reagent-added reactions, such as amine acylation, reductive amination, and Wittig olefination. The first involved a DNA-linked *N*-hydrosuccinimide (NHS) and the *in situ* formation of DNA-NHS-linked acylating reagents, which allowed the sequence-controlled acylation of DNA-linked amino groups. The templated amide bond formation (Figure 4.1.5B) proceeded in 75% yield. The scope of this reaction was shown in several applications, such as the DNA walker [65] and templated multistep synthesis to build up libraries of macrocycles [66].

Abe and Ito demonstrated a template-directed nucleophilic aromatic substitution (S_NAr) reaction to unmask a non-fluorescence dye leading to its fluorescence emission. An electrophilic non-fluorescent dinitrobenzene coumarine derivative incorporated at the 3′-end of one DNA strand was substituted with the help of nucleophilic phosphorothiate at the 5′-end of another adjacently annealed DNA strand (Figure 4.1.5C). The reaction was performed on templates originating from 23S rRNA of *Escherichia coli* [67].

A templated transfer reaction has been used to enable the *de novo* synthesis of a fluorophore. Seitz introduced a templated Wittig reaction between a DNA-linked cyanobenzyl phosphonium salt and a DNA-linked benzaldehyde (Figure 4.1.5D) [68]. The transfer of the benzylidene unit yielded a stilbene. The addition of α-cyclodextrin increased the emission at 380 nm. The reaction provided 10^2-fold enhancement of fluorescence on a 0.1 equiv template and was performed on PCR-DNA.

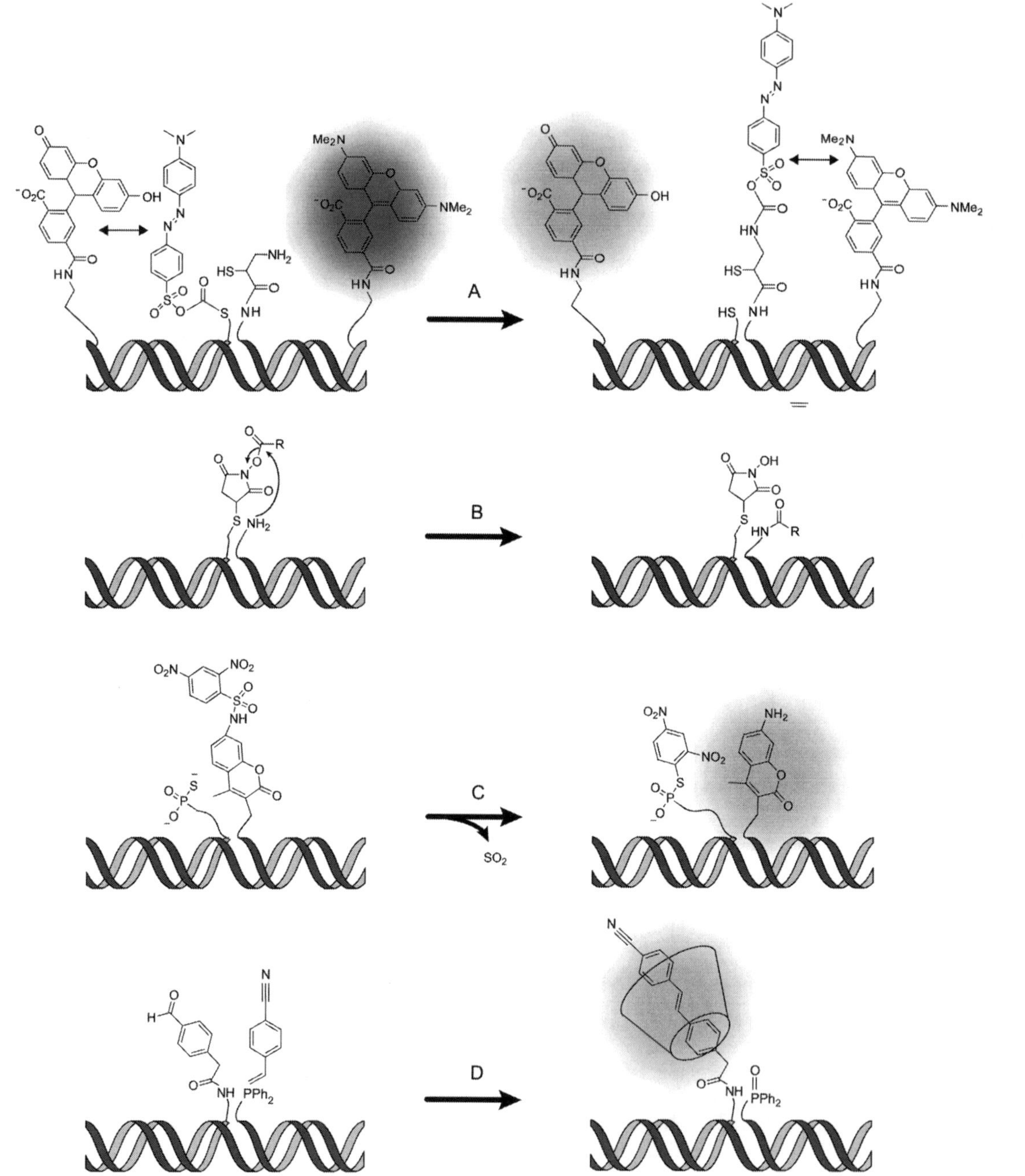

Figure 4.1.5 *Illustration of transfer reactions yielding fluorescence signals; A and B based on native chemical ligation reactions. A, Transfer of quencher (dabsyl) leading to switch-on of the donor PNA strand and simultaneous switch-off of the acceptor PNA strand; B, amine acylation of an activated acyl donor; C, nucleophilic aromatic substitution accompanied by switch-on of a fluorescent probe; D, transfer reaction based on a Wittig reaction resulting in α-cyclodextrine stabilized fluorescence* trans*-stilbene*

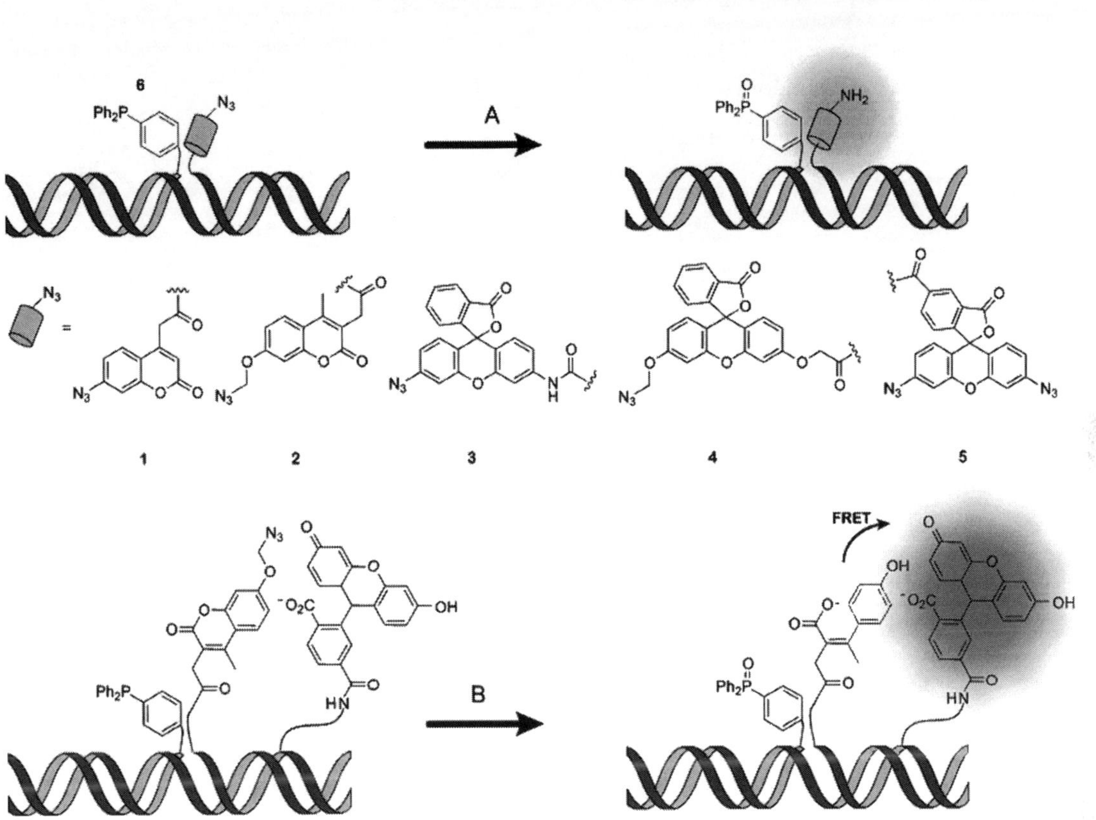

Figure 4.1.6 *Design of conversion probes: A, Staudinger reduction of azido-quenched fluorophore yielding fluorescence enhancement; and B, design of a FRET system based on conversion probes*

Nucleic acid templated reactions that lead to the *interconversion* of functional groups beyond transfer and ligation are usually based on the Staudinger chemistry. Two approaches turned out to be useful: (i) template-directed reduction of an azide-quenched fluorophore (Figure 4.1.6A and B) and (ii) the release of a linker subsequent to azide reduction (Figure 4.1.7A and B). Winssinger used an azidocoumarin PNA conjugate, which was reduced by a second PNA strand bearing a *C*-terminal triphenylphosphine group yielding a 22-fold increase of fluorescence (Figure 4.1.6A) [69]. The Staudinger reduction was tested with different fluorophores (**1–5**). Kool developed a DNA-based reaction system, which involved azidomethoxycoumarin-derived reactive probes (Figure 4.1.6A, 2) [70]. The reaction led to a 62-fold fluorescence increase within 115 min at a stoichiometric template load [71]. A two-color reporter system based on FRET was used for reactions on intracellular RNA templates (Figure 4.1.5B) [72–79].

Winssinger introduced a linker in which azide reduction triggers a fragmentation reaction (Figure 4.1.7A). The reaction resulted in the release of functional units such as rhodamine and bioactive molecules (estradiol and doxorubicin). The Staudinger reduction-induced cleavage proceeded with 30% conversion within 30 min [80]. Kool developed the quenched Staudinger triggered azidoether release (Q-STAR) probes. The reaction involves an azidoether-linked dabsyl unit, which quenches the emission of an appended fluorophor (Figure 4.1.7B). The reduction of the azidoether linker triggers the release of a quencher unit and thereby restores fluorescence. The method was adapted to a two-color readout (Figure 4.1.7C). The reaction system enabled 90% conversion within 30 min and furnished a 61-fold enhancement of the fluorescence signal after 2 h [71]. The Q-STAR probes were used to detect the presence of RNA templates within living cells (see also Figure 4.1.8) [71, 78, 81].

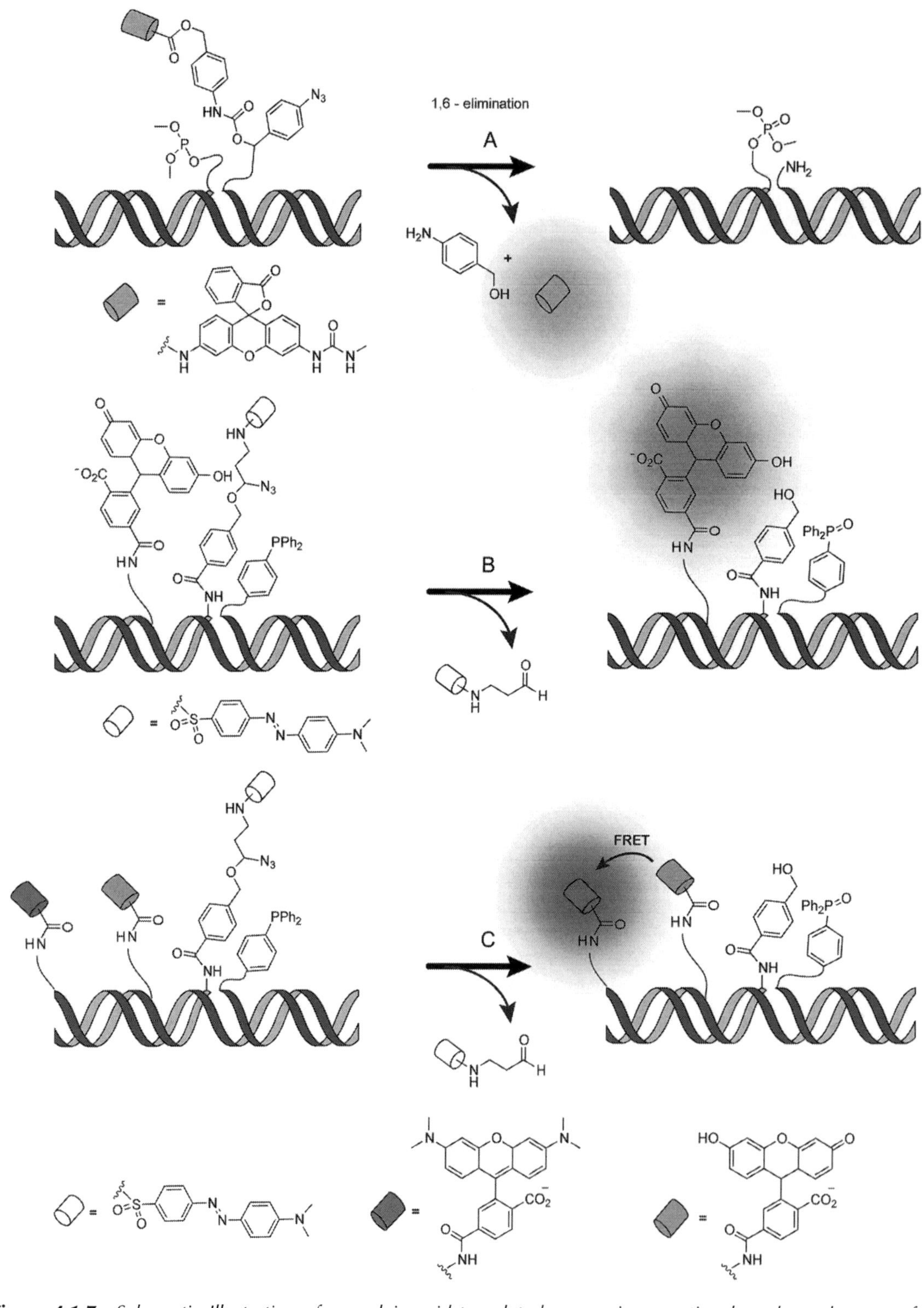

Figure 4.1.7 *Schematic Illustration of a nucleic acid templated conversion reaction based on cleavage of a linker, either release of an unquenched fluorophore (A) or release of a quencher, leading to switch-on fluorescence (B) or enabling FRET (C)*

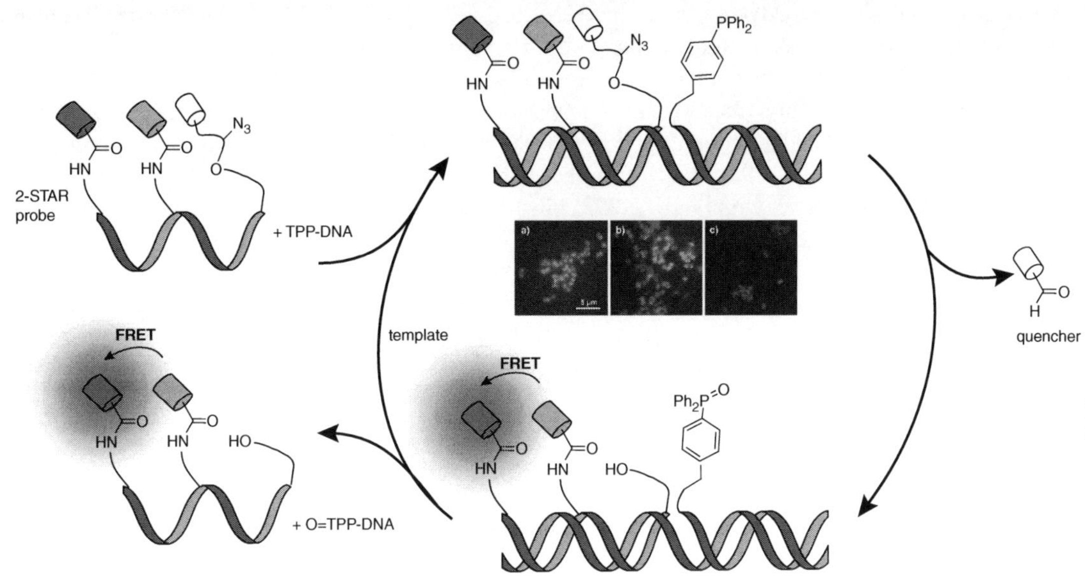

Figure 4.1.8 *Schematic illustration of FRET-Q-STAR probes detecting a 16S rRNA target; two-color detection of* E. coli *and* S. enterica *16S rRNA differing in single nucleotide polymorphism: (a)* E. coli *cells; (b)* E. coli *and* S. enterica *cells; and (c)* S. enterica *cells [71]. Adapted with permission from [71]. Copyright (2009) American Chemical Society*

The programmable nature of the nucleic acid templated reactions provides opportunities for the design of reaction systems in which a specific nucleic acid sequence is translated into assemblies of molecules and functional output with applications ranging from small-molecule synthesis, bioanalytics, and material sciences. Here we describe some representative examples and highlight challenging applications to the use of nucleic acid templated chemical reactions.

4.1.3 Applications

4.1.3.1 Reactions on intracellular RNA

The emerging role of RNA as a key regulator of biological processes has led to intensified research efforts aimed at the detection and imaging of intracellular RNA within living cells. RNA imaging has been the domain of fluorescent hybridization probes. However, the promise of high target specificity and signal amplification has inspired chemists to develop fluorogenic reactions that proceed within the cellular environment.

Kool and coworkers applied the QUAL probes for the detection of rRNA within *E. coli*. The bacterial cells were fixed with paraformaldehyde. The 5′-dabsyl containing DNA probe was labeled with fluorescein or TMR (rhodamine) and incubated with phosphorothioate DNA strands (Figure 4.1.3) [82–85]. The rRNA target triggered the generation of a green or red fluorescence signal. Later, two differently colored QUAL probe designs were used to distinguish between *E. coli* and *Salmonella enterica* 16S rRNA sequences, which are closely related and differ by only a few nucleotides [84, 86, 87]. The Q-STAR probes provide signal amplification via turnover in a template (*vide supra*). This facilitated measurements of rRNA in *E. coli* and *S. enterica* (Figure 4.1.8) [71, 78, 79, 81].

Winssinger and colleagues used probes based on GPNA to detect cellular mRNA encoding *O*-6-methylguanine-DNA-methyltransferase, which is highly expressed in HEK293 cells. GPNA oligomers are comprised of a peptide nucleic acid scaffold, which includes arginine side chains [88, 89]. The positively charged guanidino groups confer cell permeability [88, 90, 91]. The reaction system took advantage of an azido-quenched fluorophore (rhodamine, **3** and a terminal triphenylphosphine **6** (TPP-DNA), which triggers the reduction of the azido group (Figure 4.1.6A) [92]. Reactive probes that were complementary to the *O*-6-methylguanine–DNA–methyltransferase mRNA target in HEK293 cells produced a fluorescence signal whereas mismatched probes did not (Figure 4.1.9A). A reaction system that involved a bis-azidorhodamine (Figure 4.1.6A, **5**) enabled an improvement of the fold fluorescence enhancement (from 30-fold to 120-fold). The reactive probes were used to detect miRNA21, which is frequently overexpressed by solid tumors. The reactions were performed in different cell lines (MCF-7, BT474, HeLa, JIMT-1, and HEK293T). Impressively, the magnitude of the fluorescence generated upon the templated reaction correlated with the known miRNA21 expression levels (Figure 4.1.9B).

The group working with Abe and Ito demonstrated that 23S rRNA in paraformaldehyde fixed *E. coli* cells triggered the fluorogenic reaction between triphenylphosphine-containing oligonucleotide probes and azido-quenched rhodamine probes (Figure 4.1.6A) [72]. The use of azidomethyl-caged fluorescein allowed for improvements to the fold fluorescence increase [93]. The reaction was targeted to 28S rRNA and β-actin mRNA in intact human leukemia HL60 cells. To enable cellular delivery, cells were incubated with streptolysin-O (SLO). The fluorescent product was imaged by fluorescence microscopy and detected by flow cytometry. Recently, they additionally developed a fluorogenic reaction for time gated measurements in living *E. coli* cells (Figure 4.1.10) [94]. The reaction system involved an aromatic azide, which formed a phenanthridinone derivative upon RNA-templated Staudinger reduction. The phenanthridinone

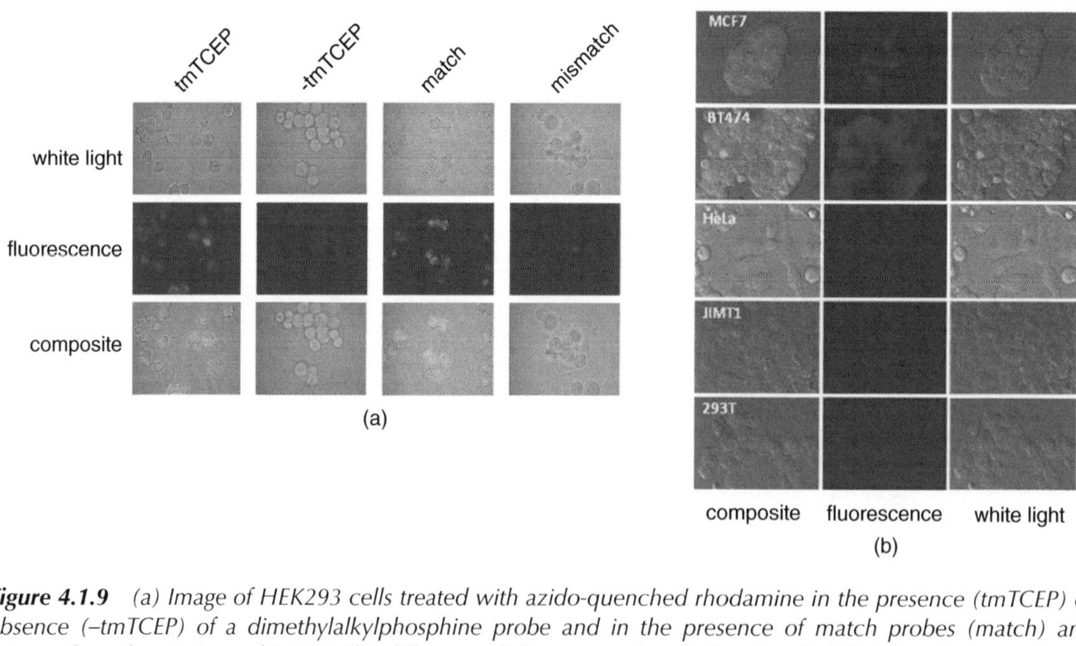

Figure 4.1.9 *(a) Image of HEK293 cells treated with azido-quenched rhodamine in the presence (tmTCEP) or absence (–tmTCEP) of a dimethylalkylphosphine probe and in the presence of match probes (match) and mismatch probes (mismatch) [92]; (b) different cell lines treated with the dimethylalkylphosphine probe and bis-azidorhodamine probes matching the miRNA21 sequence [75]. Adapted with permission from [75]. Copyright (2011) Royal Society of Chemistry (See color figure in color plate section)*

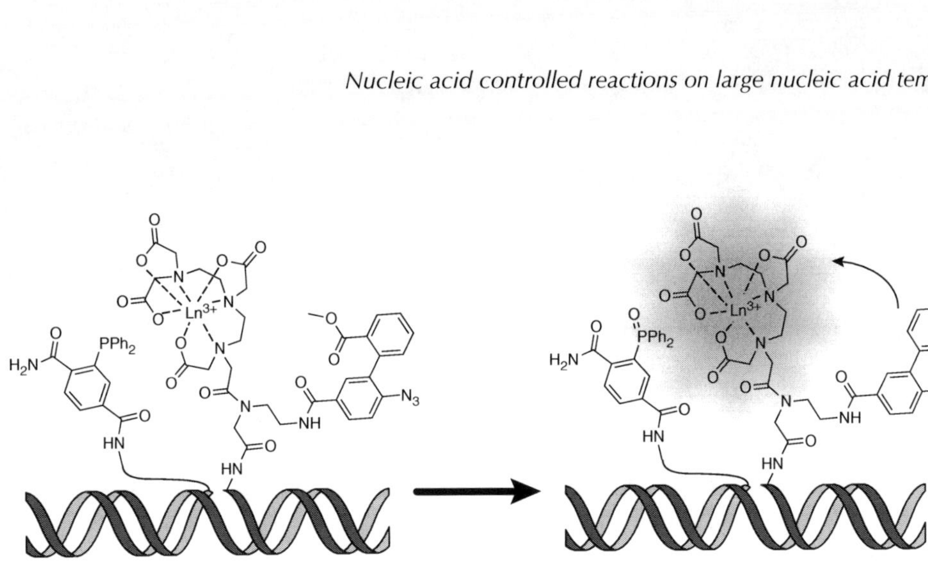

Figure 4.1.10 *Schematic illustration of antenna probes to turn on luminescence after hybridization to RNA target in living* E. coli *cells*

served as an antenna for transfer of excitation energy to a lanthanide complex. This provided a long-lived (0.6–1.8 ms) luminescence signal. The antenna probes enabled direct detection of rRNA in living *E. coli* cells through measuring crude samples [94].

4.1.3.2 Reactions on large biogenic DNA and RNA templates *in vitro*

Recently, nucleic acid templated chemistry was interfaced with molecular biology methodology. Kool and colleagues developed a chemoenzymatic method for the detection of specific miRNA molecules, which play an important role in cell proliferation processes and apoptosis. The method included two steps (Figure 4.1.11) [95]. In the first step, the miRNA template triggered a cyclization reaction of an oligonucleotide bearing a 5′-iodinated and 3′-phosphorthioate termini (see also Figure 4.1.2A). The cyclized product served as a template in a rolling circle amplification (RCA) executed by Φ29 DNA polymerase. The resulting DNA single strand contained many repeats of a sequence, which were targeted in the second step by Q-STAR probes (see also Figure 4.1.7B). To avoid degradation by the DNA polymerase the Q-STAR probes were modified by incorporation of 2′-*O*-methyl nucleosides at the 3′-end. A concentration of 0.2–2 nM miRNA was sufficient to trigger a fluorescence signal above the background [95].

Roloff and Seitz performed a native chemical PNA ligation (see Figure 4.1.2E) while the DNA template was produced in a polymerase chain reaction (PCR) [42, 96]. The PCR protocol typically involves repeated cycles of denaturation at 95 °C, primer annealing at 50 °C, and primer extension at 72 °C. The challenging task was to maintain the chemoselectivity of the reactive PNA probes in the presence of proteins and nucleic acids at temperature up to 95 °C, while enabling fast rates of templated ligation to achieve product formation within the short time available during primer annealing (20 s). The PNA probes were equipped with FAM and TAMRA fluorophores to allow FRET measurements (Figure 4.1.12). Intensive reactivity tuning lead to a reaction system comprised of a cysteinyl–PNA and a PNA probe bearing a *C*-terminal β-alanine thioester. The fluorescence signal was detected *in situ* by using a real-time thermocycler. The human *BRaf*-gene served as template in an asymmetric PCR. In the absence of *BRaf* DNA or in the presence of mismatched DNA no FRET signal was observed. A 2.5 aM concentration of the human genomic template was sufficient to trigger a detectable fluorescence signal [42].

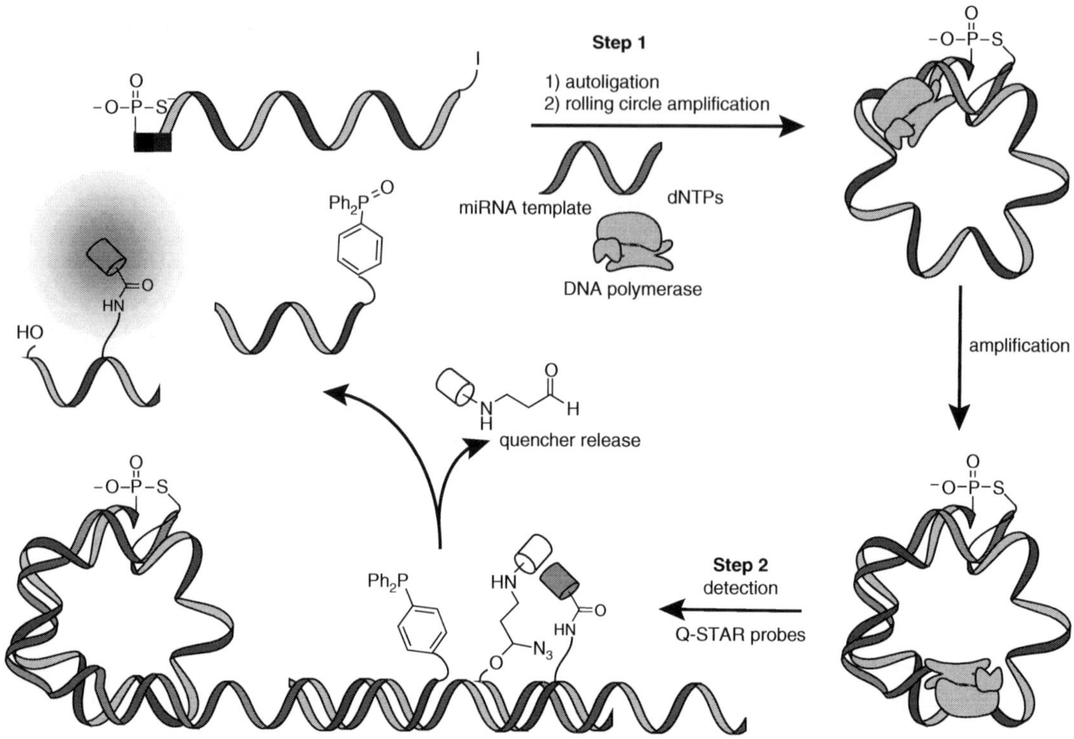

Figure 4.1.11 Illustration of miRNA detection with a DNA precursor, which is cyclized using autoligation of phosphorothioate-iodide substitution, enzymatic amplification (RCA) using DNA polymerase and detection using Q-STAR probes, enhancing a fluorescence signal after reductive quencher release

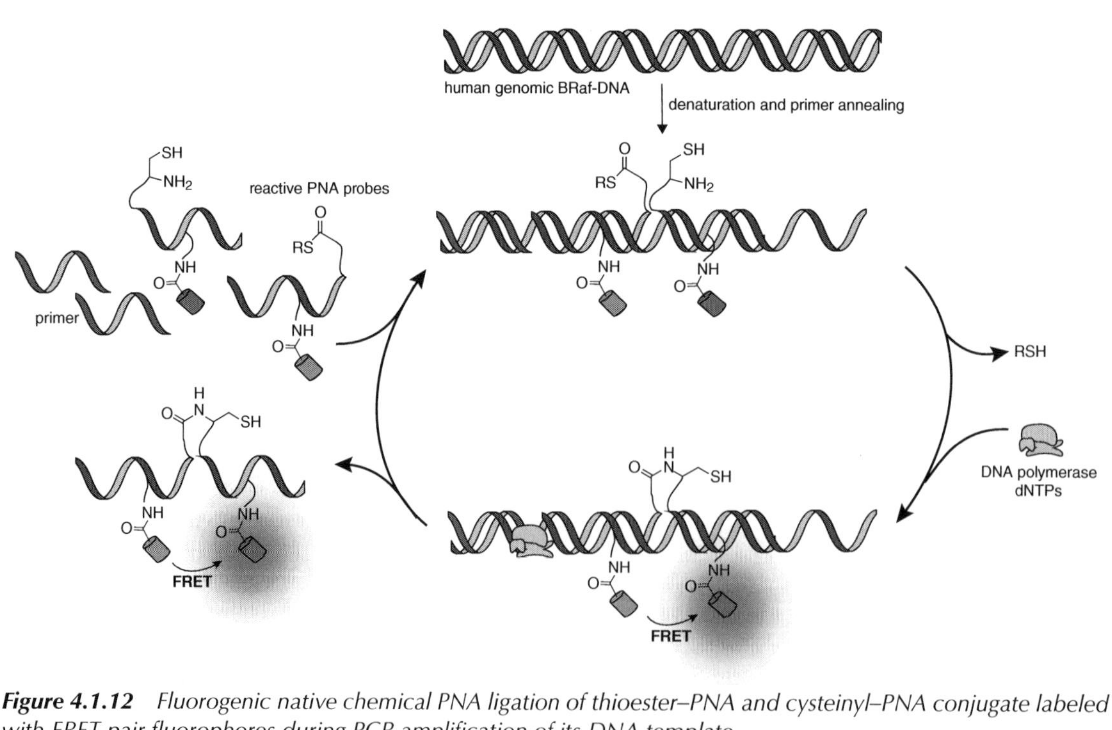

Figure 4.1.12 Fluorogenic native chemical PNA ligation of thioester–PNA and cysteinyl–PNA conjugate labeled with FRET pair fluorophores during PCR amplification of its DNA template

The authors of this book chapter explored the native chemical PNA ligation on long DNA templates encompassing the human *IT15* gene. This gene is linked to Huntington Disease, which is caused by an excessive number of triplet repeats. We developed a reaction system that provides a product when the number of triplet repeats exceeds a certain threshold (Figure 4.1.13). The targeted repeat sequences have a tendency to fold into structures that are difficult to invade by complementary probes. The use of lysine-modified PNA probes provided high template affinity and, at the same time, reduced background reactions triggered by the partial complementarity of PNA-probes containing the $(cag)_n$ motif. Auxiliary oligonucleotides **7** were targeted to the 3′-region of the repeat target sequence and blocked access to short templates. The reaction system was tested on amplified human genomic DNA (pathological, HD, and healthy, WT). It was shown that only the HD DNA triggered the formation of a ligation product detected by rapid ultra performance liquid chromatography (UPLC) measurement. This enabled a clear yes/no answer to the DNA diagnosis question [97].

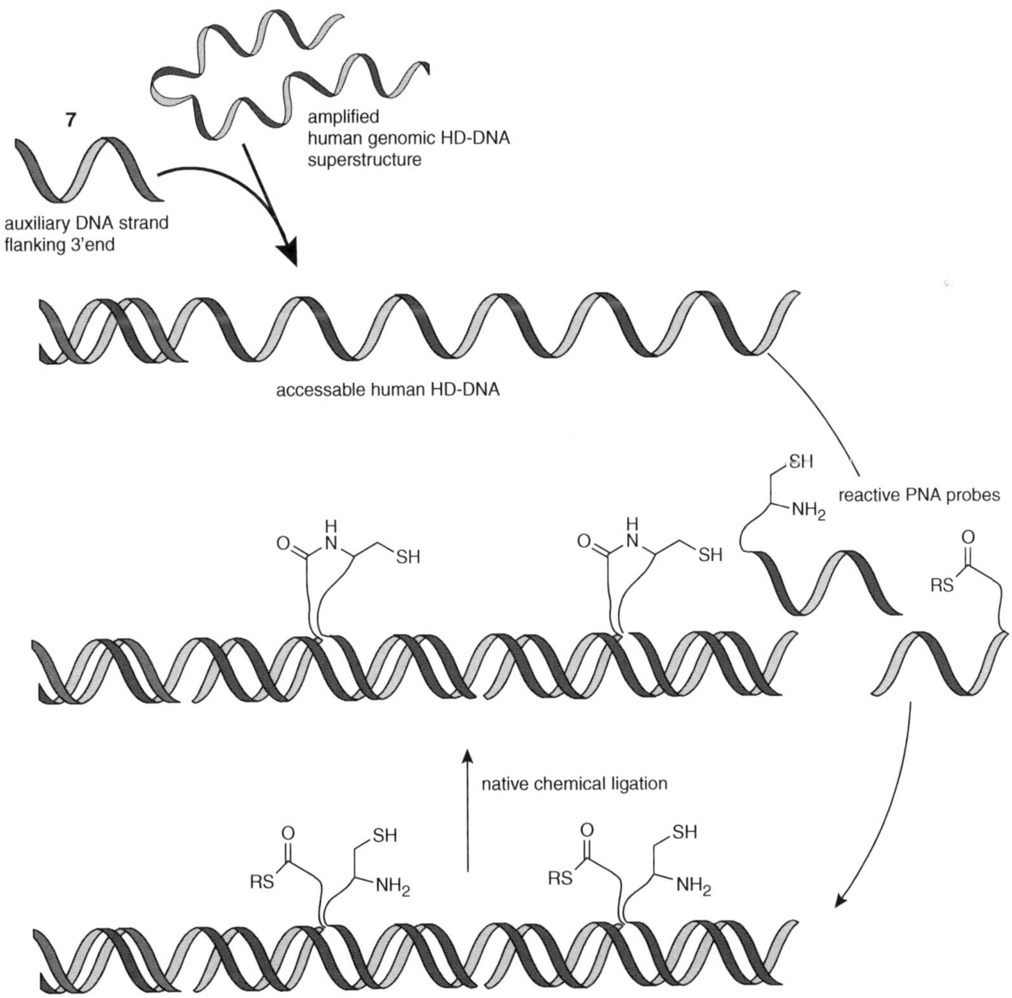

Figure 4.1.13 *Chromatographic detection of amplified human genomic HD-DNA with native chemical ligation of PNA–thioester probes and cysteinyl–PNA conjugates*

4.1.3.3 Drug screening

Based on nucleic acid templated chemical reactions (described earlier) DNA-encoded libraries were developed that allow the synthesis and screening of a huge collection of small-molecules with desired properties [98, 99]. In principle, chemical compounds (building blocks, **8**) were conjugated to DNA-fragments **9**, which serve as identification barcodes and facilitate nucleic acid templated chemical reactions. These reactions enable the formation of small-molecules (translation) linked to a specific DNA-sequence **10**. A subsequent separation step (*in vitro* selection) allows the enrichment of the DNA-conjugate bearing the library member with the desired properties **11**. Subsequently, this purified DNA-conjugate is amplified by PCR (amplification) and could be either reintroduced to another iterated cycle of translation, *in vitro* selection and amplification (diversification), or sequenced to receive information about the identity of the synthetic molecule with the desired properties (sequencing) (Figure 4.1.14).

Gartner and coworkers first demonstrated the application of DNA-directed synthesis with a model library containing 1025 members, **12**. This DNA-encoded library was allowed to react with 1025 thiol-containing reagents, **13**, in which only one reagent bears a biotine-tag, thus enabling the streptavidin binding. Owing to the nucleic-acid templated reaction (maleimide–thiol ligation, see also Figure 4.1.2D), 1025 products were obtained (instead of $>10^6$ possible products in a non-templated manner). After *in vitro* selection with streptavidin, the biotine-bearing product **14** was 1000-fold enriched (Figure 4.1.15) [21]. This principle was later extended to multistep synthesis using subsequent amine acylation (see also Figure 4.1.5B) chemistries [22].

Liu and coworkers showed the diversification of this approach by creating a DNA-encoded library with small-molecules, wherein every small-molecule binds/targets another protein with high affinity and specificity for streptavidin, carbonic anhydrase, papain, trypsin, and chymotrypsin, as well as glutathione-*S*-transferase, and thereby, demonstrated the ability of parallelization of *in vitro* selection steps [100]. Furthermore, they improved this approach by developing DNA-encoded library architectures, allowing branched small-molecule synthesis

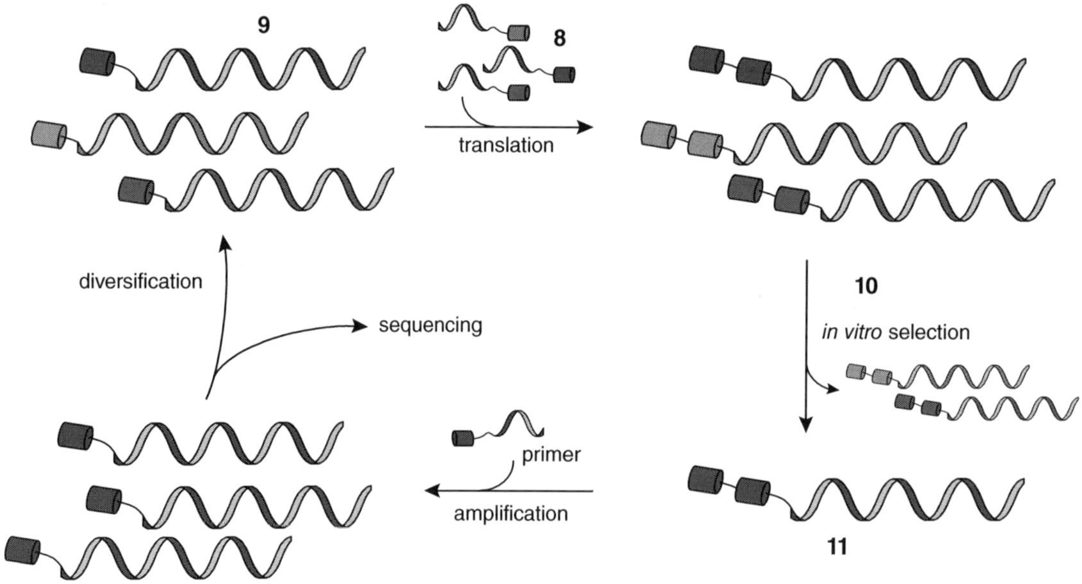

Figure 4.1.14 *Schematic illustration of DNA-encoded library technology for synthesis and screening of collections of small compounds; consisting of iterated cycles of translation, in vitro selection, amplification, and either diversification or sequencing*

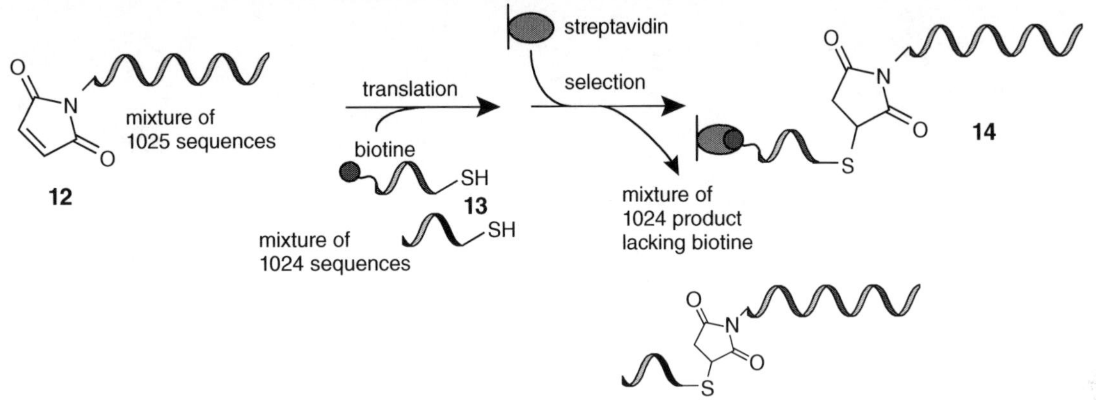

Figure 4.1.15 *Schematic illustration of DNA-encoded library principle for translation and in vitro selection of one out of 1025 products, allowing 1000-fold enrichment*

[101–103], enabling the transformation of reactive groups [104], introducing DNA-conjugates with cleavable linkers facilitating self-elution from streptavidin beads after selection of the synthesized product [101], and temperature-dependent reactions allowing the need for a temperature gradient to control the chemical reactions [105]. These amendments opened the field for impressive multistep synthesis of complex macrocyclic fumaramides by four successive DNA-templated reactions: three amine acylation steps and one Wittig olefination, giving macrocyclic fumaramide (Figure 4.1.16). Of note, the Wittig reaction also served the purpose of enabling the release of the macrocycle from the streptavidin beads [66].

The library of macrocycles obtained was selected against carbonic anhydrase affinity yielding one enriched member [100]. After PCR amplification of the template the sequence was characterized and identified as the only macrocyclic fumaramide bearing a phenol sulfonamide group, which is known for its high affinity to carbonic anhydrase [66]. In recent years, the scope of DNA-templated synthesis has been extended to the construction of a library containing >13 000 DNA-linked macrocycles [37] or unrelated polymers [36, 106]. The amplification step was improved by introducing reaction-dependent PCR for *in vitro* selection during amplification of DNA sequence encoded molecules [107, 108].

Hansen and coworkers introduced the YoctoReactor® [109–117]. This drug discovery tool is based on the molecular evolution principle described earlier (see caption to Figure 4.1.14). Building blocks were conjugated with a unique DNA sequence either with cleavable or non-cleavable linker (**15**, Figure 4.1.17). Four reactive DNA strands were designed to form a four-way junction (**16**) in which the reactive groups are arranged within a yoctoliter volume at the crossover. After assembling the four-way junction (**16**) the building blocks (colored dots) are arranged in proximity, facilitating chemical reactions such as amine acylation, reductive amination or S_NAr. Subsequent to the chemical reaction the DNA ends were ligated to form a nicked four-way junction. After cleavage of the labile linker between the building block and its DNA conjugate, the YoctoReactor is dismantled by PCR to yield a dsDNA with the product (dots) exposed on its surface (**17**). This is called the display product. The display product with the exposed molecule enables a selection step for the desired properties and affinities. The affinity-selected product is amplified by PCR facilitating the enrichment of sequences encoding for compounds that bind the target of interest. To retrieve information about the identity of the affinity selected compound, the DNA is analyzed by DNA sequencing and bioinformatics analysis.

Turberfield et al. synthesized macromolecules with a high degree of polymerization in remarkable yields (23–31% overall yield) [118]. The synthesis was initiated by a templated olefin bond forming reaction

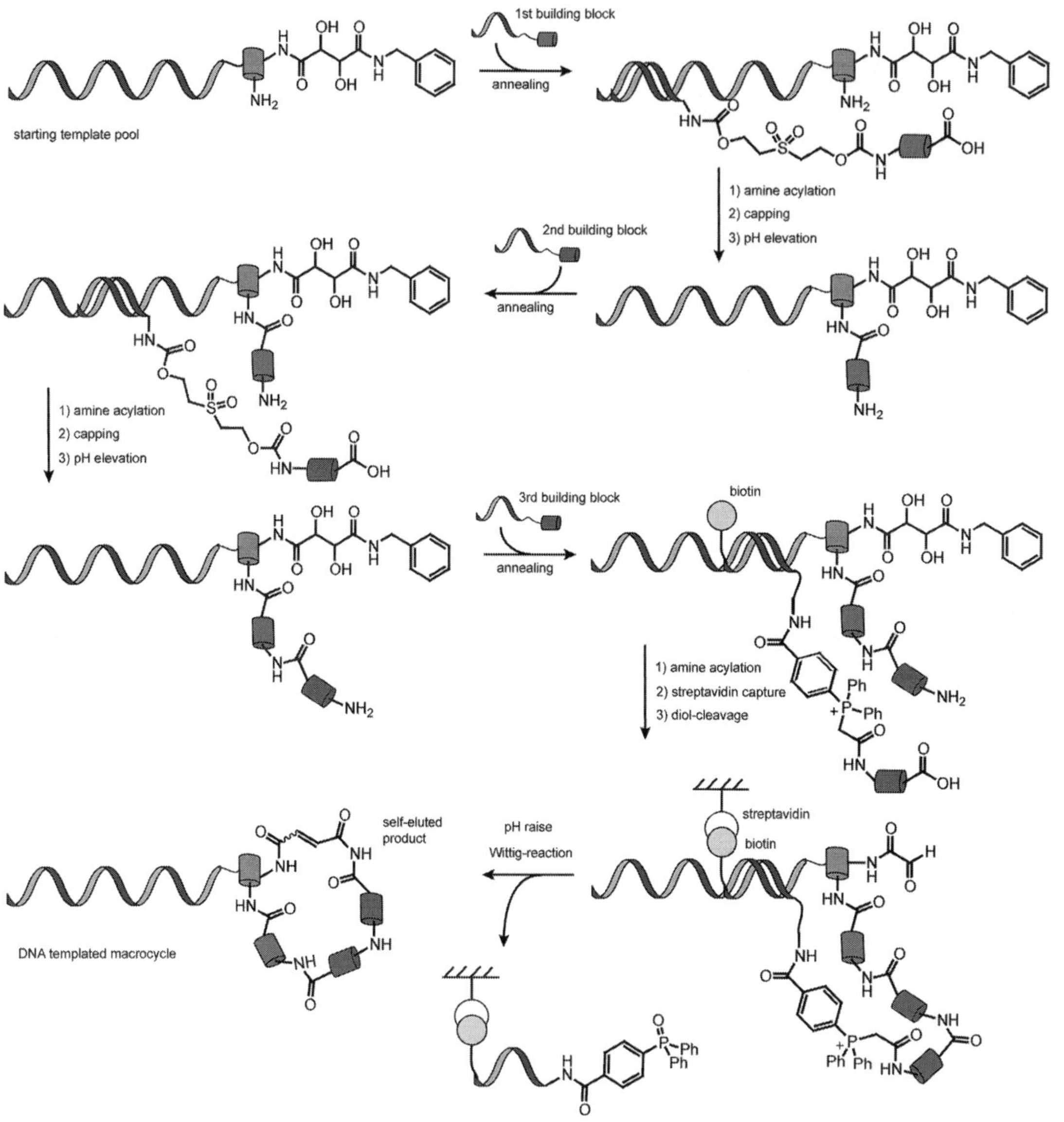

Figure 4.1.16 *Illustration of translation,* in vitro *selection strategy to generate a library of macrocycles. The library was generated by DNA-templated amine acylation reactions and one Wittig macrocyclization reaction. The DNA templated macrocycle enables the amplification by PCR, which in turn serves as starting template for iterated cycles*

between a DNA conjugate bearing an aldehyde (**19**) and a second conjugate containing an ylide **18** (Figure 4.1.18). In a Wittig reaction an alkylidene group was transferred from the ylide to the aldehyde moiety of the other reaction partner, followed by cleavage of the DNA adapter, which was released by adding a full complementary unmodified DNA strand. This heterofunctionalized product was allowed to react in a subsequent Wittig reaction with another heterofunctionalized building block, **20**. For termination of the synthesis a DNA conjugate bearing an aldehyde without an ylide linker **21** was added, which enabled the formation of the final product **22**. Later, Turberfield applied this concept to T-type template architectures [119].

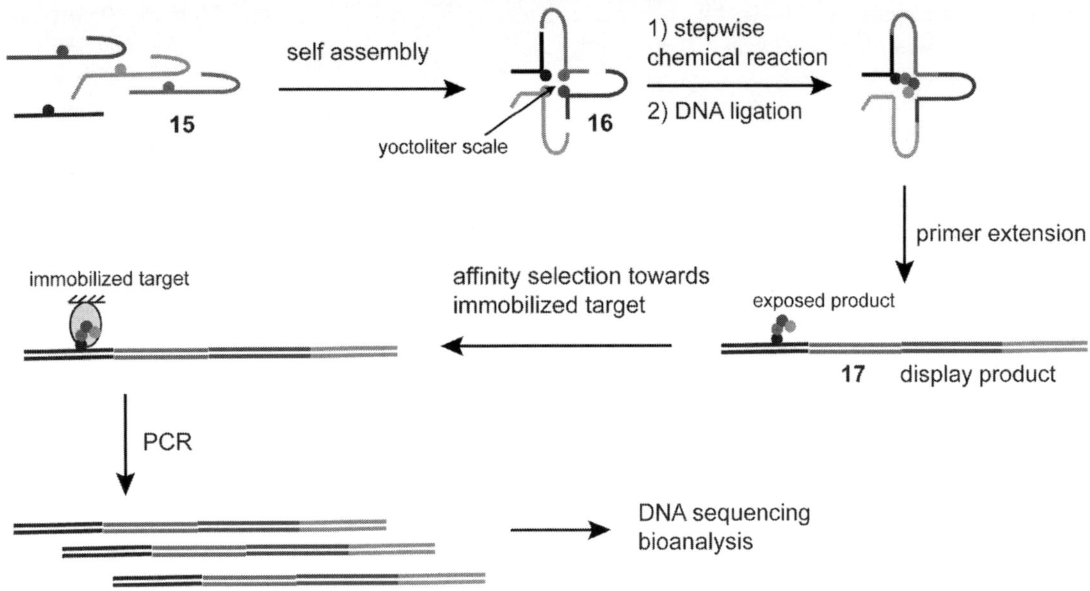

Figure 4.1.17 *Schematic principle of the YoctoReactor: annealing of ssDNA conjugates (**15**) to assemble the reaction center (**16**). The stepwise templated reaction resulting in transfer of the building block (colored dots) is facilitated as a result of increased molar concentration within the reaction center. After DNA ligation of the DNA ends, the reactor architecture is dismantled with PCR (primer extension) giving the display product (**17**), which bears the exposed product at its surface. The display product (**17**) is enriched by affinity selection towards an immobilized target. The survivors were amplified by PCR and sequenced to obtain information about the identity of the display product (See color figure in color plate section)*

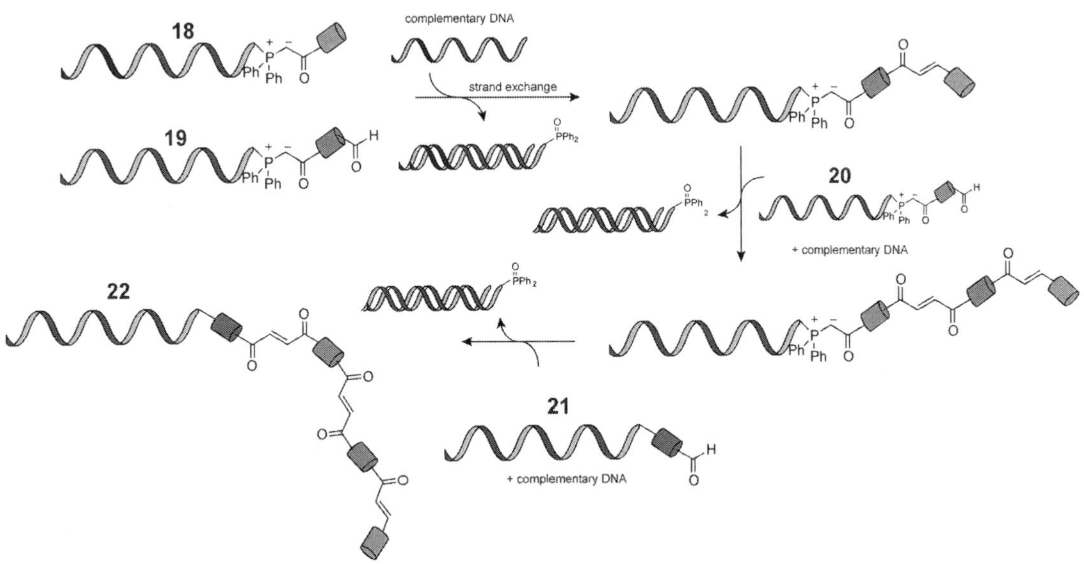

Figure 4.1.18 *DNA-templated synthesis of **22** by iterative Wittig olefination reactions*

Alternative strategies for drug discovery rely on encoded self-assembly of chemical libraries (ESAC) and lack nucleic acid templated chemical reactions and will, therefore, not be treated in this overview. For more information the reader is guided to several publications [120–124].

4.1.3.4 Materials science

The growing toolbox of nucleic acid templated reactions has brought applications within the field of materials science into reach. The aim is to develop methods for the DNA/RNA-sequence instructed construction of molecules with designed properties. For example, it is envisioned that nucleotide sequences can be translated into sequence-defined monodisperse polymer materials and nucleic acids may be able to program the formation of molecular wires and nanosized circuits.

Rosenbaum and Liu reported a nucleic acid templated oligomerization reaction based on reductive amination of PNA aldehydes on a 40-mer DNA template (Figure 4.1.19) [125]. The challenge was to enable successive chemical reactions without intermediate purification. The first PNA aldehyde (4-mer) was covalently attached to the DNA template. With this system a DNA sequence was translated into an artificial polymer (PNA). The templated reaction of side-chain functionalized PNA aldehydes was proposed as a method for laboratory evolution of synthetic polymers [126].

Based on work from von Kiedrowski's group [127], Gothelf and coworkers demonstrated DNA-templated manganese–salen bond forming reactions between linear and tripodal oligonucleotide-functionalized molecules (LOM, **23**; TOM, **24** in Figure 4.1.20, see also Figure 4.1.2H) [128, 129]. With this repertoire of DNA- functionalized building blocks they were able to build linear and angled branched products, as well as three-way and five-way junctions. It is believed that this approach enables the construction of two-dimensional

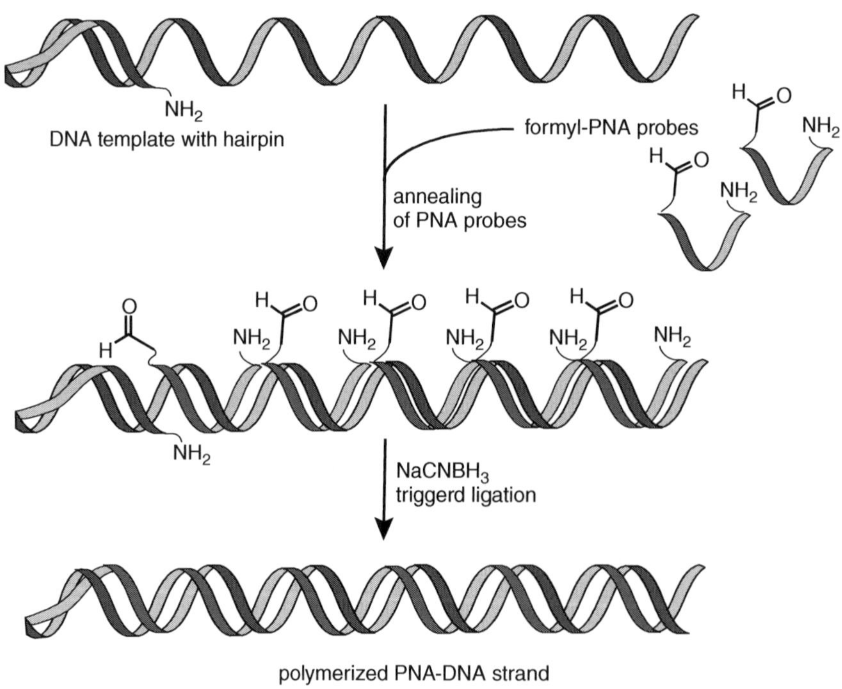

Figure 4.1.19 *DNA-templated formyl-PNA oligomerization with 5′-amino modified hairpin DNA template triggered by $NaCNBH_3$ addition*

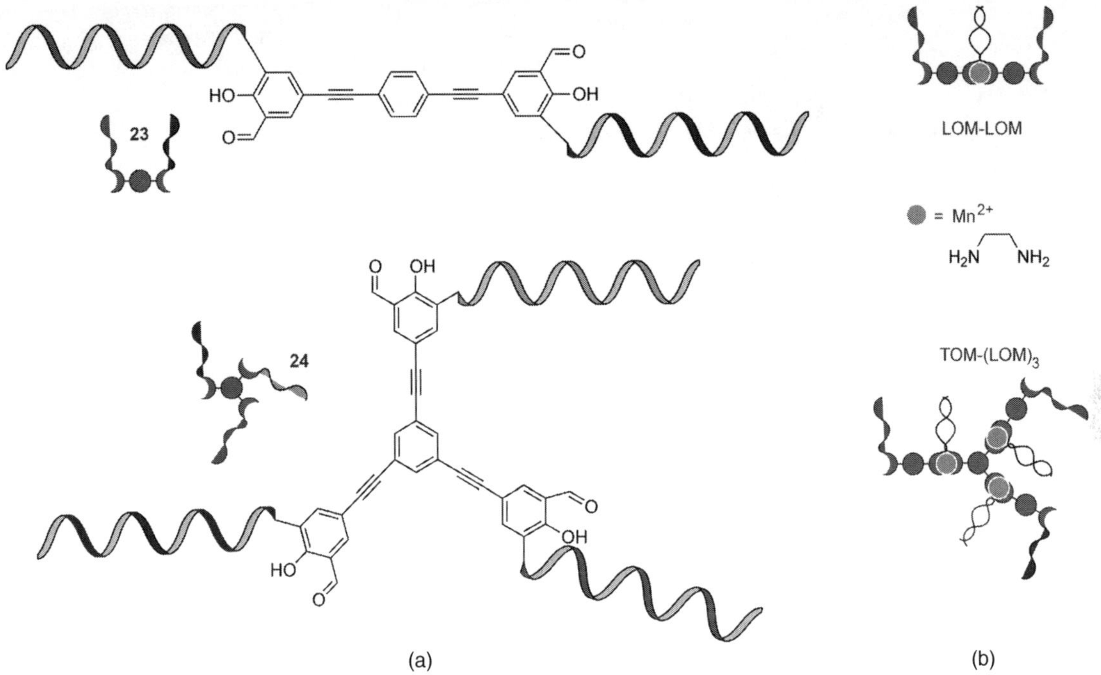

Figure 4.1.20 *(a) LOM and TOM structures introduced by Brown and (b) examples of their assembly to built-up two-dimensional nanostructures by forming manganese–salen complexes*

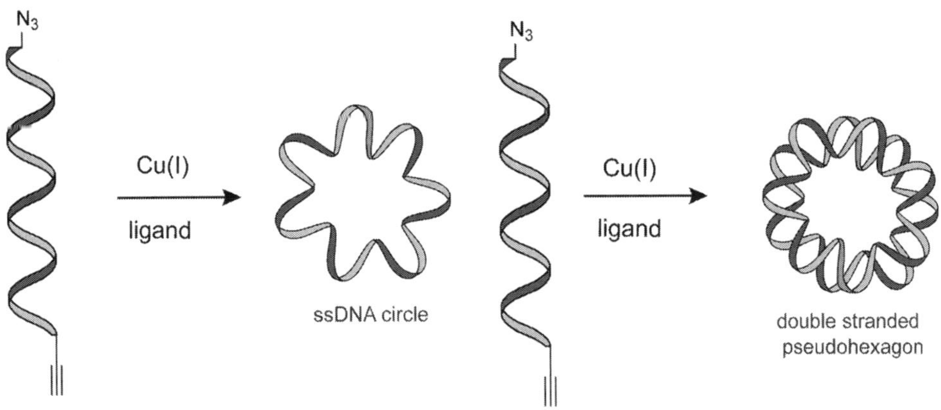

Figure 4.1.21 *Cu(I)-mediated click chemistry to form covalently closed ssDNA circle and double stranded pseudohexagon structures*

nanomatrices. Later these scientists improved the concept of manganese–salen complex formation to build potentially conducting nanomaterials [130]. The ability to design nanosized, three-dimensional objects via sequence-instructed DNA assembly is driving the field of DNA-nanotechnology. The objects are formed by means of non-covalent interactions and are therefore not within the scope of this review. Several reviews are available for further information [131–135].

Brown demonstrated the use of Cu(I)-mediated click chemistry for the synthesis of covalently closed ssDNA circle and dsDNA pseudohexagon (catenane) structures (Figure 4.1.21) [47].

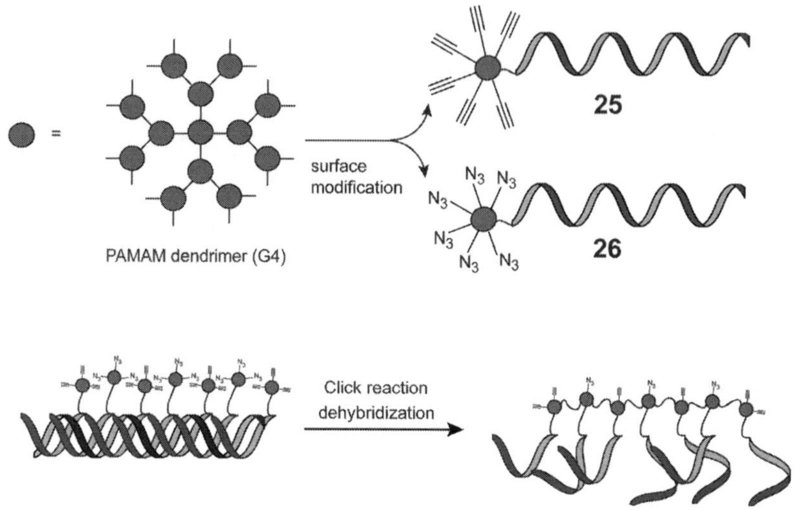

27

Figure 4.1.22 *Cu(I)-mediated click chemistry to build linear and branched polymer structures with functionalized dendrimer DNA conjugates*

Gothelf investigated templated Cu(I)-mediated click reactions in the construction of dendrimer oligomers with precise geometries and function. A polyamidoamine (PAMAM) dendrimer was either azide- (**25**) or alkyne- (**26**) functionalized at its surface and conjugated to a DNA strand. Another DNA strand served as a template, which directed the alternate assembly of the alkyne- and azide-modified dendrimer-DNA. Addition of cupper(I) triggered the coupling of the dendrimer units yielding covalently bound dimers, trimers, tetramers, and polymers (**27**, Figure 4.1.22) [136].

Gothelf and colleagues prepared functionalized nanostructures by means of the Glaser–Eglington coupling [137–139]. Copper click chemistry was used to attach phenyl acetylene units to DNA. In the presence of an amine and Cu(I) as well as Cu(II) two phenylacetylene units can be joined in a templated reaction (Figure 4.1.23a). They expanded the scope to DNA-directed heterocouplings of oligo(phenylene ethylene) building blocks to form molecular rods of predetermined length ranging from 4 to 8 nm (Figure 4.1.23b). This investigation suggests a DNA-templated route to potentially conductive materials. It was found, however, that yields decreased with the increasing level of oligomerization [140].

4.1.4 Conclusions

In the early years, nucleic acid templated reactions were explored in the context of origin-of-life studies and as a tool for the construction of synthetic genes. Later, chemists recognized the potential for DNA diagnostics. DNA/RNA-templated reactions can proceed with high sequence fidelity because adjacent hybridization of two functionalized nucleic acid strands is required to drive bond formation. Furthermore, one DNA/RNA-template molecule may trigger the synthesis of many product molecules, which provides the basis for an amplification mechanism. However, biologically occurring nucleic acid targets are structured (e.g., double-stranded DNA or folded RNA), which complicates access of the targeted segment. Furthermore, biogenic templates reside within complex matrices such as tissues, cells, cell lysates or enzyme mixtures used in

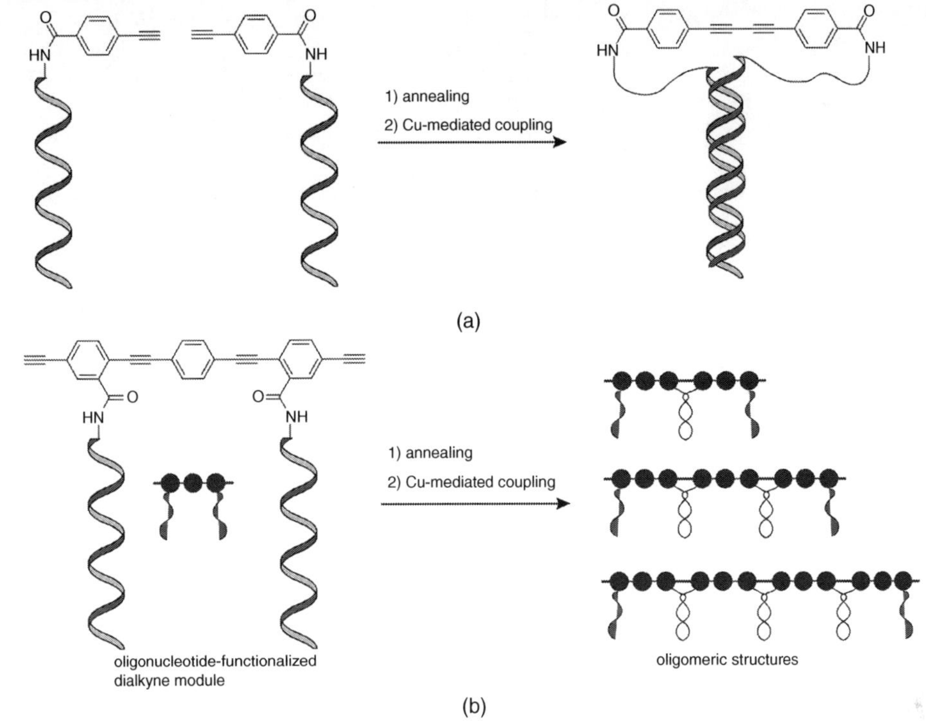

Figure 4.1.23 *(a) Nucleic acid directed Glaser–Eglington heterocoupling promoted by Cu(I) and Cu(II) and an amine base; and (b) schematic illustration of Glaser–Eglington oligomerization of phenylene ethylene DNA conjugate building blocks*

molecular biology. These challenges have led to the development of robust and biocompatible nucleic acid templated chemistries. We are now at a stage where nucleic acid directed chemistry can be used to visualize native size RNA molecules (rRNA, mRNA, miRNA) within living cells. However, limits of detection are still not sufficient to detect low copy number targets. In addition, templated chemistry has been perfected to an extent that it can be performed *in situ* while the large template is being enzymatically amplified during rolling circle amplification (RCA) of polymerase chain reactions (PCR). This may provide new opportunities for molecular diagnosis and the construction of modifiable PCR libraries.

While research into diagnostically useful NA-templated reactions has been performed within the last 20 years, the focus on applications within drug synthesis and drug screening is more recent. The extension of the repertoire of templated chemistries has been instrumental in the recent success stories of DNA-encoded synthesis of drug-like molecules. This line of research has led to the development of sequence-encoded and amplifiable drug libraries, wherein the nucleotide part of a DNA–drug conjugate not only contains the instruction for the drug part but also enables the amplification and identification. Several start-up companies have highlighted the potential usefulness of these approaches.

Applications of NA-templated reactions for the bottom-up construction of designer materials are only beginning to emerge. The first publications only date back to the 1990s and the field is still in its infancy. Yet, recent reports have demonstrated the potential of NA-templated synthesis in materials sciences. For example, the DNA-instructed synthesis of polyphenylacetyles shows that the bottom-up assembly of charge conducting circuits may be within reach.

References

[1] X. Li and D. R. Liu, DNA-templated organic synthesis: Nature's strategy for controlling chemical reactivity applied to synthetic molecules, *Angew. Chem., Int. Ed.,* **43**, 4848–4870 (2004).

[2] A. P. Silverman and E. T. Kool, Detecting RNA and DNA with templated chemical reactions, *Chem. Rev.*, **106**, 3775–3789 (2006).

[3] A. Shibata, H. Abe, and Y. Ito, Oligonucleotide-templated reactions for sensing nucleic acids, *Molecules*, **17**, 2446–2463 (2012).

[4] K. Gorska and N. Winssinger, Reactions templated by nucleic acids: more ways to translate oligonucleotide-based instructions into emerging function, *Angew. Chem., Int. Ed.*, **52**, 6820–6843 (2013).

[5] J. Michaelis, A. Roloff, and O. Seitz, Amplification by nucleic acid-templated reactions, *Org. Biomol. Chem.*, **12**, 2821–2833 (2014).

[6] R. Naylor and P. T. Gilham, Studies on some interactions and reactions of oligonucleotides in aqueous solution, *Biochemistry*, **5**, 2722–2728 (1966).

[7] Z. A. Shabarova, I. N. Merenkova, T. S. Oretskaya, N. I. Sokolova, E. A. Skripkin, E. V. Alexeyeva, A. G. Balakin, and A. A. Bogdanov, Chemical ligation of DNA: the first non-enzymatic assembly of a biologically active gene, *Nucleic Acids Res.*, **19**, 4247–4251 (1991).

[8] L. E. Orgel, Unnatural selection in chemical systems, *Acc. Chem. Res.*, **28**, 109–118 (1995).

[9] S. M. Gryaznov, R. Schultz, S. K. Chaturvedi, and R. L. Letsinger, Enhancement of selectivity in recognition of nucleic acids via chemical autoligation, *Nucleic Acid Res.*, **22**, 2366–2369 (1994).

[10] M. K. Herrlein, J. S. Nelson, and R. L. Letsinger, A covalent lock for self-assembled oligonucleotide conjugates, *J. Am. Chem. Soc.*, **117**, 10151–10152 (1995).

[11] F. D. Lewis, T. Wu, E. L. Burch, D. M. Bassani, J.-S. Yang, S. Schneider, W. Jaegerand and R. L. Letsinger, Hybrid oligonucleotides containing stilbene units. Excimer fluorescence and photodimerization, *J. Am. Chem. Soc.*, **117**, 8785–8792 (1995).

[12] L. E. Orgel, Unnatural selection in chemical systems, *Acc. Chem. Res.*, **28**, 109–118 (1995).

[13] E. Rubin, S. Rumney, S. Wang, and E. T. Kool, Convergent DNA synthesis: a non-enzymatic dimerization approach to circular oligodeoxynucleotides, *Nucleic Acid Res.*, **23**, 3547–3553 (1995).

[14] B. G. Bag and G. Von Kiedrowski, Templates, autocatalysis and molecular replication, *Pure Appl. Chem.*, **68**, 2145–2152 (1996).

[15] R. Rohatgi, D. P. Bartel, and J. W. Szostak, Nonenzymatic, template-directed ligation of oligoribonucleotides is highly regioselective for the formation of 3′-5′ phosphodiester bonds, *J. Am. Chem. Soc.*, **118**, 3340–3344 (1996).

[16] Y. Xu and E. T. Kool, Rapid and selective selenium-mediated autoligation of DNA strands, *J. Am. Chem. Soc.*, **122**, 9040–9041 (2000).

[17] Y. Xu, N. B. Karalkar, and E. T. Kool, Nonenzymatic autoligation in direct three-color detection of RNA and DNA point mutations, *Nat. Biotechnol.*, **19**, 148–152 (2001).

[18] H. Abe and E. T. Kool, Destabilizing universal linkers for signal amplification in self-ligating probes for RNA, *J. Am. Chem. Soc.*, **126**, 13980–13986 (2004).

[19] S. Wang and E. T. Kool, Circular RNA oligonucleotides. Synthesis, nucleic acid binding properties, and a comparison with circular DNAs, *Nucleic Acid Res.*, **22**, 2326–2333 (1994).

[20] J. G. Schmidt, L. Christensen, P. E. Nielsen, and L. E. Orgel, Information transfer from DNA to peptide nucleic acids by template-directed syntheses, *Nucleic Acid Res.*, **25**, 4792–4796 (1997).

[21] Z. J. Gartner and D. R. Liu, The generality of DNA-templated synthesis as a basis for evolving non-natural small molecules, *J. Am. Chem. Soc.*, **123**, 6961–6963 (2001).

[22] Z. J. Gartner, M. W. Kanan, and D. R. Liu, Multistep small-molecule synthesis programmed by DNA templates, *J. Am. Chem. Soc.*, **124**, 10304–10306 (2002).

[23] X. Li, Z.-Y. J. Zhan, R. Knipe, and D. G. Lynn, DNA-catalyzed polymerization, *J. Am. Chem. Soc.*, **124**, 746–747 (2002).

[24] Z. J. Gartner, R. Grubina, C. T. Calderone, and D. R. Liu, Two enabling architectures for DNA – templated organic synthesis, *Angew. Chem., Int. Ed.*, **42**, 1370–1375 (2003).

[25] C. Dose and O. Seitz, New isocysteine building blocks and chemoselective peptide ligation, *Org. Biomol. Chem.*, **2**, 59–65 (2004).

[26] S. Ficht, A. Mattes, and O. Seitz, Single-nucleotide-specific PNA–peptide ligation on synthetic and PCR DNA templates, *J. Am. Chem. Soc.*, **126**, 9970–9981 (2004).

[27] R. Kumar, A. El-Sagheer, J. Tumpane, P. Lincoln, L. M. Wilhelmsson, and T. Brown, Template-directed oligonucleotide strand ligation, covalent intramolecular DNA circularization and catenation using click chemistry, *J. Am. Chem. Soc.*, **129**, 6859–6864 (2007).

[28] J. L. Czlapinski and T. L. Sheppard, Nucleic acid template-directed assembly of metallosalen–DNA conjugates, *J. Am. Chem. Soc.*, **123**, 8618–8619 (2001).

[29] J. T. Goodwin and D. G. Lynn, Template-directed synthesis: use of a reversible reaction, *J. Am. Chem. Soc.*, **114**, 9197–9198 (1992).

[30] P. Luo, J. C. Leitzel, and Z. Zhan, Analysis of the structure and sStability of a backbone-modified oligonucleotide: Implications for avoiding product inhibition in catalytic template-directed synthesis, *J. Am. Chem. Soc.*, **120**, 3019–3031 (1998).

[31] J. C. Leitzel and D. G. Lynn, Template-directed ligation: From DNA towards different versatile templates, *Chem. Rec.*, **1**, 53–62 (2001).

[32] T. N. Grossmann, A. Strohbach, and O. Seitz, Achieving turnover in DNA-templated reactions, *ChemBioChem*, **9**, 2185–2192 (2008).

[33] Z. J. Gartner, M. W. Kanan, and D. R. Liu, Expanding the reaction scope of DNA – templated synthesis, *Angew. Chem., Int. Ed.*, **41**, 1796–1800 (2002).

[34] T. M. Snyder and D. R. Liu, Ordered multistep synthesis in a single solution directed by DNA templates, *Angew. Chem., Int. Ed.*, **44**, 7379–7382 (2005).

[35] Y. He and D. R. Liu, A sequential strand-displacement strategy enables efficient six-step DNA-templated synthesis, *J. Am. Chem. Soc.*, **133**, 9972–9975 (2011).

[36] J. Niu, R. Hili, and D. R. Liu, Enzyme-free translation of DNA into sequence-defined synthetic polymers structurally unrelated to nucleic acids, *Nature Chem.*, **5**, 282–292 (2013).

[37] B. T. Tse, T. M. Snyder, Y. Shen, and D. R. Liu, Translation of DNA into a library of 13000 synthetic small-molecule macrocycles suitable for *in vitro* selection, *J. Am. Chem. Soc.*, **130**, 15611–15626 (2008).

[38] R. K. Bruick, P. E. Dawson, S. B. Kent, N. Usman, and G. F. Joyce, Template-directed ligation of peptides to oligonucleotides., *Chem. Appl. Biol.*, **3**, 49–56 (1996).

[39] C. Dose, S. Ficht, and O. Seitz, Reducing product inhibition in DNA-template-controlled ligation reactions, *Angew. Chem., Int. Ed.*, **45**, 5369–5373 (2006).

[40] S. Ficht, C. Dose, and O. Seitz, As fast and selective as enzymatic ligations. Unpaired nucleobases increase the selectivity of DNA-controlled native chemical PNA ligation, *ChemBioChem*, **6**, 2098–2103 (2005).

[41] C. Dose and O. Seitz, Single nucleotide specific detection of DNA by native chemical ligation of fluorescence labeled PNA-probes, *Bioorg. Med. Chem.*, **16**, 65–77 (2008).

[42] A. Roloff and O. Seitz, Bioorthogonal reactions challenged: DNA templated native chemical ligation during PCR, *Chem. Sci.*, **4**, 432–436 (2012).

[43] R. Kumar, A. El-Sagheer, J. Tumpane, P. Lincoln, L. M. Wilhelmsson, and T. Brown, Template-directed oligonucleotide strand ligation, covalent intramolecular DNA circularization and catenation using click chemistry, *J. Am. Chem. Soc.*, **129**, 6859–6864 (2007).

[44] A. H. El-Sagheer and T. Brown, Click chemistry with DNA, *Chem. Soc. Rev.*, **39**, 1388–1405 (2010).

[45] P. Kocalka, A. H. El-Sagheer, and T. Brown, Rapid and efficient DNA strand cross-linking by click chemistry, *ChemBioChem*, **9**, 1280–1285 (2008).

[46] A. H. El-Sagheer, R. Kumar, S. Findlow, J. M. Werner, A. N. Lane, and T. Brown, A very stable cyclic DNA miniduplex with just two base pairs, *ChemBioChem*, **9**, 50–52 (2008).

[47] R. Kumar, A. H. El-Sagheer, J. Tumpane, P. Lincoln, L. M. Wilhelmsson, and T. Brown, Template-directed oligonucleotide strand ligation, covalent intramolecular DNA circularization and catenation using click chemistry, *J. Am. Chem. Soc.*, **129**, 6859–6864 (2007).

[48] J. E. Moses and A. D. Moorhouse, The growing applications of click chemistry, *Chem. Soc Rev.*, **36**, 1249–1262 (2007).

[49] J. C. Jewett and C. R. Bertozzi, Cu-free click cycloaddition reactions in chemical biology, *Chem. Soc Rev.*, **39**, 1272–1279 (2010).

[50] K. D. Hänni and D. A. Leigh, The application of CuAAC click chemistry to catenane and rotaxane synthesis, *Chem. Soc Rev.*, **39**, 1240–1251 (2010).

[51] S. K. Mamidyala and M. G. Finn, *In situ* click chemistry: probing the binding landscapes of biological molecules, *Chem. Soc Rev.*, **39**, 1252–1261 (2010).

[52] E. Paredes and S. R. Das, Click chemistry for rapid labeling and ligation of RNA, *ChemBioChem*, **12**, 125–131 (2011).

[53] K. Fujimoto, S. Matsuda, N. Takahashi and I. Saito, Template-directed photoreversible ligation of deoxyoligonucleotides via 5-vinyldeoxyuridine, *J. Am. Chem. Soc.*, **122**, 5646–5647 (2000).

[54] K. Fujimoto, S. Matsuda, M. Hayashi, and I. Saito, Reversible DNA photocircularization on triple helix: effect of vinyl substituent on base stacking, *Tetrahedron Lett.*, **41**, 7897–7900 (2000).

[55] K. Fujimoto, S. Matsuda, N. Ogawa, M. Hayashi, and I. Saito, Template-directed reversible photocircularization of DNA via 5-vinyldeoxycytidine, *Tetrahedron Lett.*, **41**, 6451–6454 (2000).

[56] K. Fujimoto, N. Ogawa, M. Hayashi, S. Matsuda, and I. Saito, Template directed photochemical synthesis of branched oligodeoxynucleotides via 5-carboxyvinyldeoxyuridine, *Tetrahedron Lett.*, **41**, 9437–9440 (2000).

[57] K. V. Gothelf, A. Thomsen, M. Nielsen, E. Cló, and R. S. Brown, Modular DNA-programmed assembly of linear and branched conjugated nanostructures, *J. Am. Chem. Soc.*, **126**, 1044–1046 (2004).

[58] T. N. Grossmann and O. Seitz, DNA-catalyzed transfer of a reporter group, *J. Am. Chem. Soc.*, **128**, 15596–15597 (2006).

[59] T. N. Grossmann, L. Röglin, and O. Seitz, Target-catalyzed transfer reactions for the amplified detection of RNA, *Angew. Chem., Int. Ed.*, **47**, 7119–7122 (2008).

[60] F. Altenbrunn, T. N. Grossmann, C. Haase, F. Mende, L. Röglin, S. Thurley, and O. Seitz, Chemical control of biomolecular interaction modules, *Pure Appl. Chem.*, **81**, 273–284 (2009).

[61] T. N. Grossmann and O. Seitz, Nucleic acid templated reactions: consequences of probe reactivity and readout strategy for amplified signaling and sequence selectivity, *Chem. Eur. J.*, **15**, 6723–6730 (2009).

[62] J. Michaelis, A. Maruyama and O. Seitz, Promoting strand exchange in a DNA-templated transfer reaction, *Chem. Commun.*, **49**, 618–620 (2013).

[63] J. Michaelis, G. J. v. d. H. v. Noort, and O. Seitz, DNA-triggered dye transfer on a quantum dot, *Bioconjugate Chem.*, **25**, 18–23 (2014).

[64] M. L. McKee, A. C. Evans, S. R. Gerrard, R. K. O'Reilly, A. J. Turberfield, and E. Stulz, Peptidomimetic bond formation by DNA-templated acyl transfer, *Org. Biomol. Chem.*, **9**, 1661–1666 (2011).

[65] Y. He and D. R. Liu, Autonomous multistep organic synthesis in a single isothermal solution mediated by a DNA walker, *Nature Nanotechnol.*, **5**, 778–782 (2010).

[66] Z. J. Gartner, DNA-templated organic synthesis and selection of a library of macrocycles, *Science*, **305**, 1601–1605 (2004).

[67] A. Shibata, H. Abe, M. Ito, Y. Kondo, S. Shimizu, K. Aikawa, and Y. Ito, DNA templated nucleophilic aromatic substitution reactions for fluorogenic sensing of oligonucleotides, *Chem. Commun.*, **43**, 6586–6588 (2009).

[68] X.-H. Chen, A. Roloff, and O. Seitz, Consecutive signal amplification for DNA detection based on *de novo* fluorophore synthesis and host-guest chemistry, *Angew. Chem., Int. Ed.*, **51**, 4479–4483 (2012).

[69] Z. L. Pianowski and N. Winssinger, Fluorescence-based detection of single nucleotide permutation in DNA via catalytically templated reaction, *Chem. Commun.*, **37**, 3820–3822 (2007).

[70] R. M. Franzini and E. T. Kool, 7-Azidomethoxy-coumarins as profluorophores for templated nucleic acid detection, *ChemBioChem*, **9**, 2981–2988 (2008).

[71] R. M. Franzini and E. T. Kool, Efficient nucleic acid detection by templated reductive quencher release, *J. Am. Chem. Soc.*, **131**, 16021–16023 (2009).

[72] H. Abe, J. Wang, K. Furukawa, K. Oki, M. Uda, S. Tsuneda, and Y. Ito, A reduction-triggered fluorescence probe for sensing nucleic acids, *Bioconjugate Chem.*, **19**, 1219–1226 (2008).

[73] K. Furukawa, H. Abe, J. Wang, M. Uda, H. Koshino, S. Tsuneda, and Y. Ito, Reduction-triggered red fluorescent probes for dual-color detection of oligonucleotide sequences, *Org. Biomol. Chem.*, **7**, 671–677 (2009).

[74] Z. Pianowski, K. Gorska, L. Oswald, C. A. Merten, and N. Winssinger, Imaging of mRNA in live cells using nucleic acid-templated reduction of azidorhodamine probes, *J. Am. Chem. Soc.*, **131**, 6492–6497 (2009).

[75] K. Gorska, I. Keklikoglou, U. Tschulena, and N. Winssinger, Rapid fluorescence imaging of miRNAs in human cells using templated Staudinger reaction, *Chem. Sci.*, **2**, 1969–1975 (2011).

[76] K. Furukawa, H. Abe, J. Wang, M. Uda, H. Koshino, S. Tsuneda, and Y. Ito, Reduction-triggered red fluorescent probes for dual-color detection of oligonucleotide sequences, *Org. Biomol. Chem.*, **7**, 671–677 (2009).

[77] K. Furukawa, H. Abe, Y. Tamura, R. Yoshimoto, M. Yoshida, S. Tsuneda, and Y. Ito, Fluorescence detection of intron lariat RNA with reduction-triggered fluorescent probes, *Angew. Chem., Int. Ed.*, **50**, 12020–12023 (2011).

[78] R. M. Franzini and E. T. Kool, Improved templated fluorogenic probes enhance the analysis of closely related pathogenic bacteria by microscopy and flow cytometry, *Bioconjugate Chem.*, **22**, 1869–1877 (2011).

[79] R. M. Franzini and E. T. Kool, Two successive reactions on a DNA template: A strategy for improving background fluorescence and specificity in nucleic acid detection, *Chem. Eur. J.*, **17**, 2168–2175 (2011).

[80] K. Gorska, A. Manicardi, S. Barluenga, and N. Winssinger, DNA-templated release of functional molecules with an azide-reduction-triggered immolative linker, *Chem. Commun.*, **47**, 4364 (2011).

[81] D. J. Kleinbaum, G. P. Miller, and E. T. Kool, Double displacement: an improved bioorthogonal reaction strategy for templated nucleic acid detection, *Bioconjugate Chem.*, **21**, 1115–1120 (2010).

[82] S. Sando and E. T. Kool, Nonenzymatic DNA ligation in Escherichia coli cells, *Nucleic Acid Res. Suppl.*, 121–122 (2002).

[83] S. Sando, H. Abe, and E. T. Kool, Quenched auto-ligating DNAs: multicolor identification of nucleic acids at single nucleotide resolution, *J. Am. Chem. Soc.*, **126**, 1081–1087 (2004).

[84] A. P. Silverman and E. T. Kool, Quenched probes for highly specific detection of cellular RNAs, *Trends Biotechnol.*, **23**, 225–230 (2005).

[85] S. Sando and E. T. Kool, Imaging of RNA in bacteria with self-ligating quenched probes, *J. Am. Chem. Soc.*, **124**, 9686–9687 (2002).

[86] A. P. Silverman and E. T. Kool, RNA-templated chemistry in cells: Discrimination of *Escherichia*, *Shigella* and *Salmonella* bacterial strains with a new two-color FRET strategy, *ChemBioChem*, **7**, 1890–1894 (2006).

[87] A. P. Silverman and E. T. Kool, Detecting RNA and DNA with templated chemical reactions., *Chem. Rev.*, **106**, 3775–3789 (2006).

[88] P. Zhou, M. Wang, L. Du, G. W. Fisher, A. Waggoner, and D. H. Ly, Novel binding and efficient cellular uptake of guanidine-based peptide nucleic acids (GPNA), *J. Am. Chem. Soc.*, **125**, 6878–6879 (2003).

[89] P. Zhou, A. Dragulescu-Andrasi, B. Bhattacharya, H. O'Keefe, P. Vatta, J. J. Hyldig-Nielsen, and D. H. Ly, Synthesis of cell-permeable peptide nucleic acids and characterization of their hybridization and uptake properties, *Bioorg. Med. Chem. Lett.*, **16**, 4931–4935 (2006).

[90] A. Dragulescu-Andrasi, P. Zhou, G. He, and D. H. Ly, Cell-permeable GPNA with appropriate backbone stereochemistry and spacing binds sequence-specifically to RNA, *Chem. Commun.*, 244–246 (2005).

[91] A. Manicardi, E. Fabbri, T. Tedeschi, S. Sforza, N. Bianchi, E. Brognara, R. Gambari, R. Marchelli, and R. Conradi, Cellular uptakes, biostabilities and anti-miR-210 activities of chiral arginine-PNAs in leukaemic K562 cells, *ChemBioChem*, **13**, 1327–1337 (2012).

[92] Z. Pianowski, K. Gorska, L. Oswald, C. A. Merten, and N. Winssinger, Imaging of mRNA in live cells using nucleic acid-templated reduction of azidorhodamine probes, *J. Am. Chem. Soc.*, **131**, 6492–6497 (2009).

[93] K. Furukawa, H. Abe, K. Hibino, Y. Sako, S. Tsuneda, and Y. Ito, Reduction-triggered fluorescent amplification probe for the detection of endogenous RNAs in living human cells, *Bioconjugate Chem.*, **20**, 1026–1036 (2009).

[94] H. Saneyoshi, Y. Ito, and H. Abe, Long-lived luminogenic probe for detection of RNA in a crude solution of living bacterial cells, *J. Am. Chem. Soc.*, **135**, 13632–13635 (2013).

[95] E. M. Harcourt and E. T. Kool, Amplified microRNA detection by templated chemistry, *Nucleic Acid Res.*, **40**, e65 (2012).

[96] A. Roloff and O. Seitz, The role of reactivity in DNA templated native chemical PNA ligation during PCR, *Bioorg. Med. Chem.*, **21**, 3458–3464 (2013).

[97] A. Kern and O. Oliver, Template-directed ligation on repetitive DNA sequences: a chemical method to probe the length of Huntington DNA., *Chem. Sci.*, **6**, 724–728 (2015).

[98] S. Brenner and R. Lerner, Encoded combinatorial chemistry, *Proc. Natl. Acad. Sci. U.S.A.*, **89**, 5381–5383 (1992).

[99] J. Nielsen, S. Brenner, and K. D. Janda, Synthetic methods for the implementation of encoded combinatorial chemistry, *J. Am. Chem. Soc.*, **115**, 9812–9813 (1993).

[100] J. B. Doyon, T. M. Snyder, and D. R. Liu, Highly sensitive *in vitro* selections for DNA-linked synthetic small molecules with protein binding affinity and specificity, *J. Am. Chem. Soc.*, **125**, 12372–12373 (2003).

[101] X. Li, Z. J. Gartner, B. N. Tse, and D. R. Liu, Translation of DNA into synthetic N-acyloxazolidines, *J. Am. Chem. Soc.*, **126**, 5090–5092 (2004).

[102] C. T. Calderone and D. R. Liu, Nucleic-acid-templated synthesis as a model system for ancient translation, *Curr. Opin. Chem. Biol.*, **8**, 645–653 (2004).

[103] C. T. Calderone and D. R. Liu, Small-molecule diversification from iterated branching reaction pathways enabled by DNA-templated synthesis, *Angew. Chem., Int. Ed.*, **44**, 7383–7386 (2005).

[104] K. Sakurai, T. M. Snyder, and D. R. Liu, DNA-templated functional group transformations enable sequence-programmed synthesis using small-molecule reagents, *J. Am. Chem. Soc.*, **127**, 1660–1661 (2005).

[105] T. M. Snyder and D. R. Liu, Ordered multistep synthesis in a single solution directed by DNA templates, *Angew. Chem., Int. Ed.*, **44**, 7379–7382 (2005).

[106] Y. Brudno, M. E. Birnbaum, R. E. Kleiner, and D. R. Liu, An *in vitro* translation, selection and amplification system for peptide nucleic acids, *Nat. Chem. Biol.*, **6**, 148–155 (2010).

[107] D. J. Gorin, A. S. Kamlet, and D. R. Liu, Reactivity-dependent pcr: direct, solution-phase *in vitro* selection for bond formation, *J. Am. Chem. Soc.*, **131**, 9189–9191 (2009).

[108] L. M. McGregor, D. J. Gorin, C. E. Dumelin and D. R. Liu, Interaction-dependent PCR: Identification of ligand–target pairs from libraries of ligands and libraries of targets in a single solution-phase experiment, *J. Am. Chem. Soc.*, **132**, 15522–15524 (2010).

[109] M. H. Hansen, P. Blakskjær, L. K. Petersen, T. H. Hansen, J. W. Højfeldt, K. V. Gothelf, and N. J. V. Hansen, A yoctoliter-scale DNA reactor for small-molecule evolution., *J. Am. Chem. Soc.*, **131**, 1322–1327 (2009).

[110] T. R. Heitner and N. J. V. Hansen, Streamlining hit discovery and optimization with a yoctoliter scale DNA reactor, *Expert Opin. Drug Discov.*, **4**, 1201–1213 (2009).

[111] D. R. Halpin and P. B. Harbury, DNA display II. Genetic manipulation of combinatorial chemistry libraries for small-molecule evolution, *PLoS Biol.*, **7**, e174 (2004).

[112] D. R. Halpin, J. A. Lee, S. J. Wrenn, and P. B. Harbury, DNA display III. Solid-phase organic synthesis on unprotected DNA, *PLoS Biol.*, **7**, e175 (2004).

[113] D. R. Halpin and P. B. Harbury, DNA display I. Sequence-encoded routing of DNA populations, *PLoS Biol.*, **7**, e173 (2004).

[114] D. R. Liu, Translating DNA into synthetic molecules, *PLoS Biol.*, **7**, e223 (2004).

[115] T. Nicholas, Methods for the synthesis of macrocycle libraries for drug discovery, *Drug Discovery today: Technol.*, **7**, e97–e104 (2010).

[116] E. M. Driggers, S. P. Hale, J. Lee, and N. K. Terret, The exploration of macrocycles for drug discovery — an underexploited structural class, *Nat. Rev. Drug Discov.*, **7**, 608–624 (2008).

[117] Y. Huang and J. Coull, Diamine catalyzed hemicyanine dye formation from nonfluorescent precursors through DNA programmed chemistry, *J. Am. Chem. Soc.*, **130**, 3238–3239 (2008).

[118] P. J. Milnes, M. L. McKee, J. Bath, L. Song, E. Stulz, A. J. Turberfield, and R. K. O'Reilly, Sequence-specific synthesis macromolecules using DNA-templated chemistry, *Chem. Commun.*, **48**, 5614–5616 (2012).

[119] M. L. McKee, P. J. Milnes, J. Bath, E. Stulz, R. K. O'Reilly, and A. J. Turberfield, Programmable one-pot multistep organic synthesis using DNA junction, *J. Am. Chem. Soc.*, **134**, 1446–1449 (2012).

[120] S. Melkko, J. Scheuermann, C. E. Dumelin, and D. Neri, Encoded self-assembling chemical libraries, *Nat. Biotechnol.*, **22**, 568–574 (2004).

[121] M. M. Rozenman, M. W. Kanan, and D. R. Liu, Development and initial application of a hybridization-independent, DNA-encoded reaction discovery system compatible with organic solvents, *J. Am. Chem. Soc.*, **129**, 14933–14938 (2007).

[122] M. M. Rozenman, B. R. McNaughton, and D. R. Liu, Solving chemical problems through the application of evolutionary principles, *Curr. Opin. Chem. Biol.*, **11**, 259–268 (2007).

[123] C. E. Dumelin, S. Trüssel, F. Buller, E. Trachsel, F. Bootz, Y. Zhang, L. Mannocci, S. E. Beck, M. Drumea-Mirancea, M. W. Seeliger, C. Baltes, T. Müggler, F. Kranz, M. Rudin, S. Melkko, J. Scheuermann, and D. Neri, A portable albumin binder from a DNA-encoded chemical library, *Angew. Chem., Int. Ed.*, **47**, 3196–3201 (2008).

[124] R. E. Kleiner, C. E. Dumelin, and D. R. Liu, Small-molecule discovery from DNA-encoded chemical libraries, *Chem. Soc Rev.*, **40**, 5707–5717 (2011).

[125] D. M. Rosenbaum and D. R. Liu, Efficient and sequence-specific DNA-templated polymerization of peptide nucleic acid aldehydes, *J. Am. Chem. Soc.*, **125**, 13924–13925 (2003).
[126] R. E. Kleiner, Y. Brudno, M. E. Birnbaum, and D. R. Liu, DNA-templated polymerization of side-chain-functionalized peptide nucleic acid aldehydes, *J. Am. Chem. Soc.*, **130**, 4646–4659 (2008).
[127] L. H. Eckardt, K. Naumann, W. N. Pankau, M. Rein, M. Schweitzer, N. Windhab, and G. v. Kiedrowski, DNA nanotechnology: Chemical copying of connectivity, *Nature*, **420**, 286–286 (2002).
[128] K. V. Gothelf, A. Thomson, M. Nielsen, E. Cló, and R. S. Brown, Modular DNA-programmed assembly of linear and branched conjugated nanostructures, *J. Am. Chem. Soc.*, **126**, 1044–1046 (2004).
[129] K. V. Gothelf and R. S. Brown, A modular approach to DNA-programmed self-assembly of macromolecular nanostructures, *Chem. Eur. J.*, **11**, 1062–1069 (2005).
[130] C. S. Andersen, H. Yan, and K. V. Gothelf, Bridging one helical turn in double-stranded DNA by templated dimerization of molecular rods, *Angew. Chem., Int. Ed.*, **47**, 5569–5572 (2008).
[131] Y. Fu, D. Zheng, J. Chao, Y. Jin, Z. Zhang, H. Liu, D. Li, H. Ma, Q. Huang, K. V. Gothelf, and C. Fang, Single-step rapid assembly of DNA origami nanostructures for addressable nanoscale bioreactors, *J. Am. Chem. Soc.*, **135**, 696-702 (2013).
[132] K. Busuttil, A. Rotaru, M. Dong, F. Besenbacher and K. V. Gothelf, Transfer of a protein pattern from self-assembled DNA origami to a functionalized substrate, *Chem. Commun*, **49**, 1927–1929 (2013).
[133] Z. Zhang, J. Song, F. Besenbacher, M. Dong, and K. V. Gothelf, Self-assembly of DNA origami and single-stranded tile structures at room temperature, *Angew. Chem., Int. Ed.*, **52**, 9219–9223 (2013).
[134] P. W. K. Rothemund, Folding DNA to create nanoscale shapes and patterns, *Nature*, **440**, 297–302 (2006).
[135] T. Torring, N. V. Voigt, J. Nangreave, J. Yan, and K. V. Gothelf, DNA origami: a quantum leap for self-assembly of complex structures, *Chem. Soc. Rev.*, **40**, 5636–5646 (2011).
[136] H. Liu, T. Torring, M. Dong, C. B. Rosen, F. Besenbacher, and K. V. Gothelf, DNA-templated covalent coupling of G4 PAMAM dendrimers, *J. Am. Chem. Soc.*, **132**, 18054–18056 (2010).
[137] C. Glaser, Beiträge zur Kenntniss des Acetenylbenzols, *Ber. Dtsch. Chem. Ges.*, **2**, 422–424 (1869).
[138] C. Glaser, Untersuchungen über einige Derivate der Zimtsäure, *Justus Liebigs Ann. Chem.*, **154**, 137–171 (1870).
[139] G. Eglinton and A. R. Galbraith, Cyclic diynes, *Chem. Ind.*, 737–738 (1956).
[140] J. B. Ravnsboek, M. F. Jacobsen, C. B. Rosen, N. V. Voigt, and K. V. Gothelf, DNA-programmed glaser-eglinton reactions for the synthesis of conjugated molecular wires, *Angew. Chem., Int. Ed.*, **50**, 10851–10854 (2011).

4.2

Lipid Oligonucleotide Bioconjugates: Applications in Medicinal Chemistry

Amit Patwa, Arnaud Gissot, Khalid Oumzil, and Philippe Barthélémy

Université Victor Segalen Bordeaux 2, UFR Sciences de la Vie; and INSERM U869, Bordeaux, France

4.2.1 Introduction

Crossbreeding is often referred to as the art, or science, of modulating the traits of biological species (animals, plants) by mixing two different populations in order to obtain desired characteristics. Interestingly, this idea can be transferred to biomolecules. In theory, the synthetic combination of two chemical species would lead to chimerical structures featuring new characteristics belonging to both molecular species. Nucleic acids and lipids are the fundamental molecules in biology. While the former store and propagate genetic information, the latter, as the structural components of cell membranes, act as boundaries and allow for compartmentalization. Thus, the combinations of lipids with oligonucleotides have been realized and new bioconjugates termed Lipid OligoNucleotides (LONs) have been synthesized and studied [1]. These amphiphiles are currently attracting a considerable degree of interest owing to their unique physicochemical and biological properties. These hybrid compounds feature the molecular code of the nucleic acids and the self-assembly properties that lead to aggregates such as micelles [2], liposomes [3], and nanoparticles [4]. Excitingly, LONs have been investigated for biomedicinal applications [4], for example, in the design of new therapeutic strategies.

In this chapter we highlight recent advances in the area of LONs with an emphasis on molecular and biomedical applications. In the first section, we focus on the design and the synthesis of LONs. This will include several examples of synthetic oligonucleotide based amphiphiles. In the next section, we describe recent biomedical applications involving LONs.

DNA in Supramolecular Chemistry and Nanotechnology, First Edition.
Edited by Eugen Stulz and Guido H. Clever.

4.2.2 Chemical approach to the synthesis of lipid–oligonucleotide conjugates

The revolutionary discoveries on gene expression [5–7] and the pioneering work on the chemistry and biochemistry of nonionic analogues [8–11] have inspired synthetic chemists/biochemists in the advancement of the therapeutic use of oligonucleotides (ONs). In this context, extensive research has been done to improve the properties of ONs by covalently linking bio-inspired reporter groups (e.g., peptides, carbohydrates, lipids, polyamines) and numerous other small molecules to ONs. Recent developments in the field of oligonucleotide conjugates have been reviewed by several groups [12–14]. To overcome the barrier due to the poor affinity of ONs with the cell membrane and to offer efficient delivery, conjugation of neutral lipids with ONs renders an interesting approach, along with the use of viral vectors as well as non-viral delivery systems. Recent reviews highlight [1, 15, 16] the advances made in the field of lipid–oligonucleotide conjugates (LONs). In this section, however, we will focus solely on the chemistry part of the LONs.

Based on location, there are three different positions for lipid conjugation to ONs: (i) 3′-terminal, (ii) 5′-terminal, and (iii) an intra-chain position (anywhere in the ON sequence except for the 3′- or 5′-termini). At each location, the lipid conjugates can typically be incorporated at various sites that is, at the phosphate backbone (either anchoring on a solid support through a linker or between two nucleotides), at a sugar unit (with 2′-, 3′- or 5′-conjugation, an LNA derivative, etc.), and/or at a nucleobase unit (at various positions or through conjugation of modified mimics of the nucleobase). The specific site/sites of the lipid conjugation has/have an impact on the biological and physical properties of the LONs. Therefore, depending on its application, the conjugation site/sites has/have to be carefully chosen so that the function of the LONs is facilitated without compromising its binding properties (hybridization with complementary ONs).

The synthetic methods used for the incorporation of lipid moieties anywhere in the sequence of the oligonucleotide, for the preparation of LONs, fall into two major categories: a solid-phase approach (or pre-synthetic approach) and a solution-phase approach (or post-synthetic approach). Both methods have their own benefits and limitations.

4.2.2.1 Solid-phase (or pre-synthetic) approach

In the pre-synthetic approach, the modified derivative (carrying the desired lipidic moiety, with or without a suitable protecting group) is incorporated prior to, during or after the stepwise process of oligonucleotide synthesis on a single solid support. Solid-phase synthesis has the advantage of high coupling efficiency and is a facile purification method. However, the wealth of molecules compatible with solid phase synthesis (acidic and oxidative conditions) and deprotection (strong alkaline and nucleophilic conditions) is limited as this method excludes the use of chemical functionalities that are not stable enough to survive in these conditions. In this context, custom synthesis of desired modified derivative is required unless it is commercially available.

4.2.2.1.1 Lipid conjugation at the 3′-terminal

Letsinger and coworkers developed the first solid-phase synthesis to functionalize the 3′-terminal of ONs with cholesterol (Scheme 4.2.1A). Firstly, they synthesized support-bound dinucleoside H-phosphonate derivative (internucleoside H-phosphonate) and then a cholesteryl group (in the form of cholester-3-yloxycarbonylaminoethylamine) was conjugated by oxidative phosphoramidation [17]. D'Onofrio and coworkers utilized a similar principle to tether a cholesterol derivative. They employed aminoalkyl supported acylate with 3-chloro-4-hydroxyphenylacetic acid as an alternative to a conventional succinyl linker, and standard phosphoramidite chemistry was utilized to couple commercially available 3′-terminal nucleoside phosphoramidite to the hydroxyl functionality [18] (Scheme 4.2.1B). Other methods to prepare 3′-conjugates are the use of solid supports attached to a branched alcohol having orthogonally protected amino and hydroxyl groups through a

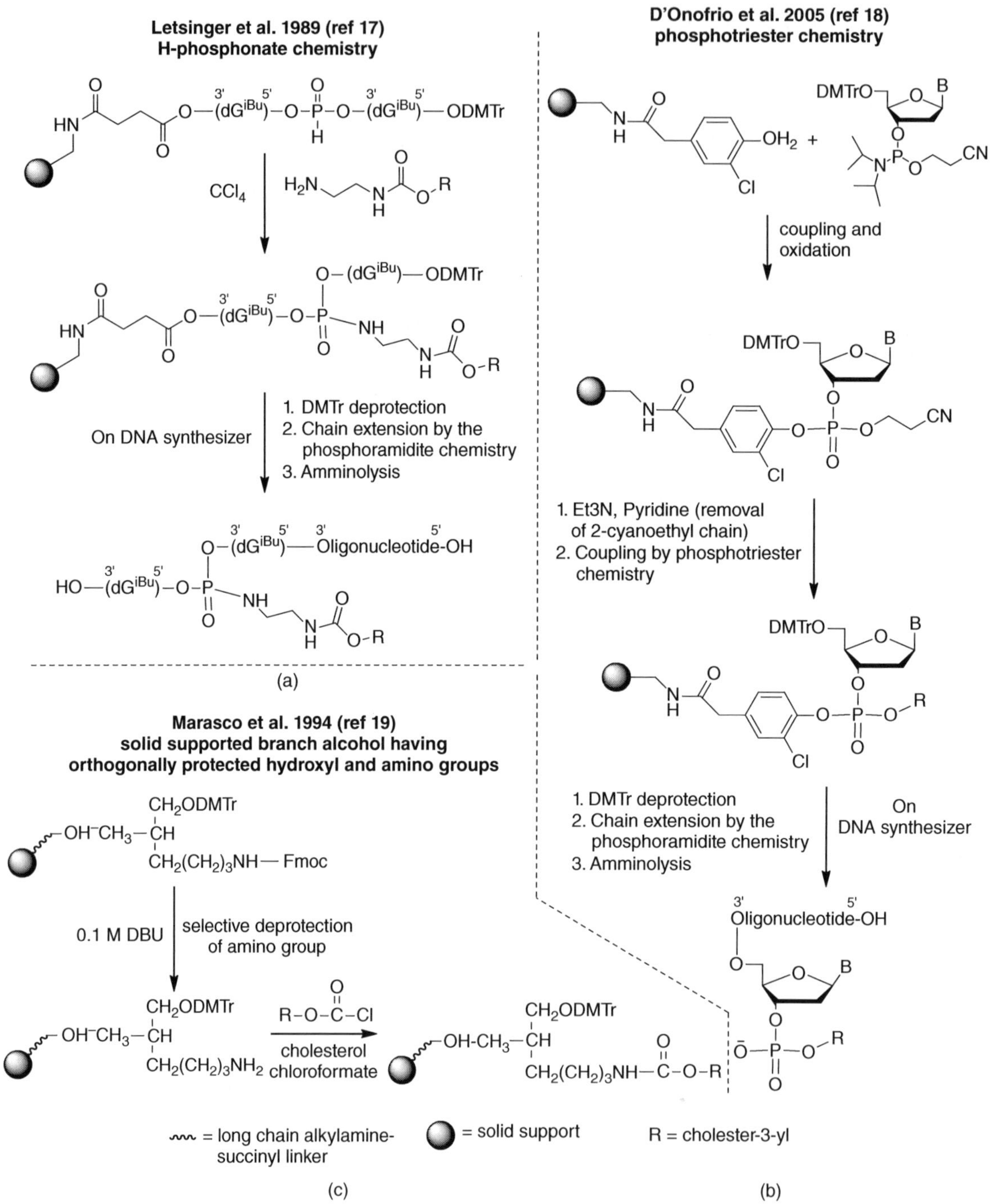

Scheme 4.2.1 *Selective solid-phase approaches for lipid conjugation at the 3′-terminal of an ON sequence*

succinyl linker. The amino group is first selectively deprotected and coupled with the hydroxyl functionality of the conjugate group as a chloroformate ester. Here all the synthesis steps were carried out on the solid support (Scheme 4.2.1C). Subsequently, stepwise solid phase oligonucleotide synthesis was carried out to obtain LONs [19, 20]. There are several other reports where linker–lipid conjugates were first synthesized and then coupled with a solid support [21–32]. Among these, Ueno and coworkers utilized a glycerol derivative (protected with DMTr and TBDMS at the primary hydroxyl group) as a branched alcohol (Scheme 4.2.2A). Aminobutylcarbonyl

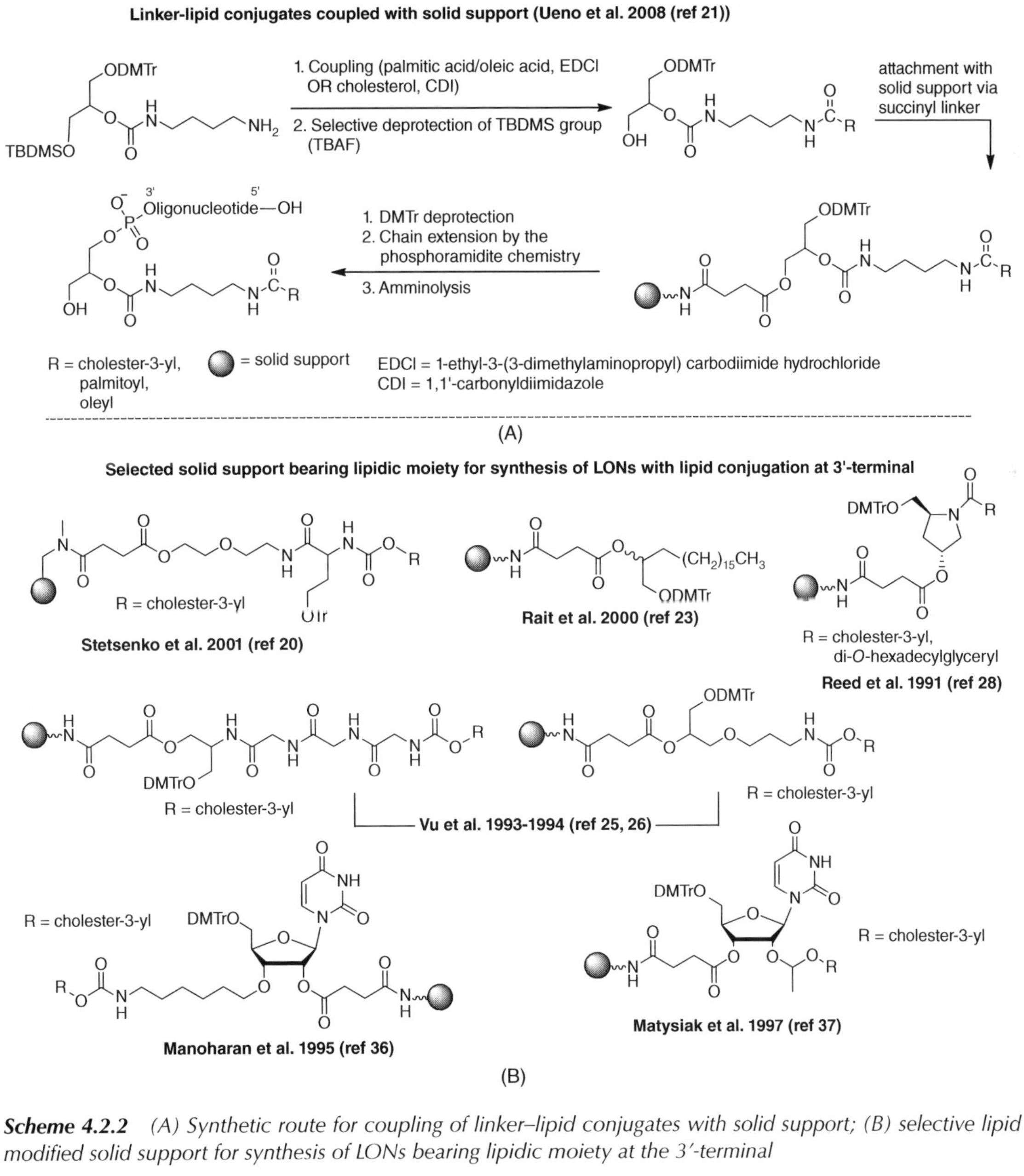

Scheme 4.2.2 *(A) Synthetic route for coupling of linker–lipid conjugates with solid support; (B) selective lipid modified solid support for synthesis of LONs bearing lipidic moiety at the 3′-terminal*

was then obtained by converting the glycerol derivative into its *O*-carbonylimidazolide and reacted with 1,4-diaminobutane. Various lipidic moieties were then coupled using appropriate coupling agents. Finally, after deprotection of the TBDMS group with TBAF, lipid modified glycerol derivatives were linked to a CPG (controlled pore glass) support through a succinyl linker [21]. Grijalvo and coworkers also utilized a glycerol derivative as a branch alcohol linker for lipid conjugation at the 3′-position [33]. Reed *et al.* [28, 34] and Soutschek *et al.* [35] used *trans*-4-hydroxy-L-prolinol as an amino–diol linker to conjugate the lipidic moiety at the 3′-position in the LONs (Scheme 4.2.2B). In a completely different approach, Manoharan and coworkers choose uridine itself as a replacement for a branch alcohol and introduced a lipidic conjugate at the 3′-hydroxy function through an *O*-aminohexyl linker. 5′-*O*-hydroxyl was protected with DMTr and the remaining secondary hydroxyl group (2′-OH) was succinylated to couple with the CPG solid support [36] (Scheme 4.2.2B). In another report by Matysiak and coworkers, 2′-*O*-cholesteryluridine building units were coupled to a polystyrene solid support via a 3′-*O*-succinate-bridge [37] (Scheme 4.2.2B).

4.2.2.1.2 Lipid conjugation at 5′-terminal

As classical ON chemical synthesis proceeds in the 3′–5′ direction, the lipid conjugation at the 3′-terminal requires significant additional efforts (in the context of chemical synthesis) [36, 38] and, therefore, it is more facile to incorporate the lipid conjugate at the 5′-terminal compared with the 3′-terminal. There are various methods reported in the literature to incorporate lipid conjugate at the 5′-terminal of ON sequences. Krieg and coworkers [39] employed H-phosphonate chemistry (described by Letsinger *et al.* [17]) to introduce cholesterol at the 5′-position. Interestingly, Shea *et al.* coupled triethylammonium 1,2-di-*O*-hexadecyl-*rac*-glycero-3-hydrogenphosphonate to the 5′-terminal of ON using H-phosphonate chemistry (Scheme 4.2.3). This was the first report on the utilization of phospholipid (non-nucleosidic) for the preparation of oligonucleotide hybrids [40]. MacKellar *et al.* employed phosphoramidite derivatives to synthesize ONs bearing 5′-lipid conjugates [38] (Scheme 4.2.3).

Recently our group has reported the synthesis of LONs featuring a double lipidic chain (with a nucleosidic phosphoramidite reagent) and a triple lipidic chain (with non-nucleosidic phosphoramidite reagent) at the 5′-position of ONs. These LONs are capable of forming micellar aggregates (self-assembly), which provide a suitable reservoir for the hydrophobic drug, and represent responsive nanomaterials for drug delivery [41] (Scheme 4.2.3). Several other recent reports from our group have shown the utilization of Huisgen 1,3-dipolar cycloaddition of alkynes and azides [42], the so-called "click" reaction [43], in the synthesis of LONs [15, 44, 45]. One of our reports demonstrated the construction of an LON-functionalized quantum dots (LON-QDs) nanoplatform for recognition and detection of miRNAs [45] (Scheme 4.2.3). These reports explained the use of nucleosidic phosphoramidite reagents for incorporation of a lipidic moiety at the 5′-position of an ON sequence. It was also revealed that the reduction of the hydrophilic character of ONs (to facilitate cellular uptake) and binding with lipoproteins and particles of LDL (low-density lipoproteins) are not, however, the only motive for lipid conjugation to ONs, as they also find applications in the drug delivery and bio-imaging field. Similarly there are numerous other H-phosphonate and phosphoramidite derivatives (both nucleosidic and non-nucleosidic) (Scheme 4.2.3) that have been developed by several research groups to conjugate lipidic moieties to the 5′-terminus of the ON chain [2, 26, 27, 29, 34, 37, 38, 40, 46–51]. Moschos *et al.* conjugated a cholesterol phosphoramidite derivative at the 5′-end of sense RNA through a disulfide linker (–(CH_2)–S–S–(CH_2)) [52].

However, Polushin and Cohen used an ester linkage to tether a lipidic group (palmitoyl) at the 5′-position of the ON chain. Since the ester functionality is not stable under the strong basic conditions (ammonia treatment) required for deprotection of the nucleobases and cleavage of ON from the solid support, they employed *tert*-butyl phenoxyacetyl (*t*-BPA) as a base labile amino protecting group and an easily cleavable oxalyl linker along with ethanolamine as a deprotection reagent, while keeping the 5′-ester linkage intact [53]. The rationale was to synthesize LONs with a biodegradable ester linkage (easily hydrolyzed by intracellular esterases) and to provide efficient accessibility of the antisense sequences for binding to their targets [53]. Very recently,

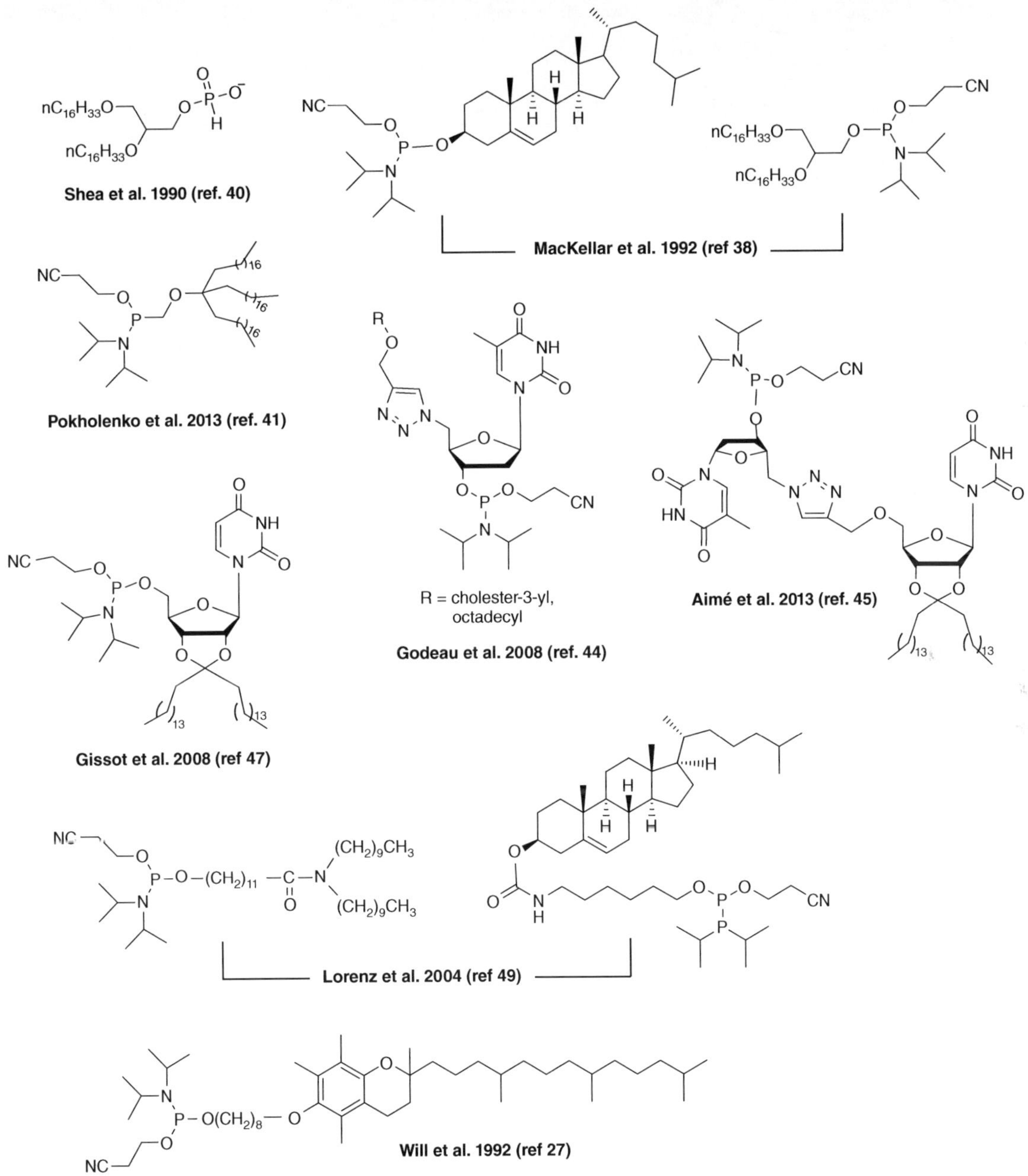

Scheme 4.2.3 *Selective lipid modified H-phosphonate and phosphoramidite derivative for incorporation of lipidic moiety at the 5′-terminal of an ON sequence via solid-phase synthesis*

Chillemi *et al.* used a 1,2-di-*O*-acyl-*sn*-glycero-3-phosphoryl derivative to afford LONs. It is worth mentioning here that this was the first report where a phosphatidyl group (carrying an ester funtionality) had been used as a lipidic moiety (Scheme 4.2.4). The use of 1-hydroxybenzotriazole (HOBt) as an activator for each phosphoramidite coupling step lead to the formation of bulky phosphate-type intermediates, which did not react with the primary amino groups of the nucleobases, and a silyl derivative as a linker (which can be

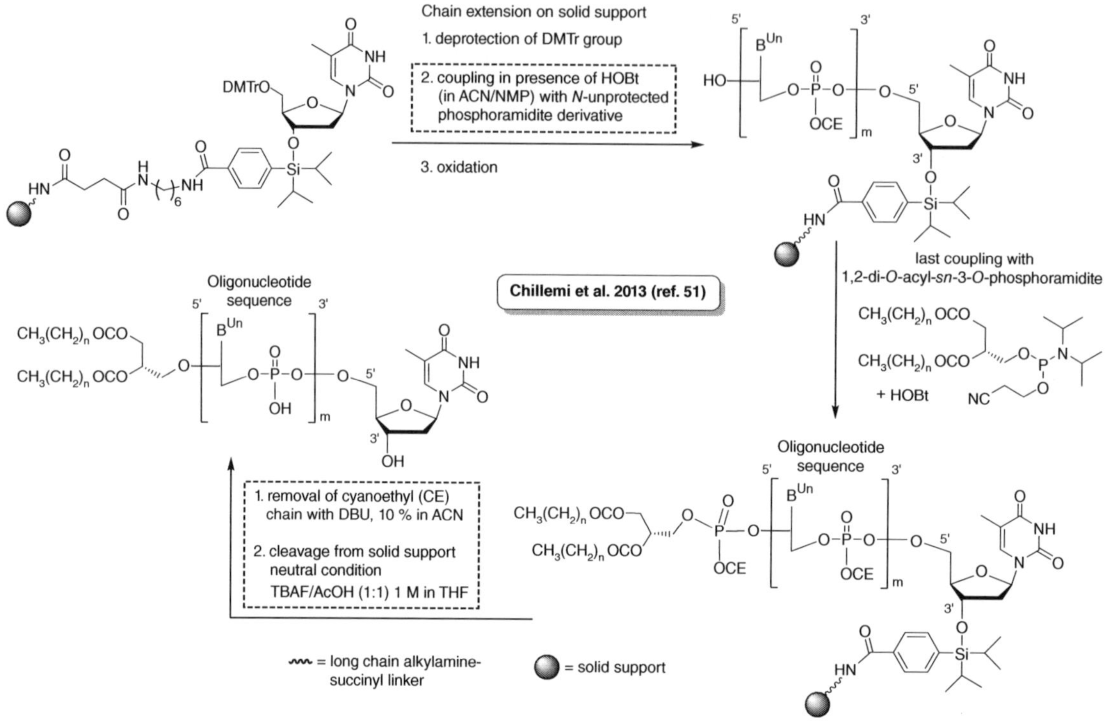

Scheme 4.2.4 Post-synthetic approach (solid-phase synthesis) for incorporation of a phosphatidyl group (1,2-di-O-acyl-sn-glycero-3-phosphoryl) at the 5′-terminal of an ON sequence

cleaved under neutral conditions) allowed the solid-phase synthesis of LONs with a phosphatidyl group, while keeping ester functionality intact [51].

An additional method for the preparation of 5′-conjugates was described by Guzaev and Manoharan [54]. Initially they anchored a phosphoramidite derivative of *N*-chloroacetyl-6-aminohexanol at the 5′-position of an ON sequence through standard solid phase ON synthesis and then coupled it with an amino (dodecylamine) or thiol (1-octadecanethiol, thiocholesterol) derivative of lipidic moieties to afford the desired LONs [54]. Kobylańska *et al.* introduced a lipidic moiety at the 5′-position of ONs through a phosphorothioate linkage. In this instance, they first synthesized 2-alkoxy-2-thiono-1,3,2-oxathiaphospholane (from various lipophilic alcohols, such as cholesterol, borneol, menthol, and heptadecanol) and then reacted this with 5′-hydroxyl groups of support bound ON in the presence of a catalytic amount of DBU [55]. Additionally, Gosse *et al.* prepared amino or iodo fluorescent lipid derivatives. The amino lipid derivative was then coupled with an activated phosphate group of the ON chain at the 5′-position. As an alternative, they introduced a terminal sulfhydryl group at the 5′-phosphate of the ON sequence and then coupled this with an iodo lipid derivative to afford the desired LONs [56].

4.2.2.1.3 Lipid conjugation at internal position (intra-chain)

Lipid conjugation at the internal position of an ON sequence could be achieved by incorporation of prefabricated nucleosidic or non-nucleosidic building blocks carrying a lipidic group. In this context a nucleobase unit and a 2′-position of the ribose sugar are preferable sites for lipid conjugation to prepare prefabricated nucleosidic phosphoramidite building blocks. A lipidic moiety could also be introduced at the internal position of an ON chain by incorporation of non-nucleosidic phosphoramidite building blocks bearing a lipidic moiety. The strategy developed by Manoharan *et al.* [36] and Matysiak *et al.* [37] for the 3′-lipid conjugation, is also applicable for the preparation of uridine based building blocks bearing a lipidic moiety at the 2′-position, which in turn can be incorporated at intra-chain positions. One of the recent reports from our group demonstrated the utilization of a Staudinger–Vilarrasa reaction for the synthesis of 2′-lipid–amido uridine phosphoramidite for the insertion of lipidic moieties within the ON sequences [57] (Scheme 4.2.5). Bryld and Lomholt employed an amino–LNA derivative and a cholesterol moiety was then introduced at a secondary amino through a C6 linker [58] (Scheme 4.2.5).

Kurz *et al.* introduced a lipidic moiety at a nucleobase unit. They first coupled a 5-iodocytidine with the *O*-propargyl derivative of tocopherol via Sonogashira coupling followed by 5′-DMTr protection, and then converted this into a 3′-phosphoramidte derivative to incorporate it into ON sequences by conventional solid-phase synthesis [59] (Scheme 4.2.5). Similar approaches have been used by Gambinossi *et al.* for the preparation of LONs bearing a cholesterol moiety at the intra-chain position [4]. Tomkins *et al.* introduced an amino functionality via a thioether linkage at the C5 position of pyrimidine nucleosides. This primary amine functionality was then reacted with lipid modified *N*-hydroxysucinimide esters to afford a nucleoside with a lipidic moiety. In an another approach, 3-(2-pyridyldithio)propionic acid *N*-hydroxysuccinimide ester was

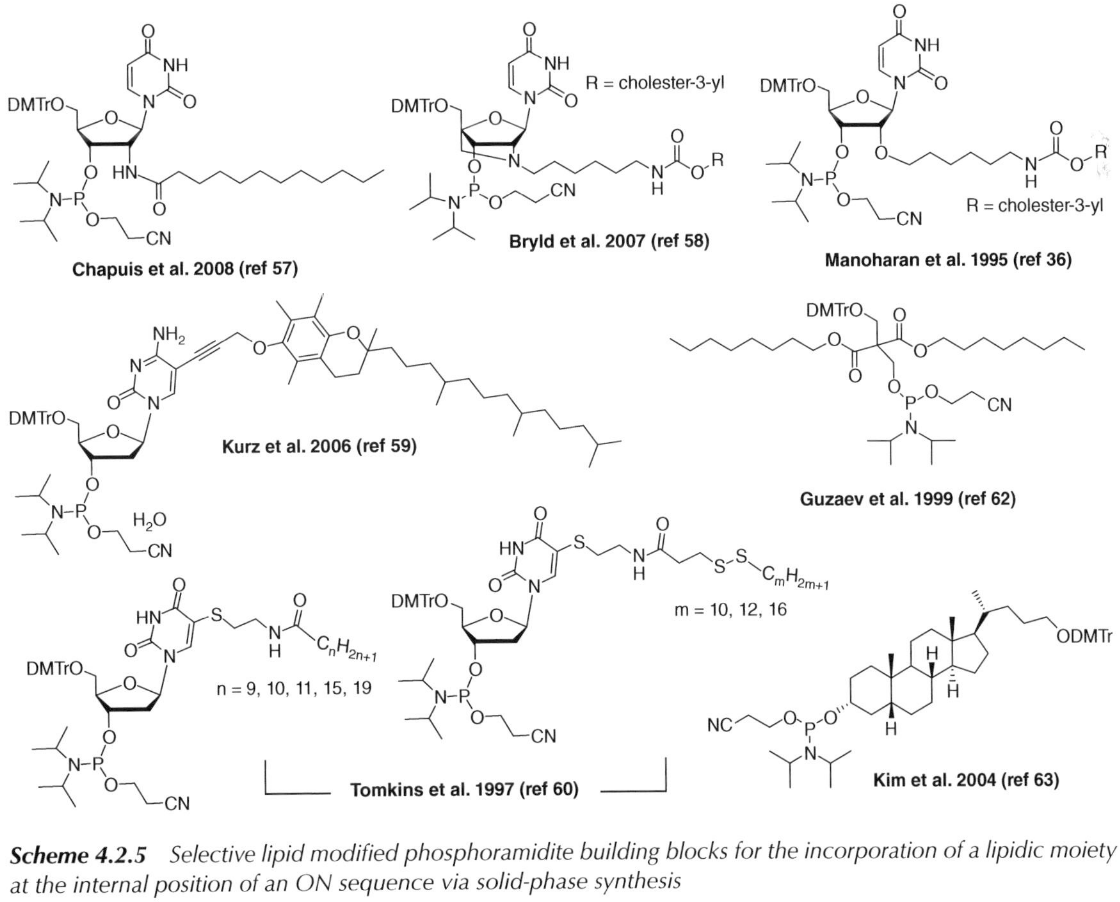

Scheme 4.2.5 *Selective lipid modified phosphoramidite building blocks for the incorporation of a lipidic moiety at the internal position of an ON sequence via solid-phase synthesis*

reacted with a primary amine followed by a thiol derivative of a lipidic moiety to introduce a disulfide linkage between the nucleobase and the lipid group [60] (Scheme 4.2.5). Several other approaches have been reported for preparing lipid conjugated phosphoramidite building blocks of derivatized heterocyclic bases [61].

For example, Guzaev and Lönnberg synthesized novel non-nucleosidic phosphoramidite building blocks from dioctyl malonate. Initially they introduced two hydroxymethyl functionalities (in dioctyl malonate) and then selectively protected one with DMTr, followed by conversion of the second hydroxyl methyl group into phosphoramidite using standard reagents and procedures (Scheme 4.2.5). It is worth mentioning that the lipidic moieties are coupled via biodegradable ester linkages and are stable enough under the standard condition of DNA deprotection, thus allowing efficient conjugation of ONs with lipidic moieties [62] (Scheme 4.2.5). Another non-nucleosidic phosphoramidite building block was synthesized by Kim *et al.* They first reduced the carboxyl group of lithocholic acid to afford cholane-3,24-diol (3α,5β). Selective protection of the primary hydroxyl group followed by reaction with chloro-(2-cyanoethyxy)-*N,N*-diisopropylaminophosphine yielded the desired phosphoramidite building block. This phosphoramidite was synthesized in such a way that it can be incorporated into ON sequences as a lipidic segment, which enabled the formation of a mimic of natural DNA hairpin structures [63] (Scheme 4.2.5).

4.2.2.2 Solution-phase (or post-synthetic) approach

The solution-phase, or post-synthetic, approach is an alternative technique to synthesizing LONs. As for solid-phase synthesis, in the solution-phase (post-synthetic) approach, ON is first synthesized on a solid support with the appropriate functional moieties at the desired positions. This in turn is deprotected and cleaved from the solid support and conjugated in the solution phase with the desired lipidic moieties to afford the LONs. Solution-phase coupling has the advantage of a choice of a variety of molecules and chemistries, which are incompatible with solid phase synthesis. Moreover, both of the components, that is, the ON sequence and lipidic derivative, can be synthesized separately employing the most efficient chemistry. Purification and characterization of both the fragments are possible before conjugation. The main drawback of the solution-phase approach is low coupling efficiency because of poor solubility of the lipidic derivative in the aqueous solution and the exhaustive purification procedure. Despite these limitations, several groups have reported the synthesis of LONs through a solution-phase approach. Vinogradove *et al.* described the preparation of LONs by coupling an ON bearing activated dithiol group with thiol phospholipids in pyridine–chloroform via a sulfur exchange reaction [64] (Scheme 4.2.6). In addition, Acedo *et al.* illustrated the synthesis of LONs having dexamethasone (a glucocorticoid with lipophilic character similar to cholesterol) as a lipidic moiety. They initially synthesized an ON sequence bearing a primary amino group at the 5′-terminus. In parallel, they prepared an *N*-hydroxysuccinimide ester derivative of dexamethasone at the primary hydroxyl group. Finally, they coupled these two fragments in 1:1 DMF–sodium carbonate buffer to yield the LONs [65] (Scheme 4.2.6).

Kabanov *et al.* used a surfactant-based reversed micelles approach in an organic solvent to synthesize the ON sequences having a 5′-phosphate group with a fatty alcohol. The coupling reaction between the fatty alcohol and the ON sequences can take place in the inner cavity (aqueous phase) of reversed micelles. This protocol seems to be advantageous for the modification of natural oligonucleotides bearing a 5′-phosphate group in the solution phase [66]. To facilitate less interference with nucleobase pairing and stacking interactions, Manoharana *et al.* described post synthetic lipid conjugation of ON sequences at the 2′-position. They synthesized 2′-*O*-hexylthiotrityl adenosine phosphoramidite and incorporated it into ON sequences at the desired position using conventional solid phase ON synthesis. The trityl deprotection was then carried out to generate free thiol groups in TEAA buffer followed by coupling with maleimide activated phospholipid derivatives in the phosphate buffer to obtain the desired LONs [67] (Scheme 4.2.6). Recently, Raouane *et al.* reported a microwave assisted coupling protocol for the synthesis of the LONs in solution phase. First they

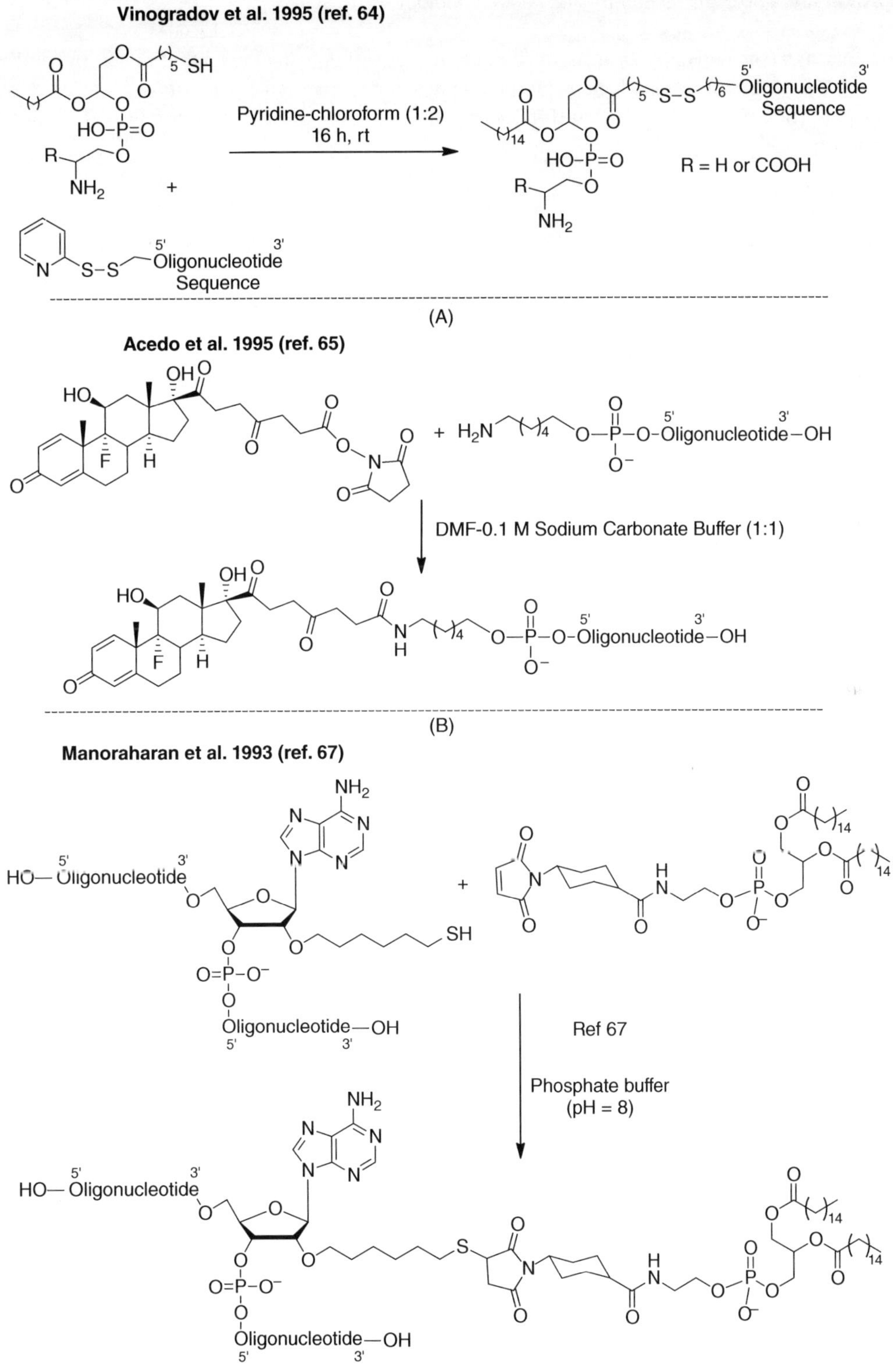

Scheme 4.2.6 *Various strategies for the synthesis of LONs via the post-synthetic route (solution-phase approach)*

synthesized a squalene derivative bearing maleimide moiety through an ethoxyethyl linker. In parallel, they prepared an ON sequence having a 3-mercaptopropyl phosphate group at the 3′-terminal. Finally, the coupling reaction was carried out in a DMF–MeOH–PBS mixture under microwave irradiation to afford the desired LONs [68].

4.2.3 Biomedical applications

4.2.3.1 LONs as efficient delivery vehicles in gene therapy

As non-coding natural RNAs are so effective at regulating gene expression *in vivo*, many investigations are aimed at developing artificial analogs capable of restoring the correct genotype of a defecting gene. Restoring the correct gene expression by means of the addition of artificial oligonucleotides (arON) constitutes the basis of gene therapy. Oligonucleotide therapeutics are no longer just laboratory curiosities and several arONs are in clinical trials or are already on the market. In this context, (i) the selectivity, (ii) the stability in biological fluids, and (iii) the ability to cross the lipidic membranes and reach the targeted oligonucleotide are among the more important features of an effective arON. Considering the amazing number of different nucleic acid sequences present in the genome and transcriptome of each cell, targeting of a unique defective gene or RNA is virtually impossible. Off-target effects were in fact often noticed in the early reports on antisense oligonucleotides. However, one has to bear in mind that regulatory non-coding RNAs such as small RNAs, mi-RNAs, si-RNAs, sn-RNAs, etc., were not known at this time. The lack of selectivity observed in the past may be ascribed (at least in part) to the targeting of non-coding RNAs. Besides, it is now well established that arONs may elicit an immune response in a sequence selective manner via their binding to toll-like receptors (TLR) or others located in the outer leaflet of the membrane [69]. As a result, the arON does not necessarily need to penetrate inside the cell to trigger the transduction of a sequence-aspecific and unwanted biological signal.

Moreover, chemists can nowadays play around with chemical modifications of the nucleic acid at will, to impart designated properties and pharmacodynamics/kinetics to the arON. These points have been addressed in a recent review and will not be covered here [14]. Oligonucleotides are high molecular weight, polar, and polyanionic molecules with a poor ability to cross apolar lipidic bilayers. Many strategies have been developed to overcome this problem. Most of them rely on a vector capable of forming a complex with the arON that is further able to diffuse across the membrane. Natural (viruses) as well as artificial (cationic polymer, lipids, etc.) vectors are used in that regard. Nevertheless, the efficacy of artificial vectors is usually low *in vivo*, they are toxic, and the complexes still tend to remain trapped in endosomal compartments. Lipid nanoparticles (LNP) containing a mixture of lipids with at least one ionizable amino group to complex the arON show great promise as a delivery vehicle for therapeutic siRNA [70]. The covalent attachment of a lipid to an arON to form an LON with inherent, “built-in”, transfection capabilities constitutes another interesting alternative. In fact, cellular membranes are mainly composed of lipids and the lipidic segment of the LON is expected to anchor the arON into the membrane, thus ultimately favoring internalization of the arON. Ideally, adsorption of the LON onto the membrane would trigger the internalization of the LON in a monomeric form, thus avoiding endosomal entrapment issues.

Given the simplicity of this hypothesis for the improvement of arON delivery, it is not surprising that the first LONs were described soon after the emergence of the antisense strategy, 20 years ago. Yet, and in spite of some reported successes [31, 63, 71], it appeared that the insertion of the LON into the membrane clearly did not guarantee efficient internalization of the arON [21]. The lipid does not serve simply as an aspecific glue that will anchor the LON to any lipidic domain at the surface of the cell [72], and the anionic nature of some membrane surfaces may limit ONA adsorption [73]. The type and length of the linker in the LON used to attach the lipid to the oligonucleotide may therefore impact the properties of the LON by decreasing electrostatic repulsions with longer linkers [74], or by influencing the nature of the LON aggregates in aqueous solutions [75]. Furthermore, if the anchoring of the arON into the membrane was the limiting step, better biological response should be correlated with an increased membrane affinity of the LON, but this clearly proved to be wrong [76, 77]. Definitely, the arON must somehow interact with the membrane first (LONs having a clear advantage over non-modified analogs in this regard). However, different mechanisms of internalization have been described with each cellular type having its own pattern of response [16]. Consequently, the derivatization strategy of the oligonucleotide must be finely tuned for an efficient delivery of the arON across plasma membranes.

A survey of the recent literature shows that the efficient delivery of arON *in vitro* and *in vivo* has usually been achieved through receptor-mediated internalization. siRNA conjugates are by far the most widely studied LON for medical applications [78]. For example, the high- (and low-) density lipoproteins (HDL and LDL) efficiently take up cholesterol molecules. Conjugation of cholesterol to the 3′-end of the passenger strand of si-RNA targeting apolipoprotein B (apoB) resulted in the suppression of the apoB mRNA *in vivo* in the liver after a systemic injection of the siRNA in mice. There is a clear correlation between the penetration of the cholesterol-conjugated siRNA and the presence of HDL (not LDL or albumin) receptors in the targeted cell membranes.

Furthermore: (i) the *in vivo* efficacy and the affinity for HDL and LDL of the different lipids conjugated to the siRNA were correlated – aliphatic C_{18} or longer alkyl chains lipid exhibit better affinities toward HDL and superior *in vivo* efficacies compared with shorter chains [79]; and (ii) the preassembly between cholesterol–siRNA and HDL was shown to cause an 8–15 times more effective apoB gene silencing *in vivo* compared with free cholesterol siRNA [79]. Recently, the silencing efficiency was found to be limited by the receptor-mediated entrapment of the LDL–chol–siRNA nanoparticles in endolysosomal compartments. Photochemical irradiation led to the release of the siRNAs from the endolysosomes and greatly improved gene silencing [80]. Interestingly, the preassembly can be constructed with mimetic lipoprotein particles prepared from recombinant apolipoprotein E3 as a delivery vehicle for chol–siRNAs. The delivery of chol–siRNA to the mouse liver is very efficient even in the presence of high endogenous levels of plasma LDL [81]. Altogether, these results suggest that the uptake of cholesterol–siRNA strongly supports the involvement of a lipoprotein receptor mediated endocytosis for cholesterol–LON cell penetration.

The detection of LDL receptors in the brain prompted the development of cholesterol-conjugated siRNA dedicated to penetrate neurons *in vivo*. The first one concerns a neurodegenerative disorder, Huntington disease, which can be addressed via RNAi. A local injection in murine neurons demonstrates the uptake and silencing efficiency of cholesterol–siRNA [82]. A second study describes the inhibition of an oligodendrocyte specific gene in the rat as well as in non-human primate central nervous systems (CNS) after a direct local injection [83]. The third investigation is also related to the silencing of a gene specifically expressed in the CNS by oligodendrocytes [84]. This study also addressed the issue of the covalent attachment of the lipid to the arON. Improved *in vivo* inhibition was observed after a direct intraparenchymal CNS infusion of the chol–siRNA conjugate modified with a cleavable disulfide linker. This shows that part of the LON may remain unproductively trapped into intracellular lipophilic compartments [2].

Interestingly, other serum proteins can be targeted to load and deliver the LONs. Yokota and coworkers prepared an siRNA whose siRNA guide strand was covalently attached to the lipophilic vitamin E (a-tocopherol). The siRNA was designed once again to silence apoB gene expression [85]. Very significantly, apoB expression was silenced after an intravenous injection of 2 mg kg^{-1} in mice whereas the same siRNA conjugated to cholesterol administered with the same dosage was inactive. The differences in efficiency between cholesterol and tocopherol siRNA can be attributed to their different affinities for serum proteins, tocopherol interacting preferentially with tocopherol-associated proteins, such as SEC14L2, SEC14L3, SEC14L4, and afamin. Overall, the targeting of serum proteins with an LON for the delivery of arONs shows great potential *in vivo* provided the targeted tissues express the receptor for the serum protein. However, care must be taken as serum proteins have also been shown to inhibit antisense LON decorated with different lipids *in vitro* [77].

Couvreur and colleagus synthesized a squalene LON siRNA that controls tumor growth *in vivo* (but not *in vitro*) [68]. Squalene is a biocompatible, inert lipid. It should be noted that no specific receptor was identified or targeted in that case.

LONs have also been targeted against microRNAs (miRNAs). miRNAs are an abundant class of small endogenous non-coding RNAs that post-transcriptionally regulate gene expression [86]. The inhibition of overexpressed miRNA with a cholesterol–LON whose sequence is complementary to the miRNA (the so-called antagomirs), has been achieved by the group working with Manoharan [87, 88]. Indded, the antagomir was not only capable of trapping the targeted miRNA but it also induced its degradation by a mechanism independent of the RNAi machinery [88].

4.2.3.2 Other biomedical applications of LONs

One can take advantage of the oligonucleotide sequence of the LON to pair with a targeting agent attached to the complementary DNA [89]. Micelles of DNA block copolymer (polypropylene) decorated with folate residues attached to the cDNA of the LON efficiently targeted *in vitro* human colon adenocarcinoma cells (caco-2), which overexpress folate receptor. The size of the micelle and the ability to trap the anticancer drug within the core of the micelle were clear advantages of this design.

In another report, LON-based micelles loaded with lipophilic anticancer drugs were shown to selectively release part of the drug payload of the micellar aggregates upon binding the complementary oligonucleotide [41]. Provided the latter is locally overexpressed, the drug could be transported and selectively released at the desired location. Moreover, surprisingly the loading capacity of the micellar aggregates were found to be dependent on the oligonucleotide sequence.

4.2.4 Conclusions

Despite the great promise of ON based therapeutic approaches, the development of novel ON drugs has been hampered by the lack of stability and the delivery of ON sequences to the target tissue. Although the resistance to degradation has been partly circumvented by the synthesis of modified ON, a hurdle remains to the delivery of the ON. This issue still remains as the main limitation for this type of molecules.

Based on the considerations discussed here, the chances are that the lipid oligonucleotide field will evolve to give diverse biomedical applications, including, for example, the self-delivery of ON sequences, the design of LONs featuring specific hydrophobic moieties targeting specific receptors, and the use of LON-based aggregates loaded with lipophilic drugs. Today, synthetic approaches to the LON chemistry are now ready to provide a large variety of molecular structures. Most of the automated synthesizers allow for the incorporation of hydrophobic modified phosphoramidites at the 3′/5′-extremities and/or within the ON sequences.

The current intense research and development activities in the field of modified ONs should be noted. Several arrangements and/or collaborations between laboratories working on the development of therapeutic ONs and those proposing new ON delivery options have been achieved. Within this framework, future tailor-made LONs should offer efficient options to address the stability and delivery issues. It is likely that the implementation of new LON based drugs should lead to novel options in the struggle against cancers or other inherited diseases. We will see how this area evolves and whether the fruits of the LON technology can be harnessed. If this should be the case, we could witness important changes to oligonucleotide based drug development.

Acknowledgements

P.B. acknowledges financial support from the Army Research Office. This work has been supported by the French National Agency (ANR) within the framework of its programmes PNANO (project NANAN, no ANR-08-NANO-028), and Emergence (project TCTP-LASOCaP, ANR-11-EMMA-0022).

References

[1] A. Patwa, A. Gissot, I. Bestel, and P. Barthélémy, Hybrid lipid oligonucleotide conjugates: Synthesis, self-assemblies and biomedical applications, *Chem. Soc. Rev.*, **40**, 5844–5854 (2011).

[2] H. Liu Z. Zhu, H. Kang, Y. Wu, K. Sefan, and W. Tan, DNA-Based micelles: Synthesis, micellar properties and size-dependent cell permeability, *Chem. - Eur. J.*, **16**, 3791–3797 (2010).

[3] M. P. Thompson, M.-P. Chien, T.-H. Ku, A. M. Rush, and N. C. Gianneschi, Smart lipids for programmable nanomaterials, *Nano Lett.*, **10**, 2690–2693 (2010).

[4] F. Gambinossi,M. Banchelli, A. Durand, D. Berti, T. Brown, G. Caminati, and P. Baglioni, Modulation of density and orientation of amphiphilic DNA anchored to phospholipid membranes. I. Supported lipid bilayers, *J. Phys. Chem. B*, **114**, 7338–7347 (2010).

[5] B. M. Paterson, B. E. Roberts, and E. L. Kuff, Structural gene identification and mapping by DNA.mRNA hybrid-arrested cell-free translation, *Proc. Natl. Acad. Sci. U.S.A.*, **74**, 4370–4374 (1977).

[6] P. C. Zamecnik and M. L. Stephenson, Inhibition of Rous sarcoma virus replication and cell transformation by a specific oligodeoxynucleotide, *Proc. Natl. Acad. Sci. U.S.A.*, **75**, 280–284 (1978).

[7] M. L. Stephenson and P. C. Zamecnik, Inhibition of Rous sarcoma viral RNA translation by a specific oligodeoxy-ribonucleotide, *Proc. Natl. Acad. Sci. U.S.A.*, **75**, 285–288 (1978).

[8] P. S. Miller, J. C. Barrett, and P. O. P. Ts'o, Alkyl phosphotriesters of dinucleotides and oligonucleotides. 4. Synthesis of oligodeoxyribonucleotide ethyl phosphotriesters and their specific complex formation with transfer ribonucleic acid, *Biochemistry (Mosc.)*, **13**, 4887–4896 (1974).

[9] J. C. Barrett, P. S. Miller, and P. O. P. Ts'o, Alkyl phosphotriesters of dinucleotides and oligonucleotides. 5. Inhibitory effect of complex formation with oligodeoxyribonucleotide ethyl phosphotriesters on transfer ribonucleic acid aminoacylation, *Biochemistry (Mosc.)*, **13**, 4897–4906 (1974).

[10] K. Jayaraman, K. McParland, P. Miller, and P. O. Ts'O, Selective inhibition of *Escherichia coli* protein synthesis and growth by nonionic oligonucleotides complementary to the 3′ end of 16S rRNA, *Proc. Natl. Acad. Sci. U.S.A.*, **78**, 1537–1541 (1981).

[11] P. S. Miller, K. B. McParland, K. Jayaraman, and P. O. P. Ts'o, Biochemical and biological effects of nonionic nucleic acid methylphosphonates, *Biochemistry (Mosc.)*, **20**, 1874–1880 (1981).

[12] H. Lönnberg, Solid-phase synthesis of oligonucleotide conjugates useful for delivery and targeting of potential nucleic acid therapeutics, *Bioconjug. Chem.*, **20**, 1065–1094 (2009).

[13] Y. Singh, P. Murat, and E. Defrancq, Recent developments in oligonucleotide conjugation, *Chem. Soc. Rev.*, **39**, 2054–2070 (2010).

[14] R. L. Juliano, X. Ming, and O. Nakagawa, The chemistry and biology of oligonucleotide conjugates, *Acc. Chem. Res.*, **45**, 1067–1076 (2012).

[15] A. Gissot, M. Camplo, M. W. Grinstaff, and P. Barthélémy, Nucleoside, nucleotide and oligonucleotide based amphiphiles: A successful marriage of nucleic acids with lipids, *Org. Biomol. Chem.*, **6**, 1324–1333 (2008).

[16] M. Raouane, D. Desmaële, G. Urbinati, L. Massaad-Massade, and P. Couvreur, Lipid conjugated oligonucleotides: A useful strategy for delivery, *Bioconjug. Chem.*, **23**, 1091–1104 (2012).

[17] R. L. Letsinger, G. Zhang, D. K. Sun, T. Ikeuchi, and P. S. Sarin, Cholesteryl-conjugated oligonucleotides: Synthesis, properties, and activity as inhibitors of replication of human immunodeficiency virus in cell culture, *Proc. Natl. Acad. Sci. U.S.A.*, **86**, 6553–6556 (1989).

[18] J. D'Onofrio, D. Montesarchio, L. De Napoli, and G. Di Fabio, An efficient and versatile solid-phase synthesis of 5′- and 3′-conjugated oligonucleotides, *Org. Lett.*, **7**, 4927–4930 (2005).

[19] C. J. Marasco Jr., N. J. Angelino, B. Paul, and B. J. Dolnick, A simplified synthesis of acridine and/or lipid containing oligodeoxynucleotides, *Tetrahedron Lett.*, **35**, 3029–3032 (1994).

[20] D. A. Stetsenko and M. J. Gait, A convenient solid-phase method for synthesis of 3′-conjugates of oligonucleotides, *Bioconjug. Chem.*, **12**, 576–586 (2001).

[21] Y. Ueno, K. Kawada, T. Naito, A. Shibata, K. Yoshikawa, H.-S. Kim, Y. Wataya, and Y. Kitade, Synthesis and silencing properties of siRNAs possessing lipophilic groups at their 3′-termini, *Bioorg. Med. Chem.*, **16**, 7698–7704 (2008).

[22] M. Manoharan, Oligonucleotide conjugates as potential antisense drugs with improved uptake, biodistribution, targeted delivery, and mechanism of action, *Antisense Nucleic Acid Drug Dev.*, **12**, 103–128 (2002).

[23] A. Rait, K. Pirollo, D. W. Will, A. Peyman, V. Rait, E. Uhlmann, and E. H. Chang, 3′-End conjugates of minimally phosphorothioate-protected oligonucleotides with 1-O-hexadecylglycerol: Synthesis and anti-ras activity in radiation-resistant cells, *Bioconjug. Chem.*, **11**, 153–160 (2000).

[24] I. Habus, Q. Zhao, and S. Agrawal, Synthesis, hybridization properties, nuclease stability, and cellular uptake of the oligonucleotide-amino-β-cyclodextrins and adamantane conjugates, *Bioconjug. Chem.*, **6**, 327–331 (1995).

[25] H. Vu, T. S. Hill, and K. Jayaraman, Synthesis and properties of cholesteryl-modified triple-helix forming oligonucleotides containing a triglycyl linker, *Bioconjug. Chem.*, **5**, 666–668 (1994).

[26] H. Vu, P. Singh, L. Lewis, J. G. Zendegui, and K. Jayaraman, Synthesis of cholesteryl supports and phosphoramidite for automated DNA synthesis of triple-helix forming oligonucleotides (TFOs), *Nucleosides Nucleotides*, **12**, 853–864 (1993).

[27] D. W. Will and T. Brown, Attachment of vitamin E derivatives to oligonucleotides during solid-phase synthesis, *Tetrahedron Lett.*, **33**, 2729–2732 (1992).

[28] M. W. Reed, A. D. Adams, J. S. Nelson, and R. B. Meyer Jr., Acridine- and cholesterol-derivatized solid supports for improved synthesis of 3′-modified oligonucleotides, *Bioconjug. Chem.*, **2**, 217–225 (1991).

[29] S. M. Gryaznov and D. H. Lloyd, Modulation of oligonucleotide duplex and triplex stability via hydrophobic interactions, *Nucleic Acids Res.*, **21**, 5909–5915 (1993).

[30] W. H. Gmeiner, W. Luo, R. T. Pon, and J. W. Lown, Development of an efficient oligonucleotide derivatization protocol, *Bioorg. Med. Chem. Lett.*, **1**, 487–490 (1991).

[31] A. S. Boutorin, L. V. Guskova, E. M. Ivanova, N. D. Kobetz, V. F. Zarytova, A. S. Ryte, L. V. Yurchenko, and V. V. Vlassov, Syntheses of alkylating oligonucleotide derivatives containing cholesterol or phenazinium residues at their 3′-terminus and their interaction with DNA within mammalian cells, *FEBS Lett.*, **254**, 129–132 (1989).

[32] B. Oberhauser and E. Wagner, Effective incorporation of 2′-O-methyl-oligoribonucleotides into liposomes and enhanced cell association through modification with thiocholesterol, *Nucleic Acids Res.*, **20**, 533–538 (1992).

[33] S. Grijalvo, S. M. Ocampo, J. C. Perales, and R. Eritja, Synthesis of oligonucleotides carrying amino lipid groups at the 3′-end for RNA interference studies, *J. Org. Chem.*, **75**, 6806–6813 (2010).

[34] M. W. Reed, E. A. Lukhtanov, V. V. Gorn, D. D. Lucas, J. H. Zhou, S. B. Pai, Y.-C. Cheng, and R. B. Meyer Jr., Structure - activity relationships of cytotoxic cholesterol-modified DNA duplexes, *J. Med. Chem.*, **38**, 4587–4596 (1995).

[35] J. Soutschek, A. Akinc, B. Bramlage, K. Charisse, R. Constien, M. Donoghue, S. Elbashir, A. Gelck, P. Hadwiger, J. Harborth, M. John, V. Kesavan, G. Lavine, R. K. Pandey, T. Racie, K. S. Rajeev, I. Röhl, I. Toudjarska, G. Wang, S. Wuschko, D. Bumcrot, V. Kotellansky, S. Limmer, M. Manoharan, and H.-P. Vornlocher, Therapeutic silencing of an endogenous gene by systemic administration of modified siRNAs, *Nature*, **432**, 173–178 (2004).

[36] M. Manoharan, K. L. Tivel, and P. D. Cook, Lipidic nucleic acids, *Tetrahedron Lett.*, **36**, 3651–3654 (1995).

[37] S. Matysiak, R. Frank, and W. Pfleiderer, Acetal oligonucleotide conjugates in antisense strategy, *Nucleosides Nucleotides*, **16**, 855–861 (1997).

[38] C. MacKellar, D. Graham, D. W. Will, S. Burgess, and T. Brown, Synthesis and physical properties of anti-HIV antisense oligonucleotides bearing terminal lipophilic groups, *Nucleic Acids Res.*, **20**, 3411–3417 (1992).

[39] A. M. Krieg, J. Tonkinson, S. Matson, Q. Zhao, M. Saxon, L.-M. Zhang, U. Bhanja, L. Yakubov, and C. A. Stein, Modification of antisense phosphodiester oligodeoxynucleotides by a 5′ cholesteryl moiety increases cellular association and improves efficacy, *Proc. Natl. Acad. Sci. U S.A.*, **90**, 1048–1052 (1993).

[40] R. G. Shea, J. C. Marsters, and N. Bischofberger, Synthesis, hybridization properties and antiviral activity of lipid-oligodeoxynucleotide conjugates, *Nucleic Acids Res.*, **18**, 3777–3783 (1990).

[41] O. Pokholenko, A. Gissot, B. Vialet, K. Bathany, A. Thiéry, and P. Barthélémy, Lipid oligonucleotide conjugates as responsive nanomaterials for drug delivery, *J. Mater. Chem. B*, **1**, 5329–5334 (2013).

[42] R. Huisgen, in *1,3-Dipolar Cycloaddition Chemistry*, A. Padwa (Ed.), John Wiley & Sons, Inc, Hobekon, 1984.

[43] V. V. Rostovtsev, L. G. Green, V. V. Fokin, and K. B. Sharpless, A stepwise huisgen cycloaddition process: Copper(I)-catalyzed regioselective "ligation" of azides and terminal alkynes, *Angew. Chem., Int. Ed.*, **41**, 2596–2599 (2002).

[44] G. Godeau, C. Staedel, and P. Barthélémy, Lipid-conjugated oligonucleotides via "click chemistry" efficiently inhibit hepatitis C virus translation, *J. Med. Chem.*, **51**, 4374–4376 (2008).

[45] A. Aimé, N. Beztsinna, A. Patwa, A. Pokolenko, I. Bestel, and P. Barthélémy, Quantum dot lipid oligonucleotide bioconjugates: Toward a new anti-microRNA nanoplatform, *Bioconjug. Chem.*, **24**, 1345–1355 (2013).

[46] T. J. Lehmann and J. W. Engels, Synthesis and properties of bile acid phosphoramidites 5′-tethered to antisense oligodeoxynucleotides against HCV, *Bioorg. Med. Chem.*, **9**, 1827–1835 (2001).

[47] A. Gissot, C. Di Primo, I. Bestel, G. Giannone, H. Chapuis, and P. Barthélémy, Sensitive liposomes encoded with oligonucleotide amphiphiles: A biocompatible switch, *Chem. Commun.*, 5550–5552 (2008).

[48] Y.-H. M. Chan, B. van Lengerich, and S. G. Boxer, Lipid-anchored DNA mediates vesicle fusion as observed by lipid and content mixing, *Biointerphases*, **3**, FA17–FA21 (2008).

[49] C. Lorenz, P. Hadwiger, M. John, H.-P. Vornlocher, and C. Unverzagt, Steroid and lipid conjugates of siRNAs to enhance cellular uptake and gene silencing in liver cells, *Bioorg. Med. Chem. Lett.*, **14**, 4975–4977 (2004).

[50] M. Banchelli, F. Betti, D. Berti, G. Caminati, F. B. Bombelli, T. Brown, L. Marcus Wilhelmsson, B. Norden, and P. Baglioni, Phospholipid membranes decorated by cholesterol-based oligonucleotides as soft hybrid nanostructures, *J. Phys. Chem. B*, **112**, 10942–10952 (2008).

[51] R. Chillemi, V. Greco, V. G. Nicoletti, and S. Sciuto, Oligonucleotides conjugated to natural lipids: Synthesis of phosphatidyl-anchored antisense oligonucleotides, *Bioconjug. Chem.*, **24**, 648–657 (2013).

[52] S. A. Moschos, S. W. Jones, M. M. Perry, A. E. Williams, J. S. Erjefalt, J. J. Turner, P. J. Barnes, B. S. Sproat, M. J. Gait, and M. A. Lindsay, Lung delivery studies using siRNA conjugated to TAT(48-60) and penetratin reveal peptide induced reduction in gene expression and induction of innate immunity, *Bioconjug. Chem.*, **18**, 1450–1459 (2007).

[53] N. N. Polushin and J. S. Cohen, Antisense pro-drugs: 5′-ester oligodeoxynucleotides, *Nucleic Acids Res.*, **22**, 5492–5496 (1994).

[54] A. Guzaev and M. Manoharan, Conjugation of oligonucleotides via an electrophilic tether: N- chloroacetamidohexyl phosphoramidite reagent, *Bioorg. Med. Chem. Lett.*, **8**, 3671–3676 (1998).

[55] A. Kobylańska, A. Okruszek, and W. J. Stec, Application of oxathiaphospholane method for the synthesis of oligodeoxyribonucleotide 5′-O-conjugates, *Nucleosides Nucleotides*, **17**, 1977–1982 (1998).

[56] C. Gosse, A. Boutorine, I. Aujard, M. Chami, A. Kononov, E. Cogné-Laage, J.-F. Allemand, and J. Li, L. Jullien, Micelles of lipid-oligonucleotide conjugates: Implications for membrane anchoring and base pairing, *J. Phys. Chem. B*, **108**, 6485–6497 (2004).

[57] H. Chapuis, L. Bui, I. Bestel, and P. Barthélémy, 2′-Lipid-modified oligonucleotides via a "Staudinger-Vilarrasa" reaction, *Tetrahedron Lett.*, **49**, 6838–6840 (2008).

[58] T. Bryld and C. Lomholt, Attachment of cholesterol to amino-LNA: Synthesis and hybridization properties, *Nucleosides Nucleotides Nucleic Acids*, **26**, 1645–1647 (2007).

[59] A. Kurz, A. Bunge, A.-K. Windeck, M. Rost, W. Flasche, A. Arbuzova, D. Strohbach, S. Müller, J. Liebscher, D. Huster, and A. Herrmann, Lipid-anchored oligonucleotides for stable double-helix formation in distinct membrane domains, *Angew. Chem., Int. Ed.*, **45**, 4440–4444 (2006).

[60] J. M. Tomkins, K. J. Barnes, A. J. Blacker, W. J. Watkins, and C. Abell, Lipophilic modification of oligonucleotides, *Tetrahedron Lett.*, **38**, 691–694 (1997).

[61] A. Durand and T. Brown, Synthesis and properties of oligonucleotides containing a cholesterol thymidine monomer, *Nucleosides Nucleotides Nucleic Acids*, **26**, 785–794 (2007).

[62] A. Guzaev and H. Lönnberg, Solid support synthesis of ester linked hydrophobic conjugates of oligonucleotides, *Tetrahedron*, **55**, 9101–9116 (1999).

[63] S. J. Kim, E.-K. Bang, H. J. Kwon, J. S. Shim, and B. H. Kim, Modified oligonucleotides containing lithocholic acid in their backbones: Their enhanced cellular uptake and their mimicking of hairpin structures, *ChemBioChem*, **5**, 1517–1522 (2004).

[64] S. V. Vinogradov, T. Le Doan, and C. Hélène, Synthesis of phospholipid-oligodeoxyribonucleotide conjugates, *Tetrahedron Lett.*, **36**, 2493–2496 (1995).

[65] M. Acedo, G. Tarrason, J. Piulats, M. Mann, M. Wilm, and R. Eritja, Preparation of oligonucleotide-dexamethasone conjugates, *Bioorg. Med. Chem. Lett.*, **5**, 1577–1580 (1995).

[66] A. V. Kabanov, S. V. Vinogradov, A. V. Ovcharenko, A. V. Krivonos, N. S. Melik-Nubarov, V. I. Kiselev, and E. S. Severin, A new class of antivirals: antisense oligonucleotides combined with a hydrophobic substituent effectively inhibit influenza virus reproduction and synthesis of virus-specific proteins in MDCK cells, *FEBS Lett.*, **259**, 327–330 (1990).

[67] M. Manoharan, L. K. Johnson, K. L. Tivel, R. H. Springer, and P. D. Cook, Introduction of a lipophilic thioether tether in the minor groove of nucleic acids for antisense applications, *Bioorg. Med. Chem. Lett.*, **3**, 2765–2770 (1993).

[68] M. Raouane, D. Desmaele, M. Gilbert-Sirieix, C. Gueutin, F. Zouhiri, C. Bourgaux, E. Lepeltier, R. Gref, R. Ben Salah, G. Clayman, L. Massaad-Massade, P. Couvreur, Synthesis, characterization, and in vivo delivery of siRNA-squalene nanoparticles targeting fusion oncogene in papillary thyroid carcinoma, *J. Med. Chem.*, **54**, 4067–4076 (2011).

[69] M. Avci-Adali, H. Steinle, T. Michel, C. Schlensak, and H. P. Wendel, Potential capacity of aptamers to trigger immune activation in human blood, *PLoS ONE*, **8** doi: 10.1371/journal.pone.0068810 (2013).

[70] R. L. Rungta, H. B. Choi, P. J. Lin, R. W. Ko, D. Ashby, J. Nair, M. Manoharan, P. R. Cullis, B. A. Macvicar, Lipid Nanoparticle delivery of siRNA to silence neuronal gene expression in the brain, *Mol. Ther. Nucleic Acids*, **2**, DOI: 10.1038/mtna.2013.65 (2013).

[71] A. S. Boutorine and E. V. Kostina, Reversible covalent attachment of cholesterol to oligodeoxyribonucleotides for studies of the mechanisms of their penetration into eucaryotic cells, *Biochimie*, **75**, 35–41 (1993).

[72] M. Loew, R. Springer, S. Scolari, F. Altenbrunn, O. Seitz, J. Liebscher, D. Huster, A. Herrmann, and A. Arbuzova, Lipid Domain Specific Recruitment of Lipophilic Nucleic Acids: A Key for Switchable Functionalization of Membranes, *J. Am. Chem. Soc.* **132**, 16066–16072 (2010).

[73] M. J. Palte and R. T. Raines, Interaction of nucleic acids with the glycocalyx, *J. Am. Chem. Soc.*, **134**, 6218–6223 (2012).

[74] N. S. Petrova, I. V. Chernikov, M. I. Meschaninova, Ii. S. Dovydenko, A. G. Venyaminova, M. A. Zenkova, V. V. Vlassov, and E. L. Chernolovskaya, Carrier-free cellular uptake and the gene-silencing activity of the lipophilic siRNAs is strongly affected by the length of the linker between siRNA and lipophilic group, *Nucleic Acids Res.*, **40**, 2330–2344 (2012).

[75] T. R. Pearce, B. Waybrant, and E. Kokkoli, The role of spacers on the self-assembly of DNA aptamer-amphiphiles into micelles and nanotapes, *Chem. Commun.*, **50**, 210–212 (2014).

[76] M. Manoharan, K. L. Tivel, T. P. Condon, L. K. Andrade, I. Barber-Peoch, G. Inamati, S. Shah, V. Mohan, M. J. Graham, C. F. Bennett, S. T. Crooke, and P. D. Cook, Conjugated antisense oligonucleotides, *Nucleosides Nucleotides*, **16**, 1129–1138 (1997).

[77] A. E. Felber, N. Bayó-Puxan, G. F. Deleavey, B. Castagner, M. J. Damha, and J.-C. Leroux, The interactions of amphiphilic antisense oligonucleotides with serum proteins and their effects on in vitro silencing activity, *Biomaterials*, **33**, 5955–5965 (2012).

[78] J. H. Jeong, H. Mok, Y.-K. Oh, and T. G. Park, siRNA Conjugate delivery systems, *Bioconjug. Chem.*, **20**, 5–14 (2009).

[79] C. Wolfrum S. Shi, K. N. Jayaprakash, M. Jayaraman, G. Wang, R. K. Pandey, K. G. Rajeev, T. Nakayama, K. Charrise, E. M. Ndungo, T. Zimmermann, V. Koteliansky, M. Manoharan, and M. Stoffel, Mechanisms and optimization of in vivo delivery of lipophilic siRNAs, *Nat. Biotechnol.*, **25**, 1149–1157 (2007).

[80] H. Jin *et al.*, Mechanistic insights into LDL nanoparticle-mediated siRNA delivery, *Bioconjug. Chem.*, **23**, 33–41 (2012).

[81] T. Nakayama, J. S. Butler, A. Sehgal, M. Severgnini, T. Racie, J. Sharman, F. Ding, S. S. Morskaya, J. Brodsky, L. Tchangov, V. Kosovrasti, M. Meys, L. Nechev, G. Wang, C. G. Peng, Y. Fang, M. Maier, K. G. Rajeev, R. Li, J. Hettinger, S. Barros, V. Clausen, X. Zhang, Q. Wang, R. Hutabarat, N. V. Dokholyan, C. Wolfrum, M. Manoharan, V. Kotelianski, M. Stoffel, and D. W. Sah, Harnessing a physiologic mechanism for siRNA delivery with mimetic lipoprotein particles, *Mol. Ther. J. Am. Soc. Gene Ther.*, **20**, 1582–1589 (2012).

[82] M. DiFiglia, M. Sena-Esteves, K. Chase, E. Sapp, E. Pfister, M. Sass, J. Yoder, P. Reeves, R. K. Pandey, K. G. Rajeev, M. Manoharan, D. W. Y. Sah, P. D. Zamore, N. Aronin, Therapeutic silencing of mutant huntingtin with siRNA attenuates striatal and cortical neuropathology and behavioral deficits, *Proc. Natl. Acad. Sci. U.S.A.*, **104**, 17204–17209 (2007).

[83] W. Querbes, P. Ge, W. Zhang, Y. Fan, J. Costigan, K. Charisse, M. Maier, L. Nechev, M. Manoharan, V. Kotelianski, and D. W. Y. Sah, Direct CNS delivery of siRNA mediates robust silencing in oligodendrocytes, *Oligonucleotides*, **19**, 23–29 (2009).

[84] Q. Chen, D. Butler, W. Querbes, R. K. Pandey, P. Ge, M. A. Maier, L. Zhang, K. G. Rajeev, L. Nechev, V. Kotelianski, M. Manoharan, and D. W. Y. Sah, Lipophilic siRNAs mediate efficient gene silencing in oligodendrocytes with direct CNS delivery, *J. Control. Release*, **144**, 227–232 (2010).

[85] K. Nishina, T. Unno, Y. Uno, T. Kubodera, T. Kanouchi, H. Mizusawa, and T. Yokota, Efficient *in vivo* delivery of siRNA to the liver by conjugation of α-tocopherol, *Mol. Ther.*, **16**, 734–740 (2008).

[86] J. Stenvang, A. Petri, M. Lindow, S. Obad, and S. Kauppinen, Inhibition of microRNA function by antimiR oligonucleotides, *Silence*, **3**, 1–17 (2012).

[87] J. Krützfeldt, N. Rajewsky, R. Braich, K.G. Rajeev, T. Tuschl, M. Manoharan, and M. Stoffel, Silencing of microRNAs in vivo with "antagomirs," *Nature*, **438**, 685–689 (2005).

[88] J. Krützfeldt, S. Kuwajima, R. Braich, K.G. Rajeev, J. Pena, T. Tuschl, M. Manoharan, and M. Stoffel, Specificity, duplex degradation and subcellular localization of antagomirs, *Nucleic Acids Res.*, **35**, 2885–2892 (2007).

[89] F. E. Alemdaroglu, N. C. Alemdaroglu, P. Langguth, and A. Herrmann, DNA block copolymer micelles – A combinatorial tool for cancer nanotechnology, *Adv. Mater.*, **20**, 899–902 (2008).

4.3
Amphiphilic Peptidyl–RNA

Peter Strazewski

Institut de Chimie et Biochimie Moléculaires et Supramoléculaires (UMR 5246), Université de Lyon, Lyon, France

4.3.1 Introduction

This is not a review on synthetic peptide–RNA conjugates: many excellent reviews on synthetic DNA, RNA or PNA covalently linked to synthetic peptides are available in the literature [1]. Instead, let me give you a personal view, in a nutshell, of our own work on synthetic amphiphilic peptidyl–RNA and address certain aspects of the chemistry, physico-chemical properties and the potential applications of such macromolecules. As far as the chemistry is concerned, it will be the, to date, unpublished problems, strategic drawbacks, comparison with similar published approaches and suggestions of how to resolve the problems that will be discussed. With respect to the physico-chemical properties, the importance of the amphiphilic character of our objects will be emphasised. The potential applications are mostly speculative and are entirely open to further discussion.

Hopefully, this chapter will inspire the interested reader to go ahead and establish his or her own projects where amphiphilic peptidyl–RNA is given the main rôle. The author set out to experiment on amphiphilic peptidyl–RNA in one particular and conceptually adventurous way, which will be introduced briefly in the following section. Fortunately, the chemical space of synthetic peptidyl–RNA is extremely large, the build-up of the conjugates highly modular and the number of possible variants so incredibly vast that a large number of researchers should find a fundamental or an applicative interest in this class of synthetic conjugate compounds. The author is just one pawn, one that is somewhat hard-bitten, who is spearheading the search in a fastidious, and for some perhaps swashbuckling, direction: towards the synthetic living cell. When the journey is long and the dynamics are ruled by a nonlinear conduct, the number of spontaneous bifurcations is always high. This book chapter is devoted to those bifurcations, past and future, which the author will no longer step into, not more than he already has done. But, then again, never say never.

DNA in Supramolecular Chemistry and Nanotechnology, First Edition.
Edited by Eugen Stulz and Guido H. Clever.

4.3.2 Three souls alas! are dwelling in my breast [2]

Typically, the 'collective soul' of an archetypal organic chemist is split into three (Figure 1 in reference [3]). The deepest 'vegetative' part over which we have the least control – meaning that most of us cannot help loving its achievements – feeds on the challenges of synthesis. It is best satisfied by elegant pathways leading to complicated molecular objects, cleverly designed reagents, or brilliant catalysts that give access to new covalent structures or much sought-after high-energy molecules, preferably through environmentally sustainable routes. It is Marcellin Berthelot's glorious heritage, and this is how we usually grow:

> Chemistry creates its own object. [4]

An even higher level of complexity calls for the 'designer' in us. It is our desire to construct higher-order molecular structures, termed supramolecules. They are based on our control over weak interactions through knowledgeably predisposed strong (covalent) bonds. Pre-organisation is the catchword. The design and exploitation of supramolecules, whether static, as in staple DNA, for example, or dynamic, as in dynamic combinatorial libraries, are based on thermodynamic control over their formation and Le Châtelier's Principle, in other words, on thermodynamically convergent systems [5]. 'Designers' are fishing in deep waters, so to speak, where everything must assemble with minimal free energies.

The third part of our divided soul is the one that enjoys observations more than synthesis. It veers away from Berthelot's joy of free creation:

> This creative quality, resembling that of art itself, distinguishes it essentially from natural and historical sciences. The latter have an object given in advance and independent of the will and action of the scientist. [4]

Mind you, it is a chemist's soul that we are thinking about, not a biologist's, geologist's or other natural or historical scientist's soul – no. Yes, we chemists do create our own object, rather than 'have an object given in advance', but the main motivation for its creation (synthesis) may be driven by our curiosity at observing how this object 'behaves'. Molecular objects (molecules) do not 'behave' very unpredictably, unless we are not sufficiently well educated or our model happens do be flawed. Supramolecular objects bear the potential for unexpected 'behaviour' but, as it usually turns out, the convergent nature of their creation will set natural limits on this unexpectedness. Therefore, one should really replace 'behaviour' with property.

Chemical systems, or more precisely, molecular systems that are kept far away from thermodynamic equilibrium (off-equilibrium systems), and biological systems (cells, organelles, organs, organisms, similar populations, biospheres), present the most promising potential for unexpected behaviour. This is because such systems are inherently divergent [5]. With energy input, they spontaneously produce variants and varieties of themselves, until they are outcompeted by others. Self-evolvability is the catchword. The principle that rules their behaviour is much less understood than Le Châtelier's Principle. What has emerged from the past few decades of research are autocatalytic and cross-catalytic replication reactions, and whole reaction networks, in which the permanent or recurrent formation of chemically related and compartmented compounds mean they persist despite their relative thermodynamic instability.

'Observers' prefer the top of a mountain to deep waters, as it were, where everything is at high free energy, with quite a short existence, and an appearance that is changing permanently. Only from further away from such a 'systemic object' will its guise be revealed. Imagine yourself watching the tip of Geneva's water fountain through a telescope [5], then zooming out and then being able to recognise it for what it is. This is the observer's joy. Imagine now, that you are the engineer who built Geneva's fountain and are testing whether it works. This is the observer's task – heavily dependent on the designer's output. So envisage that the fountain is spontaneously producing baby fountains whilst it runs: this is the observer's ultimate goal.

4.3.3 Why RNA? Why peptides?

To link RNA with peptides was an early choice for two out of three motivations, but the third incentive will occupy us most in this exposé. First of all, it is challenging to synthesise, step-by-step, an amphiphilic RNA–peptide conjugate from its monomers. RNA monomers are extremely densely functionalised and need a high degree of sophistication of orthogonal protection for their oligomerisation. Amino acids are extremely variant rich and, depending on the actual choice, require still more orthogonal protection for the synthesis of the target amphiphile. The purification and isolation of macromolecular amphiphiles remains a problem, which has relatively idiosyncratic solutions.

The amphiphilicity of peptidyl–RNA relies on the presence of fairly densely localised polar hydroxy groups and negative charges on one side of the macromolecule, and hydrophobic residues in a long stretch on the remaining part. There are many ways to create a macromolecular amphiphile that will provide these characteristics. The 'vegetative' needs of a good organic chemist may well be more ambitious than simply to reproduce in the laboratory what Nature has long accomplished and explored. Where is the adventurous 'designer'? Indeed, but there is another *raison d'être*, the one called forth by the 'observer'.

RNA primary sequences are obviously well replicated. In living systems fragments of RNA molecules are reproduced (and degraded) constantly in every cell during their transcription from DNA. In chemical systems, *ergo* in the strict absence of any added natural enzymes (biotically evolved catalysts), a mixture of different RNA molecules – phosphodiester polymers essentially – can very readily be prepared under aqueous conditions. Depending on the pH and the salts present, such polymers can persist for considerable amounts of time. For this rudimentary technique of preparing combinatorial RNA libraries, where neither precise length nor specific primary sequence are under the chemist's control, we have to compromise on isomeric purity. We have to accept the presence of 2′- and 3′-phosphodiester isomers, and we can only shrug our shoulders with the formation of nitrogen-branched RNA oligomers, and so forth.

At first sight, lipid–RNA conjugates could be an alternative for amphiphilic peptidyl–RNA [1c, d, 6]. However, it is much easier to prepare peptide libraries under aqueous conditions than libraries of congeneric fatty acids, lipids, terpenes or steroids. The last class of hydrophobic molecules is built from carbon–carbon bonds – not exactly an obvious undertaking in water. Peptides, being made up from a mixture of different amino acids, should be an ideal combinatorial pool for many slightly (and also significantly) different molecular and supramolecular properties. Amide bonds are chemically persistent and can be formed in water under only moderately reactive conditions, provided that a large excess of (permanently hydrolysing) condensing reagent is furnished. The 'simple' peptides that the author is primarily interested in are those that present no chemical function in their side chains, or only a poor variety of few polar functions, to make sure that the peptide will be reasonably hydrophobic.

In summary, the main motivation for preparing macromolecular amphiphilic peptidyl–RNA is to set the stage for – and then observe! – an off-equilibrium chemical system supplied with combinatorial libraries of synthetic amphiphilic peptidyl–RNA esters. Of particular interest will be the fate and perhaps evolution of amphiphilic 2′-/3′-peptidyl–RNA esters. Such conjugates are natural 'active esters', that is, peptide–RNA conjugates that are intermediates for the ribosomal protein biosynthesis. In cells they appear only as 3′-peptidyl transfer RNA in the P-site of ribosomes and act as peptide donors for the nucleophilic attack of 3′-aminoacyl transfer RNAs that have found their way into the neighbouring ribosomal A-site. In systemic feeding experiments, where 'simple' amino acids, hydrophobic peptides, ribonucleosides, RNA and condensing agents are brought together, RNA-assisted peptide elongations and peptide- or protein-catalysed RNA elongations are expected to emerge and evolve concomitantly.

All such experiments are initially addressing a chemical system's self-evolvability. The author feels, however, that the reader of this book is more likely to be interested in the controlled synthesis of a designed macromolecular amphiphile, and less so in the observer's imaginings and piloting preparations. If this is not the case, please take the time to read another book chapter or related article written by the same author [7]; if, however, this is the case, please carry on.

4.3.4 Hydrolysis-resistant amphiphilic 3′-peptidyl–RNA

For the reasons mentioned previously (and in reference [7]), structural mimics of natural peptidyl transfer RNA fragments were envisaged as targets for a stepwise solid support synthesis. The connection between the *C*-terminus of the peptide and the RNA had to be realised at its 2′,3′-end leaving the peptide's *N*-terminus 'free' (cationic at neutral pH) and the RNA's 5′-end. In addition, the target compounds needed to carry an amphiphile's 'polar head' that was comparable in size (spatial length) to its single 'lipophilic tail'. The head should be sufficiently broad (wide) and the tail tip have a relatively thin average (hydrodynamic) diameter. A head-to-tail tip width ratio of 2.5–3.0:1 would afford the amphiphile a pronounced overall conical shape, and thus render it detergent-like and prone to self-assembly in micelles.

Theoretically, such macromolecular detergents should be endowed with some influence over transgressing the impermeability of lipid bilayers or similar membranes. Micelles, on the other hand, would be a useful aggregate form for 'feeding' experiments, where some other, perhaps lamellar aggregate phase composed of peptidyl-RNA, could be rapidly enriched with more material when provided with fresh peptidyl–RNA micelles.

The head was an RNA hairpin 22-mer with a well known solution structure [8] and folding thermodynamics [9]. The RNA stem–loop hairpin disposed of 21 negative charges over a double-helical object of roughly 5.0 nm length and 2.0–2.2 nm width (Figure 4.3.1). The tail had to be any nominally hydrophobic 'simple' peptide consisting of about 20 amino acids, and thus be long enough to cross a lipid bilayer approximately 4.5 nm thick, whatever the stretched (non-looped) conformation – α-helix, 3_{10}-helix, π-helix, β-strand, disordered 'random coil'.

In the peptidyl–RNA models depicted in Figure 4.3.1, the tilt between the helical axis of the A-RNA and the peptide backbone axis was set to 37° in the α-helical and 52° in the β-stranded conjugate. This spatial arrangement between the RNA and the peptide is by no means the only one, since the single-stranded part of the RNA is fairly flexible and the chosen Cα–CO torsional angle ψ in the *C*-terminal, RNA-bound alanine moiety is only one of many possible angles: here $\psi_{\text{C-terminal}} = 165°$ in the α-helical and 132° in the β-stranded conjugate. Therefore, all types of RNA–peptide hinge angles are imaginable in the unaggregated molecule, including perpendicular or coaxial geometries. Note that the negative end of the macro dipole of the modelled α-helix points, owing to the unidirectional arrangement of the intrahelical CO···HN hydrogen bonds, towards the RNA hairpin, and thus, is expected to reinforce the amphiphilicity of the all-helical conjugate isomer.

Peptides can be designed through a specific primary sequence to become lipid-like [10]. We were refraining from too much design, however, because our future peptide libraries were to be produced in a reversible and off-equilibrium fashion. So, for the first model test series, our choice was narrowed down to oligoalanine peptides. However, pure oligoalanines, or even those oligoalanines that are flanked by more efficiently helix-stabilising amino acids, are not as lipophilic as perhaps would be expected, they have a strong preference to partition into water when faced with phospholipids [11]. Hence, certain alanines in our peptides were replaced with leucines so as to enhance the peptide's lipophilicity, and in some cases with (negatively charged) glutamates, to reduce the chances that the peptide would somehow associate with 'its own' RNA through electrostatic interactions. The presence of any cationic amino acid side chain was avoided for the same reason.

Both the peptide and the RNA had to be joined to one another through a reasonably stable covalent bond, in order to be able to observe the conjugates for long enough time periods, during which the connections must remain intact. Natural 2′- or 3′-*O*-(α-amino)acyl or peptidyl esters are not resistant against isomerisation or

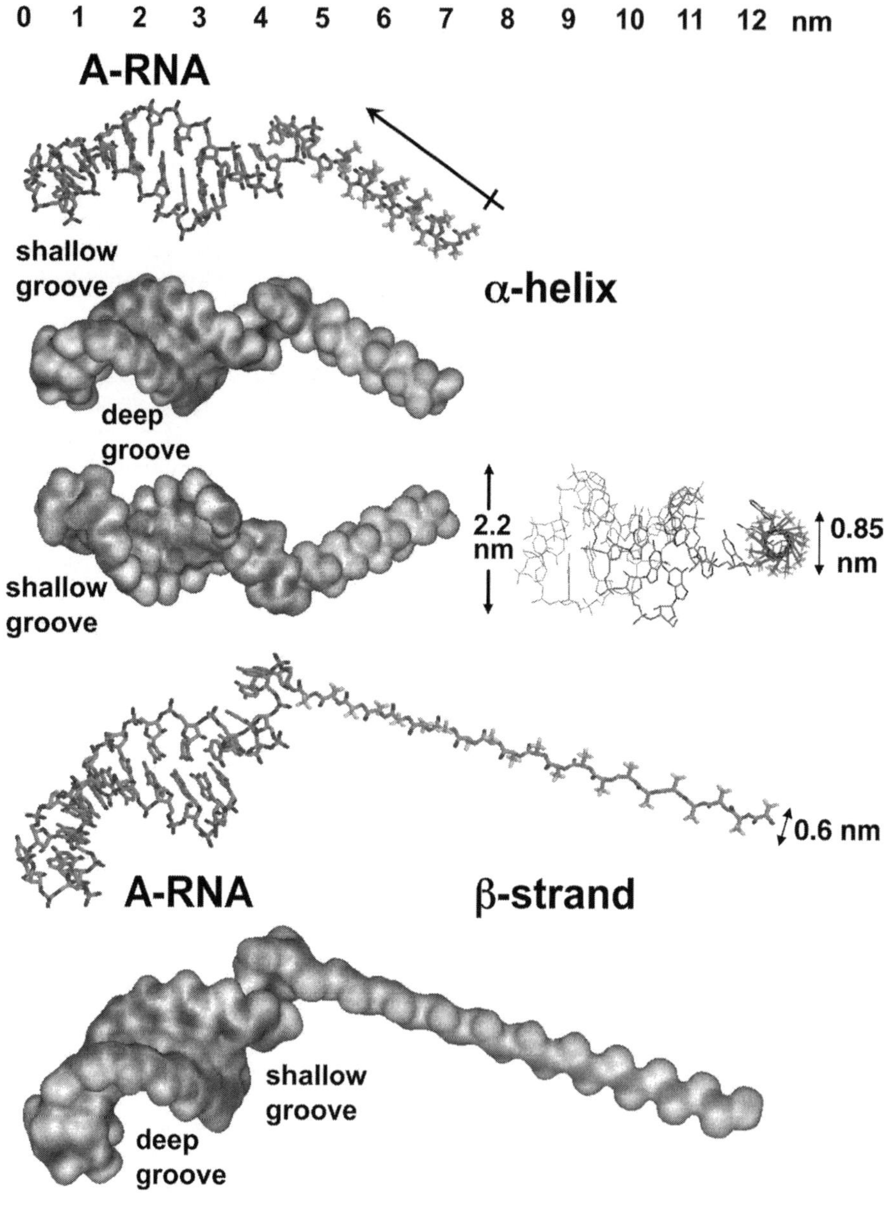

Figure 4.3.1 Shapes and sizes of a 3′-peptidyl–RNA model. RNA: solution structure of the synthesised 22-mer hairpin (model 4 in 1IKD from the Brookhaven database) [8]. Peptide: Ala_{21} modelled by PyMOL as an α-helix and β-strand. Top line = length scale in nanometers; crossed arrow (tilted) = direction of the α-helix macro dipole (See colour figure in colour plate section)

hydrolysis, as such esters show half-lives of minutes to hours at neutral pH. The connection between the peptide and the RNA had to be isosteric to natural 2′-/3′-peptidyl transfer RNA. Therefore, we decided to prepare and test *C*-terminal 3′-amide-linked peptidyl–RNA. This necessitated the synthesis of 3′-amino-3′-deoxyadenosine (= N^6-*bis*-demethyl puromycin aminonucleoside) as the 3′-terminal nucleoside. To cut a long story short, much of our effort was devoted to developing useful synthetic routes to puromycin analogues [12]. We elaborated an efficient way of immobilising a fully orthogonally protected puromycin analogue [13], and prepared a number of hydrolysis-resistant peptidyl–RNAs [14]. Two studies on the supramolecular properties of these amphiphilic conjugates have since followed [15].

4.3.5 Synthetic strategy

One can imagine several different ways of synthesising 3′-peptidyl–RNA. The first decision was whether it would be prepared on separate solid supports, one for the peptide and the other for the RNA, to be followed by de-immobilisation, partial deprotection of both and then joining the fragments in solution – thus, through the fragment (segment) or post-synthetic condensation approach [1c] – or else through fully stepwise synthesis on one and the same solid support. The choice quickly fell on the latter, for a seemingly obvious reason. We envisaged the synthesis of highly hydrophobic peptides that were expected to be barely soluble. We were counting on a significant effect of the RNA on the solubility of the peptide.

Another peril would be the growing hydrophobic main peptide chain. At the usually high functional group loadings of commercial solid supports (0.4–3.0 mmol g^{-1} resin), a ‘difficult’ peptide – in the sense of difficult to synthesise and isolate pure because it is highly hydrophobic – it can fold upon itself and also aggregate while still bound to the solid support, which could jeopardise an efficient oligomerisation after only five or six peptide couplings [16].

The 3′-terminal nucleoside offers a welcome 2′-hydroxyl group as an attachment point for the solid support. This 2′-*O*-ester function would be cleavable under relatively harsh aminolytic conditions, which should allow for the use of the highly base-labile and very practical Fmoc–amino acid protection. A classical C→N build-up of the peptide would precede an equally classical 3′→5′ build-up of the RNA strand, as summarised in Figure 4.3.2. The methodology of oligomerising Fmoc–amino acids, using the allyl ester side chain protection for glutamate, to avoid a final aminolysis into glutamine residues (see step 6a), and 2′-*O*-silylated RNA monomers on 2′-*O*-support-linked puromycin analogue **1** complements several other techniques that were being explored at the same time or developed later [17].

Our strategy and methodology were quite successful and were soon adopted by the Micura group [18]. In addition, Micura and colleagues developed an alternative, using the same overall strategy and monomer coupling methods but starting from a 2′-deoxypuromycin analogue being attached to the solid support through its adenine base. While we were focussing on the synthesis of unmodified middle-length RNA being linked

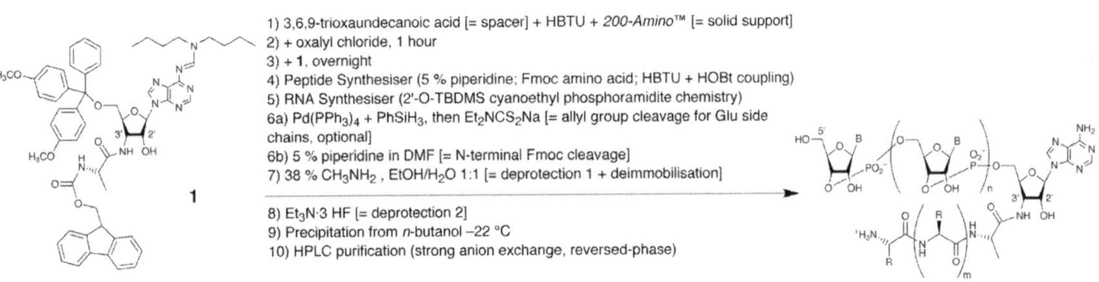

Figure 4.3.2 *Solid support synthesis of 3′-peptidyl-RNA. HBTU = 2-(1H-benzotriazol-1-yl)-1,1,3,3-tetramethyluronium hexafluorophosphate, Fmoc = fluorenylmethyloxycarbonyl, HOBt = 1-hydroxybenzotriazole, TBDMS =* tert-*butyldimethylsilyl, DMF = dimethylformamide, R = amino acid side chain, B = ribonucleobase,* m *= 19–20,* n *= 20*

to comparatively long hydrophobic peptides, Micura and colleagues concentrated on naturally modified long RNA connected to short mixed-sequence peptides. Their success was based in part on the impressive DNAzyme-catalysed ligation of synthetic 3′-tetrapeptidyl–RNA conjugates to fully modified but 3′-truncated natural transfer RNAs, to give fully modified full-length 3′-tetrapeptidyl transfer RNAs [18b].

We have synthesised the following peptides on pyromycin derivative **1** (sequence from N to C terminus): Ala$_8$, GluAla$_{10}$ (EA$_{10}$), Ala$_{16}$, Ala$_{20}$, Ala$_{21}$, pyroGluGlu$_2$Ala$_{18}$ (pEE$_2$A$_{18}$), GluAla$_4$GluAla$_7$GluAla$_7$ (EA$_4$EA$_7$EA$_7$ 21-mer), GluLeu$_{18}$Ala (EL$_{18}$A), Glu$_3$Leu$_{18}$Ala (E$_3$L$_{18}$A) [19a], and LeuAla$_3$LeuAla$_3$LeuAla$_3$LeuAla$_3$LeuAla$_3$ (LA$_3$ 20-mer), LeuAla$_3$LeuAla$_2$LeuAla$_3$LeuAla$_2$LeuAla$_3$LeuAla (LA$_3$LA$_2$ 20-mer), LeuAla$_2$Leu$_2$Ala$_2$LeuAla$_2$ Leu$_2$Ala$_2$LeuAla$_2$Leu$_2$Ala (LA$_2$L$_2$A$_2$ 20-mer), LeuGluAlaLeuAlaGluLeu$_2$GluAlaLeuAlaGluLeu$_2$GluAlaLeu AlaGluLeuAla (LEALAEL 22-mer), AlaGluLeu$_2$GluAlaLeuAlaGluLeu$_2$GluAlaLeuAlaGluLeu$_2$GluAla LeuAla (AELLEAL 22-mer) [19b].

The RNA sequences synthesised so far were 5′-GG*G*GCUC[UUXG]GAGC*U*CCACCA-3′-NH-Y, where Y = Ala, Glu or Pro, and X = C or T(FAM), the latter being a commercial fluorescein-linked thymidine 2′-deoxyribonucleotide. The [square-bracketed tetranucleotide] folds into a stable loop; the underligned tetranucleotide is single-stranded like the 'CCA terminus' of natural transfer RNA; the *nucleotides G3 and U16 shown in italics* form a G·U wobble base pair, the remaining nucleotides are all Watson–Crick paired.

With respect to the analytics, high-resolution mass spectrometry (HRMS) is the only way of precisely identifying the macromolecular conjugate target compounds. Our MALDI-MS attempts failed too often (MALDI = matrix-assisted laser desorption ionisation). Sometimes the unusually strong affinity of some of our peptides – not RNA ! – with the sodium ions produced, and in spite of strong deionising pre-treatments of the final target peptidyl–RNAs, MALDI-MS signals with too low an intensity and mass resolution. Once we replicated Micura and colleague's HPLC-coupled electrospray ionisation (HPLC-ESI) conditions, we were able to reliably identify any examined amphiphilic peptidyl–RNA conjugate. The basic HPLC eluant containing triethylamine and hexafluoroisopropanol at pH 8.0 apparently provided conditions that were strong enough to get rid of all sodium ions during reversed-phase HPLC and delivered clean ESI-MS signals. Interestingly, the abundance distribution of the negatively charged states accumulated best between –6 and –13 with a maximal abundance for ten negative charges, which showed that about one half of the 21 phosphodiesters in the RNA hairpin of the conjugate were readily neutralised by protons under the applied ESI conditions.

4.3.6 Pros'n cons

On the whole our strategy was satisfactory, which does not mean that it could not be done better. Several practical problems keep us steadfastly looking for alternatives. The first one is easy to resolve. We should simply drop our own immobilisation method and adopt the one published by the Micura group. Both our methods profit from the high loading of functionalisable amino groups on the water-repellent 'Custom Primer Support' *200-Amino*™ available from GE Healthcare (not yet been used by us in 2003) [20]. The nominal 200 µmol amino groups per gram of resin can be charged with a PEG-like spacer (3,6,9-trioxaundecanoic diacid), then treated with oxalyl chloride followed by puromycin derivative **1** (Figure 4.3.2) [13b], to give 150 µmol DMT$^+$ (*p,p*-dimethoxytrityl) release per gram of resin, measured colorimetrically [15b]. This compares with a loading of 31–46 µmol DMT$^+$ per gram of *200-Amino* resin when a similar puromycin derivative was linked through a 2′-*O*-adipic acid pentafluorophenyl ester, and to 115–130 µmol DMT$^+$ per gram of *200-Amino* resin when a corresponding 2′-deoxypuromycin derivative was linked through its base using a similar carboxylic pentafluorophenyl ester [18a]. In our case, the added excess of valuable **1** could be readily recovered and a loading of 150 µmol DMT$^+$ g^{-1} was first seen as advantageous, but eventually it turned out to have been something of a Pandora's box.

All our attempts to optimise the Fmoc–amino acid oligomerisations on *200-Amino* charged with **1** ended in a final (after 19 or so peptide couplings) approximate 40 μmol DMT^+ per gram of resin, which is a useful loading for the subsequent RNA oligomerisation. We assumed that this was owing to the partial aminolysis of the PEG spacer-2′-nucleoside ester bond, thus, the 'bleeding out' from the resin of the somewhat base-labile 3′-(peptidyl)amino-3′-deoxyadenosine (pept-3′-NHdA) fragments in the course of the repeated Fmoc deprotections – even though we used as low as 5% piperidine in dimethylformamide. Such bleeding out would not be detected by HPLC of the crude target pept-3′-NHdA, and finding the fragments in the piperidine washings appeared too fastidious, if not impossible. The purities of the crude full-length pept-3′-NHdA were impeccable, so the strong drop in DMT^+ release appeared to us to be not much more than an exotic conundrum.

However, the final HPLC analyses of the target peptidyl–RNAs taught us better. Unexpectedly large amounts of unpeptidylated, only 3′-aminoacylated RNA, needed to be separated out first, in order to obtain the pure full-length peptidyl–RNA conjugate (unpublished work). The chemical yield of the target compounds with respect to the initial loading of the solid support was not so important, since in those studies we worked with fluorescent material that was easy to detect. However, we caught the javelin before full impact: it is better to charge the support through a mild pentafluorophenyl ester activation method and be satisfied with an initial 40 μmol DMT^+ g^{-1} loading, than ignore an unsolved riddle and take the final HPLC pain.

The *200-Amino* polymer is a functionalised polystyrene support of chemically proprietary identity (unknown, not published). The Kaiser (ninhydrin) test on the untreated commercial polymer resulted in yellow solutions (unpublished) consistent with the presence of secondary, not primary, amino groups or, for that matter, no amino groups altogether. Apparently, our super reactive activation method produced a loaded solid support where approximately two-thirds of bound **1** would not be suitable for peptide synthesis (or Fmoc deprotection), only for RNA synthesis.

A problem that was grimmer to resolve was the RNA coupling yields in the presence of the full-length peptide. In our hands, on our machine and without prior peptide synthesis, we were satisfied with 96.0–98.9% average stepwise RNA coupling yields on *200-Amino* loaded with **1**. However, after the synthesis of a full-length Fmoc–peptide 20-mer, the subsequent average stepwise RNA coupling yields consistently dropped to 92–95%, despite all attempts at optimisations of the synthesis cycle and reagents. The mere presence of a covalently bound peptide icosamer kept the RNA coupling yields lower for no real obvious reasons. As long as the stepwise yield did not drop below 92% we could live with it: $0.92^{19} = 20.5\%$ theoretical overall RNA yield.

The third inconvenience of this chemistry was the use of 2′-*O*-TBDMS RNA monomers (*tert*-butyldimethylsilyl), or more precisely, the necessity to treat, after de-immobilisation of the target compound with methylamine, the fully silylated peptidyl–RNA conjugate with triethylamine *tris*-hydrofluoride (step 8 in Figure 4.3.2). Usually, the desilylation reaction worked very well, as did the subsequent precipitation from *n*-butanol. However, we could only isolate sufficient amounts of pure target conjugate if the HPLC purification of the fully deprotected product was performed on the same day, within a few hours of the desilylation. The residual fluoride ions in the crude precipitate were revealed to be harmful over prolonged time periods.

This severely strained the nerves and any spontaneous inspiration of the person(s) who was (were) hands-on. In our case this meant: begin the automated peptide synthesis on day *j* and run it over night, install the peptidylated solid support on the RNA synthesiser (usually on two columns) the next morning, carry out the RNA synthesis (two in parallel) before noon, condition the ion-exchange or reversed-phase semi-prep HPLC column (numerous blank runs) while synthesising, deimmobilising, vacuum-centrifuging and desilylating the crude target conjugate, precipitate it over lunch at –22 °C, centrifuge and dissolve it in buffer A, purify it using a slowly eluting buffer B gradient, then concentrate the desired fractions as fast and as effectively as possible, to be able to start lyophilising the pure product, hopefully before midnight of day *j + 1*. Of course, day *j + 2* is committed to a second HPLC purification, or other such task.

4.3.7 Alternative methods and strategies

The compassionate reader might have some other chemistry or a different protective group methodology for the RNA oligomerisation on the tip of his or her tongue. We have tried a potentially attractive new protocol that uses 2′-*O*-(1,1-dioxo-1λ^6-thiomorpholine-4-carbothioate) RNA monomers [21]. TC-RNA synthesis does not require any separate 2′-*O*-deprotection step, but it relies on a firm non-covalent association of the deprotected (oligoanionic) target RNA with the solid support during a crucial washing step. Deprotected amphiphilic peptidyl–RNA does not associate firmly enough, the lipophilic nature of the peptide spoils the chances of the TC method of being successful.

One other card – as far as we know, that is – has still not been played. Why not prepare the peptide (as pept-3′-NHdA) and the rest of the RNA on two separate supports after all? The RNA could be synthesised with superior stepwise coupling yields according to an inverse 5′→3′ build-up protocol [22]. The peptide synthesis works well on a solid support and gives chemically pure products. Full-length pept-3′-NHdA would be deimmobilised without being fully deprotected, transformed into a 5′-*O*-phosphoramidite and coupled (in a molar excess) as a 3′-terminal peptidyl–'RNA monomer' to 3′-OH deprotected RNA, which is still bound to the solid support through its 5′-terminus. The future will tell whether this joker card leads anywhere useful.

4.3.8 Molecular properties

Temperature-dependent ultraviolet (UV) and circular dichroism (CD) spectroscopy of the amphiphilic peptidyl–RNA, and separate analyses of the RNA, Ala-3′-NH-RNA and pept-3′-NHdA, provided evidence for the anticipated folding properties of the peptides, as well as the RNA hairpin. Briefly, the UV-detected denaturation temperature of the RNA hairpin was not influenced by the presence of any examined peptide to any great extent (87.6–88.8 °C, Figure S9 in reference [14]). If any significant differences could be pointed out, these would be a slight stabilisation rather than destabilisation of the RNA hairpin folding by any of the peptides, and much more tilted baselines in the melting curves of peptidyl–RNA indicating enhanced aggregation (thus molecularity) at low molar concentrations.

With some surprise came the observation that, after subtracting the CD spectrum of the RNA hairpin from the CD spectra of the peptidyl–RNA conjugates, the residual ellipticity $\Delta\Theta$ due to RNA alone (> 245 nm) withstood heat denaturation between 0 and 60 °C better than that of any examined peptide (< 240 nm) [14], suggesting that the single-stranded 3′-terminal ACCA tetranucleotide was more rigid in its conformation when bound to a peptide. Such a conformational stabilisation could be due to the self-aggregation of amphiphilic peptidyl–RNA even at low molar concentrations, as confirmed by dynamic light scattering (see the next section). Most peptides showed the presence of an α-helical conformation, with a notable exception: the preferred random coil for the LA_3LA_2 20-mer [15b).

4.3.9 Supramolecular properties

Dynamic light scattering (DLS), atomic force microscopy (AFM) and confocal laser scanning fluorescence microscopy (CFM) helped examine the supramolecular properties of amphiphilic peptidyl–RNAs, on their own and in association with lipidic giant vesicles (GVs). DLS revealed a high polydispersity of a mixture of Ala_{20-22}-3′-NH-RNA that became less disperse when pure Ala_{21}-3′-NH-RNA was present in solution [14]. The hydrodynamic diameters of the Ala_{20-22}-3′-NH-RNA aggregates ranged between 60 nm and 2 μm, and peaked at 400–500 nm.

The salt-free solution of a peptidyl–RNA was deposited onto an optical glass slide and dried at 40 °C during the 24 h prior to AFM scanning in the non-contact mode [15a]. The AFM images of Ala_{21}-3′-NH-RNA revealed three types of solid phases (Figure 4.3.3, A, B and C): a central spheroidal or vesicular area (B), surrounded by planar fractal type structures (C) and an interface between the two, a circular rim (A).

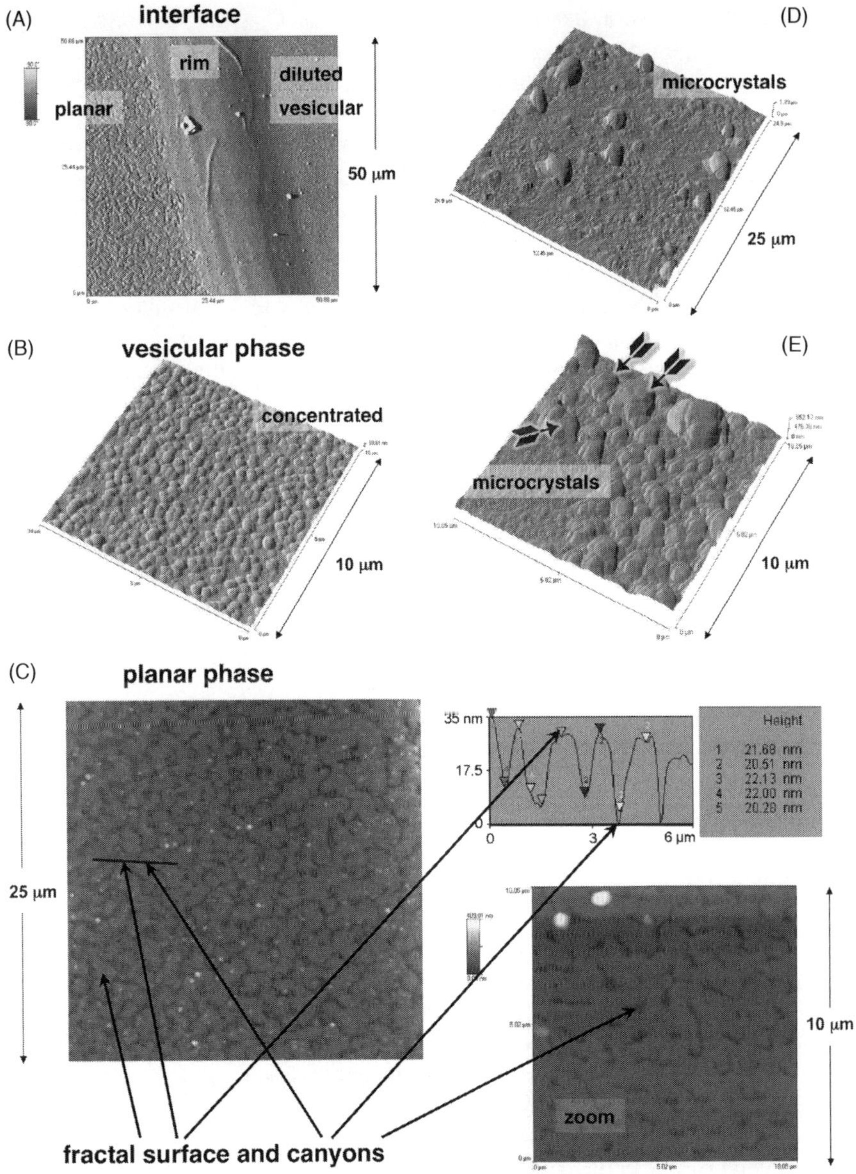

Figure 4.3.3 AFM images of amphiphilic 3′-peptidyl-RNA deposited on glass and dried. In the non-contact mode, the distance between the tip and the sample is >1 nm. A, B, C = Ala_{21}-3′-NH-RNA; D = pGluGlu$_2$Ala$_{18}$-3′-NH-RNA; E = GluLeu$_{18}$Ala-3′-NH-RNA. Dart arrows point at well recognisable hexagonal faces. Bright spots in C are the residual vesicles on top of the surface

The vesicular Ala_{21}-3′-NH-RNA aggregates consisted of flattened spheroids of 400–650 nm diameter and 100–200 nm height, which are of similar average size to the aggregates observed by DLS.

The planar phase outside the circular rim appeared, due to strong shearing forces, in the outer parts of the liquid droplet. It was covered with raft-like fractal structures, holes and canyons of 20–22 nm depth (Figure 4.3.3C). This is consistent with the formation of supported bilayers (SB) of Ala_{21}-3′-NH-RNA, where the upper and lower SB surfaces were composed of layers of RNA hairpins. In between the polar RNA surfaces there was sufficient space for the hydrophobic peptide in an aggregated disordered or β-sheeted form ($2 \times <12$ nm $\simeq$ 20–22 nm, Figure 4.3.1), not unlike the molecular organisation of lipid bilayers. In the planar zones of higher conjugate concentrations, cliffs of the same 20–22 nm depth revealed a local collection of multiple SB sheets stratified on top of one another (not shown). No further evidence for the supramolecular organisation of such SB self-assembly of amphiphilic peptidyl–RNA exists.

The supramolecular organisation of the spheroidal 'vesicular' peptidyl–RNA phase is unknown. Since peptidyl–RNA is oligoanionic, added counter cations or RNA ligands are expected to exert a visible influence on the self-assembly. For example, the addition of spermidine to deposited vesicles of Ala_{20-22}-3′-NH-RNA changed their shape, size and polydispersity very significantly [19a]. On the other hand, AFM images of deposited amphiphilic peptidyl–RNA, in which the peptide contained *N*-terminal glutamates (pGluGlu$_2$Ala$_{18}$-3′-RNA) and much higher amounts of leucine (GluLeu$_{18}$Ala-3′-RNA), showed no vesicular or SB phase, but up to 2 μm sized microcrystals instead (Figure 4.3.3D and E), the latter presenting hexagonal faces (marked with dart arrows). The RNA hairpin with no peptide did not form any deposited aggregates under the same conditions.

The interaction between amphiphilic peptidyl–RNA with lipidic GVs visualised by CFM is an on-going study, which deserves a separate detailed description [15b]. It is suffice to state here that the affinity between peptidyl–RNA and GVs composed of phospholipids, be it in the absence or presence of glycolipids, is at least as high or superior to the affinity between the GVs and the hydrophobic peptide without the RNA. To visualise this in pictures, enjoy Figure 4.3.4. This shows green peptidyl–RNA that has been added to a red mixed-lipid GV at an initial 300-fold molar excess of lipid over peptidyl–RNA. The different sliced views through the GV demonstrate an inhomogeneous lipid phase interacting, through its outer surface, with the even more unevenly distributed conjugate. The GV-anchored peptidyl–RNA aggregates were stabilised through firm peptide–peptide interactions.

4.3.10 Conclusions and perspectives

Amphiphilic 3′-peptidyl–RNA are macromolecular objects that are challenging to synthesise, possess exciting supramolecular properties and are of high scientific value. Their detergent-like molecular shape, pronouced amphiphilicity, plasticity with respect to the conformation of the 'hydrophobic tail', and the size of their highly charged 'polar head' make them unique among amphiphiles. In addition, amphiphilic peptidyl–RNAs combine highly modular shape and (as yet unexamined) functional characteristics of the peptide with a capacity for information storage and potential catalytic function of the RNA. The supramolecular properties of the conjugates, that is, the spontaneous formation of micelles, vesicular microaggregates, microcrystals, supported bilayers upon shearing, or anchoring to lipidic vesicles, depend to a great extent on the actual identity of the peptide. These generally RNA-immobilising properties – self-assembly, assembly on solid mineral or lipidic supports – blend well with the replicative and catalytic properties of RNA alone [7, 23].

After considering the studies on designed, chemically more resistant model conjugates described here, future experiments on the self-evolution of chemical systems that are off-equilibrium, where libraries of hydrophobic peptides are being reversibly joined with RNA libraries and brought in contact with lipidic

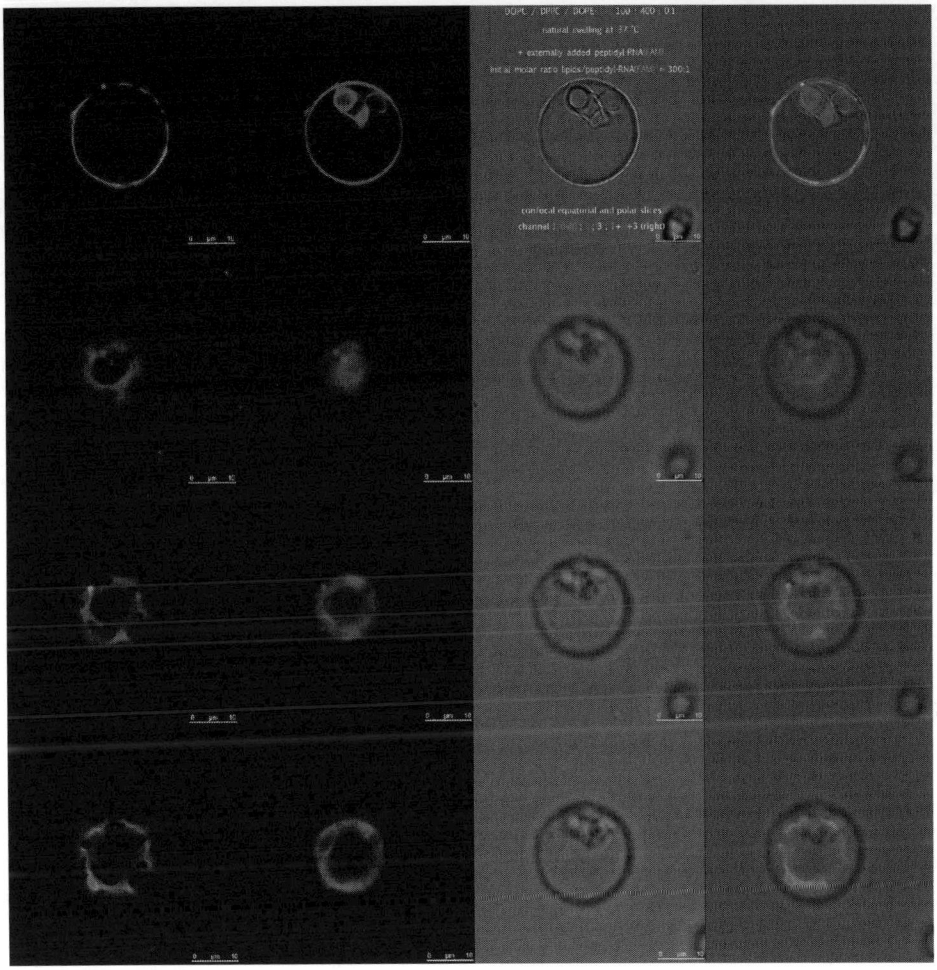

Figure 4.3.4 *Confocal fluorescent images of amphiphilic 3′-peptidyl–RNA interacting with mixed-lipid giant vesicles. The RNA hairpin part of the conjugate contains green fluorescent T(FAM), the peptide is the LA_3LA_2 20-mer (see text). The 20 μm GV was prepared from the slow hydration of a dry 1:4 mixture of 1,2-dioleoyl-sn-glycero-3-phosphatidylcholine (DOPC) and 1,2-dipalmitoyl-sn-glycero-3-phosphatidylcholine (DPPC) containing 0.1% red fluorescent 1,2-dioleoyl-sn-glycero-3-phosphatidylethanolamine-N-lissamine rhodamine B sulfonate (DOPE-Rh). From left to right: green, red, bright light, and all three channels superimposed. Confocal 1 μm thin slices: first row = equatorial, second row = polar, third and fourth rows = in between equatorial and polar depths of the same multivesicular GV (See colour figure in colour plate section)*

vesicles [7b], should be less problematic to be set up, and their outcome easier to interpret. Applications for well-defined hydrolysis-resistant synthetic peptidyl–RNA conjugates could, for example, be the pathogen-specific delivery of small interfering RNA, or more generally, a designed peptide-assisted nucleic acid invasion into liposomes and cells. The peptide part would recognise specific targets (target proteins, cell receptors), whereas the oligonucleotide part could be double-stranded instead of a hairpin, and release the hydridised strand after invasion.

Acknowledgements

I should like to thank two teams for their development of the synthetic procedure to prepare amphiphilic 3′-peptidyl–RNA. The Basel team of pioneers, above all Silvia Terenzi [19a], together with Nhat Quang Nguyen Trung [13b] and Ewa Biała [14], and the Lyon team, who began using a peptide synthesiser, Alexandra Le Chevalier Isaad and Krishnakumar KS [15b,19b]. I am indebted to Anthony Coleman (Lyon), whose immense experience with generating and interpreting atomic force microscopy images lead to the discovery of the supramolecular properties of pure amphiphilic peptidyl–RNA [15a]. Most of the funding for the development of the synthesis was provided by the Swiss National Science Foundation.

References

[1] (a) M. Éthève-Quelquejeu and M. Arthur, Covalent functionalization of nucleic acids, *in Chemistry of Organo-Hybrids: Synthesis and Characterization of Functional Nano-Objects*, E. Lacôte, B. Charleux, and C. Copéret (Eds), Wiley-VCH, Weinheim (2014); (b) P. Järver, T. Coursindel, S. E. L. Andaloussi, C. Godfrey, M. J. A. Wood, and M. J. Gait, Peptide-mediated cell and in vivo delivery of antisense oligonucleotides and siRNA, *Mol. Ther.-Nucleic Acids*, **1**, e27 (2011); (c) K. Lu, Q.-P. Duan, L. Ma, and D.-X. Zhao, Chemical strategies for the synthesis of peptide-oligonucleotide conjugates, *Bioconjugate Chem.*, **21**, 187–202 (2010); (d) H. Lönnberg, Solid-phase synthesis of oligonucleotide conjugates useful for delivery and targeting of potential nucleic acid therapeutics, *Bioconjugate Chem.*, **20**, 1065–1094 (2009); (e) J. H. Jeong, H. Mok, Y.-K. Oh, and T. G. Park, siRNA conjugate delivery systems, *Bioconjugate Chem.*, **20**, 5–14 (2009); (f) N. Venkatesan and B. H. Kim, Peptide conjugates of oligonucleotides: Synthesis and applications, *Chem. Rev.*, **106**, 3712–3761 (2006).

[2] H. Faust, freely adapted from: *Faust. Eine Tragödie von Goethe*, J. W. Goethe (author), Cotta′sche Verlagsbuchhandlung, Tübingen (1808).

[3] P. Strazewski, Braucht es ein Plädoyer für Chemie? *Chimia*, **51**, 69–75 (1997). English version, A Pleading for Chemistry (1999), www.chemie.unibas.ch/~strazi/abstracts/Pleading.html; Polish version: W obronie chemii, *Nauka*, **1**, 109–125, (2000).

[4] Translated from M. Berthelot, *Chimie organique fondée sur la synthèse*, Vol. **2**, Mallet-Bachelier, Paris, p. 811 (1860).

[5] A. Pross, *What Is Life? How Chemistry Becomes Biology*, Oxford University Press (2012).

[6] M. Raouane, D. Desmaële, G. Urbinati, L. Massaad-Massade, and P. Couvreur, Lipid conjugated oligonucleotides: A useful strategy for delivery, *Bioconjugate Chem.*, **23**, 1091–1104 (2012).

[7] (a) P. Strazewski, RNA as major components in chemical evolvable systems, in *Chemical Biology of Nucleic Acids: Fundamentals and Clinical Applications*, V. A. Erdmann, W. T. Markiewicz, and J. Barciszewski (Eds), Series*: RNA Technologies*. V. A. Erdmann and J. Barciszewski (series Eds), Springer (2014), chapt. 1, pp. 1–24: (b) P. Strazewski. *Omne vivum ex vivo … omne*? How to feed an inanimate evolvable chemical system so as to let it self-evolve into increased complexity and life-like behaviour, *Isr. J. Chem.*, DOI: 10.1002/ijch.201400175 (2015).

[8] A. Ramos and G. Varani, Structure of the acceptor stem of *Escherichia coli* tRNAAla: role of the G3·U70 base pair in synthetase recognition, *Nucleic Acids Res.*, **25**, 2083–2090 (1997).

[9] E. Biała and P. Strazewski, Internally mismatched RNA: pH and solvent dependence of the thermal unfolding of $tRNA^{Ala}$ acceptor stem microhairpins, *J. Am. Chem. Soc.*, **124**, 3540–3545 (2002).

[10] S. Zhang, Lipid-like self-assembling peptides, *Acc. Chem. Res.*, **45**, 2142–2150 (2012).

[11] R. Lewis, Y. P. Zhang, R. S. Hodges, W. K. Subczyński, A. Kusumi, C. R. Flach, R. Mendelsohn, and R. N. McElhaney, A polyalanine-based peptide cannot form a stable transmembrane α-helix in fully hydrated phospholipid bilayers, *Biochemistry*, **40**, 12103–12111 (2001).

[12] (a) O. Botta, E. Moyroud, C. Lobato, and P. Strazewski, Synthesis of 3′-azido- and 3′-amino-3′-deoxyadenosine in both enantiomeric forms, *Tetrahedron*, **54**, 13529–13546 (1998); (b) N.Q. Nguyen Trung, O. Botta, S. Terenzi, and P. Strazewski, A practical route to 3′-amino-3′-deoxyadenosine derivatives and puromycin analogs, *J. Org. Chem.*, **68**, 2038–2041 (2003); (c) H. Chapuis and, P. Strazewski, Shorter puromycin analog synthesis by means of an efficient Staudinger-Vilarrasa coupling, *Tetrahedron*, **62**, 12108–12115 (2006); (d) A. Charafeddine, W. Dayoub, H. Chapuis, and P. Strazewski, First synthesis of 2′-deoxyfluoropuromycin analogues: Experimental insight into

the mechanism of the Staudinger reaction, *Chem. Eur. J.*, **13**, 5566–5584 (2007); (e) B. Y. Michel, K. S. Krishnakumar, and P. Strazewski, Synthesis of a *xylo*-puromycin analogue, *Synlett*, 2461–2464 (2008); (f) H. Saneyoshi, B. Y. Michel, Y. Choi, P. Strazewski, and V. E. Marquez, Synthesis of conformationally locked versions of puromycin analogues, *J. Org. Chem.*, **73**, 9435–9438 (2008); (g) B. Y. Michel and P. Strazewski, Total syntheses of a conformationally locked North-type methanocarba puromycin analogue and its dinucleotide derivative, *Chem. Eur. J.*, **15**, 6244–6257 (2009); (h) K. S. Krishnakumar, S. Goudedranche, D. Bouchu, and P. Strazewski, The shortest synthetic route to puromycin analogues using a modified Robins approach, *J. Org. Chem.*, **76**, 2253–2256 (2011).

[13] (a) S. Gunzenhauser, E. Biała, and P. Strazewski, Tetraethylene glycol-derived spacer for oligonucleotide synthesis, *Tetrahedron Lett.*, **39**, 6277–6280 (1998); (b) N.Q. Nguyen Trung, S. Terenzi, G. Scherer, and P. Strazewski, High yield immobilisation of a puromycin analogue for the solid support synthesis of aminoacyl-tRNA fragments, *Org. Lett.*, **5**, 2603–2606 (2003).

[14] S. Terenzi, E. Biała, N. Q. Nguyen-Trung, and P. Strazewski, Amphiphilic 3′-peptidyl-RNA conjugates, *Angew. Chem.*, **42**, 3015–3018 (2003); *Angew. Chem., Int. Ed.*, **42**, 2909–2912 (2003).

[15] (a) A. Coleman, A. Lazar, S. Terenzi, and P. Strazewski, Observation of the formation of supported bilayers by amphiphilic peptidyl-RNAs, *Chem. Commun.*, 63—64 (2006); (b) A. Le Chevalier Isaad, P. Carrara, P. Stano, K. S. Krishnakumar, D. Lafont, A. Zamboulis, R. Buchet, D. Bouchu, F. Albrieux, and P. Strazewski, A hydrophobic disordered peptide spontaneously anchors a covalently bound RNA hairpin to giant lipidic vesicles, *Org. Biomol. Chem.*, **12**, 6363–6373 (2014).

[16] R. Warrass, J.-M. Wieruszeski, C. Boutillon, G. Lippens, High-resolution Magic Angle Spinning NMR study of resin-bound polyalanine peptides, *J. Am. Chem. Soc.*, **122**, 1789–1795 (2000).

[17] (a) L. Debéthune, V. Marchán, G. Fàbregas, E. Pedroso, and A. Grandas, Towards nucleopeptides containing any trifunctional amino acid (II), *Tetrahedron*, **58**, 6965–6978 (2002); (b) D. A. Stetsenko, A. D. Malakhov, and M. J. Gait, Total stepwise solid-phase synthesis of oligonucleotide-(3′→N)-peptide conjugates, *Org. Lett.*, **4**, 3259–3262 (2002); (c) S. Zaramella, E. Yeheskiely, and R. Strömberg, A method for solid-phase synthesis of oligonucleotide 5′-peptide-conjugates using acid-labile α-amino protections, *J. Am. Chem. Soc.*, **126**, 14029–14035 (2004); (d) S. M. Ocampo, F. Albericio, I. Fernández, M. Vilaseca, and R. Eritja, A straightforward synthesis of 5′-peptide oligonucleotide conjugates using N^{α}-Fmoc-protected amino acids, *Org. Lett.*, **7**, 4349–4352 (2005); (e) M. Fonvielle, D. Mellal, D. Patin, M. Lecerf, D. Blanot, A. Bouhss, M. Santarem, D. Mengin-Lecreulx, M. Sollogoub, M. Arthur, and M. Éthève-Quelquejeu, Efficient access to peptidyl-RNA conjugates for picomolar inhibition of non-ribosomal FemXWv aminoacyl transferase, *Chem. Eur. J.*, **19**, 1357–1363 (2013).

[18] (a) H. Moroder, J. Steger, D. Graber, K. Fauster, K. Trappl, V. Marquez, N. Polacek, D. N. Wilson, and R. Micura, Non-hydrolyzable RNA-peptide conjugates: a powerful advance in the synthesis of mimics for 3′-peptidyl tRNA termini, *Angew. Chem., Int. Ed.*, **48**, 4056–4060 (2009); (b) D. Graber, H. Moroder, J. Steger, K. Trappl, N. Polacek, and R. Micura, Reliable semi-synthesis of hydrolysis-resistant 3′-peptidyl-tRNA conjugates containing genuine tRNA modifications, *Nucleic Acids Res.*, **38**, 6796–6802 (2010).

[19] Described in (a) S. Terenzi, Synthesis and characterisation of amphiphilic 3′-peptidyl-RNA conjugates, Dissertation, University of Basel (2003); (b) K. S. Krishnakumar, Synthesis and analysis of puromycin analogues and amphiphilic peptidyl-RNA conjugates, Doctoral Thesis, Université Claude Bernard Lyon 1 (2010).

[20] J. Zimmermann, M. P. J. Cebulla, S. Mönninghoff, and G. von Kiedrowski, Self-assembly of a DNA dodecahedron from 20 trisoligonucleotides with C3h linkers, *Angew. Chem., Int. Ed.*, **47**, 1–6 (2008).

[21] D. Dellinger, Z. Timár, J. Myerson, A. B. Sierzchała, J. Turner, F. Ferreira, Z. Kupihár, G. Dellinger, K. W. Hill, J. A. Powell, J. R. Sampson, and M. H. Caruthers, Streamlined process for the chemical synthesis of RNA using 2′-O-thionocarbamate-protected nucleoside Streamlined process for the chemical synthesis of RNA using 2′-O-thionocarbamate-protected nucleoside phosphoramidites in the solid phase, *J. Am. Chem. Soc.*, **133**, 11540–11556 (2011).

[22] S. C. Srivastava, D. Pandey, N. P. Srivastava, and S. P. Bajpai, RNA synthesis by reverse direction process: phosphoramidites and high purity RNAs and introduction of ligands, chromophores, and modifications at 3′-end, *Curr. Protoc. Nucleic Acid Chem.*, **45**:3.20:3.20.1–3.20.39 (2011).

[23] P. Strazewski, Adding to Hans Kuhn's thesis on the emergence of the genetic apparatus: Of the Darwinian advantage to be neither too soluble, nor too insoluble, neither too solid, nor completely liquid, *Colloid Surf B: Biointerfaces*, **74**, 419–425 (2009).

4.4

Oligonucleotide-Stabilized Silver Nanoclusters

Alfonso Latorre and Álvaro Somoza

Instituto Madrileño de Estudios Avanzados en Nanociencia (IMDEA Nanociencia) and CNB-CSIC-IMDEA Nanociencia Associated Unit "Unidad de Nanobiotecnología", Madrid, Spain

4.4.1 Introduction

Silver nanoclusters (AgNCs) stabilized by oligonucleotides are fluorescent structures, whose properties are attracting researchers from various areas. In particular, the two main characteristics that have prompted the development of projects involving AgNCs are: (i) their ease of preparation and (ii) their tunable fluorescent properties, which can be modulated simply by the selection of the oligonucleotide employed. In this chapter the properties and applications of AgNCs are summarized.

Silver nanoclusters are composed of a few atoms of metallic silver (2–10 atoms) and have a reduced size (<2 nm). Owing to their small size, they have different properties compared with silver nanoparticles, specifically, these nanostructures are fluorescent [1]. This new property appears because the electrons are confined in a reduced space, close to the Fermi wavelength of an electron (~0.7 nm), leading to discrete quantum-confined electronic transitions. The interaction with light through these electronic transitions between different energy levels, leads to a strong photoluminescence [2, 3].

Silver nanoclusters need to be stabilized to prevent their oxidation and to keep their fluorescent properties. In this sense several structures have been employed as ligands to protect the nanoclusters, such as polymers [4–7], dendrimers [8, 9], peptides [10], and microgels [11–13].

In addition to these ligands, it is also possible to obtain fluorescent materials upon reduction of the silver cations bound to the oligonucleotides (DNA or RNA) [14], yielding silver nanoclusters. This was first

DNA in Supramolecular Chemistry and Nanotechnology, First Edition.
Edited by Eugen Stulz and Guido H. Clever.

observed by Braun and coworkers [15] in 1998, but Dickson and coworkers were the ones who studied the DNA stabilized silver nanoclusters in detail and initiated this area of research [16–19].

The main interest in this material comes from its fluorescent properties, which may be useful in multiple applications. Fluorescent materials are valuable tools in research, since they allow the visualization of processes in a very convenient way. They have been particularly useful in bioscience, where the localization of specific biomolecules within a cell or even in animals has allowed the study of processes at the molecular level and the development of sensors for diagnosis.

The most common fluorescent materials used in bioscience are fluorescent proteins, such as green fluorescent protein (GFP), organic dyes, such as fluorescein or cyanine based dyes, and semiconductor nanoparticles, such as quantum dots. All of these have excellent fluorescent properties, but also some drawbacks that limit their use. For instance, fluorescent proteins are big (the molecular mass of GFP is around 27 kDa) and in some cases can interfere with the study [20]. Organic dyes are prone to quenching and are toxic materials. Quantum dots are relatively large (10–20 nm) and are composed of toxic metals such as cadmium and nickel, thus, their use *in vivo* is raising some concerns [21].

For these reasons the development of new fluorescent materials is of great interest, particularly for their application in bioscience. In this regard DNA stabilized silver nanoclusters (DNA–AgNCs) have emerged as new fluorescent materials with interesting properties [22–27]. Compared with quantum dots, DNA–AgNCs are less toxic. There are several reports where their toxicity has been evaluated using different cell lines and it seems that this nanomaterial is indeed safe [28]. The easy preparation of DNA–AgNCs is clearly an advantage compared with the use of organic dyes, whose incorporation into biomolecules can be time-consuming and expensive.

The synthesis of DNA–AgNCs only requires the incubation of an oligonucleotide with a silver salt (usually $AgNO_3$) for few minutes to allow the Ag^+ to interact with the oligonucleotide. The reduction to obtain the metallic silver nanoclusters is usually performed by addition of $NaBH_4$. Even though the oxidation of silver nanoclusters leads to a decrease in their fluorescence, the presence of oxygen seems to be critical to obtain fluorescent nanoclusters [29].

Structurally, DNA–AgNCs are composed of an oligonucleotide and few atoms of silver (2–10). The size of these structures is small, around 1–2 nm. These clusters are quite stable compared with those prepared with other polymers in water, because they usually have lower affinity for silver, and are either cationic or metallic, leading to the formation of nanoparticles. Unlike these nanoparticles, DNA–AgNCs have discrete electronic energy bands allowing the radiative relaxation to produce fluorescence with high quantum yields (QY) and excellent photostability. The standard sequence employed in most of the reports, a poly deoxycytidine (polydC) of 12 nucleotides, has a QY of 17% [30]. However, there are sequences with QY higher than 50% [31]. Regarding the photostability, a recent report has described the preparation of a DNA–AgNCs where 31% of the initial emission was maintained after ten months [32].

The stability of DNA–AgNCs is mainly provided by the interaction of cytosines through the N3 [33] and to a lesser extent by guanines through the N7, as determined by NMR and calculations [34]. However, mass spectrometry studies suggest that cytosines and guanines stabilize AgNCs at the same level [35] and that even thymines can stabilize them [36, 37]. The interaction of nucleobases with silver ions can be modulated by changing the pH of the solution. Thus, at high pH, the N3 of thymine is deprotonated, favoring its binding affinity by Ag^+ [29].

The pH can also change the secondary structure of the DNA strands bound to AgNCs, and therefore the emission properties of the fluorescent AgNCs. This is the case for an i-motif reported by Petty and coworkers, which is able to stabilize red and green emitters at pH 6 and 9, respectively [38]. The different emissions could be explained by the secondary structures of the oligonucleotide obtained at each pH, which yield a different emitter. The structure favored at pH 6 is an i-motif, whereas at pH 9 the structure is different.

Although DNA–AgNCs are normally generated with single-stranded oligonucleotides, other structures such as duplexes [39] or quadruplexes [40] that are rich in guanines can yield the fluorescent nanomaterials as well.

Especially interesting is the case of triplexes [41]. A DNA triplex is a motif where an oligonucleotide is able to bind a DNA duplex at the major groove in a sequence-specific manner through Hoogsteen or reverse Hoogsteen base pairing. In this case the triplet CG·C is able to bind silver and form the complex CG·CAg^0 after reduction with $NaBH_4$. Using this approach the number of clusters obtained can be controlled as well as their position. What is more, the as-prepared AgNCs are better protected inside the triplex, affording very stable fluorescent clusters, which did not decay even in a medium with a high concentration of NaCl (200 mM).

Modified oligonucleotides have also been utilized to generate AgNCs. In an example reported by Somoza and coworkers, three polydC of 12 nucleotides were placed in alternate positions of a benzene ring and used as a template for AgNCs. These trimers showed a fluorescence intensity 60 times higher than the corresponding monomer. Probably, the globular structure of the trimer allows for a cooperative effect between the oligonucleotides and a better protection of AgNCs, favoring their generation and preventing their quenching [42].

One of the features that make DNA–AgNCs so interesting is that their fluorescent properties depend greatly on the sequence of the oligonucleotide. Actually, it is possible to tune their emissions from blue to near infrared just by changing the oligonucleotide sequence or length (Table 4.4.1).

Table 4.4.1 *The emission of AgNCs can be tuned by the selection of the oligonucleotide sequence. Selected examples are shown to illustrate this feature*

Entry [ref.]	Sequence	Exc (nm)	Emi (nm)	Emitting color
1 [29]	$T_4C_4T_4$	370	475	Blue
2 [17]	$C_3T_3A_2C_4$	340	485	Blue
3 [29]	$C_4T_4C_4$	340	495	Blue
4 [43]	(CCGTAA)3CCGTA	439	495	Blue
5 [43]	(GGCTAA)3GGCTA	453	515	Green
6 [17]	$C_3TCT_2A_2C3$	425	520	Green
7 [29]	T_{12}	350	540	Green
8 [31]	$TGACTA_5C_3T_2A_2TC_4$	460	550	Green
9 [43]	(CGCTAA)3CGCTA	466	554	Yellow
10 [43]	$(CCCTAA)_3CCCTA$	468	560	Yellow
11 [44]	$G_3T_2AG_3TC_6AC_3T_2AC_3$	494	570	Yellow
12 [17]	$C_3T_2A_2TC_4$	480	572	Yellow
13 [44]	$G_3TG_3TC_6AC_3AC_3$	512	582	Yellow
14 [31]	$AGTCAC_4A_2C_2TGC_3TAC_2ACG_2ACT$	530	600	Orange
15 [45]	$A_2T_2C_{12}A_2T_2$	550	610	Orange
16 [35]	$TATC_2GTG_5ACG_2ATA$	544	614	Orange
17 [32]	$AC_3GA_2C_2TG_3CTAC_2AC_3T_2A_2TC_4$	535	615	Orange
18 [17]	$C_2TC_2T_2C_2TC_2$	525	620	Red
19 [43]	$(GCGTAA)_3GCGTA$	548	623	Red
20 [38]	(C4A2)3C4	560	625	Red
21 [43]	(GGGTAA)3GGGTA	560	626	Red
22 [46]	$AG_2TCGC_2GC_3$	540	629	Red
23 [47]	C_{12}	560	630	Red
24 [48]	C_{24}	580	640	Red
25 [33]	C_{12}	580	665	Red
26 [18]	C_{12}	650	700	NIR
27 [31]	$AGTC_2GTG_2TAG_3CAG_2T_2G_4TGACTA_5C_3T_2A_2TC_4$	640	700	NIR
28 [17]	$C_3TA_2CTC_4$	650	705	NIR
29 [19]	$(C_3A)_2C_3TC_3A$	750	810	NIR

Currently, there is no way to predict the fluorescent properties of AgNCs from a given oligonucleotide sequence. However, several research groups are investigating the fundamental origins of color emissions of DNA–AgNCs, which are needed to develop better probes and signaling systems [49–51]. In this sense, the different fluorescence that is obtained, which depends on the oligonucleotide employed, can be explained by the length and shape of the nanoclusters. Gwinn and coworkers have recently observed that nanoclusters seem to have a rod shape and the color emission is closely related to the length of each nanocluster [52].

Wang and coworkers have studied the preparation of AgNCs using just the nucleobases. The results showed that clusters prepared with thymine, adenine, and guanine were not fluorescent. On the other hand, an abundant fluorescence was observed when cytosine was employed. In this work they also carried out some calculations and found that the fluorescence was due to the fifteenth exited state of the dC–Ag_9 complex [53]. Although the problem is more complex in DNA, this is a good approach to try to shed some light on the fluorescent properties of DNA stabilized AgNCs.

Despite the fact that the structure of DNA–AgNCs is not completely known and their fluorescent properties cannot be anticipated, many researchers are working in this area. Most of the contributions reported refer to various applications of these fluorescent materials, particularly sensors.

4.4.2 Sensors

The fluorescence of DNA–AgNCs is greatly influenced by their environment, and for this reason most of the applications reported for these systems refer to their use as sensors. Different analytes can be detected using DNA–AgNCs, such as small molecules, metallic ions, proteins, but above all, sensors of nucleic acids are the most common applications.

4.4.2.1 Metallic sensors

Oligonucleotides have a great affinity for different metallic cations, such as Cu^{2+}, Hg^{2+} or Pb^{2+}, among others. For this reason, when DNA–AgNCs are incubated in the presence of these types of cations there is a change in the fluorescence, which can be used to build a sensor. The fluorescence can be increased or quenched depending on the cation and DNA–AgNCs employed.

For example, a sensor for Hg^{2+} based on DNA–AgNCs can be easily prepared since it quenches the fluorescence due to the oxidation of AgNCs by Hg^{2+} [54].

On the other hand, Cu^{2+} can be detected using a turn-on strategy. In this case, the interaction between the copper ions and metallic silver led to a 9.4-fold increase in the fluorescence [55]. What is more, this observation has allowed the preparation of AgNCs doped with Cu^{2+} showing excellent QY (51.2%) [56]. These results could be due to the generation of more rigid bimetallic nanoclusters, which may be more stable against quenching effects.

Wang and coworkers applied a different strategy for the detection of Hg^{2+}. They combined the production of oligonucleotides using a DNA machine with the preparation of DNA–AgNCs. The machine uses two enzymes, a polymerase (KSF^-) and an endonucleoase (Nb BbvC I), the required nucleotide triphosphates, and an oligonucleotide strand. The strand has three different regions: one encodes for a sequence that is able to form AgNCs, another encodes for the nicking site of the endonuclease, and the last one is able to bind Hg^{2+} and fold the oligonucleotide to yield the initiator duplex required for the polymerase to work. Only in the presence of Hg^{2+} the polymerase will be able to complete the duplex introducing the required sequence to prepare AgNCs. Once it is completed, the endonuclease can break the duplex, releasing the fragment required to stabilize the AgNCs. The remaining DNA structure is intact and the polymerase can work again

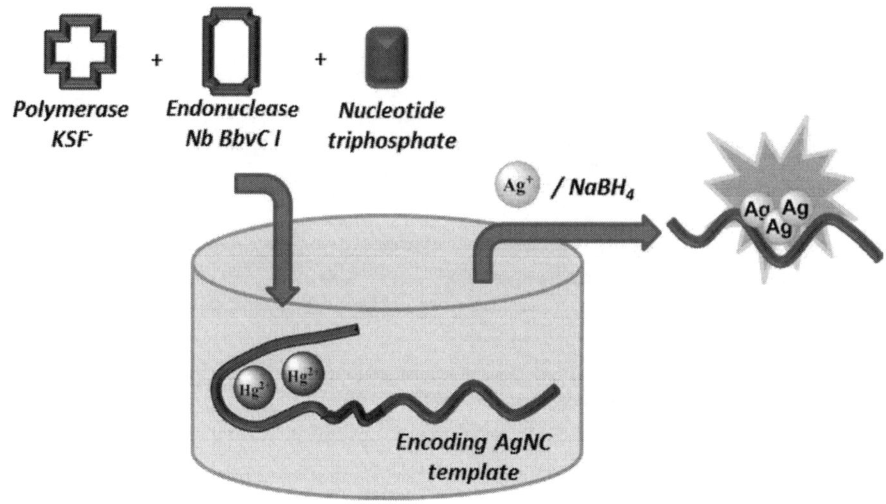

Figure 4.4.1 Schematic representation of the DNA machine employed to detect Hg^{2+}. The metallic cation folds the oligonucleotide into a duplex, which can be recognized by the polymerase. Once the synthesis has finished, an endonuclease releases the fragment, which can be used in the preparation of AgNCs

producing, in this way, multiple copies of the oligonucleotide strand required in the preparation of AgNCs. Then, the oligonucleotide is purified by an NAP-5 column and used in the preparation of AgNCs under standard conditions (Figure 4.4.1). Using this machine these workers were able to detect Hg^{2+} at 80 pM [57].

In addition to the fluorescence of DNA–AgNCs, these structures have other interesting properties. For example, AgNCs are able to reduce metallic salts, particularly cooper(II) to copper(0), leading to bimetallic nanoparticles [58]. The reaction also works with Hg^{2+} and Pb^{2+} and it is a convenient method for the preparation of alloys at the nanoscale.

4.4.2.2 Small molecule sensors

The sensors developed for small molecules require there to be a strong interaction with the compound in order to modulate the fluorescent properties of the system. For this reason, most of the molecules employed contain thiols that strongly interact with silver clusters, quenching their fluorescence. A wide variety of molecules have been detected by AgNCs stabilized with oligonucleotides, such as, glutathione, cysteine, homocysteine [45], and *N*-acetylcysteine [59].

Most of the systems are based on fluorescence quenching (turn-off approach), although an increase in the fluorescence is preferred for sensing [60]. In this regard, Li and coworkers found a DNA sequence that can be used for the preparation of fluorescent AgNCs whose fluorescence is increased in the presence of cysteine [43].

The use of aptamers in combination with AgNCs has allowed the detection of different structures, including small molecules. Aptamers are short sequences of oligonucleotides that can be designed to bind different structures with high affinity and selectivity. Based on this approach, Zhou and Dong developed a method to detect cocaine at a concentration of 0.1 μM. The sensor consisted of two domains, one able to generate AgNCs and an aptamer specific for cocaine binding. The treatment of this sequence with $AgNO_3$ and the subsequent reduction yield the fluorescent AgNCs with moderate intensity. However, the binding between the aptamer and cocaine promotes a conformational rearrangement, inducing a fluorescence increase of threefold (Figure 4.4.2). The same approach was employed to detect ATP using the corresponding aptamer [61].

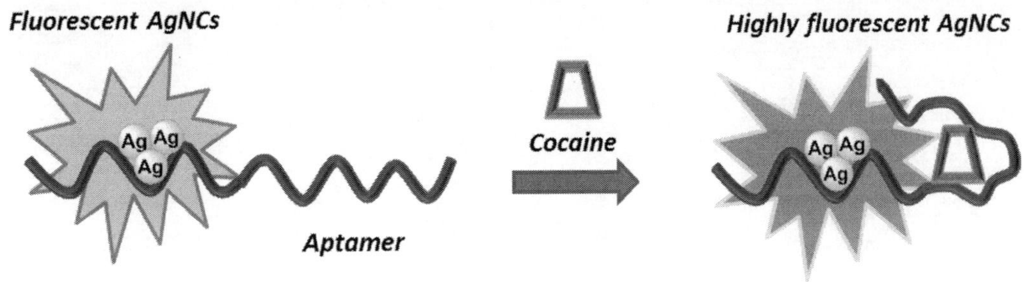

Figure 4.4.2 Schematic representation of the use of aptamers and AgNCs as sensors. The aptamer domain recognizes the ligand (cocaine) and binding leads to an increase in the fluorescence intensity

4.4.2.3 Protein sensors

The sensors developed for proteins are based on two different strategies. One requires the activity of the protein (enzymes) to trigger a change in the fluorescence of the DNA–AgNCs. The other one is more general and can be applied to other structures, as in the previous section. In this case a targeting molecule, such as an aptamer, is used to bind the desired protein. This interaction changes the fluorescent properties of AgNCs, allowing the detection of the protein.

Martinez and coworkers reported a thrombin sensor based on this approach that allowed its detection at 0.1 nM [62]. The oligonucleotide employed was formed by two domains, an AgNCs template and the corresponding thrombin aptamer. The interaction with the target protein induced a rearrangement in the structure that quenched the fluorescent signal of the DNA–AgNCs. Three different proteins, PDBGF, BSA, and streptavidin, were tested as negative controls showing the specificity of the system.

Thrombin has also been detected in a turn-on strategy using two different oligonucleotides with three different domains [63]. One domain was an aptamer, which was different in each strand and each one could bind distinct regions of the protein in a very specific manner. Also, both oligonucleotides had complementary regions that were used to promote their assembly. The third domain was different in the two oligonucleotides, one strand contained a template region for AgNCs, and the other one a guanine-rich sequence. In this system, thrombin is recognized by both aptamers, whose interaction with thrombin allows the hybridization of the complementary fragments. Such rearrangement places the guanine-rich region close to the AgNCs, leading to an increase in the fluorescence intensity, allowing the thrombin detection at a concentration of 1 nM of concentration (Figure 4.4.3). Although this method requires the use of two different probes, it is highly specific.

Yang and coworkers developed another turn-on system to detect the PDBGF-BB protein, where two probes were also required [64]. In this case, one probe is a folded aptamer, which becomes unfolded upon target binding. The rearrangement releases a domain complementary to a fragment of a second folded probe. The hybridization between these two oligonucleotides unfolds the second probe, which releases a fragment able to stabilize AgNCs. Therefore, the treatment of these two folded probes with $AgNO_3$ and $NaBH_4$ allows the generation of bright AgNCs just in the presence of the target protein (Figure 4.4.4). Using this strategy, PDBGF-BB protein was detected in saliva at 5 nM.

The other approach that is employed in the detection of proteins is only used with enzymes since it exploits their activity to turn on or off the fluorescence of the sensor. A basic system that detects the activity of an enzyme was reported by Zhang and coworkers [65] for the detection of the endonuclease EcoRI. They used the nuclease activity to cut the oligonucleotides employed in the preparation of fluorescent AgNCs, which contain the recognition sequence for the enzyme. In this way, the fluorescent system disassembles leading to a decrease in fluorescence. The same workers developed a turn-on sensor for a different enzyme,

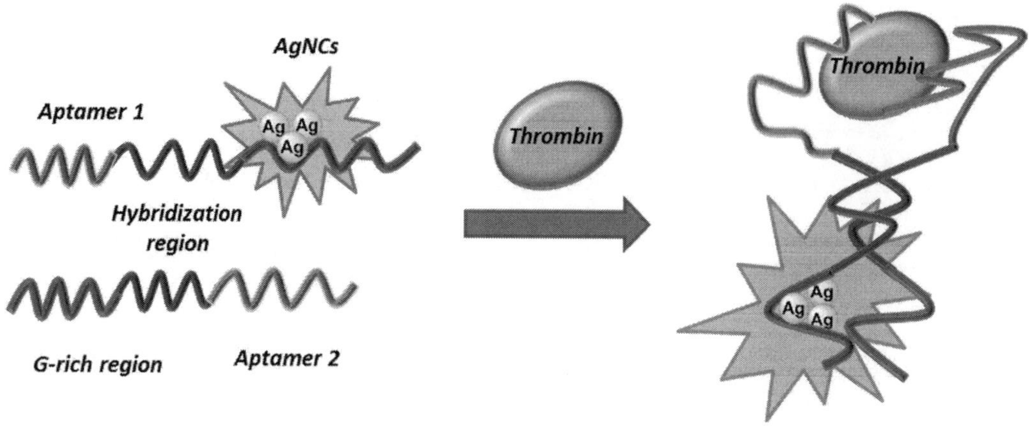

Figure 4.4.3 Detection of thrombin using two different oligonucleotides. Both strands recognize thrombin but at different sites. Once they are bound to the thrombin, the complementary regions hybridize and place the AgNCs and the G-rich domain in close contact, leading to an increase in fluorescence

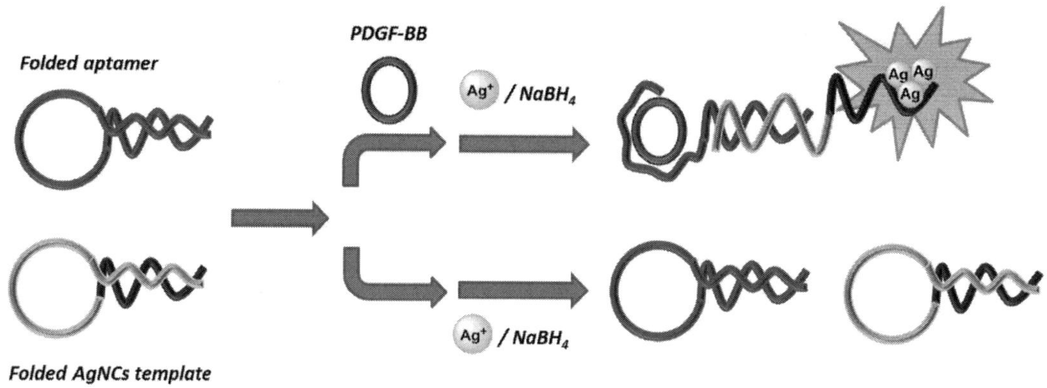

Figure 4.4.4 Detection of PDGF-BB protein using two different oligonucleotides. In the presence of the protein, both oligonucleotide unfold and a region for the preparation of AgNCs can be used (See color figure in color plate section)

the acetylcholinesterase, which is able to remove the acetyl group of acetylthiocholine [66]. The resulting thiocholine reacts with AgNCs present in the solution, increasing their fluorescence. The sensing capabilities of this system were comparable to the standard assay for acetylcholinesterase, known as the Ellman method.

A related system was developed to detect the glutathione reductase. Glutathione (GSH) is present mainly in its reduced form, and its oxidized derivative is associated with some diseases such as diabetes, Parkinson's disease or cancer. The glutathione reductase transforms oxidized glutathione (GSSG) to its reduced form (GSH), which interacts with AgNCs increasing their fluorescence. In this way, the enzymatic activity can be detected by an increase in the fluorescence intensity (Figure 4.4.5) [67].

Other strategies are based on the oxidation of AgNCs, which can be done easily in the presence of hydrogen peroxide (H_2O_2), leading to a fluorescence quenching. This can be used to develop sensors for oxidases or their substrates. This approach has been applied to the detection of glucose, which, in the presence of glucose

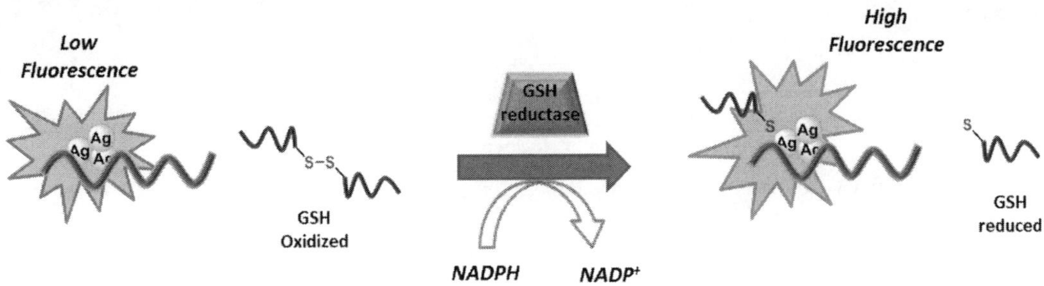

Figure 4.4.5 Sensor of GSH reductase using DNA–AgNCs and the reduced form of glutathione. GSH reductase reduces the disulfide bond present in the GSH dimer. The reduced GSH activates the fluorescence of the AgNCs

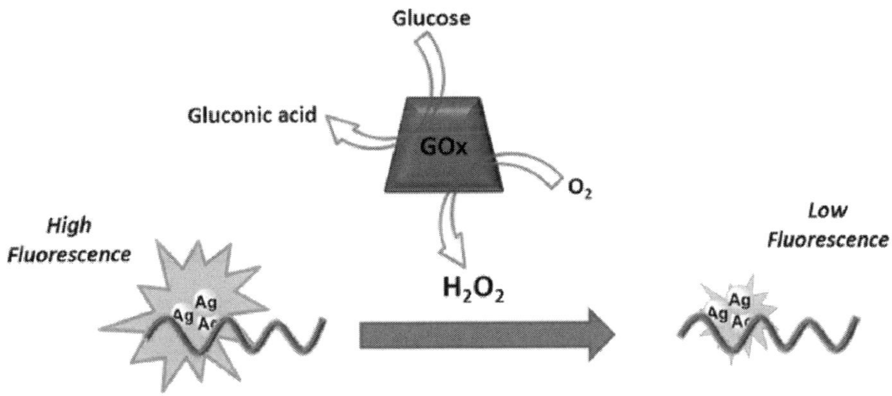

Figure 4.4.6 Glucose sensor. The glucose is oxidized in the presence of glucose oxidase. During this process the molecular oxygen is transformed into hydrogen peroxide, which oxidizes the AgNCs quenching their fluorescence

oxidase (GOx) and O_2, produces gluconic acid and H_2O_2. The reaction is carried out in the presence of fluorescent DNA stabilized AgNCs, whose fluorescence is quenched by the hydroperoxide (Figure 4.4.6). This method has allowed the detection of glucose at 150 nM [68]. The same approach was employed in the detection of cholesterol using the cholesterol oxidase in the generation of H_2O_2 [69].

DNA–AgNCs could be also used in the detection of the activity of tyrosinase, which is found at high concentrations in melanoma cancer cells. In this case the enzyme is able to oxidize phenols to quinones, which can quench the fluorescence of AgNCs. The substrates of this enzyme are diverse, such as tyramine, dopamine or L-tyrosine. In the presence of tyrosinase these compounds are transformed into their corresponding quinones, which are able to oxidize the AgNCs and, therefore, quench their fluorescence.

Liu and coworkers studied Föster resonance energy transfer (FRET) between organic dyes and DNA–AgNCs and used this phenomenon to develop a nuclease sensor. In this case, when the enzyme is not present a fluorescent signal arising from the DNA–AgNCs can be observed upon excitation of a dye close to the clusters. However, when the enzyme is present the DNA is cleaved and the FRET cannot be observed (Figure 4.4.7) [70].

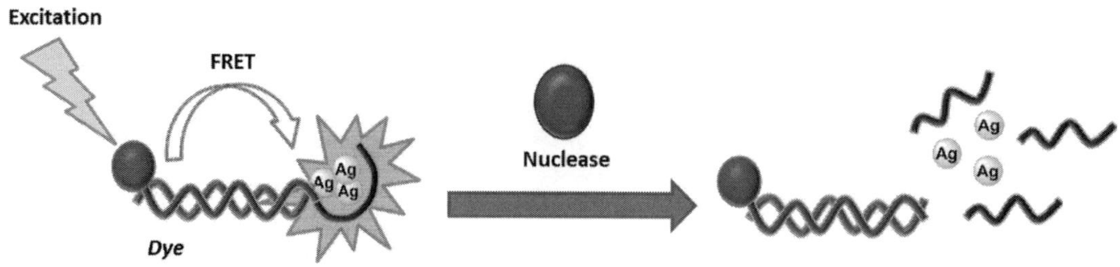

Figure 4.4.7 Nuclease sensor. The fluorescence of AgNCs can be activated through a FRET process when the nuclease is not present. However, when the enzyme is in the media it cleaves the duplex, reducing the FRET process

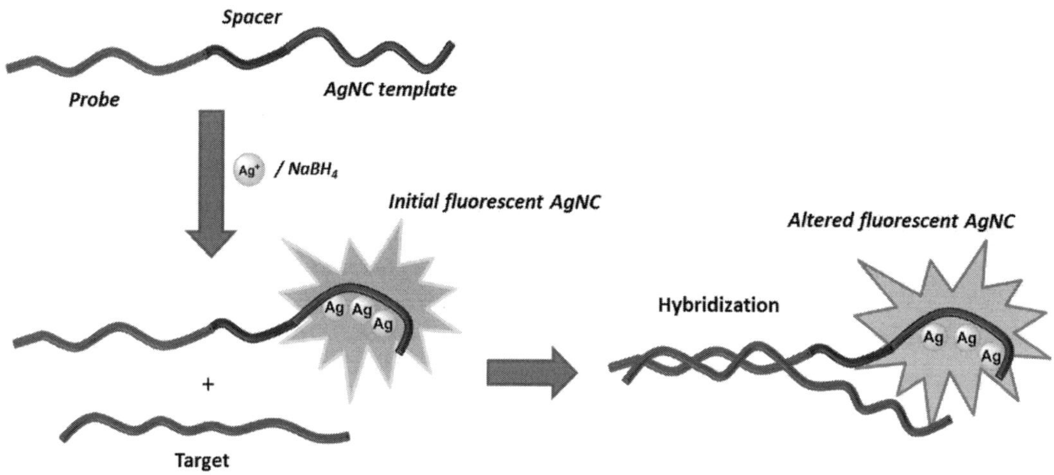

Figure 4.4.8 Nucleic acid sensor based on DNA–AgNCs. A single strand with two domains, one for the preparation of AgNCs and the other complementary to the target nucleic acid can be used to prepare the sensor. Once the AgNCs are generated, the incubation with the target leads to changes in the original fluorescence intensity due to the hybridization of both sequences

4.4.2.4 Nucleic acid sensors

The detection of nucleic acids can be performed easily using DNA–AgNCs. It requires the preparation of an oligonucleotide with at least two different domains. One should have the complementary sequence of the target nucleic acid and the other a sequence that will be used to generate AgNCs. Both domains can be placed consecutively although using a spacer sequence seems to give better results [71]. Therefore, the desired probe can be obtained from a simple DNA strand, after incubation with silver nitrate and reduction with sodium borohydride to generate AgNCs. The fluorescence of the probe is very sensitive to the environment and its initial intensity is altered upon hybridization with the corresponding target (Figure 4.4.8). This probe is much cheaper than the standard ones that require the use of organic dyes covalently attached to the DNA strand. On the other hand, probes made from AgNCs can be less stable and the combination of some targeting and clustering sequences might be not compatible, since AgNCs can be generated in both domains. This type of sensor is very versatile and can be used to detect DNA, mRNA, microRNAs, and even the human telomerase RNA template [72].

4.4.2.4.1 DNA sensors

An example of this type of sensor was reported by Chang and coworkers [47], where the single point mutation of a gene involved in tyrosinemia type I was detected with high sensitivity and selectivity. Upon hybridization, a ninefold increase in fluorescence was observed, probably due to the formation of more rigid duplex structures where the AgNCs were more protected against the quenching effects.

Another strategy employed to detect genes was based on the fluorescence enhancement of DNA–AgNCs by guanine-rich oligonucleotides. The system requires two oligonucleotides that are able to bind the desired target at different consecutive regions. One of the oligonucleotides should contain a sequence to stabilize AgNCs and the other should have a guanine rich domain. Upon hybridization of both oligonucleotides with their corresponding target regions, the AgNCs and the guanine rich fragment are arranged in close contact, inducing a drastic enhancement of the fluorescence intensity, as well as a red shift of the wavelength emission. The system allowed the detection of the *Braf* oncogene at 10 nM, and was named by these workers as NanoClusterBeacon due to its similarity with the classical molecular beacons [73].

An analogous strategy was used to prepare a sensor that distinguishes the three possible mutations in a gene. The system, termed Chamaleon Nanocluster Beacons [74], works on the basis that different colors can be obtained by repositioning the guanine-rich sequence with respect to the AgNCs. Furthermore, the fluorescence intensity could be used to quantify such mutations. Using this sensor, KRAS mutations could be characterized using real clinical samples from patients with ovarian serous borderline tumors.

As commented in the introduction of this chapter, duplexes are able to stabilize AgNCs and this feature has been exploited in the preparation of sensors. One example was reported by Wang and coworkers where single-point mutations of a gene responsible for sickle cell disease were detected [75]. In that report, the probe employed was an oligonucleotide complementary to the target sequence, with a loop of six cytosines close to the single-point mutation. Once the duplex was hybridized, the C6 loop was able to form bright yellow AgNCs using the standard procedure.

Another strategy employed to detect genes using duplexes takes advantage of the capability of abasic sites to host silver ions. An oligonucleotide with an abasic site is an oligonucleotide where a nucleobase has been removed, leaving a deoxyribose residue in the structure. These types of damage can occur by endogenous or exogenous processes spontaneously. Shao and coworkers managed to detect the single-point mutation of p53 gene using this methodology [76]. The probe employed was an oligonucleotide complementary to the target gene, which has an abasic nucleotide at the mutation site. The mutated base of p53 gene is a cytosine, which is able to bind Ag^+ more efficiently than A, T or G, and therefore yields higher fluorescence (18 folds) after AgNCs preparation.

Kim and coworkers employed a strand-displacement process to detect sequences that contain single-nucleotide polymorphisms (SNP) [77]. In this method, a DNA duplex is prepared by combining a strand that contains the complementary sequence of the gene with the SNP and a region for the generation of AgNCs with a strand that partially matches the recognition sequence. In the presence of the perfect match, the former strand is replaced by the new strand, which contains a fluorescence enhancer domain. Once the new duplex, is formed an increase in fluorescence can be observed (Figure 4.4.9).

A remarkable approach was reported by Willner, where the quenching properties of graphene oxide (GO) and the selective binding properties of single-stranded oligonucleotides were employed to prepare a gene sensor. Single-stranded oligonucleotides interact strongly with the GO through π–π staking interactions between the hexagonal graphitic units and the nucleobases. When a duplex is formed this interaction decreases and the oligonucleotides detach from the GO. In this approach an oligonucleotide with two domains, one for the generation of AgNCs and the other to bind the target sequence, was bound to GO through π–π interactions. At this stage, the fluorescence of AgNCs is quenched by the GO, but when the complementary sequence is added it forms the corresponding duplex with the probe and it detaches from the GO leading to an increase

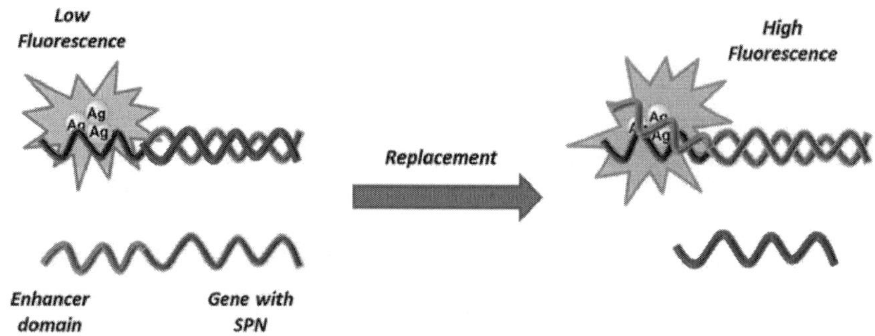

Figure 4.4.9 Sensor of SNP based on strand displacement. A DNA duplex composed of sequences with partial complementarity and AgNCs is incubated with a sequence complementary to the one that holds AgNCs. This sequence replaces the sequence with partial complementarity and the enhancer domain increase the fluorescence of AgNCs

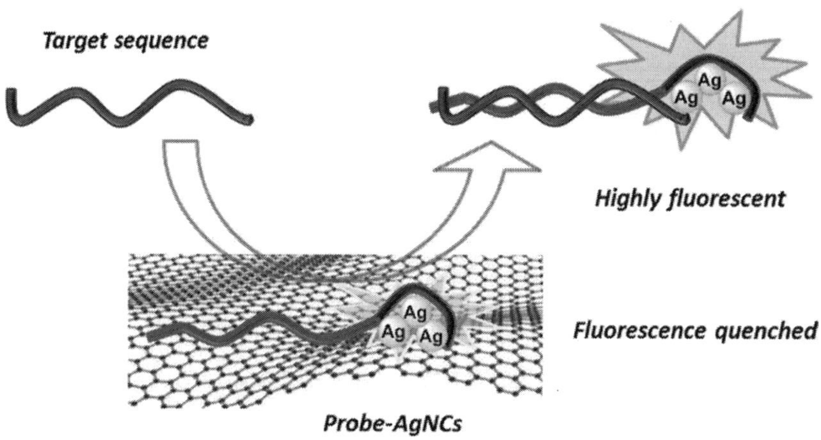

Figure 4.4.10 A gene sensor using GO and DNA–AgNCs. The fluorescence is quenched by the GO, but after duplex formation with the complementary sequence the probe detaches and the fluorescence increases

in fluorescence (Figure 4.4.10) [78]. The use of GO in combination with DNA–AgNCs is fairly versatile and has allowed the multiplex analysis of two different genes as well as the detection of ATP and thrombin using their corresponding aptamers.

Other quenching structures employed in the preparation of sensors are G-quadruplexes, particularly a parallel G-quadruplex–hemin complex. This structure efficiently quenches the fluorescence of AgNCs through a photoinduced electron transfer process (PET) [79]. An oligonucleotide with three different domains was necessary to construct this type of sensor; the first domain was a fragment able to stabilize AgNCs, the second was a G-rich fragment and the last one contained the complementary sequence of the target nucleic acid. In the absence of the target, the last fragment formed a stable loop with part of the G-rich fragment, leading to highly fluorescent AgNCs. However, upon target annealing, the loop was unfolded and the formation of G-quadruplexes–hemin complex was allowed. As a consequence the AgNCs were quenched (Figure 4.4.11). Using this approach a single-point mutation was detected at 0.6 nM, making this system the

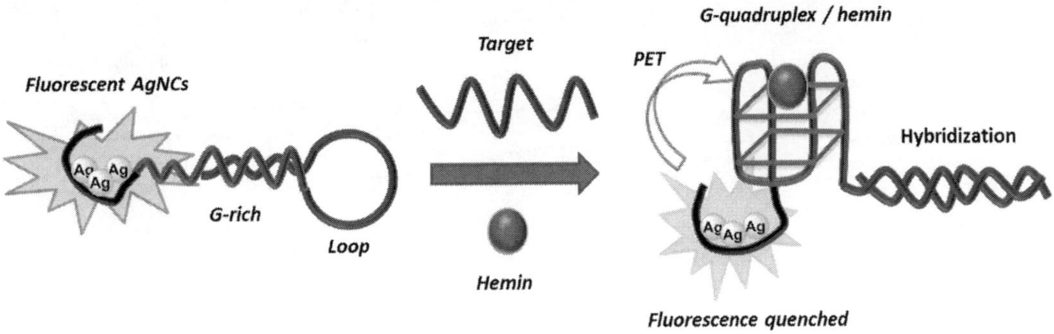

Figure 4.4.11 *Gene sensor prepared with an oligonucleotide with three domains. The probe is folded and the system is fluorescent, but in the presence of the target sequence it unfolds and a G-quadruplex is formed. A hemin molecule binds the G-quadruplex and the fluorescence of AgNCs is quenched (See color figure in color plate section)*

one with the lowest detection limit for DNA based on DNA–AgNCs. Furthermore, the sensor could be adapted to sense ATP with excellent selectivity and sensitivity just by adjusting the probe sequence.

4.4.2.4.2 MiRNAs sensors

The detection of miRNAs is of great interest, and particularly the detection of multiple miRNAs at the same time. These regulatory RNAs are involved in multiple processes and their dysregulation is a clear fingerprint of disease. The method currently employed to detect these noncoding RNAs uses quantitative reverse transcriptase PCR (qRT-PCR), which has some drawbacks such as the cost of the equipment and reagents as well as the requirement of trained personnel to perform the assays. For these reasons, there is a pronounced interest in the development of new systems for the simultaneous detection of different miRNAs.

Similarly to the gene detection discussed here, one strategy consists of the use of DNA–AgNCs where the oligonucleotide sequence has a fragment complementary to miRNA target. The hybridization leads to a quenching of the fluorescence of AgNCs, allowing the detection of miRNAs with high selectivity. In this sense, Vosch and coworkers managed to detect miRNAs from plant extracts at a concentration 82 nM using this approach [80, 81].

Regarding the multiplexed detection of miRNAs, Ye and coworkers have used silver nanoclusters to simultaneously detect two different miRNAs (miR-21 and miR-141). Owing to the low concentration of these RNAs, they used a target-triggered isothermal exponential amplification reaction (TIEAR), which allowed the detection of the miRNAs at picomolar concentrations. The method requires a template sequence that is complementary to the target and encodes for an oligonucleotide, which can be used in the preparation of AgNCs. The template also encodes sites for nicking enzymes between the different regions. Once the template binds the target miRNA, a polymerase elongates the miRNA using the information encoded in the template. After that, nicking enzymes can cleave the new strands affording the oligonucleotides required for the preparation of AgNCs and the targets, which can start the process again. After several cycles, the oligonucleotides obtained can be used for the preparation of AgNCs that allow the detection and the quantification of the target miRNAs [82]. As mentioned earlier, this method could be applied for simultaneous detection of two miRNAs, since the sequences used could be adjusted to generate AgNCs with different emission wavelengths (Figure 4.4.12).

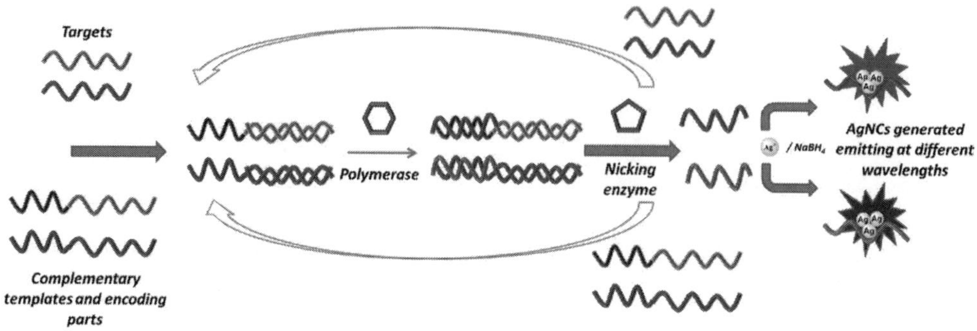

Figure 4.4.12 Multiplexed detection of miRNAs based on TIEAR and AgNCs. Complementary sequences of miRNAs that contain a coding region for the formation of AgNCs are incubated with the miRNAs in the presence of enzymes. The system produces several copies of oligonucleotides that can be used in the preparation of AgNCs. The emitting colors can be encoded specifically for each miRNA (See color figure in color plate section)

4.4.2.5 Cells

The use of DNA–AgNCs is not limited to *ex vivo* samples, and they can also be used inside the cells to detect different structures in combination with targeting molecules. In this sense, the use of aptamers is the most straightforward approach, since oligonucleotides can be prepared with two different domains, one to stabilize AgNCs and the other to recognize the target structure (aptamer).

An example of this strategy was reported by Gao and coworkers, who employed an oligonucleotide that contains a poly-cytosine domain, for the preparation of AgNCs and the sgc8 aptamer. After the preparation of AgNCs with this oligonucleotide, the probe was employed to target the nucleus of CCRF-CEM cells, allowing its visualization by fluorescence microscopy [83]. The use of a spacer between the two domains seems to be beneficial, as mentioned before, and the fluorescence increases [84]. The sgc8 aptamer was designed to target endosomes of CCRF-CEM cells [85], and it has been shown that it specifically targets the human protein tyrosine kinase 7. Its translocation inside the cell is mediated by receptors and subsequently it reaches the endosomes [86]. However, these workers observed that the clusters ended in the nucleus of live cells after incubation for 2 h. They suggest that the structure of the aptamer may have been modified after the formation of AgNCs thus being able to recognize the nucleus. Other aptamers, and even antibodies, bound to AgNCs have shown a similar behavior ending in the nucleus of cells as well. This could suggest that additional factors are involved in the translocation of DNA–AgNCs to the nucleus.

In this regard, the aptamer AS1411 was used to direct the AgNCs to the nucleus of MCF-7 cells [87]. Hence, the aptamer and a poly-cytosine region were connected through a spacer of five thymidines. AS1411 interacts strongly with nucleolin, a protein overexpressed in the membrane of MCF-7 cells. This aptamer has shown cytotoxic effects in cancer cells and is currently in phase II clinical trials [88]. Interestingly, the toxic effect of AS1411 was not influenced by the presence of AgNCs in the same oligonucleotide strand.

Aptamers labeled with DNA–AgNCs have been employed not only to detect membrane receptors but also in the staining of proteins inside the cells. For example, Huang and coworkers reported the use of an aptamer to detect prion protein (PrPc) labeled with DNA–AgNCs in SK–N–SH cells [89].

Other targeting molecules, such as antibodies, have also been employed, but their preparation is more complex since it requires the conjugation between the oligonucleotide and the targeting molecule.

An example of this approach was reported by Dickson and coworkers. They used antibodies of the heparin sulfate polysaccharide (SH receptor) with DNA–AgNCs to detect NIH 3T3 cells that overexpress the SH receptor. The incubation of the labeled antibodies with the cells at 4 °C allows the visualization of the cell membrane. Once the temperature is raised to 37 °C, the fluorescent AgNCs are able to enter inside the cell and even the nucleus [90].

Besides antibodies, proteins with strong binding capabilities such as avidin have also been employed in cells. In this case a poly-cytidine was conjugated with avidin, allowing the detection of cells that were previously incubated with biotin. The strong affinity between avidin and biotin promotes the accumulation of the fluorescent AgNCs in the cells treated [84].

4.4.3 DNA computing (logic gates)

A different use of fluorescent DNA–AgNCs is in DNA computing, particularly in the elaboration of logic gates. In DNA computing, a structure made from DNA should change under some stimuli, mimicking logic arithmetic operations that are performed by semiconductors. The signal transducers can be organic dyes or quantum dots, but the use of AgNCs eases the preparation of these systems and renders them available at lower cost.

The first logic gate based on AgNCs was reported by Wang and coworkers and it was able to perform three logic operations: NOR, NOT and AND [44]. It was made of a single DNA strand with three different domains. One domain contained a sequence able to stabilize two types of AgNCs, one emitting at 570 and other one at 640 nm. This domain was inserted between a domain rich in cytosines and another one rich in guanines. Once prepared, the fluorescent properties of the system could be modulated by two inputs, the addition of K^+ or a low pH (H^+), as well as the combination of both. When none of the inputs were applied, the original fluorescence was maintained (NOR). When the pH was low, the C-rich fragment rearranged into an i-motif, leading to a non-fluorescent system (NOT). When both inputs, K^+ and H^+, were applied to the logic gate, an emission at 601 nm was observed (AND). This is a basic logic gate but it shows the potential of DNA–AgNCs in this area of research, due to the ease of preparation and the modulation of their fluorescent properties by external stimuli.

A more complex example of a logic gate was reported by the same group a few years later, which was able to perform the following operations: OR, NOT, INHABIT, XNOR, and IMPLICATION (Figure 4.4.13).

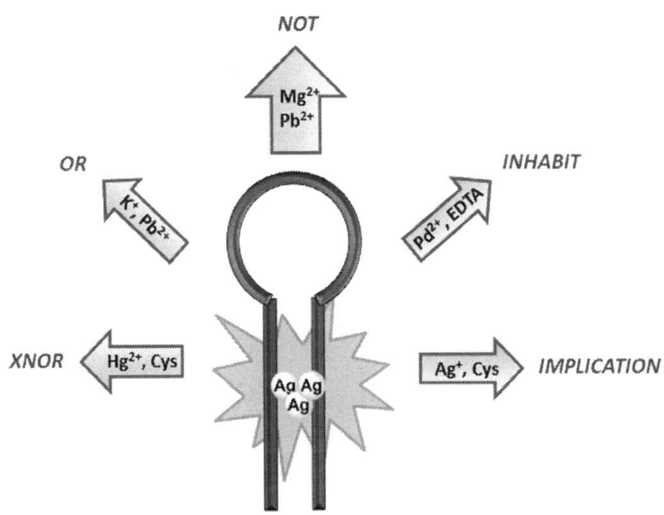

Figure 4.4.13 Logic gate prepared with DNA–AgNCs. The combination of inputs allows the mimicking of several logic operations

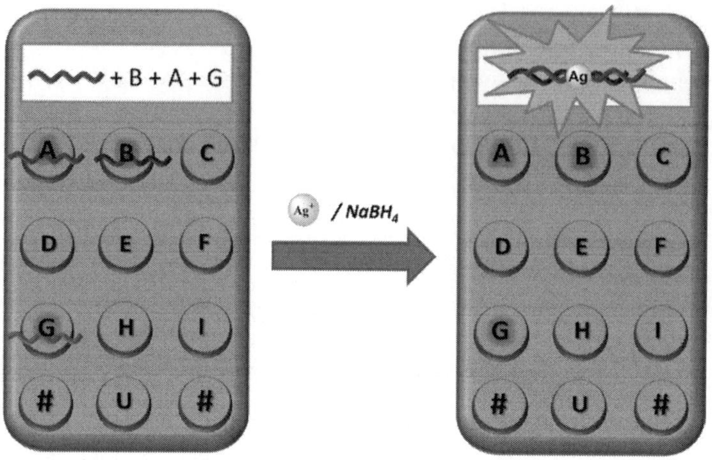

Figure 4.4.14 Representation of a keypad lock. The addition to one oligonucleotide of the inputs in the correct order (B + A + G) allows the generation of fluorescent AgNCs

In this case the oligonucleotide employed had a hairpin structure where the loop was composed of thymidines, whereas the stem was mainly composed by deoxycytidines and deoxyguanosines [91].

The fluorescence of the AgNCs stabilized by the DNA hairpin was modulated by the presence of different cations. Particularly, the OR logic operation was obtained using K^+ and Pb^{2+}; when these cations were absent, the fluorescence was low. Upon the addition of either input (K^+ or Pb^{2+}) the hairpin structure changed to a G-quadruplex and the fluorescence increased. The NOT operation was achieved using Mg^{2+} and Pb^{2+}. In this case the presence of Mg^{2+} (alone or with Pb^{2+}) reduced the original fluorescence. In a similar way the INHABIT operation was obtained using Pd^{2+} and EDTA as inputs. When Pd^{2+} was added an increase in fluorescence was clearly observed, but it was prevented in the presence of EDTA. Hg^{2+} and cysteine (Cys) were used for the construction of the XNOR gate. In this system a fluorescent signal was obtained when both inputs were present or absent, but the fluorescence was quenched when either of them were present. The last logic gate (IMPLICATION) was prepared using Ag^+ and Cys where the fluorescence was intense in all combinations except when Cys was added alone.

Another example of the use of AgNCS in DNA computing is the preparation of a keypad lock system where the fluorescent AgNCs are used as signal transducers. The fundamental difference between a molecular logic gate and a molecular keypad lock is that in the former the output signal is obtained not only by the presence of the proper combination of inputs, but it is also required that the inputs are introduced in the right order. Dong and coworkers described the use of a DNA strand as the keypad lock and three different sequences of DNA as the inputs. When the sequences were added in the order B, A, and G and the resulting mixture was incubated with $AgNO_3$ and treated with $NaBH_4$, a fluorescent signal was obtained. When a different order was applied to the keypad lock the fluorescence obtained was significantly less than that obtained with the correct combination (Figure 4.4.14). What is more, the keypad lock could be reused after treating the mixtures with exonuclease III [92].

4.4.4 Assorted examples

Besides the applications described earlier, DNA–AgNCs can be used to label other structures, such as DNA nanotubes. In a recent report a DNA nanotube was prepared with hairpin protrusions, where the loops of these hairpins were composed of cytosines. Thus, the generation of AgNCs can be directed to those positions.

Fygenson and coworkers managed to prepare a DNA nanotube labeled with AgNCs, proving that these fluorescent clusters can be arranged in a specific way along several micrometers [93]. This strategy, or other related ones, can be extremely useful in the spatial arrangement of interesting optical materials, such as AgNCs, for further photophysical studies.

In addition to the applications of DNA–AgNCs in sensing or imaging, they can be used as delivery systems of oligonucleotides in cells. Particularly, a sequence commonly present in virus and bacteria with an immunostimulatory activity, the CpG motif, has been efficiently delivered to cells. This was achieved using a strand containing a C12 region and a CpG motif in the preparation of AgNCs. The new material was efficiently internalized, as confirmed by fluorescent microscopy and FACS and it was able to induce the immune response better than well-established immunostimulatory lipopolysaccharide (LPS) [94].

4.4.5 Conclusions

As discussed throughout this chapter, the fluorescent properties of DNA–AgNCs have attracted the interest of several research groups. Specifically, their extraordinary sensibility to the environment makes them very useful materials for the development of sensors. For this reason, most of the applications reported have been directed at this area. So far, there have been systems described for the detection of a wide variety of structures, such as small molecules, cations, proteins, enzymes, and nucleic acids.

Thanks to the versatility of oligonucleotides and the responsiveness of AgNCs, DNA–AgNCs can be employed to build complex sensing devices with high selectivity and sensibility as well as nano devices that perform basic logic operations.

The application of fluorescent DNA–AgNCs in cell cultures has been demonstrated, allowing the visualization of the organelles and proteins. However, more studies are required to assert that AgNCs are a real alternative to the standard dyes currently used in cell culture and *in vivo*.

Besides the fluorescent properties of DNA–AgNCs, their ease of preparation as well as their reduced cost will further promote their use in the near future. Some of their applications may possibly become routine practice in research laboratories.

References

[1] I. Díez, and R. H. A. Ras, Fluorescent silver nanoclusters, *Nanoscale*, **3**, 1963–1970 (2011).

[2] W. A. de Heer, The physics of simple metal clusters: experimental aspects and simple models, *Rev. Mod. Phys.*, **65**, 611–676 (1993).

[3] J. Zheng, C. Zhang, and R. M. Dickson, Highly fluorescent, water-soluble, size-tunable gold quantum dots, *Phys. Rev. Lett.*, **93**, 077402/1–077402/4 (2004).

[4] B. L. Li, J. R. Chen, H. Q. Luo, and N. B. Li, Electrocatalytic activity of polymer-stabilized silver nanoclusters for hydrogen peroxide reduction, *J. Electroanal. Chem.*, **706**, 706, 64–68 (2013).

[5] L. Li, Q. Guo, J. Li, W. Yan, C. Leng, H. Tang, Q. Lu, and B. Tan, Design of a polymer ligand for the one-step preparation of highly stable fluorescent Ag5 clusters for tissue labeling, *J. Mater. Chem. B*, **1**, 3999–4004 (2013).

[6] H. Zhang, X. Huang, L. Li, G. Zhang, I. Hussain, Z. Li, and B. Tan, Photoreductive synthesis of water-soluble fluorescent metal nanoclusters, *Chem. Commun.*, 567–569 (2012).

[7] X. Wang, S. Xua, and W. Xu, Synthesis of highly stable fluorescent Ag nanocluster @ polymer nanoparticles in aqueous solution, *Nanoscale*, **3**, 4670–4675 (2011).

[8] X. Sun, S. Dong, and E. Wang, One-step preparation and characterization of poly(propyleneimine) dendrimer-protected silver nanoclusters, *Macromolecules*, **37**, 7105–7108 (2004).

[9] J. Zheng and R. M. Dickson, Individual water-soluble dendrimer-encapsulated silver nanodot fluorescence, *J. Am. Chem. Soc.*, **124**, 13982–13983 (2002).

[10] J. Yu, S. A. Patel, and R. M. Dickson, *In vitro* and intracellular production of peptide-encapsulated fluorescent silver nanoclusters, *Angew. Chem., Int. Ed.*, **46**, 2028–2030 (2007).

[11] S. Saha, A. Pal, S. Pande, S. Sarkar, S. Panigrahi, and T. Pal, Alginate gel-mediated photochemical growth of mono- and bimetallic gold and silver nanoclusters and their application to surface-enhanced Raman scattering, *J. Phys. Chem. C*, **113**, 7553–7560 (2009).

[12] F. Bouchama, M. B. Thathagar, G. Rothenberg, D. H. Turkenburg, and E. Eiser, Self-assembly of a hexagonal phase of wormlike micelles containing metal nanoclusters, *Langmuir*, **20**, 477–483 (2004).

[13] E. Eiser, F. Bouchama, M. B. Thathagar, and G. Rothenberg, Trapping metal nanoclusters in "soap and water" soft crystals, *ChemPhysChem*, **4**, 526–528 (2003).

[14] D. Schultz and E. Gwinn, Stabilization of fluorescent silver clusters by RNA homopolymers and their DNA analogs: C,G versus A,T(U) dichotomy, *Chem. Commun.*, 4715–4717 (2011).

[15] E. Braun, Y. Eichen, U. Sivan, and G. Ben-Yoseph, DNA-templated assembly and electrode attachment of a conducting silver wire, *Nature*, **391**, 775–778 (1998).

[16] J. T. Petty, J. Zheng, N. V. Hud, and R. M. Dickson, DNA-templated Ag nanocluster formation, *J. Am. Chem. Soc.*, **126**, 5207– 5212 (2004).

[17] C. I. Richards, S. Choi, J.-C Hsiang, Y. Antoku, T. Vosch, A. Bongiorno, Y.-L. Tzeng, and R. M. Dickson, Oligonucleotide-stabilized Ag nanocluster fluorophores, *J. Am. Chem. Soc.*, **130**, 5038–5039 (2008).

[18] T. Vosch, Y. Antoku, J.-C. Hsiang, C. I. Richards, J. I. Gonzalez, and R. M. Dickson, Strongly emissive individual DNA-encapsulated Ag nanoclusters as single-molecule fluorophores, *Proc. Natl. Acad. Sci. U.S.A.*, **104**, 12616 –12621 (2007).

[19] J. T. Petty, C. Fan, S. P. Story, B. Sengupta, A. St. John Iyer, Z. Prudowsky, and R. M. Dickson, DNA encapsulation of 10 silver atoms producing a bright, modulatable, near-infrared-emitting cluster, *J. Phys. Chem. Lett.*, **1**, 2524–2529 (2010).

[20] A. Bavec, Constructing glucagon like peptide-1 receptor fused with derivatives of GFP for visualizing protein-protein interaction in living cells, *Mol. Biol. Rep.*, **37**, 2749–2755 (2010).

[21] U. Resch-Genger, M. Grabolle, S.Cavaliere-Jaricot, R. Nitschke, and T. Nann, Quantum dots versus organic dyes as fluorescent labels, *Nat. Methods*, **5**, 763–775 (2008).

[22] B. Han and E. Wang, DNA-templated fluorescent silver nanoclusters, *Anal. Bioanal. Chem.*, **402**, 129–138 (2012).

[23] A. Latorre and Á. Somoza, DNA-mediated silver nanoclusters: synthesis, properties and applications, *ChemBioChem,* **13**, 951–958 (2012).

[24] A. Latorre, R. Lorca, and Á. Somoza, Fluorescent DNA stabilized silver nanoclusters as biosensors, *J. Chem.*, 2013, 631421/0–631421/6 (**2013**).

[25] J. M. Obliosca, C. Liua, and H.-C. Yeh, Fluorescent silver nanoclusters as DNA probes, *Nanoscale*, **5**, 8443–8461 (2013).

[26] J. T. Petty, S. P. Story, J.-C. Hsiang, and R. M. Dickson, DNA-Templated Molecular Silver Fluorophores, *J. Phys. Chem. Lett.*, **4**, 1148–1155 (2013).

[27] J. M. Obliosca, C. Liu, R. A. Batson, M. C. Babin, J. H. Werner, and H.-C. Yeh, DNA/RNA detection using DNA-templated few-atom silver nanoclusters, *Biosensors*, **3**, 185–200 (2013).

[28] Y. Antoku, J.-I. Hotta, H. Mizuno, R. M. Dickson, J. Hofkens, and T. Vosch, Transfection of living HeLa cells with fluorescent poly-cytosine encapsulated Ag nanoclusters, *Photochem. Photobiol. Sci.*, **9**, 716–721 (2010).

[29] B. Sengupta, C. M. Ritchie, J. G. Buckman, K. R. Johnsen, P. M. Goodwin and J. T. Petty, Base-directed formation of fluorescent silver clusters, *J. Phys. Chem. C*, **112**, 18776–18782 (2008).

[30] S. A. Patel, C. I. Richards, J.-C. Hsiang, and R. M. Dickson, Water-soluble Ag nanoclusters exhibit strong two-photon-induced fluorescence, *J. Am. Chem. Soc.*, **130**, 11602–11603 (2008).

[31] J. Sharma, H.-C. Yeh, H. Yoo, J. H. Werner, and J. S. Martinez, A complementary palette of fluorescent silver nanoclusters, *Chem. Commun.*, 3280–3282 (2010).

[32] J. Sharma, R. C. Rocha, M. L. Phipps, H.-C. Yeh, K. A. Balatsky,D. M. Vu, A. P. Shreve, J. H. Wernera, and J. S. Martinez, A DNA-templated fluorescent silver nanocluster with enhanced stability, *Nanoscale*, **4**, 4107–4110 (2012).

[33] C. M. Ritchie, K. R. Johnsen, J. R. Kiser, Y. Antoku, R. M. Dickson, and J. T. Petty, Ag nanocluster formation using a cytosine oligonucleotide template, *J. Phys. Chem. C.*, **111**, 175–181 (2007).

[34] V. Soto-Verdugo, H. Metiu, and E. Gwinn, The properties of small Ag clusters bound to DNA bases, *J. Chem. Phys.*, **132**, 195102/1–195102/10 (2010).

[35] E. G. Gwinn, P. O'Neill, A. J. Guerrero, D. Bouwmeester, and D. K. Fygenson, Sequence-dependent fluorescence of DNA-hosted silver nanoclusters, *Adv. Mater.*, **20**, 279–283 (2008).

[36] H.-C. Yeh, J. Sharma, J. J. Han, J. S. Martinez, and J. H. Werner, A DNA-silver nanocluster probe that fluoresces upon hybridization, *Nano Lett.*, **10**, 3106–3110 (2010).

[37] H.-C. Yeh, J. Sharma, I.-M. Shih, D. M. Vu, J. S. Martinez, and J. H. Werner, A fluorescence light-up Ag nanocluster probe that discriminates single-nucleotide variants by emission color, *J. Am. Chem. Soc.*, **134**, 11550–11558 (2012).

[38] B. Sengupta, K. Springer, J. G. Buckman, S. P. Story, O. H. Abe, Z. W. Hasan, Z. D. Prudowsky, S. E. Rudisill, N. N. Degtyareva, and J. T. Petty, DNA templates for fluorescent silver clusters and i-motif folding, *J. Phys. Chem. C*, **113**, 19518–19524 (2009).

[39] Y. Fu, J. Zhang, X. Chen, T. Huang, X. Duan, W. Li, and J. Wang, Silver nanomaterials regulated by structural competition of G-/C-rich oligonucleotides, *J. Phys. Chem. C*, **115**, 10370–10379 (2011).

[40] J. Ai, W. Guo, B. Li, T. Li, D. Li, and E. Wang, DNA G-quadruplex-templated formation of the fluorescent silver nanocluster and its application to bioimaging, *Talanta*, **88**, 450–455 (2012).

[41] L. Feng, Z. Huang, J. Ren, and X. Qu, Toward site-specific, homogeneous and highly stable fluorescent silver nanoclusters fabrication on triplex DNA scaffolds, *Nucleic Acids Res.*, **40**, e122 (2012).

[42] A. Latorre, R. Lorca, F. Zamora, and Á. Somoza, Enhanced fluorescence of silver nanoclusters stabilized with branched oligonucleotides, *Chem. Commun.*, 4950–4952 (2013).

[43] G. Liu, D.-Q. Feng, X. Mu, W. Zheng, T. Chen, L. Qi, and D. Li, DNA-functionalized silver nanoclusters as a chemopalette: tunable fluorescence for turn-on detection of cysteine, *J. Mater. Chem. B*, **1**, 2128–2131 (2013).

[44] T. Li, L. Zhang, J. Ai, S. Dong, and E. Wang, Ion-tuned DNA/Ag fluorescent nanoclusters as versatile logic device, *ACS Nano*, **5**, 6334–6338 (2011).

[45] B. Han and E. Wang, Oligonucleotide-stabilized fluorescent silver nanoclusters for sensitive detection of biothiols in biological fluids, *Biosens. Bioelectron.*, **26**, 2585–2589 (2011).

[46] J. Yu, S. Choi and R. M. Dickson, Shuttle-based fluorogenic silver-cluster biolabels, *Angew. Chem., Int. Ed.*, **48**, 318–320 (2009).

[47] G.-Y. Lan, W.-Y. Chen, and H.-T. Chang, One-pot synthesis of fluorescent oligonucleotide Ag nanoclustersfor specific and sensitive detection of DNA, *Biosens. Bioelectron.*, **26**, 2431–2435 (2011).

[48] C. I. Richards, J.-C. Hsiang, D. Senapati, S. Patel, J. Yu, T. Vosch, and R. M. Dickson, Optically modulated fluorophores for selective fluorescence signal recovery, *J. Am. Chem. Soc.*, **131**, 4619–4621 (2009).

[49] M. L. Neidig, J. Sharma, H.-C. Yeh, J. S. Martinez, S. D. Conradson, and A. P. Shreve, Ag K-edge EXAFS analysis of DNA-templated fluorescent silver by DNA sequence variations, *J. Am. Chem. Soc.*, **133**, 11837–11839 (2011).

[50] J. Wu, Y. Fu, Z. He, Y. Han, L. Zheng, J. Zhang, and W. Li, Growth mechanisms of fluorescent silver clusters regulated by polymorphic DNA templates: a DFT study, *J. Phys. Chem. B*, **116**, 1655–1665 (2012).

[51] R. R. Ramazanov and A. I. Kononov, Excitation spectra argue for threadlike shape of DNA-stabilized silver fluorescent clusters, *J. Phys. Chem. C*, **117**, 18681–18687 (2013).

[52] D. Schultz, K. Gardner, S. S. R. Oemrawsingh, N. Markešević, K. Olsson, M. Debord, D. Bouwmeester, and E. Gwinn, Evidence for rod-shaped DNA-stabilized silver nanocluster emitters, *Adv. Mater.*, **25**, 2797–2803, (2013).

[53] X. Yang, L. Gan, L. Han, E. Wang, and J. Wang, High-yield synthesis of silver nanoclusters protected by DNA monomers and DFT prediction of their photoluminescence properties, *Angew. Chem., Int. Ed. Engl.*, **52**, 2022–2026 (2013).

[54] K. Morishita, J. L. MacLean, B. Liu, H. Jiang, and J. Liu, Correlation of photobleaching, oxidation and metal induced fluorescence quenching of DNA-templated silver nanoclusters, *Nanoscale*, **5**, 2840–2849 (2013).

[55] G.-Y. Lan, C.-C. Huang, and H.-T. Chang, Silver nanoclusters as fluorescent probes for selective and sensitive detection of copper ions, *Chem. Commun.*, 1257–1259 (2010).

[56] G.-Y. Lan, W.-Y. Chen, and H.-T. Chang, Characterization and application to the detection of single-stranded DNA binding protein of fluorescent DNA-templated copper/silver nanoclusters, *Analyst*, **136**, 3623–3628 (2011).

[57] J. Yin, X. He, X. Jia, K. Wang, and F. Xu, Highly sensitive label-free fluorescent detection of $Hg2^+$ ions by DNA molecular machine-based Ag nanoclusters, *Analyst*, **138**, 2350–2356 (2013).

[58] G. Liu, D.-Q. Feng, W. Zheng, T. Chen, and D. Li, An anti-galvanic replacement reaction of DNA templated silver nanoclusters monitored by the light-scattering technique, *Chem. Commun.*, 7941–7943 (2013).

[59] X. Wang, R. Lin, Z. Xu, H. Huang, L. Li, F. Liu, N. Li, and X. Yang, N-acetylcysteine induced quenching of red fluorescent oligonucleotide-stabilized silver nanoclusters and the application in pharmaceutical detection, *Anal. Chim. Acta*, **793**, 79–85 (2013).

[60] Z. Huang, F. Pu, Y. Lin, J. Ren, and X. Qu, Modulating DNA-templated silver nanoclusters for fluorescence turn-on detection of thiol compounds, *Chem. Commun.*, **47**, 3487–3489 (2011).

[61] Z. Zhou, Y. Du, and S. Dong, DNA–Ag nanoclusters as fluorescence probe for turn-on aptamer sensor of small molecules, *Biosens. Bioelectron.*, **28**, 33–37 (2011).

[62] J. Sharma, H.-C. Yeh, H. Yoo, J. H. Werner, and J. S. Martinez, Silver nanocluster aptamers: in situ generation of intrinsically fluorescent recognition ligands for protein detection, *Chem. Commun.*, **47**, 2294–2296 (2011).

[63] J. Li, X. Zhong, H. Zhang, X. C. Le, and J.-J. Zhu, Binding-induced fluorescence turn-on assay using aptamer-functionalized silver nanocluster DNA probes, *Anal. Chem.*, **84**, 5170–5174 (2012).

[64] J.-J. Liu, X.-R. Song, Y.-W. Wang, A.-X. Zheng, G.-N. Chen, and H.-H. Yang, Label-free and fluorescence turn-on aptasensor for protein detection via target-induced silver nanoclusters formation, *Anal. Chim. Acta*, **749**, 70–74 (2012).

[65] Y. Qian, Y. Zhang, L. Lu and Y. Cai, A label-free DNA-templated silver nanocluster probe for fluorescence on–off detection of endonuclease activity and inhibition, *Biosens. Bioelectron.*, **51**, 408–412 (2014).

[66] Y. Zhang, Y. Cai, Z. Qi, L. Lu, and Y. Qian, DNA-templated silver nanoclusters for fluorescence turn-on assay of acetylcholinesterase activity, *Anal. Chem.*, **85**, 8455–8461 (2013).

[67] S. Zhu, X. Zhao, W. Zhang, Z. Liu, W. Qi, S. Anjum, and G. Xu, Fluorescence detection of glutathione reductase activity based on deoxyribonucleic acid-templated silver nanoclusters, *Anal. Chim. Acta*, **786**, 111–115 (2013).

[68] X. Liu, F. Wang, A. Niazov-Elkan, W. Guo, and I. Willner, Probing biocatalytic transformations with luminescent DNA/silver nanoclusters, *Nano Lett.*, **13**, 309–314 (2013).

[69] M. Duan, Y. Peng, L. Zhang, X. Wang, J. Ge, J. Jiang, and R. Yu, DNA-stabilized silver nanoclusters with guanine-enhanced fluorescence as a novel indicator for enzymatic detection of cholesterol, *Anal. Methods*, **5**, 2182–2187 (2013).

[70] Y. Xiao, F. Shu, K. Wong, and Z. Liu, Förster resonance energy transfer-based biosensing platform with ultrasmall silver nanoclusters as energy acceptors, *Anal. Chem.*, **85**, 8493–8497 (2013).

[71] J. Yin, X. He, K. Wang, Z. Qing, X. Wu, H. Shi, and X. Yang, One-step engineering of silver nanoclusters–aptamer assemblies as luminescent labels to target tumor cells, *Nanoscale*, **4**, 110–112 (2012).

[72] Y. Wei, R. Liu, Z. Sun, Y. Wang, Y. Cui, Y. Zhao, Z. Cai, and X. Gao, *Analyst*, **138**, 1338–1341 (2013).

[73] H.-C. Yeh, J. Sharma, J. J. Han, J. S. Martinez, and J. H. Werner, A DNA–silver nanocluster probe that fluoresces upon hybridization, *Nano Lett.*, **10**, 3106–311 (2010).

[74] H.-C. Yeh, J. Sharma, I.-M. Shih, D. M. Vu, J. S. Martinez, and J. H. Werner, A fluorescence light-up Ag nanocluster probe that discriminates single-nucleotide variants by emission color, *J. Am. Chem. Soc.*, **134**, 11550–11558 (2012).

[75] W. Guo, J. Yuan, Q. Dong, and E. Wang, Highly sequence-dependent formation of fluorescent silver nanoclusters in hybridized DNA duplexes for single nucleotide mutation identification, *J. Am. Chem. Soc.*, **132**, 932–934 (2010).

[76] K. Ma, Q. Cui, G. Liu, F. Wu, S. Xu, and Y. Shao, DNA abasic site-directed formation of fluorescent silver nanoclusters for selective nucleobase recognition, *Nanotechnology*, **22**, DOI: 10.1088/0957-4484/22/30/305502 (2011).

[77] J. Park, J. Lee, C. Ban, and W. J. Kim, An approach toward SNP detection by modulating the fluorescence of DNA-templated silver nanoclusters, *Biosens. Bioelectron.*, **43**, 419–424 (2013).

[78] X. Liu, F. Wang, R. Aizen, O. Yehezkeli, and I. Willner, Graphene oxide/nucleic-acid-stabilized silver nanoclusters: functional hybrid materials for optical aptamer sensing and multiplexed analysis of pathogenic DNAs, *J. Am. Chem. Soc.*, **135**, 11832–11839 (2013).

[79] L. Zhang, J. Zhu, S. Guo, T. Li, J. Li, and E. Wang, Photoinduced electron transfer of DNA/Ag nanoclusters modulated by G-quadruplex/hemin complex for the construction of versatile biosensors, *J. Am. Chem. Soc.*, **135**, 2403–2406 (2013).

[80] S. W. Yang and T. Vosch, Rapid detection of microRNA by a silver nanocluster DNA probe, *Anal. Chem.*, **83**, 6935–6939 (2011).

[81] P. Shah, A. Rørvig-Lund, S. B. Chaabane, P. W. Thulstrup, H. G. Kjaergaard, E. Fron, J. Hofkens, S. W. Yang, and T. Vosch, Design aspects of bright red emissive silver nanoclusters/DNA probes for microRNA detection, *ACS Nano*, **6**, 8803–8814 (2012).

[82] M. Zhang, Y.-Q. Liu, C.-Y. Yu, B.-C. Yin, and B.-C. Ye, Multiplexed detection of microRNAs by tuning DNA-scaffolded silver nanoclusters, *Analyst*, **138**, 4812–4817 (2013).

[83] Z. Sun, Y. Wang, Y. Wei, R. Liu, H. Zhu, Y. Cui, Y. Zhao, and X. Gao, Ag cluster–aptamer hybrid: specifically marking the nucleus of live cells, *Chem. Commun.*, 11960–11962 (2011).

[84] J. Yin, X. He, K. Wang, Z. Qing, X. Wu, H. Shi, and X. Yang, One-step engineering of silver nanoclusters–aptamer assemblies as luminescent labels to target tumor cells, *Nanoscale*, **4**, 110–112 (2012).

[85] D. Shangguan, Z. Tang, P. Mallikaratchy, Z. Xiao, and W. Tan, Optimization and modifications of aptamers selected from live cancer cell lines, *ChemBioChem*, **8**, 603–606 (2007).

[86] Y.-F. Huang, D. Shangguan, H. Liu, J. A. Phillips, X. Zhang, Y. Chen, and W. Tan, Molecular assembly of an aptamer-drug conjugate for targeted drug delivery to tumor cells, *ChemBioChem*, **10**, 862–868 (2009).

[87] J. Li, X. Zhong, H. Zhang, X. C. Le, and J.-J. Zhu. One-pot synthesis of aptamer-functionalized silver nanoclusters for cell-type-specific imaging, *Anal. Chem.*, **84**, 4140–4146 (2012).

[88] E. M. Reyes-Reyes, Y. Teng, and P. J. Bates, A new paradigm for aptamer therapeutic AS1411 action: uptake by macropinocytosis and its stimulation by a nucleolin-dependent mechanism, *Cancer Res.*, **70**, 8617–8629 (2010).

[89] Y. W. Zhou, C. M. Li, Y. Liu, and C. Z. Huang, Effective detection and cell imaging of prion protein with new prepared targetable yellow-emission silver nanoclusters, *Analyst*, **138**, 873–878 (2013).

[90] J. Yu, S. Choi, C. I. Richards, Y. Antoku, and R. M. Dickson, Live cell surface labeling with fluorescent Ag nanocluster conjugates, *Photochem. Photobiol.*, **84**, 1435–1439 (2008).

[91] J. Li, X. Jia, D. Li, J. Ren, Y. Han, Y. Xia, and E. Wang, Stem-directed growth of highly fluorescent silver nanoclusters for versatile logic devices, *Nanoscale*, **5**, 6131–6138 (2013).

[92] Z. Zhou, Y. Liu, and S. Dong, DNA-templated Ag nanoclusters as signal transducers for a label-free and resettable keypad lock, *Chem. Commun.*, 3107–3109 (2013).

[93] P. R. O'Neill, K. Young, D. Schiffels, and D. K. Fygenson, Few-atom fluorescent silver clusters assemble at programmed sites on DNA nanotubes, *Nano Lett.*, **12**, 5464–5469 (2012).

[94] Y. Tao, Z. Li, E. Ju, J. Ren, and X. Qu, One-step DNA-programmed growth of CpG conjugated silver nanoclusters: a potential platform for simultaneous enhanced immune response and cell imaging, *Chem. Commun.*, **49**, 6918–6920 (2013).

Part V

Alternative DNA Structures, Switches and Nanomachines

5.1

Structure and Stabilization of CGC$^+$ Triplex DNA

Fang Pu and Jinsong Ren

Laboratory of Chemical Biology and State Key Laboratory of Rare Earth Resources Utilization, Changchun Institute of Applied Chemistry, Chinese Academy of Sciences, Changchun, China

5.1.1 Introduction

Nucleic acids exist in a variety of secondary structures in addition to the canonical B-form DNA structure, depending on the sequence, ionic environment, temperature, solvent, and the presence of ligands [1–3]. Conformational transition of nucleic acids under different conditions has received much attention due to its influence on the functionality of the nucleic acids. In 1957, Felsenfeld and coworkers found that a chain of polyriboadenylic acid (polyrA) and two chains of polyribouridylic acid (polyrU) could form a three-stranded structure in the presence of Mg^{2+} ions [4, 5]. During the three decades after the initial discovery, studies focused on the structures and physico-chemical properties of triplex structures composed of various polynucleotides. In 1987, it was demonstrated that a short oligonucleotide could bind sequence specifically to an oligopurine–oligopyrimidine double helix to form a triple-helical structure through hydrogen bonds [6, 7]. The hydrogen bonds involved in triplex formation are different from those in Watson–Crick base pairing in duplex nucleic acids, and are referred to as Hoogsteen hydrogen bonds, or reverse Hoogsteen hydrogen bonds. Since then, tremendous progress has been made. Sequence-specific recognition of duplex DNA by forming triplex DNA structure renders the third strands as potentially useful in the inhibition of gene expression, in site-directed DNA cleavage and repair, and as tools in biotechnology [8–10]. Herein, a focus on the structure and stabilization of triplex DNA is described.

DNA in Supramolecular Chemistry and Nanotechnology, First Edition.
Edited by Eugen Stulz and Guido H. Clever.

5.1.2 Classification of DNA triplets

The canonical DNA triplex structure is generally formed between an oligonucleotide and a homopurine–homopyrimidine duplex. The third strand (triplex-forming oligonucleotide, TFO) binds in the major groove of the duplex and forms hydrogen bonds with the hydrogen bond donor and acceptor groups available on the major groove edge of the purine bases. On the basis of base sequence composition and the relative orientation of the backbone of the third strand to that of the purine strand of the duplex, the triple helices can be separated into three motifs [11–13]. (1) In the pyrimidine motif (TC triplex), a TFO binds parallel to the purine strand of the duplex, forming T•A*T and C•G*C triplets in the Hoogsteen configuration (the first two bases form the Watson–Crick pair, the third base refers to the Hoogsteen strand, and the symbols "dot" and "asterisk" refer to Watson–Crick and non-Watson–Crick pairings, respectively). Generally, the formation of a C•G*C triplet requires protonation of the N3 atoms of the third strand's cytosines under acidic conditions. Thus, the triplet can be written as CGC^+. (2) In the purine motif (GA triplex), a TFO binds antiparallel to the purine strand of the duplex and forms C•G*G and T•A*A triplets via reverse Hoogsteen bonds. (3) In the purine–pyrimidine motif (GT triplex), C•G*G and T•A*T triads can form in either the Hoogsteen or reverse Hoogsteen configuration. The hydrogen bonding schemes found in these triplets are depicted in Figure 5.1.1. Furthermore, some triplexes are hybrids between these different classes. For example, a TCG motif contains T•A*T, C•G*C, and C•G*G triplets in the same structure in the Hoogsteen configuration [14]. C•G*G, T•A*A, and T•A*T triplets can also form a mixed GAT motif in the reverse Hoogsteen configuration.

According to where the TFO originates from, triplexes can be classified as intramolecular and intermolecular triplexes. In the intramolecular triplex, the third strand can come from an endogenous mirror repeat sequence of the same duplex DNA molecule. For example, one half of the pyrimidine strand folds back and binds to the half of the purine strand in the Watson–Crick duplex in a parallel fashion to form TAT and CGC triplets. The half of the displaced complementary purine strand remains single stranded. Owing to the requirement of H^+ to protonate the cytosines in the third strand, this structure is called H-DNA. H*-DNA, consisting of C•G*G and T•A*A triplets in the reverse Hoogsteen configuration, can be formed at neutral pH and in the presence of divalent cations such as Mg^{2+} and Zn^{2+}. A schematic description of the formation of H-DNA and H*-DNA is shown in Figure 5.1.2. A minimum of 15 base pairs of purine–pyrimidine are necessary to form an intramolecular triplex [15, 16]. The energy driving the formation of an intramolecular triplex is the negative superhelical tension [17]. Naturally occurring mirror repeat sequences capable of adopting intramolecular triplex structures are frequently found in mammalian genomes [18]. Intramolecular triplexes may play important roles in the processes of replication, genetic recombination, or gene expression, and are thus receiving significant attention [10, 17]. Moreover, owing to the correct stoichiometry of the three strands, the defined orientation of these strands with respect to each other, and the unfavorable formation of competing structures, intramolecular folded sequences are beneficial to the stability study of triplex structures.

The third strand in the triplex can also result from an exogenously applied molecule to form an intermolecular triplex structure. At least 12 base pairs of purine–pyrimidine runs are required to form a stable triple helix, while shorter triplexes are not sufficiently stable [15, 19]. Intermolecular triplexes have attracted much attention because of their potential therapeutic application in inhibiting the expression of genes involved in cancer and other human diseases.

5.1.3 Structure of triplexes

The structural features of nucleic acid triplexes have been studied using different techniques. X-ray fiber diffraction was first employed to deduce the structure of poly(A)•2poly(U) and poly(dA)•2poly(dT) [20, 21]. The data obtained demonstrated the formation of TAT triplets with an axial rise per residue of 3.26 Å and 12 residues per turn [20]. An A-type model with all the sugars in the C3′-endo conformation was

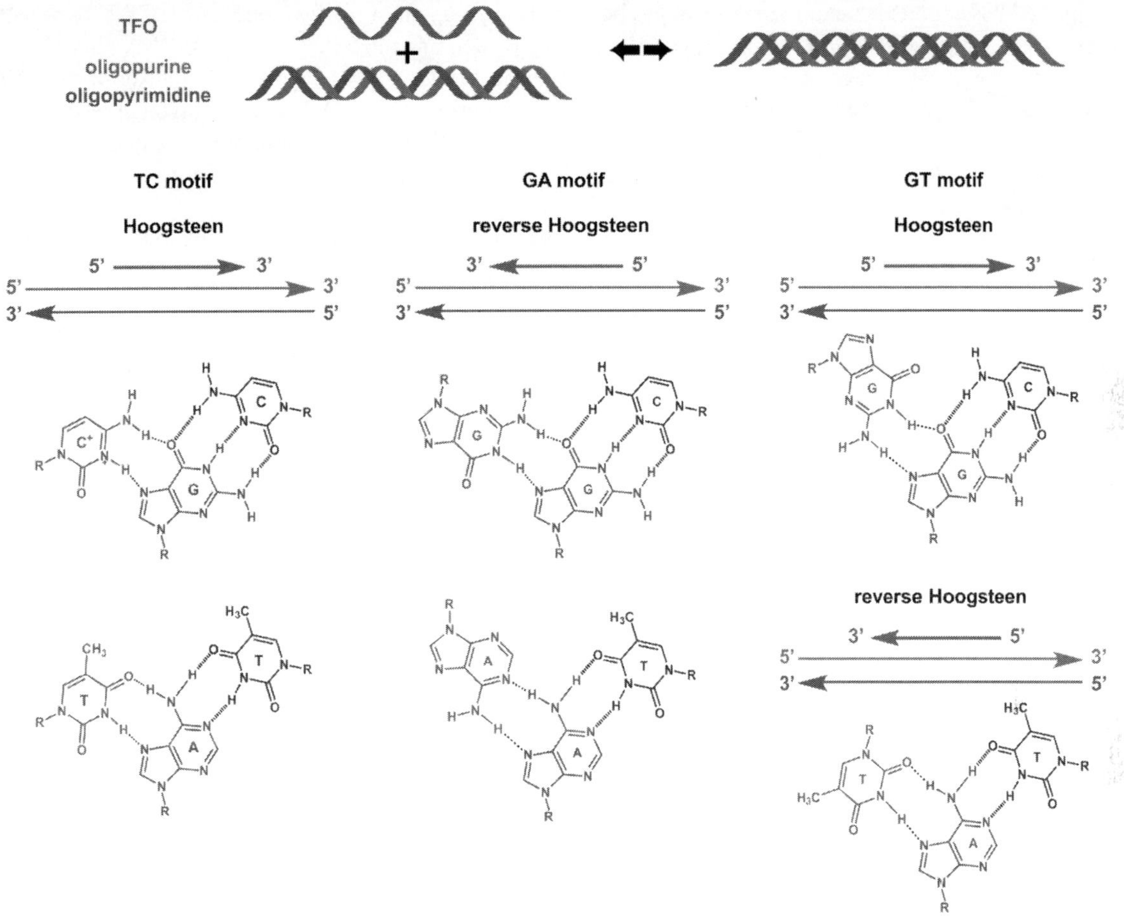

Figure 5.1.1 *Representative scheme of the triplet motifs involved in the triplex structures (See color figure in color plate section)*

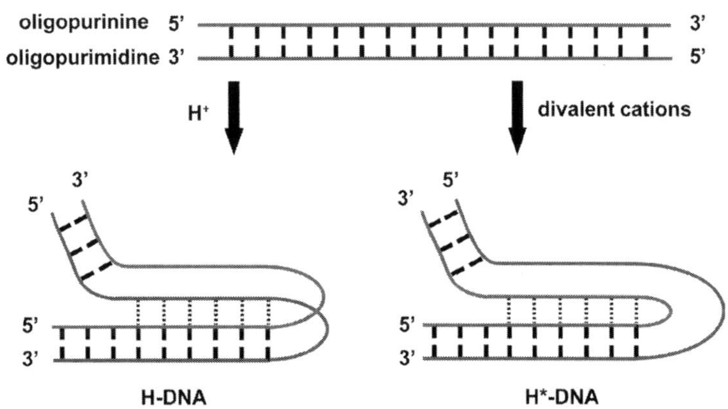

Figure 5.1.2 *Schematic illustration of the formation of intramolecular triple helices*

suggested based on these data. This model has been widely accepted in the literature. However, subsequent stereochemical modeling and infrared spectroscopy studies demonstrated that the triplex DNA structure is structurally and conformationally similar to the double helical B-form DNA with a sugar pucker in the C2′-endo region [22, 23]. The triplex structure can be readily determined by NMR at the individual base-triple level with distinct markers differentiating between Watson–Crick and Hoogsteen pairing. A triplex structure that is closer to B-DNA than A-DNA was suggested [24]. The protonation of cytosines in triplexes has been unambiguously observed [25–30].

In 1997, Nunn *et al.* observed a single CGC base triplet in the crystal structure of an oligonucleotide–drug complex between the minor-groove drug netropsin and the decanucleotide $d(CGCAATTGCG)_2$ [31]. Based on this experimental observation, they calculated the structure of a 10-mer triplex, which has the B-type family characteristics for the target duplex, since its Watson–Crick base pairs are perpendicular to the helix axis and all the sugar groups of three strands have C2′-endo puckers. Meanwhile, the third strand of the triplex has a backbone conformation closer to the A-type family with the C^+ bases in the third strand tilted out of the plane of the GC base pairs. The model implies that the structures of triplexes cannot be simply described as classic A- or B-type families, and they could even show features of both.

Single crystals of $d(CTCCT_SCCGCGCG)$•d(CGCGCGGAG) (containing one phosphorothioate modification in the backbone) have been grown by the vapor-diffusion method [32]. The overall triplex dimensions differ somewhat from those of B-form DNA, but the general topology remains similar [33]. The presence of CGC triplets causes sequence-specific changes in the groove width of the Crick–Hoogsteen strands by the electrostatic interactions between the C^+ bases and the phosphate groups of the Crick strand. These conformational changes restrict the access of water to the minor groove and may be unique structural features for the recognition of triplex DNA by protein molecules.

Sugimoto *et al.* employed circular dichroism (CD) to characterize the conformation of a pyrimidine-purine-pyrimidine triplex [34]. The results indicate that the Hoogsteen strand would be located right in the center of the major groove of the host duplex to form the parallel triplex. Meanwhile, the addition of the Hoogsteen strand needs only small readjustments of the duplex to adapt to this conformational change without significantly distorting the duplex structure. Similar observations in CD studies were obtained by other research teams [35–37].

Soliva *et al.* reported the data of molecular dynamics simulations on the triplexes containing CGC triplets in aqueous solution [38]. The third strand divides the major groove of the target duplex into two grooves with different sizes: the major part (MM) and the minor part (mM) of the major groove. The MM groove is very wide (around 15–16 Å for the shortest interstrand phosphate–phosphate distance), and is only slightly less (around 1–2 Å) than that of a TAT triplex found previously [39]. The width of the MM groove alone is similar to that of the major groove of B-DNA, which means that the binding of the third strand to duplex DNA significantly widens the major groove of the duplex. Thus, the MM groove is wide enough to function as a recognition site for molecules. The mM groove is narrow (around 9 Å) and fairly shallow. The width and depth of the minor groove of the triplex are almost identical to those of B-type duplex DNA. The shortest average distance between phosphates across this groove is about 11–12 Å, which gives a groove width of 5.8 Å, matching typical values for B-type duplex DNA (5.9 Å). However, the groove shapes of the B-type duplex and triplex are rather different in cross-section. The former is U-shaped, while the latter is V-shaped. The minor groove of the triplex could be a target for specific minor groove binders.

5.1.4 Triplex stabilizing factors

Among the structures mentioned previously, parallel triplexes containing TAT and CGC triplets have been studied most widely. However, the applications of these triplexes *in vitro* and *in vivo* are limited due to their relatively low thermodynamic stability and slow kinetic formation, compared with the parent duplex DNA.

This is a result of fewer hydrogen bonds being involved in the formation of triplex DNA and the unfavorable electrostatic repulsion between the negatively charged phosphate backbones. The stability of the triplexes should be dependent on the length and composition of the three strands, pH values, and the presence of stabilizing agents such as metal ions and binding ligands. For intermolecular triplexes, the formation and stabilization would also depend on the concentration and ratio of the target duplex and the TFO.

5.1.4.1 Effect of pH on the triplex

The pH of the solution is an essential parameter for the CGC^+ triplet. The formation of the CGC^+ triplet requires protonation of the N3 atom of cytosine under acidic conditions for appropriate Hoogsteen bonding with the N7 atom of guanine. The pK_a of free deoxycytidine is about 4.3, and the value is higher in the triplex state, resulting from the different local environment and the hydrogen bonds with guanine. Consequently, the CGC^+ triplet formation is pH dependent and of relatively low stability at neutral pH.

Melting studies are commonly employed to detect the thermodynamic stability of triplexes. At acidic pH, direct melting of the triplex into single strands can be observed, while at higher pH, two transitions are observed [40]. The low-temperature transition exhibits clear pH dependence, suggesting the conversion of a triplex into a duplex and a single strand. The transition at higher temperatures is either pH-independent between pH 6 and 7 [41, 42], or shows a slight reduction with decreasing pH values [34], involving the separation of the Watson–Crick DNA duplex. At pH 8.0, the low temperature transition disappears, while the high temperature transition remains.

The formation pathway of a triplex is also significantly pH-dependent. In an acidic environment, the triplex is formed by a one-step docking. Under neutral conditions, the target duplex DNA structure forms first and then accepts the third strand into its major groove [34]. The difference in thermodynamic stability between the protonated and unprotonated CGC triplets is so large that it cannot merely be attributed to the energy associated with the formation of a single hydrogen bond. The contributions from electrostatic interactions with the negatively charged phosphate backbones and the base-stacking interaction should not be ruled out [43]. Also, it has been shown that triplexes containing CGC^+ triplets are more stable than those consisting of TAT triplets below pH 7.0 [43–45]. This observation is attributed to the fact that the positive charge on C^+ can partly overcome the charge repulsion and favor disposition of this positive charge within the π-stack [46]. However, the unprotonated CGC triplet with only one hydrogen bond displays less stability compared with the TAT triplet [47, 48].

5.1.4.2 Effect of cations on the triplex

The formation of triplexes requires a reduction in the repulsion between the negatively charged phosphate groups of the three strands. Monovalent cations, including Li^+, Na^+, and K^+, can shielded the repulsion. Plum *et al.* found that 200 mM Na^+ were sufficient to induce complete formation of a pyrimidine–purine–pyrimidine intermolecular triplex at pH 6.5 [41]. In further investigations they found that the dependence of the thermodynamic stability of an intramolecular triplex on the Na^+ ion concentration was different for different pH ranges [42]. At pH 5.0, a single transition shifted to higher temperatures with the increase in the Na^+ ion concentration. At pH 6.0, a single transition broadened and then split into two transitions when the salt concentration increased. However, the low-temperature transition shifted to a lower temperature while the high-temperature transition shifted to a higher temperature. At pH 6.5 and 7.0, the shift of the biphasic melting curves was similar to that at pH 6.0 with a higher salt concentration. At pH 8.0, a monophasic melting curve shifted to higher temperatures when the salt concentration increased. Moreover, it has been suggested that the effect of ionic strength decreases with increasing CGC content [49].

Divalent metal cations, such as Mg^{2+}, Mn^{2+}, Ca^{2+}, and Ba^{2+}, are more effective in shielding the negative charges. Thus, they are more effective than Na^+ in the stabilization of triplex structures *in vitro* [50].

The cations with smaller ionic radius are expected to enhance the triplex stability. Owing to the biological significance of Mg^{2+}, its effect on triplex stabilization was investigated very thoroughly. The stability of triplexes containing TAT increases with increasing concentrations of Mg^{2+} ions at neutral pH [51, 52]. Analysis of cation effects in the pyrimidine motif is complicated by the electrostatic potential created by the protonated cytosines in the third strand. Pilch *et al.* reported that C^{+}•G and T•A Hoogsteen base pairs appear to have similar stability in the presence of Mg^{2+} ions and at low pH [35]. James *et al.* demonstrated that the effect of Mg^{2+} decreases with an increasing content of CGC^{+} triplets [46].

Sugimoto *et al.* have investigated the effect of mixtures of mono- and divalent cations (Na^{+} and Mg^{2+}) on the triplex stability [34]. In the absence of Na^{+} ions, Mg^{2+} ions favor stabilizing the duplex more than the third strand, since Mg^{2+} presents a site-specific binding and is predisposed to bind to phosphate groups. In the presence of Na^{+}, the thermodynamic stability of the duplex is almost Mg^{2+}-concentration independent, while the thermal stability of the triplex is clearly lower than that in the absence of Na^{+}. This was attributed to the competition between the two different cations when binding to the triplex. Since Na^{+} is typically non-site-specific and binds to phosphates and bases, shielding the negative charges where Mg^{2+} would bind, results in the insensitivity of the duplex stability to Mg^{2+} and reduces the stability of the Hoogsteen paired strand.

5.1.4.3 Effect of length and composition of the three strands on the triplex

Firstly, the influence of the position of the cytosine bases within the triplexes is critical to the stability. The triplex is less stable when a single CGC^{+} triplet is situated at the terminal position [43]. The stability increases markedly as the CGC^{+} triplet is moved into the center of the molecule. The lower stability of triplexes containing terminal cytosines might be attributed to the fact that the interaction of the protonated cytosine with only one nearest-neighbor is much weaker than when it stacks with two nearest-neighbors.

Secondly, the number of cytosine bases within the triplexes is another factor related to the stability of the triplex. Asensio *et al.* found that triplexes containing two CGC^{+} triplets are more stable than that containing a single CGC^{+} triplet at low pH. Moreover, similar to the single CGC^{+} triplet, the stability increases as the CGC^{+} triplets are moved into the center of the molecule [43]. Adjacent placing of two protonated cytosine residues has only a small effect on the stability of the triple helix [53]. Furthermore, triplexes containing more CGC^{+} triplets in different positions were analyzed. Keppler and Fox revealed that increasing the proportion of CGC^{+} triplets increases triplex stability under conditions in which the TFO should be protonated [44]. The stabilization results from the effects that the positively charged cytosine has by favoring the interaction with the stacked π-system or screening the charge of the phosphate groups. However, triplexes having consecutive protonated cytosines in the Hoogsteen strand are unstable, which can be explained by the electrostatic repulsion between adjacent protonated cytosines. Molecular dynamics simulations and Poisson–Boltzmann (PB) calculations suggested that Hoogsteen cytosines in triplexes containing contiguous CGC triplets are not fully protonated, but a certain percentage of the imino form exists [38]. The distance between the amino groups of protonated cytosines and the imino groups is very close, which favors a fast proton interchange, without involving water molecules. The presence of neutral Hoogsteen cytosines can screen the electrostatic repulsion between ptotonated cytosines. Therefore, triplexes can be stable when protonated cytosines are separated by at least one neutral base. The most stable triplexes contain alternating CGC^{+} and TAT triplets [46].

Thirdly, the thermodynamic stability of the target duplex also affects the stability of the triplex. Rusling *et al.* designed a variety of different duplexes containing the same 14 bp oligopurine–oligopyrimidine target sequence with different flank sequences on either or both sides [54]. At pH 5.0, they found that the triplex melting temperature increased with increasing length of the duplex, and that GC-tails produced more stable complexes than AT-tails. There is no significant difference between triplexes containing flank sequences on the 3′- or the 5′-ends. Triplexes containing GC pairs at both sides were found to be even more stable. The triplex generated from an intramolecular duplex with GC pairs at both termini presented the highest stability. However, at pH 5.8, all these triplexes displayed little difference in their melting temperatures. These data demonstrate that as the triplex becomes more stable, the melting temperatures become more dependent on the nature of the underlying duplex.

Furthermore, the thermodynamic properties of intramolecular triplexes are slightly dependent on associated junctions or loops. Sugimoto and coworkers compared three parallel pyrimidine DNA triplexes, including one intermolecular triplex, one intramolecular triplex with a 3′-junction, and one intramolecular triplex with a 5′-junction [55]. They found that the intramolecular triplexes with different junction orientations were thermodynamically more stable than the intermolecular triplex. The CD spectra showed that their peaks are similar and differ only in the intensity, suggesting that the presence of junctions would not affect the binding sites of the TFO on the target duplex, but change the groove width and backbone torsion angles. Shimizu *et al.* studied intramolecular triplexes with identical base triads in the stem but with different loop sizes (4, 6, 8, and 10 bases) in supercoiled plasmids [56]. They found that a higher supercoil energy is required to form triplexes with longer loops. Thus, the thermodynamic stability of triplexes with longer loops is lower than that of triplexes with shorter loops. Besides the loop length, the base composition of the loop region also affects the structural transition and triplex stability. Loop regions containing high GC content might require more supercoil energy for triplex formation, leading to more unstable structures compared with loops with a low GC content.

5.1.4.4 Molecular crowding

The stability of triplex structures is sensitive to their aqueous environment. Some groups have studied the molecular crowding effect on the triplex stability. They found that the Hoogsteen base pairs were stabilized in the presence of poly(ethylene glycol) (PEG), used as crowding ligand [57–60]. It was found that on the Watson–Crick and Hoogsteen base pairs the molecular crowding effects were opposite, resulting from the different behaviors of the water molecules binding to the DNA strands [61].

5.1.5 Formation of stable CGC^+ triplex DNA

The requirement for cytosine protonation limits the formation of CGC^+ triplexes in living cells since the cellular pH is usually above 7.0. To overcome this drawback, several strategies have been employed to improve the stability of CGC^+ triplexes. One strategy focuses on the design and use of non-natural nucleosides to form Hoogsteen hydrogen bonds with guanine, at physiological pH. Modification of the sugar–phosphate backbone provides another means of improving the stability of the triplex. Small molecules that can selectively bind to triplex DNA and regulate their stability were designed and synthesized. The attachment of DNA-binding ligands at the termini of TFO is an alternative for stabilizing the triplexes.

5.1.5.1 Analogues mimicking protonated cytosine

In 1984, Morgan and coworkers found that poly(pyrimidine)•poly(purine) DNA (poly[d(Tm^5C)]•poly[d(GA)]) containing 5-methylcytosine could form a triplex at pH values below 8, while the unmethylated analogue poly[d(TC)]•poly[d(GA)] only forms a triplex at pH values below 6 [62]. Similarly, replacement of 2′-deoxycytidine with 5-methyl-2′-deoxycytidine (m^5C, the structure is shown in Figure 5.1.3A) increases the oligonucleotide affinity to the target dsDNA and extends the pH range for binding [63]. This stabilization can be explained as follows: substitution of methyl for hydrogen at position 5 promotes binding of the oligonucleotide via a hydrophobic effect created by the "spine" of methyl groups in the major groove. Moreover,

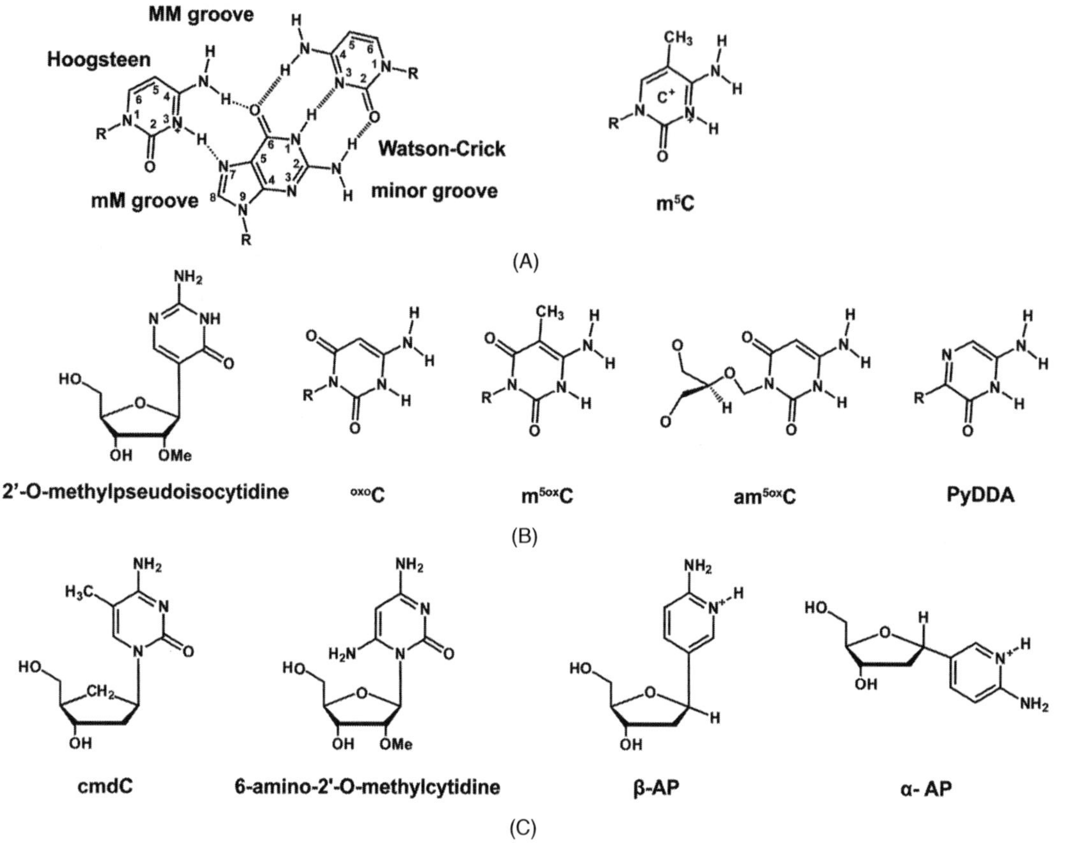

Figure 5.1.3 (A) and (B) Chemical structures of cytosine analogues that mimic protonated cytosine; (C) chemical structures of cytosine analogues with increased pK_a

owing to the increased pK_a of m^5C relative to dC, methylation leads to extension of the pH range (about 0.4 units) for triplex stability [63]. The m^5C was the first base to be shown to stabilize CGC triplexes. Since this discovery, it has frequently been used as a reference nucleobase for stability studies. However, the use of m^5C alleviates the pH limitation only partially and protonation of N3 is still required. However, for the binding of contiguous tracts of GC base pairs on dsDNA by the pyrimidine motif, over a wide range of pH, even DNA containing m^5C may be inadequate. The low stability of triplexes containing adjacent m^5C residues could result from charge–charge repulsion by the adjacent protonated residues.

Neutral pyrimidine analogues with two hydrogen bond donor groups were designed and synthesized to eliminate the requirement for protonation of cytidine and to stabilize parallel triple helices (Figure 5.1.3B). Oligonucleotide analogue containing 2-amino-5-(2-*O*-methyl-β-D-ribofuranyl)-4(1*H*)-pyrimidinone (2′-*O*-methylpseudoisocytidine) substituted for 2′-deoxycytidine has been synthesized [64, 65]. These strands were shown to cause triplex formation in neutral and basic conditions without the requirement for protonation, since this nucleoside already contains one H at the N3 position for hydrogen bonding with N7 of guanine in the Hoogsteen pair of the triad. Besides, methoxy substitution at the 2′-position of pyrimidine nucleosides stabilized the triplex formation [64]. Furthermore, since the base residue is uncharged or unprotonated, DNA containing 2′-*O*-methylpseudoisocytidine could form triplexes with dsDNA containing a homoguanine cluster [65]. Berressem *et al.* found that oligonucleotides containing 6-oxocytosine (^{oxo}C) can form stable triplexes over a pH range of 6.0–8.0 [66]. The electron-withdrawing oxo group can increase the acidity at N3H and the 4-NH_2 group to stabilize the hydrogen bonds to guanines. However, the stability of the triplexes in more acidic solutions is still inferior to oligonucleotides containing the natural cytosine bases. The more electron-poor heterocycle of the 6-oxocytosine bases causes a weaker base stacking and, thus, decreases the stability of the triple helix. This effect could not be compensated by more stable hydrogen bonds.

Another analogue 4-amino-1-(β-2-deoxy-D-erythro-pentofuranosy1)-5-methy1-2,6-[1*H*,3*H*]-pyrimidione ($m^{5ox}C$), a 6-keto derivative of m^5C, was described by Xiang and coworkers [67, 68]. This maintains the 5-methyl group that has been shown to enhance stability in the m^5C–G–C triplet. Its use in targeting dsDNA containing isolated and contiguous G–C base pairs has been studied. For duplex targets containing G–C and A–T base pairs, with each of the G–C base pairs isolated by one or more A–T base pairs, the T_m values are lower than those of the C- or m^5C-containing triplexes at acidic pH. This behavior is caused by differences in the hydrogen-bonding characteristics. C or m^5C can form one charged hydrogen bond, which enhances the triplex stability through providing a shielding effect for the three polyionic strands, while $m^{5ox}C$ forms only an uncharged hydrogen bond. However, there is virtually no variation in the T_m for the $m^{5ox}C$-containing triplex over the pH range 6.4–8.5. The C- or m^5C-containing triplexes would be destabilized with the increase of the pH value, resulting from the loss of hydrogen bonding functionality at the N3 nitrogen of the C (m^5C) residues and the loss of compensating charge effects. The $m^{5ox}C$ analog lacks any compensating positive charge effects, but with the presence of a pH-independent hydrogen bond donor at the pyrimidine N3 nitrogen, stabilized triple helices are formed equally well at pH 6.4 or 8.5.

In addition, TFOs containing $m^{5ox}C$ cannot target dsDNA containing a series of contiguous G–C base pairs. This is attributed to the ineffective base stacking between the analogue residues or the undesirable steric effects resulting from the carbonyl substituent at the 6-position [68]. A sequence containing alternating m^5C and $m^{5ox}C$ bases, however, is effective for targeting dsDNA containing contiguous G–C base pairs. Furthermore, the nucleoside derivatives of the pyrimidine bases $m^{5ox}C$ have been prepared with flexible acyclic carbohydrate linkers ($am^{5ox}C$) [69]. Replacement of the $m^{5ox}C$ residues by the corresponding acyclic derivative simply induces some added flexibility into the third strand, without altering the nature of the base–base interactions. Nevertheless, this alteration results in an increase of the thermal stability over the pH range 7.0–8.5. Vonkrosigk *et al.* reported that the pyDDA base, a pyrimidine analogue based on a pyrazine ring, presented the same hydrogen-bonding pattern as protonated cytosine. It supported triple helix formation that was independent of pH from pH 6.3 to 8.0 [70].

5.1.5.2 Analogues with increased pK_a

The previous approaches to overcome the requirement of protonation consist of the introduction of analogues that emulate the hydrogen-bonding face of an N3-protonated cytosine without the need for protonation. Another approach to increase the affinity of oligonucleotides to target dsDNA at neutral pH is the replacement of the cytidine nucleosides by analogues that, ideally, are completely protonated under physiological conditions and provide the hydrogen-bonding pattern required for triplet formation (Figure 5.1.3C). A carbocyclic analogue of 5-methyl-2′-deoxycytidine (cmdC) was designed by substitution of the furanosyl oxygen with a methylene bridge to generate a cyclopentane ring. Removal of the oxygen of the furanosyl ring has a significant effect on the basicity of the parent heterocycle. The analogue has a slightly higher pK_a than that of the furanosyl nucleoside. When incorporated into an oligopyrimidine third strand, cmdC stabilizes a triple helix relative to the furanosyl nucleoside, presumably due to the increased basicity of the heterocycle [71].

However, the increase in basicity is limited. Pudlo *et al.* synthesized DNA containing 6-amino-2′-*O*-methylcytidine, a protonated cytidine analogue for triple helix binding studies [72]. Although the analogue displayed a pK_a of 6.8, a stable triplex could not form due to unfavorable conformational properties around the nucleosidic bond in this 2,6-substituted pyrimidine nucleoside. Replacement of the nitrogen atom N1 in the pyrimidine ring by carbon and concomitant removal of the 2-oxo function should result in the more basic pyridine C-nucleoside, **P**, without perturbing the hydrogen bonding pattern relevant for triple helix formation [73–76]. The pK_a value is 6.26 for **P**, which is about two pK_a units higher than that of natural cytidine [73, 74]. Owing to its enhanced basicity, it is an excellent substitute for natural cytidine in DNA duplex recognition at physiological pH values. Furthermore, the lack of the 2-oxo function present in cytidine does not perturb third-strand binding in the major groove of DNA. Moreover, Bates *et al.* assessed the ability of the α- and β-anomers of **P** (α-AP and β-AP) to form triplexes [75]. It was found that TFOs containing α-AP had a slightly higher affinity for the target duplex than those with β-AP, since α-AP had a higher pK_a value compared with β-AP. The ODNs containing these non-natural AP nucleosides exhibited increased resistance to nucleases present in the medium, while ODNs containing cytosine were degraded quickly.

5.1.5.3 Backbone modification

In order to change the electrostatic properties and flexibility of the negative phosphodiester backbone of DNA, or to restrict the conformation of the sugar part, different synthetic modifications have been proposed, including phosphate modification, sugar modification, and full backbone replacement (Figure 5.1.4).

Phosphorothioate (PS) or methylphosphonate containing strands show reduced degradation by many nucleases. For phosphorothioate DNA, the presence of a sulfur atom leads to subtle conformational changes due to the larger size of sulfur and changes in the charge distribution around the phosphate group. Moreover, it was expected that the uncharged methylphosphonate substitution can overcome the electrostatic repulsion of the third strand. However, it was found that substitutions of phosphodiesters for phosphorothioates or methylphosphonates destabilize the pyrimidine motif [40, 49, 77]. Replacements of the O3′→P5′ phosphodiester

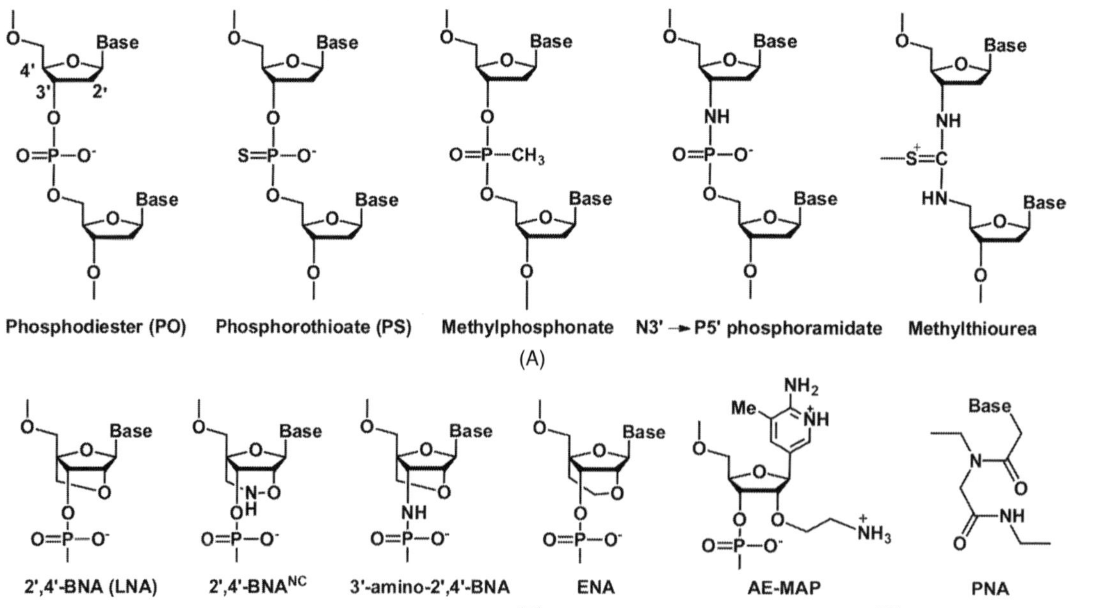

Figure 5.1.4 *Modifications used to enhance triplex stability. (A) Phosphate modifications, (B) sugar modifications, and (C) PNA*

linkage by a N3′→P5′ phosphoramidate linkage can yield more stable triplexes under physiological conditions [78, 79]. Even more drastic modifications of the phosphate group were examined. It was found that substitution of the phosphate group by a methylthiourea is favorable for triplex stability [80, 81].

The sugar modifications were also analyzed extensively. The TC motif can be stabilized by 2′-*O*-methylated TFO substitution [40]. The sugars of the 2′-OMe strand adopted a predominantly C3′-endo conformation, which may be more favorable for triplex formation. Locked nucleic acid (LNA) is an RNA derivative in which the ribose ring is constrained by a methylene linkage between the 2′-oxygen and the 4′-carbon. It was found that TFOs based on LNA could form triplexes at physiological pH values with significantly increased thermostability [82]. In addition, it was reported that 2′-*O*,4′-*C*-aminomethylene-bridged nucleic acid (2′,4′-BNA^{NC}), 3′-amino-2′-*O*,4′-*C*-methylene bridged nucleic acid (3′-amino-2′,4′-BNA), and 2′-*O*,4′-*C*-ethylene bridged nucleic acid (ENA) modifications of TFOs could promote pyrimidine motif triplex formation at physiological pH [83–87]. The phosphoramidite monomer of the C-nucleoside 2′-aminoethoxy-2-amino-3-methylpyridine (AE-MAP) has been synthesized and incorporated into TFOs [88]. It is selective for G–C base pairs.

Full backbone replacement by other polymerizing units has also been explored. Peptide nucleic acid (PNA) is a mimic of nucleic acid in which the natural sugar–phosphate backbone has been replaced by achiral *N*-(2-aminoethyl) glycine units [89]. The lack of charge in the backbone and the special plasticity of PNA make the hybrid structures more stable [90, 91].

Another strategy to enhance triplex stability is to attach a DNA-binding agent to one end of the triplex-binding oligonucleotide. An acridine derivative was covalently linked to the 5′-end of a homopyrimidine oligonucleotide [92]. The acridine intercalates at the triplex–duplex junction and strongly stabilizes the complex formed by the oligopyrimidine with its target duplex sequence. Cytosine methylation further increases the stability of the complexes. A δ-carboline derivative was attached covalently to a TFO at its 5′- or 3′-terminus, which is effective in stabilizing the triplex [93–95]. Introduction of naphthalene diimide [96] or azobenzene [97] to the terminus of the TFO also presents triplex-stabilizing effects.

5.1.5.4 Triplex binding ligands

Molecules that would bind preferentially to triple-helical DNA compared with double-stranded DNA should enhance the stability of triple-helical structures. Besides, if the formation of triple-stranded DNA structures has a biological role, compounds that bind to these structures would provide a new tool for studying their biological functions. So far it has been reported that many ligands can interact more tightly with a triplex than with a duplex, and strongly stabilize the triplex structure through intercalation or minor groove binding (Figure 5.1.5) [16, 98, 99].

The classical DNA intercalator, ethidium bromide (EB), can bind to DNA triple helices by intercalation [100, 101]. This compound stabilizes TAT triplexes slightly, but destabilizes CGC⁺ triplex DNA. This is attributed to electrostatic repulsion between positively charged EB and the protonated cytosines. It was found that a benzo[e]pyridoindole derivative (BePI) showed preferential binding to a triplex rather than to a duplex and strongly stabilized the triplex [102]. The first molecule that was described to stabilize triplexes has four rings with a crescent shape and can optimize stacking interactions with base triplets. Hydrophobic and electrostatic interactions also play important roles. Subsequently, several benzopyridoindole derivatives (BePIs and BgPIs) were synthesized and shown to efficiently stabilize triplexes [103, 104].

The sequence specificity for triplex stabilization was also investigated. It was found that BePIs and BgPIs bind preferentially to triplexes containing a stretch of adjacent TAT triplets. Their stabilization effect is much weaker on a triplex where the stretch of adjacent TAT triplets is interrupted by CGC triplets. Meanwhile, the ability of these ligands to stabilize sequences containing CGC triplets strongly depends on the structure and the charge of the ligands [104]. Benzo[f]quino[3, 4-b]quinoxaline derivatives (BQQ) with a five-membered ring bind strongly to triplexes and show a high discrimination between triplexes and duplexes [105]. A further

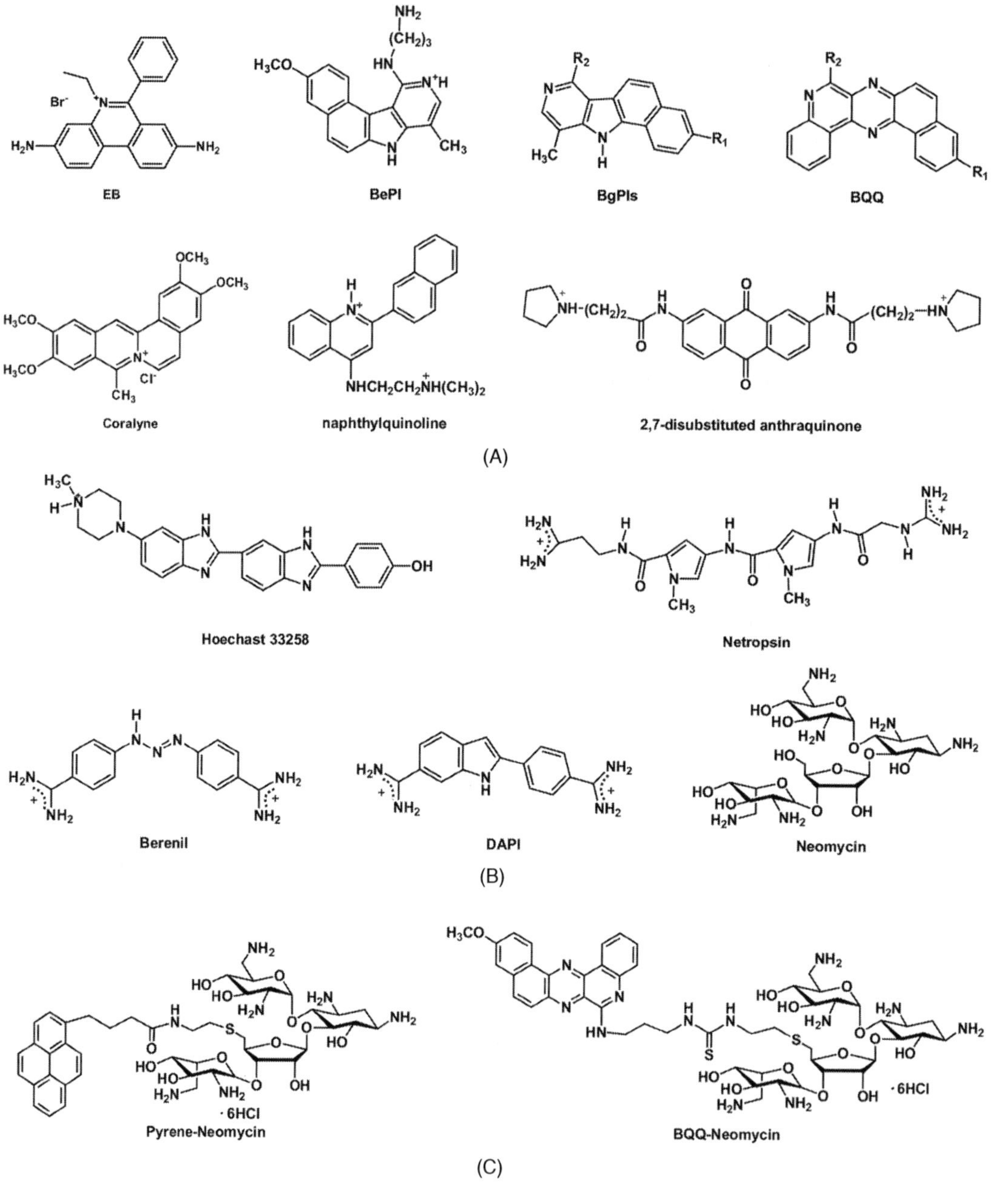

Figure 5.1.5 *Chemical structures of triplex binding ligands. (A) Intercalators, (B) groove binding ligands, and (C) conjugates of two ligands*

aromatic ring can engage in additional stacking interactions with the pyrimidine strand of the Watson–Crick duplex upon binding to the triplex.

Coralyne, a crescent-shape DNA-binding ligand, exhibits significant antitumor activity against leukemias in mice. Lee *et al.* reported that coralyne has a higher affinity for triplexes than for duplexes, especially in the presence of Mg^{2+} ions [106]. Intercalation and stacking interactions are accommodated by the triplexes. Meanwhile, coralyne has little sequence specificity, unlike EB, although coralyne is also positively charged. The large unsaturated ring system of coralyne is required. Less charge in coralyne will be delocalized onto the methoxy groups so that there is no sequence specificity. Contrary to these results, Fox's group reported that coralyne has a moderate preference for triplexes over duplexes, but has a significant binding preference for the TAT triplet over the CGC^+ triplet, even at pH 5.5 [107]. The weaker binding to CGC^+-containing triplexes results in a different conformation, in which the planar chromophore is only partly intercalated.

Quinoline derivatives can provide significant and selective stabilization of the triplex DNA [108–110]. They possess an appropriate aromatic surface, positive charge, and some flexibility in the aromatic system to match the propeller twist of the triple-base interactions in the triplex. For naphthylquinoline, there is a good stacking of the naphthylquinoline ring with the bases at a triplex intercalation site, while the ring is too large to stack optimally with base pairs at a duplex intercalation site. Although naphthylquinoline stabilizes triplexes containing blocks of TAT triplets, it has no effect on triplexes formed with third strands composed of $(TC)_n$ or $(CCT)_n$. Fox's group found that 2,7-disubstituted anthraquinone produced a stabilization of triplexes, except for those containing alternating TAT and CGC^+ triplets, although it is much more effective at stabilizing triplexes that are rich in TAT triplets [111]. The broader sequence selectivity of 2,7-disubstituted anthraquinones compared with naphthylquinolines resulted from the absence of a positive charge on the aromatic ring system. Intercalation of the uncharged anthraquinone ring system between the adjacent protonated cytosines generates a stack of alternating charged and uncharged residues, separating the charged CGC^+ triplets from each other. Triplexes containing alternating TAT and CGC^+ triplets are not stabilized by 2,7-disubstituted anthraquinones, since the ligand would interrupt the alternating pattern of charged and uncharged residues.

The previously mentioned ligands stabilize triplexes by intercalation. The influence of minor groove binding agents on triplexes was also studied (Figure 5.1.5B). Some minor groove binders decrease the thermal stability of triplexes. For example, Hoechst 33258 can bind to the triplex and destabilize the triplex whereas it stabilizes the duplex [98, 112]. Netropsin was reported to bind in the minor groove of DNA and thermally destabilize the triplex–duplex equilibrium dramatically [113]. Some minor groove binding ligands have a different behavior; for example, 1,3-bis(4′-amidinophenyl)triazene (berenil) binds to triplexes in a minor groove binding mode without displacing the major groove-bound third strand [114]. The magnitude and the direction of the effect of berenil on the thermal stability of DNA triplexes depends on both the Na^+ concentration and the drug binding density. The triplex was stabilized at lower and destabilized at higher Na^+ concentrations. 4′,6-Diamidino-2-phenylindole (DAPI) had a much larger effect on the duplex than on the TAT triplex [98]. However, DAPI destabilized the triplex containing mixed TAT and CGC^+ triplets. Neomycin is a very effective aminoglycoside in stabilizing a DNA triple helix without affecting the DNA duplex [115]. It shows a preference for stabilization of TAT triplets but can also accommodate CGC^+ triplets [116]. Binding/modeling studies show a marked preference for neomycin binding to the larger MM groove.

Another class of triplex stabilizing agents has been developed by linking two triplex binders together (Figure 5.1.5C). It was found that pyrene–neomycin conjugates can stabilize a triplex much more effectively than neomycin alone at low concentrations [117]. Computer modeling suggests that the pyrene can intercalate between the base pairs while neomycin stays bound to the MM groove. The BQQ–neomycin conjugate, consisting of a triplex-specific intercalator and a minor groove binder, is more potent in stabilizing DNA triplexes than BQQ, neomycin, or a combination of both [118]. The principle of dual recognition could be applicable to the design of more triplex stabilizing ligands [119].

Although a large number of studies have been conducted on the interaction of triplex DNA with ligands, these ligands stabilize triplexes in a sequence-dependent manner. Most of them stabilize triplexes containing consecutive TAT triplets. Some of them stabilize triplexes which also contain CGC^+ triplets, but with a relatively low efficiency.

5.1.5.5 Other stabilizing effects

5.1.5.5.1 Polyamines

In eukaryotes, spermine and spermidine (structures shown in Figure 5.1.6A) are present in millimolar concentrations and may be as high as 5 mM in the nucleus. These naturally occurring polyamines are largely protonated at physiological pH and exhibit a net positive charge close to +4 and +3, respectively. Their inherent polycationic nature and conformational mobility favor electrostatic interaction with the anionic phosphates of DNA. It was found that spermine and spermidine favor both inter- and intramolecular pyrimidine motif triplexes under physiological pH [120–122]. Firstly, they can be used as free ligands. Upon binding to DNA, they reduce the electrostatic repulsion between the phosphate backbones. The addition of spermine can also alter the relative stability of triplexes containing C, m^5C, and $m^{5ox}C$ significantly, due to its ability to shield the charged phosphates and enhance triplex stability [68]. However, the effect of spermine is more marked on triplexes containing $m^{5ox}C$ than that on triplexes containing C and m^5C because protonated residues already assist in limiting destabilization as a result of charge repulsion effects. Secondly, spermine and other polyamines could be covalently linked to the 5′-terminus of TFOs to form triplexes [123]. It was found that the appended polyamine enhanced the thermal stability of the resulting triple helix at pH 6.5. The conjugation of spermine could eliminate the requirement for free polyamine or Mg^{2+} ions in triplex formation. Barawkar *et al.* combined the properties of m^5C and spermine to design 5-Me-dC-(N^4)-spermine containing TFOs. Stable triplexes formed at physiological pH, even in the absence of Mg^{2+} ions [124, 125]. The spermine attached to the TFO is accommodated in the deep major groove of the duplex along with the TFO. The higher stability of 5-Me-dC-(N^4)-spermine containing triplexes resulted from the extra binding energy between the target duplex and the polyamine appended to the TFO. N3 protonation of C was not observed in the resulting triplex. The loss in stability due to the absence of the Hoogsteen bond between N3 of cytosine and N7 of guanine is compensated by favorable electrostatic interactions between DNA and spermine. Furthermore, 1,11-diamino-3,6,9-trioxaundecane was covalently attached to N4 of 5-Me-dC instead of spermine. The triplex could form at physiological pH through hydrophobic binding [126].

5.1.5.5.2 Basic oligopeptides

Basic oligopeptides, another class of biologically relevant polycations, can also stabilize triplex DNA. Cationic amino acid residues of basic peptides might bind to and neutralize phosphate groups of triple-helical nucleic acids [127]. Potaman and Sinden found that lysine-rich oligopeptides can substitute for other triplex-stabilizing factors including low pH, divalent metal ions, or polyamines. In the presence of lysine-rich peptides, the intermolecular triplex was stabilized up to pH 6.5 [128]. The stabilizing effect of the peptide was much greater than that of Mg^{2+} and a little weaker than that of spermine. A peptide sequence containing five cationic amino acid residues Lys-Lys-Lys-Lys-Lys has a pronounced triplex-stabilizing effect, while peptide sequences containing alternating cationic and neutral residues such as Lys-Gly-Lys-Gly-Lys and Lys-Ala-Lys-Als-Lys had less stabilizing effect. The differences resulted from the positive charge densities. For Lys-Gly-Lys-Gly-Lys and Lys-Ala-Lys-Als-Lys with the same charge densities, the reduced stabilization of the latter might be correlated with the more hydrophobic group within Ala. Since the major groove of the duplex target would be occupied by the TFO and thus is inaccessible, oligopeptides might bind in the minor groove or in the new grooves formed upon TFO binding. The study offered the possibility of tuning triplex stability by changing the sequence and fraction of amino acids residues within the peptides.

Ferdous and Torigoe and coworkers demonstrated that poly(L-lysine)–graft-dextran (PLL-g-Dex) copolymer was extremely good at promoting pyrimidine motif triplex formation at neutral pH (Figure 5.1.6B) [129, 130]. The sequence specificity of triplex formation was not affected since the hydrophilic dextran

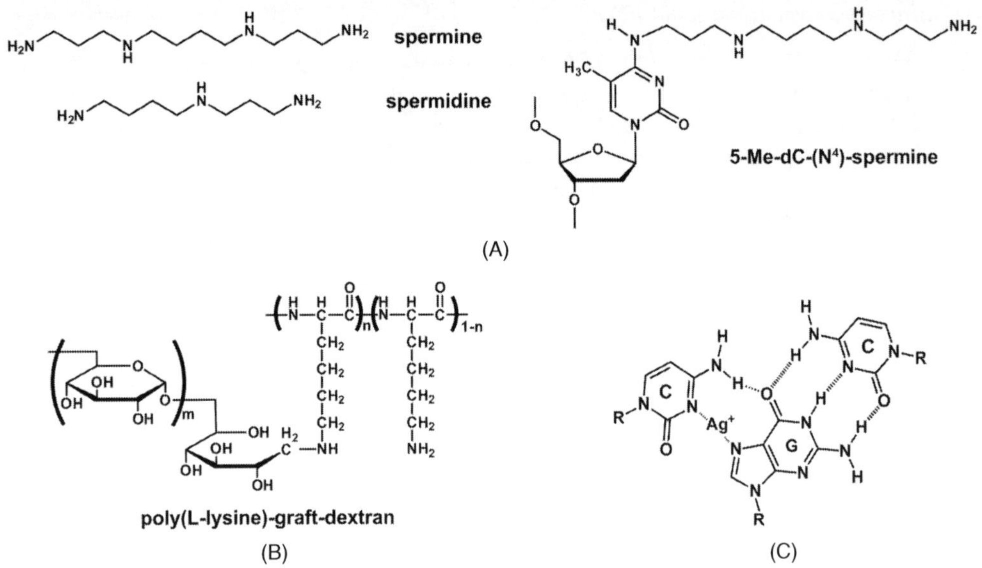

Figure 5.1.6 *(A) Chemical structures of spermine and spermidine, (B) structure of poly(L-lysine)-graft-dextran, and (C) the Ag^+-mediated base triplet CG.CAg+*

chains did not disorder the higher-order structure of the triplex. Moreover, the triplex-promoting efficiency of the copolymer was higher than that of spermine. The major contribution of the copolymer is from the considerable increase in the association rate constant, which increases with the valence of the cation. In addition, the polynucleotides associated with the copolymer may be forced to merge into the dextran-rich phase, which has a low dielectric constant. Such an environment under a low dielectric constant may enhance hydrogen bonding between the TFO and the target duplex to promote triplex formation.

5.1.5.5.3 Silver ions and silver nanoclusters

In 2009, Ihara *et al.* reported an effective alternative method for stabilization of the parallel-motif triple helix of DNA using Ag^+ ions [131]. In the duplexes, Ag^+ ions are placed between the bases to form specific C–Ag^+–C bridges. The silver ions displace an N3 proton of a cytosine in the CGC^+ base triplet to form a CG.CAg^+ (Figure 5.1.6C). This was found to stabilize parallel-motif triplexes even at neutral pH. The shapes of the circular dichroism (CD) spectra in the presence of Ag^+ were rather different from the typical spectra of DNA triplexes. The coordination distance in N–Ag^+–N was longer than that of the Hoogsteen hydrogen bonds in CGC^+. Model studies showed that the cytosines on the third strand are forced into twisting from the plane of Watson–Crick GC pairs in CG.CAg^+ triplets. This nonplanarity of CG.CAg^+ triplets seems to alter the whole structure of the triple helix.

Furthermore, Feng *et al.* found that site-specific formation of fluorescent Ag nanoclusters (AgNCs) could be realized by reducing the Ag^+ ions *in situ* in the triplet CG.CAg^+ base sites [132]. The formation of AgNCs did not destroy the triplex structure. The UV melting studies indicated that melting temperatures were increased in the presence of Ag^+ ions and AgNCs, demonstrating that the triplex retained its integrity during the synthesis of AgNCs. The influence of the number of the set of two successive CGC^+ triplets on AgNCs synthesis was investigated. Emission intensity of AgNCs increased with increasing of number of the set of two successive CGC^+ triplets while the corresponding emission wavelengths remained unchanged.

5.1.5.5.4 Single-walled carbon nanotubes

To date, the influence of intercalative drugs and minor groove binding agents on triple helices has been widely investigated. However, there is no ligand that has been reported to selectively stabilize CGC⁺ triplets rather than TAT. Some groups have turned their sights to the nanomaterials with unique structural and biological properties. Single-walled carbon nanotubes (SWNTs) are recognized as a promising modulator of DNA structures due to their one-dimensional nanoscale geometry, biocompatibility, and controllable surface properties [133–135]. Very recently, Qu's group reported that a duplex d(CT)•d(AG) can be disproportioned into triplex d(CT)•d(AG)•d(C⁺T) and single-stranded d(AG) in the presence of carboxyl-modified SWNTs (SWNTs-COOH) at pH 6.5 [136]. The single-stranded d(CT) has overhangs around the triplex sequence, which may immobilize the triplex on the surface of SWNTs and promote the triplex formation. According to this study, SWNTs act as helicases to catalyze the unwinding of dsDNA [133]. Meanwhile, the interaction between SWNTs and d(CT)•d(AG) can promote protonation of dC residues by changing their pK_a values. Besides SWNTs-COOH, negatively charged hydroxyl-SWNTs can induce triplex formation, while positively charged amino-modified SWNTs cannot induce the formation of triplexes. Electrostatic interaction plays a crucial role in the repartition of duplex d(CT)•d(AG) into a triplex and a single strand. The study suggests that SWNTs can reduce the stringency of conditions that are required for the formation of CGC⁺ triplexes.

5.1.6 Summary

In the past decades, the formation of triplex structures has attained considerable attention due to their important roles in regulating DNA metabolism and gene function. Substantial progress in understanding the structure and energetics of triplex structures has been achieved. However, owing to the instability of the Hoogsteen base pairs at physiological pH, their applications have been severely limited. Many studies have been devoted to increasing the stability of triplexes, including modifications of the base, sugar, and phosphate backbone, using triplex binding ligands, polyamines, oligopeptides, silver ions, and single-walled carbon nanotubes, or conjugation of polyamines and triple helix-specific binding agents. These efforts are valuable for the use of triplex forming oligonucleotides. However, there are still many unanswered questions that are required to be resolved through further investigations.

References

[1] Song, G.; Ren, J., Recognition and regulation of unique nucleic acid structures by small molecules. *Chem. Commun.*, **46**, 7283 (2010).

[2] Ono, A.; Torigoe, H.; Tanaka, Y.; Okamoto, I., Binding of metal ions by pyrimidine base pairs in DNA duplexes. *Chem. Soc. Rev.*, **40**, 5855 (2011).

[3] Choi, J.; Majima, T., Conformational changes of non-B DNA. *Chem. Soc. Rev.*, **40**, 5893 (2011).

[4] Felsenfeld, G.; Davies, D. R.; Rich, A., Formation of a three-stranded polynucleotide molecule. *J. Am. Chem. Soc.*, **79**, 2023 (1957).

[5] Felsenfeld, G.; Rich, A., Studies on the formation of two- and three-stranded polyribonucleotides. *Biochim. Biophys. Acta*, **26**, 457 (1957).

[6] Le Doan, T.; Perrouault, L.; Praseuth, D.; Habhoub, N.; Decout, J. L.; Thuong, N. T.; Lhomme, J.; Helene, C., Sequence-specific recognition, photocrosslinking and cleavage of the DNA double helix by an oligo-[α]-thymidylate covalently linked to an azidoproflavine derivative. *Nucleic Acids Res.*, **15**, 7749 (1987).

[7] Moser, H. E.; Dervan, P. B., Sequence-specific cleavage of double helical DNA by triple helix formation. *Science*, **238**, 645 (1987).

[8] Chan, P. P.; Glazer, P. M., Triplex DNA: fundamentals, advances, and potential applications for gene therapy. *J. Mol. Med.*, **75**, 267 (1997).

[9] Tumpane, J.; Kumar, R.; Lundberg, E. P.; Sandin, P.; Gale, N.; Nandhakumar, I. S.; Albinsson, B.; Lincoln, P.; Wilhelmsson, L. M.; Brown, T.; Norden, B., Triplex addressability as a basis for functional DNA nanostructures. *Nano Lett.*, **7**, 3832 (2007).

[10] Jain, A.; Wang, G.; Vasquez, K. M., DNA triple helices: biological consequences and therapeutic potential. *Biochimie*, **90**, 1117 (2008).

[11] Buchini, S.; Leumann, C. J., Recent improvements in antigene technology. *Curr. Opin. Chem. Biol.*, **7**, 717 (2003).

[12] Duca, M.; Vekhoff, P.; Oussedik, K.; Halby, L.; Arimondo, P. B., The triple helix: 50 years later, the outcome. *Nucleic Acids Res.*, **36**, 5123 (2008).

[13] Buske, F. A.; Bauer, D. C.; Mattick, J. S.; Bailey, T. L., Triplexator: detecting nucleic acid triple helices in genomic and transcriptomic data. *Genome Res.*, **22**, 1372 (2012).

[14] Giovannangeli, C.; Rougee, M.; Garestier, T.; Thuong, N. T.; Helene, C., Triple-helix formation by oligonucleotides containing the three bases thymine, cytosine, and guanine. *Proc. Natl. Acad. Sci. U.S.A.*, **89**, 8631 (1992).

[15] Ghosh, M. K.; Katyal, A.; Brahmachari, V.; Chandra, R., Design and structural analysis of hairpin-TFO for transcriptional activation of genes in S. cerevisiae. *J. Biomol. Struct. Dyn.*, **20**, 265 (2002).

[16] Jain, A. K.; Bhattacharya, S., Groove binding ligands for the interaction with parallel-stranded *ps*-duplex DNA and triplex DNA. *Bioconjug Chem.*, **21**, 1389 (2010).

[17] Bissler, J. J., Triplex DNA and human disease. *Front. Biosci.*, **12**, 4536 (2007).

[18] Schroth, G. P.; Ho, P. S., Occurrence of potential cruciform and H-DNA forming sequences in genomic DNA. *Nucleic Acids Res.*, **23**, 1977 (1995).

[19] Cheng, A. J.; Van Dyke, M. W., Oligodeoxyribonucleotide length and sequence effects on intermolecular purine-purine-pyrimidine triple-helix formation. *Nucleic Acids Res.*, **22**, 4742 (1994).

[20] Arnott, S.; Selsing, E., Structures for the polynucleotide complexes poly(dA)·poly (dT) and poly(dT)·poly(dA)·poly(dT). *J. Mol. Biol.*, **88**, 509 (1974).

[21] Arnott, S.; Bond, P. J.; Selsing, E.; Smith, P. J., Models of triple-stranded polynucleotides with optimised stereochemistry. *Nucleic Acids Res.*, **3**, 2459 (1976).

[22] Raghunathan, G.; Miles, H. T.; Sasisekharan, V., Symmetry and molecular structure of a DNA triple helix: $d(T)_n{\cdot}d(A)_n{\cdot}d(T)_n$. *Biochemistry*, **32**, 455 (1993).

[23] Howard, F. B.; Miles, H. T.; Liu, K.; Frazier, J.; Raghunathan, G.; Sasisekharan, V., Structure of $d(T)_n{\cdot}d(A)_n{\cdot}d(T)_n$: the DNA triple helix has B-form geometry with C2′-endo sugar pucker. *Biochemistry*, **31**, 10671 (1992).

[24] Macaya, R.; Wang, E.; Schultze, P.; Sklenar, V.; Feigon, J., Proton nuclear magnetic resonance assignments and structural characterization of an intramolecular DNA triplex. *J. Mol. Biol.*, **225**, 755 (1992).

[25] de los Santos, C.; Rosen, M.; Patel, D., NMR studies of DNA $(R+)_n{\cdot}(Y-)_n{\cdot}(Y+)_n$ triple helices in solution: imino and amino proton markers of T·A·T and C·G·C+ base-triple formation. *Biochemistry*, **28**, 7282 (1989).

[26] Rajagopal, P.; Feigon, J., Triple-strand formation in the homopurine:homopyrimidine DNA oligonucleotides $d(G\text{-}A)_4$ and $d(T\text{-}C)_4$. *Nature*, **339**, 637 (1989).

[27] Rajagopal, P.; Feigon, J., NMR studies of triple-strand formation from the homopurine-homopyrimidine deoxyribonucleotides $d(GA)_4$ and $d(TC)_4$. *Biochemistry*, **28**, 7859 (1989).

[28] Mooren, M. M.; Pulleyblank, D. E.; Wijmenga, S. S.; Blommers, M. J.; Hilbers, C. W., Polypurine/polypyrimidine hairpins form a triple helix structure at low pH. *Nucleic Acids Res.*, **18**, 6523 (1990).

[29] Sklenar, V.; Feigon, J., Formation of a stable triplex from a single DNA strand. *Nature* 1990, **345**, 836.

[30] Tarkoy, M.; Phipps, A. K.; Schultze, P.; Feigon, J., Solution structure of an intramolecular DNA triplex linked by hexakis(ethylene glycol) units: d(AGAGAGAA-$(EG)_6$-TTCTCTCT-$(EG)_6$-TCTCTCTT). *Biochemistry*, **37**, 5810 (1998).

[31] Nunn, C. M.; Trent, J. O.; Neidle, S., A model for the $[C^{+}\text{-}G{\cdot}C]_n$ triple helix derived from observation of the C^{+}-G·C base triplet in a crystal structure. *FEBS Lett.*, **416**, 86 (1997).

[32] Han, Z. J.; Rhee, S.; Liu, K.; Miles, H. T.; Davies, D. R., Crystallization and preliminary crystallographic study of triple-helical DNA. *Acta Cryst.*, **56**, 104 (2000).

[33] Rhee, S.; Han, Z.; Liu, K.; Miles, H. T.; Davies, D. R., Structure of a triple helical DNA with a triplex-duplex junction. *Biochemistry*, **38**, 16810 (1999).

[34] Sugimoto, N.; Wu, P.; Hara, H.; Kawamoto, Y., pH and cation effects on the properties of parallel pyrimidine motif DNA triplexes. *Biochemistry*, **40**, 9396 (2001).

[35] Pilch, D. S.; Brousseau, R.; Shafer, R. H., Thermodynamics of triple helix formation: spectrophotometric studies on the $d(A)_{10}{\cdot}2d(T)_{10}$ and $d(C^{+3}T4C^{+3}){\cdot}d(G_3A_4G_3){\cdot}d(C_3T_4C_3)$ triple helices. *Nucleic Acids Res.*, **18**, 5743 (1990).
[36] Manzini, G.; Xodo, L. E.; Gasparotto, D.; Quadrifoglio, F.; van der Marel, G. A.; van Boom, J. H., Triple helix formation by oligopurine-oligopyrimidine DNA fragments. Electrophoretic and thermodynamic behavior. *J. Mol. Biol.*, **213**, 833 (1990).
[37] Hashem, G. M.; Wen, J. D.; Do, Q.; Gray, D. M., Evidence from CD spectra and melting temperatures for stable Hoogsteen-paired oligomer duplexes derived from DNA and hybrid triplexes. *Nucleic Acids Res.*, **27**, 3371 (1999).
[38] Soliva, R.; Laughton, C. A.; Luque, F. J.; Orozco, M., Molecular dynamics simulations in aqueous solution of triple helices containing d(G·C·C) trios. *J. Am. Chem. Soc.*, **120**, 11226 (1998).
[39] Shields, G. C.; Laughton, C. A.; Orozco, M., Molecular dynamics simulations of the d(T·A·T) triple helix. *J. Am. Chem. Soc.*, **119**, 7463 (1997).
[40] Mills, M.; Arimondo, P. B.; Lacroix, L.; Garestier, T.; Klump, H.; Mergny, J. L., Chemical modification of the third strand: differential effects on purine and pyrimidine triple helix formation. *Biochemistry*, **41**, 357 (2002).
[41] Plum, G. E.; Park, Y. W.; Singleton, S. F.; Dervan, P. B.; Breslauer, K. J., Thermodynamic characterization of the stability and the melting behavior of a DNA triplex: a spectroscopic and calorimetric study. *Proc. Natl. Acad. Sci. U.S.A.*, **87**, 9436 (1990).
[42] Plum, G. E.; Breslauer, K. J., Thermodynamics of an intramolecular DNA triple helix: a calorimetric and spectroscopic study of the pH and salt dependence of thermally induced structural transitions. *J. Mol. Biol.*, **248**, 679 (1995).
[43] Asensio, J. L.; Lane, A. N.; Dhesi, J.; Bergqvist, S.; Brown, T., The contribution of cytosine protonation to the stability of parallel DNA triple helices. *J. Mol. Biol.*, **275**, 811 (1998).
[44] Keppler, M. D.; Fox, K. R., Relative stability of triplexes containing different numbers of T·AT and C^+·GC triplets. *Nucleic Acids Res.*, **25**, 4644 (1997).
[45] Soto, A. M.; Loo, J.; Marky, L. A., Energetic contributions for the formation of TAT/TAT, TAT/CGC^+, and CGC^+/CGC^+ base triplet stacks. *J. Am. Chem. Soc.*, **124**, 14355 (2002).
[46] James, P. L.; Brown, T.; Fox, K. R., Thermodynamic and kinetic stability of intermolecular triple helices containing different proportions of C^+·GC and T·AT triplets. *Nucleic Acids Res.*, **31**, 5598 (2003).
[47] Volker, J.; Klump, H. H., Electrostatic effects in DNA triple helices. *Biochemistry* 1994, **33**, 13502.
[48] Roberts, R. W.; Crothers, D. M., Prediction of the stability of DNA triplexes. *Proc. Natl. Acad. Sci. U.S.A.*, **93**, 4320 (1996).
[49] Latimer, L. J.; Hampel, K.; Lee, J. S., Synthetic repeating sequence DNAs containing phosphorothioates: nuclease sensitivity and triplex formation. *Nucleic Acids Res.*, **17**, 1549 (1989).
[50] Wu, P.; Kawamoto, Y.; Hara, H.; Sugimoto, N., Effect of divalent cations and cytosine protonation on thermodynamic properties of intermolecular DNA double and triple helices. *J. Inorg. Biochem.*, **91**, 277 (2002).
[51] Pilch, D. S.; Levenson, C.; Shafer, R. H., Structural analysis of the $(dA)_{10}{\cdot}2(dT)_{10}$ triple helix. *Proc. Natl. Acad. Sci. U.S.A.*, **87**, 1942 (1990).
[52] Durand, M.; Peloille, S.; Thuong, N. T.; Maurizot, J. C., Triple-helix formation by an oligonucleotide containing one $(dA)_{12}$ and two $(dT)_{12}$ sequences bridged by two hexaethylene glycol chains. *Biochemistry*, **31**, 9197 (1992).
[53] Asensio, J. L.; Brown, T.; Lane, A. N., Comparison of the solution structures of intramolecular DNA triple helices containing adjacent and non-adjacent CG·C^+ triplets. *Nucleic Acids Res.*, **26**, 3677 (1998).
[54] Rusling, D. A.; Rachwal, P. A.; Brown, T.; Fox, K. R., The stability of triplex DNA is affected by the stability of the underlying duplex. *Biophys. Chem.*, **145**, 105 (2009).
[55] Wu, P.; Hara, H.; Kawamoto, Y.; Sugimoto, N., Effect of cytosine protonation and cation on thermodynamic properties of parallel DNA triplex family. *Nucleic Acids Res. Suppl.*, 39 (2001).
[56] Shimizu, M.; Hanvey, J. C.; Wells, R. D., Intramolecular DNA triplexes in supercoiled plasmids. I. Effect of loop size on formation and stability. *J. Biol. Chem.*, **264**, 5944 (1989).
[57] Spink, C. H.; Chaires, J. B., Selective stabilization of triplex DNA by poly(ethylene glycols). *J. Am. Chem. Soc.*, **117**, 12887 (1995).
[58] Goobes, R.; Minsky, A., Thermodynamic aspects of triplex DNA formation in crowded environments. *J. Am. Chem. Soc.*, **123**, 12692 (2001).

[59] Goobes, R.; Kahana, N.; Cohen, O.; Minsky, A., Metabolic buffering exerted by macromolecular crowding on DNA-DNA interactions: origin and physiological significance. *Biochemistry*, **42**, 2431 (2003).

[60] Nakamura, K.; Karimata, H.; Ohmichi, T.; Miyoshi, D.; Sugimoto, N., Effects of cosolutes on the thermodynamic stability of parallel DNA duplex and triplex. *Nucleic Acids Symp. Ser.*, 167 (2007).

[61] Miyoshi, D.; Nakamura, K.; Tateishi-Karimata, H.; Ohmichi, T.; Sugimoto, N., Hydration of Watson-Crick base pairs and dehydration of Hoogsteen base pairs inducing structural polymorphism under molecular crowding conditions. *J. Am. Chem. Soc.*, **131**, 3522 (2009).

[62] Lee, J. S.; Woodsworth, M. L.; Latimer, L. J.; Morgan, A. R., Poly(pyrimidine)·poly(purine) synthetic DNAs containing 5-methylcytosine form stable triplexes at neutral pH. *Nucleic Acids Res.*, **12**, 6603 (1984).

[63] Povsic, T. J.; Dervan, P. B., Triple helix formation by oligonucleotides on DNA extended to the physiological pH range. *J. Am. Chem. Soc.*, **111**, 3059 (1989).

[64] Ono, A.; Tso, P. O. P.; Kan, L. S., Triplex formation of oligonucleotides containing 2′-*O*-methylpseudoisocytidine in substitution for 2′-deoxycytidine. *J. Am. Chem. Soc.*, **113**, 4032 (1991).

[65] Ono, A.; Tso, P. O. P.; Kan, L. S., Triplex formation of an oligonucleotide containing 2′-*O*-methylpseudoisocytidine with a DNA duplex at neutral pH. *J. Org. Chem.*, **57**, 3225 (1992).

[66] Berressem, R.; Engels, J. W., 6-Oxocytidine a novel protonated C-base analogue for stable triple helix formation. *Nucleic Acids Res.*, **23**, 3465 (1995).

[67] Xiang, G. B.; Soussou, W.; Mclaughlin, L. W., A new pyrimidine nucleoside ($m^{5ox}c$) for the pH-independent recognition of G-C base-pairs by oligonucleotide-directed tripler formation. *J. Am. Chem. Soc.*, **116**, 11155 (1994).

[68] Xiang, G.; Bogacki, R.; McLaughlin, L. W., Use of a pyrimidine nucleoside that functions as a bidentate hydrogen bond donor for the recognition of isolated or contiguous G-C base pairs by oligonucleotide-directed triplex formation. *Nucleic Acids Res.*, **24**, 1963 (1996).

[69] Xiang, G. B.; McLaughlin, L. W., A cytosine analogue containing a conformationally flexible acyclic linker for triplex formation at sites with contiguous G-C base pair. *Tetrahedron*, **54**, 375 (1998).

[70] Vonkrosigk, U.; Benner, S. A., pH-independent triple-helix formation by an oligonucleotide containing a pyrazine donor-donor-acceptor base. *J. Am. Chem. Soc.*, **117**, 5361 (1995).

[71] Froehler, B. C.; Ricca, D. J., Triple-helix formation by oligodeoxynucleotides containing the carbocyclic analogs of thymidine and 5-methyl-2′-deoxycytidine. *J. Am. Chem. Soc.*, **114**, 8320 (1992).

[72] Pudlo, J. S.; Wadwani, S.; Milligan, J. F.; Matteucci, M. D., Synthesis of 6-amino-2′-Q-methylcytidine, a protonated cytidine analog for triple helix binding-studies. *Bioorg. Med. Chem. Lett.*, **4**, 1025 (1994).

[73] Hildbrand, S.; Leumann, C.,Enhancing DNA triple helix stability at neutral pH by the use of oligonucleotides containing a more basic deoxycytidine analog. *Angew. Chem., Int. Ed.*, **35**, 1968 (1996).

[74] Hildbrand, S.; Blaser, A.; Parel, S. P.; Leumann, C. J., 5-substituted 2-aminopyridine *C*-nucleosides as protonated cytidine equivalents: Increasing efficiency and selectivity in DNA triple-helix formation. *J. Am. Chem. Soc.*, **119**, 5499 (1997).

[75] Bates, P. J.; Laughton, C. A.; Jenkins, T. C.; Capaldi, D. C.; Roselt, P. D.; Reese, C. B.; Neidle, S., Efficient triple helix formation by oligodeoxyribonucleotides containing α- or β-2-amino-5-(2-deoxy-D-ribofuranosyl) pyridine residues. *Nucleic Acids Res.*, **24**, 4176 (1996).

[76] Cassidy, S. A.; Slickers, P.; Trent, J. O.; Capaldi, D. C.; Roselt, P. D.; Reese, C. B.; Neidle, S.; Fox, K. R., Recognition of GC base pairs by triplex forming oligonucleotides containing nucleosides derived from 2-aminopyridine. *Nucleic Acids Res.*, **25**, 4891 (1997).

[77] Kibler-Herzog, L.; Kell, B.; Zon, G.; Shinozuka, K.; Mizan, S.; Wilson, W. D., Sequence dependent effects in methylphosphonate deoxyribonucleotide double and triple helical complexes. *Nucleic Acids Res.*, **18**, 3545 (1990).

[78] Escude, C.; Giovannangeli, C.; Sun, J. S.; Lloyd, D. H.; Chen, J. K.; Gryaznov, S. M.; Garestier, T.; Helene, C., Stable triple helices formed by oligonucleotide N3′→P5′ phosphoramidates inhibit transcription elongation. *Proc. Natl. Acad. Sci. U.S.A.*, **93**, 4365 (1996).

[79] Giovannangeli, C.; Perrouault, L.; Escude, C.; Gryaznov, S.; Helene, C., Efficient inhibition of transcription elongation in vitro by oligonucleotide phosphoramidates targeted to proviral HIV DNA. *J. Mol. Biol.*, **261**, 386 (1996).

[80] Arya, D. P.; Bruice, T. C., Triple-helix formation of DNA oligomers with methylthiourea-linked nucleosides (DNmt): a kinetic and thermodynamic analysis. *Proc. Natl. Acad. Sci. U.S.A.*, **96**, 4384 (1999).

[81] Arya, D. P.; Bruice, T. C., Fidelity of deoxynucleic S-methythiourea (DNmt) binding to DNA oligomers: Influence of C mismatches. *J. Am. Chem. Soc.*, **121**, 10680 (1999).

[82] Sorensen, J. J.; Nielsen, J. T.; Petersen, M., Solution structure of a dsDNA:LNA triplex. *Nucleic Acids Res.*, **32**, 6078 (2004).

[83] Torigoe, H.; Rahman, S. M.; Takuma, H.; Sato, N.; Imanishi, T.; Obika, S.; Sasaki, K., 2′-*O*,4′-*C*-aminomethylene-bridged nucleic acid modification with enhancement of nuclease resistance promotes pyrimidine motif triplex nucleic acid formation at physiological pH. *Chem. Eur. J.*, **17**, 2742 (2011).

[84] Torigoe, H.; Rahman, S. M.; Takuma, H.; Sato, N.; Imanishi, T.; Obika, S.; Sasaki, K., Interrupted 2′-*O*,4′-*C*-aminomethylene bridged nucleic acid modification enhances pyrimidine motif triplex-forming ability and nuclease resistance under physiological condition. *Nucleosides Nucleotides Nucleic Acids*, **30**, 63 (2011).

[85] Torigoe, H.; Nakagawa, O.; Imanishi, T.; Obika, S.; Sasaki, K., Chemical modification of triplex-forming oligonucleotide to promote pyrimidine motif triplex formation at physiological pH. *Biochimie*, **94**, 1032 (2012).

[86] Torigoe, H.; Sato, N.; Nagasawa, N., 2′-*O*,4′-*C*-ethylene bridged nucleic acid modification enhances pyrimidine motif triplex-forming ability under physiological condition. *J. Biochem.*, **152**, 17 (2012).

[87] Rahman, S. M. A.; Seki, S.; Obika, S.; Haitani, S.; Miyashita, K.; Imanishi, T., Highly stable pyrimidine-motif triplex formation at physiological pH values by a bridged nucleic acid analogue. *Angew. Chem., Int. Ed.*, **46**, 4306 (2007).

[88] Lou, C.; Shelbourne, M.; Fox, K. R.; Brown, T., 2′-Aminoethoxy-2-amino-3-methylpyridine in triplex-forming oligonucleotides: high affinity, selectivity and resistance to enzymatic degradation. *Chem. Eur. J.*, **17**, 14851 (2011).

[89] Ray, A.; Norden, B., Peptide nucleic acid (PNA): its medical and biotechnical applications and promise for the future. *FASEB J.*, **14**, 1041 (2000).

[90] Kuhn, H.; Demidov, V. V.; Nielsen, P. E.; Frank-Kamenetskii, M. D., An experimental study of mechanism and specificity of peptide nucleic acid (PNA) binding to duplex DNA. *J. Mol. Biol.*, **286**, 1337 (1999).

[91] Egholm, M.; Christensen, L.; Dueholm, K. L.; Buchardt, O.; Coull, J.; Nielsen, P. E., Efficient pH-independent sequence-specific DNA binding by pseudoisocytosine-containing bis-PNA. *Nucleic Acids Res.*, **23**, 217 (1995).

[92] Sun, J. S.; Francois, J. C.; Montenay-Garestier, T.; Saison-Behmoaras, T.; Roig, V.; Thuong, N. T.; Helene, C., Sequence-specific intercalating agents: intercalation at specific sequences on duplex DNA via major groove recognition by oligonucleotide-intercalator conjugates. *Proc. Natl. Acad. Sci. U.S.A.*, **86**, 9198 (1989).

[93] Todorovic, N.; Phuong, N. T.; Langer, P.; Weisz, K., DNA triplex stabilization by a δ-carboline derivative tethered to third strand oligonucleotides. *Bioorg. Med. Chem. Lett.*, **16**, 1647 (2006).

[94] Eick, A.; Xiao, Z.; Langer, P.; Weisz, K., Spectroscopic studies on the formation and thermal stability of DNA triplexes with a benzoannulated δ-carboline-oligonucleotide conjugate. *Bioorg. Med. Chem.*, **16**, 9106 (2008).

[95] Eick, A.; Riechert-Krause, F.; Weisz, K., Binding and NMR structural studies on indoloquinoline-oligonucleotide conjugates targeting duplex DNA. *Bioconjug Chem.*, **23**, 1127 (2012).

[96] Gianolio, D. A.; Segismundo, J. M.; McLaughlin, L. W., Tethered naphthalene diimide-based intercalators for DNA triplex stabilization. *Nucleic Acids Res.*, **28**, 2128 (2000).

[97] Asanuma, H.; Liang, X. G.; Yoshida, T.; Yamazawa, A.; Komiyama, M., Photocontrol of triple-helix formation by using azobenzene-bearing oligo(thymidine). *Angew. Chem., Int. Ed.*, **39**, 1316 (2000).

[98] Wilson, W. D.; Mizan, S.; Tanious, F. A.; Yao, S.; Zon, G., The interaction of intercalators and groove-binding agents with DNA triple-helical structures: the influence of ligand structure, DNA backbone modifications and sequence. *J. Mol. Recognit.*, **7**, 89 (1994).

[99] Escude, C.; Sun, J. S., DNA major groove binders: Triple helix-forming oligonucleotides, triple helix-specific DNA ligands and cleaving agents. *Top. Curr. Chem.*, **253**, 109 (2005).

[100] Mergny, J. L.; Collier, D.; Rougee, M.; Montenaygarestier, T.; Helene, C., Intercalation of ethidium-bromide into a triple-stranded oligonucleotide. *Nucleic Acids Res.*, **19**, 1521 (1991).

[101] Scaria, P. V.; Shafer, R. H., Binding of ethidium bromide to a DNA triple helix. Evidence for intercalation. *J. Biol. Chem.*, **266**, 5417 (1991).

[102] Mergny, J. L.; Duval-Valentin, G.; Nguyen, C. H.; Perrouault, L.; Faucon, B.; Rougee, M.; Montenay-Garestier, T.; Bisagni, E.; Helene, C., Triple helix-specific ligands. *Science*, **256**, 1681 (1992).

[103] Pilch, D. S.; Martin, M. T.; Nguyen, C. H.; Sun, J. S.; Bisagni, E.; Garestier, T.; Helene, C., Self-association and DNA-binding properties of 2 triple helix-specific ligands - comparison of a benzo[e]pyridoindole and a benzo[g] pyridoindole. *J. Am. Chem. Soc.* 1993, **115**, 9942 (1993).

[104] Escude, C.; Nguyen, C. H.; Mergny, J. L.; Sun, J. S.; Bisagni, E.; Garestier, T.; Helene, C., Selective stabilization of DNA triple helices by benzopyridoindole derivatives. *J. Am. Chem. Soc.*, **117**, 10212 (1995).
[105] Escude, C.; Nguyen, C. H.; Kukreti, S.; Janin, Y.; Sun, J. S.; Bisagni, E.; Garestier, T.; Helene, C., Rational design of a triple helix-specific intercalating ligand. *Proc. Natl. Acad. Sci. U.S.A.*, **95**, 3591 (1998).
[106] Lee, J. S.; Latimer, L. J. P.; Hampel, K. J., Coralyne binds tightly to both T·A·T- and C·G·C+-containing DNA triplexes. *Biochemistry*, **32**, 5591 (1993).
[107] MoraruAllen, A. A.; Cassidy, S.; Alvarez, J. L. A.; Fox, K. R.; Brown, T.; Lane, A. N., Coralyne has a preference for intercalation between TA·T triples in intramolecular DNA triple helices. *Nucleic Acids Res.*, **25**, 1890 (1997).
[108] Wilson, W. D.; Tanious, F. A.; Mizan, S.; Yao, S.; Kiselyov, A. S.; Zon, G.; Strekowski, L., DNA triple-helix specific intercalators as antigene enhancers: unfused aromatic cations. *Biochemistry*, **32**, 10614 (1993).
[109] Cassidy, S. A.; Strekowski, L.; Wilson, W. D.; Fox, K. R., Effect of a triplex-binding ligand on parallel and antiparallel DNA triple helixes using short unmodified and acridine-linked oligonucleotides. *Biochemistry*, **33**, 15338 (1994).
[110] Chandler, S. P.; Strekowski, L.; Wilson, W. D.; Fox, K. R., Footprinting studies on ligands which stabilize DNA triplexes: effects on stringency within a parallel triple helix. *Biochemistry*, **34**, 7234 (1995).
[111] Keppler, M. D.; James, P. L.; Neidle, S.; Brown, T.; Fox, K. R., DNA sequence specificity of triplex-binding ligands. *Eur. J. Biochem.*, **270**, 4982 (2003).
[112] Durand, M.; Thuong, N. T.; Maurizot, J. C., Interaction of Hoechst 33258 with a DNA triple helix. *Biochimie*, **76**, 181 (1994).
[113] Park, Y. W.; Breslauer, K. J., Drug binding to higher ordered DNA structures: netropsin complexation with a nucleic acid triple helix. *Proc. Natl. Acad. Sci. U.S.A.*, **89**, 6653 (1992).
[114] Pilch, D. S.; Kirolos, M. A.; Breslauer, K. J., Berenil binding to higher ordered nucleic acid structures: complexation with a DNA and RNA triple helix. *Biochemistry*, **34**, 16107 (1995).
[115] Arya, D. P.; Coffee, R. L., Jr.; Willis, B.; Abramovitch, A. I., Aminoglycoside-nucleic acid interactions: remarkable stabilization of DNA and RNA triple helices by neomycin. *J. Am. Chem. Soc.*, **123**, 5385 (2001).
[116] Arya, D. P.; Micovic, L.; Charles, I.; Coffee, R. L., Jr.; Willis, B.; Xue, L., Neomycin binding to Watson-Hoogsteen (W-H) DNA triplex groove: a model. *J. Am. Chem. Soc.*, **125**, 3733 (2003).
[117] Xue, L.; Charles, I.; Arya, D. P., Pyrene-neomycin conjugate: dual recognition of a DNA triple helix. *Chem. Commun.*, 70 (2002).
[118] Arya, D. P.; Xue, L.; Tennant, P., Combining the best in triplex recognition: synthesis and nucleic acid binding of a BQQ neomycin conjugate. *J. Am. Chem. Soc.*, **125**, 8070 (2003).
[119] Arya, D. P., New approaches toward recognition of nucleic acid triple helices. *Acc. Chem. Res.*, **44**, 134 (2011).
[120] Hampel, K. J.; Crosson, P.; Lee, J. S., Polyamines favor DNA triplex formation at neutral pH. *Biochemistry*, **30**, 4455 (1991).
[121] Thomas, T.; Thomas, T. J., Selectivity of polyamines in triplex DNA stabilization. *Biochemistry*, **32**, 14068 (1993).
[122] Thomas, T. J.; Ashley, C.; Thomas, T.; Shirahata, A.; Sigal, L. H.; Lee, J. S., Pyrimidine-purine-pyrimidine triplex DNA stabilization in the presence of tetramine and pentamine analogues of spermine. *Biochem. Cell Biol.*, **75**, 207 (1997).
[123] Tung, C. H.; Breslauer, K. J.; Stein, S., Polyamine-linked oligonucleotides for DNA triple helix formation. *Nucleic Acids Res.*, **21**, 5489 (1993).
[124] Barawkar, D. A.; Kumar, V. A.; Ganesh, K. N., Triplex formation at physiological pH by oligonucleotides incorporating 5-Me-dC-(N^4-spermine). *Biochem. Biophys. Res. Commun.*, **205**, 1665 (1994).
[125] Barawkar, D. A.; Rajeev, K. G.; Kumar, V. A.; Ganesh, K. N., Triplex formation at physiological pH by 5-Me-dC-N^4-(spermine) [X] oligodeoxynucleotides: Non protonation of N3 in X of X*G:C triad and effect of base mismatch ionic strength on triplex stabilities. *Nucleic Acids Res.*, **24**, 1229 (1996).
[126] Rajeev, K. G.; Jadhav, V. R.; Ganesh, K. N., Triplex formation at physiological pH: Comparative studies on DNA triplexes containing 5-Me-dC tethered at N^4 with spermine and tetraethyleneoxyamine. *Nucleic Acids Res.*, **25**, 4187 (1997).
[127] Potaman, V. N.; Sinden, R. R., Stabilization of intramolecular triple/single-strand structure by cationic peptides. *Biochemistry*, **37**, 12952 (1998).

[128] Potaman, V. N.; Sinden, R. R., Stabilization of triple-helical nucleic acids by basic oligopeptides. *Biochemistry* 1995, **34**, 14885 (1995).

[129] Ferdous, A.; Watanabe, H.; Akaike, T.; Maruyama, A., Poly(L-lysine)-graft-dextran copolymer: amazing effects on triplex stabilization under physiological pH and ionic conditions (*in vitro*). *Nucleic Acids Res.*, **26**, 3949 (1998).

[130] Torigoe, H.; Ferdous, A.; Watanabe, H.; Akaike, T.; Maruyama, A., Poly(L-lysine)-graft-dextran copolymer promotes pyrimidine motif triplex DNA formation at physiological pH. Thermodynamic and kinetic studies. *J. Biol. Chem.*, **274**, 6161 (1999).

[131] Ihara, T.; Ishii, T.; Araki, N.; Wilson, A. W.; Jyo, A., Silver ion unusually stabilizes the structure of a parallel-motif DNA triplex. *J. Am. Chem. Soc.*, **131**, 3826 (2009).

[132] Feng, L.; Huang, Z.; Ren, J.; Qu, X., Toward site-specific, homogeneous and highly stable fluorescent silver nanoclusters fabrication on triplex DNA scaffolds. *Nucleic Acids Res.*, **40**, e122 (2012).

[133] Li, X.; Peng, Y.; Qu, X., Carbon nanotubes selective destabilization of duplex and triplex DNA and inducing B-A transition in solution. *Nucleic Acids Res.*, **34**, 3670 (2006).

[134] Li, X.; Peng, Y.; Ren, J.; Qu, X., Carboxyl-modified single-walled carbon nanotubes selectively induce human telomeric i-motif formation. *Proc. Natl. Acad. Sci. U.S.A.*, **103**, 19658 (2006).

[135] Zhao, C.; Peng, Y.; Song, Y.; Ren, J.; Qu, X., Self-assembly of single-stranded RNA on carbon nanotube: polyadenylic acid to form a duplex structure. *Small*, **4**, 656 (2008).

[136] Song, Y.; Feng, L.; Ren, J.; Qu, X., Stabilization of unstable CGC^+ triplex DNA by single-walled carbon nanotubes under physiological conditions. *Nucleic Acids Res.*, **39**, 6835 (2011).

5.2

Synthetic Molecules as Guides for DNA Nanostructure Formation

Andrea Greschner, Fiora Rosati, and Hanadi Sleiman

Department of Chemistry, McGill University, Montreal, Quebec, Canada

5.2.1 Introduction

DNA is recognized as the molecule of life for its role in preserving and transferring our genetic code, but its unique assembly features have also made it an ideal building block for nanomaterials [1]. Because of the high fidelity and cooperativity of molecular recognition between nucleobases resulting from the specificity of hydrogen bonding, DNA forms predictable double helical structures of well defined nanoscale dimensions. The ability to construct a near infinite number of DNA sequences using facile automated synthesis has allowed this molecule to emerge as a programmable scaffold for organizing molecules and materials into one-, two-, and three-dimensional structures. One important emerging field is the use of synthetic molecules to tune and modify the stability, functionality, and assembly of these DNA-based structures. These molecules can be incorporated into the structures through one of two methods: covalent insertion or non-covalent interactions, both of which will be discussed here.

5.2.2 Covalent insertion of synthetic molecules into DNA

Even the simplest chemical insertions into DNA strands are able to alter the hybridization of nucleobases and control their self-assembly outcomes. For instance, by positioning organic molecules with specific geometries within the phosphate backbone of DNA, linear duplexes can be oriented relative to each other in a unique

DNA in Supramolecular Chemistry and Nanotechnology, First Edition.
Edited by Eugen Stulz and Guido H. Clever.

manner and lead to complex, high-order structures. Furthermore, by inserting molecules capable of coordinating metal ions, it is possible not only to modulate the assembly process and provide an enhancement of the duplex stability, but also to equip the resulting structures with entirely new functions relevant to many fields (e.g., nanoelectronics, medicine, light harvesting).

Thanks to the ease of automated synthesis and the efficiency of coupling approaches, a plethora of synthetic modifications are available.

5.2.2.1 Incorporating organic molecules into DNA

The ability to insert synthetic molecules, vertices and linkers, directly into the DNA sequence, through the use of automated synthesis and off-column coupling techniques has led to a variety of assemblies with newly embedded properties and functions.

In many cases, it is preferable to maintain all the natural characteristics of DNA while incorporating chemical modifications. The most widely used approach to modify the DNA backbone includes the design and insertion, via automated solid-phase synthesis, of artificial non-nucleoside phosphoramidites. Alternatively, organic molecules can be coupled to the termini of the DNA strands via covalent strategies that involve the use of an appropriate derivative of the molecule of choice, in solution. Thus, different research groups have explored organic/inorganic molecule–DNA hybrid structures or polymer–DNA hybrids by looking at compounds that do not require traditional phosphoramidite chemistry. For example, Lee *et al.* screened several types of phosphoramidite-free coupling reactions and determined that amide coupling with flexible synthetic linkers gave the highest yields (35%) [2]. Similarly, amide coupling has also proved feasible for creating DNA–polymer hybrids [3, 4], and attaching metal-binding ligands [5]. Other crosslinking methods include NHS-ester reactions [6], and copper(I) catalyzed alkyne–azide cycloaddition (CuAAC or "click" chemistry) [7]. Recently, Nguyen and coworkers carried out CuAAC-mediated coupling of a tetraaryl azide molecule with alkyne-modified DNA strands on a controlled pore glass solid support. Interestingly, this gave branched structures with four DNA arms in good yield (in addition to 1, 2, and 3 arms), due to the increased local concentration of DNA strands on the solid support [8]. El-Sagheer and Brown have demonstrated the CuAAC ligation of DNA and RNA strands using a complementary DNA strand that brings the azide and alkyne substituted ends into close proximity. This gives a triazole instead of a phosphate linkage, and the resulting structures could be replicated using the polymerase chain reaction [9, 10].

Non-nucleoside phosphoramidites are very adaptable in that they can be created from a variety of molecules with interesting properties. If the molecule is to be inserted within the DNA backbone, the synthetic building block is designed with two hydroxyl groups. One hydroxyl group can be protected using an acid-labile group, such as a triphenyl methyl (trityl) derivative, and mimic the 5′-end of a nucleotide for synthesis purposes. The other hydroxyl is reacted with a chloro-phosphoramidite or a phosphorodiamidite, creating the phosphoramidite necessary for automated synthesis (Figure 5.2.1).

These phosphoramidite derivatives can be designed to introduce a number of modifications to DNA, such as: minimizing steric hindrance, increasing stability, increasing or reducing flexibility, favoring the self-assembly of a specific product, and imparting new functions to the resulting structures. In this section, a series of selected examples will be discussed in more detail.

5.2.2.2 Adjusting flexibility

Flexibility in DNA assemblies has many benefits. Increased flexibility can allow for the formation of structures that would otherwise be too sterically hindered for correct assembly, whereas rigid molecules can favor certain structural shapes over others. For example, Shchepinov *et al.* reported a study on the optimal composition and length of a phosphoramidic spacer in order to reduce the steric hindrance between oligonucleotides

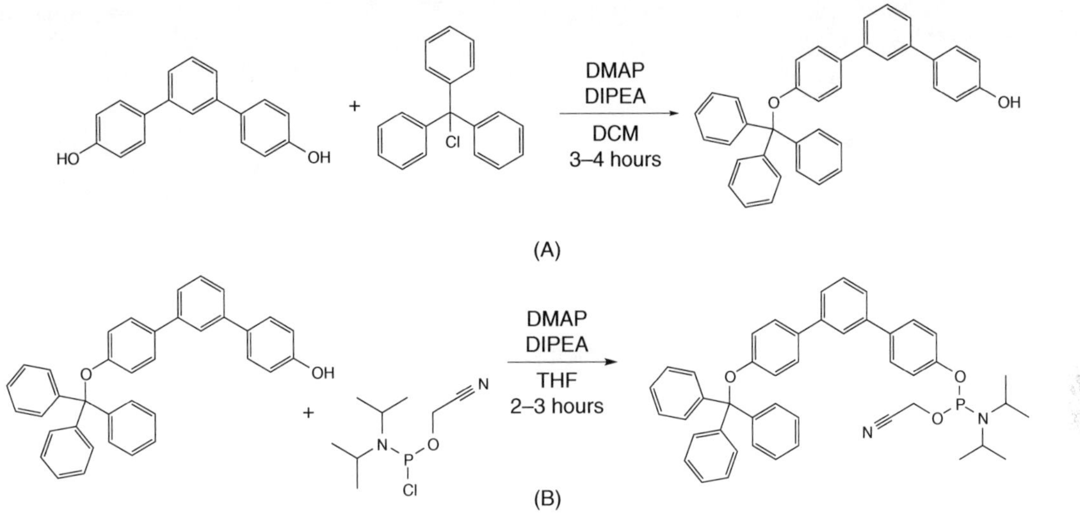

Figure 5.2.1 An example of phosphoramidite preparation using a chloro-phosphoramidite. (A) Protection of a hydroxyl via tritylation. DIPEA should be added in a dropwise fashion [11,12]. (B) Conversion into a phosphoramidite suitable for automated synthesis [12,13]. Dry conditions should be maintained throughout both reactions: DMAP = 4-dimethylaminopyridin, DIPEA = N,N-diisopropylethylamine, and DCM = dichloromethane

tethered on a polypropylene solid support [14]. These phosphoramidite spacers have a significant effect on hybridization, the most important determinant being their length and thus the distance of the oligonucleotides from the surface. Similar spacers were eventually incorporated into DNA nanoassemblies to lend flexibility to junctions [15, 16].

An early example of how DNA assembly could be manipulated by synthetic insertions is illustrated by the work of Shi and Bergstrom [17]. In this case, two identical, self-complementary, single-stranded (ss) DNA arms were convergently synthesized from a rigid *p*-(2-hydroxyethyl)phenylethynylphenyl spacer unit attached to controlled pore glass (CPG). From the resulting strands, discrete macrocycle formation occurred as a function of monomer concentration (Figure 5.2.2A). When attempting to increase the flexibility of the linkers that connected the vertex to the two arms, hybridization resulted in increased amounts of monomer and lower-order structures.

5.2.2.2.1 *Mediating DNA self-assembly*

The groups working with von Kiedrowski [18] and Sawai [19] reported three-way organic junctions on which DNA can grow by standard phosphoramidite chemistry. von Kiedrowski and coworkers constructed a trifunctional flexible linker connected to three oligonucleotides of identical sequence, which could associate, under kinetic conditions and in the presence of their complementary strands, into dimeric or tetrameric structures. However, Sawai and coworkers described the synthesis of a branched tris-DNA module in which two identical and one different oligonucleotide strands were connected by a rigid linker molecule. By hybridization to a complementary branched strand, a variety of closed cages were formed. As demonstrated by mung bean nuclease digestion, the linker effectively pre-organized the DNA strands, preferentially forming fully duplexed cage structures over oligomers.

In all these examples, however, self-complementarity results in a ladder of concentration-dependent products rather than a single assembly. Sleiman and coworkers developed an approach that used six rigid organic

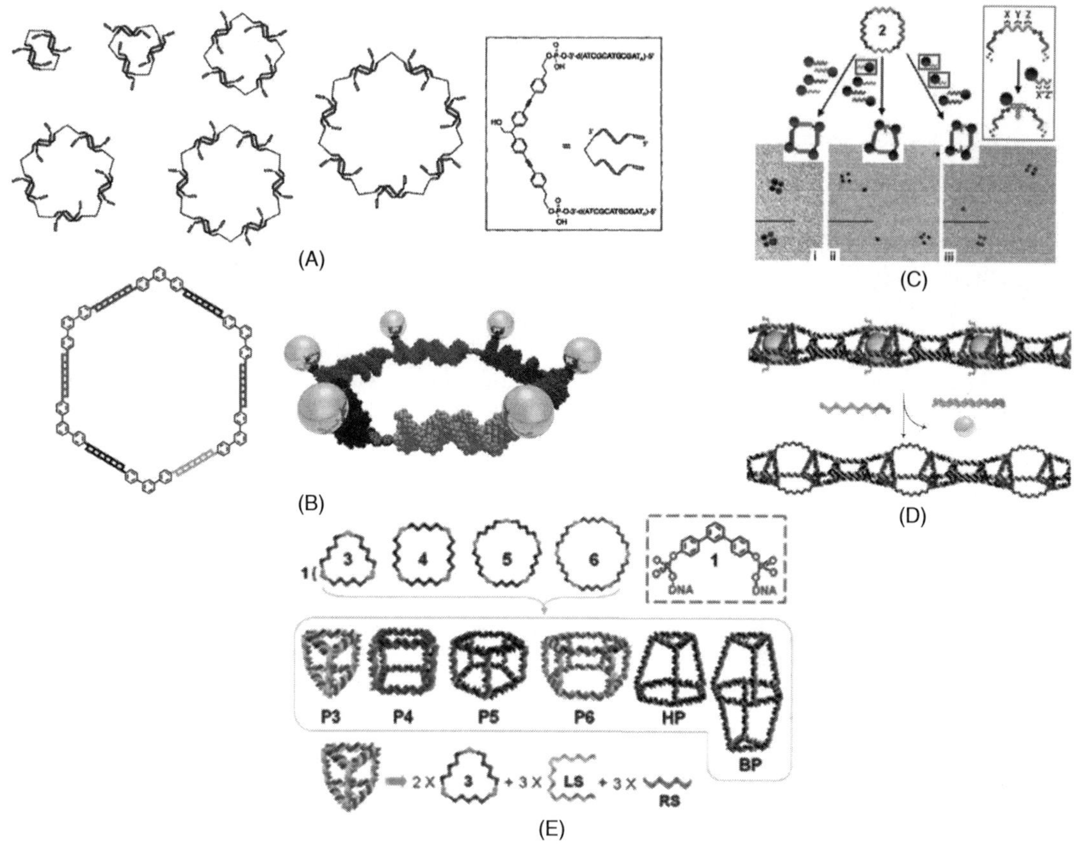

Figure 5.2.2 (A) Macrocycle formation through the use of rigid organic linkers [17]. Reproduced with permission from [17]. Copyright © 1997 WILEY-VCH Verlag GmbH & Co. KGaA, Weinheim. (B) Use of rigid linkers to create a well-defined DNA scaffold for gold nanoparticle arrangement [12]. Reprinted with permission from [12] Copyright 2006, American Chemical Society. (C) A simple square-shaped DNA template can position gold nanoparticles in three different orientations depending on which complementary strands are added: (i) square, (ii) trapezoid, and (iii) rectangle. The scale bar on the TEM images corresponds to 50 nm [21]. Reprinted with permission from [12] Copyright 2006, American Chemical Society. (D) A nanotube built from rungs containing rigid linkers can hold and release cargo [22]. Reproduced from [22]. Copyright © 2010, Rights Managed by Nature Publishing. (E) Single-stranded polygons **3–6** are constructed using a rigid linker (**1**). A face-centered approach was then used to create three-dimensional cages [23]. Reprinted with permission from [23] Copyright 2007, American Chemical Society

vertices (*m*-terphenyl based)-conjugated ssDNA arms that were not self-complementary and not identical with one another, which led to the preferential formation of a single assembly. Linker sequences were designed to ensure a very narrow product distribution, with a hexamer as the predominant product obtained through classical base pairing. This system was used as a scaffold to precisely position gold nanoparticles (AuNPs) into a discrete 2D structure (Figure 5.2.2B).

Using the same rigid triphenyl vertex, this group reported the synthesis of cyclic single stranded DNA polygons, such as squares, pentagons, and hexagons, by directed chemical ligation of their ends. Gold nanoparticles monosubstituted with a DNA strand were positioned by hybridization to their single-stranded arms, thus yielding 2D-nanoparticle assemblies with pre-designed geometries. This approach allowed the writing, erasing, and modification of structural features of gold nanoparticle DNA assemblies with externally added DNA strands (Figure 5.2.2C) [20, 21].

These studies led to the development of other small DNA assemblies that rely on a minimum number of high-affinity interactions between complementary strands to form small, distinct structures in high yield [24–26]. Chemically-introduced small molecules allow the use of a relatively small number of strands, as compared with DNA origami approaches [27, 28], resulting in "DNA-economic" nanostructures [29, 30].

Using the single-stranded DNA polygons (**3–6,** Figure 5.2.2E) Sleiman's group then showed a face-centered approach to create a number of three-dimensional DNA cages in a quantitative manner (Figure 5.2.2). Addition of strand loops and DNA overhangs resulted in cages that could be effectively "opened" and "closed" through the addition of an eraser strand complementary to the strand with an overhang [23].

A modular approach to construct DNA nanotubes with finely controlled geometry, size, and persistent length was then reported, using the polygons **3–6**. These nanotubes can be generated in single-stranded and open forms, as well as double-stranded and closed versions, of dramatically different stiffness. A DNA nanotube structure that encapsulates gold nanoparticles of specific sizes to form nanoparticle "pea-pod" lines was created. An interesting "sieving" ability was noted: only specific nanoparticle sizes that match the size of the capsules along the nanotubes could be encapsulated, and the process is very selective. When a specific DNA strand is added, it causes release of the strands that close the nanotube, and this nanoparticle cargo is released quantitatively in less than 15 min (Figure 5.2.2D) [22, 23]. A method to control the length of these nanotubes to monodisperse, pre-programmed values was reported (up to 1 μm). This method to create lines of encapsulated gold nanoparticles of all the same length shows promise for size-defined plasmonic wires [31]. von Kiedrowski and coworkers have used tris-oligonucleotide branched molecules with an aromatic linker to prepare a 3D DNA dodecahedron [32].

5.2.2.3 Direct effect of synthetic organic linkers on DNA stability and assembly

Insertion of synthetic linkers has several direct effects on the DNA duplexes. Firstly, adjacent duplexes are held in much closer proximity than would normally be accessible through Brownian motion. Secondly, the aromatic nature of many of the linkers entails the addition of a hydrophobic break in the otherwise hydrophilic phosphate backbone of DNA. Thirdly, rigid linkers constrain the ends of the duplexes in specific conformations. In order to properly predict how linkers will affect assembly, it is necessary to understand their immediate influence on DNA.

To determine how nearby duplexes might interact, Schatz and coworkers investigated the cooperative melting of small molecule DNA hybrids possessing only two closely associated, asymmetric double-stranded DNAs. Each strand contained a central linker. The aim was to facilitate the formation of cage-like dimers where two duplexes are oriented in a parallel fashion (Figure 5.2.3A) [33]. These aggregates displayed sharper melting profiles compared with unmodified DNA duplexes, consistent with cooperative melting on the two duplexes [34, 35].

A follow-up study [16] employed experimental and computational methods to elucidate the role of a hydrophobic rigid organic core in the assembly of discrete nanostructures (Figure 5.2.3B) in an aqueous environment. When sequences were designed such that the hydrophobic surfaces of linkers were unable to interact with one another (i.e., they were too far apart), a mixture of cage dimers and ill-defined networks formed. In contrast, if the DNA sequence allowed for stacking of the hydrophobic linkers, a good minimization of the hydrophobic surfaces of the cores occurred, resulting in a single highly-stable structure with a dramatic enhancement of thermal denaturation temperature of 15 °C (T_m, a measure of DNA stability). These results emphasize the significant role of hydrophobic interactions occurring between organic cores in nanostructures built from organic–DNA hybrid building blocks.

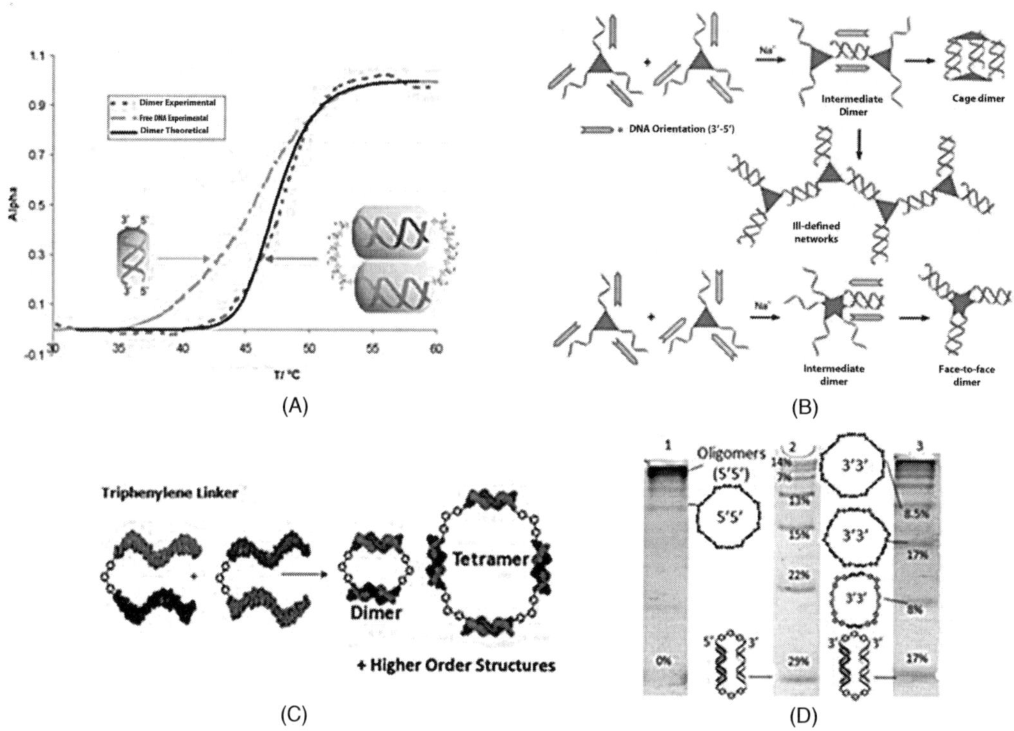

Figure 5.2.3 (A) Two parallel duplexes connected via linkers have a sharper melting curve than a single duplex [33]. Reprinted with permission from [33] Copyright 2010, American Chemical Society. (B) Hydrophobic cores stacked atop one another (bottom) have greater stability than the same core in isolation (top) [16]. Reprinted with permission from [16] Copyright 2012, American Chemical Society. (C) Two strands bisected by a rigid linker are capable of assembling into a variety of shapes. (D) By altering the DNA-linker connectivity of the strands in (C), the product distribution is greatly altered: (1) 3′-linker-3′; (2) 5′-linker-3′ (3) 5′-linker-5′ [36]. Reprinted with permission from [36] Copyright 2012, American Chemical Society

A study by Greschner *et al.* [36] further elucidated the cooperative interactions between parallel duplexes by creating two complementary strands of DNA, each bisected by a linker. These strands could then be assembled into a variety of structures (Figure 5.2.3C). It was shown that, when connecting two DNA strands equipped with synthetic linkers, factors such as linker rigidity, size, and DNA strand orientation affected both the assembly outcome and the stability of the resulting duplexes. A significant increase in thermal denaturation temperature was observed in comparison with a single DNA duplex. A short and rigid triphenylene linker gave the greatest stabilization with a 10 °C increase in T_m, and almost full cooperativity in the melting of the two duplexes was observed. Using the same DNA sequences with more flexible linkers gave a degree of stabilization greater than that of single duplexes, but lower than that obtained using the rigid linkers. Thus, the same DNA sequence can be tuned to melt in a wide range of different temperatures by correctly selecting the linker structure. In addition, changes in the DNA-to-linker connectivity (e.g., 5′-linker-3′ versus 3′-linker-3′) can lead to dramatic changes in the self-assembly behavior of the same base-set of strands, from the formation of cyclic dimers and tetramers to higher-order cycles and oligomeric assemblies (Figure 5.2.3D) [36].

The observed increases in stability with the use of synthetic linkers are especially useful in biological environments that may degrade DNA. For example, a combination of folded topology and 5′/3′ end-capping with synthetic insertions has been observed to significantly improve the serum stability of DNA cages. The greatest improvement was achieved through end-capping using hexaethylene glycol, which reduced the ability of exonucleases to recognize the oligo nucleotide 5′/3′ ends [37].

5.2.2.3.1 Supramolecular assembly guided through synthetic additions to DNA

Instead of simply replacing a nucleotide in the DNA strand, synthetic linkers also allow for additional DNA strands to be attached at a single branch point (DNA branching) [18, 23, 38, 39]. Shchepinov *et al.* [39] reported higher-order branching junctions that allow for the formation of cross-linked systems. The group working with Majima and Seeman [40] used a tetrabranched DNA structure with a porphyrin core to induce the rolling of DNA tiles into tubular structures. More recently, synthetic cores have been expanded to six DNA arms with a pseudo-octahedral structure and an adamantane derivative [41].

In addition to creating vertices with multiple DNA arms, synthetic branches can be used to introduce dendritic moieties onto the ends of DNA strands. When the dendrimer is of a different polarity than the parent DNA strand, assembly patterns similar to those seen in block copolymers emerge.

For example, Carneiro *et al.* [42] used phosphoramidite chemistry to create a series of DNA dendrons consisting of from two to eight branches of hexaethylene glycol attached to short (10–20 bases) DNA strands (Figure 5.2.4A). When assembled in a mixture of acetonitrile and aqueous buffer, these short block copolymers create distinct, long-range fibers whose assembly is mediated by the two "blocks" (hydrophilic DNA and the more acetonitrile-soluble oligoethylene glycol) [42]. The Gothelf's group reported the organization of poly(amidoamine) dendrimers (PAMAM) on a DNA strand and their covalent "click" coupling by CuAAc, thus producing DNA-templated PAMAM dimers, trimers, and oligomers [43].

Dendritic moieties have also been used to modify the properties of DNA cages. DNA cubes decorated with dendritic hydrophilic or hydrophobic chains (D-DNA) display different cellular uptake profiles when compared with naked structures. While the hydrophobic chains help with the rapid uptake of the DNA cages, the hydrophilic chains favor sustained uptake of the cages over time; these properties can be used to finely tune the cellular uptake profile of a given DNA nanostructure [44].

Furthermore, selective placement of dendritic hydrophobic residues on a 3D DNA scaffold can modulate the amphiphilic assembly properties. When four such hydrophobic units are organized on one face of the cube, the alkyl chains engage in an intermolecular "*handshake*" across two cubes, resulting exclusively in a cube dimer. When eight dendritic units are organized on the top and bottom faces of the cube, they engage in a "handshake" *inside* the cube (Figure 5.2.4B). This forms a monodisperse micelle within a DNA cage, which can encapsulate small hydrophobic molecules, and release them with added DNA strands [45].

Block copolymer assembly can also be accessed without the use of dendrons. Currently, sequence-controlled synthetic polymers are a major synthetic challenge in polymer chemistry. Recently, a stepwise, solid-phase, phosphoramidite synthesis has been employed to produce sequence-controlled polymers attached to DNA. These sequence-specific DNA amphiphiles can be produced with control over the length and pattern of the hydrophobic portion. Investigation of the amphiphilic self-assembly of these DNA conjugates has revealed that molecules with the same composition but different sequences have very different properties. This highlights the importance of sequence control for the production of well-defined materials with predictable assembly behavior (Figure 5.2.4C) [46].

5.2.2.3.2 Backbone insertion of metal binding organic molecules

The site-specific incorporation of transition metal binding molecules into the DNA backbone represents an alternative strategy to drive and tune the self-assembly of DNA nanostructures. This is because transition metal centers provide access to numerous coordination geometries and new bond angles that can be useful for

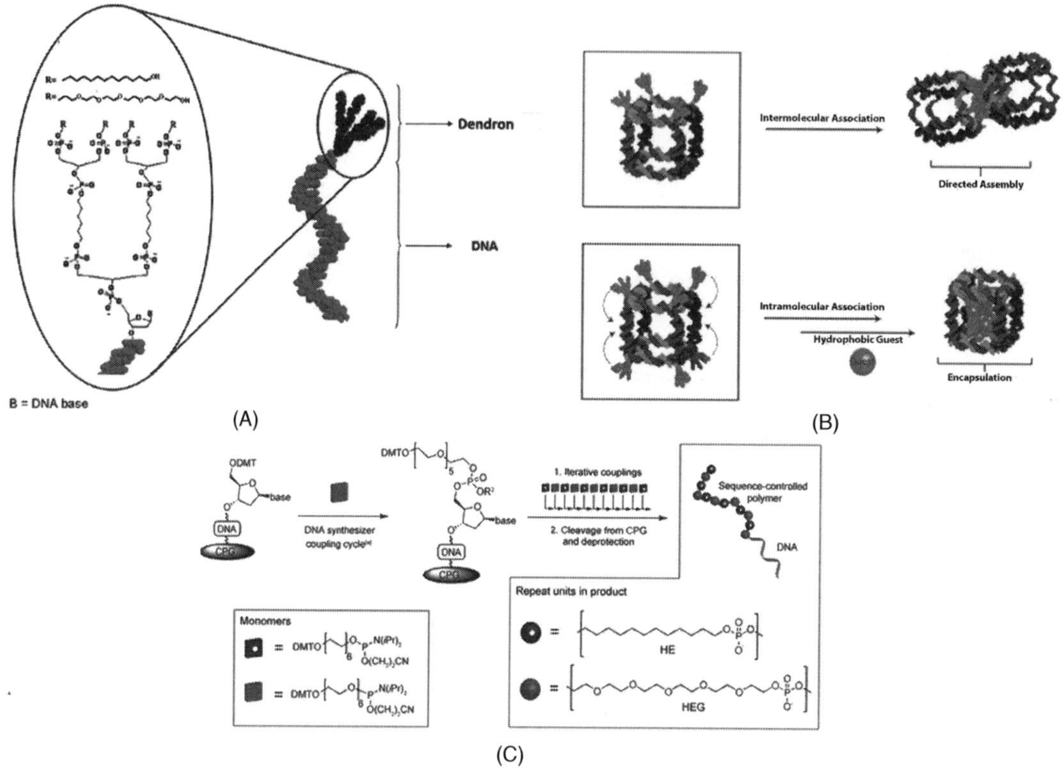

Figure 5.2.4 *(A) Structure of a DNA–ethylene glycol or DNA–dodecanol dendron using synthetic branching phosphoramidites [42]. Reprinted with permission from [42] Copyright 2010, American Chemical Society. (B) Selective orientation of hydrophobic moieties results in dimerization (top), or encapsulation (bottom) [45]. Reproduced from [45]. Copyright © 2010, Rights Managed by Nature Publishing. (C) Synthesis scheme for sequence-defined polymer–DNA amphiphilesm [46]. Reproduced with permission from [46]. Copyright © 2014, WILEY-VCH Verlag GmbH & Co. KGaA, Weinheim*

controlling the stability and reversibility of duplex formation. Moreover, they possess electronic, photochemical, and catalytic properties that can be directly transferred to the DNA assembly and are embedded as new functional features, which are useful for materials science and biotechnology applications. In addition, DNA represents a way to pattern metals in precise positions and in a programmable manner, and to generate dynamic, switchable, and molecule-responsive structures.

Early examples of direct metal–complex insertions into DNA, where the coordination environment influenced the arrangement of the strands, were reported by the groups working with McLaughlin [47, 48] and Sleiman [32, 33]. In the first case, McLaughlin's group designed a six-oligonucleotide-armed ruthenium(II) tris-(bipyridyl)-centered complex as a building module for supramolecular nanoscale assemblies. The Sleiman group reported on the synthesis of DNA junctions modified with Ru complexes and their application in nanostructure assembly [49, 50] (Figure 5.2.5). Two of these units, by hybridization with complementary sequences, led to the formation of a discrete cyclic structure, where the Ru metal center dictated the appropriate orientation of the arms.

While creating metal junctions through phosphoramidic modification of DNA strands, it should be considered that the metals must be sufficiently inert and resistant to automated DNA synthesis conditions. If more

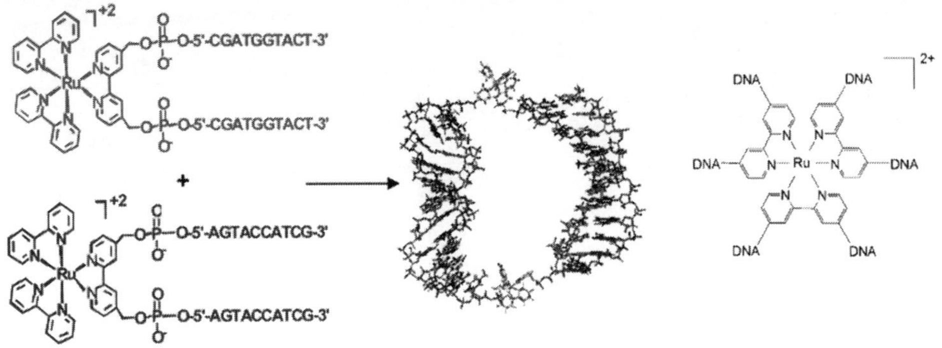

Figure 5.2.5 (A) The Ru(bpy)$_3$ corner molecule orients the attached ssDNA such that a cyclic dimer forms [51]. Reproduced from [51], with permission from Elsevier. (B) A six-armed, metal-based DNA vertex [48]. Reproduced with permission from [48]. Copyright © 2004, WILEY-VCH Verlag GmbH & Co. KGaA, Weinheim

kinetically labile metals need to be incorporated into the DNA backbone, another approach is available. This is based on the conjugation of a metal-binding ligand to the end of the purified oligonucleotides, through a phosphoramidite derivative attached to DNA during standard automated synthesis, followed by the binding of the metal to the ligand-modified DNA. The ligand phosphoramidites can differ significantly from the DNA bases, so that the metals can bind to them selectively.

Terpyridine, bipyridine, and phenanthroline have been conjugated to the DNA backbone to coordinate with a variety of metals. Han and coworkers demonstrated that ssDNA end-modified with a terpyridine ligand could be used to guide the assembly process upon addition of specific metals (Figure 5.2.6A) [52]. In the presence of Fe^{2+} ions, the ligand–metal coordination clipped together two unique strands of DNA, forming a metal-bound branched building block. The assembly resulted in the formation of a mixture of heterodimers and homodimers so that a purification step was needed. Both metal coordination and predesigned base pairing allowed control of the assembly process and programming of the outcome.

Göritz and Krämer presented a short DNA sequence modified at both termini with a terpyridine derivative. In the presence of metals such as Fe^{2+}, Zn^{2+}, and Ni^{2+}, this system preferentially cyclizes (Figure 5.2.6), displacing a complementary DNA strand [53]. Even though this metal–ssDNA unit was not incorporated into higher-ordered structures, it represents a valid example of how metal modification can control the conformation of the resulting system. In contrast, Sugimoto and coworkers used ligand insertions to assemble higher complexity structures, providing an example of a reversible, metal-mediated self-assembly method. Upon addition of Ni^{2+} ions, bipyridine modified G-quadruplexes could be switched from an anti-parallel to a parallel structure, resulting in intermolecular formation of a "G-wire". By adding EDTA, the process could be reversed [54].

Selective incorporation of a range of reactive transition metals into DNA-templated junctions was reported by the Sleiman group. Two single strands were modified with a phosphoramidite derivative of a phenanthroline molecule. This allowed close contact between the metal complex and DNA, resulting in a synergistic stabilization upon addition of the metal (Ag or Cu) and highly stable DNA–metal assemblies [55]. Besides greatly enhanced stability (e.g., thermal denaturation increases from 40 to 80 °C), the possibility to organize different transition metals in precise locations within a nanostructure is particularly appealing.

Later, the same group designed three different ligand environments – terpyridine (tpy), diphenylphanthroline (dpp) and a combination of the two – each of them capable of selectively binding to a specific transition metal ion (Figure 5.2.7B) [56]. Based on thermal denaturation (T_m) studies, it was shown that (dpp)$_2$–DNA

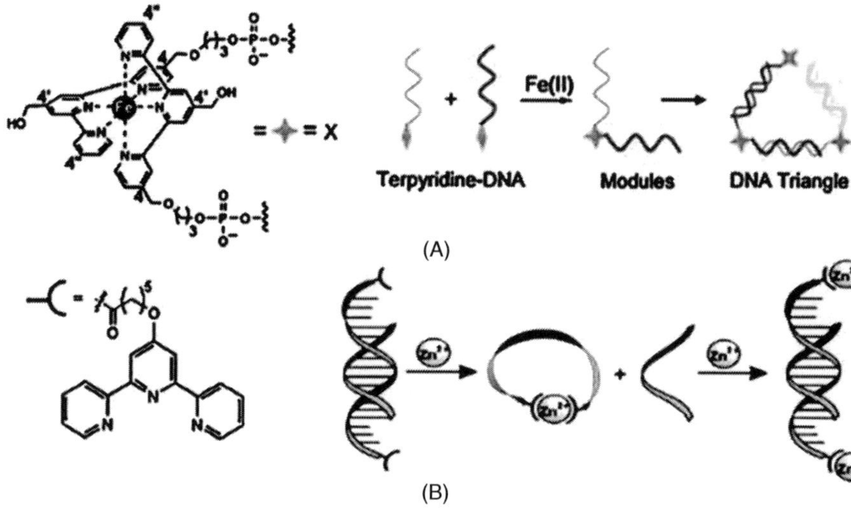

Figure 5.2.6 (A) Terpyridine-modified DNA can bind to Fe(II), connecting different DNA strands into a triangle [53]. Reprinted with permission from [53] Copyright 2005, American Chemical Society. (B) With Zn(II), a terpyridine–DNA can be switchably opened and closed [54]. Reprinted with permission from [54] Copyright 2007, American Chemical Society

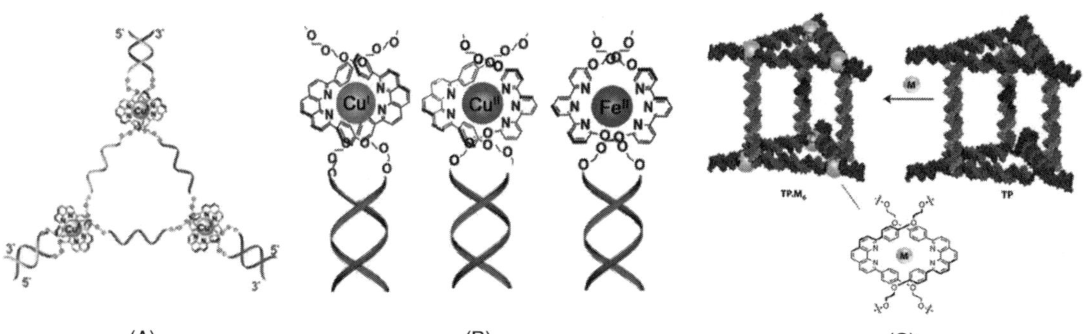

Figure 5.2.7 (A) Assembly of a stable DNA triangle with metal–ligand vertices [55]. Reproduced with permission from [55]. Copyright © 2008, WILEY-VCH Verlag GmbH & Co. KGaA, Weinheim. (B) Selective metal incorporation via ligand selection [51]. Reproduced with permission from [51]. Copyright © 2009, WILEY-VCH Verlag GmbH & Co. KGaA, Weinheim. (C) Metal-DNA cage [15]. Reproduced from [15]. Copyright © 2009, Rights Managed by Nature Publishing

bound most strongly with Cu(I), $(tpy)_2$–DNA bound strongest with Fe(II), and the mixed ligand environment tpy–dpp prefers Cu(II). The Fe(II)–$(tpy)_2$–DNA gives a T_m increase of 40 °C, one of the highest reported T_m increases for metal complex modification of a DNA duplex. Interestingly, "error checking" was observed: when an "incorrect" metal ion was placed within one of those ligand environments, either a spontaneous adjustment of the oxidation state of the metal occurred, or the metal was displaced in favor of a more stable species when multiple metals were present.

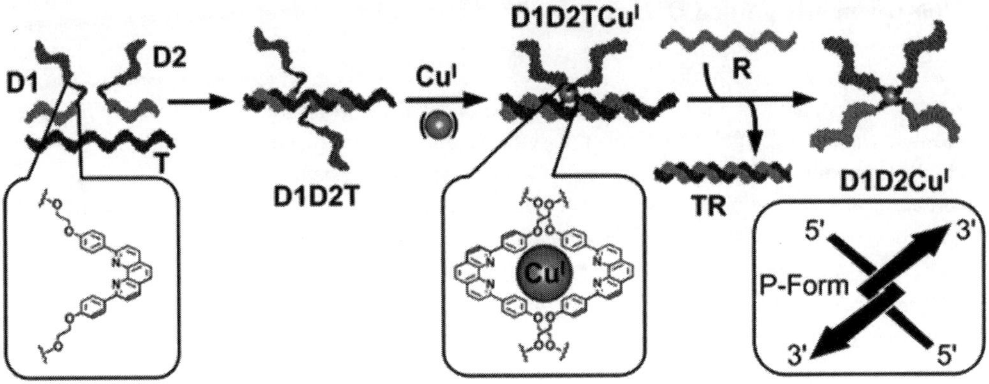

Figure 5.2.8 *Assembly of a four-way DNA junction mediated by metal–ligand interactions [73]. Reproduced from [73]. Copyright © 2009, Rights Managed by Nature Publishing*

Following this approach, the construction of a metal–DNA cage, with site-specific incorporation of transition metals in the vertices of the structure, was reported [15]. The metal coordination geometry defines the geometry of the vertex, and the DNA strands give programmable regions that can be used as building blocks for 2D and 3D structures with properties potentially similar to a unit of metal–organic frameworks (MOF), such as metal-mediated redox, photochemical, magnetic or catalytic control on encapsulated guest molecules (Figure 5.2.7C).

Complementary work by Sheppard first reported the templated synthesis of salen–metal–DNA complexes from salicylic aldehyde-modified DNA [57]. Using one DNA strand as a template, two salicylic aldehyde-modified DNA strands can be covalently linked and coordinate a metal ion, resulting in a very stable structure that resists chemical denaturation. The same group also reported that two complementary DNA strands with salicylic aldehyde modifications at the ends can also template the formation of a salen metal complex [58]. The covalently linked product has a slightly higher stability than a DNA hairpin. Gothelf and coworkers reported a series of studies on using DNA hybridization and metal complexation to join organic functional groups together to make macromolecular nanostructures [59–63]. The idea is that DNA can direct the assembly of the functional groups so that they are located at desired positions. They can then be connected together using metal–salen complex formation and the macromolecular product would be cleaved from the DNA.

It is of note that Takezawa and Shionoya [64], Schultz and coworkers [65], Carell and coworkers [66, 67], in addition to others groups [68–71], have reported the replacement of DNA bases with metal-binding ligands, and the creation of metal-mediated base pairs, a topic reviewed elsewhere. More recently, Shionoya and coworkers [72] synthesized a DNA bipyridine-modified three-way junction structure that can be specifically stabilized by the addition of Ni(II) ions (3:1 ligand–metal complexation). Although metal selectivity and coordination structure are still difficult to predict, they envisaged that the metal-locked junction system would significantly increase the programmability of DNA nanostructures whose stability and rigidity can be regulated by reversible metal coordination at the branching points. It has been shown that a diastereomeric preference could be a result of the chiral DNA environment at the junction. A DNA-templated method to create chiral metal–DNA junctions was presented by the Sleiman group [73]. This involved the use of a phenanthroline-modified strand and a minimal amount of DNA. This compact chiral four-arm junction has great potential as a building block to construct other structures (Figure 5.2.8).

5.2.3 Non-covalently guided DNA assembly

Covalently inserting synthetic molecules into the DNA backbone introduces many unique properties for DNA assembly and represents a powerful tool towards controlling structure. However, covalent insertion of a molecule is not always feasible for a given assembly. Instead, adding molecules that interact with DNA non-covalently provides an attractive alternative for modifying DNA behavior.

There are several ways by which small molecules can interact with DNA, including: hydrogen bonding [74–76], van der Waals interactions [77, 78], metal-ion coordination [79–81], and electrostatics [82, 83]. Furthermore, many molecules display a variety of specific binding modes. For the purposes of guiding DNA assembly, groove binders and intercalators have interesting properties. The application of these properties to DNA self-assembly will now be discussed.

5.2.3.1 DNA intercalators

Intercalators are small, flat, polyaromatic molecules that can insert between the base pairs of B-DNA. First described by Lerman in 1961 [84], intercalators insert into the 3.4 Å inter-base pairing site, allowing for enhanced π-stacking interactions when the flat, aromatic molecules are introduced. Ideal intercalators contain three-to-four fused rings [85] and interact with the DNA through hydrogen bonding, π-stacking, van der Waals interactions, hydrophobic interactions, and steric effects (Figure 5.2.9A), all of which contribute to an overall stabilization of the DNA duplex [86].

In addition to stabilizing the DNA duplex, intercalation has a variety of direct effects on the structure of DNA, which can be harnessed to control nanostructure assembly. These effects include lengthening and unwinding of the duplex, and occur as an intercalator inserts between the base pairs, causing local disturbances in the helical structure. As more intercalators bind, the changes to the structure multiply and become more pronounced (Figure 5.2.9B) [84, 87].

One of the most readily characterized outcomes of intercalation is helical unwinding, which has been used as a method of differentiating between groove binding and intercalation. In 1970, Michael Waring of the University of Cambridge postulated that if intercalators unwind DNA, they should have

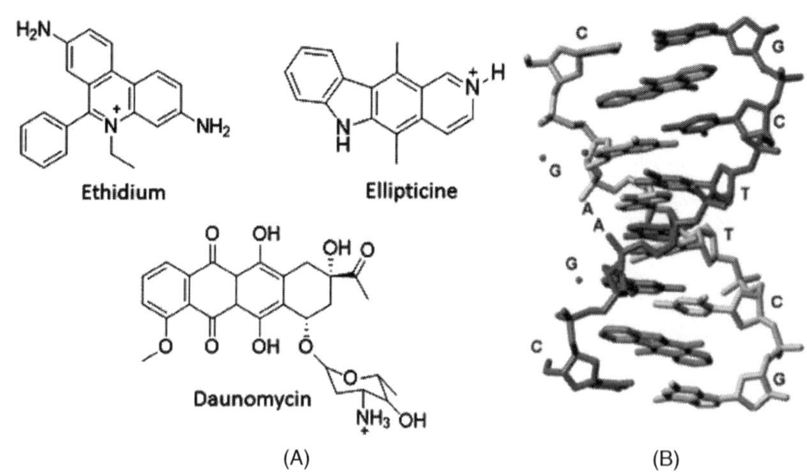

Figure 5.2.9 (A) Various intercalators. (B) Crystallographic structure of ellipticine in DNA. Inter-base pair spaces containing ellipticine are stretched [87]. Reproduced with permission from [87], IUCr

a pronounced effect on supercoiled circular DNA [88]. The degree of coiling was judged using ultracentrifugation to ascertain the sedimentation coefficient. For each of seven intercalators tested, a decrease in sedimentation coefficient was observed as the number of intercalators per nucleotide was increased, corresponding to helical unwinding. In several cases, the sedimentation coefficient reached a minimum and then began to increase at greater intercalator concentrations. This was interpreted as the helix reaching a point where it was fully unwound (the sedimentation coefficient minimum). A similar test with known groove binders gave no change to the sedimentation coefficient. From this study, it was concluded that not only did intercalators unwind DNA, but different intercalators unwound DNA to different extents. Similar results were later obtained using the methods of buoyant density [89] and viscometric titration measurements [90].

The first determination of the precise degree of helical unwinding per intercalator that stood the test of time was performed by Wang in 1974. By ligating DNA in the presence of known amounts of intercalator (ethidium bromide, in this case), Wang was able to generate circular DNA of varying superhelicity. Subsequently, he performed an alkaline titration to determine the point at which each DNA lost all helicity. From this he was able to calculate the degree of unwinding of ethidium bromide as –26°, the value still in use today [91].

The stabilizing and helical unwinding properties of intercalators were harnessed for structural DNA nanoassembly by Greschner, Bujold, and Sleiman [92]. They used a strained two-dimensional (2D) system that could potentially form a wide variety of structures including: those that were fully duplexed (containing even numbers of strands hybridized into cycles) and those with single-stranded ends (ssDNA, those with odd numbers of strands or uncyclized even numbers of strands, Figure 5.2.10A). Without the

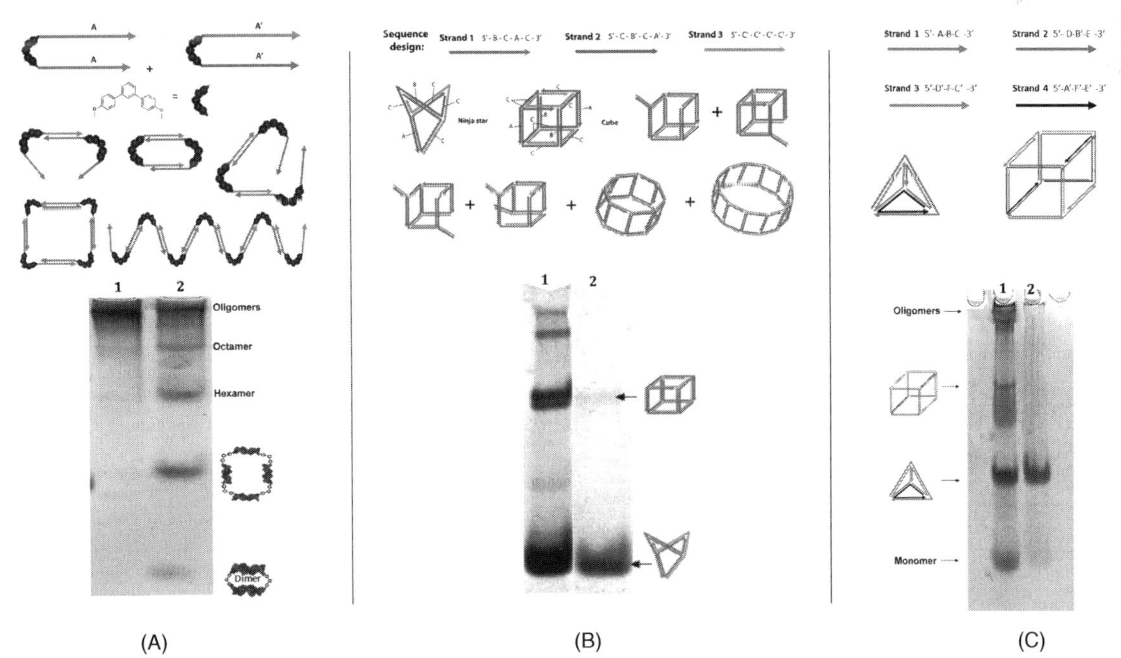

Figure 5.2.10 *(A) A strained 2D system that can assemble into many possible structures. Upon addition of EtBr, (lane 2), only cyclic products form. (B) A 3D "ninja star" system. The symmetrical nature of the strands provides many possible products, but addition of EtBr results in one product. (C) A second 3D system shows a similar change in product distribution [92]. Reprinted with permission from [92] Copyright 2013, American Chemical Society*

addition of an intercalator, this system formed undefined oligomers. With the addition of an intercalator prior to annealing, well-defined, fully cyclized structures emerged. This experiment was successful with several different intercalators, including ethidium bromide (EtBr), acridine orange, methylene blue, and proflavine.

With EtBr, the authors were able to use the defined unwinding angle of –26° to model and explain different product distributions among the cyclized structures. Without an intercalator, the alignment of DNA strand-ends – rigidly defined by a restrictive organic linker [36] – is unfavorable for the formation of cycles, resulting in oligomers. Addition of EtBr alters the strand-end alignment such that even small cycles, such as dimers and tetramers, can be formed, allowing for the production of a variety of nanoassemblies from the same initial DNA strands.

The presence of only fully duplexed structures was explained through the stabilizing effects of intercalators on DNA. Favorable π-interactions, and electrostatic interactions between the positively-charged intercalator and the negatively-charged phosphate backbone of DNA, energetically favor double-stranded DNA (cyclic DNA, in this case) over ssDNA.

The group then designed a 3D system using repeating strands, such that many products were possible. When assembled in the presence of EtBr, only one product (the smallest, fully duplexed "ninja star" product) formed (Figure 5.2.10B). The same result was observed using a nanostructure from the literature [25] – an assembly with many products was reduced to the single, smallest, fully-duplexed tetrahedron product (Figure 5.2.10C). Finally, it was demonstrated that the intercalator could be removed post-assembly without disturbing the final product distributions. As such, intercalators can stabilize and isolate a single product from many possible products, increasing the yield and simplifying the design process of DNA nanoassemblies.

Lengthening of the DNA helix is another inherent property of intercalators. As an intercalator inserts into the duplex the distance between base pairs is enlarged to better accommodate this molecule. The lengthening of the inter-base pair distance is achieved by straightening the backbone of DNA, allowing the base pairs to be further apart without changing the inter-phosphate distance [93]. Although straightening the backbone of DNA creates extra space for intercalator binding, it also strongly distorts the phosphate backbone that is directly flanking the intercalator. As such, the inter-base pair spaces directly above and below the intercalation site become distorted and inaccessible to a second intercalator. Therefore, in many intercalated systems, the neighbor exclusion principle applies, limiting intercalation events to a maximum of one out of every two inter-base pair sites [93–95].

Duplex length can play an important role in DNA nanostructure design. Adding additional bases to structures not only changes the overall helical twist of a structure, but also provides increased flexibility. In 2009, Shih and coworkers demonstrated that in tightly-packed DNA origami, adding bases to obtain an underwound 11 bp/turn for each crossover strand resulted in a higher yield than using the conventional 10.5 bp/turn [96]. They hypothesized that this might be due to the adjacent duplexes being more flexible, allowing them to bow away from one another and reducing electrostatic repulsion between the phosphate backbones.

In 2012, members of the same laboratory went a step further. By adding intercalators to underwound DNA origami structures, they were able to stabilize the duplexes and increase the yield of the final assemblies from around 14% to 24–30%, depending on the initial degree of unwinding (Figure 5.2.11A) [97]. Furthermore, these workers demonstrated that in addition to aiding in assembly, intercalators could be used to non-covalently introduce functionality to DNA nanostructures. This was done using pegylated *tris*-acridine. When the *tris*-acridine intercalated into the underwound duplexes, the PEG chains (polyethylene glycol) were secured to the nanoassembly, coating its surface and presenting an easy, viable method of surface functionalization (Figure 5.2.11B). Intercalation-mediated DNA functionalization with gold nanoparticles [98, 99] and streptavidin [100] has also been shown.

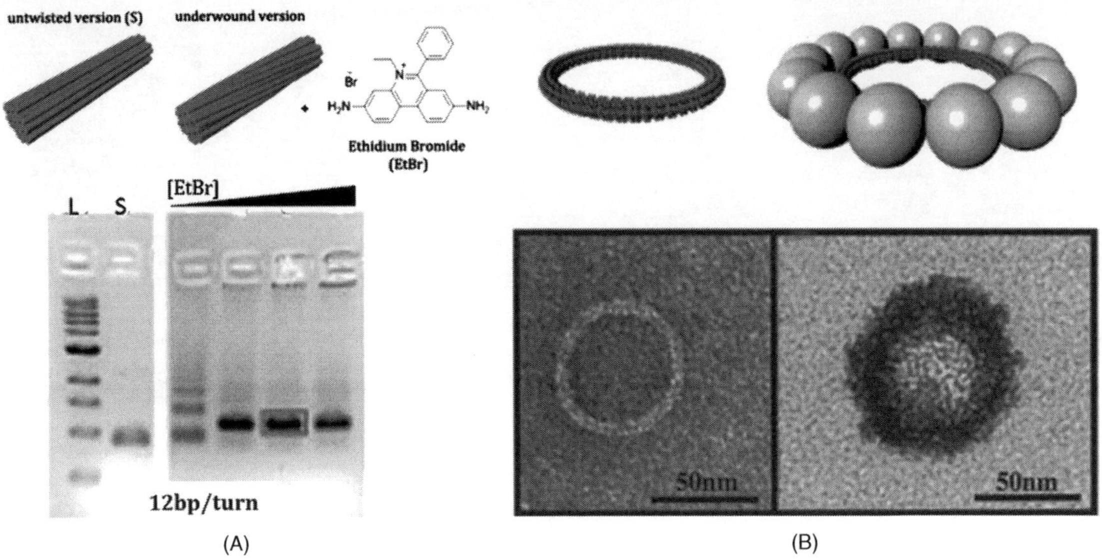

Figure 5.2.11 (A) Addition of an intercalator stabilizes underwound DNA origami structures: L is a reference ladder and S is a purified, untwisted cylinder. The gel on the right shows the underwound cylinder with increasing amounts of EtBr. The best yield is shown by the box in the middle. (B) Left, DNA origami ring. Right, PEGylated intercalator inserts between the base pairs of the ring [97]. Reproduced from [97] with permission from The Royal Society of Chemistry

5.2.3.2 Groove binding

In contrast to intercalators, groove binders have not been as well explored in DNA nanostructure assembly. However, they do possess an interesting characteristic, which is less common in intercalators – sequence specificity.

Groove binders can target either the major or minor groove of DNA. Minor groove binders include a variety of anticancer drugs such as distamycin [101], and netropsins (Figure 5.2.12) [94, 102]. Many minor groove binders consist of several ring structures linked in an arched shape that fits the concavity of the minor groove. Specificity for the minor groove over the major groove comes from hydrogen bonding, electrostatic interactions, and van der Waals forces (Figure 5.2.12) [74, 103–105]. The group working with Dervan have developed pyrrole–imidazole polyamides that bind to the minor groove of DNA and target specific DNA sequences [106, 107]. They showed similar sequence-binding specificity to DNA crossover tiles, and were able to create an ordered streptavidin array by attaching biotin units to these minor groove binders [108].

Although the early discovery of natural minor groove binders directed research towards this class of molecules [109, 110], major groove binding has also had some successes. Some proteins such as the herpes simplex virus [111] and enzymes like group II intron endonucleases [112] are known major groove binders. Both of these naturally occurring molecules bind through interactions with specific bases and phosphate groups. As such, they are sequence specific to 8 and 44 base segments, respectively.

Sequence specificity is often a goal in designing intercalators for therapeutic use. By targeting the intercalator drugs to different DNA sequences, it should be possible to decrease their cytotoxicity. In the field of DNA nanoassembly, a promising approach would be to combine the sequence specificity of groove binders with the stabilizing and duplex-modifying properties of intercalators. Early advances in sequence-specific intercalators were seen in drug therapeutics. Connecting two intercalators that have a G–C or A–T preference

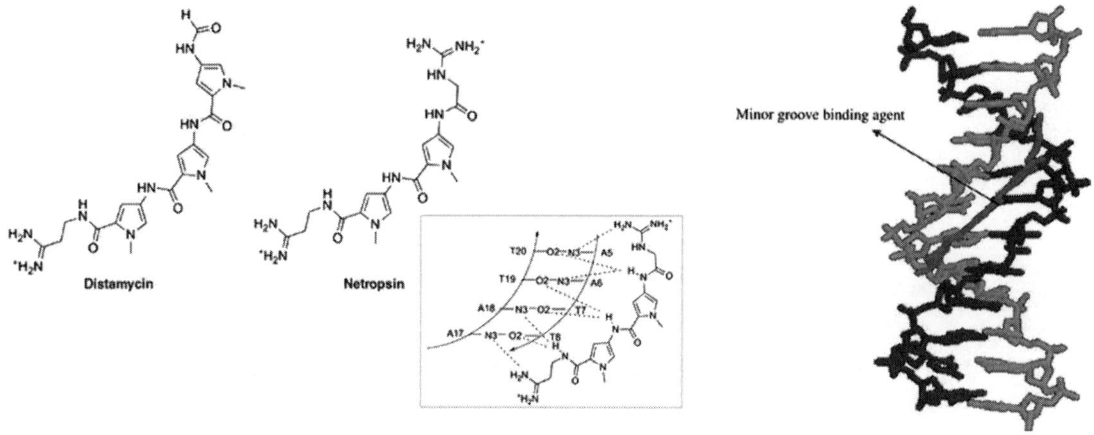

(A) (B)

Figure 5.2.12 *(A) Arc-shaped groove binders can bind specifically to certain functional groups in DNA base pairs [94]. Reproduced with permission from [94] Copyright 2011, John Wiley & Sons, Inc. (B) A model of how a groove binder sits in the DNA minor groove [103]. Reproduced from [103] with permission from Elsevier*

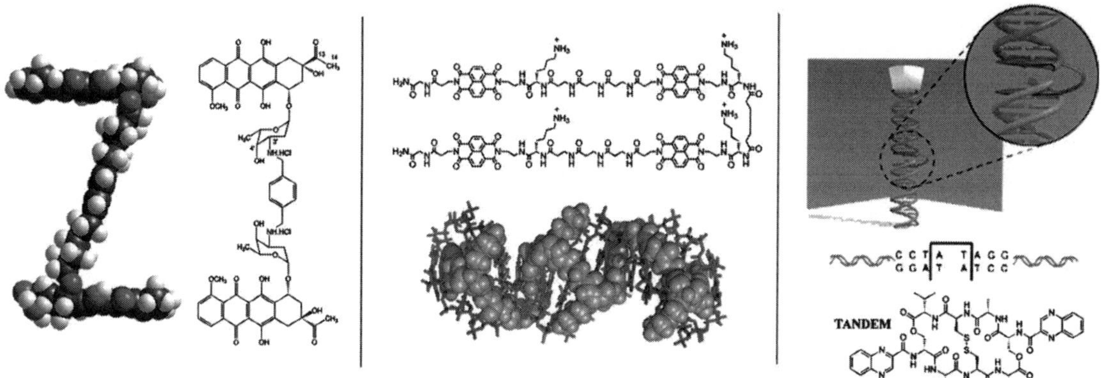

(A) (B) (C)

Figure 5.2.13 *(A) bis-intercalator WP631 [114]. Reprinted with permission from [114] Copyright 1998, American Chemical Society. (B) Threading tetra-intercalator [116]. Reproduced from [116]. Copyright © 2011, Rights Managed by Nature Publishing. (C) The intercalator TANDEM can join two duplexes non-covalently in an end-to-end fashion [117]. Reproduced from [117] with permission from The Royal Society of Chemistry*

should result in an overall increase in specificity. One example is a *bis*-daunorubicin connected by a *p*-xylyl linker [113, 114]. This *bis*-intercalator (Figure 5.2.13) was found to bind to a specific six-base sequence, with the xylyl linker residing in the DNA minor groove [115].

More recently, a tetraintercalator was introduced that is specific to a 14 bp target sequence. In addition to intercalating at four different sites, the molecule threads itself through the DNA strand, such that the links between intercalators move from the minor groove to the major groove and back to the minor groove (Figure 5.2.13B). The large number of intercalator–DNA interactions make the resulting complex both specific and very stable, resulting in a halflife of 16 days [116].

Finally, the stabilizing abilities of sequence specific *bis*-intercalators have been used to create duplexes that can be connected via their blunt ends. By designing duplexes with the correct base sequences at their terminal ends, Rackham *et al.* were able to use echinomycin to bridge between the two duplexes such that their blunt ends abutted (Figure 5.2.13C). Force measurements revealed that the connection was as strong as a normal single bond, indicating that the intercalators were well anchored [117]. It is feasible that *bis*-intercalators might become a tool for assembling nanostructures, bringing blunt ends together.

5.2.4 Conclusions

All these approaches can be exploited to incorporate a broad range of organic/inorganic molecules into the DNA structure, allowing for the control and modulation of its stability and self-assembly properties. The features and functions of the resulting DNA assemblies might be tuned at will by appropriate design of the added synthetic complexes or, for instance, by external stimuli that affect dynamic aspects of coordination bonding. These materials could be easily envisioned for DNA molecular hosts, nanomachines and nanodevices, and for several applications in drug delivery, optoelectronics, magnetic materials, light harvesting, diagnostics, and bioimaging.

With the area of DNA nanotechnology still very much under development, non-covalent methods of controlling assembly are a burgeoning field. Intercalators and groove binders have demonstrated the ability to stabilize fully duplexed structures, modify assembly outcomes, increase yields, functionalize assemblies, and connect blunt-ended duplexes. These techniques will simplify the sequence-design requirements for many DNA assemblies, ultimately leading to the formation of larger, more complex, and functional DNA nanomaterials.

References

[1] F. A. Aldaye, A. L. Palmer, and H. F. Sleiman, Assembling materials with DNA as the guide, *Science*, **321**, 1795–1799 (2008).

[2] J. K. Lee, Y. H. Jung, J. B. H. Tok and Z. Bao, Syntheses of organic molecule–DNA hybrid structures, *ACS Nano*, **5**, 2067–2074 (2011).

[3] M. Kwak and A. Herrmann, Nucleic acid/organic polymer hybrid materials: Synthesis, superstructures, and applications, *Angew. Chem., Int. Ed.*, **49**, 8574–8587 (2010).

[4] J. Chen and N. C. Seeman, Synthesis from DNA of a molecule with the connectivity of a cube, *Nature*, **350**, 631–633 (1991).

[5] J. Irvoas, A. Noirot, N. Chouini-Lalanne, O. Reynes, J.-C. Garrigues, and V. Sartor, Programmable multimetallic linear nanoassemblies of ruthenium-DNA conjugates, *R. Soc. Chem. Adv.*, **2**, 9538–9542 (2012).

[6] N. S. Oltra and G. Roelfes, Modular assembly of novel DNA–based catalysts, *Chem. Commun.*, 6039–6041 (2008).

[7] J. Gierlich, G. A. Burley, P. M. E. Gramlich, D. M. Hammond, and T. Carell, Click chemistry as a reliable method for the high-density postsynthetic functionalization of alkyne-modified DNA, *Org. Lett.*, **8**, 3639–3642 (2006).

[8] R. V. Thaner, I. Eryazici, O. K. Farha, C. A. Mirkin, and S. T. Nguyen, Facile one-step solid-phase synthesis of multitopic organic-DNA hybrids via "click" chemistry, *Chem. Sci.*, **5**, 1091–1096 (2014).

[9] A. H. El-Sagheer and T. Brown, Click nucleic acid ligation: Applications in biology and nanotechnology, *Acc. Chem. Res.*, **45**, 1258–1267 (2012).

[10] H. Isobe, T. Fujino, N. Yamazaki, M. Guillot-Nieckowski, and E. Nakamura, Triazole-linked analogue of deoxyribonucleic acid (TLDNA): Design, synthesis, and double-strand formation with natural DNA, *Org. Lett.*, **10**, 3729–3732 (2008).

[11] S. Rele and S. K. Nayak, Low-valent titanium mediated reductive cleavage of o/n-trityl bonds via free radical pathway, *Synth. Commun.*, **32**, 3533–3540 (2002).

[12] F. A. Aldaye and H. F. Sleiman, Sequential self-assembly of a DNA hexagon as a template for the organization of gold nanoparticles, *Angew. Chem.*, **45**, 2204–2209 (2006).

[13] K. K. Ogilvie and M. J. Nemer, The synthesis of phosphite analogues of ribonucleotides, *Tetrahedron Lett.*, **21**, 4145–4148 (1980).

[14] M. S. Shchepinov, S. C. Case-Green, and E. M. Southern, Steric factors influencing hybridisation of nucleic acids to oligonucleotide arrays, *Nucleic Acids Res.*, **25**, 1155–1161 (1997).

[15] H. Yang, C. K. McLaughlin, F. A. Aldaye, G. D. Hamblin, A. Z. Rys, I. Rouiller, and H. F. Sleiman, Metal-nucleic acid cages, *Nat. Chem.*, **1**, 390–396 (2009).

[16] I. Eryazici, I. Yildirim, G. C. Schatz, and S. T. Nguyen, Enhancing the melting properties of small molecule-DNA hybrids through designed hydrophobic interactions—an experimental-computational study, *J. Am. Chem. Soc.*, **143**, 7450–7458 (2012).

[17] J. Shi and D. E. Bergstrom, Assembly of novel DNA cycles with rigid tetrahedral linkers, *Angew. Chem., Int. Ed.*, **36**, 111–113 (1997).

[18] M. Scheffler, A. Dorenbeck, S. Jordan, M. Wüstefeld, and G. von Kiedrowski, Self-assembly of trisoligonucleotidyls: The case for nano-acetylene and nano-cyclobutadiene. *Angew. Chem., Int. Ed.*, **38**, 3311–3315 (1999).

[19] T. Kuroda, Y. Sakurai, Y. Suzuki, A. O. Nakamura, M. Kuwahara, H. Ozaki, and H. Sawai, Assembly of DNA nanostructures with branched Tris-DNA, *Chem. – Asian J.*, **1**, 575–580 (2006).

[20] F. A. Aldaye and H. F. Sleiman, Sequential self-assembly of a DNA hexagon as a template for the organization of gold nanoparticles, *Angew. Chem., Int. Ed.*, **45**, 2204–2209 (2006).

[21] F. A. Aldaye and H. F. Sleiman, Dynamic DNA templates for discrete gold nanoparticle assemblies: control of geometry, modularity, write/erase and structural switching, *J. Am. Chem. Soc.*, **129**, 4130–4131 (2007).

[22] P. K. Lo, P. Karam, F. A. Aldaye, C. K. McLaughlin, G. D. Hamblin, G. Cosa, and H. F. Sleiman, Loading and selective release of cargo in DNA nanotubes with longitudinal variation, *Nat. Chem.*, **2**, 319–328 (2010).

[23] F. A. Aldaye and H. F. Sleiman, Modular access to structurally switchable 3D discrete DNA assemblies, *J. Am. Chem. Soc.*, **129**, 13376–13377 (2007).

[24] J. Tumpane, P. Sandin, R. Kumar, V. E. C. Powers, E. P. Lundberg, N. Gale, P. Baglioni, J.-M. Lehn, B. Albinsson, P. Lincoln, L. M. Wilhelmsson, T. Brown, and B. Nordén, Addressable high-information-density DNA nanostructures, *Chem. Phys. Lett.*, **440**, 125–129 (2007).

[25] R. P. Goodman, R. M. Berry, and A. J. Turberfield, The single-step synthesis of a DNA tetrahedron, *Chem. Commun.*, 1372–1373 (2004).

[26] G. D. Hamblin, A. A. Hariri, K. M. M. Carneiro, K. L. Lau, G. Cosa, and H. F. Sleiman, Simple design for DNA nanotubes from a minimal set of unmodified strands: Rapid, room-temperature assembly and readily tunable structure, *ACS Nano*, **7**, 3022–3028 (2013).

[27] P. W. K. Rothemund, Folding DNA to create nanoscale shapes and patterns, *Nature*, **440**, 297–302 (2006).

[28] Á. Somoza, Evolution of DNA origami, *Angew. Chem., Int. Ed.*, **48**, 9406–9408 (2009).

[29] C. K. McLaughlin, G. D. Hamblin, K. D. Hänni, J. W. Conway, M. K. Nayak, K. M. M. Carneiro, H. S. Bazzi, and H. F. Sleiman, Three-dimensional organization of block copolymers on "DNA-minimal" scaffolds, *J. Am. Chem. Soc.*, **134**, 4280–4286 (2012).

[30] C. K. McLaughlin, G. D. Hamblin, and H. F. Sleiman. Supramolecular DNA assembly, *Chem. Soc. Rev.*, **40**, 5647–5656 (2011).

[31] P. K. Lo, F. Altvater, and H. F. Sleiman, Templated synthesis of DNA nanotubes with controlled, predetermined lengths, *J. Am. Chem. Soc.*, **132**, 10212–10214 (2010).

[32] J. Zimmermann, M. P. J. Cebulla, S. Mönninghoff, and G. von Kiedrowski, Self-assembly of a DNA dodecahedron from 20 trisoligonucleotides with C3h linkers, *Angew. Chem., Int. Ed.*, **47**, 3626–3630 (2008).

[33] I. Eryazici, T. R. Prytkova, G. C. Schatz, and S. T. Nguyen, Cooperative melting in caged dimers with only two DNA duplexes, *J. Am. Chem. Soc.*, **132**, 17068–17070 (2010).

[34] R. Jin, G. Wu, Z. Li, C. A. Mirkin, and G. C. Schatz, What controls the melting properties of DNA-linked gold nanoparticle assemblies? *J. Am. Chem. Soc.*, **125**, 1643–1654 (2003).

[35] J. M. Gibbs-Davis, G. C. Schatz, and S. T. Nguyen, Sharp melting transitions in DNA hybrids without aggregate dissolution: Proof of neighboring-duplex cooperativity, *J. Am. Chem. Soc.*, **129**, 15535–15540 (2007).

[36] A. A. Gręschner, V. Toader, and H. F. Sleiman, The role of organic linkers in directing DNA self-assembly and significantly stabilizing DNA duplexes, *J. Am. Chem. Soc.*, **134**, 14382–14389 (2012).

[37] J. W. Conway, C. K. McLaughlin, K. J. Castor, and H. Sleiman, DNA nanostructure serum stability: greater than the sum of its parts, *Chem. Commun.*, **49**, 1172–1174 (2013).

[38] F. A. Aldaye, P. K. Lo, P. Karam, C. K. McLaughlin, G. Cosa, and H. F. Sleiman, Modular construction of DNA nanotubes of tunable geometry and single- or double-stranded character, *Nat. Nanotechnol.*, **4**, 349–352 (2009).

[39] M. S. Shchepinov, K. U. Mir, J. K. Elder, M. D. Frank-Kamenetskii, and E. M. Southern, Oligonucleotide dendrimers: stable nano-structures, *Nucleic Acids Res.*, **27**, 3035–3041 (1999).

[40] M. Endo, N. C. Seeman, and T. Majima, DNA tube structures controlled by a four-way-branched DNA connector, *Angew. Chem., Int. Ed.*, **44**, 6074–6077 (2005).

[41] A. Singh, M. Tolev, M. Meng, K. Klenin, O. Plietzsch, C. I. Schilling, T. Muller, M. Nieger, S. Bräse, W. Wenzel, and C. Richert, Branched DNA that forms a solid at 95 °C, *Angew. Chem., Int. Ed.*, **50**, 3227–3231 (2011).

[42] K. M. Carneiro, F. A. Aldaye, and H. F. Sleiman, Long-range assembly of DNA into nanofibers and highly ordered networks using a block copolymer approach, *J. Am. Chem. Soc.*, **132**, 679–685 (2010).

[43] H. Liu, T. Tørring, M. Dong, C. B. Rosen, F. Besenbacher, and K. V. Gothelf, DNA-templated covalent coupling of G4 PAMAM dendrimers, *J. Am. Chem. Soc.*, **132**, 18054–18056 (2010).

[44] K. E. Bujold, J. Fakhoury, T. G. W. Edwardson, K. M. M. Carneiro, J. N. Briard, A. G. Godin, L. Amrein, G. D. Hamblin, L. C. Panasci, P. W. Wiseman, and H. F. Sleiman, Sequence-responsive unzipping DNA cubes with tunable cellular uptake profiles, *Chem. Sci.*, **5**, 2449–2455 (2014).

[45] T. G. W. Edwardson, K. M. M. Carneiro, C. K. McLaughlin, C. J. Serpell, and H. F. Sleiman, Site-specific positioning of dendritic alkyl chains on DNA cages enables their geometry-dependent self-assembly, *Nat. Chem.*, **5**, 868–875 (2013).

[46] T. G. W. Edwardson, K. M. M. Carneiro, C. J. Serpell, and H. F. Sleiman, An efficient and modular route to sequence-defined polymers appended to DNA, *Angew. Chem., Int. Ed.*, **53**, 4567–4571 (2014).

[47] K. M. Stewart and L. W. McLaughlin, Four-arm oligonucleotide Ni(II)–cyclam-centered complexes as precursors for the generation of supramolecular periodic assemblies, *J. Am. Chem. Soc.*, **126**, 2050–2057 (2004).

[48] K. M. Stewart, J. Rojo, and L. W. McLaughlin, Ru(II) tris(bipyridyl) complexes with six oligonucleotide arms as precursors for the generation of supramolecular assemblies, *Angew. Chem., Int. Ed.*, **43**, 5808–5811 (2004).

[49] D. Mitra, N. Di Cesare, and H. F. Sleiman, Self-assembly of cyclic metal-DNA nanostructures using ruthenium tris(bipyridine)-branched oligonucleotides, *Angew. Chem.*, **43**, 5804–5808 (2004).

[50] I. Vargas-Baca, D. Mitra, H. J. Zulyniak, J. Banerjee, and H. F. Sleiman, Solid-phase synthesis of transition metal linked, branched oligonucleotides, *Angew. Chem., Int. Ed.* **40**, 4629–4632 (2001).

[51] H. Yang, K. L. Metera, and H. F. Sleiman, DNA modified with metal complexes: Applications in the construction of higher order metal–DNA nanostructures, *Coord. Chem. Rev.*, **254**, 2403–2415 (2010).

[52] J. S. Choi, C. W. Kang, K. Jung, J. W. Yang, Y.-G. Kim, and H. Han, Synthesis of DNA triangles with vertexes of bis(terpyridine)iron(II) complexes, *J. Am. Chem. Soc.*, **126**, 8606–8607 (2004).

[53] M. Göritz and R. Krämer, Allosteric control of oligonucleotide hybridization by metal-induced cyclization, *J. Am. Chem. Soc.*, **127**, 18016–18017 (2005).

[54] D. Miyoshi, H. Karimata, Z.-M. Wang, K. Koumoto, and N. Sugimoto, Artificial G-wire switch with 2,2′-bipyridine units responsive to divalent metal ions, *J. Am. Chem. Soc.*, **129**, 5919–5925 (2007).

[55] H. Yang and H. F. Sleiman, Templated synthesis of highly stable, electroactive, and dynamic metal-DNA branched junctions, *Angew. Chem.*, **47**, 2443–2446 (2008).

[56] H. Yang, A. Z. Rys, C. K. McLaughlin, and H. F. Sleiman, Templated ligand environments for the selective incorporation of different metals into DNA, *Angew. Chem.*, **48**, 9919–9923 (2009).

[57] J. L. Czlapinski and T. L. Sheppard, Nucleic acid template-directed assembly of metallosalen–DNA conjugates, *J. Am. Chem. Soc.*, **123**, 8618–8619 (2001).

[58] J. L. Czlapinski and T. L. Sheppard, Template-directed assembly of metallosalen–DNA hairpin conjugates, *ChemBioChem*, **5**, 127–129 (2004).

[59] K. V. Gothelf and R. S. Brown, A modular approach to DNA-programmed self-assembly of macromolecular nanostructures, *Chem. –Eur. J.*, **11**, 1062–1069 (2005).

[60] M. Nielsen, A. H. Thomsen, E. Cló, F. Kirpekar, and K. V. Gothelf, Synthesis of linear and tripoidal oligo(phenylene ethynylene)-based building blocks for application in modular DNA-programmed assembly, *J. Org. Chem.*, **69**, 2240–2250 (2004).

[61] K. V. Gothelf, A. Thomsen, M. Nielsen, E. Cló, and R. S. Brown, Modular DNA-programmed assembly of linear and branched conjugated nanostructures, *J. Am. Chem. Soc.*, **126**, 1044–1046 (2004).

[62] R. S. Brown, M. Nielsen, and K. V. Gothelf, Self-assembly of aluminium-salen coupled nanostructures from encoded modules with cleavable disulfide DNA-linkers, *Chem. Commun.*, 1464–1465 (2004).

[63] M. Nielsen, V. Dauksaite, J. Kjems, and K. V. Gothelf, DNA-directed coupling of organic modules by multiple parallel reductive aminations and subsequent cleavage of selected DNA sequences, *Bioconjugate Chem.*, **16**, 981–985 (2005).

[64] Y. Takezawa and M. Shionoya, Metal-mediated DNA base pairing: Alternatives to hydrogen-bonded Watson–Crick base pairs, *Acc. Chem. Res.*, **45**, 2066–2076 (2012).

[65] E. Meggers, P. L. Holland, W. B. Tolman, F. E. Romesberg, and P. G. Schultz, A novel copper-mediated DNA base pair, *J. Am. Chem. Soc.*, **122**, 10714–10715 (2000).

[66] G. H. Clever, C. Kaul, and T. Carell, DNA–metal base pairs, *Angew. Chem., Int. Ed.*, **46**, 6226–6236 (2007).

[67] G. H. Clever, S. J. Reitmeier, T. Carell, and O. Schiemann, Antiferromagnetic coupling of stacked CuII–salen complexes in DNA, *Angew. Chem., Int. Ed.*, **49**, 4927–4929 (2010).

[68] H. Weizman and Y. Tor. 2,2′-Bipyridine ligandoside: A novel building block for modifying DNA with intra-duplex metal complexes, *J. Am. Chem. Soc.*, **123**, 3375–3376 (2001).

[69] B. D. Heuberger, D. Shin, and C. Switzer, Two Watson–Crick-like metallo base-pairs, *Org. Lett.*, **10**, 1091–1094 (2008).

[70] F.-A. Polonius and J. Müller, An artificial base pair, mediated by hydrogen bonding and metal-ion binding, *Angew. Chem., Int. Ed.*, **46**, 5602–5604 (2007).

[71] T. Richters, O. Krug, J. Kösters, A. Hepp, and J. Müller, A family of "click" nucleosides for metal-mediated base pairing: Unravelling the principles of highly stabilizing metal-mediated base pairs, *Chem. –Eur. J.*, **20**, 781–7818 (2014).

[72] J.-L. H. A. Duprey, Y. Takezawa, and M. Shionoya, Metal-locked DNA three-way junction. *Angew. Chem.*, **125**, 1250–1254 (2013).

[73] H. Yang, F. Altvater, A. D. de Bruijn, C. K. McLaughlin, P. K. Lo, and H. F. Sleiman, Chiral metal–DNA four-arm junctions and metalated nanotubular structures, *Angew. Chem., Int. Ed.*, **50**, 4620–4623 (2011).

[74] C. L. Kielkopf and S. White, A structural basis for recognition of A… and T… base pairs in the minor groove of B-DNA, *Science*, **282**, 111 (1998).

[75] C. L. Kielkopf, E. E. Baird, P. B. Dervan, and D. C. Rees, Structural basis for G[bull]C recognition in the DNA minor groove, *Nat. Struct. Mol. Biol.*, **5**, 104–109 (1998).

[76] B. Geierstanger, M. Mrksich, P. Dervan, and D. Wemmer, Design of a G.C-specific DNA minor groove-binding peptide, *Science*, **266**, 646–650 (1994).

[77] A. Sitlani and J. K. Barton, Sequence-specific recognition of DNA by phenanthrenequinone diimine complexes of rhodium(III): Importance of steric and van der Waals interactions, *Biochemistry*, **33**, 12100–12108 (1994).

[78] F. Leng, J. B. Chaires, and M. J. Waring, Energetics of echinomycin binding to DNA, *Nucleic Acids Res.* **31**, 6191–6197 (2003).

[79] H. Cho, Y. Guo, D. E. Sosnovik, and L. Josephson, Imaging DNA with fluorochrome bearing metals, *Inorg. Chem.*, (2013).

[80] G. Barone, A. Terenzi, A. Lauria, A. M. Almerico, J. M. Leal, N. Busto, and B. García, DNA-binding of nickel(II), copper(II) and zinc(II) complexes: Structure–affinity relationships, *Coord. Chem. Rev.*, **257**, 2848–2862 (2013).

[81] S. Yang, C. Wu, H. Tan, Y. Wu, S. Liao, Z. Wu, G. Shen, and R. Yu, Label–free liquid crystal biosensor based on specific oligonucleotide probes for heavy metal ions, *Anal. Chem.*, **85**, 14–18 (2012).

[82] X. Jiang and X. Lin, Voltammetry of the interaction of metronidazole with DNA and its analytical applications, *Bioelectrochemistry*, **68**, 206–212 (2006).

[83] I. Rouzina and V. A. Bloomfield, Influence of ligand spatial organization on competitive electrostatic binding to DNA, *J. Phys. Chem.*, **100**, 4305–4313 (1996).

[84] L. S. Lerman, Structural considerations in the interaction of DNA and acridines, *J. Mol. Biol.*, **3**, 18–30 (1961).

[85] W. D. Wilson, in *Comprehensive Natural Products Chemistry* (O. Meth-Cohn, D. Barton, and K. Nakanishi (Eds-in-Chief), Pergamon, pp. 427–476 (1999).

[86] B. Neto and A. Lapis, Recent developments in the chemistry of deoxyribonucleic acid (DNA) intercalators: Principles, design, synthesis, applications and trends, *Molecules*, **14**, 1725–1746 (2009).

[87] M. Purciolas, A. Canals, M. Coll, and J. Aymamí, The anticancer agent ellipticine unwinds DNA by intercalative binding in an orientation parallel to base pairs, *Acta Crystallogr., Sect. D: Biol. Crystallogr.*, **61**, 1009–1012 (2005).

[88] M. Waring, Variation of the supercoils in closed circular DNA by binding of antibiotics and drugs: Evidence for molecular models involving intercalation, *J. Mol. Biol.*, **54**, 247–279 (1970).

[89] B. M. J. Révet, M. Schmir, and J. Vinograd, Direct determination of the superhelix density of closed circular DNA by viscometric titration, *Nature*, **229**, 10–13 (1971).

[90] J. C. Wang, Helical repeat of DNA in solution, *Proc. Natl. Acad. Sci. U.S.A.*, **76**, 200–203 (1979).

[91] J. G. Wang, The degree of unwinding of the DNA helix by ethidium: I. Titration of twisted PM2 DNA molecules in alkaline cesium chloride density gradients, *J. Mol. Biol.*, **89**, 783–801 (1974).

[92] A. A. Greschner, K. E. Bujold, and H. F. Sleiman, Intercalators as molecular chaperones in DNA self-assembly, *J. Am. Chem. Soc.*, **135**, 11283–11288 (2013).

[93] L. D. Williams, M. Egli, Q. Gao, and A. Rich, in *Structure and Function*, Vol. **1**: Nucleic Acids (R. H. Sarmaand and M. H Sarma, Eds), Adenine Press, pp. 107–125 (1992).

[94] S. Li and Z. Xi, in *Medicinal Chemistry of Nucleic Acids*, John Wiley & Sons, Inc., pp. 164–205 (2011).

[95] D. M. Chothers, Calculation of binding isotherms for heterogeneous polymers, *Biopolymers*, **6**, 575–584 (1968).

[96] H. Dietz, S. M. Douglas, and W. M. Shih, Folding DNA into twisted and curved nanoscale shapes, *Science*, **325**, 725–730 (2009).

[97] Y. Ke, G. Bellot, N. V. Voigt, E. Fradkov, and W. M. Shih, Two design strategies for enhancement of multilayer-DNA-origami folding: underwinding for specific intercalator rescue and staple-break positioning, *Chem. Sci.*, **3**, 2587–2597 (2012).

[98] J. P. Renault, A. Bernard, D. Juncker, B. Michel, H. R. Bosshard, and E. Delamarche, Fabricating microarrays of functional proteins using affinity contact printing, *Angew. Chem., Int. Ed.*, **41**, 2320–2323 (2002).

[99] M. Mehrabi and R. Wilson, Intercalating gold nanoparticles as universal labels for DNA detection, *Small*, **3**, 1491–1495 (2007).

[100] M. Slim, N. Durisic, P. Grutter, and H. F. Sleiman, DNA-protein noncovalent cross-linking: ruthenium dipyridophenazine biotin complex for the assembly of proteins and gold nanoparticles on DNA templates, *Chembiochem: Eur. J. Chem. Biol.*, **8**, 804–812 (2007).

[101] T. C. Bruice, H. Y. Mei, G. X. He, and V. Lopez, Rational design of substituted tripyrrole peptides that complex with DNA by both selective minor-groove binding and electrostatic interaction with the phosphate backbone, *Proc. Natl. Acad. Sci. U.S.A.*, **89**, 1700–1704 (1992).

[102] R. Zhang, X. Wu, L. J. Guziec, F. S. Guziec, G.-L. Chee, J. C. Yalowich, and B. B. Hasinoff, Design, synthesis and biological evaluation of a novel series of anthrapyrazoles linked with netropsin-like oligopyrrole carboxamides as anticancer agents, *Biorg. Med. Chem.*, **18**, 3974–3984 (2010).

[103] G. S. Khan, A. Shah, R. Zia-ur-Rehman, and D. Barker, Chemistry of DNA minor groove binding agents, *J. Photochem. Photobiol., B*, **115**, 105–118 (2012).

[104] K. A. Muzikar, J. L. Meier, D. A. Gubler, J. A. Raskatov, and P. B. Dervan, Expanding the repertoire of natural product-inspired ring pairs for molecular recognition of DNA, *Org. Lett.*, **13**, 5612–5615 (2011).

[105] J. A. Raskatov, J. L. Meier, J. W. Puckett, F. Yang, P. Ramakrishnan, and P. B. Dervan, Modulation of NF-κB-dependent gene transcription using programmable DNA minor groove binders, *Proc. Natl. Acad. Sci. U.S.A.*, **109**, 1023–1028 (2012).

[106] P. B. Dervan and R. W. Bürli, Sequence-specific DNA recognition by polyamides, *Curr. Opin. Chem. Biol.*, **3**, 688–693 (1999).

[107] J. S. Kang, J. L. Meier, and P. B. Dervan, Design of sequence-specific DNA binding molecules for DNA methyltransferase inhibition, *J. Am. Chem. Soc.*, **136**, 3687–3694 (2014).

[108] J. D. Cohen, J. P. Sadowski, and P. B. Dervan, Addressing single molecules on DNA nanostructures, *Angew. Chem., Int. Ed.*, **46**, 7956–7959 (2007).

[109] P. B. Dervan, Molecular recognition of DNA by small molecules, *Biorg. Med. Chem.*, **9**, 2215–2235 (2001).

[110] P. L. Hamilton and D. P. Arya, Natural product DNA major groove binders, *Nat. Prod. Rep.*, **29**, 134–143 (2012).

[111] S. Simonsson, T. Samuelsson, and P. Elias, The herpes simplex virus type 1 origin binding protein: specific recognition of phosphates and methyl groups defines the interacting surface for a monomeric DNA binding domain in the major groove of DNA, *J. Biol. Chem.*, **273**, 24633–24639 (1998).
[112] N. N. Singh and A. M. Lambowitz, Interaction of a group II intron ribonucleoprotein endonuclease with its DNA target site investigated by DNA footprinting and modification interference, *J. Mol. Biol.*, **309**, 361–386 (2001).
[113] J. B. Chaires, F. Leng, T. Przewloka, I. Fokt, Y.-H. Ling, R. Perez-Soler, and W. Priebe, Structure-based design of a new bisintercalating anthracycline antibiotic, *J. Med. Chem.*, **40**, 261–266 (1997).
[114] F. Leng, W. Priebe, and J. B. Chaires, Ultratight DNA binding of a new bisintercalating anthracycline antibiotic, *Biochemistry*, **37**, 1743–1753 (1998).
[115] K. R. Fox, R. Webster, R. J. Phelps, I. Fokt, and W. Priebe, Sequence selective binding of bis-daunorubicin WP631 to DNA, *Eur. J. Biochem.*, **271**, 3556–3566 (2004).
[116] G. G. Holman, M. Zewail-Foote, A. R. Smith, K. A. Johnson, and B. L. Iverson, A sequence-specific threading tetra-intercalator with an extremely slow dissociation rate constant, *Nat. Chem.*, **3**, 875–881 (2011).
[117] B. D. Rackham, L. A. Howell, A. N. Round, and M. Searcey, Non-covalent duplex to duplex crosslinking of DNA in solution revealed by single molecule force spectroscopy, *Org. Biomol. Chem.*, **11**, 8340–8347 (2013).

5.3

DNA-Based Nanostructuring with Branched Oligonucleotide Hybrids

Alexander Schwenger, Helmut Griesser, and Clemens Richert

Institute for Organic Chemistry, University of Stuttgart, Stuttgart, Germany

5.3.1 Introduction

Oligodeoxynucleotides are fascinating building materials. They are readily synthesized in automatic fashion in lengths of up to 100–50 nucleotides, and they form antiparallel duplex structures, based on the base-paring rules of Watson and Crick [1]. Duplexes of DNA strands are approximately 2 nm in diameter and have a typical persistence length in aqueous buffer of over 100 base pairs. When combined with branching points, such as Holliday junctions, bends or loops, diverse shapes and structures can be designed and built through hybridization [2]. A Holliday junction is a biologically relevant structure that helps with the exchange of genetic information between two homologous or non-homologous DNA strands [3], and its use in nanoconstruction is one of many examples in this field of new, non-natural functions for the structural elements found in a cell. The structural diversity of DNA nanostructures can be further increased by incorporating non-duplex motifs, such as triplexes or G-quadruplexes, or by including RNA strands in the design [4].

Initially, DNA-based nanoconstruction was focused on finite nano-objects, such as Ned Seeman's cube [5] or Turberfield's tetrahedron [6]. However, Rothemund's origami technique that uses a long single-stranded template, folded by more than 100 scaffold strands, moved the field to a new level [7]. Origami structures created by Shih, Dietz, Liedl, and others, then included delicate shapes with curved surfaces and dazzling levels of structural engineering [8–10]. Designed origami structures now include an entirely artificial ion channel [11], and a three-dimensional object whose structural details were confirmed in a high-resolution

DNA in Supramolecular Chemistry and Nanotechnology, First Edition.
Edited by Eugen Stulz and Guido H. Clever.

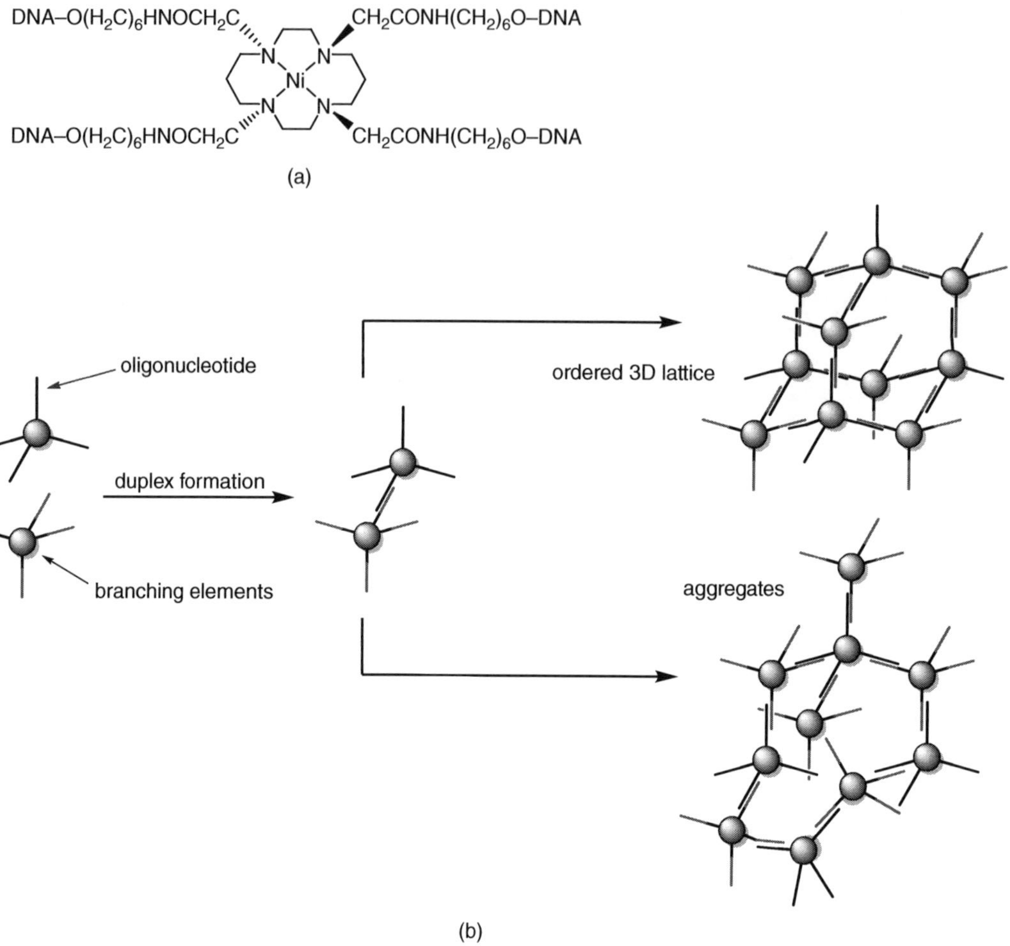

Figure 5.3.1 *DNA hybridization-driven building of 3D lattices or aggregates*

structure [12]. While most groups in the field focus on structures of ever-increasing size, we, and others, have studied ways of building small origami structures that can be prepared from less than 25 staple strands, so that new synthetic methods can be readily tested in affordable experiments [13].

Building infinite networks, rather than finite objects, from DNA is more challenging. Porous three-dimensional networks with cavities with a size sufficient to bind proteins were the initial goal of the pioneering work of Seeman, but it took more than 25 years from the early phases of the project [14] to the design of the first DNA crystals with tuneable unit cell dimensions [15]. Although small proteins can be included in DNA lattices [16], it remains unclear whether X-ray crystal structures of high resolution can be obtained from such inclusion complexes.

Ten years ago we asked whether designed, porous three-dimensional lattices, held together by DNA duplexes, could be obtained by allowing branched oligonucleotide hybrids to assemble, based on hybridization of their DNA arms. The hybrids were designed to consist of a pre-organizing organic core element and short oligodeoxynucleotides as 'sticky ends'. At the time, Stewart and McLaughlin had just published work on branched structures with transition metal complexes as cores (Figure 5.3.1a) [17]. Complexes with six

DNA arms were also tested by this group, but it remained unclear whether the assemblies formed had an ordered geometry [18].

Even though there was already a rich chemistry of one- and two-dimensional arrays, based on DNA [19], a review published in 2005 still considered periodic matter constructed from DNA to be a long-term goal of this field [20]. At the time, the breakthrough for DNA-coated gold nanoparticles [21] (Figure 5.3.1b) forming lattices with a high level of crystallinity [22–24] was still a long way off, as was the design of the first crystal of the triangle folding motif [15]. It was unclear whether DNA duplex formation could be used as the guiding principle for assembling ordered three-dimensional lattices. It seemed quite likely that aggregation would be the predominant outcome of hybridization experiments involving branched oligonucleotide constructs, as shown schematically in the lower part of Figure 5.3.1b for colloids.

5.3.2 Branched oligonucleotides

We felt that it would be worthwhile pursuing a different approach from those using inorganic branching elements (including gold colloids) or DNA folding motifs. Instead, we decided to synthesize hybrids of rigid hydrocarbon 'cores' with terminal hydroxy groups and oligonucleotide chains. The term 'hybrid' is meant to indicate that these species are devoid of bifunctional linkers of the type commonly used in bioconjugation chemistry [25]. We reasoned that the cores would pre-organize the DNA arms, so as to induce stable and possibly periodically ordered structures during hybridization. The hydrocarbon cores should also be chemically more resistant than inorganic complexes, facilitating syntheses. Further, the hybrids should be monodisperse, pure compounds, avoiding the complications that can arise from preparing statistically coated nanoparticles. The appropriate geometric form should then ensure that all DNA arms could, at least in principle pair with a complementary strand in the final assembly.

Firstly, we focused on hybrids with four DNA arms. The initial synthesis used a commercial phosphoramidite that gave hybrids such as **I** with a flexible core and self-complementary DNA arms (Figure 5.3.2). The short-hand used for **I** is $(AGGCCT)_4$Trebler, where 'Trebler' represents the core. Sequences in this and all other hybrids mentioned in short-hand in the text are given in the 5′- to 3′-direction. The synthesis was carried out on a solid support using a combination of 3′ and 5′-phosphoramidites of the natural 2′-deoxynucleosides. Automated DNA synthesis of the first strand was followed by manual coupling of the pentaerythritol-based phosphoramidite trebler and subsequent coupling cycles on a DNA synthesizer. Two hybrids with octamer DNA arms, such as $(5'\text{-CAGGCCTG-}3')_4$Trebler (**II**) with self-complementary arms and $(5'\text{-CGCACGTG-}3')_4$Trebler (**III**) with non-self-complementary sequences, were also prepared.

Next, UV-melting curves of the hybrids were measured in 10 mM phosphate buffer at hybrid concentrations of 1–2 μM and a salt concentration of 150 mM NaCl, that is, under conditions that are common for UV-melting curves experiments with linear DNA. When allowed to hybridize and subjected to melting, an increase in the UV-melting point of >40 °C and a significant hysteresis between heating and cooling curves was found for **I** when compared with the melting curve of the linear hexamer duplex [26]. This suggested that large assemblies were forming, whose assembly and disassembly processes were too slow to achieve equilibrium at the given heating and cooling rates of 1 °C. Further, the hyperchromicity reading for the melting transition was only half that for the linear control duplex of the same hexamer sequence. The subsequent cooling curve showed even less intense hypochromicity than the reduced hyperchromicity observed upon heating. Together, these suggested that only a fraction of the DNA arms engaged in duplex formation, most probably because of slow kinetics and/or the inability to adopt a more optimal structure for maximum base pairing. Electrophoresis using PAGE gels revealed that the hexamer hybrid **I** did not form assemblies large enough (or stable enough) to prevent migration into the gel, while self-complementary octamer hybrid **II** formed some structures incapable of migrating out of the loading pocket. Control octamer hybrid **III** did

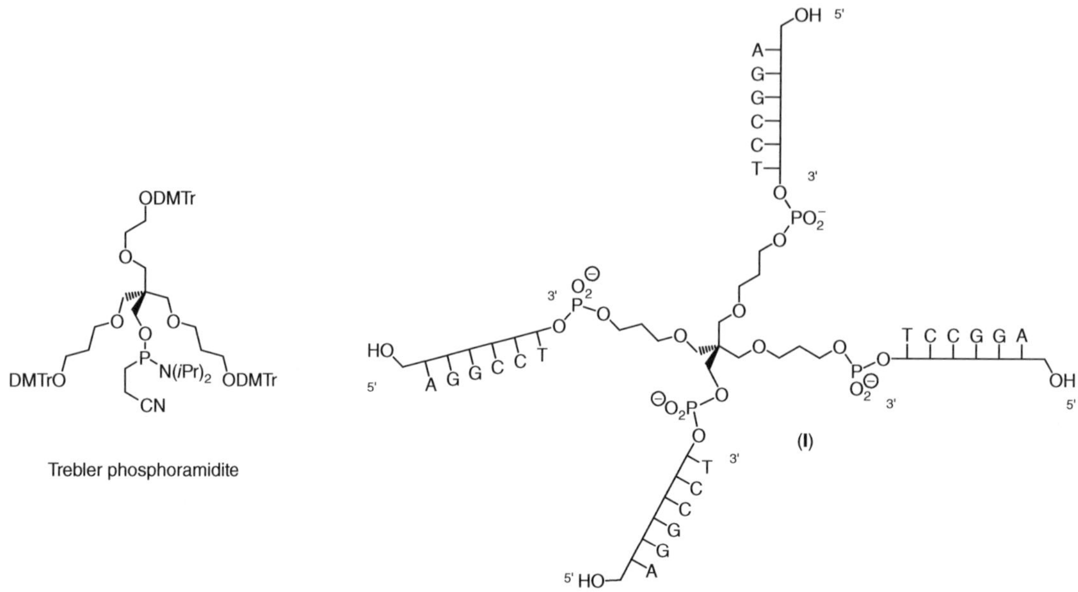

Figure 5.3.2 *Branched oligonucleotide **I**, prepared by solid-phase synthesis using the commercial 'Trebler phosphoramidite' shown on the left*

migrate into the gel, but the band was significantly broadened compared with that of a linear control. Finally, AFM images showed small assemblies for **I** but much larger, mushroom-type structures for assemblies of **II**, with lateral dimensions greater than 1 μm [27]. Taken together, these results confirmed that branched oligonucleotides form larger assemblies that are thermally more stable than linear duplexes with the same oligonucleotide length. The small hyperchromicity accompanying melting and the irregular shapes observed by AFM suggested that the flexible core did not provide a strong pre-organizing effect.

5.3.3 Hybrids with rigid cores

Based on the results of the studies with the non-symmetrical and flexible core, it was decided to employ core molecules with a higher degree of symmetry and rigidity. Further, since binding was strong, the length of the DNA arms was further reduced. Consequently, a symmetrical, tetrahedral DNA hybrid was synthesized, using tetrakis(*p*-hydroxyphenyl)methane (TPM) as the organic core (Figure 5.3.3), which was initially provided by Stefan Bräse and coworkers as part of a collaborative project [28, 29].

The initial hybrids were constructed via solid-phase synthesis (Figure 5.3.4). The synthetic approach is similar to that known from studies by the groups of McLaughlin [17, 30] and Sleiman [31], in that both 3′- and 5′-phosphoramidites are used as building blocks and two stages of chain assembly are involved. The initial stage used so-called 'reverse' DNA synthesis to construct the first oligonucleotide arm. This began with 5′-phosphoramidites and long-chain alkylamine controlled-pore glass (LCAA cpg) as the support, to which the first nucleoside is immobilized. After three coupling cycles with 5′-amidites that produced the first tetramer chain, the oligonucleotide terminus was phosphitylated, followed by immediate coupling of the core and oxidation. A chloroamidite was used as phosphitylating agent, and the TPM core was added in a solution containing tetrazole as the activator. Microwave-assisted couplings [32] gave fewer side products than the corresponding reactions at ambient temperature.

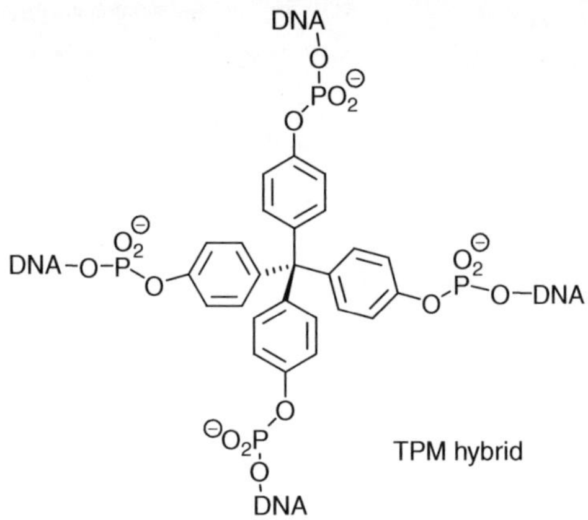

Figure 5.3.3 *Second generation DNA hybrid with stiffer tetrakis(p-hydroxyphenyl)methane (TPM) as the organic core*

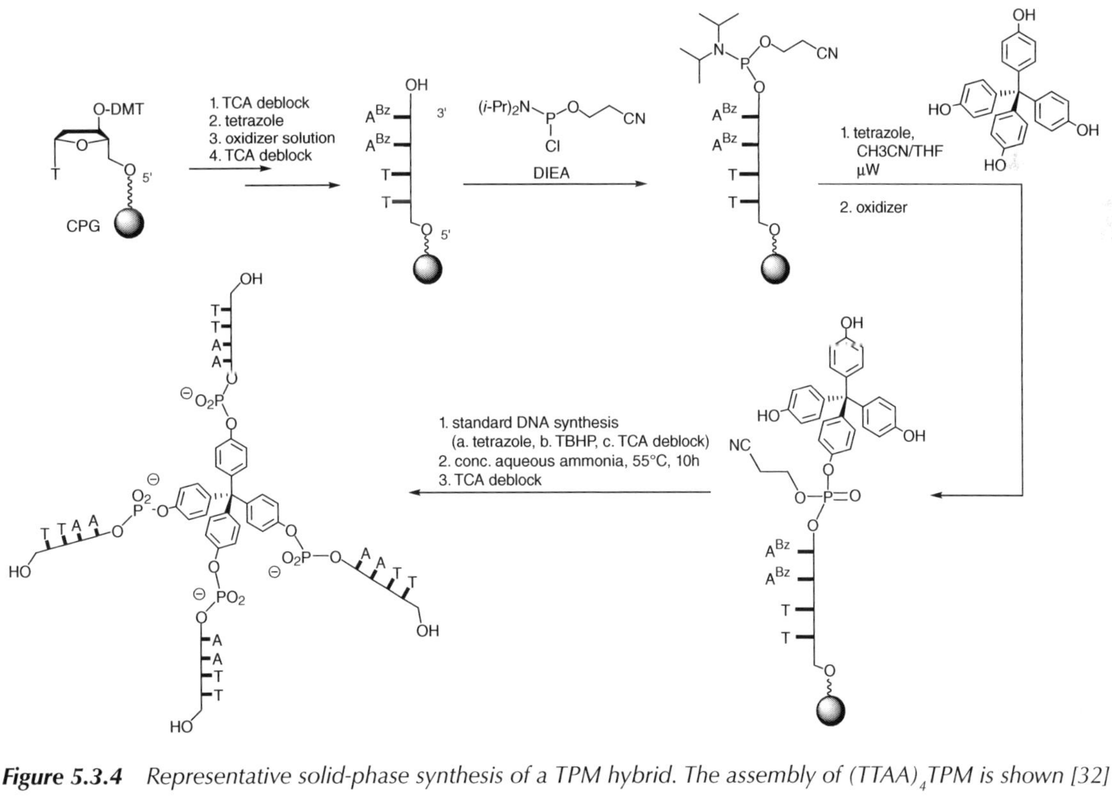

Figure 5.3.4 *Representative solid-phase synthesis of a TPM hybrid. The assembly of* $(TTAA)_4TPM$ *is shown [32]*

During the last phase of the solid-phase synthesis, the remaining three oligonucleotide arms were built simultaneously on the remaining three hydroxyl groups of the TPM core via conventional DNA coupling cycles, using 3′-phosphoramidites. To achieve high yields, more equivalents of the phosphoramidites were used and the coupling time was extended, as compared to the standard protocol. After completion of the chain

Table 5.3.1 *UV-melting points (°C) of complexes of DNA hybrids and controls at a hybrid concentration of 4 μM and a strand concentration of 16 μM for linear controls [26, 33]*

Compound	Buffer only[a]	+NaCl[b]	$+MgCl_2$[c]
AGGCCT	18	24	27
$(AGGCCT)_4TPM$	62	70	70
$(GTCCAG)_4TPM$	n.t.[d]	n.t.	n.t.
TTAA	n.t.	n.t.	<15
$(TTAA)_4TPM$	19	26	40
$(CAAA)_4TPM$	n.t.	n.t.	n.t.
$(CG)_4TPM$[e]	n.t.	<15	solid[f]
$(CC)_4TPM$[e]	n.t.	n.t.	n.t.

[a] 10 mM triethylammonium acetate (TEAA) buffer, pH 7.0.
[b] Plus150 mM NaCl.
[c] Plus100 mM $MgCl_2$.
[d] No cooperative transition.
[e] Hybrid concentration 15 μM.
[f] Precipitation observed at <15 °C.

assembly, DNA hybrids were cleaved from the solid support using aqueous ammonia [26]. Purification was challenging, often requiring a two-step chromatographic protocol based on reversed-phase and/or ion-exchange stationary phases. Using the solid-phase approach, several TPM hybrids with self-complementary DNA arms, such as $(AGGCCT)_4TPM$, $(TTAA)_4TPM$ and $(CG)_4TPM$ were obtained. Further, we also synthesized control hybrids with non-self-complementary DNA sequences, including $(GTCCAG)_4TPM$, $(CAAA)_4TPM$ and $(CC)_4TPM$. Yields of purified TPM hybrids ranged from 1 to 15% [33].

Hybrids in hand we then proceeded to studying their hybridization properties. The results of the measurements are shown in Table 5.3.1. UV-melting curve experiments showed that the TPM core, when decorated with four hexamer DNA strands, gives very stable assemblies. The melting point at an ionic strength similar to that of physiological salt concentration was recorded as 70 °C, which is 46 °C higher than that of the linear control duplex $(5'\text{-AGGCCT-}3')_2$, whose duplex gave a melting point of 25 °C under the chosen experimental conditions, confirming that rigid and highly symmetrical cores favour stable assemblies. Surprisingly, DNA arms as short as dimers were found to induce hybridization in low micromolar solutions of hybrids, if (and only if) the DNA sequences were self-complementary. The predicted UV-melting point of the duplex of linear CG dimers is well below –20 °C, demonstrating how strong the effect of multivalency is in the hybridization of hybrids with stiff, symmetrical cores.

While the assemblies of hybrids remained soluble in aqueous buffers devoid of divalent cations, addition of magnesium chloride led to the formation of solids that slowly precipitated upon cooling [26]. Again, this unusual property (the ability to form solids from micromolar solutions), was not found for linear oligonucleotides and was limited to TPM hybrids bearing self-complementary DNA arms. Control hybrids with non-self-complementary sequences did not produce solids, nor did they give cooperative transitions in UV-melting experiments. Taken together, these results showed that symmetrically branched oligonucleotides with rigid, four-arm branching elements hybridize to form more stable assemblies than their counterparts with flexible cores, and that CG dimers suffice as 'sticky ends'. In fact, the term 'CG zippers' was coined in these and subsequent studies involving such hybrids.

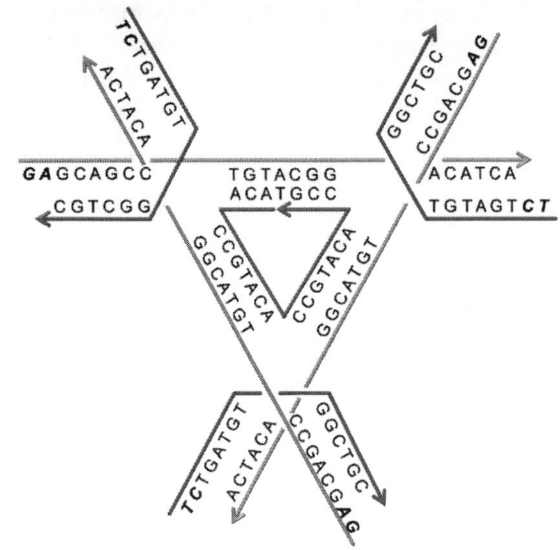

Figure 5.3.5 *Tensegrity triangle of Seeman and colleagues formed by three different strands, with six dimer sticky ends. This motif forms designed DNA crystals via hybridization of the dimer ends [15] (See colour figure in colour plate section)*

The melting curves of the assemblies formed by the dimer hybrid $(CG)_4$TPM were broad, a feature typical for very short duplexes. In contrast, the assemblies of DNA-coated gold nanoparticles gave very sharp melting curves [34]. We concluded that the electrostatic situation must be very different between the 'hybrid case' and the 'nanoparticle case' with its much longer DNA arms, many of which most probably do not engage in duplex formation, but rather produce an unusual electrostatic environment around the strands that hybridize. Perhaps, because the strands are short and held tightly by the rigid core, there is no longer a counter ion cloud that acts cooperatively. Further, the UV-melting studies on the hybrids showed a steep salt dependence for the formation of the assemblies. We assume that this is because a tight packing of the negatively charged backbones occurs. The unusually strong pairing of the DNA arms of our branched oligonucleotide hybrids also led us to assume that the lack of crystallinity in the assemblies of earlier branched constructs, with long arms [17, 30], may have been due to irreversible hybridization that prevented adopting a thermodynamically more stable, ordered state [24].

Shortly after our publication on assemblies of hybrids with dimers as DNA arms [26], Seeman and coworkers succeeded in crystallizing DNA folding motifs into lattices with designed crystal structure [15]. They used dimer sticky ends. The motif carrying the dimers was a so-called tensegrity triangle (Figure 5.3.5), and the resolution of the crystal structures reached 4 Å. The tensegrity triangle consists of a rigid C_3-symmetric all-DNA core, which is formed through hybridization of seven oligonucleotide strands. Crystal growth was probably favoured by the high degree of symmetry and rigidity of the folding motif. The short, dimer sticky ends at the corners of the triangle probably hybridized with the necessary degree of reversibility during crystal growth. Apparently, kinetically accessible alternative structures that would have led to less ordered lattices were sufficiently unstable to dissociate before the growing assembly enclosed the defect. The fact that non-self-complementary sticky ends were used and that a rigid, six-arm motif was the one that crystallized was not unexpected, given the results with the organic hybrids, mentioned earlier. Further confirmation for the hypothesis that much shorter sequences support crystal growth than are needed for hybridization of linear duplexes came from the studies on DNA-coated nanoparticles, where a combination of stiff duplex regions, 'flexor' regions, and short sticky ends (as short as tetramers) were shown to favour crystallographic order [23].

5.3.4 Second-generation hybrids with a rigid core

In order to develop the next generation of hybrids, we sought a collaboration with theoreticians. Wolfgang Wenzel and coworkers took it upon themselves to develop a system for simulating the assembly of DNA hybrids, using a coarse-grained model. The rigidity of the hybrids, and the geometry of the core, including the number of DNA arms were varied, and the extent to which periodic lattices formed upon simulated adiabatic cooling was measured using an artificially defined 'order parameter' [35]. Even though the coarse-grained model could not accurately reproduce the intramolecular conformation, salt concentration and solvent effects, interesting insights were gained. The first was that hybrids have a tendency to form assemblies with a range of different ring sizes, particularly when self-complementary sequences and tetrahedral geometries are involved. This tendency reduces the likelihood of obtaining crystals through cooling of hybrid solutions. As long as there is a similar level of base pairing, order had little energetic significance in these systems. Secondly, more ordered model assemblies were observed in the simulations when octahedral, six-arm models of hybrids were studied. Thirdly, and perhaps unsurprisingly, an increase in rigidity was found to favour crystallinity. Finally, the association strength was predicted to increase with a decrease in the number of like charges in the backbone of the sticky ends. This effect was again expected, based on conventional Coulombic effects. For three representative hybrid models, the simulations predicted the following order of crystallinity: tetrahedral four-arm with a flexible core < tetrahedral four-arm with rigid core < octahedral six-arm hybrid. The predicted order in the thermal stability of the assemblies was: tetrahedral four-arm with a flexible core ≈ tetrahedral four-arm with rigid core << octahedral six-arm hybrid.

Encouraged by the theoretical studies, we then designed DNA hybrids with new organic cores. The structures of these hybrids are shown in Figure 5.3.6. Two of the second-generation hybrids feature adamantane-based cores. The TBA core contains stiff biphenyl linkers to further increase rigidity. The TPPA core is based on the same design principle, in that the adamantane bears four biaryl-arms, but the linkage between the first nucleoside and the core itself is a triazole moiety. This linkage not only contains a smaller number of bonds that are able to rotate, it is also uncharged, so that $(CG)_4$TPPA had no more than four negative charges. Unlike the phenol-based cores, which were assembled using phosphoramidite chemistry, the triazole linker required a 1,3-dipolar cycloaddition ('click reaction') to connect the DNA arms to the core. This was the first hybrid to be fully assembled in solution phase, rather than on a solid support. Details of solution-phase syntheses are discussed later (see Section 5.3.5).

The third of the second-generation hybrids was $(CG)_6$HPX. The HPX core is pseudo-octahedral, and was as close as we and our collaborators got to truly octahedral structures using our established approach of building cores by conventional organic synthesis. Like the TPM core, it features *para*-hydroxyphenyl arms to whose distal alcohol functions DNA arms were attached by solid-phase DNA syntheses, as described in Figure 5.3.4. A crystal structure of the core by itself [35] confirmed its propensity to readily form ordered lattices. Owing to the steric crowding, a higher level of structural rigidity was expected for the HPX hybrids, when compared with the TPM counterparts. However, the yields were also lower, and the purification and characterization was more challenging. We also had to find the correct conditions for MALDI-TOF (matrix-assisted laser desorption ionisation time of flight) mass spectrometry to detect both the final $(CG)_6$HPX hybrid and partially deprotected intermediates of the syntheses. After successful synthesis and purification, UV-melting curve experiments were again conducted for the hybrids with CG zippers as DNA arms for all three new cores [35]. Table 5.3.2 gives an overview of the results.

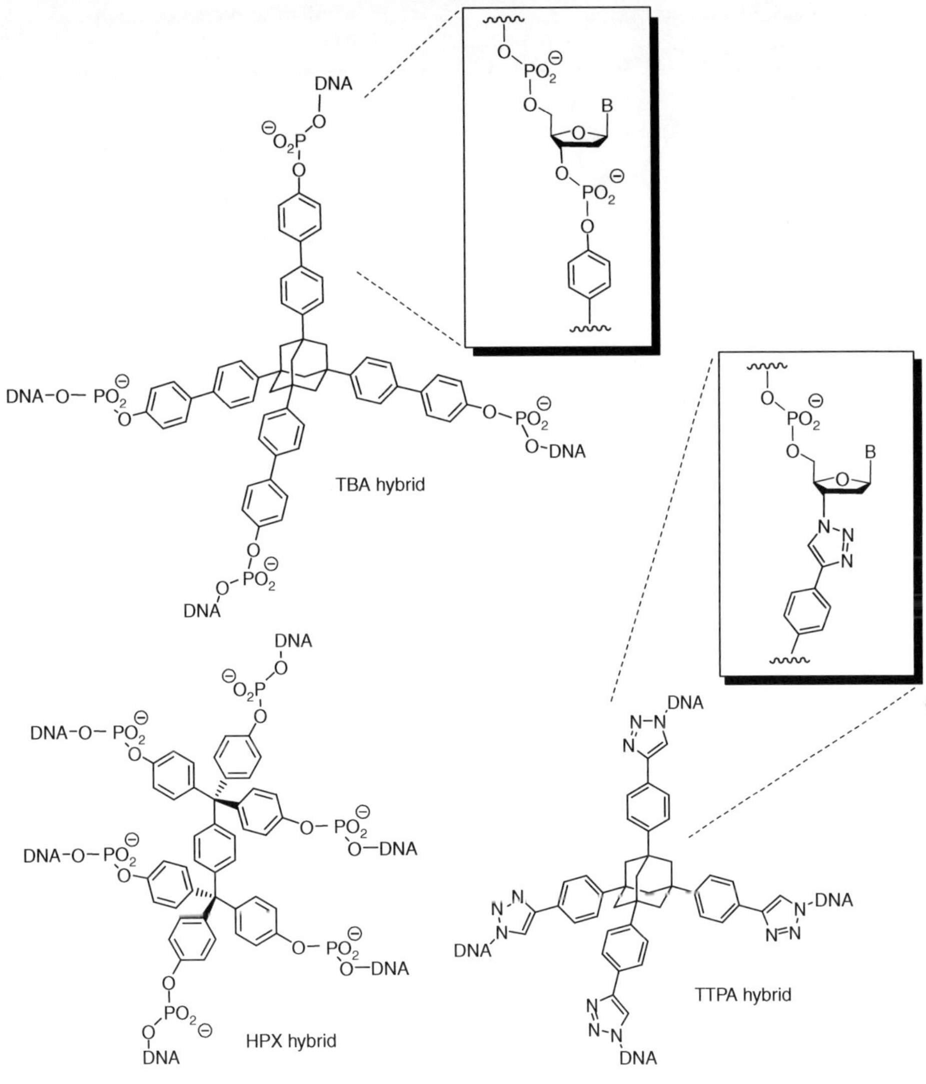

Figure 5.3.6 *Second generation hybrids with rigid organic cores: TBA = tetrakis(*p*-hydroxybiphenyl)adamantane core; TTPA = tetrakis(triazolylphenyl)adamantane core and HPX = hexakis(*p*-hydroxyphenyl)xylene core*

The strength of the association of the second-generation hybrids warranted a special approach for melting curves. To ensure that at the beginning the assay solution contained monomeric hybrids, strongly denaturing conditions were employed. For this, a sample containing the hybrid and 10 mM NaOH was heated to 95 °C, followed by neutralization with AcOH in the heat, buffering, dilution and initiation of the first cooling curve. Even in the absence of other salts (other than NaOAc and 10 mM TEAA) a transition was observed for pseudo octahedral $(CG)_6$HPX, but not for the less rigid four-arm hybrid $(CG)_4$TPM. For rigid tetra-anionic $(CG)_4$TTPA, the 'assembly point' was already above 60 °C, despite the low ionic strength. Addition of a near-physiological concentration of NaCl (150 mM) had a very significant effect on the ability of the hybrids

Table 5.3.2 *UV-melting points (°C) of assemblies formed by branched DNA hybrids at a hybrid concentration of 10 μM [35, 36]*

Compound	Buffer only[a]	+NaCl[b]	$+MgCl_2$[c]
$(CG)_4TPM$	n.t.[d]	14 ± 0.8	solid[e]
$(CC)_4TPM$	n.t.	n.t.	n.t.
$(CG)_6HPX$	18.7 ± 0.9	25.1 ± 1.3	solid[f]
$(TC)_6HPX$	n.t.	n.t.	n.t.
$(CG)_4TBA$	65	over 95[g]	solid[f]
$(TC)_4TBA$	n.t.	n.t.	n.t.
$(CG)_4TTPA$	61.5 ± 0.7	solid[h]	solid[h]
$(CC)_4TTPA$	n.t.	n.t.	n.t.

[a] 10 mM TEAA buffer, 1.5 mM NaOAc, pH 7.
[b] Plus 150 mM NaCl.
[c] Plus 100 mM $MgCl_2$.
[d] No cooperative transition.
[e] Precipitation at <10 °C.
[f] Precipitation at <25 °C.
[g] The T_m could not be calculated because of the limited experimental temperature range.
[h] Precipitation at <95 °C.

to assemble. For the TPM hybrid, a cooperative transition was now observable, for the HPX hybrid, the transition point shifted to 25 °C, and for $(CG)_4TTPA$, a solid formed, even in the heat.

Following the addition of divalent cations (100 mM $MgCl_2$), all three hybrids with CG zippers formed solids. For the least rigid four-arm core (TPM), the onset of precipitation occurred at approximately 10 °C, for its six-arm counterpart HPX, precipitation was visible at 25 °C, whereas the tetratriazolide $(CG)_4TTPA$ began to form a solid immediately after neutralization and $MgCl_2$ addition at 95 °C. The assemblies melted slowly, if at all, during subsequent heating, as expected for extended networks. The material formed by $(CG)_4TTPA$ did not fully dissolve again, even at 95 °C. Control experiments with hybrids featuring non-self-complementary DNA arms did not show precipitation, suggesting that the assembly was indeed induced by base pairing.

At the end of the cooling and heating cycles, MALDI-TOF mass spectrometry yielded signals for pure DNA hybrids, without discernable decomposition. Electron microscopy of samples of $(CG)_6HPX$ and $(CG)_4TTPA$ showed rectangular crystals of approximately 100–200 nm length that diffracted the electron beam in the diffraction mode (*vide infra*) [35]. Further, diffusion of intercalators into pre-formed materials indicated that nanoporous materials had formed. Again, the intercalation was limited to the materials found for hybrids with self-complementary DNA arms, but not the solutions of control hybrids, indicating that Watson–Crick base pairing was the molecular basis of assembly. For ethidium bromide, a stoichiometry of 1:3 between the intercalator and the DNA hybrid was found in the solid soaked in ethidium bromide solution. Elemental analysis of the neat material formed by $(CG)_4TTPA$ indicated that 24 equivalents of $MgCl_2$ and 40% water had been included in the materials.

Overall, the results from the study on the second-generation hybrids confirmed the predictions made by the theoretical study. More rigid hybrids have a greater propensity to assemble into materials, as also confirmed by the properties of $(CG)_4TBA$, which was studied in more detail later [36, 37], and the six-arm pseudo-octahedral core geometry of $(CG)_6HPX$ that led to more avid assembly than the tetrahedral geometry of $(CG)_4TPM$. The step from octaanionic $(CG)_4TPM$ to tetraanionic $(CG)_4TTPA$ gave the most dramatic effect, with the formation of nanoporous materials from low-millimolar hybrid solutions in aqueous buffer, even at temperatures as high as 95 °C.

5.3.5 Solution-phase syntheses: Synthetic challenges

The interesting properties of the material formed by the second-generation hybrids prompted us to develop solution-phase syntheses of these compounds, so that larger amounts could be produced. The main drawbacks of the solid-phase syntheses used thus far had been the cost of the supports, the use of a large excess of phosphoramidites and other reagents, the difficulty of scaling up the syntheses with the available instrumentation, low yields and challenging purifications. To be viable industrially, we also wished to avoid chromatography in the newly developed solution-phase routes. We focused on hybrids with HPX or TBA as the core. Further, we opted for a 'block condensation' strategy, involving dimer syntheses, and coupling of dimers to the core, rather than mononucleotide-based conventional assembly of the DNA arms. When pure dimer building blocks are coupled to cores, no products lacking a single nucleotide residue should be formed, so that purification should be easier. Further, the number of steps with large and polar intermediates is reduced, facilitating handling. The more convergent approach was also expected to be more versatile and cheaper. Finally, we decided to prepare all cores ourselves, rather than relying on our (fruitful and enjoyable) collaborations. Figure 5.3.7 shows typical syntheses of cores, that of TBA and that of the acetylide used to construct the TTPA hybrids.

Starting from 1-bromoadamantane, 1,3,5,7-tetraphenyladamantane is obtained in a variation of a Friedel–Crafts alkylation [38, 39], followed by four-fold iodination to 1,3,5,7-tetrakis(4-iodophenyl)adamantine [40]. The latter can be subjected to a four-fold Suzuki coupling with the boronic acid derivative of anisole to give the tetraether precursor of the TBA core [41]. Ether cleavage with BBr_3 then yields TBA itself in 84% overall yield [29]. Alternatively, the tetra-iodide can be employed in a four-fold Sonogashira coupling with ethynyltrimethylsilane, and 1,3,5,7-tetrakis(4-ethynylphenyl)adamantane can then be released with potassium fluoride [41]. The latter is suitable for 1,3-dipolar cycloadditions to azidonucleosides.

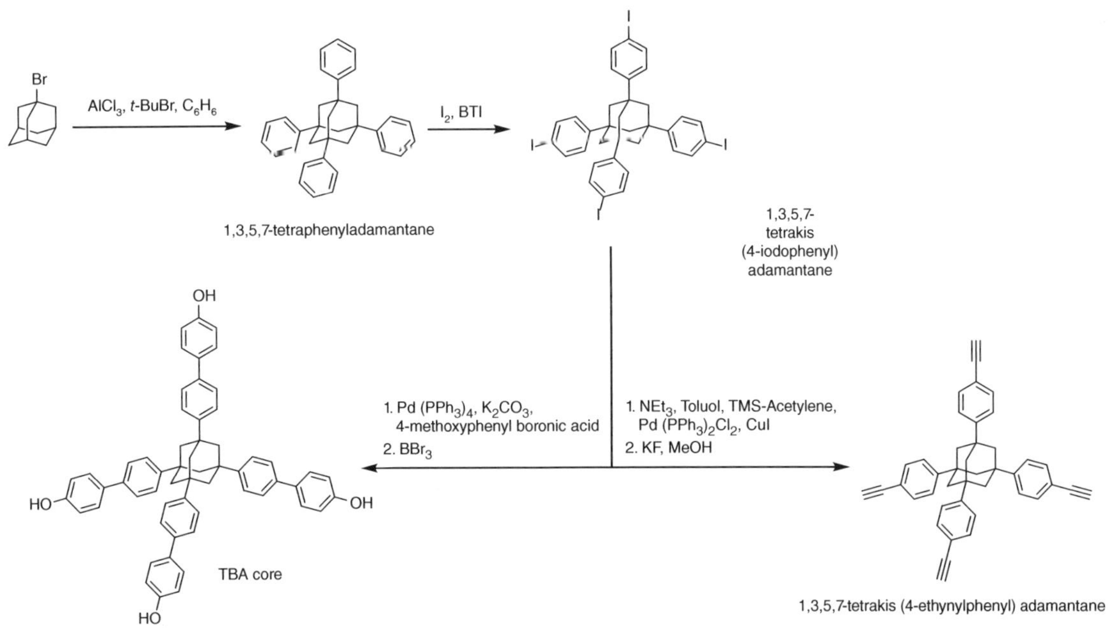

Figure 5.3.7 *Synthesis of 1,3,5,7-tetrakis(4-ethynylphenyl)adamantane and tetrakis(p-hydroxybiphenyl)adamantane (TBA) from bromoadamantane as starting material [38, 39]. The iodination reagent is bis(trifluoroacetoxy) iodobenzene (BTI) [40]*

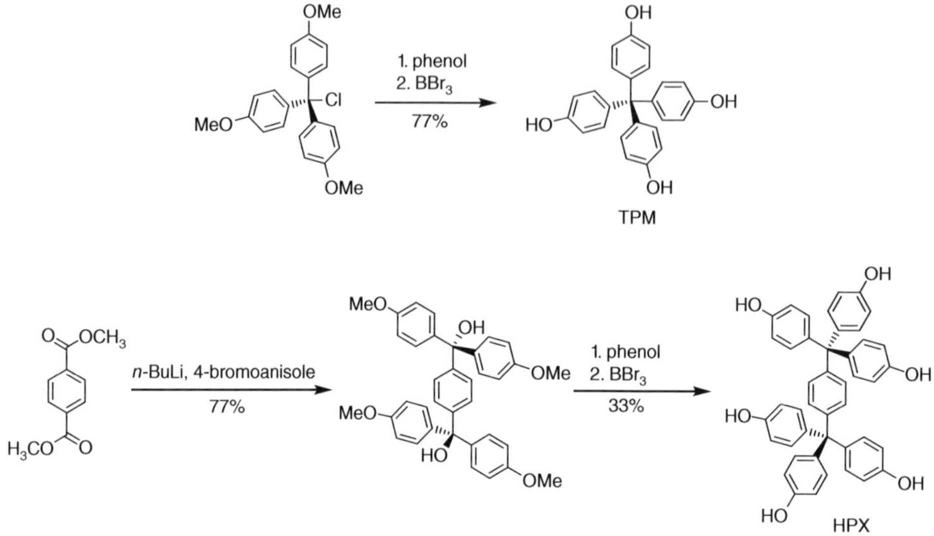

Figure 5.3.8 *Synthesis of the tetrakis(*p*-hydroxyphenyl)methane (TPM) and hexakis(*p*-hydroxyphenyl)xylene (HPX) cores [35]*

The syntheses of the TPM and HPX cores are shown in Figure 5.3.8. The earliest synthesis of tetrakis(4-hydroxyphenyl)methane had been reported by Wuest and coworkers [42]. Their three-step route started from trityl chloride and gave a total yield of 50%. Our current route starting from 4,4′,4″-trimethoxytrityl chloride gives an overall yield of 77% [26]. The HPX core was prepared from dimethyl terephthalate via 1,4-phenylenebis[bis(4′-methoxyphenyl)methanol] to give the desired hexakis(*p*-hydroxyphenyl)xylene in 36% overall yield [43]. In the crystal structure mentioned earlier, HPX showed the expected pseudo-octahedral geometry. In the solid state, HPX forms a two-dimensional network of intermolecular O–H···O hydrogen and extensive channels [28].

With the desired core in hand, the assembly of the actual hybrid can be undertaken. When we first turned to solution-phase syntheses, we were faced with low yields. Some of the difficulties were due to the dendrimer-like structure of hybrids, others were due to the difficulty of handling and purifying richly functional polar molecules. Figure 5.3.9 gives a graphical overview for phenolic cores.

The first challenge of the block condensation-based assembly was to produce the dimer or oligomer building blocks cheaply from commercial starting materials. In order to be attractive, the building block assembly has to occur in solution and chromatography should be avoided. Secondly, because cores are highly symmetrical, rigid molecules, they can form stable crystals and tend to be poorly soluble in many of the more common solvents. Thirdly, and perhaps most importantly, during the initial phase of the project, coupling conditions had to be found that give high yields without the massive excess of phosphoramidites or *H*-phosphonates that is common in solid-phase syntheses. With the coupling going to completion, a fourth challenge arose from the fact that polar and richly functionalized molecules had to be isolated and (pre)purified. Though straightforward conceptually, the subsequent two- or three-step deprotection became the fifth challenge and the toughest nut to crack during the optimization phase of the project, as decomposition reactions were difficult to suppress. The final, sixth, major challenge was to prevent premature assembly of hybrids with self-complementary DNA arms during the final purification steps.

Several methods for synthesizing dimer building blocks were considered that differ in the protecting groups, coupling chemistry and the monomers used [44–46]. The method of Moroney and coworkers

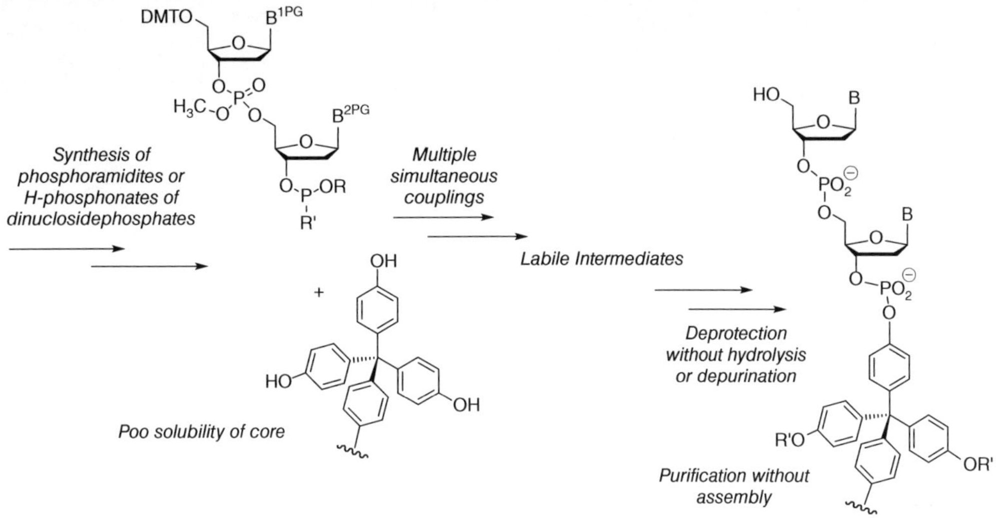

Figure 5.3.9 *Graphical representation of some of the issues making solution-phase syntheses of branched oligonucleotide hybrids challenging; PG = protecting group*

originally developed for trimer phosphoramidites [47], gave satisfactory yields for dimer phosphoramidites. It uses methyl protecting groups for the phosphodiester, which are more stable than a cyanoethyl group. Figure 5.3.10 shows the first successful solution-phase route to the $(TC)_6HPX$ hybrid using dimer phosphoramidites and subsequent coupling to phenolic cores [48]. Dimer construction started from commercially available phosphoramidites and a previously 3′-protected nucleoside. A phosphoramidite was then coupled to the 5′-alcohol, and the resulting phosphite triesters were oxidized *in situ* to give fully protected dimers that could be chromatographed. The 3′-terminal phenoxyacetyl (PAC) protecting group was removed, followed by phosphitylation of this position. After this time-consuming and difficult construction of the dimer, the building block was coupled to the HPX core, using tetrazole as the catalyst, followed by oxidation with *tert*-butylhydroperoxide (TBHP). Diisoproylethylamine and 1-thionaphthol were added to the fully protected hybrid to cleave the methyl groups, followed by detritylation with Dowex MAC3 resin or trichloroacetic acid (TCA). Base labile protecting groups were then removed with ammonium hydroxide and methylamine [48].

A detailed study of the coupling reaction showed that over time decomposition or hydrolysis of the initial phosphite occurred, lowering the yield. The low solubility of the core further complicated the coupling. Lowering the concentration of the tetrazole catalyst, improving the drying protocol for the core and the dimer, and an unusual temperature protocol (first slightly elevated to get the core into solution, and then cooling to drive coupling to completion) improved yields. Satisfactory conversions could finally be achieved by using two successive coupling cycles, driving coupling to 85–99% yield [48].

Still, the route described above was long and not free from chromatographic steps. Therefore, an alternative approach was developed that uses *H*-phosphonates of dimers as building blocks. This approach is an extension of the work of Jones and coworkers on cyclic diguanosine monophosphate [49], in that it uses a combination of phosphoramidite and *H*-phosphonate chemistry. With the adapted route (see Figure 5.3.11 for a TBA hybrid), our DNA dimer building blocks became accessible in gram quantities without chromatography in near-quantitative yield in less than three days of laboratory work, starting exclusively from affordable commercial compounds.

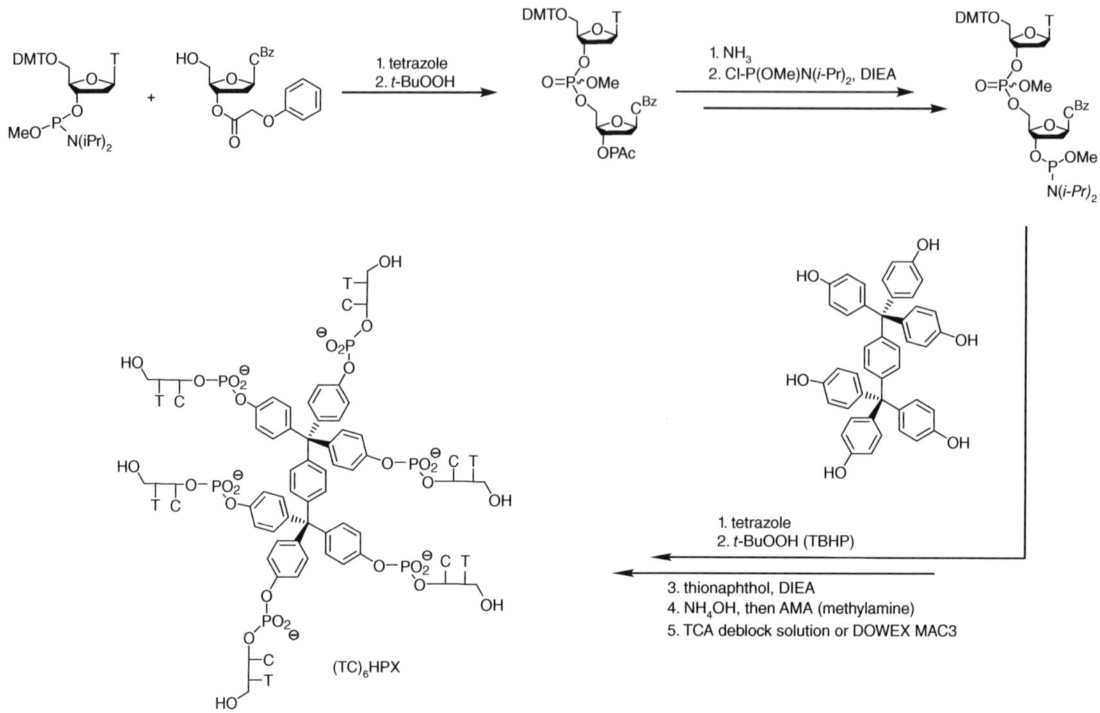

Figure 5.3.10 *Solution-phase synthesis of the (TC)$_6$HPX hybrid using dimer phosphoramidites as building blocks [35–37, 48]*

The synthetic route begins with commercially available phosphoramidites that are hydrolysed, followed by removal of the cyanoethyl groups with *tert*-butylamine and detritylation with dichloroacetic acid. The resulting *H*-phosphonate is then coupled at its 5′-position with a second phosphoramidite. Oxidation with a peroxide gave the *H*-phosphonate dimer in 44% yield. Similarly, other dimers, such as 5′-CG-3′, 5′-TC-3′ and 5′-GA-3′ were also obtained [36]. The assembly of the hybrids started with an *H*-phosphonate coupling employing diphenylchlorophosphate (DPCP) as the condensing agent [50]. At –40 °C, a temperature previously recommended by Reese and Song for other *H*-phosphonate couplings [51], near-quantitative conversion was observed. The crude product was then oxidized by treatment with I_2 and subsequent addition of water [50]. The fully protected DNA hybrid was detritylated with 6% DCA solution, followed by purification through precipitation. To remove the cyanoethyl groups and the protecting groups of the nucleobases, the hybrid was treated with ammonium hydroxide. After lyophilisation, the crude product can be purified by extraction and precipitation steps, combined with reversed-phase chromatography, where necessary [36, 37].

In the laboratory, the *H*-phosphonate route is now more popular, as it produces the dimer building blocks more quickly and from inexpensive starting materials. The *H*-phosphonate group of the 3′-terminal nucleoside is unreactive during the coupling at the 5′-position, and the subsequent oxidation, thereby eliminating the need for a protecting group at the 3′-position. Though GA-containing sequences continue to be challenging, the *H*-phosphonate route has proven reliable and attractive.

This account would not be complete without mentioning hybrid synthesis via cycloaddition or the 'click reaction'. Figure 5.3.12 shows a representative route. Starting from 1,3,5,7-tetrakis(4-ethynylphenyl)adamantane and a 3′-azido-2′,3′-dideoxynucleoside, the first four nucleosides are attached via Cu(I)-catalysed

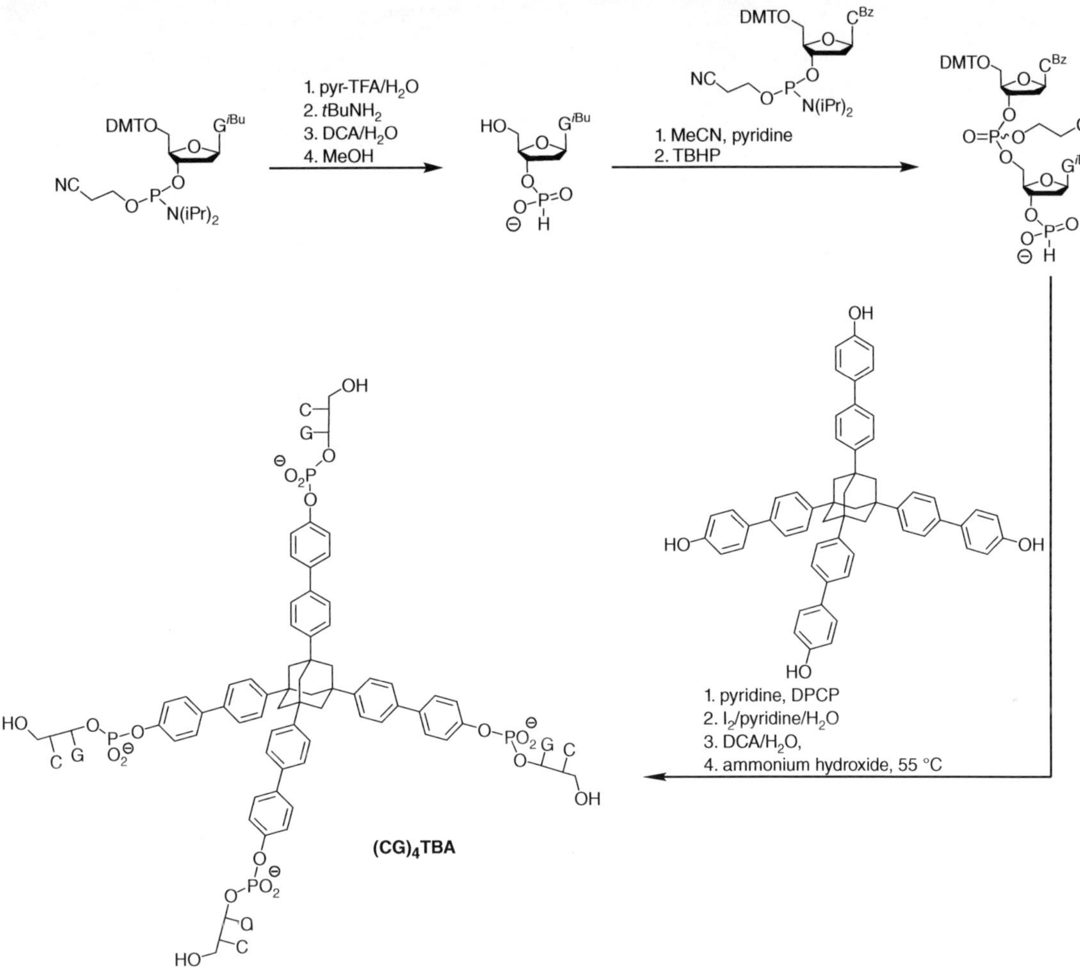

Figure 5.3.11 *Solution-phase synthesis of $(CG)_4$TBA employing a* H*-phosphonate block condensation coupling [36]*

Huisgen cycloaddition [52]. Use of an equimolar amount of copper and a reducing agent led to a short reaction time of just 1 h, instead of the 10 h of earlier protocols [28]. Addition of a 5% aqueous EDTA solution at the end of the reaction produces the $(Gi^{Bu})_4$TTPA hybrid as a colourless precipitate that can be easily separated from all other components of the reaction mixture by centrifugation [35, 36. 48, 53]. After this first coupling, either *H*-phosphonate or phosphoramidite chemistry can be used to append the terminal nucleoside to the $(Gi^{Bu})_4$TTPA intermediate to then give, after deprotection, tetra-anionic $(CG)_4$TTPA.

5.3.6 Hybrid materials

Ultimately, branched oligonucleotide hybrids with rigid cores will remain interesting only if they show their usefulness in practical applications. Testing material properties requires routine access to gram, if not kilogram amounts. We have recently begun to scale-up the syntheses of hybrids with a phenolic core, using the

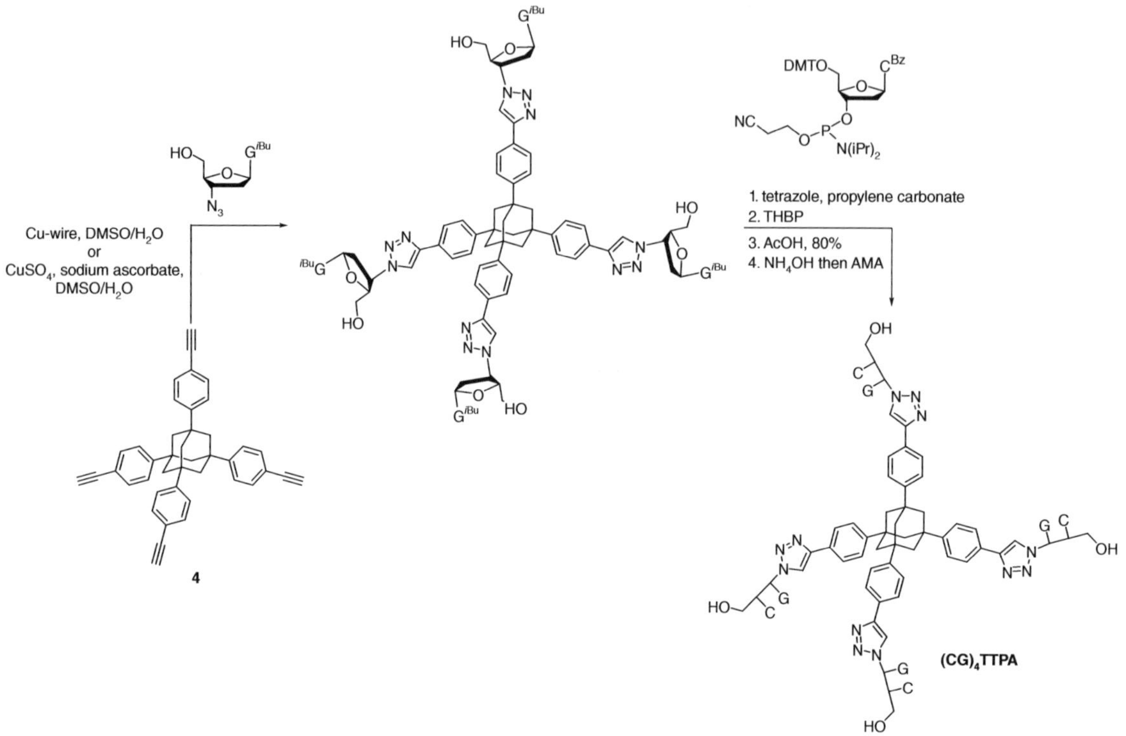

Figure 5.3.12 Solution-phase synthesis of $(CG)_4$TTPA via cycloaddition and subsequent phosphoramidite coupling [36, 48, 53]

solution-phase approach outlined here, so that tests on the ability to store and release host molecules can be run in earnest. While oligonucleotides are usually produced in microscopic quantities, with a 50 nanomole scale being standard for commercial oligonucleotide providers, a single run of our modestly scaled-up protocol readily produces 10 μmoles of pure $(CG)_4$TBA within less than a week and in excellent purity, without the need for HPLC (Figure 5.3.13).

The hybridization of the zipper arms of hybrids can be detected in UV-melting experiments, but the proof of assembly is the visible formation of a material. The branched oligonucleotide hybrids presented here produce materials upon addition of millimolar concentrations of magnesium chloride [26, 35]. When the precipitate is isolated and treated with solutions of intercalators, stoichiometric quantities of the intercalators are taken up without any need to disassemble and reassemble the materials. This suggests that the materials are nanoporous. Usually, one equivalent of intercalator is taken up per CG dimer duplex formed [35, 36, 48]. Figure 5.3.14 shows images of samples of $(CG)_4$TBA, together with corresponding samples of $(TC)_4$TBA as a control compound that does not form a solid. The fluorescence images show that despite the high loading density, some intercalators are fluorescent when taken up into the materials of $(CG)_4$TBA. The reversible uptake and release of small molecules and the light harvesting properties are among those that are currently being studied in our laboratories.

Certainly, one of the intellectually most interesting questions is whether the assemblies of branched hybrids can be induced to grow into macroscopic crystals suitable for X-ray crystallography. We presume that obtaining large crystals is largely a kinetic problem, and that finding conditions that prevent the formation of a large number of seed crystals at a concentration high enough to produce crystals quickly is not trivial. We note that

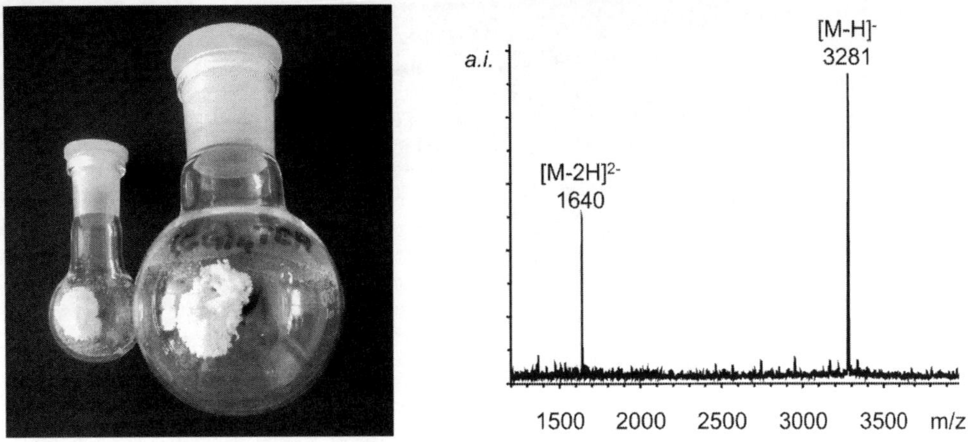

(a) (b)

Figure 5.3.13 *Scaled-up solution-phases synthesis of a hybrid. (a) Photograph of flasks with product fractions of a run of the current solution-phase synthesis of $(CG)_4TBA$, after lyophilization, as obtained after a single cartridge purification step, that is, without HPLC purification. (b) MALDI-TOF mass spectrum of the product in the 100 mL flask shown in (a). Calculated for $C_{134}H_{144}N_{32}O_{52}P_8$ $[M-H]^-$ 3281, found 3281; calculated for $[M-2H]^{2-}$ 1640, found 1640*

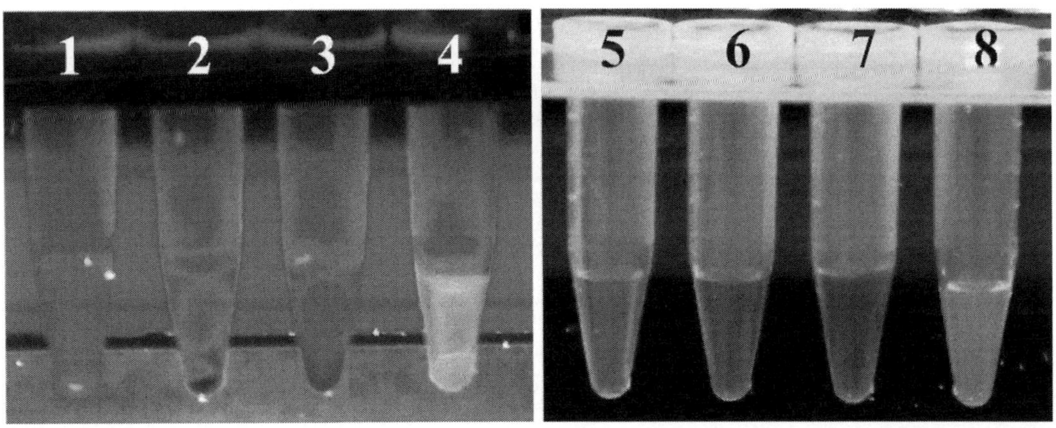

Figure 5.3.14 *Fluorescence images from intercalation experiments with two different hybrids under UV light [37]. On the left-hand side, materials formed from aqueous solution of (CG)4TBA are shown. The materials were produced by treating micromolar solutions of the hybrid with NaCl (150 mM) and $MgCl_2$ (100 mM), cooling to 4 °C, and isolation of precipitates, followed by addition of the respective dye. On the right-hand side, corresponding images of samples of the (TC)4TBA hybrid as control compound, treated in the same fashion, are shown. The dyes added are ethidium bromide (samples 1 and 5), methylene blue (samples 2 and 6), tetrakis(N-methyl-4-pyridinium)porphyrin tetra(p-toluenesulfonate) (samples 3 and 7), and YOYO-1 iodide (samples 4 and 8). The luminescence images were taken at λ_{ex} = 366 nm (See colour figure in colour plate section)*

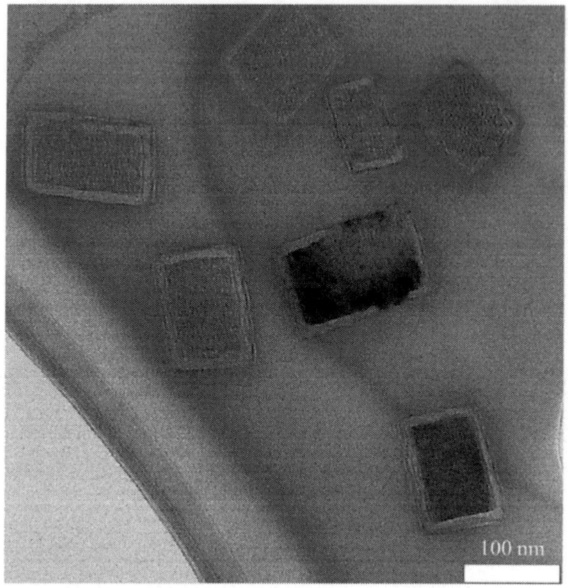

Figure 5.3.15 *Transmission electron micrograph of a sample of $(CG)_6$HPX, produced by cooling an aqueous solution (100 μM), free of added salts or buffers, and keeping the sample at 10 °C for one day [33]. The image was taken after 5 min of TEM imaging, leading to visible decomposition on the surface of the crystallites, indicating that they are indeed made up of an organic material, not an inorganic salt. The TEM image was taken at room temperature and 120 keV. The length bar in the lower right-hand corner is 100 nm*

for DNA-coated gold nanoparticles, obtaining large single crystals remains challenging [54]. Growing crystals can also be complicated by the formation of salt crystals made up of components of the buffered aqueous solutions of samples. Figure 5.3.15 shows an electron microscopy image of a sample of $(CG)_6$HPX that was allowed to assemble from a pure aqueous solution, devoid of any added salts or buffer components.

5.3.7 Outlook

Porous materials are attractive as catalysts. Zeolites are well-established a well-known example of a porous catalyst [55, 56]. Their pore size is in the sub-nanometre range. Producing nanoporous catalysts is a challenge. It is interesting to ask whether DNA assemblies can act as such catalysts. Roelfes and colleagues elegantly showed that DNA is an interesting chiral ligand in catalytic systems [57]. Among the reactions studied by this group are asymmetric Diels–Alder reactions and addition reactions to various substrates that rely on copper complexes bound to duplexes. We have recently begun to explore the synthesis of branched oligonucleotide hybrids that feature transition metal complexes. Figure 5.3.16 shows the ruthenium complex $(CG)_6$HPB-$Ru(biPy)_2$ that was prepared via solution-phase synthesis, starting from a literature-known bipyridyl core [58].

Bipyridyl units are strong bidentate ligands, and bipyridyl complexes have been studied previously in the context of DNA [30, 59]. Moreover, DNA nanostructures containing transition metal complexes are being studied for their photophysical, redox, catalytic and magnetic properties [60]. Hybrid $(CG)_6$HPB-$Ru(biPy)_2$ with its 2,2′-bipyridine metal complex core was obtained in pure form after stringent HPLC purification and

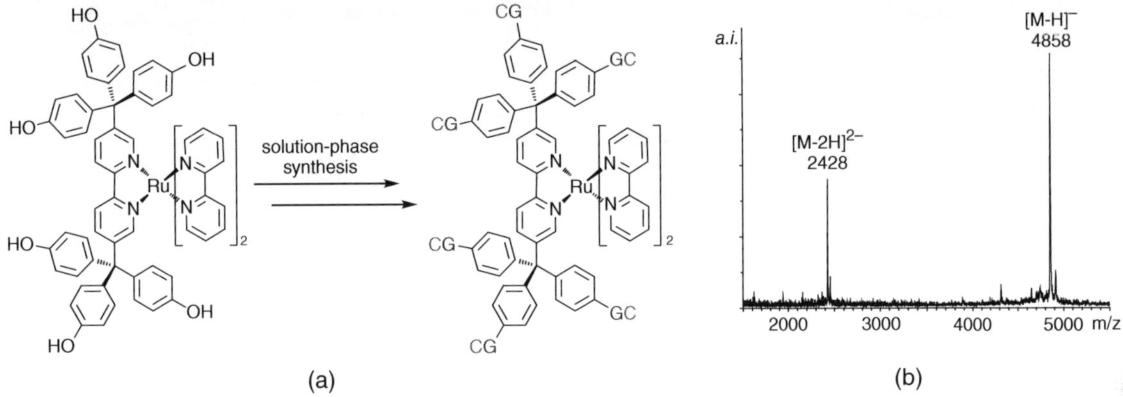

Figure 5.3.16 *Transition metal-containing six-arm hybrid $(CG)_6$HPB-Ru(biPy)$_2$. (a) Structures of core and hybrid and (b) MALDI-TOF mass spectrum after HPLC purification, calculated for $C_{182}H_{196}N_{54}O_{78}P_{12}Ru$: $[M-H]^-$ 4859, found 4858; calculated for $[M-2H]^{2-}$ 2428, found 2428*

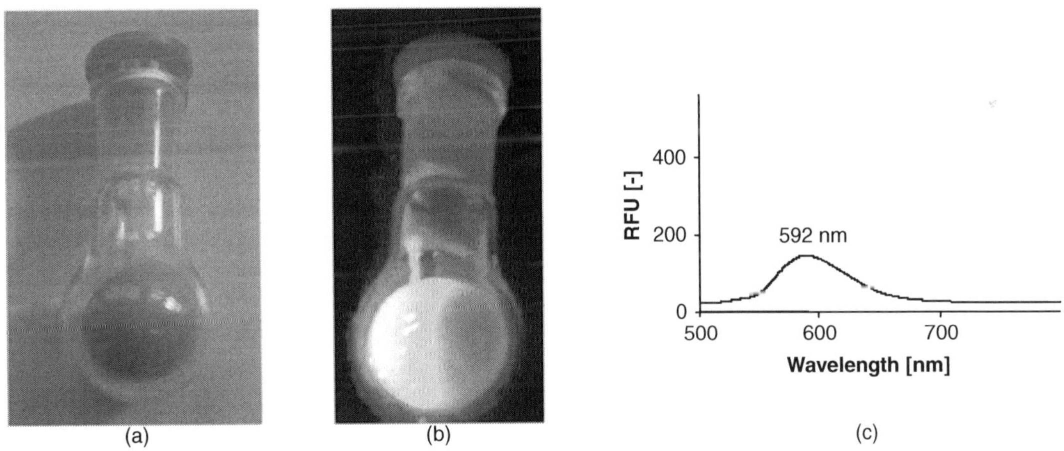

Figure 5.3.17 *Optical properties of hybrid $(CG)_6$HPB-Ru(biPy)$_2$. (a) Photograph of a sample of 8 mg (1.6 μmol) in daylight, (b) the same sample under UV light (366 nm), (c) fluorescence spectrum of a 1 μM solution of $(CG)_6$HPB-Ru(biPy)$_2$ in 10 mM NH_4Ac buffer, λ_{ex} = 455 nm (See colour figure in colour plate section)*

showed strong luminescence, even in solid form (Figure 5.3.17). The photochemical stability of the material is currently under investigation. From the initial results, it seems likely that photochemical reactions can indeed be induced, similar to what was observed by us with porphyrin-containing branched oligonucleotides [61].

We are currently preparing metal-free versions of the hybrid with a bipyridyl core, assuming that different metal ions can be inserted post-synthesis, as demonstrated for other DNA constructs [62]. Another area of particular interest to us is to combine hybrids with DNA sequences that form nucleotide- and cofactor-binding motifs [63, 64]. In addition, we are synthesizing new hybrids with core molecules of different geometries and increasing number of DNA arms. The progress achieved along these lines will be published in due course.

5.3.8 Conclusions

Branched, symmetrical oligonucleotides have properties very different from their linear counterparts. They hybridize to form duplex-mediated superstructures based on sticky ends that are much shorter than are required for linear oligonucleotides forming duplexes at micromolar concentrations in aqueous buffers. Upon addition of divalent cations, such as magnesium ions, they form macroscopically visible materials that take up intercalators, salts and other guest substances. Thus, they are attractive as nanoporous storage media. Significant advances have been made in the synthesis of branched oligonucleotide hybrids with rigid core molecules, most importantly by developing solution-phase approaches. Solution-phase syntheses that are rapid, inexpensive and robust now produce hybrid materials routinely in ever-increasing quantities, opening the door to the exploration of their fascinating properties.

Acknowledgements

The authors are grateful to Th. Sabirov and D. Göhringer for expert technical assistance, Ch. Rothe, H. Weischedel and Drs. M. Meng, A. Singh and M. Tolev for sharing results and discussions, and C. Gerlach and Dr. B. Claasen for spectroscopic measurements. The work of the authors on DNA nanostructuring is supported by DFG (grants No. RI 1063/13-1 and RI 1063/15-1 to C.R.).

References

[1] (a) J. D. Watson and F. H. Crick, *Nature*, **171**, 737–738 (1953); (b) J. D. Watson and F. H. Crick, *Nature*, **171**, 964–967 (1953).
[2] N. C. Seeman, *Nature*, **421**, 427–431 (2003).
[3] T. Fu, Y. Tse-Dinh, and N. C. Seeman, *J. Mol. Biol.*, **236**, 91–105 (1994).
[4] M. Endo, S. Yamamoto, K. Tatsumi, T. Emura, K. Hidaka, and H. Sugiyama, *Chem. Commun.*, **49**, 2879–2881 (2013).
[5] N. C. Seeman, *Nature*, **421**, 427–431 (2003).
[6] R. P Goodman, I. A. T. Schaap, C. F. Tardin, C. M. Erben, R. M. Berry, C. F. Schmidt, and A. J. Turberfield, *Science*, **310**, 1661–1665 (2005).
[7] P. W. K. Rothemund, *Nature*, **440**, 297 (2006).
[8] S. M. Douglas, H. Dietz, T. Liedl, B. Högberg, F. Graf, and W. M. Shih, *Nature*, **459**, 414–418 (2009).
[9] H. Dietz, S. M. Douglas, and W. M. Shih, *Science*, **325**, 725–730 (2009).
[10] A. Kuzyk, R. Schreiber, Z. Fan, G. Pardatscher, E.-M. Roller, A. Högele, F. C. Simmel, A. O. Govorov, and T. Liedl, *Nature*, **483**, 311–314 (2012).
[11] M. Langecker, V. Arnaut, T. G. Martin, J. List, S. Renner, M. Mayer, H. Dietz, and F. C. Simmel, *Science* **338**, 932–936 (2012).
[12] X. Bai, T. Martin, S. Scheres, and H. Dietz, *Proc. Natl. Acad. Sci. U.S.A.*, **109**, 20012–20017 (2012).
[13] H. Said, V. Schueller, F. Eber, C. Wege, T. Liedl, and C. Richert, *Nanoscale*, **5**, 284–290 (2013).
[14] N. C. Seeman, *J. Theor. Biol.*, **99**, 237 (1982).
[15] J. Zheng, J. J. Birktoft, Y. Chen, T. Wang, R. Sha, P. E. Constantinou, S. L. Ginnell, C. Mao, and N. C. Seeman, *Nature*, **461**, 74–77 (2009).
[16] P. J. Paukstelis, *J. Am. Chem. Soc.*, **128**, 6794–6795 (2006).
[17] K. M. Stewart and L. W. McLaughlin, *J. Am. Chem. Soc.*, **126**, 2050–2057 (2004).
[18] K. M. Stewart, J. Rojo, and L. W. McLaughlin, *Angew. Chem.*, **116**, 5932–5935 (2004).
[19] H. Yan, S. H. Park, G. Finkelstein, J. H. Reif, and T.H. LaBean, *Science*, **301**, 1882–1884 (2003).
[20] K. V. Gothelf and T. H. LaBean, *Org. Biomol. Chem.*, **3**, 4023–4037 (2005).
[21] C. A. Mirkin, R. L. Letsinger, R. C. Mucic, and J. J. Storhoff, *Nature*, **382**, 607–609 (1996).

[22] D. Nykypanchuk , M. M. Maye , D. van der Lelie, and O. Gang, *Nature*, **451**, 549–552 (2008).
[23] S. Y. Park, A. K. R. Lytton-Jean, B. Lee, S. Weigand, G. C. Schatz, and C. A. Mirkin, *Nature*, **451**, 553–556 (2008).
[24] C. Richert, M. Meng, K. Müller, and K. Heimann, *Small*, **4**, 1040–1042 (2008).
[25] G. T. Hermanson, *Bioconjugate Techniques*, Academic Press: San Diego, pp. 639–671 (1995).
[26] M. Meng, C. Ahlborn, M. Bauer, O. Plietzsch, S. A. Soomro, A. Singh, T. Muller, W. Wenzel, S. Bräse, and C. Richert, *ChemBioChem*, **10**, 1335–1339 (2009).
[27] C. I. Y. Ahlborn, Ph.D. Thesis, University of Karlsruhe (2007).
[28] O. Plietzsch, C. I. Schilling, M. Tolev, M. Nieger, C. Richert, T. Muller, and S. Bräse, *Org. Biomol. Chem.*, **7**, 4734–4743 (2009).
[29] C. Schilling, Ph.D. Thesis, Karlsruhe Institute of Technology (2010).
[30] (a) K. M. Stewart, J. Rojo, and L. W. McLaughlin, *Chem. Commun.*, **2934–2935** (2003); (b) K. M. Stewart, J. Rojo, and L. W. McLaughlin, *Angew. Chem.*, **116**, 5932–5935 (2004); *Angew. Chem., Int. Ed.*, **43**, 5808–5811 (2004).
[31] F. A. Aldaye and H. F. Sleiman, *Angew. Chem.*, **118**, 2262–2267 (2006).
[32] P. Gruenefeld and C. Richert, *Nucleosides Nucleotides Nucl. Acids*, **25**, 815–821 (2006).
[33] M. Meng, Ph.D. Thesis, University of Karlsruhe (2010).
[34] J. Rongchao, W. Guosheng, L. Zhi, C. A. Mirkin, and G. C. Schatz, *J. Am. Chem. Soc.*, **125**, 1643–1654 (2003).
[35] A. Singh, M. Tolev, M. Meng, K. Klenin, O. Plietzsch, C. I. Schilling, T. Muller, M. Nieger, S. Bräse, W. Wenzel, and C. Richert, *Angew. Chem., Int. Ed.*, **50**, 3227–3231 (2011).
[36] A. Singh, M. Tolev, C. I. Schilling, S. Bräse, H. Griesser, and C. Richert, *J. Org. Chem.*, **77**, 2718–2728 (2012).
[37] A. Singh, Ph.D. Thesis, Institute of Organic Chemistry University of Stuttgart (2011).
[38] H. Newman, *Synthesis*, 692–693 (1972).
[39] V. Reichert and L. Mathias, *Macromolecules*, **27**, 1030–1034 (1994).
[40] E.B. Merkushev, *Synthesis*, **6**, 486–487 (1980).
[41] C. I. Schilling, O. Plietzsch, M. Nieger, T. Muller, and S. Bräse, *Eur. J. Org. Chem.*, 1743–1754 (2011).
[42] J.-H. Fournier, X. Wang, and J. D. Wuest, *Can. J. Chem.*, **81**(5), 376–380 (2003).
[43] T. Hatano and T. Kato, *Tetrahedron*, **64**, 8368–8380 (2008).
[44] G. Kumar and M. S. Poonian, *J. Org. Chem.*, **49**, 4905–4912 (1984).
[45] R. Wörl and H. Köster, *Tetrahedron*, **55**, 2957–2972 (1999).
[46] J. Yanez, M. Arguello, J. Osuma, X. Soberon, and P. Gaytan, *Nucleic Acids Res.*, **32**, e158 (2004).
[47] B. Virnekäs, L. Ge, A. Plückthun, K. C. Schneider, G. Wellnhofer, and S. E. Moroney, *Nucleic Acids Res.*, **22**, 5600–5607 (1994).
[48] H. Griesser, M. Tolev, A. Singh, T. Sabirov, C. Gerlach, and C. Richert, *J. Org. Chem.*, **77**, 2703–2717 (2012).
[49] B. L. Gaffney, E. Veliath, J. Zhao, and R. A. Jones, *Org. Lett.*, **12**, 3269–3271 (2010).
[50] J. Cieślak, J. Jankowska, M. Sobkowski, M. Wenska, J. Stawiński, and A. J. Kraszewski, *J. Chem. Soc., Perkin Trans.* **1**, 31–37 (2002).
[51] C. B. Reese and Q. Song, *Bioorg. Med. Chem. Lett.*, **7**, 2787–2792 (1997).
[52] (a) C. W. Tornoe, C. Christensen, and M. Meldal, *J. Org. Chem.*, **66**, 3057–3064 (2002); (b) V. V. Rostovtsev, L. G. Green, V. V. Fokin, and K. B. Sharpless, *Angew. Chem., Int. Ed.*, **41**, 2596–2599 (2002).
[53] M. Tolev, Ph. D. Thesis, Institute of Organic Chemistry University of Stuttgart (2012).
[54] E. Auyeung, T. I. N. G. Li, A. J. Senesi, A. L. Schmucker, B. C. Pals, M. Olvera de la Cruz, and C. A. Mirkin, *Nature*, **505**, 73–77 (2014).
[55] (a) X. Li and V. Ramamurthy, *J. Am. Chem. Soc.*, **118**, 10666–10667 (1996); (b) R. J. Robbins and V. Ramamurthy, *J. Chem. Soc., Chem. Commun.*, 1071–1072 (1997);(c) V. Ramamurthy, P. Lakshminarasimhan, C. P. Grey, and L. J. Johnston, *J. Chem. Soc., Chem. Commun.*, 2411–2424 (1998).
[56] (a) E. L. Clennan and J. P. Sram, *Tetrahedron Lett.*, **40**, 5275–5278 (1999); (b) M. Stratakis and G. Froudakis, *Org. Lett.*, **2**, 1369–1372 (2000); (c) E. L. Clennan and J. P. Sram, *Tetrahedron*, **56**, 6945–6950 (2000).
[57] G. Roelfes and B. L. Feringa, *Angew. Chem.*, **117**, 3294–3296 (2005).
[58] T. Hatano and T. Kato, *Tetrahedron*, **64**, 8368–8380 (2008).
[59] (a) A. C. Komor and J. K. Barton, *Chem. Commun.*, **49**, 3617–3630 (2013); (b) D. Mitra, N. Di Cesare, and H. F. Sleiman, *Angew. Chem., Int. Ed.*, **43**, 5804–5808 (2004).

[60] H. Yang, F. Altvater, A. Dowine de Bruijn, Ch. K. McLaughlin, P. K. Lo, and H. F. Sleiman, *Angew. Chem.*, **123**, 4716–4719 (2011).
[61] R. Haug, H. Griesser, Th. Sabirov, and C. Richert, *J. Porphyrins Phthalocyanines*, **16**, 488–498 (2012).
[62] J.-L. H. A. Duprey, Y. Takezawa, and M. Shionoya, *Angew. Chem.*, **125**, 1250–1254 (2013).
[63] C. Kroener, M. Roethlingshoefer, and C. Richert, *J. Org. Chem.*, **76**, 2933–2936 (2011).
[64] C. Kröner, A. Göckel, W. Liu, and C. Richert, *Chem. Eur. J.*, **19**, 15879–15887 (2013).

5.4

DNA-Controlled Assembly of Soft Nanoparticles

Stefan Vogel

Department of Physics, Chemistry and Pharmacy, University of Southern Denmark, Biomolecular Nanoscale Engineering Center (BioNEC), Denmark

5.4.1 Introduction

Controlled assembly of soft nanoparticles requires a recognition event to trigger the formation of a non-covalent assembly [1]. Among known supramolecular structures such as proteins and oligonucleotides, DNA is by far the most versatile building block to perform such tasks [2]. The main advantage, besides; the predictable hybridization and sequence design, is the highly automated and cost-effective synthesis of DNA [3] and the ever increasing number of available chemistries to extend the natural toolbox of DNA with unusual properties, such as lipids, dyes, additional charges and conformational restricted monomers (LNA), unlocked nucleic acids (UNA) and uncharged nucleic acids, such as peptide nucleic acids (PNAs) [4, 5]. The DNA controlled assembly of solid nanoparticles is well established for a number of solid nanoparticles despite the often tedious and not generally applicable surface chemistry [6–8], but much less has been reported on soft nanoparticles [1, 9–11].

A large number of soft nanoparticles are known, (e.g., dendrimer-, polymer- or lipid-based nanoparticles) and among the most important class of natural soft nanoparticles are vesicles [12, 13]. Vesicles are lipid nanoparticles based on natural lipids and are of particular interest due to their occurrence in nature as intra- or extracellular transport vehicles (e.g., endosomes and exosomes). The equivalent type of soft particles based on synthetic lipids are liposomes, which are equivalent to vesicles in all aspects except for properties that are linked to the particular nature of the synthetic lipid or lipid mixture used. Lipid based soft nanoparticles often form lipid bilayers, that allow surface modification through attachment of other molecules by simple membrane anchoring, a process which is entirely non-covalent and driven by the hydrophobic effect. Membrane

---

DNA in Supramolecular Chemistry and Nanotechnology, First Edition.
Edited by Eugen Stulz and Guido H. Clever.

anchoring is highly predictable for lipid nanoparticles in aqueous solution. Depending on the nature of the lipid-membrane anchor a wide range of anchoring strengths is available and can easily be adjusted to the required level of anchoring strength by variation of the anchor length or number of anchors for each anchored molecular entity [14–16]. DNA-controlled assembly does require lipid-modified oligonucleotides (e.g., DNA or RNA) with a number of DNA design considerations (see Figure 5.4.1). Immobilization of DNA (encoding)

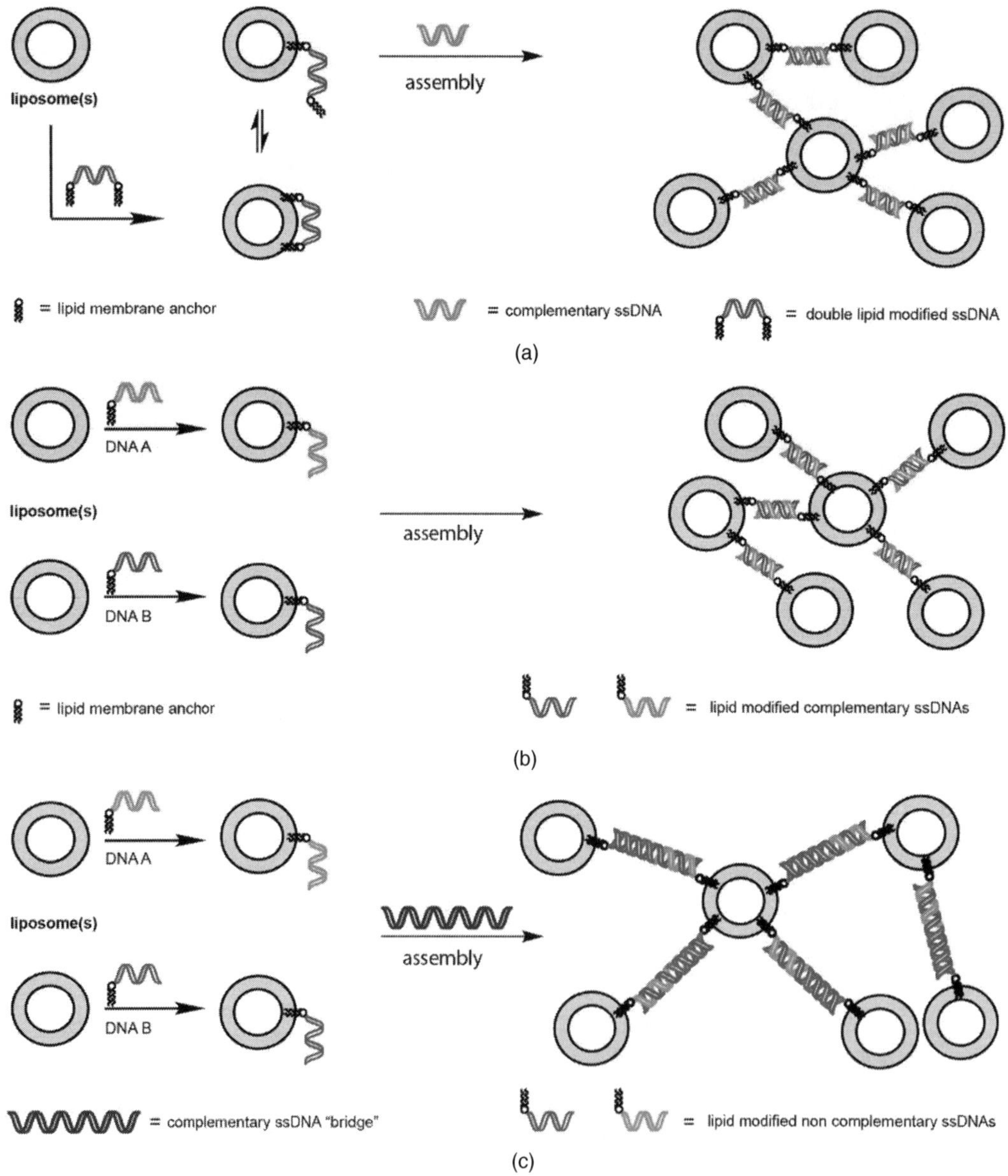

Figure 5.4.1 *DNA-sequence design. Schematic representation of liposome aggregation upon duplex formation between lipid-modified DNA strand(s) and an unmodified (a, c) or modified complementary DNA strand (b). Liposomes and DNA strands are not drawn to scale*

on solid nanoparticles requires surface chemistry, which is well established for a limited number of materials such as gold (via sulfur chemistry) or silica (silanol derivatization), but for most other materials no established chemistry is available and the surface anisotropy found for most solid materials requires tedious procedures to control surface chemistry and uniform surface coverage (e.g., encoding with DNA).

However, soft nanoparticles based on lipid bilayers (e.g., vesicles or liposomes) possess a universal mode for encoding surfaces by membrane anchoring of lipid modified DNA. In light of the technically demanding procedures for solid nanoparticles, non-covalent attachment of DNA to lipid bilayer surfaces becomes a very attractive technology [1, 17].

5.4.2 Sequence design

Incorporation of one or more lipid-membrane anchors into DNA leads to different sequence designs, which in general all allow assembly of liposomes. However, working with multiple sets of encoded liposomes or applications in molecular diagnostics with unmodified DNA target sequences requires a different sequence design (see Section 5.4.2).

5.4.2.1 Double membrane anchor ssDNA design

In applications using just one type of liposomes, all of the DNA design suggestions (see Figure 5.4.1) are feasible. However subtle differences during the DNA-controlled assembly of liposomes are a direct result of different DNA designs. For a single DNA strand that is modified at both ends (Figure 5.4.1a), the DNA is tightly but reversibly anchored to the lipid bilayer of the liposome. The attached DNA is free floating on the surface of the liposome with full lateral mobility and restricted mobility for upwards movement (partial lipid membrane anchor release) [11, 18]. This dynamic process allows hybridization at one end, which subsequently exerts an increasing pull from the DNA duplex formed. This process continues until one end of the DNA is no longer anchored but is exposed to the aqueous bulk phase. Owing to the energetically unfavorable disturbance of the hydrogen bonding network of the solvent, a strong hydrophobic effect forces the free anchor to anchor into another lipid bilayer surface. The DNA will therefore attach to the next available liposome in the proximity and DNA-controlled assembly will proceed until all free DNA has hybridized to the complementary membrane-anchored DNA [1]. It is important to note that re-anchoring of the free lipid-membrane anchor of the resulting DNA duplex into the same liposome is energetically very unfavorable. Anchoring of both ends of the formed DNA duplex to the same liposome would require bending of the rigid double stranded DNA and also require partial desolvation of the DNA duplex caused by the very close contact to the lipid surface. The assembly process can be tuned by adjustment of the lipid anchor (see Section 5.4.3) and length of the DNA.

5.4.2.2 Single membrane anchor multiple ssDNA design

In order to assemble different batches of liposomes, a number of alternative designs are available. A simple design consists of two different complementary ssDNAs, each modified with a single membrane anchor. Both strands have to be modified either at the 3′-end or at the 5′-end, as sequence designs with lipid membrane anchors juxtapositioned at the same end may promote hemifusion instead of reversible assembly (see subsequent section). This design does work for liposomes encoded with complementary DNA strands and allows assembly of more than two different liposomes, each encoded with one or more membrane anchored ssDNAs. For applications in molecular diagnostics with DNA targets from a biological source a design with a bridging DNA (see Figure 5.4.1c) or a double-modified ssDNA (see Figure 5.4.1a) may be used.

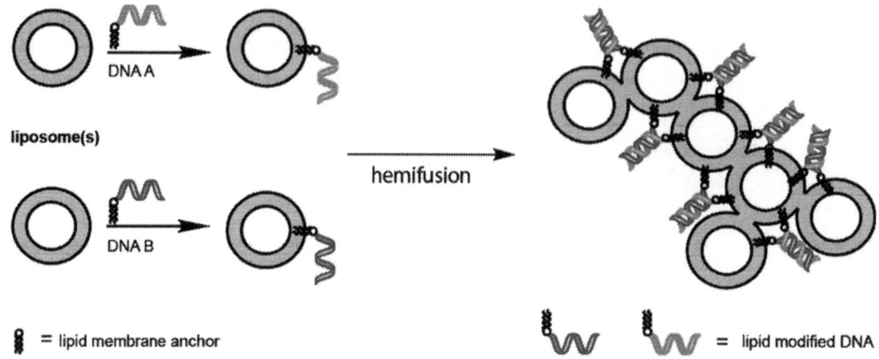

Figure 5.4.2 DNA-sequence design. Schematic representation of irreversible liposome aggregation (hemifusion/fusion) upon duplex formation between a two lipid-modified DNA strands. Liposomes and DNA strands are not drawn to scale

5.4.2.3 Single membrane anchor multiple ssDNA design for irreversible assembly (fusion/hemifusion)

An irreversible DNA controlled assembly will enable hemifusion or fusion of liposomes and requires a design that reduces the distance between two liposomes to an extent that the resulting destabilization of the lipid bilayer membrane is sufficient to induce hemifusion (fusion of the outer lipid layer, see Figure 5.4.2) or eventually full fusion (fusion of the outer and inner lipid layer). The required DNA design will, in contrast to designs shown in Figure 5.4.1, have both lipid anchors on the same end of the DNA duplex to force the liposomes together in a 'zipper-fashion' rather than forming inter-liposomal spacers. Not all aspects of the DNA controlled hemifusion/fusion mechanism are currently well understood and the process is dependent on a number of parameters, such as DNA sequence design, number and structure of the membrane anchor(s) and DNA concentration [19, 20].

5.4.3 Lipid membrane anchors

The non-covalent anchoring of DNA to the surface of liposomes requires a membrane anchoring unit that keeps the DNA attached to the liposome, and the obvious choice are lipophilic moieties. In principle, natural or synthetic lipids are equally applicable. A number of lipid modifications have been described in the literature, including modifications on the nucleobase, terminal tocopherol and cholesterol-derived building blocks [21–23]. However, terminal modifications with lipids often lead to DNA-conjugates with strong detergent properties and, depending on the concentrations, a strong tendency for self-aggregation. To avoid self-aggregation and to increase flexibility in the DNA design, a generic key intermediate (see Scheme 5.4.1) based on an aza macrocycle has been developed [23–25], which allows attachment of different lipid moieties and subsequent conversion into the corresponding DMT-protected phosphoramidite building block for automated solid phase synthesis [3].

The resulting lipid membrane anchor building blocks allow multiple incorporations into the DNA and are not limited to terminal modifications (see Scheme 5.4.1). An appropriate lipid membrane anchor is crucial for efficient DNA controlled assembly of liposomes, and not only the lipid anchor structure and length but also the positioning in the DNA strand is important. A general solution to circumvent self-aggregation is to add a number of additional nucleobases (usually 1–3) on the terminal end relative to the position of the membrane

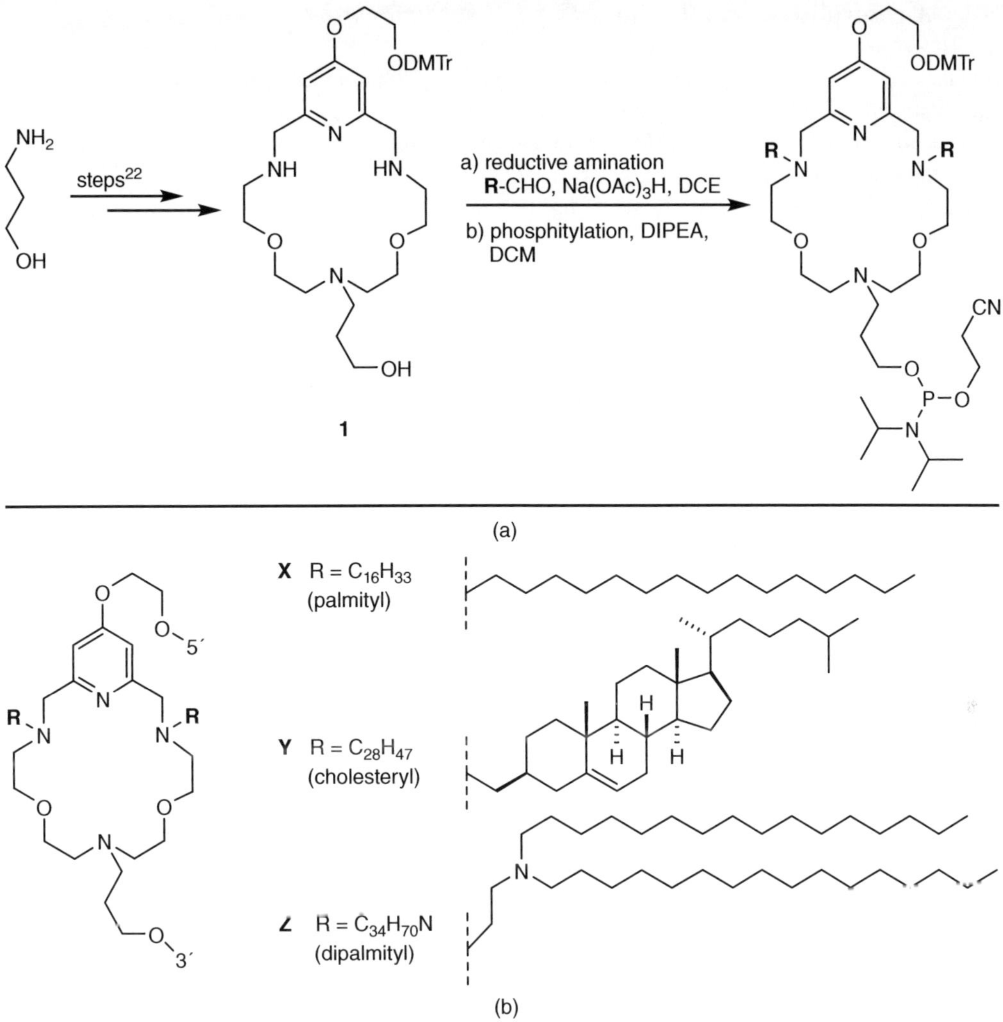

Scheme 5.4.1 *Synthesis of aza crown ether based lipid membrane anchor amidite building blocks (a) and key building blocks incorporated into DNA (denoted* ***X, Y, Z****, see (b))*

anchor(s). This approach is sufficient to suppress self-aggregation and development of detergent properties of the resulting DNA–lipid conjugate. Membrane anchoring requires at least two lipid membrane anchors of sufficient lipophilicity to anchor the DNA–lipid conjugate without significant partitioning between the liposome surface and the aqueous solution [26], and the minimal length of the lipid membrane anchor is around 10–12 carbon chains (unsaturated linear lipid chains). This estimation has been validated for DNA sequences of 9–30 nucleobases, reliable anchoring is achieved with 14–16 carbons whereas longer anchors (>20 carbons) lead to very strong anchoring. In particular, DNA designs with ssDNAs modified with two membrane anchor building blocks with 20 carbons incorporated at each end of the DNA lead to permanent irreversible anchoring and no assembly is observed. This is of course of no consequence for single lipid modified DNAs, which are membrane anchored just stronger than is necessary for efficient liposome assembly (Jakobsen, U. and Vogel, S., unpublished results).

5.4.4 DNA-controlled assembly studied by UV spectroscopy

Ultraviolet spectroscopy (UV) is a very convenient and powerful method for monitoring the process of DNA controlled liposome assembly and disassembly [27]. The assembly process can be followed for different wavelengths, with signals decreasing at longer wavelengths, in full agreement with the Stokes–Einstein equation where the scattering of light decreases with increasing wavelength. The observed change in apparent absorbance upon thermal cycling is a result of light scattering by particle aggregates formed during particle assembly and aggregates dissolved during disassembly, rather than the absorption of light by the individual molecular components (e.g., DNA nucleobases at 260 nm).

The apparent absorbance is also a measure of the reversibility of the assembly and disassembly process. Measurement of the same level of apparent absorbance after a full thermal cycle will indicate that the same number and average size (polydispersity) of aggregates has been formed and dissolved during thermal cycling [1]. Ultraviolet spectroscopy in the presence of liposomes allows monitoring of DNA controlled assembly processes based on double or triple helix formation. More complex DNA structures, such as triple helices, are often difficult to monitor by conventional thermal denaturation studies (in the absence of liposomes), since competing hybridization events from self-association of single strands will often obscure measurements, but thermal denaturation in the presence of liposomes will distinguish between the different possible DNA structures based on the level of liposome assembly, as expressed by the change in apparent absorbance and the observed T_m value [28]. The light scattered away from light path by individual liposomes or liposome aggregates will reduce the amount of light that reaches the detector and the result is an increased apparent absorption upon liposome assembly and disassembly.

The recorded curves for thermal cycling in aqueous buffer solutions are, therefore, fundamentally different from regular thermal denaturation profiles of hybridized oligonucleotides measured at 260 nm (absorption maximum for DNA nucleobases). Regular thermal denaturation measurements lead to a rise in absorbance with increased temperature due to onset of denaturation of hybridized structures to free ssDNA

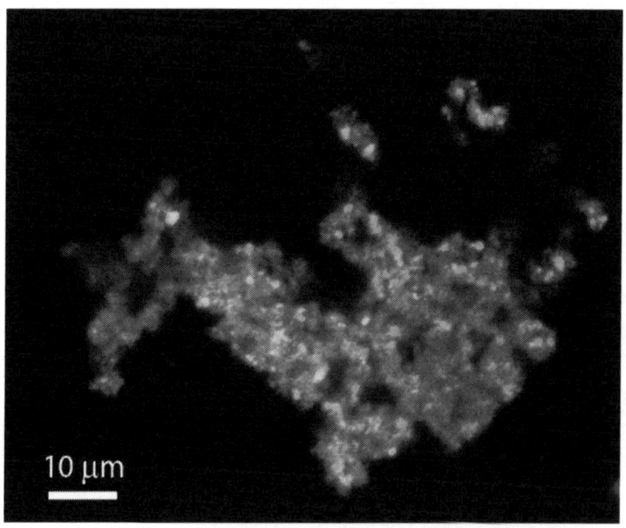

Figure 5.4.3 *Confocal image of aggregates formed by DNA controlled assembly with a pair of complementary lipid modified DNA strands and two colored batches of liposomes (5′-TTT-**X**-GTG-ATA-TGC-**X**-TTT : 3′-TTT-**X**-CAC-TAT-ACG-**X**-TTT, 1 μM DNA, POPC liposomes Ø = 100 nM, pH 7) (Bagatolli, L., Jakobsen, U. and Vogel, S., unpublished results) (See color figure in color plate section)*

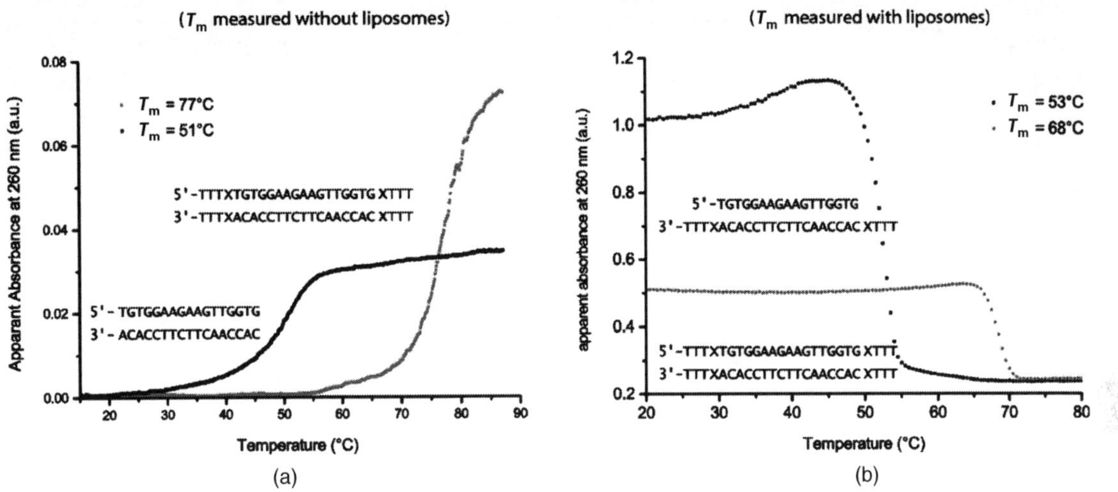

Figure 5.4.4 *Thermal denaturation data for complementary lipid-modified DNA strands in the absence (a) and presence of liposomes (b), (DNA concentration for experiment (a) is 0.5 μM and for (b) is 0.125 μM, POPC liposomes Ø = 100 nm, pH 7)*

with typically broad transitions covering a 10–20 °C thermal window. In strong contrast to regular measurements, DNA controlled liposome assembly shows inverted curves with a much narrower thermal transition window of 2–4 °C. The resulting aggregates vary in size (see Figure 5.4.3) depending on the available DNA concentration and the initial size of the individual liposomes [1].

A comparison of thermal denaturation results from complementary lipid-modified DNA strands (see Figure 5.4.4a and b) shows that lipid modifications (**X**) positioned opposite to each other result in a significantly increased thermal stability ($\Delta T_m = +26$ °C). This is also observed, but to a lesser degree, for the same DNA pair in the presence of liposomes ($\Delta T_m = +15$ °C), but it also shows that even a combination of two lipid modified ssDNA strands with two incorporated lipid membrane anchors at each end allow DNA controlled liposome assembly (with reduced efficiency in direct comparison, see Figure 5.4.4b). The reason for this finding is attributed to the much stronger hydrophobic interactions between the lipid modifications in the absence of liposomes than in the presence of liposomes, as the liposomes compete for the interaction with the lipid moieties. In the presence of liposomes, most of the hydrophobic interactions will be directed towards anchoring into the lipid bilayer of liposomes rather than DNA interstrand interactions [17, 23].

Another feature of the complex DNA controlled liposome assembly and disassembly is the reversibility of the aggregate formation. Sometimes the first thermal cycle is associated with a slightly lower absolute value for the apparent absorbance compared with subsequent thermal cycles. It is assumed that not all membrane-anchored olgonucleotides are are able to partcipate in the assembly process, but in the next cycle an equilibrium is reached and all subsequent thermal cycles show virtually superimposed curves (see Figure 5.4.5b). In addition, an important observation is the difference in signal intensity. Regular thermal denaturation measurements in the absence of liposomes lead to only minor changes in apparent absorbance (Δabs ~ 0.08 a.u.), whereas the same measurement in the presence of liposomes and four times lower DNA concentration yields a ten times increased signal (Δabs ~ 0.8 a.u., see Figure 5.4.4).

The thermal transition temperatures are highly reproducible and have, in combination with the remarkable sharp thermal transition (see Figure 5.4.5a), potential diagnostic applications (e.g., SNP analysis and post-PCR diagnostic methods) [6, 29].

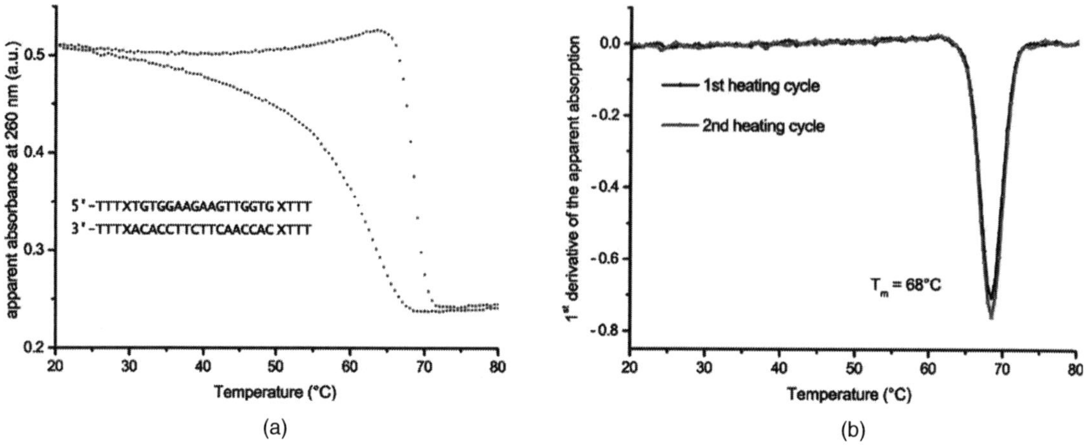

Figure 5.4.5 *Thermal denaturation data for complementary lipid-modified DNA strands showing a full thermal cycle (a) and an overlay of the first derivative of two consecutive heating cycles (b) in the presence of liposomes (POPC liposomes, Ø = 65 nm, 0.125 μM DNA concentration, pH 7)*

The remarkable sharp thermal transitions have been described earlier for the DNA-controlled assembly of gold nanoparticles [6] and can be explained by a cooperative behavior of DNA-linked liposomal aggregates. Upon heating of DNA-linked liposomes, the DNA linker structures are increasingly destabilized until the thermal denaturation point is reached. Each individual liposome is presumably linked by a number of membrane anchored duplex structures (interliposomal DNA linker) Also, each DNA linker denaturation event is associated with a decrease in local ion strength, which results in a subsequent destabilization of the remaining hybridized DNA linkers. This process leads to a rapid destabilization and disassembly of the liposomal aggregates and results in a remarkably sharp thermal transition compared with regular unmodified duplex DNA in the absence of liposomes [30].

A special case of DNA controlled liposome assembly is the irreversible assembly. This requires a DNA design as shown in Figure 5.4.2 with lipid-modified DNA strands that position both membrane anchors at one end of the corresponding DNA duplex. Two batches of liposomes are encoded by complementary DNA anchored to the liposome surface at one end and will trigger assembly in a 'zipper-fashion' upon mixing. The liposomes are presumably forced very close together during the assembly process and the DNA linker exerts a considerable mechanical stress on the outer layer of the lipid membrane [19, 20]. At one point the outer lipid layer of two juxtapositioned liposomes will fuse and the whole assembly process will ultimately lead to hemifusion, and also potentially to mixing of the inner lipid layers with subsequent fusion. This hemifusion process is not reversible and each thermal cycle will only lead to additional hemifusion of free liposomes in solution, depending on the DNA concentration. The resulting apparent absorbance will increase for each cycle and will not return to the value observed for free liposomes (see Figure 5.4.6).

Examples of DNA controlled hemifusion/fusion have been reported in the literature, but so far only with very limited fusion and content mixing (<5%) [19, 20]. However, the DNA programmed approach to fusion of lipid bilayers is an extremely elegant application of DNA-hybrid materials and a very advanced example for DNA based supramolecular chemistry.

5.4.4.1 Thermal stability of lipid-modified DNA conjugates

The assembly of liposomes requires lipid-modified DNA strands. Such DNA-hybrid material has, to the best of our knowledge, no equivalent in nature but does possess unique properties.

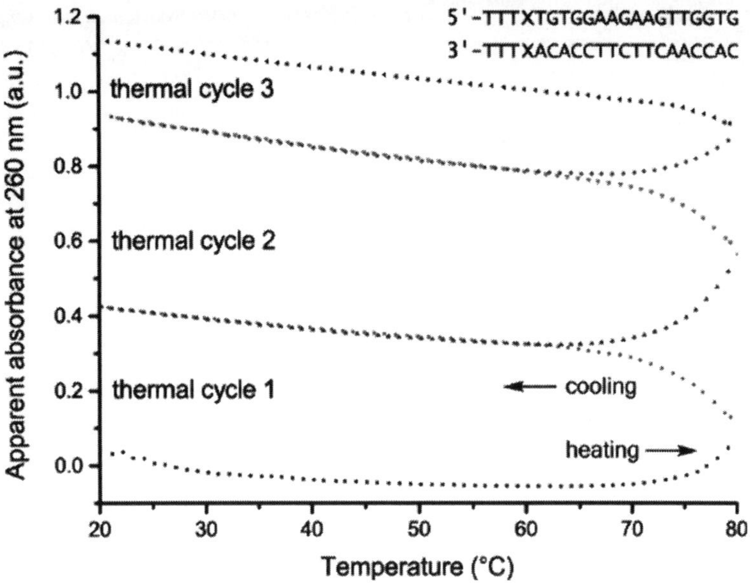

Figure 5.4.6 *Thermal denaturation data for irreversible hemifusion/fusion of liposomes by DNA-controlled assembly (POPC liposomes, Ø = 100 nM, 0.2 μM DNA concentration, pH 7)*

It becomes clear from thermal denaturation data of lipid-modified DNA strands that positioning of the corresponding lipid membrane anchor building block is crucial for DNA duplex stability (see Table 5.4.1) and biophysical properties such as surface activity. Two juxtapositioned monomers in each DNA strand, as dangling ends, stabilize the duplex considerably ($\Delta T_m \geq 20$ °C, Table 5.4.1, entries 2, 7, and 8), whereas juxtapositioned incorporations in the middle of the DNA duplex lead to a strong stabilization for palmityl modified DNA (**X**, $\Delta T_m = 8$ °C, Table 5.4.1, entry 4) and an even stronger stabilization for cholesteryl-modified DNA (**Y**, $\Delta T_m \geq 20$ °C, Table 5.4.1, entry 5). This is remarkable in light of only two incorporations opposite to each other for either palmityl or cholesteryl derived membrane anchors, which is caused by strong hydrophobic forces that induce dense packing of the lipids placed opposite to each other. The stabilization by lipid-modified building blocks in the middle of a DNA duplex correlates with the relative lipophilicity of the substituents (Table 5.4.1). The results are remarkable considering the steric demand (compared with a natural T-nucleotide) and the structure of the 'abasic' non-nucleosidic building blocks, which introduce a large abasic site in the DNA duplex [17, 23, 24].

5.4.4.2 DNA mismatch discrimination studies

The remarkably sharp thermal transitions during DNA-controlled assembly and disassembly are very valuable in applications for detection of single nucleotide polymorphism analysis (SNP analysis), as even weakly discriminated mismatches are easily distinguished from the fully complementary target by the simplified analysis of non-overlapping thermal transitions. The discrimination of single mismatches (ΔT_m) in lipid-modified oligonucleotide sequences is otherwise comparable to unmodified DNA and allows traditional probe sequences design [1].

Comparison of the thermal denaturation data for the same sequence with or without liposomes results in surprisingly similar T_m values despite the much lower DNA concentrations used for experiments with liposomes (0.5 μM versus 0.062 μM, see Table 5.4.2). This indicates an additional stabilization of the DNA duplex as part of the interliposomal DNA linker bundles connecting the individual liposomes to larger aggregates.

Table 5.4.1 *Influence of aza crown ether monomers **X**, **Y** on thermal DNA duplex stability*[a,b]

Entry	Duplex	T_m (°C)	ΔT_m (°C)
1	5′-TTTGTGATATGCTTT 3′-TTTCACTATACGTTT	31.0	—
2	5′-TTT**X**GTGATATGC**X**TTT 3′-TTT**X**CACTATACG**X**TTT	60	29
3	5′-TGTGGAAGAAGTTGGTG 3′-ACACCTTCTTCAACCAC	56.0	—
4	5′-TGTGGAAG**X**AGTTGGTG 3′- ACACCTTC**X**TCAACCAC	64.0	8.0
5	5′-TGTGGAAG**Y**AGTTGGTG 3′- ACACCTTC**Y**TCAACCAC	77.5	21.5
6	5′-TTTTGTGGAAGAAGTTGGTGTTT 3′-TTTACACCTTCTTCAACCACTTT	58	—
7	5′-TTT**X**TGTGGAAGAAGTTGGTG**X**TTT 3′-TTT**X**ACACCTTCTTCAACCAC**X**TTT	77.0	21.0
8	5′-TTT**Y**TGTGGAAGAAGTTGGTG**Y**TTT 3′-TTT**Y**ACACCTTCTTCAACCAC**Y**TTT	>86	>28

[a] **X, Y** denote the aza crown ether monomer (see Scheme 5.4.1).
[b] Conditions: T_m values (°C) and ΔT_m = change in T_m value calculated relative to the DNA : DNA reference duplex, measured as the maximum of the first derivative of the melting curve (A_{260} versus temperature) recorded in medium salt buffer (110 mM Na^+, pH 7.0, 100 mM NaCl, 10 mM NaH_2PO_4/Na_2HPO_4), using 0.5 μM concentrations of the two complementary strands. Exp. error: ±0.5 °C.

Table 5.4.2 *Mismatch discrimination data from thermal denaturation experiments*[a,b]

Entry	Duplex	T_m (°C)	ΔT_m (°C)	T_m (°C)	ΔT_m (°C)
	3´-TTT**X**ACACCTTCTTCAACCAC**X**TTT	With liposomes[c]		Without liposomes[d]	
1	5´-TGTG**G**AAGAAGTTGGTG	48	—	47	—
2	5´- TGTG**T**AAGAAGTTGGTG	36	−12	37	−10
3	5´ -TGTG**A**AAGAAGTTGGTG	38	−10	38	−11
4	5´ -TGTG**C**AAGAAGTTGGTG	36	−12	37	−11

[a] **X** denotes the palmityl substituted aza crown ether base surrogate (see Scheme 5.4.1).
[b] Conditions: T_m values (°C) and ΔT_m = change in T_m value calculated relative to the DNA : DNA reference duplex measured as the maximum of the first derivative of the melting curve (A_{260} versus temperature) recorded in medium salt buffer (10 mM HEPES, 110 mM Na^+, pH 7.0).
[c] 62 nM concentrations of the two complementary strands in presence of liposomes.
[d] 1 μM concentrations of the two complementary strands in the absence of liposomes. Exp. error: ±1 °C.

5.4.5 Assembly on solid support

Earlier chapters focused on DNA-controlled assembly in solution, but DNA tethering to solid supported membranes is possible using the same DNA sequence design as described in Section 5.4.2. DNA controlled tethering of liposomes to solid supported membranes has been described with our dipalmityl lipid membrane

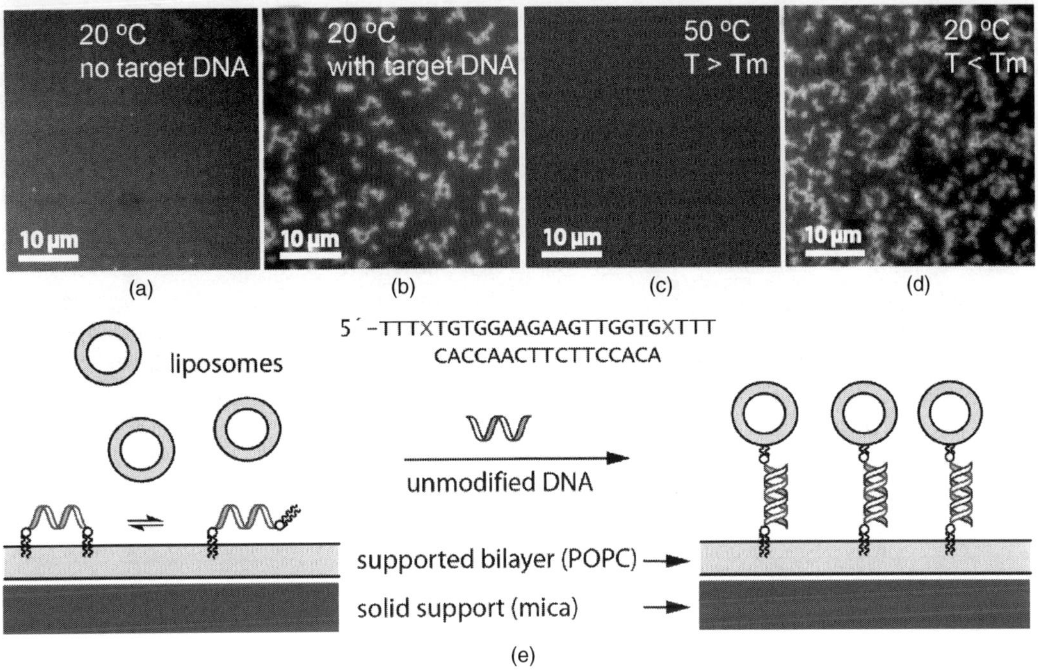

Figure 5.4.7 Fluorescence microscopy data for the reversible DNA controlled liposome assembly on solid supported POPC bilayers (a) to (b) and the subsequent disassembly and reattachment of liposome aggregates after thermal denaturation (c) to (d); (e) schematic representation of DNA-controlled assembly of POPC liposomes on the surface of a solid supported POPC bilayer. Adapted from [1]. Reproduced with permission from [1]. Copyright © 2008, American Chemical Society

anchors [1] and earlier using terminal cholesteryl [26] incorporated into two individual DNA strands bearing one cholesteryl membrane anchor each. This results in a more complex DNA design with four DNA strands for the assembly as well as in another important study on terminal dialkyllipid-modified DNA [20] describing tethering and movement of liposomes in an electrical field.

The reversible attachment of liposomes as shown in Figure 5.4.7 involves only a single DNA strand, which contains two dipalmityl based membrane anchors at each end of the DNA. This allows control of the assembly process by unmodified DNA and may even be used for analytical or diagnostic applications on biological DNA samples with the assembly process as a powerful read out for hybridization to a biological target DNA sequence [1].

Thermal denaturation of the surface bound liposome aggregates occurs by heating of the system beyond the thermal denaturation temperature (T_m = 46 °C in Figure 5.4.7). Heating dissolves all aggregates, which reform upon cooling. The process is highly reversible and can be visualized directly by fluorescence microscopy (see Figure 5.4.7). Multiple heating–cooling cycles lead to similar surface coverage by liposome aggregates despite a large excess of the POPC liposome surface in solution (>600 times) [1]. This surprising finding shows that DNA (23-mer) anchored by dipalmityl membrane anchors does not partition significantly into the solution and is probably bound stronger to the surface of the solid supported POPC lipid bilayer than to the free floating POPC liposomes. It is speculated that the more ordered and densely packed POPC bilayer on the solid support is able to bind the lipid-modified DNA stronger than free liposomes

[1, 31]. The lateral diffusion of the liposomes monitored by fluorescence microscopy shows fast diffusing small aggregates and much slower diffusing larger aggregates, as expected from a solid supported fluid membrane [1, 11, 18].

5.4.6 Assembly of giant unilamellar liposomes (GUVs)

Most studies have been performed in solution with nanoparticles between 50 and 200 nm in size, but DNA controlled assembly is also highly versatile for larger particles or extended lipid bilayer surfaces. Attempts to transfer the concepts from nanoparticles to larger macroscopic lipid particles have led to studies on giant liposomes (~50 μm) using the same lipid-modified DNA strands employed in the liposome assembly on solid supported lipid bilayer surfaces (see Section 5.4.5) [32]. The almost perfectly spherical giant liposomes (Figure 5.4.8a) perform unaltered as long as lipid-modified DNA is only anchored to the surface. As soon as a complementary DNA strand is added to the solution, a major change is observed and the spherical shapes of the liposome group from Figure 5.4.8a changes very significantly. DNA controlled assembly, as seen for rapidly diffusing nanometer sized liposomes, is not possible for large GUVs and we observe, therefore, a large scale (square micrometer) 'glue-effect'. The interliposomal DNA linkers force larger areas to attach to the neighboring liposome, altering their shape in the process. The observed process is relatively fast given the nanomolar concentrations of DNA and the enormous size of the giant liposomes used (Jakobsen, U., Bagatolli L. and Vogel, S., unpublished results). Most of the interliposomal surface attachment occurs within a few minutes. However, the process is dependent on the sequence complementarity between the lipid-modified DNA and the complementary DNA. Introduction of a single mismatch in the complementary DNA strand slows the process considerably, down to approximately 45 min

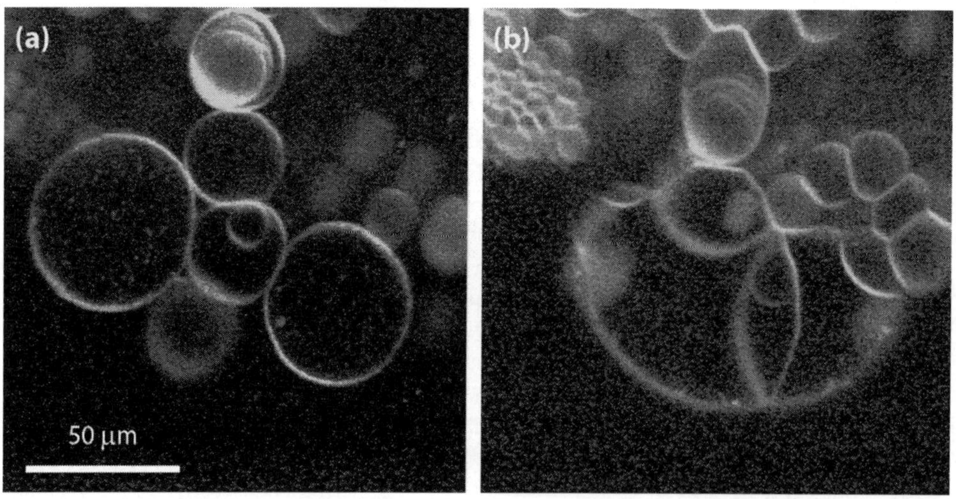

Figure 5.4.8 *Images from confocal microscopy studies showing giant unilamellar liposomes (POPC, 0.5% DiI dye, attached lipid modified DNA–5′-TTT**X**CACCAACTTCTTCCACA**X**TTT, 100 nM) followed over time, at time point 0 min (a) complementary DNA is added and the next snapshot is taken after 15 min (b). The observed red hue is due to the addition of 4′,6-diamidino-2-phenylindole (DAPI) at time point 5 min (however, intercalation of DAPI into the formed DNA duplex could not be distinguished from the DAPI staining of the membrane) (Jakobsen, U., Bagatolli L., and Vogel, S., unpublished results) (See color figure in color plate section)*

compared with <5 min for the fully complementary strand. The extended 'glue-effect' shows the generic power of non-covalent interactions based on hybridization of lipid membrane anchored DNA in the context of larger particles, such as GUVs.

5.4.7 Conclusions

The DNA controlled assembly of liposomes is applicable to particles spanning sizes in the range of three orders of magnitude from 50 to 50000 nm (GUVs). This is another striking example of the power of DNA in supramolecular chemistry. It also exemplifies the inherent DNA programmability and the enormous molecular flexibility of DNA-hybrid materials synthesized by automated solid-phase synthesis, in combination with a large toolbox of non-natural chemistries available for modification and enrichment of DNA with new functionalities in a complex setting, combining design on a nanometer scale (a DNA-hybrid material) and triggering of macroscopic effects (liposome aggregates of >50 μm). The combination of DNA with natural or synthetic lipids leads to a new class of unusual DNA-hybrid material. This material combines the DNA programmability through hybridization and the hydrophobic effect as a strong driving force for controlled membrane anchoring and subsequent controlled assembly of soft nanoparticles such as liposomes. Controlled adhesion as well as bilayer fusion in natural processes, such as endosome and exosome formation, is of fundamental interest for the understanding of key processes in molecular biology (e.g., cell–cell communication by exosomes or transport of encapsulated biological material within endosomes). DNA controlled assembly of model nanoparticles such as liposomes may shed light on the basic biophysical requirements for controlled fusion of natural lipid bilayers and at the same allows new applications within the field of drug delivery (e.g., artificial exosomes for delivery of oligonucleotides) and molecular diagnostics (e.g., PCR free detection of DNA). Future systematic studies with new DNA–lipid conjugates will allow fine-tuning of important properties and functionalities needed for assembly, hemifusion, and fusion of lipid bilayers and extend the range of useful DNA-hybrid materials even further.

Acknowledgements

This work has been supported by the Nucleic Acid Center funded by the Danish National Research Foundation and BioNEC, a centre of Excellence funded by THE VILLUM FOUNDATION for studies on biomolecular nanoscale engineering.

References

[1] Jakobsen, U., Simonsen, A. C., and Vogel, S., DNA-controlled assembly of soft nanoparticles, *J. Am. Chem. Soc.*, **130**, 10462–10463 (2008).

[2] McLaughlin C. K., Hamblin G. D., and Sleiman H. F., Supramolecular DNA assembly, *Chem. Soc. Rev.*, **40**, 5647–5656 (2011).

[3] Caruthers, M. H., Chemical synthesis of DNA and DNA analogs, *Acc. Chem. Res.*, **24**, 278–284 (1991).

[4] Campbell M. A. and Wengel J., Locked vs. unlocked nucleic acids (LNA vs. UNA): contrasting structures, work towards common therapeutic goals, *Chem. Soc. Rev.*, **40**, 5680–5689 (2011).

[5] Egholm, M., Buchardt, O., Christensen, L., Behrens, C., Freier, S. M., Driver, D. A., Berg, R. H., Kim, S. K., Norden, B., and Nielsen, P. E., PNA hybridizes to complementary oligonucleotides obeying the Watson–Crick hydrogen-bonding rules, *Nature*, **365**, 566–568 (1993).

[6] Mirkin, C. A., Letsinger, R. L., Mucic, R. C., and Storhoff, J. J., A DNA-based method for rationally assembling nanoparticles into macroscopic materials, *Nature*, **382**, 607–609 (1996).

[7] Li, Z., Jin, R. C., Mirkin, C. A., and Letsinger, R. L., Multiple thiol-anchor capped DNA-gold nanoparticle conjugates. *Nucleic Acids Res.*, **30**, 1558–1562 (2002).

[8] Taton, T. A., Mucic, R. C., Mirkin, C. A., and Letsinger, R. L., The DNA-mediated formation of supramolecular mono- and multilayered nanoparticle structures, *J. Am. Chem. Soc.*, **122**, 6305–6306 (2000).

[9] Pfeiffer, I. and Höök, F., Bivalent cholesterol-based coupling of oligonucletides to lipid membrane assemblies, *J. Am. Chem. Soc.*, **126**, 10224–10225 (2004).

[10] Zhang, G. R., Farooqui, F., Kinstler, O., Letsinger, R. L., Informational liposomes: Complexes derived from cholesteryl-conjugated oligonucleotides and liposomes, *Tetrahedron Lett.*, **37**, 6243–6246 (1996).

[11] Yoshina-Ishii, C. and Boxer, S. G., Arrays of mobile tethered vesicles on supported lipid bilayers, *J. Am. Chem. Soc.*, **125**, 3696–3697 (2003).

[12] Nayak S. and Lyon A. L., Soft nanotechnology with soft nanoparticles, *Angew. Chem., Int. Ed.*, **44**, 7686–7708 (2005).

[13] (a) Bangham, A. D. and Horne, R. W., Negative staining of phospholipids and their structural modification by surface-active agents as observed in the electron microscope, *J. Mol. Biol.*, **8**(5), 660–668 (1964); (b) Sessa, G. and Weissmann, G., Phospholipid spherules (liposomes) as a model for biological membranes, *J. Lipid Res.*, **9**(3), 310–318 (1968).

[14] Chandra, R. A., Douglas, E. S., Mathies, R. A., Bertozzi, C. R., and Francis, M. B., Programmable cell adhesion encoded by DNA hybridization, *Angew. Chem., Int. Ed.*, **45**, 896–901 (2006).

[15] Vogel S., Stembera K., Hennig L., Findeisen M., Giesa S., Welzel P., Tillier C., Bonhomme C., and Lampilas M., Moenomycin analogs with long-chain amine lipid parts from reductive amination, *Tetrahedron*, **57**, 4147–4160 (2001).

[16] Vogel S., Stembera K., Hennig L., Findeisen M., Giesa S., Welzel P., and Lampilas M., Moenomycin analogues with modified lipid side chains from indium-mediated Barbier-type reactions, *Tetrahedron*, **57**, 4139–4146 (2001).

[17] Jakobsen, U. and Vogel, S. DNA-controlled assembly of liposomes in diagnostics, *Methods Enzymol.*, **464**, 233–248 (2009).

[18] Benkoski, J. J. and Höök, F., Lateral mobility of tethered vesicle - DNA assemblies, *J. Phys. Chem. B*, **109**, 9773–9779 (2005).

[19] Stengel, G., Zahn, R., and Höök, F., DNA-induced programmable fusion of phospholipid vesicles, *J. Am. Chem. Soc.*, **129**, 9584–9585 (2007).

[20] Chan Y.-H. M., van Lengerich B., and Boxer S. G., Effects of linker sequences on vesicle fusion mediated by lipid-anchored DNA oligonucleotides, *Proc. Natl. Acad. Sci. U.S.A.*, **106**, 4, 979–984 (2009).

[21] MacKellar, C., Graham, D., Will, D. W., Burgess, S., and Brown T., Synthesis and physical properties of anti-HIV antisense oligonucleotides bearing terminal lipophilic groups, *Nucleic Acid Res.*, **20**, 3411–3417 (1992).

[22] Letsinger R. L., Zhang G. R., Sun D. K., Ikeuchi T., and Sarin P. S., Cholesteryl-conjugated oligonucleotides: synthesis, properties, and activity as inhibitors of replication of human immunodeficiency virus in cell culture, *Proc. Natl. Acad. Sci. U.S.A.*, **86**, 17, 6553–6556 (1989).

[23] Rohr, K. and Vogel, S., Polyaza crown ethers as non-nucleosidic building blocks in DNA conjugates: Synthesis and remarkable stabilization of dsDNA, *ChemBioChem*, **7**, 463–470 (2006).

[24] Jakobsen, U., Rohr, K., Madsen, R. K., and Vogel, S., Polyaza crown ether as non-nucleosidic building blocks in DNA-conjugates, *Nucleosides Nucleotides Nucl. Acids*, **26**, 1221–1224 (2007).

[25] Jakobsen, U., Rohr, K., and Vogel, S., Toward a catalytic site in DNA: Polyaza crown ether as non-nucleosidic building blocks in DNA conjugates, *Nucleosides Nucleotides Nucl. Acids*, **26**, 1419–1422 (2007).

[26] Pfeiffer, I. and Höök, F., Bivalent Cholesterol-based coupling of oligonucleotides to lipid membrane assemblies, *J. Am. Chem. Soc.*, **126**, 10224–10225 (2004).

[27] Mergny, J.-L. and Lacroix, L. Analysis of thermal melting curves, *Oligonucleotides*, **13**, 515–537 (2003).

[28] Jakobsen, U. and Vogel, S., Assembly of liposomes controlled by triple helix formation, *Bioconj. Chem.*, **24**, 1485–1495 (2013).

[29] Rosi, N. L. and Mirkin, C. A., Nanostructures in biodiagnostics, *Chem. Rev.*, **105**, 1547–1562 (2005).

[30] Jin, R. C., Wu, G. S., Li, Z., Mirkin, C. A., and Schatz, G. C., What controls the melting properties of DNA-linked gold nanoparticle assemblies? *J. Am. Chem. Soc.*, **125**, 1643–1654 (2003).

[31] Simonsen, A. C. and Bagatolli L. A., Structure of spin-coated lipid films and domain formation in supported membranes formed by hydration. *Langmuir*, **20**, 9720 (2004).

[32] Montes, L.-R., Alonso, A., Goñi, F. M.. and Bagatolli, L. A., Giant unilamellar vesicles electroformed from native membranes and organic lipid mixtures under physiological conditions. *Biophys. J.*, **93**(10), 3548–3554 (2007).

5.5

Metal Ions in Ribozymes and Riboswitches

Anastasia Musiari, Magdalena Rowinska-Zyrek, Sofia Gallo, and Roland K. O. Sigel

Department of Chemistry, University of Zurich, Zurich, Switzerland

5.5.1 Introduction

In the last decades, a vast number of studies have shown that RNA is much more than a passive, single-stranded carrier of genetic information. Similar to proteins, RNA can form sophisticated tertiary structures that have various "enzymatic" and regulatory functions. RNA enzymes, that is, ribozymes, use similar catalytic strategies as their protein counterparts. General catalytic strategies employed by ribozymes include metal ion, acid–base, and electrostatic catalysis or a combination thereof. Interestingly, RNA is able to carry out its functions as it has a much-reduced variety in "side chains" (i.e., nucleobases) and functional groups, as compared with proteins. Other RNAs are involved in regulatory processes, riboswitches being one example. These RNAs bind small metabolites with high affinity, changing their three-dimensional structure upon complex formation. This binding equilibrium is used in the regulation of protein expression of the downstream gene.

When single-stranded RNA folds into large and compact structures, its negatively charged phosphate groups are brought together in close proximity, but positive cations are required to compensate for the charge repulsion. Metal ions are thus needed in the first place for simple charge compensation, and hence for the folding of RNA into stable tertiary structures. In addition, being omnipresent they also participate in and influence directly the catalytic activity of all ribozymes. Consequently, also the function of riboswitches is crucially linked to metal ions, as only in their presence the active fold prone to bind the ligand with high affinity can be achieved. As such, metal ions are also key factors in the riboswitch-based regulation mechanism or in the efficient drug targeting of RNA.

DNA in Supramolecular Chemistry and Nanotechnology, First Edition.
Edited by Eugen Stulz and Guido H. Clever.

Within the scope of this chapter, we will provide a concise overview of the delicate relationship between biologically relevant metal ions or metal complexes and the two classes of naturally occurring functional RNAs: ribozymes or riboswitches.

The chapter is divided into three main parts. The first introduces the most important types of metal ion–RNA interactions, concentrating on the biologically most significant metal ions and discussing the various coordination modes and most frequent specific binding sites. Secondly, we will present the world of both small and large ribozymes giving special attention to the influence of cations on the structure, folding, and function. In the third part, we will explain how metal ions and metal complexes are involved in the correct structure formation and functioning of riboswitches.

5.5.2 Coordination chemistry of RNA

As a rough rule, monovalent ions are necessary for the formation of secondary structures, while divalent metal ions are needed to establish tertiary interactions within RNA (Figure 5.5.1). In small RNAs, monovalent ions may substitute divalent ions, but only at unphysiologically high, that is, molar, concentrations, which is concurrent with a major loss in activity [1]. The two biologically most relevant cations that interact with nucleic acids are K^+ and Mg^{2+}. Their concentrations in the cytosol are, at 1 mM (Mg^{2+}) and 140 mM (K^+), the highest among all monovalent and divalent metal ions [2]. The intracellular concentration of Na^+ does not usually exceed 12 mM, therefore its influence on RNA is much smaller compared with K^+ [3]. The most important role of K^+ ions is to compensate the negative charge of the phosphate backbone, which will, in turn, allow closer distances between the phosphate groups in the folded secondary structure of the RNA. Nevertheless, the binding mode of K^+ can be inner-sphere. For example, typical structures that are supported by K^+ ions are G-quadruplexes – specific four-stranded structures formed by guanine-rich nucleic acids (for recent reviews, see [4, 5]).

Divalent ions, apart from being crucial for the mediation and stabilization of tertiary interactions, are also directly involved in the biological activity of ribozymes under physiological conditions. The usual divalent key player, Mg^{2+}, is a hard metal ion that binds six ligands in an octahedral geometry and prefers hard ligands, such as oxygen, rather than nitrogen and sulfur atoms. Studying the specificity of Mg^{2+} binding to RNA is particularly challenging, since this ion is spectroscopically silent and difficult to distinguish from Na^+ and H_2O in crystal structures [2]. Moreover, its binding is quite weak, making it a difficult target for thermodynamic studies. For these reasons, it is often difficult to show its specific binding site and, frequently, in experimental studies, Mg^{2+} needs to be replaced by Zn^{2+}, Cd^{2+}, Pb^{2+} or lanthanides as mimics [2, 7, 8]. Other metal ions are also used as mimics, for example Ca^{2+} has similar ligand preferences as Mg^{2+}, but its complexes

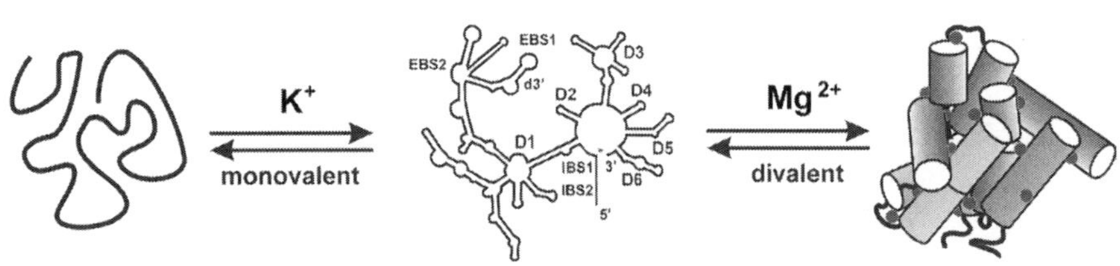

Figure 5.5.1 Schematic and simplified view of RNA folding. Monovalent ions are crucial for charge-screening and secondary-structure formation. Divalent ions (dark spheres) are needed to form tertiary interactions essential for a catalytically active architecture. Adapted from [6]. Reproduced with permission from [6]. Copyright © 2005 WILEY-VCH Verlag GmbH & Co. KGaA, Weinheim

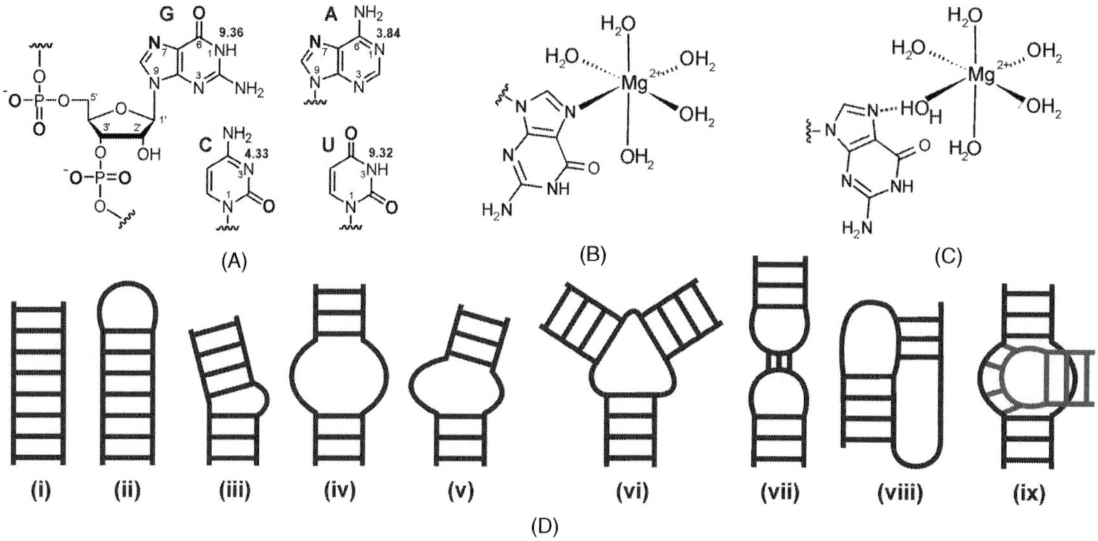

Figure 5.5.2 Metal ion coordination and RNA structures. (A) Most important metal ion coordinating atoms of RNA nucleobases (in bold) and important pK_a values (adapted from references [1, 17]. (B) Inner-sphere binding of Mg^{2+} to N7 of a guanosine. The remaining sites of the hexacoordinated Mg^{2+} are occupied by H_2O. (C) Outer-sphere coordination of Mg^{2+} to N7 of a guanosine mediated by a water molecule [18]. (D) Schematic representation of specific secondary and tertiary RNA motifs: (i) double helix, (ii) hairpin, (iii) bulge, (iv) symmetric internal loop, (v) asymmetric internal loop, (vi) 3-way junction, (vii) kissing loops, (viii) pseudoknot, and (ix) tetraloop receptor. Other tertiary structural motifs not described here, for simplicity, include the triple-stranded helix, 4-way and multihelical junctions as well as quadruplex structures

with monophosphate groups are 0.75 times weaker than those with Mg^{2+}. Nevertheless, even low concentrations of Ca^{2+} are able to substitute Mg^{2+} at specific places, inhibiting for example the cleavage of group II introns [9, 19]. Also, Mn^{2+} is a good example of a common mimic of inner-sphere bound Mg^{2+} [11].

Potential sites of metal ion–RNA interactions include the oxygen atoms of each phosphodiester linkage, the N7 and N3 atoms of purines, the N1 of adenine, the N3 of cytosine, as well as the carbonyl oxygens at C6 of guanosine, and at C2 and C4 of pyrimidines (Figure 5.5.2A) [12]. These sites are unprotonated and are therefore accessible for metal–ion binding at physiological pH. It is important to keep in mind that all nitrogen ligands mentioned are less basic than imidazole, and therefore have quite a low affinity towards most metal ions [13]. However, several exceptions to this rule exist – examples of nucleobase sites with pK_a values shifted towards neutrality are known and will be mentioned later in the text. The other sites such as guanine N1, uracil N3, and the exocyclic amino groups are usually protonated at physiological pH and are therefore unable to participate in metal coordination [14]. While considering potential metal binding sites in RNA, it is worth remembering, that RNA is not a simple single stranded random coil; metal ions have a variety of binding sites to choose from the RNA's specific secondary and tertiary RNA motifs (Figure 5.5.2D). The most important motifs are summarized in Figure 5.5.2D (see also recent reviews [15, 16]).

Metal ions can bind in at least four different ways to their potential binding sites [14, 19]:

i. *Nonspecific charge screening ions.* Before RNA folds into compact forms, the charge of the phosphate backbone, which will be brought into close proximity, must be neutralized. This is done by numerous weakly and unspecifically bound monovalent and divalent metal ions [20].

ii. *Electrostatically localized metal ions.* Higher order RNA structures frequently contain well-defined sites that have a higher affinity towards metal ions, often associated with an accumulation of negative electrostatic potential. Such binding is driven by electrostatics, and is not metal-specific, therefore various metal ions or even protonated organic polyamines can be bound [20].
iii. *Inner-sphere site-bound metal ions.* These types of binding sites are strongly metal-specific, and involve a direct interaction between a metal ion and a specific RNA binding site (Figure 5.5.2B). A common example of such binding is observed for K^+, which fills almost every coordination site through direct interactions with phosphoryl oxygens and/or base heteroatoms [1]. Inner-sphere bound Mg^{2+} are also commonly observed: the vast majority of site-bound Mg^{2+} ions are partially dehydrated with direct RNA contacts [21, 22].
iv. *Outer-sphere site-bound metal ions.* These interactions occur through a "shell" of water molecules that are coordinated to the metal ion and that mediate the metal–RNA interactions. Usually, an extensive network of hydrogen bonds is formed between RNA and the water molecules of the first and second hydration sphere of the metal ion [2]. Every H_2O molecule can form up to three hydrogen bonds to RNA sites (Figure 5.5.2C), reaching relatively high thermodynamic stabilities. Mg^{2+} is prone to form outer-sphere interactions because of its small ionic radius and high charge density, leading to a high hydration energy [14].

Several web tools are available for analyzing metal ion–RNA interactions in the PDB and other databases, such as MeRNA [23], MINAS [18], and SwS [24].

5.5.3 Ribozymes

5.5.3.1 Overview of the ribozyme world

The term ribozyme describes the composition of the main building block (ribonucleotides) and the activity (enzymatic) of this molecule. A ribozyme is thus an RNA or an RNA–protein complex capable of catalyzing specific biochemical reactions. Such functions, beyond the mere carriage of information, were first described at the end of the 1970s by Sidney Altman, who showed that the RNA moiety in RNase P (an RNA–protein complex) is responsible for catalytic activity of RNA cleavage [25]. A few years later, Thomas Cech described the catalytic activity of RNA in the absence of any protein [26]. Altman and Cech were rewarded for their groundbreaking discovery with the 1989 Nobel Prize.

Nowadays, ribozymes are found in the genomes of species from all the kingdoms of life and, based on their size, can roughly be divided into five classes, comprising small and large ribozymes. All of these employ metal ions within their structural elements, and most also have been shown to employ them directly for catalysis [14]. Not having a wide variety of building blocks (21 amino acids in proteins versus four nucleotides in RNA), and not having a residue with a pK_a close to neutral pH, RNAs learned to cope with this situation, taking advantage of those metal ions that are most ubiquitous in the cytoplasm.

Being both genetic materials and biological catalysts, ribozymes contributed to the RNA world hypothesis, which suggests that RNA may have been important in the evolution of prebiotic self-replicating systems. Although most ribozymes are quite rare in the cell, their roles are essential to life. The structure, function, and relations with metal ions of naturally occurring small and large ribozymes will now be discussed.

5.5.3.2 Small ribozymes

In the last 30 years, several small (40–200 nucleotide), naturally occurring ribozymes have been discovered. According to their secondary structure and cleavage site, they can be divided into five classes: the hammerhead [27], hairpin [28], hepatitis delta virus (HDV) [29], *glmS* [30] (Figure 5.5.3A), and Varkud satellite (VS,

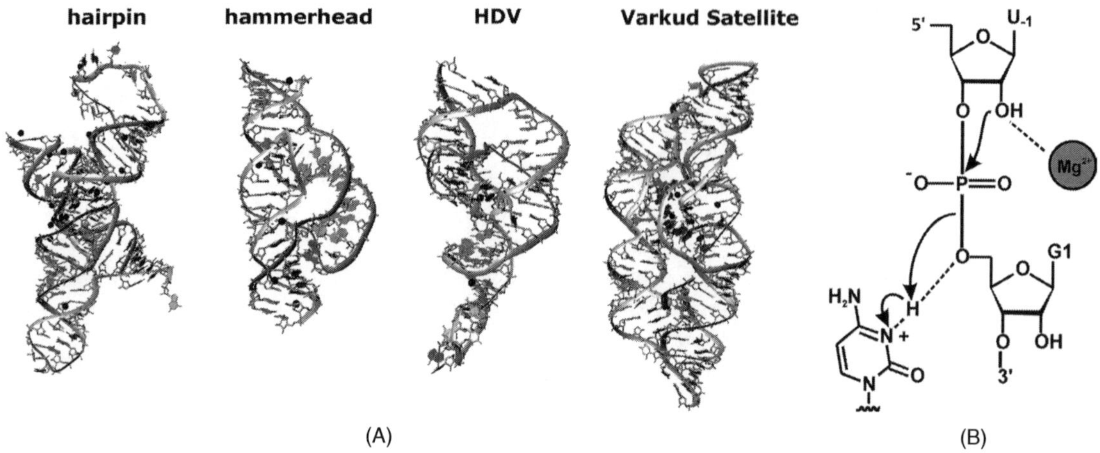

Figure 5.5.3 *Small ribozymes. (A) Three-dimensional structures and metal ion binding sites of the natural small ribozymes: the hairpin ribozyme in the presence of Ca^{2+} (PDB ID 1M5K), the hammerhead ribozyme with Mn^{2+} (2OEU), the HDV ribozyme with Mg^{2+} (1SJ3) and the glmS ribozyme with Mg^{2+} (2NZ4). Divalent M^{2+} ions are shown as black spheres. (B) Mg^{2+} stabilizes the pre-cleavage state of the HDV ribozyme. Panels have been drawn with MOLMOL [45]*

structure not yet published) ribozymes. All five form different types of higher order structures, in which separate regions form mostly a pseudoknot (Figure 5.5.2D). Nevertheless, these small ribozymes share a common general mechanism of self-excision.

Until recently, small ribozymes were known to occur only in the genomes of viruses, prokaryotes, and lower eukaryotes: The hairpin, hammerhead, and the HDV ribozymes play an important role in virus and viroid genome replication but are also found in transcripts of lower eukaryotes. The *glmS* ribozyme is found in *Bacillus subitilis* and the VS ribozyme is located in mitochondrial transcripts of the mould *Neurospora crassa* [31]. Only recently, small hammerhead [32] and HDV [33] related ribozymes were found in several mammalian genomes: the CoTC motif in the 3′-untranslated region of the beta-globin gene [34]; a discontinuous hammerhead ribozyme in the 3′-UTR of C-type lectin type 2 genes [32]; a hammerhead ribozyme in an intron of a tumor suppressor gene [34]; and the CPEB3 ribozyme in an intron of the *cpeb3* gene, which encodes a cytoplasmic polyadenylation element binding protein that is involved in the elongation of the polyadenine tail of messenger RNA, mediation of germ cell development, synaptic plasticity, learning, and memory [35].

To the best of our knowledge, all known natural small ribozymes catalyze the same internal phosphodiester isomerization reaction, yielding a 2′,3′-cyclic phosphate and a 5′-hydroxyl terminus (Figure 5.5.3B). The first step of this reaction is the activation of the 2′-OH nucleophile by a general base, followed by an attack of this "oxygen anion" on the adjacent phosphate, and a protonation of the 5′-oxygen leaving group by a general acid [19, 36]. Metal ions are unequivocally linked to conformational changes and catalysis within the hairpin, HDV, and VS ribozymes. In the hairpin ribozyme, catalysis requires docking of two loops from separate helical stems [37–39]. An analogous docking is observed in the VS ribozyme, and, in this case, a crucial change in the base-pairing pattern is needed before catalysis [40]. Both reactions require the presence of mono- and divalent metal ions, and usually, during docking interactions, Na^+ and Mg^{2+} ions are being taken up [41]. For catalysis of the HDV ribozyme, a Mg^{2+} ion needs to be present in the C75 active site, and it dissociates directly after self-cleavage [33, 42–44].

All ribozymes require the presence of metal ions for proper folding. The concentrations of mono-, di-, and trivalent metal ions required are specific to the individual ribozyme. As an example, under *in vitro* conditions, the hairpin ribozyme requires 1 M monovalent salts, 10 mM Mg^{2+} or 0.5 mM $Co(NH_3)_6^{3+}$ to achieve similar folding [46]. A recent study on the HDV ribozyme suggests that monovalent metal ions might help the ribozyme to find the correct minimum-energy native structure by destabilizing alternative misfolds, while divalent and trivalent metal ions are a better charge compensation in specific negative pockets [36].

In general, the function of metal ions in small ribozymes, is diverse [14]. As far as catalysis is concerned, Mg^{2+} ions have a role in: (i) organizing the active site, (ii) organizing a network of water molecules that facilitates proton transfer during the cleavage reaction, (iii) inducing the shift of pK_a values of nucleobases towards neutrality (a significant shift of pK_a values of adenine in the case of the hairpin ribozyme [47] or cytosine in the case of the HDV ribozyme [43]), or (iv) direct participation in catalysis, which would include the direct activation of the 2′-OH nucleophile, electrostatic stabilization of the newly formed transition state, and the protonation of the 5′-oxygen leaving group [41, 48, 49]. However, the last roles, usually employed by the large ribozymes, have not yet been directly demonstrated to be used by small ribozymes.

The type of coordination is crucial for catalysis. All divalent metal ions found in crystal structures of the hammerhead, hairpin, *glmS*, and HDV ribozymes remain at least partially hydrated [41]. $Co(NH_3)_6^{3+}$ is a popular mimic of Mg^{2+} for complete outer-sphere binding because its ammonia ligands are exchange-inert [50]. In the case of the hammerhead ribozyme [51], the presence of only $Co(NH_3)_6^{3+}$ impedes catalysis, proving that inner-sphere contacts are necessary for its biological activity. This agrees well with the observation that the nature of the M^{2+} ion has a large effect on the catalytic rate, Mn^{2+} being the most proficient M^{2+} ion [52, 53]. Thus the trend observed proves a direct phosphate interaction of the catalytically crucial ion [1]. In contrast, outer-sphere coordination is sufficient for catalytic activity of the hairpin, *glmS* and HDV ribozymes [54, 55].

Alongside natural ribozymes, selected engineered metal-dependent ribozymes have also been developed *in vitro*, such as the lead-dependent ribozyme (a small self-cleaving ribozyme that catalyzes its cleavage in the presence of Pb^{2+}) [56], the Diels–Alder ribozyme (able to catalyze reactions such as the acylation, alkylation, and formation of glycosidic bonds) [57], or allosteric ribozymes that respond to specific divalent metal ions [58]. For more details, please refer to recent review papers [46, 59].

5.5.3.3 Large ribozymes

Large ribozymes consist either solely of RNA (group I and II introns, bacterial RNase P) or are ribonucleoprotein complexes (the spliceosome, ribosome, and RNase P). All have in common a strict requirement for Mg^{2+} ions to achieve their structure and function [19, 49, 60–63]. We will now briefly mention all five classes and later focus on group I and II introns, two ribozyme classes composed only of RNA.

The well-known **ribosome** [64], a large ribonucleoprotein with an RNA-based active site, catalyzes the formation of peptide bonds during translation. The crystal structure of the *Haloarcula marismortui* large ribosomal subunit showed 88 monovalent and 116 Mg^{2+} ions bound mostly to the rRNA [22]. Nine out of the 116 Mg^{2+} ions are fully outer-sphere bound, while the remaining 107 show at least one inner-sphere coordination, mostly to a non-bridging phosphate oxygen. Only one Mg^{2+} ion seems to be fully dehydrated [21, 22]. A recent structural and computational analysis of the *Thermus thermophilus* large ribosomal subunit shows that the role of RNA-bound Mg^{2+} goes far beyond the neutralization of electrostatic repulsion and direct coordination of RNA functional groups. Mg^{2+} takes part in localizing and polarizing water molecules and in increasing the specificity of the interaction with the partner proteins [65]. The general role of metal ions in the ribosome is their contribution to structure and substrate recognition. A direct participation of metal ions in peptide bound formation has not yet been shown.

The **spliceosome** is a massive complex of five RNAs and numerous proteins that catalyzes the removal of non-self-splicing introns from precursor mRNA in eukaryotes. Its catalytic core consists of three small nuclear RNAs (U2, U5, and U6) involved in substrate positioning and catalysis [61, 66, 67]. The spliceosome is evolutionally closely related to group II introns, having a similar splicing mechanism and a reminiscent structure of the active site. Both ribozymes utilize two Mg^{2+} ions to stabilize the transition state [68, 69]. The most likely candidates for Mg^{2+} coordination are two regions (termed U2 and U6) that pair with each other and with the pre-mRNA substrate [70, 71]. Structural [72, 73] and NMR studies [74, 75] showed that a specific Mg^{2+} coordinating residue is the conserved U80 in U6.

The catalytic mechanism of the eukaryotic spliceosome is closely related to that of other large ribozymes – group I and group II introns and ribonuclease P. All four groups employ a classical S_N2 mechanism to cleave and ligate their phosphodiester backbones in order to be excised from the remaining exons [76, 77]. The cleavage results in the formation of a 5′-phosphate and a 3′-hydroxyl group and is based on the attack of an exogenous nucleophile. Cleavage further requires an appropriately oriented active site and an activated scissile phosphate. To fulfill both requirements, the presence of M^{2+} in close proximity to the active site is necessary [61].

RNase P is a ribonucleoprotein that catalyzes the removal of the 5′-leader sequence from precursor tRNA [78]. Its catalytic core consists only of RNA and bacterial RNase P has also been shown to function in the absence of its protein unit [62].

Several crystal structures of the protein and/or RNA subunit of RNase P exist [79–82], but to date, no high resolution structural data are available. A 3.3 Å resolution structure of RNase P from *Staphylococcus aureus* is known [80]. Osmium(III) hexammine, used for phasing, was found in several regions of this enzyme. Two of the complexes bind in direct proximity to the active site, in the so-called P4 region. Also, in *Bacillus subtilis* this region was shown to bind Zn^{2+} and cobalt(III) hexamine, which were used to mimic Mg^{2+} binding [83].

Group I and II introns are distinct classes of self-splicing ribozyme introns, intervening RNA sequences that need to be removed from the primary transcripts to obtain the functional RNA. Both introns catalyze their own excision from pre-mRNA, tRNA, and rRNA precursors and subsequently ligate the flanking exons to yield mature RNAs [60, 63, 84].

Group I introns have an average length of 400 nt (nucleotide) and share a characteristic secondary structure, which consists of nine paired regions folded into two domains [85]. The first group I ribozyme to be discovered [26] and shown to directly require a Mg^{2+} ion to promote catalysis [86] was the group I intron ribozyme from *Tetrahymena thermophila*e. Since then, more than 2000 other sequences have been identified. A large amount of interest has been given to the folding pathway of group I introns; for example, the previously mentioned *Tetrahymena* folds very slowly from an unfolded state in the presence of only Mg^{2+} ions, at low ionic strength, but, if pre-folded in the presence of Na^+, then Mg^{2+} mediated refolding proceeds very fast [87].

The stabilization of this ribozyme's tertiary structure is influenced by the charge density of the bound cation – the higher the charge density, the better the stabilization [88]. *Azoarcus* group I intron folds in a two-step reaction [89, 90], while the correct folding of the *Tetrahymena* ribozyme is a three-state reaction, in which the second step of folding leads to metastable misfolded intermediates [90, 91]. This middle step is affected by both charge and size of the metal ions present: correct folding is more likely to occur in the presence of Mg^{2+} and Ca^{2+} rather than Sr^{2+} and Ba^{2+}.

Crystal structures of the *Azoarcus* group I intron reveal 18 specifically bound K^+ (via inner-sphere contacts) and Mg^{2+} ions (both, inner- and outer-sphere-bound), five of which are located in the center of the ribozyme, within 12 Å from the scissile phosphate [92, 93]. Two of those five are Mg^{2+} ions that form five inner-sphere contacts and are crucial for catalytic activity, since they activate the reaction, stabilize the transition state, and organize the structure of the active site. The remaining metal ions are responsible for tertiary contacts within the intron as well as contribute to the neutralization of the close-by phosphate backbone-negative charges.

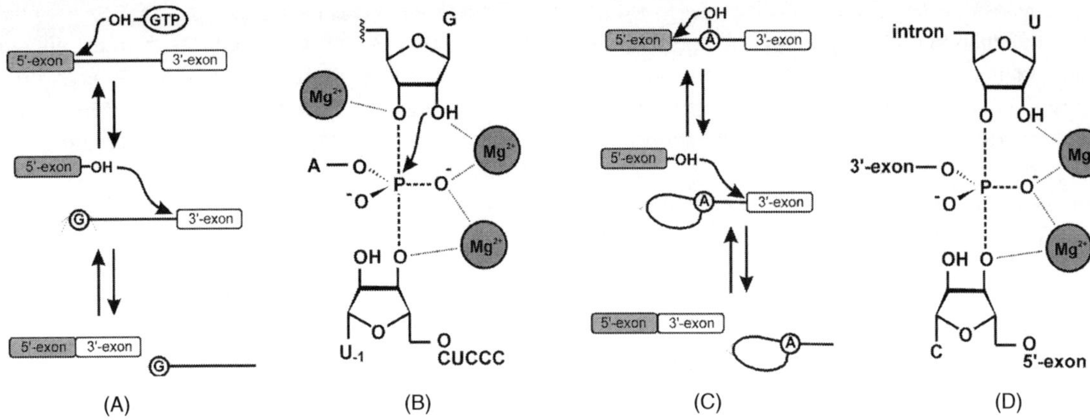

Figure 5.5.4 *Splicing pathways of group I and II introns. (A) In the first step of group I intron splicing, the 3′-OH of an exogenous GTP (or guanosine) attacks the backbone phosphate at the 5′-splice site, thus cleaving off the 5′-exon. In the second step, the 3′-OH of the 5′-exon attacks the 3′-splice site. (B) Transition-state model of the first step of splicing of group I introns, based on the* Tetrahymena *ribozyme [60, 87]. (C) In group II intron splicing, first the 2′-OH of the intronic branch point adenosine attacks the 5′-splice site. The 5′-exon is released and the intron forms a lariat structure. The second step joins the exons proceeding as in (A), but releasing the intron as a lariat. All steps of group I and II intron splicing are reversible. (D) Transition-state model of the second step of splicing of group II introns, based on a crystal structure of the* Oceanobacillus iheyensis *intron [102]. Adapted from ref [60].*

Group I introns require an exogenous guanosine to carry out the two ester-transfer reactions [94]. Upon docking to the active site, the guanosines nucleophilic 3′-OH attacks the phosphodiester bond at the 5′-splice site resulting in a free upstream exon and the intron elongated by a G. In the second step, the 3′-terminal intronic G now occupies the G-binding site and the free 3′-OH group of the upstream exon attacks the 3′-splice site, ligating the exons and releasing the linear intron, extended by a 5′-G (Figure 5.5.4A) (for details, see e.g., references [95–97].

Group I introns employ a two metal ion mechanism with a third metal ion present in the active site: one activates the exogenous guanine, the second neutralizes the charge of the leaving group, and the third stabilizes the geometry of the leaving group (Figure 5.5.4B) [87]. The natural cofactor Mg^{2+} can be substituted by Mn^{2+} in the case of *Tetrahymena* group I intron, while in the presence of Ca^{2+}, the ribozyme is able to fold properly, but its catalytical activity is blocked [98]. However, both the *Thetrahymena* [99, 100] and *Azoarcus* group I ribozymes [101] can be engineered in a way that they specifically require Ca^{2+} ions for splicing.

Group II intron ribozymes are significantly larger (600–2500 nt long) than group I introns with a conserved secondary structure with six domains arranged around a central wheel (Figure 5.5.1). These RNAs are best known to catalyze their self-excision, but they also perform a variety of other reactions including RNA and DNA hydrolysis as well as intron mobility into RNA and DNA substrates. First identified over 30 years ago [26], the primary sequence of group II introns is not well conserved, however the secondary and tertiary contacts are [103]. The domains are independent folding units that require the presence of an appropriate amount of metal ions to fold into an active tertiary structure [100]. The folding of group II introns is strongly dependent on metal ions and the necessary concentration of ions varies between group II introns from various species: it can reach values up to 2 M K^+ [104] and between 0.1 and 100 mM Mg^{2+} [6, 105–107].

Out of the six domains, only domains 1 (D1) and 5 (D5) are indispensible. D2 stabilizes the ribozyme fold, D3 increases the rate of catalysis, D4 can contain an open reading frame that encodes a maturase protein, and D6 contains the conserved bulged adenosine that serves as a nucleophile during splicing (the so-called branch point adenosine) [103]. One of the best-described group II intron constructs is D135, a shortened version of the yeast mitochondrial Sc.ai5γ intron, composed only of domains D1, D3, and D5 [106]. D135 folds without kinetic traps in the presence of Mg^{2+} [108], the folding being rapid at 100 mM Mg^{2+} but very slow at physiological Mg^{2+} concentration [109, 110]. Recent single molecule FRET (Förster Resonance Energy Transfer) studies [111] described details of D135 folding, confirming the absence of kinetic traps and identifying the presence of a particularly stable near-native state. The presence of Ca^{2+} perturbs the global architecture and leads to two distinct subpopulations that do not interchange [9]. The rate-limiting step of group II intron folding is the compaction of D1, followed by rapid docking of other domains to this scaffold. Recently, within D1, specific binding of Mg^{2+} to a central region was identified, supposedly the first micro-step of D1 folding [108, 112].

The specificity of Mg^{2+} is also obvious from the catalytic activity: Mg^{2+} can partially be substituted with Mn^{2+} [113], but other metal ions seem to have major inhibitory effects: only 50% activity of Sc.ai5γ was reached in the presence of Ca^{2+} and 100-fold excess of Mg^{2+} [10]. The presence of 10 mM Ca^{2+}, Mn^{2+}, Ni^{2+}, Zn^{2+}, Cd^{2+}, Pb^{2+} or $[Co(NH_3)_6]^{3+}$, with 90 mM Mg^{2+} again causes a large decrease in splicing rate and in some cases RNA degradation [10].

Solution structures and accompanying studies of several domains and substructures of the Sc.ai5γ intron reveal multiple specific metal ion binding sites: the catalytic core domain 5 harbors Mg^{2+} ions at the bulge, the catalytic triad and the tetraloop; the branch domain 6 binds at least four Mg^{2+} ions [114], all in close proximity to the highly conserved branch adenosine [115]; the so-called κ-ζ region in D1, a complex three-way junction being involved in the first as well as the last step of folding, is only structured upon Mg^{2+} binding; and the 5′-splice site recognition complex is only stable in the presence of Mg^{2+} ions [116, 117], even more when a DNA substrate is bound [118]. Intrinsic affinity constants for M^{2+} binding could be obtained for D6 [119, 120] and estimated for the other domains. Single molecule FRET studies revealed that splice site formation is highly sensitive to the M^{2+}, the on and off rates of RNA binding being strongly influenced by the type of M^{2+} [121, 122].

Recently, the first crystal structures of five constructs of a *Oceanobacillus iheyensis* group II intron were published [123]. The 17 structures of those five constructs [107, 124–127] visualize different stages of splicing, that is, the pre-catalytic state [102, 127], the conformational rearrangements between the first and second splicing steps [101], the post-catalytic state [106], the ligand-free, linear form of the intron [102], and the retrotransposable form of the intron after target substrate binding [102]. For a detailed review see reference [123]. Numerous metal ions including Mg^{2+} and K^+, Rb^+, Tl^+, Cs^+, NH_4^+, Na^+ and Ba^{2+} and K^+ have been identified and described in detail [102]. Most importantly, these structures confirm earlier biochemical studies on direct M^{2+} involvement in catalysis [68]. In the first step of splicing, the nucleophilic adenosine 2′-OH within D6 attacks the scissile phosphate; two Mg^{2+} ions stabilize the transition state – one activates the nucleophile and the other stabilizes the leaving group. In the second step, the 5′-phosphate and 3′-OH termini of the exons are ligated and the intron simultaneously released as a lariat (Figure 5.5.4, C and D). Both steps of this reaction are highly reversible – the free lariat group II introns can catalytically insert themselves into the same or other regions of RNA and DNA [84]. The metal ions involved in such a two-metal ion mechanism have been putatively identified in the crystal structures [126]. Generally, group II introns offer a wealth of information on the role of metal ions in folding, structure, and catalysis, due to their size and complexity. For more detailed overviews on group II introns, please refer to references [1, 60, 84, 123].

5.5.4 Riboswitches

5.5.4.1 Overview of the riboswitch world

Little more than ten years have passed since a new mechanism of gene regulation was discovered in riboswitches [128–130]. These conserved RNA sequences are able to control gene expression through a highly specific interaction with a particular metabolite [131, 132]. They are involved in gene regulation mainly in

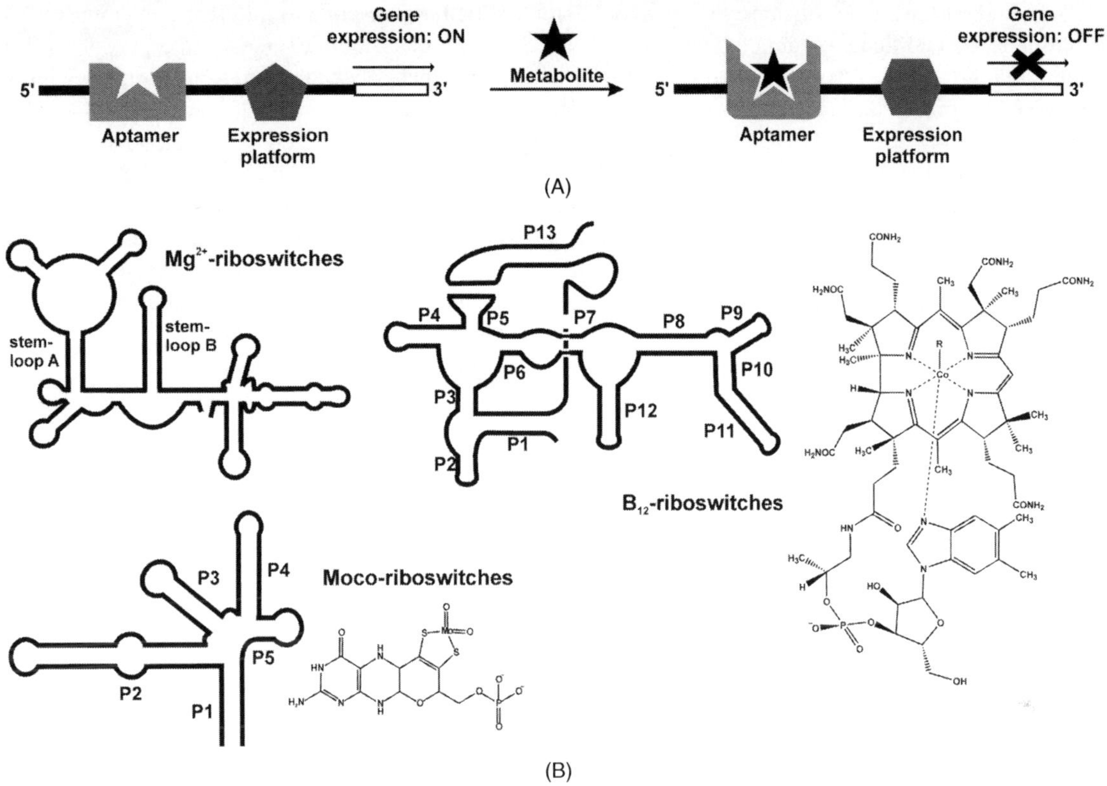

Figure 5.5.5 *Schematic representation of (A) the general mechanism of the riboswitch-based gene-regulation and (B) of the secondary structures of the riboswitches and their ligands described in this chapter. (A) Binding of the metabolite (shown as a star) to the aptamer region of the riboswitch provokes not only a structural rearrangement of the aptamer but also of the downstream situated expression platform, finally leading to an altered gene expression. (B) The riboswitches shown in this figure are the Mg^{2+}-binding* mgtA *riboswitch of* S. enterica *in its bound form [139], the coenzyme B_{12}-binding* btuB *riboswitch from* E. coli *in its bound form [140], 171 and the general consensus secondary structure of the common molybdenum cofactor responding riboswitches in its unbound form [110]*

bacteria, but also in plants and fungi [133]. They are mostly found in the 5′-untranslated region (UTR) of bacterial mRNAs where they modulate gene expression either at the transcriptional or translational level by a metabolite-dependent rearrangement [134, 135], through an intron splicing mechanism [136], a self-cleavage activation [30], or due to *trans* regulation [137]. As in the case of ribozymes, in riboswitches also, metal ions, in particular Mg^{2+}, greatly influence the function and folding.

In general, riboswitches are composed of two domains, the aptamer and the expression platform [138]. The aptamer is a highly conserved region and is involved in the recognition and interaction with the ligand. The second domain varies more across species and contains gene control elements. Ligand binding causes a structural rearrangement in the aptamer region that leads to an altered expression of the downstream genes that usually encode proteins involved in the biosynthetic pathway or in the transport of the ligand itself (Figure 5.5.5A). For example, in *Escherichia coli* an adenosyl cobalamin (B_{12}) sensing riboswitch regulates the expression of the *btuB* gene that encodes an outer membrane receptor responsible for cobalamin's transport into the periplasmic space [129].

To date, about 20 classes of riboswitches have been identified that respond to different metabolites, such as purine-related compounds, amino acids, coenzymes, Mg^{2+} cations, fluoride anions and a phosphorilated amino sugar [142, 143]. Moreover, many so-called *orphan* riboswitches exist that are RNA sequences found in bacterial genomes that have peculiar riboswitch features, but whose metabolite is still unknown [144–146]. Since riboswitches are mostly found in bacteria and they are able to control gene expression by binding a particular metabolite with high affinity and specificity, they are an ideal target for antimicrobial drugs [147, 148]. As seen for ribozymes, riboswitches also require metal cations to neutralize the strong electrostatic repulsions that would occur during folding. However, metals take part in the riboswitch-based gene regulation mechanism not only for folding but in many other different modes. We will next describe in detail how metal ions can influence the riboswitch folding process and the ligand recognition by the aptamer. Moreover, significant attention is given to those riboswitches classes whose metabolite is or contains a metal ion. These are the Mg^{2+}-sensing riboswitches, the B_{12}-riboswitches, and the MoCo/TuCo-riboswitches.

5.5.4.2 Metal ions assisting riboswitch folding and ligand recognition

In general, metal ions are also essential in riboswitch processes, mainly due to the electrostatic interactions with the polycharged backbone. Thus various biophysical and biochemical techniques have been utilized to elucidate the influence of metal ions, in particular Mg^{2+}, on the riboswitch conformational rearrangement of different riboswitch classes, such as the riboswitches sensing *S*-adenosylmethionine (SAM) [149–153], thiamine pyrophosphate (TPP) [154, 155], coenzyme B_{12} [156], and cyclic diguanylate (c-di-GMP) [157].

The SAM riboswitch class has been studied the most, probably due to the vast occurrence in the bacterial kingdom and due to its further classification into five different SAM riboswitch groups, each able to bind the metabolite in a different way [158]. Single molecule FRET studies performed with the SAM-I riboswitch in the absence/presence of both Mg^{2+} and the ligand highlight the presence of three RNA conformational states [153]. The low FRET state, predominant in the absence of both Mg^{2+} and SAM, corresponds to a relatively open conformation. The presence of Mg^{2+} induces the formation of a new conformational state, characterized by helical stacking and a pseudoknot formation. Finally, a third population is observed at high Mg^{2+} concentrations, which is further strongly stabilized upon SAM addition. This last conformation corresponds to the native folded state, characterized by additional helical stacking. Solid experimental evidence demonstrates how both Mg^{2+} and SAM are essential for the global gene regulation process, with the metal ion involved in the pre-organization of the ligand binding site and in the stabilization of the final native state [153]. Recently, strong cooperativity of Mg^{2+} and SAM has been shown in driving the SAM-I riboswitch to the folded state and in stabilizing it [84]. Further FRET studies were performed with the SAM-II class riboswitch, which is known to modulate the gene expression upon a metabolite-induced pseudoknot formation [144, 159]. These experiments focused on the dynamics of the pseudoknot formation and revealed the presence of three FRET states [150], comparable to those observed for the SAM-I riboswitch [153], however with Mg^{2+} allowing the formation of more compact species. Furthermore, SAXS (Small Angle X-ray Scattering) data suggest that both Mg^{2+} and SAM alone are able to pre-organize the RNA, even if both are necessary to lead the riboswitch to the native state [149]. NMR analyses have also been performed in order to elucidate the features of the secondary and tertiary interactions occurring in the presence of Mg^{2+} and the ligand, showing a hairpin undergoing a reorganization in the presence of SAM to form the pseudoknot determinant for the translation termination [149, 150].

Similar structural investigations have been conducted on the TPP riboswitch and on the c-di-GMP riboswitch [155, 157]. In the TPP riboswitch the ligand is bound by two helix "arms", each recognizing a different functional moiety of the ligand [160]. The dynamics of these structural components have been investigated through smFRET studies, showing that one of the helices is predominantly in a folded conformation in the presence of Mg^{2+} alone, even if TPP provides further stabilization [155]. However, the ligand and the metal ion are both required to reach the native folded state.

As far as the c-di-GMP riboswitch is concerned, four different structural folds were identified by FRET, with the stable bound population requiring the presence of both Mg^{2+} and ligand [157]. The riboswitch is not able to reach the final bound conformation in the absence of Mg^{2+}, as demonstrated through experiments performed in the presence of only monovalent ions and a saturating amount of ligand [157].

The impact of metal ions on the folding of another important riboswitch, the B_{12}-sensing *btuB* riboswitch, has recently been investigated through biochemical assays, such as in-line probing [156]. Mg^{2+} has been proven to be determinant to fold the *btuB* riboswitch in the correct binding competent conformation that is essential to bind coenzyme B_{12} [156]. Since some regions of this RNA show different K_D values towards Mg^{2+}, a possible stepwise folding process was also proposed. Metal cations mediate the ligand recognition by riboswitches, especially when metabolites carry negatively charged moieties, such as phosphate or carboxylate groups. In the TPP riboswitch, Mg^{2+} ions are not only the determinant for correct folding, but they also mediate the ligand recognition. The TPP riboswitch recognizes different moieties of its ligand by two distinct subdomains, and the pyrophosphate group is situated in a pocket where one or more Mg^{2+} ions bind the ligand via inner- or outer-sphere coordination [161–163]. Co-crystallization experiments show that Mg^{2+} can be replaced by Mn^{2+}, Ca^{2+} or Ba^{2+} to coordinate the pyrophosphate of TPP [162].

Similarly, in the flavin mononucletide (FMN) riboswitch, a non-bridging oxygen of the phosphate group of the ligand is directly coordinated to a Mg^{2+} ion [164, 165]. In this case, Mg^{2+} can successfully be substituted by Ca^{2+}, Ba^{2+} or Mn^{2+}, but not by Cs^{+} [164]. Riboflavin, a similar molecule lacking the phosphate group, binds the FMN riboswitch with a 1000-fold lower affinity, indicating the essential role of the phosphate group for the ligand recognition [164]. However, not only Mg^{2+} ions are able to significantly assist ligand recognition. For example it was found that the affinity of the lysine riboswitch to its ligand is 50–100 times weaker when Na^{+} or Mg^{2+} are present instead of K^{+} [166, 167], as lysine is located in the binding site with a K^{+} ion directly bound to its carboxylate group [166].

The importance of Mg^{2+} and other metal ions is certainly evident for the correct function and folding of riboswitches. However, what has to be kept in mind is that the real intracellular environment differs from the conditions commonly used for *in vitro* studies, as shown by SHAPE experiments on the adenine riboswitch [168]. The structural conformation that RNA adopts *in vivo* is different from that *in vitro*, indicating a great impact of the cellular environment on the compaction of the riboswitches [168]. Although structural *in vitro* studies provide great insights in the regulation mechanism of these RNAs, it is important to remember that a different behavior could occur *in vivo*.

5.5.4.3 Riboswitch ligands containing a metal

In the introductory part on riboswitches, we already listed the great variety of metabolites that these regulatory RNA sequences are able to sense. We will now focus on those riboswitches, which depend either on a metal ion (Mg^{2+}), or require a metal-containing cofactor as a natural ligand (coenzyme B_{12} with its cobalt center, the molybdenum cofactor (MoCo) and the tungsten cofactor (TuCo, also abbreviated as WCo)).

5.5.4.3.1 Mg^{2+}-sensing riboswitches

Earlier we extensively discussed the key role of Mg^{2+} ions for both riboswitch folding and assistance in ligand recognition. However, the importance of Mg^{2+} is not limited to these two cases; its role is even more pronounced in two different types of riboswitches, which have been found to be directly regulated by this metal ion.

In 2006, the first example of a metal-sensing riboswitch was found in *Salmonella enterica,* regulating the expression of the *mgtA* gene, which is furthermore regulated by the PhoP protein (Figure 5.5.5B) [169]. Recently, it has been proven that the *mgtA* riboswitch follows a new alternative mechanism of gene regulation, a Rho-dependent transcription termination [139]. Biochemical assays demonstrated that high Mg^{2+} concentrations allow the *mgtA* riboswitch to reach a structural conformation, which favors the binding of Rho, inducing premature transcription termination [170]. Moreover, NMR studies on the anti-terminator loop of the *mgtA* riboswitch of *Yersinia enterocolitica* highlighted the presence of specific Mg^{2+} binding sites that are probably widespread in the whole riboswitch. Interestingly, a Mg^{2+} binding site described is located in an AU-rich region [171].

Among the various existing orphan riboswitch classes, one was designated as an Mg^{2+}-sensing type and named the M-box. The first experiments with the M-box riboswitch revealed the ability of Mg^{2+} to induce transcription termination. However, differently from the *mgtA* riboswitch, the M-box carries an intrinsic transcription terminator sequence. The formation of this terminator is primarily responsible for the regulatory mechanism [172]. The X-ray crystal structure of the M-box riboswitch from *mgtE* highlighted the presence of six Mg^{2+} ions exhibiting many inner- and outer-sphere interactions in a specific recognition site of the aptamer [172]. Despite the presence of further four K^+ ions in this crystal structure, it has been demonstrated that the M-box RNA is able to reach the correct folded conformation even in the absence of monovalent ions, while the presence of Mg^{2+} is strictly required [172].

Riboswitches usually bind only a single metabolite to undergo the structural rearrangement required for gene regulation. The presence of more Mg^{2+} ions bound to the M-box represents a peculiar feature of this riboswitch class and could be associated with a cooperative influence exercised by the sensed metabolite. To further investigate the characteristics of these binding pockets the crystal structure of the M-box in the presence of Mn^{2+} ions has been solved (it was demonstrated that the presence of Mn^{2+} does not influence the overall RNA folding) [173]. This new crystal structure shows the presence of eight binding sites, six of them being the same as those observed in the Mg^{2+}-structure and the other two having already been suggested to be Mg^{2+}-binding sites by previous biochemical experiments [173].

The *mgtA* and the M-box riboswitches are well-defined examples of metal-sensing RNA. If we make a comparison with the "protein world" and remember that proteins are able to respond to different metal ions, it should be considered that the existence of other riboswitch classes recognizing different metal ions is highly probable.

5.5.4.3.2 B_{12}-riboswitches

The B_{12}-riboswitch class is rather well investigated and in recent years, numerous biochemical studies have been conducted in order to elucidate the features of these mRNAs in different bacterial families, such as *Escherichia coli, Thermotoga lettingae, Desulfitobacterium hafniense* [140, 156, 174–177]. However, it was not until 2012 that the first X-ray structures of these riboswitches were published. The B_{12}-riboswitches are the largest and most complex known riboswitches and coenzyme B_{12} (AdoCbl) is the most complex of all known vitamins and one of the few organometallic compounds present in nature (Figure 5.5.5B).

AdoCbl is constituted of a cobalt center surrounded by the large corrin ring. The coordination sphere is completed by two apical ligands, an adenosyl group and a dimethylbenzimidazole moiety. Cobalamins are essential cofactors for enzymatic reactions in all living organism [178]. The B_{12}-riboswitches are abundant across bacterial classes and control gene expression of proteins involved in cobalamin biosynthesis and well as transport at the level of both transcription termination and translation initiation [129]. In 2012, two papers were published providing detailed molecular insight into the cobalamin–riboswitch system, thanks to the high-resolution X-ray structures obtained [179, 180]. Batey and coworkers introduced an interesting classification of the cobalamin-binding riboswitches in two different groups, based on the effective B_{12}-derivative bound to the RNA. Although the natural ligand associated to the B_{12}-riboswitches is traditionally AdoCbl, biochemical assays proved that some riboswitches interact with methylcobalamin (MeCbl) and aquocobalamin (AqCbl) with a 500-fold higher affinity [180].

Looking at the secondary structures of the riboswitches tested it is clear that the absence of the P6-extension allows the interaction only with compounds with a small upper apical ligand, such as MeCbl and AqCbl.

This finding was clarified by the solved structures of the *env8* riboswitch bound to AqCbl, the *Thermoanaerobacter tengcongensis* B_{12}-riboswitch bound to AdoCbl and the *Symbiobacterium thermophilum* B_{12}-riboswitch bound to AdoCbl [179, 180]. The selectivity of these riboswitches derives from the conformational differences in the peripheral extensions that determine the ability of the binding core to dissect between the large and the small apical ligand of B_{12} metabolites.

Interestingly, in all the mentioned B_{12}-riboswitches, the major contributing binding forces derive from hydrophobic packing and electrostatic interactions, while the great hydrogen-bonding potential of the ligand is not fully exploited. Moreover, it was demonstrated the importance of metal ions for the B_{12}-riboswitch to reach the active binding conformation, as already discussed in the Section 5.4.2 [156].

5.5.4.3.3 *MoCo/TuCo-riboswitches*

In 2007, through a computational method known as CM finder comparative genomics pipeline, 22 novel structured RNA sequences were identified in bacteria [145]. Between them, a riboswitch candidate was shown to sense the molybdenum cofactor (MoCo) (Figure 5.5.5B) [181].

MoCo is constituted of a molybdenum center covalently bound to the two thiolate groups of a tricyclic pyranopterin. Mo-dependent enzymes mostly use the metal in the form of MoCo and usually catalyze key redox reactions in the carbon, nitrogen, and sulfur cycles [141]. Early bioinformatic studies found 176 examples of this MoCo–RNA motif as a highly conserved aptamer region; moreover, these aptamer regions are followed by sequences that can form different expression platforms commonly found in other riboswitches [145]. The *moaA* RNA of *E. coli* is the representative of the MoCo–riboswitch class most studied in biochemical assays [181, 182]. The most recent studies highlight a complex scenery for the *moaA*-dependent gene regulation, also involving the so-called CsrA (carbon storage regulator) protein [182]. This is the first example of an RNA-aptamer to be recognized by a protein (CsrA) and a small metabolite (MoCo) in an independent manner. Further studies showed that the P2 stem and the junctions undergo a structural rearrangement that could lead to the formation of the metabolite-binding site. Unfortunately, the high chemical instability of MoCo does not allow direct testing of the RNA-ligand interactions.

An analogus tungsten cofactor (TuCo) is found in some species of Clostridia and thermophilic archaean [183]. In these organisms, tungsten, instead of molybdenum, is incorporated in the molybdopterin (MPT) compound to form the corresponding cofactor. Hence, the question arises as to whether MoCo–RNA motifs are able to differentiate between these two such similar cofactors. The *moaA* riboswitch of *E. coli* does not respond to the presence of TuCo, thus indicating the presence of features that allow MoCo, but not TuCo, to interact with the RNA.

From the sequence alignment of MoCo–-RNA motifs, a classification in two groups can be made, based on the presence or absence of the P3 stem [181]. Interestingly, the RNA sequences carrying the P3 stem are only associated with genes encoding proteins related to MPT or MoCo, while 17 of the 19 RNA motifs lacking the P3 stem are associated with genes related to MPT or TuCo. Therefore, from the structural point of view, the selectivity towards MoCo and TuCo seems to derive from the presence or absence of the P3 stem. Concerning the regulation mechanism, the binding of MoCo to the *moaA* riboswitch leads to a structural rearrangement that influences the transcription of the whole *moaABCDE* operon [181, 184].

5.5.5 Summary

In this chapter, we provide a short overview of the world of metal ions and RNA, an exciting and relatively unexplored field of studies. The complicated relationship between these two partners goes far beyond simple electrostatics, with metal ions being an absolute necessity for proper RNA folding, recognition, and catalysis.

Although often underestimated, Mg^{2+} ions in particular form important and complex associations with RNA. Ubiquitous in the cytoplasm, Mg^{2+} has a distinct coordination chemistry but stabilizes an unknown wealth of different RNA motifs and structures [12].

In general, the formation of RNA tertiary structures depends on four parameters: (i) RNA sequence, (ii) metal-ion identity, (iii) metal-ion concentration, and (iv) the presence of RNA binding proteins and polyamines. Two out of these four requirements are metal-ion related. Most certainly, the world of riboswitches and ribozymes would not be complete without their metallic partners.

Acknowledgement

Financial support by a Marie Curie fellowship (No. PIEF-GA-2012-329700 to MRZ), the Swiss National Science Foundation (to RKOS), within the COST Action CM1105, and the University of Zurich is gratefully acknowledged. RKOS is holder of an ERC starting grant (MIRNA).

References

[1] J. Schnabl and R. K. O. Sigel, Controlling ribozyme activity by metal ions, *Curr. Opin. Chem. Biol.*, **14**, 269–275 (2010).

[2] M. C. Erat and R. K. O. Sigel, Methods to detect and characterize metal ion binding sites in RNA, *Met. Ions Life Sci.*, **9**, 37–100 (2011).

[3] M. C. Erat and R. K. O. Sigel, Structure determination of catalytic RNAs and investigations of their metal ion-binding properties, *Chimia*, **59**, 817–821 (2005).

[4] A. Bugaut and S. Balasubramanian, 5 ′-UTR RNA G-quadruplexes: translation regulation and targeting, *Nucleic Acids Res.*, **40**, 4727–4741 (2012).

[5] S. Millevoi, H. Moine, and S. Vagner, G-quadruplexes in RNA biology, *Wiley Interdiscip. Reviews: RNA*, **3**, 495–507 (2012).

[6] R. K. O. Sigel,Group II intron Ribozymes and metal ions - a delicate relationship, *Eur. J. Inorg. Chem.*, 2281–2292 (2005).

[7] M. Pechlaner and R. K. O. Sigel, Characterization of metal ion-nucleic acid interactions in solution, *Met. Ions Life Sci.*, **10**, 1–42 (2012).

[8] J. K. Frederiksen and J. A. Piccirilli, Identification of catalytic metal ion ligands in ribozymes, *Methods*, **49**, 148–166 (2009).

[9] M. Steiner, D. Rueda, and R. K. O. Sigel, Ca^{2+} Induces the formation of two distinct subpopulations of group II intron molecules, *Angew. Chem., Int. Ed.*, **48**, 9739–9742 (2009).

[10] M. C. Erat and R. K. O. Sigel, Divalent metal ions tune the self-splicing reaction of the yeast mitochondrial group II intron Sc.ai5 gamma, *J. Biol. Inorg. Chem.*, **13**, 1025–1036 (2008).

[11] P. Auffinger, N. Grover, and E. Westhof, Metal ion binding to RNA, *Met. Ions Life Sci.*, **9**, 1–35 (2011).

[12] R. K. O. Sigel and H. Sigel, Metal ion interactions with nucleic acids and their constituents, *Compr. Inorg. Chem. II*, **3**, 623–660 (2013).

[13] L. E. Kapinos, B. Song, and H. Sigel, Acid-base and metal-ion-coordinating properties of benzimidazole and derivatives (=1,3-dideazapurines) in aqueous solution: Interrelation between complex stability and ligand basicity, *Chem. Eur. J.*, **5**, 1794–1802 (1999).

[14] R. K. O. Sigel and A. M. Pyle, Alternative roles for metal ions in enzyme catalysis and the implications for ribozyme chemistry, *Chem. Rev.*, **107**, 97–113 (2007).

[15] A. I. Petrov, C. L. Zirbel, and N. B. Leontis, Automated classification of RNA 3D motifs and the RNA 3D Motif Atlas, *RNA*, **19**, 1327–1340 (2013).

[16] D. K. Hendrix, S. E. Brenner, and S. R. Holbrook, RNA structural motifs: Building blocks of a modular biomolecule, *Q. Rev. Biophys.*, **38**, 221–243 (2005).

[17] A. Dominguez-Martin, S. Johannsen, A. Sigel, B. P. Operschall, B. Song, H. Sigel, A. Okruszek, J. M. Gonzalez-Perez, J. Niclos-Gutierrez, and R. K. O. Sigel, Intrinsic acid-base properties of a hexa-2-deoxynucleoside pentaphosphate, d(ApGpGpCpCpT): Neighboring effects and isomeric equilibria, *Chem. Eur. J.*, **19**, 8163–8181 (2013).

[18] J. Schnabl, P. Suter, and R. K. O. Sigel, MINAS - a database of metal ions in nucleic acids, *Nucleic Acids Res.*, **40**, 434–438 (2012).

[19] J. C. Bowman, T. K. Lenz, N. V. Hud, and L. D. Williams, Cations in charge: Magnesium ions in RNA folding and catalysis, *Curr. Opin. Struct. Biol.*, **22**, 262–272 (2012).

[20] V. K. Misra and D. E. Draper, A thermodynamic framework for Mg^{2+} binding to RNA, *Proc. Natl. Acad. Sci. U.S.A.*, **98**, 12456–12461 (2001).

[21] E. Freisinger and R. K. O. Sigel, From nucleotides to ribozymes – a comparison of their metal ion binding properties, *Coord. Chem. Rev.*, **251**, 1834–1851 (2007).

[22] D. J. Klein, P. B. Moore, and T. A. Steitz, The contribution of metal ions to the structural stability of the large ribosomal subunit, *RNA*, **10**, 1366–1379 (2004).

[23] L. R. Stefan, R. Zhang, A. G. Levitan, D. K. Hendrix, S. E. Brenner, and S. R. Holbrook, MeRNA: A database of metal ion binding sites in RNA structures, *Nucleic Acids Res.*, **34**, 131–134 (2006).

[24] P. Auffinger and Y. Hashem, SwS: A solvation web service for nucleic acids, *Bioinformatics*, **23**, 1035–1037 (2007).

[25] B. C. Stark, R. Kole, E. J. Bowman, and S. Altman, Ribonuclease-P - enzyme with an essential RNA component, *Proc. Natl. Acad. Sci. U.S.A.*, **75**, 3717–3721 (1978).

[26] K. Kruger, P. J. Grabowski, A. J. Zaug, J. Sands, D. E. Gottschling, and T. R. Cech, Self-splicing RNA - auto-excision and auto-cyclization of the ribosomal-RNA intervening sequence of tetrahymena, *Cell*, **31**, 147–157 (1982).

[27] G. A. Prody, J. T. Bakos, J. M. Buzayan, I. R. Schneider, and G. Bruening, Autolytic processing of dimeric plant-virus satellite RNA, *Science*, **231**, 1577–1580 (1986).

[28] A. Hampel, R. Tritz, M. Hicks, and P. Cruz, Hairpin catalytic RNA model - evidence for helices and sequence requirement for substrate RNA, *Nucleic Acids Res.*, **18**, 299–304 (1990).

[29] H. N. Wu, Y. J. Lin, F. P. Lin, S. Makino, M. F. Chang, and M. M. C. Lai, Human Hepatitis-Delta Virus – RNA subfragments contain an autocleavage activity, *Proc. Natl. Acad. Sci. U.S.A.*, **86**, 1831–1835 (1989).

[30] W. C. Winkler, A. Nahvi, A. Roth, J. A. Collins, and R. R. Breaker, Control of gene expression by a natural metabolite-responsive ribozyme, *Nature*, **428**, 281–286 (2004).

[31] C. Hammann, R. K. Hartmann, and A. Marchfelder, 25 years of catalytic RNA: Looking younger than ever! *Biol. Chem.*, **388**, 659–660 (2007).

[32] M. Martick, L. H. Horan, H. F. Noller, and W. G. Scott, A discontinuous hammerhead ribozyme embedded in a mammalian messenger RNA, *Nature*, **454**, 899–U57 (2008).

[33] C. H. T. Webb and A. Luptak, HDV-like self-cleaving ribozymes, *RNA Biol.*, **8**, 719–727 (2011).

[34] M. de la Pena and I. Garcia-Robles, Intronic hammerhead ribozymes are ultraconserved in the human genome, *EMBO Reports*, **11**, 711–716 (2010).

[35] J. D. Richter, CPEB: a life in translation, *Trends Biochem. Sci.*, **32**, 279–285 (2007).

[36] N. Veeraraghavan, A. Ganguly, B. L. Golden, P. C. Bevilacqua, and S. Hammes-Schiffer, Mechanistic strategies in the HDV ribozyme: Chelated and diffuse metal ion interactions and active site protonation, *J. Phys. Chem. B*, **115**, 8346–8357 (2011).

[37] N. G. Walter, K. J. Hampel, K. M. Brown, and J. M. Burke, Tertiary structure formation in the hairpin ribozyme monitored by fluorescence resonance energy transfer, *EMBO J.*, **17**, 2378–2391 (1998).

[38] P. B. Rupert and A. R. Ferre-D'Amare, Crystal structure of a hairpin ribozyme-inhibitor complex with implications for catalysis, *Nature*, **410**, 780–786 (2001).

[39] N. G. Walter, P. A. Chan, K. J. Hampel, D. P. Millar, and J. M. Burke, A base change in the catalytic core of the hairpin ribozyme perturbs function but not domain docking, *Biochemistry*, **40**, 2580–2587 (2001).

[40] M. J. B. Pereira, E. N. Nikolova, S. L. Hiley, D. Jaikaran, R. A. Collins, and N. G. Walter, Single VS ribozyme molecules reveal dynamic and hierarchical folding toward catalysis, *J. Mol. Biol.*, **382**, 496–509 (2008).

[41] J. C. Cochrane and S. A. Strobel, Catalytic strategies of self-cleaving ribozymes, *Acc. Chem. Res.*, **41**, 1027–1035 (2008).

[42] R. A. Tinsley, D. A. Harris, and N. G. Walter, Magnesium dependence of the amplified conformational switch in the trans-acting hepatitis delta virus ribozyme, *Biochemistry*, **43**, 8935–8945 (2004).

[43] B. Gong, J. H. Chen, E. Chase, D. M. Chadalavada, R. Yajima, B. L. Golden, P. C. Bevilacqua, and P. R. Carey, Direct measurement of a pK_a near neutrality for the catalytic cytosine in the genomic HDV ribozyme using Raman crystallography, *J. Am. Chem. Soc.*, **129**, 13335–13342 (2007).

[44] J. H. Chen, B. Gong, P. C. Bevilacqua, P. R. Carey, and B. L. Golden, A catalytic metal ion interacts with the cleavage site GU wobble in the HDV ribozyme, *Biochemistry*, **48**, 1498–1507 (2009).

[45] R. Koradi, M. Billeter, and K. Wüthrich, MOLMOL: A program for display and analysis of macromolecular structures, *J. Mol. Graph. Model.*, **14**, 29–55 (1996).

[46] A. E. Johnson-Buck, S. E. McDowell, and N. G. Walter, Metal ions: Supporting actors in the playbook of small ribozymes, *Met. Ions Life Sci.*, **9**, 175–196 (2011).

[47] M. Guo, R. C. Spitale, R. Volpini, J. Krucinska, G. Cristalli, P. R. Carey, and J. E. Wedekind, Direct Raman measurement of an elevated base pK(a) in the active site of a small ribozyme in a precatalytic conformation, *J. Am. Chem. Soc.*, **131**, 12908–12909 (2009).

[48] J.-H. Chen, R. Yajima, D. M. Chadalavada, E. Chase, P. C. Bevilacqua, and B. L. Golden, A 1.9 Å crystal structure of the HDV ribozyme precleavage suggests both Lewis acid and general acid mechanisms contribute to phosphodiester cleavage, *Biochemistry*, **49**, 6508–6518 (2010).

[49] S. A. Strobel and J. C. Cochrane, RNA catalysis: Ribozymes, ribosomes, and riboswitches, *Curr. Opin. Chem. Biol.*, **11**, 636–643 (2007).

[50] M. Rowinska-Zyrek, M. Skilandat, and R. K. O. Sigel, Hexaamminecobalt(III) – Probing metal ion binding sites in nucleic acids by NMR spectroscopy, *Z. Anorg. Allg. Chem.*, **639**, 1313–1320 (2013).

[51] T. E. Horton and V. J. DeRose, Cobalt hexammine inhibition of the hammerhead ribozyme, *Biochemistry*, **39**, 11408–11416 (2000).

[52] J. L. Boots, M. D. Canny, E. Azimi, and A. Pardi, Metal ion specificities for folding and cleavage activity in the Schistosoma hammerhead ribozyme, *RNA*, **14**, 2212–2222 (2008).

[53] M. Roychowdhury-Saha and D. H. Burke, Extraordinary rates of transition metal ion-mediated ribozyme catalysis, *RNA*, **12**, 1846–1852 (2006).

[54] A. Roth, A. Nahvi, M. Lee, I. Jona, and R. R. Breaker, Characteristics of the *glmS* ribozyme suggest only structural roles for divalent metal ions, *RNA*, **12**, 607–619 (2006).

[55] M. J. Fedor, The role of metal ions in RNA catalysis, *Curr. Opin. Struct. Biol.*, **12**, 289–295 (2002).

[56] T. Pan, B. Dichtl, and O. C. Uhlenbeck, Properties of an in-vitro selected Pb^{2+} cleavage motif, *Biochemistry*, **33**, 9561–9565 (1994).

[57] J. J. Agresti, B. T. Kelly, A. Jaschke, and A. D. Griffiths, Selection of ribozymes that catalyse multiple-turnover Diels-Alder cycloadditions by using in vitro compartmentalization, *Proc. Natl. Acad. Sci. U.S.A.*, **102**, 16170–16175 (2005).

[58] M. Zivarts, Y. Liu, and R. R. Breaker, Engineered allosteric ribozymes that respond to specific divalent metal ions, *Nucleic Acids Res.*, **33**, 622–631 (2005).

[59] M. A. Vorobjeva, A. S. Davydova, and A. G. Venyaminova, Artificial hammerhead ribozymes: Engineering and applications, *Russ. Chem. Rev.*, **80**, 127–143 (2011).

[60] D. Donghi and J. Schnabl, Multiple roles of metal ions in large ribozymes, *Met. Ions Life Sci.*, **9**, 197–234 (2011).

[61] S. E. Butcher, The spliceosome and its metal ions, *Met. Ions Life Sci.*, **9**, 235–251 (2011).

[62] N. J. Reiter, A. Osterman, A. Torres-Larios, K. K. Swinger, T. Pan, and A. Mondragon, Structure of a bacterial ribonuclease P holoenzyme in complex with tRNA, *Nature*, **468**, 784–789 (2010).

[63] H. Nielsen and S. D. Johansen, Group I introns moving in new directions, *RNA Biol.*, **6**, 375–383 (2009).

[64] C. L. Shenvi, K. C. Dong, E. M. Friedman, J. A. Hanson, and J. H. D. Cate, Accessibility of 18S rRNA in human 40S subunits and 80S ribosomes at physiological magnesium ion concentrations - implications for the study of ribosome dynamics, *RNA*, **11**, 1898–1908 (2005).

[65] A. S. Petrov, C. R. Bernier, C. L. Hsiao, C. D. Okafor, E. Tannenbaum, J. Stern, E. Gaucher, D. Schneider, N. V. Hud, S. C. Harvey, and L. D. Williams, RNA-magnesium-protein interactions in large ribosomal subunit, *J. Phys. Chem. B*, **116**, 8113–8120 (2012).

[66] D. A. Brow, Allosteric cascade of spliceosome activation, *Annu. Rev. Genet.*, **36**, 333–360 (2002).

[67] M. C. Wahl, C. L. Will, and R. Luhrmann, The spliceosome: design principles of a dynamic RNP machine, *Cell*, **136**, 701–718 (2009).
[68] P. M. Gordon, E. J. Sontheimer, and J. A. Piccirilli, Metal ion catalysis during the exon-ligation step of nuclear pre-mRNA splicing: Extending the parallels between the spliceosome and group II introns, *RNA*, **6**, 199–205 (2000).
[69] E. J. Sontheimer, S. G. Sun, and J. A. Piccirilli, Metal ion catalysis during splicing of premessenger RNA, *Nature*, **388**, 801–805 (1997).
[70] S. E. Butcher and D. A. Brow, Towards understanding the catalytic core structure of the spliceosome, *Biochem. Soc. Trans.*, **33**, 447–449 (2005).
[71] T. Villa, J. A. Pleiss, and C. Guthrie, Spliceosomal snRNAs: Mg^{2+}-dependent chemistry at the catalytic core? *Cell*, **109**, 149–152 (2002).
[72] N. J. Reiter, L. J. Nikstad, A. M. Allmann, R. J. Johnson, and S. E. Butcher, Structure of the U6 RNA intramolecular stem-loop harboring an S-P-phosphorothioate modification, *RNA*, **9**, 533–542 (2003).
[73] A. Huppler, L. J. Nikstad, A. M. Allmann, D. A. Brow, and S. E. Butcher, Metal binding and base ionization in the U6 RNA intramolecular stem-loop structure, *Nature Struct. Biol.*, **9**, 431–435 (2002).
[74] V. Venditti, L. Clos, N. Niccolai, and S. E. Butcher, Minimum-energy path for a U6 RNA conformational change involving protonation, base-pair rearrangement and base flipping, *J. Mol. Biol.*, **391**, 894–905 (2009).
[75] V. Venditti, N. Niccolai, and S. E. Butcher, Measuring the dynamic surface accessibility of RNA with the small paramagnetic molecule TEMPOL, *Nucleic Acids Res.*, **36**, 20–30 (2008).
[76] T. A. Steitz and J. A. Steitz, A general 2-metal-ion mechanism for catalytic RNA, *Proc. Natl. Acad. Sci. U.S.A.*, **90**, 6498–6502 (1993).
[77] D. M. J. Lilley, The origins of RNA catalysis in ribozymes, *Trends Biochem. Sci.*, **28**, 495–501 (2003).
[78] V. Gopalan, Uniformity amid diversity in RNase P, *Proc. Natl. Acad. Sci. U.S.A.*, **104**, 2031–2032 (2007).
[79] A. V. Kazantsev, A. A. Krivenko, and N. R. Pace, Mapping metal-binding sites in the catalytic domain of bacterial RNase P RNA, *RNA*, **15**, 266–276 (2009).
[80] A. V. Kazantsev, A. A. Krivenko, D. J. Harrington, S. R. Holbrook, P. D. Adams, and N. R. Pace, Crystal structure of a bacterial ribonuclease P RNA, *Proc. Natl. Acad. Sci. U.S.A.*, **102**, 13392–13397 (2005).
[81] A. Torres-Larios, K. K. Swinger, A. S. Krasilnikov, T. Pan, and A. Mondragon, Crystal structure of the RNA component of bacterial ribonuclease P, *Nature*, **437**, 584–587 (2005).
[82] T. Persson, S. Cuzic, and R. K. Hartmann, Catalysis by RNase P RNA - Unique features and unprecedented active site plasticity, *J. Biol. Chem.*, **278**, 43394–43401 (2003).
[83] K. S. Koutmou, A. Casiano-Negroni, M. M. Getz, S. Pazicni, A. J. Andrews, J. E. Penner-Hahn, H. M. Al-Hashimi, and C. A. Fireke, NMR and XAS reveal an inner-sphere metal binding site in the P4 helix of the metallo-ribozyme ribonuclease P, *Proc. Natl. Acad. Sci. U.S.A.*, **107**, 2479–2484 (2010).
[84] O. Fedorova and N. Zingler, Group II introns: Structure, folding and splicing mechanism, *Biol. Chem.*, **388**, 665–678 (2007).
[85] Q. Vicens and T. R. Cech, Atomic level architecture of group I introns revealed, *Trends Biochem. Sci.*, **31**, 41–51 (2006).
[86] J. A. Piccirilli, J. S. Vyle, M. H. Caruthers, and T. R. Cech, Metal-ion catalysis in the *Tetrahymena* ribozyme reaction, *Nature*, **361**, 85–88 (1993).
[87] S. Shan, A. V. Kravchuk, J. A. Piccirilli, and D. Herschlag, Defining the catalytic metal ion interactions in the *Tetrahymena* ribozyme reaction, *Biochemistry*, **40**, 5161–5171 (2001).
[88] E. Koculi, C. Hyeon, D. Thirumalai, and S. A. Woodson, Charge density of divalent metal cations determines RNA stability, *J. Am. Chem. Soc.*, **129**, 2676–2682 (2007).
[89] P. Rangan, B. Masquida, E. Westhof, and S. A. Woodson, Architecture and folding mechanism of the Azoarcus group I pre-tRNA, *J. Mol. Biol.*, **339**, 41–51 (2004).
[90] S. Moghaddam, G. Caliskan, S. Chauhan, C. Hyeon, R. M. Briber, D. Thirumalai, and S. A. Woodson, Metal ion dependence of cooperative collapse transitions in RNA, *J. Mol. Biol.*, **393**, 753–764 (2009).
[91] J. Pan and S. A. Woodson, Folding intermediates of a self-splicing RNA: Mispairing of the catalytic core, *J. Mol. Biol.*, **280**, 597–609 (1998).
[92] P. L. Adams, M. R. Stahley, A. B. Kosek, J. M. Wang, and S. A. Strobel, Crystal structure of a self-splicing group I intron with both exons, *Nature*, **430**, 45–50 (2004).

[93] M. R. Stahley and S. A. Strobel, Structural evidence for a two-metal-ion mechanism of group I intron splicing, *Science*, **309**, 1587–1590 (2005).

[94] J. H. Cate, A. R. Gooding, E. Podell, K. H. Zhou, B. L. Golden, C. E. Kundrot, T. R. Cech, and J. A. Doudna, Crystal structure of a group I ribozyme domain: Principles of RNA packing, *Science*, **273**, 1678–1685 (1996).

[95] M. Forconi, T. Benz-Moy, K. R. Gleitsman, E. Ruben, C. Metz, and D. Herschlag, Exploring purine N7 interactions via atomic mutagenesis: the group I ribozyme as a case study, *RNA*, **18**, 1222–1229 (2012).

[96] M. Forconi, R. N. Sengupta, J. A. Piccirilli, and D. Herschlag, A rearrangement of the guanosine-binding site establishes an extended network of functional interactions in the T*etrahymena* group I ribozyme active site, *Biochemistry*, **49**, 2753–2762 (2010).

[97] M. Forconi, R. N. Sengupta, M. C. Liu, A. C. Sartorelli, J. A. Piccirilli, and D. Herschlag,Structure and function converge to identify a hydrogen bond in a group I ribozyme active site, *Angew. Chem., Int. Ed.*, **48**, 7171–7175 (2009).

[98] D. W. Celander and T. R. Cech, Visualizing the higher-order folding of a catalytic RNA molecule, *Science*, **251**, 401–407 (1991).

[99] N. Lehman and G. F. Joyce, Evolution *in vitro* - analysis of a lineage of ribozymes, *Curr. Biol.*, **3**, 723–734 (1993).

[100] N. Lehman and G. F. Joyce, Evolution *in vitro* of an RNA enzyme with altered metal dependence, *Nature*, **361**, 182–185 (1993).

[101] A. S. Burton and N. Lehman, Calcium(II)-dependent catalytic activity of the *Azoarcus* ribozyme: testing the limits of resolution for in vitro selection, *Biochimie*, **88**, 819–825 (2006).

[102] (a)M. Marcia and A. M. Pyle, Visualizing group II intron catalysis through the stages of splicing, *Cell*, **151**, 497–507 (2012);(b)M. Marcia and A. M. Pyle, Principles of ion recognition in RNA: Insights from the group II intron structures, *RNA*, **20**, 516–527 (2014).

[103] P. Z. Qin and A. M. Pyle, The architectural organization and mechanistic function of group II intron structural elements, *Curr. Opin. Struct. Biol.*, **8**, 301–308 (1998).

[104] H. R. Huang, M. Y. Chao, B. Armstrong, Y. Wang, A. M. Lambowitz, and P. S. Perlman, The DIVa maturase binding site in the Yeast group II intron aI2 is essential for intron homing but not for in vivo splicing, *Mol. Cell. Biol.*, **23**, 8809–8819 (2003).

[105] C. Adamidi, O. Fedorova, and A. M. Pyle, A group II intron inserted into a bacterial heat-shock operon shows autocatalytic activity and unusual thermostability, *Biochemistry*, **42**, 3409–3418 (2003).

[106] J. F. Swisher, L. H. J. Su, M. Brenowitz, V. E. Anderson, and A. M. Pyle, Productive folding to the native state by a group II intron ribozyme, *J. Mol. Biol.*, **315**, 297–310 (2002).

[107] N. Toor, K. S. Keating, S. D. Taylor, and A. M. Pyle, Crystal structure of a self-spliced group II intron, *Science*, **320**, 77–82 (2008).

[108] C. Waldsich and A. M. Pyle, A folding control element for tertiary collapse of a group II intron ribozyme, *Nature Struct. Mol. Biol.*, **14**, 37–44 (2007).

[109] L. J. Su, C. Waldsich, and A. M. Pyle, An obligate intermediate along the slow folding pathway of a group II intron ribozyme, *Nucleic Acids Res.*, **33**, 6674–6687 (2005).

[110] O. Fedorova, C. Waldsich, and A. M. Pyle, Group II intron folding under near-physiological conditions: Collapsing to the near-native state, *J. Mol. Biol.*, **366**, 1099–1114 (2007).

[111] M. Steiner, K. S. Karunatilaka, R. K. O. Sigel, and D. Rueda, Single-molecule studies of group II intron ribozymes, *Proc. Natl. Acad. Sci. U.S.A.*, **105**, 18071–18071 (2008).

[112] D. Donghi, M. Pechlaner, C. Finazzo, B. Knobloch, and R. K. O. Sigel, The structural stabilization of the kappa three-way junction by Mg^{2+} represents the first step in the folding of a group II intron, *Nucleic Acids Res.*, **41**, 2489–2504 (2013).

[113] K. Chin and A. M. Pyle, Branch-point attack in group II introns is a highly reversible transesterification, providing a potential proofreading mechanism for 5′-splice site selection, *RNA*, **1**, 391–406 (1995).

[114] M. C. Erat, O. Zerbe, T. Fox, and R. K. O. Sigel, Solution structure of domain 6 from a self-splicing group II intron ribozyme: A Mg^{2+} binding site is located close to the stacked branch adenosine, *ChemBioChem*, **8**, 306–314 (2007).

[115] M. C. Erat, H. Kovacs, and R. K. O. Sigel, Metal ion-N7 coordination in a ribozyme branch domain by NMR, *J. Inorg. Biochem.*, **104**, 611–613 (2010).

[116] D. Kruschel, M. Skilandat, and R. K. O. Sigel, NMR structure of the 5′-splice site in the group IIB intron Sc.ai5γ – conformational requirements for positioning of the exon-intron junction, *RNA*, **20**, 295–307 (2014).

[117] D. Kruschel and R. K. O. Sigel, Divalent metal ions promote the formation of the 5 ′-splice site recognition complex in a self-splicing group II intron, *J. Inorg. Biochem.*, **102**, 2147–2154 (2008).

[118] M. Skilandat and R. K. O. Sigel, The role of magnesium(II) for DNA cleavage site recognition in Group II intron ribozymes - Solution structure and metal ion binding sites of the RNA·DNA complex, *J. Biol. Chem.*, **289**, 20650–20663 (2014).

[119] M. C. Erat, J. Coles, C. Finazzo, B. Knobloch, and R. K. O. Sigel, Accurate analysis of Mg^{2+} binding to RNA: From classical methods to a novel iterative calculation procedure, *Coord. Chem. Rev.*, **256**, 279–288 (2012).

[120] M. C. Erat and R. K. O. Sigel, Determination of the intrinsic affinities of multiple site-specific Mg^{2+} ions coordinated to domain 6 of a group II intron ribozyme, *Inorg. Chem.*, **46**, 11224–11234 (2007).

[121] S. L. B. Koenig, D. Kowerko, M. Khier, M. Hadzic, and R. K. O. Sigel, Cation-promoted RNA tertiary structure formation dissected by single-molecule fluorescence, submitted for publication.

[122] D. Kowerko, S. L. B. König, M. Skilandat, M. C. A. S. Hadzic, L. Cardo, and R. K. O. Sigel, Cation-induced kinetic heterogeneity of the intron-exon recognition in single group II introns, *Proc. Natl. Acad. Sci. U.S.A.*, in the press.

[123] M. Marcia, S. Somarowthu, and A. M. Pyle, Now on display: A gallery of group II intron structures at different stages of catalysis, *Mobile DNA*, **4**:14, (2013).

[124] N. Toor, K. S. Keating, O. Fedorova, K. Rajashankar, J. M. Wang, and A. M. Pyle, Tertiary architecture of the oceanobacillus iheyensis group II intron, *RNA*, **16**, 57–69 (2010).

[125] K. S. Keating, N. Toor, P. S. Perlman, and A. M. Pyle, A structural analysis of the group II intron active site and implications for the spliceosome, *RNA*, **16**, 1–9 (2010).

[126] N. Toor, K. Rajashankar, K. S. Keating, and A. M. Pyle, Structural basis for exon recognition by a group II intron, *Nature Struct. Mol. Biol.*, **15**, 1221–1222 (2008).

[127] R. T. Chan, A. R. Robart, K. R. Rajashankar, A. M. Pyle, and N. Toor, Crystal structure of a group II intron in the pre-catalytic state, *Nature Struct. Mol. Biol.*, **19**, 555–557 (2012).

[128] A. S. Mironov, I. Gusarov, R. Rafikov, L. E. Lopez, K. Shatalin, R. A. Kreneva, D. A. Perumov, and E. Nudler, Sensing small molecules by nascent RNA: A mechanism to control transcription in bacteria, *Cell*, **111**, 747–756 (2002).

[129] A. Nahvi, N. Sudarsan, M. S. Ebert, X. Zou, K. L. Brown, and R. R. Breaker, Genetic control by a metabolite binding mRNA, *Chem. Biol.*, **9**, 1043–1049 (2002).

[130] W. Winkler, A. Nahvi, and R. R. Breaker, Thiamine derivatives bind messenger RNAs directly to regulate bacterial gene expression, *Nature*, **419**, 952–956 (2002).

[131] E. Nudler and A. S. Mironov, The riboswitch control of bacterial metabolism, *Trends Biochem. Sci.*, **29**, 11–17 (2004).

[132] W. C. Winkler and R. R. Breaker, Regulation of bacterial gene expression by riboswitches, *Annu. Rev. Microbiol.*, **59**, 487–517 (2005).

[133] J. E. Barrick and R. R. Breaker, The distributions, mechanisms, and structures of metabolite-binding riboswitches, *Genome Biol.*, **8**, 239–246 (2007).

[134] W. C. Winkler and R. R. Breaker, Genetic control by metabolite-binding riboswitches, *ChemBioChem*, **4**, 1024–1032 (2003).

[135] J. K. Soukup and G. A. Soukup, Riboswitches exert genetic control through metabolite-induced conformational change, *Curr. Opin. Struct. Biol.*, **14**, 344–349 (2004).

[136] T. Kubodera, M. Watanabe, K. Yoshiuchi, N. Yamashita, A. Nishimura, S. Nakai, K. Gomi, and H. Hanamoto, Thiamine-regulated gene expression of Aspergillus oryzae thiA requires splicing of the intron containing a riboswitch-like domain in the 5′-UTR, *FEBS Lett.*, **555**, 516–520 (2003).

[137] E. Loh, O. Dussurget, J. Gripenland, K. Vaitkevicius, T. Tiensuu, P. Mandin, F. Repoila, C. Buchrieser, P. Cossart, and J. Johansson, A trans-acting riboswitch controls expression of the virulence regulator PrfA in Listeria monocytogenes, *Cell*, **139**, 770–779 (2009).

[138] A. Roth and R. R. Breaker, The structural and functional diversity of metabolite-binding riboswitches, *Annu. Rev. Biochem.*, **78**, 305–334 (2009).

[139] K. Hollands, S. Proshkin, S. Sklyarova, V. Epshtein, A. Mironov, E. Nudler, and E. A. Groisman, Riboswitch control of Rho-dependent transcription termination, *Proc. Natl. Acad. Sci. U.S.A.*, **109**, 5376–5381 (2012).

[140] S. Gallo, M. Oberhuber, R. K. O. Sigel, and B. Kräutler, The corrin moiety of coenzyme B-12 is the determinant for switching the *btuB* riboswitch of *E. coli*, *ChemBioChem*, **9**, 1408–1414 (2008).

[141] G. Schwarz and R. R. Mendel, Molybdenum cofactor biosynthesis and molybdenum enzymes, *Annu. Rev. Plant Biol.*, **57**, 623–647 (2006).

[142] A. Serganov and D. J. Patel, Metabolite recognition principles and molecular mechanisms underlying riboswitch function, *Annu. Rev. Biophys.*, **41**, 343–370 (2012).

[143] J. L. Baker, N. Sudarsan, Z. Weinberg, A. Roth, R. B. Stockbridge, and R. R. Breaker, Widespread genetic switches and toxicity resistance proteins for fluoride, *Science*, **335**, 233–235 (2012).

[144] K. A. Corbino, J. E. Barrick, J. Lim, R. Welz, B. J. Tucker, I. Puskarz, M. Mandal, N. D. Rudnick, and R. R. Breaker, Evidence for a second class of S-adenosylmethionine riboswitches and other regulatory RNA motifs in alpha-proteobacteria, *Genome Biol.*, **6**, 70–78 (2005).

[145] Z. Weinberg, J. E. Barrick, Z. Yao, A. Roth, J. N. Kim, J. Gore, J. X. Wang, E. R. Lee, K. F. Block, N. Sudarsan, S. Neph, M. Tompa, W. L. Ruzzo, and R. R. Breaker, Identification of 22 candidate structured RNAs in bacteria using the CMfinder comparative genomics pipeline, *Nucleic Acids Res.*, **35**, 4809–4819 (2007).

[146] Z. Weinberg, J. X. Wang, J. Bogue, J. Yang, K. Corbino, R. H. Moy, and R. R. Breaker, Comparative genomics reveals 104 candidate structured RNAs from bacteria, archaea, and their metagenomes, *Genome Biol.*, **11**, 31–47 (2010).

[147] R. R. Breaker, Riboswitches: From ancient gene-control systems to modern drug targets, *Future Microbiol.*, **4**, 771–773 (2009).

[148] K. E. Deigan and A. R. Ferre-D'Amare, Riboswitches: discovery of drugs that target bacterial gene-regulatory RNAs, *Acc. Chem. Res.*, **44**, 1329–1338 (2011).

[149] B. Chen, X. Zuo, Y. X. Wang, and T. K. Dayie, Multiple conformations of SAM-II riboswitch detected with SAXS and NMR spectroscopy, *Nucleic Acids Res.*, **40**, 3117–3130 (2012).

[150] A. Haller, U. Rieder, M. Aigner, S. C. Blanchard, and R. Micura, Conformational capture of the SAM-II riboswitch, *Nature Chem. Biol.*, **7**, 393–400 (2011).

[151] R. L. Hayes, J. K. Noel, U. Mohanty, P. C. Whitford, S. P. Hennelly, J. N. Onuchic, and K. Y. Sanbonmatsu, Magnesium fluctuations modulate RNA dynamics in the SAM-I riboswitch, *J. Am. Chem. Soc.*, **134**, 12043–12053 (2012).

[152] S. P. Hennelly, I. V. Novikova, and K. Y. Sanbonmatsu, The expression platform and the aptamer: cooperativity between Mg^{2+} and ligand in the SAM-I riboswitch, *Nucleic Acids Res.*, **41**, 1922–1935 (2013).

[153] B. Heppell, S. Blouin, A. M. Dussault, J. Mulhbacher, E. Ennifar, J. C. Penedo, and D. A. Lafontaine, Molecular insights into the ligand-controlled organization of the SAM-I riboswitch, *Nature Chem. Biol.*, **7**, 384–392 (2011).

[154] P. C. Anthony, C. F. Perez, C. Garcia-Garcia and S. M. Block, Folding energy landscape of the thiamine pyrophosphate riboswitch aptamer, *Proc. Natl. Acad. Sci. U.S.A.*, **109**, 1485–1489 (2012).

[155] A. Haller, R. B. Altman, M. F. Souliere, S. C. Blanchard, and R. Micura, Folding and ligand recognition of the TPP riboswitch aptamer at single-molecule resolution, *Proc. Natl. Acad. Sci. U.S.A.*, **110**, 4188–4193 (2013).

[156] P. K. Choudhary and R. K. O. Sigel, Mg^{2+}-induced conformational changes in the btuB riboswitch from E. coli, *RNA*, **20**, 36–45 (2014).

[157] S. Wood, A. R. Ferre-D'Amare, and D. Rueda, Allosteric tertiary interactions preorganize the c-di-GMP riboswitch and accelerate ligand binding, *ACS Chem. Biol.*, **7**, 920–927 (2012).

[158] J. X. Wang, E. R. Lee, D. R. Morales, J. Lim and R. R. Breaker, Riboswitches that sense S-adenosylhomocysteine and activate genes involved in coenzyme recycling, *Mol. Cell*, **29**, 691–702 (2008).

[159] S. D. Gilbert, R. P. Rambo, D. Van Tyne and R. T. Batey, Structure of the SAM-II riboswitch bound to S-adenosylmethionine, *Nature Struct. Mol. Biol.*, **15**, 177–182 (2008).

[160] K. Lang, R. Rieder, and R. Micura, Ligand-induced folding of the thiM TPP riboswitch investigated by a structure-based fluorescence spectroscopic approach, *Nucleic Acids Res.*, **35**, 5370–5378 (2007).

[161] A. Serganov, The long and the short of riboswitches, *Curr. Opin. Struct. Biol.*, **19**, 251–259 (2009).

[162] T. E. Edwards and A. R. Ferre-D'Amare, Crystal structures of the thi-box riboswitch bound to thiamine pyrophosphate analogs reveal adaptive RNA-small molecule recognition, *Structure*, **14**, 1459–1468 (2006).

[163] S. Thore, M. Leibundgut, and N. Ban, Structure of the eukaryotic thiamine pyrophosphate riboswitch with its regulatory ligand, *Science*, **312**, 1208–1211 (2006).

[164] A. Serganov, L. Huang, and D. J. Patel, Coenzyme recognition and gene regulation by a flavin mononucleotide riboswitch, *Nature*, **458**, 233–237 (2009).

[165] Q. Vicens, E. Mondragon, and R. T. Batey, Molecular sensing by the aptamer domain of the FMN riboswitch: a general model for ligand binding by conformational selection, *Nucleic Acids Res.*, **39**, 8586–8598 (2011).

[166] A. Serganov, L. L. Huang, and D. J. Patel, Structural insights into amino acid binding and gene control by a lysine riboswitch, *Nature*, **455**, 1263–1276 (2008).

[167] A. D. Garst, E. B. Porter, and R. T. Batey, Insights into the regulatory landscape of the lysine riboswitch, *J. Mol. Biol.*, **423**, 17–33 (2012).

[168] J. Tyrrell, J. L. McGinnis, K. M. Weeks, and G. J. Pielak, The cellular environment stabilizes adenine riboswitch RNA structure, *Biochemistry*, **52**, 8777–8785 (2013).

[169] M. J. Cromie, Y. Shi, T. Latifi, and E. A. Groisman, An RNA sensor for intracellular Mg^{2+}, *Cell*, **125**, 71–84 (2006).

[170] J. M. Peters, A. D. Vangeloff, and R. Landick, Bacterial transcription terminators: The RNA 3′-end chronicles, *J. Mol. Biol.*, **412**, 793–813 (2011).

[171] M. M. Korth and R. K. O. Sigel, Unusually high-affinity Mg^{2+} binding at the AU-rich sequence within the antiterminator hairpin of a Mg^{2+} riboswitch, *Chem. Biod.*, **9**, 2035–2049 (2012).

[172] C. E. Dann, 3rd, C. A. Wakeman, C. L. Sieling, S. C. Baker, I. Irnov, and W. C. Winkler, Structure and mechanism of a metal-sensing regulatory RNA, *Cell*, **130**, 878–892 (2007).

[173] A. Ramesh, C. A. Wakeman, and W. C. Winkler, Insights into metalloregulation by M-box riboswitch RNAs via structural analysis of manganese-bound complexes, *J. Mol. Biol.*, **407**, 556–570 (2011).

[174] S. Gallo, S. Mundwiler, R. Alberto, and R. K. O. Sigel, The change of corrin-amides to carboxylates leads to altered structures of the B_{12}-responding btuB riboswitch, *Chem. Commun.*, 403–405 (2011).

[175] P. K. Choudhary, A. Duret, E. Rohrbach-Brandt, C. Holliger, R. K. O. Sigel, and J. Maillard, Diversity of cobalamin riboswitches in the corrinoid-producing organohalide respirer *Desulfitobacterium hafniense*, *J. Bacteriol.*, **195**, 5186–5195 (2013).

[176] N. C. Butzin, M. A. Secinaro, K. S. Swithers, J. P. Gogarten, and K. M. Noll, *Thermotoga lettingae* can salvage cobinamide to synthesize vitamin B_{12}, *Appl. Environ. Microbiol.*, **79**, 7006–7012 (2013).

[177] G. A. Perdrizet, 2nd, I. Artsimovitch, R. Furman, T. R. Sosnick, and T. Pan, Transcriptional pausing coordinates folding of the aptamer domain and the expression platform of a riboswitch, *Proc. Natl. Acad. Sci. U.S.A.*, **109**, 3323–3328 (2012).

[178] J. R. Roth, J. G. Lawrence, and T. A. Bobik, Cobalamin (coenzyme B_{12}): Synthesis and biological significance, *Annu. Rev. Microbiol.*, **50**, 137–181 (1996).

[179] A. Peselis and A. Serganov, Structural insights into ligand binding and gene expression control by an adenosylcobalamin riboswitch, *Nature Struct. Mol. Biol.*, **19**, 1182–1184 (2012).

[180] R. T. Batey, Structure and mechanism of purine-binding riboswitches, *Quart. Rev. Biophys.*, **45**, 345–381 (2012).

[181] E. E. Regulski, R. H. Moy, Z. Weinberg, J. E. Barrick, Z. Yao, W. L. Ruzzo, and R. R. Breaker, A widespread riboswitch candidate that controls bacterial genes involved in molybdenum cofactor and tungsten cofactor metabolism, *Mol. Microbiol.*, **68**, 918–932 (2008).

[182] L. M. Patterson-Fortin, C. A. Vakulskas, H. Yakhnin, P. Babitzke, and T. Romeo, Dual posttranscriptional regulation via a cofactor-responsive mRNA leader, *J. Mol. Biol.*, **425**, 3662–3677 (2013).

[183] J. L. Johnson, K. V. Rajagopalan, S. Mukund, and M. W. W. Adams, Identification of molybdopterin as the organic component of the tungsten cofactor in 4 enzymes from hyperthermophilic archaea, *J. Biol. Chem.*, **268**, 4848–4852 (1993).

[184] L. A. Anderson, E. McNairn, T. Lubke, R. N. Pau, and D. H. Boxer, ModE-dependent molybdate regulation of the molybdenum cofactor operon moa in *Escherichia coli*, *J. Bacteriol.*, **182**, 7035–7043 (2000).

5.6

DNA Switches and Machines

Fuan Wang, Chun-Hua Lu, and Itamar Willner

Institute of Chemistry, The Center for Nanoscience and Nanotechnology, The Hebrew University of Jerusalem, Jerusalem, Israel

5.6.1 Introduction

DNA nanotechnology is a rapidly developing research area that implements nucleic acids as functional building materials for the assembly of nanostructures, nanodevices, and nano-composites of programmed properties and functions [1–5]. The unique structural and functional properties of DNA originate from the information encoded in the base sequences of the biopolymers. Structural features of DNA include the formation of duplex nucleic acid via adenine–thymine (A–T) and guanine–cytosine (G–C) base pairing, ion-driven stabilization of guanosine (G)-quadruplexes [6–9], pH-stimulated proton-stabilized cytosine (C)-quadruplex (i-motif) structures [10–13], or ion-stimulated, and cooperatively-stabilized duplex DNA structures through the formation of T–Hg^{2+}–T [14–16] or C–Ag^{+}–C bridges [17–19], Figure 5.6.1(A). Also, sequence-specific DNA structures dictate the binding of proteins to the nucleic acids [20–23], or provide instructive information for enzymes that manipulate DNA, for example, the cleavage of DNA by endonucleases or nicking enzymes [24–27]. Functional information encoded by the base sequence in the nucleic acid structures includes the sequence-specific recognition of low molecular weight substrates or macromolecules (aptamers) [28–32], or the catalytic functions of nucleic acids (DNAzymes or ribozymes) [33–37]. For example, Figure 5.6.1(B) depicts the aptamer complexes with cocaine [38, 39] or thrombin [40, 41], and Figure 5.6.1(C) exemplifies the hemin–G-quadruplex horseradish peroxidase (HRP)-mimicking DNAzyme [42–46] or the metal-ion dependent hydrolytic DNAzyme [47–49].

The structures of DNA are controlled by the energy associated with the respective assemblies, and the interactions of the DNA structures with external triggers might reconfigure the DNA nanostructures, and even introduce new reactivity patterns controlled by the energy stabilizing the different states. For example, the stability of duplex DNA is controlled by the number of base pairs and the nature of bases (G–C

DNA in Supramolecular Chemistry and Nanotechnology, First Edition.
Edited by Eugen Stulz and Guido H. Clever.

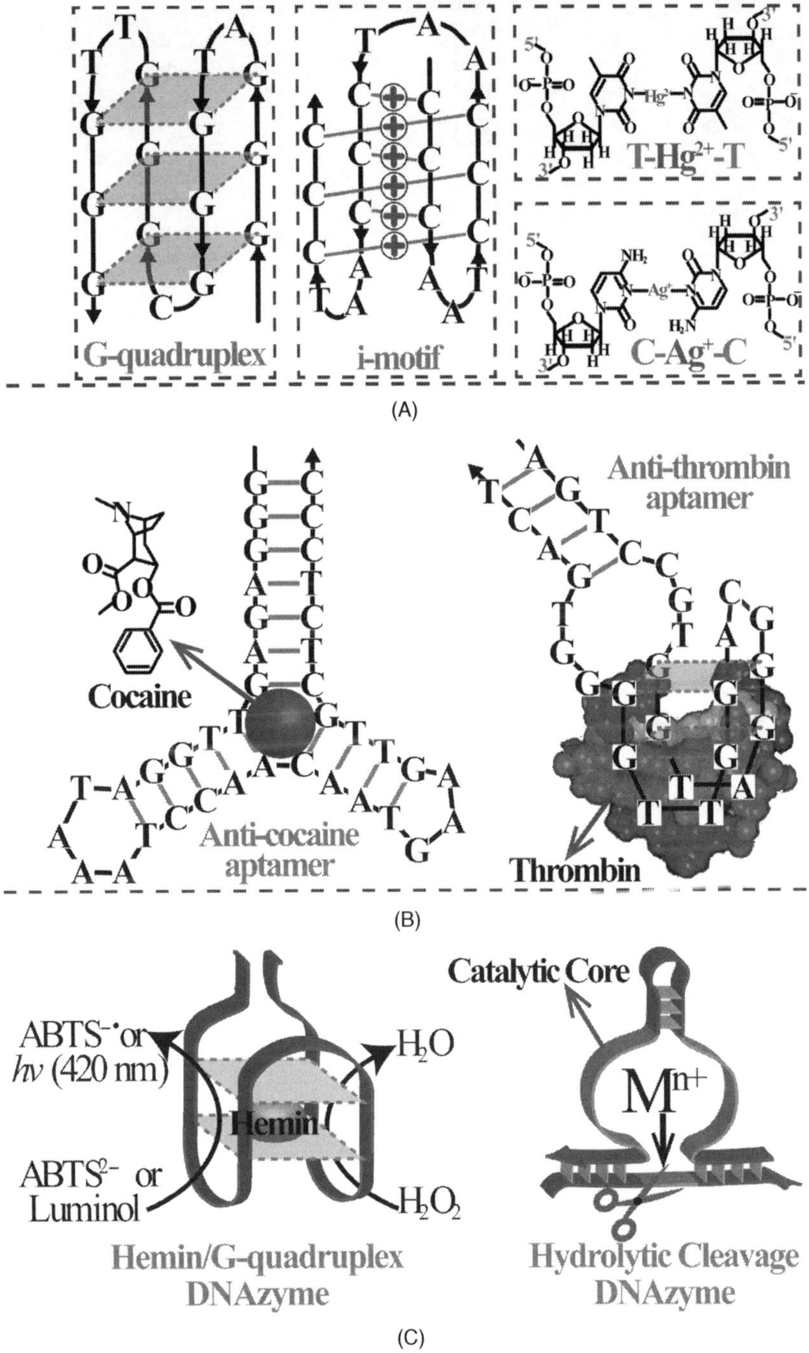

Figure 5.6.1 *(A) Self-assembly of G-quadruplex, i-motif, and metal-ion bridged DNA nanostructures. (B) Examples of aptamer–cocaine and aptamer–thrombin complexes. (C) Examples of the hemin–G-quadruplex horseradish peroxidase (HRP)-mimicking DNAzyme and the metal-ion dependent hydrolytic cleavage DNAzymes*

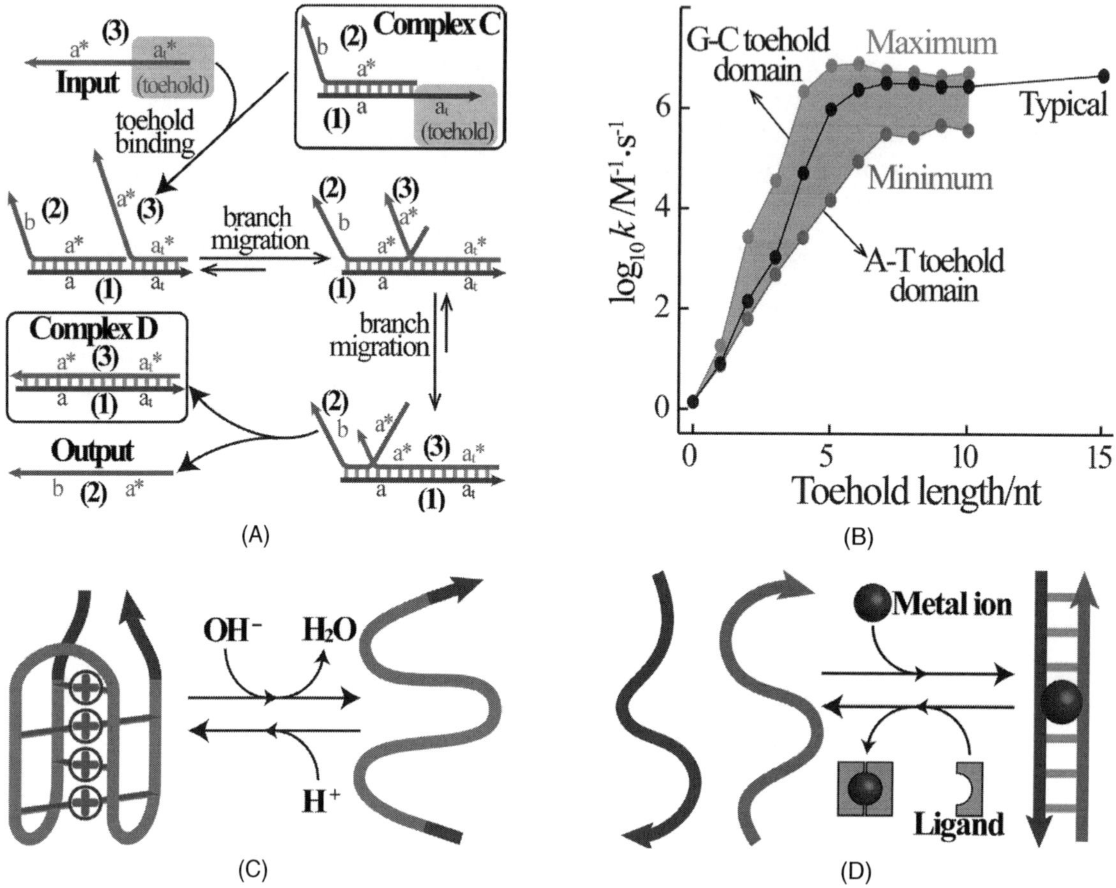

Figure 5.6.2 (A) The strand-displacement principle. (B) Theoretical modeling of the kinetics of DNA strand displacement as a function of base-pairing numbers and base-pair compositions in the toehold domain. Reproduced with permission from [50]. Copyright 2009 American Chemical Society. (C) pH-switchable formation and dissociation of an i-motif DNA nanostructure. (D) Metal-ion-stimulated formation of a duplex DNA nanostructure and its dissociation by ligand–metal-ion affinity interactions

base pairing is energetically favored over A–T base pairing). As a result, a duplex DNA (**1–2**) that includes a single-stranded toehold undergoes strand displacement in the presence of an auxiliary strand (**3**), which reveals enhanced stability [50], as outlined in Figure 5.6.2(A). The rate of strand-displacement is, similarly, dictated by the degree of base-pairing stability, Figure 5.6.2(B). Likewise, pH-driven transitions between i-motif structures and random-coil configurations, Figure 5.6.2(C), or metal-ion-assisted formation of duplex DNA structures and their separation by ligands that bind the ions, Figure 5.6.2(D), represent dynamic transitions triggered by external stimuli. These unique structural and functional features of DNA provide a "rich" toolbox for using nucleic acids as attractive materials in bio-nanotechnology. Indeed, recent advances have addressed the use of nucleic acids as building blocks to construct one-dimensional (1D) [51–53], two-dimensional (2D) [54–56], and three-dimensional (3D) [57–59] nanostructures, to implement the DNA nanostructures as scaffolds for the organization of enzymes [60–64] or nanoparticles [65–68], for applying nucleic acids for sensing [69–72], for using DNA–nanoparticle conjugates for nano-medicine [73–77], for

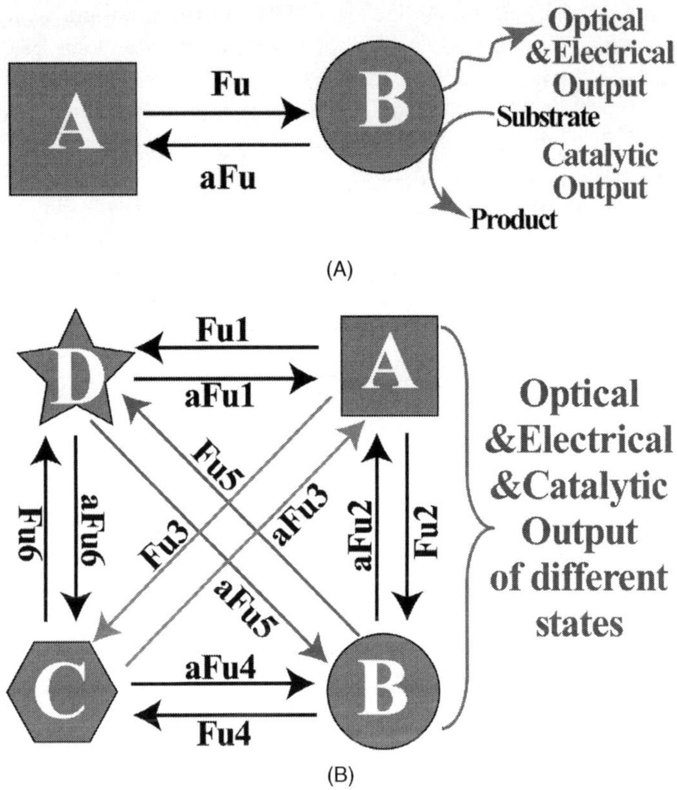

Figure 5.6.3 *(A) Switchable, cyclic, transitions between two states using fuel and anti-fuel stimuli. (B) Multiplexed fuel and anti-fuel-driven transitions between different states (states A to D)*

developing DNA computing circuits [78–82], and for constructing nanoscale devices by using DNAs as templates [83–87], for the synthesis of nanomaterials, for example, metallic nanowires [88, 89].

A major research area in DNA nanotechnology involves the development of DNA switching systems and DNA machines [90–94]. DNA switches and machines include several fundamental features. (i) The DNA switches or machines duplicate functions of macroscopic devices. They undergo cyclic "ON/OFF" operations, or they stimulate mechanical transitions such as rotation or translocations. (ii) The DNA switches or devices are triggered by external stimuli acting as "fuels". The reverse operation of the devices requires the addition of a stimulus that opposes the fuel trigger, and thus it is often termed, as "anti-fuel". (iii) The utilization of the fuel–anti-fuel often generates a "waste" product, similar to the utilization of the fuels in macroscopic machines. (iv) The operation of the DNA devices is energy-driven, and the different "mechanical" or "switchable" operations are energetically downhill, $\Delta G < 0$. A DNA switch represents a mechanical device that reconfigures reversibly between two states, A and B, in the presence of fuel and anti-fuel components, respectively, Figure 5.6.3(A). The switched process exhibiting different DNA structures, or eventually the structures, might lead to "ON/OFF" switchable functionalities such as electrical, optical, and plasmonic properties, or catalytic functions. DNA machines or devices exhibit fuel–anti-fuel-driven switchable operations across several different states, Figure 5.6.3(B). Such molecular devices may reveal energetically controlled configurations or functions along the operation paths.

Beyond the intellectual curiosity in the fabrication of DNA devices, the practical applications of DNA machines hold great promise. These include the use of DNA machines as logic gate devices [95–98], the application of DNA machines for nano-medicine (drug carriers and release systems) [99–103], for controlled and programmed synthesis [104–107], and for the use of DNA machines as functional materials to control optical properties [108–110]. Substantial progress in designing DNA machines has been demonstrated, and the advances in the development of DNA machines have been addressed in several reviews [90–94]. The present chapter aims to exemplify several principles to assemble DNA switching devices and DNA machines, and to discuss the potential applications. Specifically, we wish to highlight the existing structural and functional "tool-boxes" of nucleic acids to assemble DNA switches and machines, and to highlight the perspectives of the area within the broad topic of DNA nanotechnology.

5.6.2 Ion-stimulated and photonic/electrical-triggered DNA switches

The ion-stimulated control of DNA nanostructures has been widely implemented to develop DNA switches. The pH-induced transitions of random-coil to i-motif nanostructures, the metal-ion-induced conversion of random-coil to G-quadruplex assemblies, and the metal-ions bridging of nucleotide bases (C–Ag^+–C or T–Hg^{2+}–T), Figure 5.6.1(A), represent the means to reversibly switch DNA structures between two different states. Similarly, physical stimuli, such as optical or electrical signals, can be used to trigger the reversible switching of DNA devices. For example, photoisomerizable azobenzene units may control the stability of DNA duplexes [111–113], Figure 5.6.4. While the intercalation of *trans*-azobenzene units into duplex DNAs cooperatively stabilizes the double-stranded structure, the photoisomerized *cis*-azobenzene isomer lacks affinity for the duplex DNA, and this might separate the duplex structure due to insufficient base-pairing stability. Similarly, an electrical trigger might either reduce metal ions to the metals or reversibly oxidize the metals to metal ions. This provides a general route to deplete or regenerate metal ions from the DNA surroundings. As a result the DNA nanostructures may be switched between ion-stabilized or ion-lacking configurations. This chapter will exemplify the tailoring of DNA switches by ions or physical triggers such as photonic or electrical signals.

5.6.2.1 Ion-stimulated DNA switches

The pH-stimulated switchable transitions between a hairpin structure, state I, and a hybrid DNAzyme–i-motif nanostructure, state II, is exemplified [114] in Figure 5.6.5(A). The hairpin structure (**4a**), includes, in domain ***a***, the C-rich sequence that might lead at an acidic pH to the energetically-stabilized i-motif structure. Domain ***b*** of the hairpin consists of the G-rich sequence that may lead to the self-assembly of the G-quadruplex. At neutral pH, the base pairing between G–C bases of domains ***a*** and ***b*** leads to the

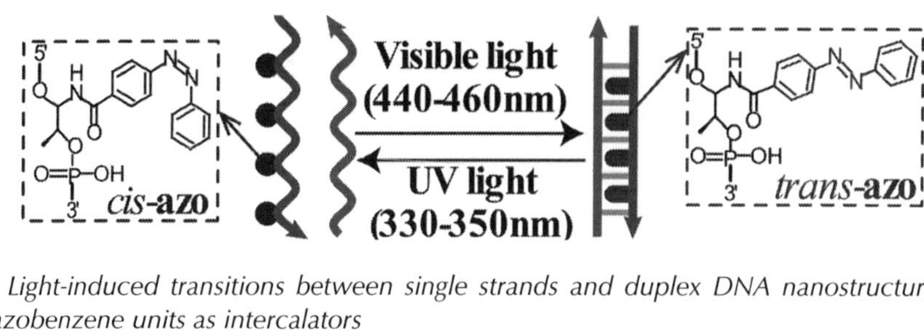

Figure 5.6.4 *Light-induced transitions between single strands and duplex DNA nanostructures using photo-isomerizable azobenzene units as intercalators*

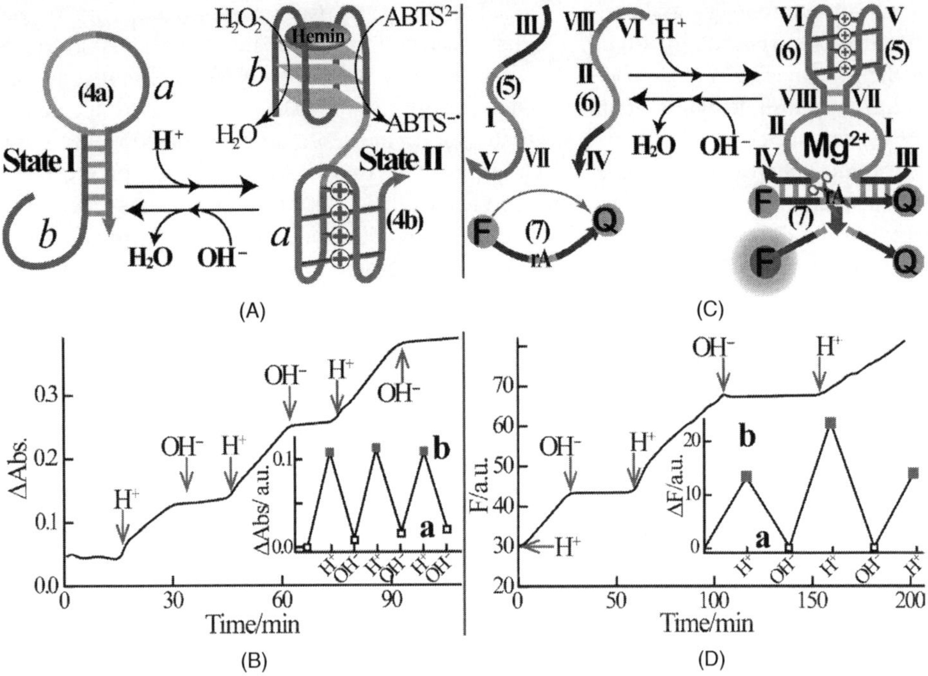

Figure 5.6.5 (A) Cyclic pH-triggered switchable transitions of a functional hairpin structure, state I, into an i-motif-hemin–G-quadruplex DNAzyme conjugate, state II, and the reverse process. (B) Absorbance changes upon the H^+-triggered formation of the i-motif-hemin–G-quadruplex DNAzyme conjugate, state II, that catalyzes the H_2O_2-mediated oxidation of $ABTS^{2-}$ to $ABTS^{\bullet -}$, switched "ON", and the OH^--induced switched "OFF" system, state I. Inset: Cyclic absorbance changes upon switching the system between states I (a) and II (b), respectively. Reproduced with permission from [114] The Royal Society of Chemistry. (C) Cyclic pH-induced switchable assembly and separation of the Mg^{2+}-dependent DNAzyme using an i-motif bridging unit. The activity of the DNAzyme is followed by the fragmentation of the fluorophore–quencher (F–Q)-modified substrate, and the generation of fluorescence. (D) Fluorescence changes upon the cyclic switchable activation (H^+) and deactivation (OH^-) of the Mg^{2+}-dependent DNAzyme. Inset: Switchable fluorescence changes upon the pH-stimulated "OFF" (a) and "ON" (b) activation of the DNAzyme. Reproduced with permission from [115] The Royal Society of Chemistry

stabilization of the hairpin structure (**4a**). At acidic pH, pH = 5.2, domain ***a*** of (**4a**) self-assembles into the i-motif structure, leading to the separation of the hairpin to the hybrid (**4b**) composed of the i-motif and the G-quadruplex units. The association of hemin to the G-quadruplex yields the horseradish peroxidase mimicking DNAzyme that catalyzes the H_2O_2-mediated oxidation of 2,2′-azino-bis(3-ethylbenzothiazoline-6-sulfonic acid), $ABTS^{2-}$, to the colored product, $ABTS^{\bullet -}$. The subsequent neutralization of the system, pH = 7.0, separates the i-motif structure, and this dissociates the hemin–G-quadruplex counter component by regenerating the energetically favored hairpin structure (**4a**). By subjecting the system to cyclic pH changes between pH = 7.0 and pH = 5.2 the system was switched between the hairpin structure (**4a**) and the catalytically active hemin–G-quadruplex–i-motif hybrid nanostructure (**4b**), Figure 5.6.5(B). A related system has implemented the i-motif structure for the switchable activation of the Mg^{2+}-dependent DNAzyme [115], Figure 5.6.5(C). The single strands (**5**) and (**6**) include, in domains I and II, the base sequences corresponding to the Mg^{2+}-dependent DNAzyme, and in domains III and IV, the binding sequences for the association of the DNAzyme substrate (**7**). The sequences V and VI in (**5**) and (**6**), respectively, included the sequences corresponding to the i-motif nanostructure. At neutral pH, 7.2, the subunits (**5**) and (**6**) exist as separate components, with no catalytic function. At pH = 5.8, the i-motif structure formed between the respective subunits yields the loop domain of the Mg^{2+}-dependent DNAzyme that cooperatively binds the substrate (**7**) to form the supramolecular DNAzyme–substrate hybrid. In the presence of Mg^{2+} ions, the DNAzyme is activated, resulting in the cleavage of the ribonucleobase-containing substrate (**7**). As the substrate (**7**) is functionalized at its 3′- and 5′-ends with a fluorophore–quencher pair, which results in the quenching of the fluorophore, the cleavage of the substrate results in a fluorophore-functionalized nucleic acid that provides a fluorescence readout signal for the DNAzyme functions. The neutralization of the system dissociates the i-motif structure, leading to the separation of the subunits (**5**) and (**6**). By the cyclic pH-stimulated formation and dissociation of the i-motif unit, the system was switched between a catalytically active structure and a catalytically inactive system, respectively, Figure 5.6.5(D).

Different metal ions, such as K^+, Pb^{2+} or Tb^{3+}, are known to stabilize G-quadruplex structures [116, 117]. The elimination of the ions from the quadruplexes by appropriate affinity ligands can then be used to switch the G-quadruplex into a random coil structure. By the secondary association of stabilizing components to the G-quadruplex structures, the switchable transitions between the G-quadruplex and random coil configurations may then lead to switchable functionalities (such as catalytic functions) of the systems. Figure 5.6.6(A) depicts the Pb^{2+} ion-stimulated assembly and dissociation of a G-quadruplex structure and the switchable control of the fluorescence properties of a fluorophore interacting with the reconfigured states [118]. The strand (**8**) includes the G-quadruplex sequence and this is partially blocked by the complementary strand (**9**) through the formation of a duplex structure. In the presence of Pb^{2+} ions the duplex is separated by forming the energetically stabilized Pb^{2+}–G-quadruplex. Subjecting the system to the 1,4,7,10-tetraazacyclododecane-1,4,7,10-tetraacetic acid (DOTA) ligand eliminates the Pb^{2+} ions, through the formation of the Pb^{2+}–DOTA complexes, leading to the dissociation of the G-quadruplex. The secondary association of Zn(II)-protoporphyrin to the G-quadruplex results in an enhanced fluorescence of the fluorophore. According to the treatment of the duplex DNA structure (**8**)–(**9**) with Pb^{2+} and the subsequent DOTA-stimulated release of Pb^{2+} ions from the Pb^{2+}-stabilized G-quadruplex lead to the cyclic formation and dissociation of the G-quadruplex. This results in the switchable binding and dissociation of the Zn(II)-protoporphyrin (ZnPPIX) fluorophore to the G-quadruplex and leads to high and low fluorescence properties of the system, respectively, Figure 5.6.6(B). A related system has implemented the switchable formation and dissociation of a K^+-stabilized G-quadruplex to cycle the catalytic functions of the Mg^{2+}-dependent DNAzyme [119], Figure 5.6.6(C). The strands (**10**) and (**11**) include the domains I and II, corresponding to the G-quadruplex subunits, and the domains III and IV include the sequences corresponding to the Mg^{2+}-dependent DNAzyme. In the absence of K^+ ions, the strands (**10**) and (**11**) are separated and the system lacks catalytic functions. In the presence of K^+ ions, the G-quadruplex structure combines the strands (**10**) and (**11**) resulting in a stable loop sequence of the Mg^{2+}-dependent DNAzyme that cooperatively binds the ribonucleobase-containing DNAzyme substrate (**12**). In this structure, the Mg^{2+}-dependent DNAzyme is activated, leading to the cleavage of the substrate. Subjecting the system to 18-crown-6-ether leads to the elimination of K^+ ions from the G-quadruplex and to the separation of the nanostructure to a catalytically inactive mixture of the individual components. By the cyclic addition of K^+ ions and 18-crown-6-ether, the catalytic functions of the system were switched between "ON" and "OFF" states, respectively. By labeling of the substrate (**12**) with a fluorophore–quencher pair, the catalytic cleavage of the substrate by the Mg^{2+}-dependent DNAzyme could be followed by the fluorescence changes generated by the fluorophore-labeled fragmented product. Figure 5.6.6(D) depicts the fluorescence changes corresponding to the switchable "ON/OFF" activation and deactivation of the catalytic functions of the system in the presence of the K^+ and 18-crown-6-ether, respectively.

The further metal-ion-stimulated reconfiguration of a duplex DNA to a G-quadruplex structure on electrode surfaces is exemplified [120] in Figure 5.6.7(A). The cyclic structural transitions were followed by

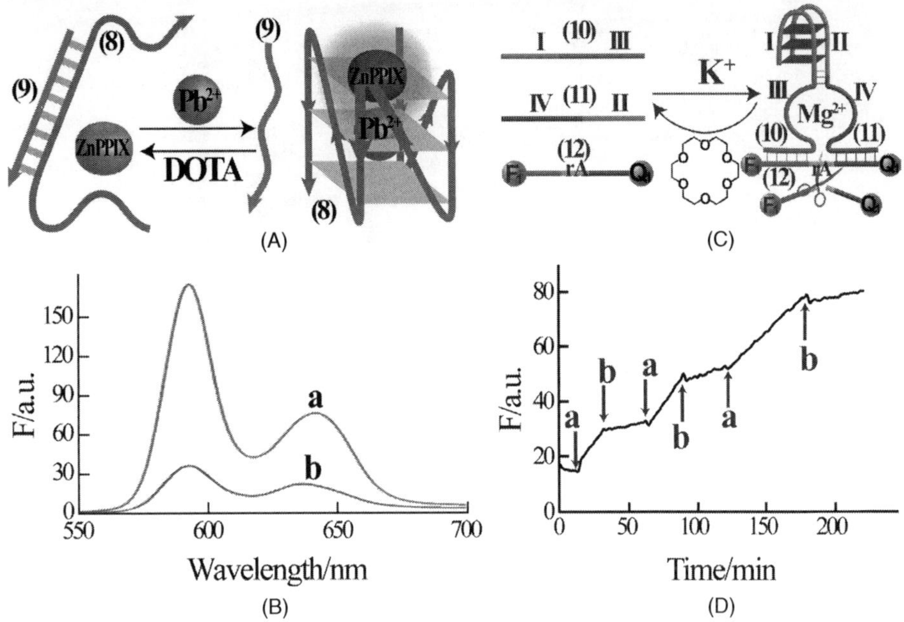

Figure 5.6.6 (A) Pb^{2+}-stimulated switchable reconfiguration of a duplex DNA structure into the Pb^{2+}-stabilized G-quadruplex structure, and the reverse separation of the G-quadruplex by the DOTA-assisted elimination of the Pb^{2+} ions. The formation and disassociation of the G-quadruplex are followed by the binding of Zn(II)-protoporphyrin IX (ZnPPIX) to the G-quadruplex (high fluorescence) and by the depletion of the fluorescence of ZnPPIX upon dissociation of the G-quadruplex. (B) Fluorescence spectra corresponding to: (a) the ZnPPIX bound to the Pb^{2+}-ion-stabilized G-quadruplex and (b) the free ZnPPIX formed upon the DOTA-stimulated separation of the G-quadruplex. Reproduced with permission from [118]. Copyright 2010, American Chemical Society. (C) K^+ ion-stimulated switchable "ON/OFF" activation and deactivation of the Mg^{2+}-dependent DNAzyme. The DNAzyme is formed by the K^+ ion-stimulated bridging of the DNAzyme subunits through the formation of a G-quadruplex. The Mg^{2+}-dependent DNAzyme is separated into its subunits by the 18-crown-6-ether (CE)-induced elimination of the K^+ ions from the G-quadruplex. The switchable supramolecular system is followed by the Mg^{2+}-dependent DNAzyme cleavage of a fluorophore–quencher (F–Q)-functionalized substrate. (D) Time-dependent fluorescence changes upon (a) the K^+ ion-stimulated activation of the Mg^{2+}-dependent DNAzyme and (b) the subsequent CE-stimulated elimination of K^+ ions and switched "OFF" the DNAzyme.Reproduced with permission from [119]. Copyright © 2014, WILEY-VCH Verlag GmbH & Co. KGaA, Weinheim

electrochemical means. The thiolated nucleic acid, (**13**), was assembled on an Au electrode, and the ferrocene-labeled nucleic acid, (**14**), exhibiting partial complementarity to (**13**) was hybridized with (**13**) to yield the duplex structure (**13**)–(**14**), which includes two loop regions I and II that are composed of the G-quadruplex subunits. In the presence of K^+ ions, the duplex structure reconfigured into the K^+-stabilized G-quadruplex. The subsequent treatment of the G-quadruplex structure with 18-crown-6-ether eliminates the K^+ ions, resulting in the duplex structure. The ferrocene redox label associated with the different nanostructures exhibits different spatial orientations. While in the duplex structure, the ferrocene redox labels are spatially separated from the electrode, giving rise to a low voltammetric response, Figure 5.6.7(B), curve (a), and the ferrocene units associated with the reconfigured G-quadruplex are in close proximity to the electrode surface, leading to a high voltammetric signal, Figure 5.6.7(B), curve (b). The cyclic, switchable transitions between the duplex configuration and G-quadruplex structure of (**13**)–(**14**) were then followed by the electrical responses of the ferrocene redox labels, Figure 5.6.7(B), inset.

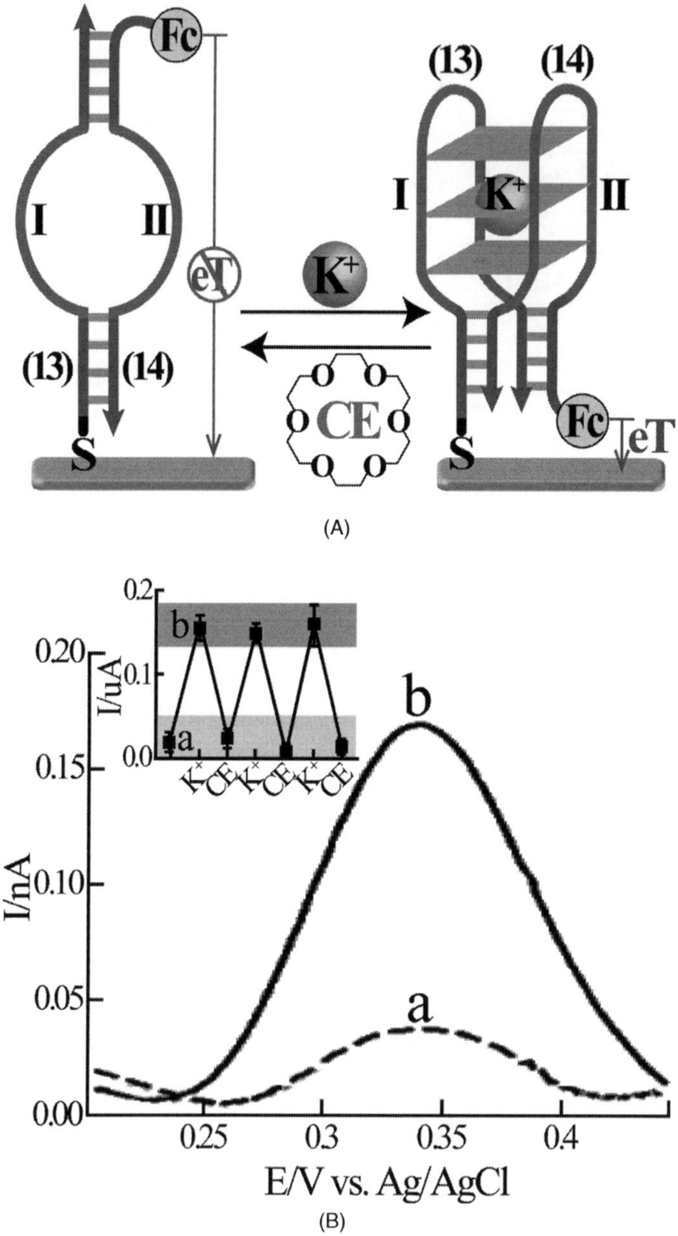

Figure 5.6.7 *(A) Electrochemical transduction of the cyclic transitions between a ferrocene-labeled double-loop duplex DNA nanostructure and a K^+ ion-stabilized G-quadruplex structure associated with the electrode. The G-quadruplex is formed in the presence of K^+ ions whereas the G-quadruplex reconfigures into the double-loop duplex nanostructure in the presence of 18-crown-6-ether (CE). (B) Voltammetric responses of the ferrocene units in: (a) the double-loop duplex structure and (b) the G-quadruplex DNA structure. Inset: Cyclic voltammetric responses upon treatment of the modified electrode in the presence of CE (a) and K^+ ions (b). Reproduced with permission from [120]. Copyright © 2010, WILEY-VCH Verlag GmbH & Co. KGaA, Weinheim*

Metal ions such as Ag^+ or Hg^{2+} are known to form complexes with cytosine (C) or thymine (T) bases, respectively (C–Ag^+–C or T–Hg^{2+}–T) [14–19]. Such complexes might cooperatively stabilize the formation of duplex DNA structures. The ligand-induced elimination of metal ions from the duplex assemblies may then lead to the separation of the strands. By the appropriate design of nucleic acid strands, the metal-ion-induced formation or dissociation of functional DNA nanostructures was demonstrated. Figure 5.6.8(A) depicts the design of two strands (**15**) and (**16**) that include the sequences which can assemble in domains I and II of the strands, under appropriate conditions, into the G-quadruplex structure. The binding of hemin to the resulting G-quadruplex yields the horseradish peroxidase (HRP)-mimicking DNAzyme [121]. The domains III and IV include partial complementarities and C–C mismatches. The complementarity of domains III and IV provides insufficient stabilization energy to form a stable duplex structure with the concomitant assembly of the G-quadruplex. However, in the presence of Ag^+ ions, the bridging C–Ag^+–C complexes stabilize the generated duplex structure between domains III and IV, leading to the co-stabilization of the G-quadruplex unit. The association of hemin to the G-quadruplex then yields the hemin–G-quadruplex HRP-mimicking DNAzyme, which catalyzes the oxidation of $ABTS^{2-}$ by H_2O_2 to the colored product $ABTS^{\bullet-}$. Subjecting the catalytically active nanostructure to the cysteine (Cys) ligand, eliminated Ag^+ ions from the duplex nanostructures, resulting in the separation of the strands (**15**) and (**16**), and to the depletion of the catalytically active properties of the system. That is, by the cyclic treatment of the system with Ag^+ ions and cysteine, the switchable "ON" and "OFF" activation and deactivation of the catalytic functions of the system were achieved. Similarly, Figure 5.6.8(B) shows the implementation of Hg^{2+} ions, and the cysteine ligand, to control the switchable catalytic functions of the hemin–G-quadruplex HRP-mimicking DNAzyme [122]. The strands (**17**) and (**18**) include the G-rich sequences in domains I and II, providing the subunits to assemble the G-quadruplex. The tethers III and IV include partial complementarities to the auxiliary strand (**19**) and T–T mismatches between the tethers III, IV, and strand (**19**). Under appropriate conditions, and in the absence of Hg^{2+} ions, the G-quadruplex consisting of (**17**)–(**18**) is energetically stabilized. The binding of hemin to the G-quadruplex yields the HRP-mimicking DNAzyme that catalyzes the H_2O_2-mediated oxidation of luminol to yield chemiluminescence as the readout signal. Treatment of the system with Hg^{2+} ions separates the G-quadruplex structure while reconfiguring the system into a catalytically inactive duplex composed of the scaffold (**19**) that binds the subunits (**17**) and (**18**) by cooperative base pairing and T–Hg^{2+}–T bridges, resulting in the stabilization of the hybrid (**19**)–(**17**)+(**18**). Addition of cysteine to the resulting structure eliminated the co-stabilizing Hg^{2+} ions, resulting in the separation of strands (**17**) and (**18**) from the scaffold (**19**), and to their reassembly into the catalytically active hemin–G-quadruplex HRP-mimicking DNAzyme nanostructure. By the cyclic treatment of the system with Hg^{2+} ions and cysteine, the nucleic acid structures were reversibly switched between a catalytically inactive configuration of (**19**)–(**17**)+(**18**) and a catalytically active nanostructure of the hemin–G-quadruplex DNAzyme, respectively, Figure 5.6.8(C).

5.6.2.2 Photonic and electrical triggering of DNA switches

Reversible photoisomerizable chromophores may interact differently with duplex DNA structures, thus leading to the cooperative photochemical control of the stabilities of nucleic acid duplexes [123]. Specifically, photoisomerizable azobenzene derivatives exhibit light-controlled affinities toward double-stranded nucleic acids. While *trans*-azobenzene intercalates into duplex DNA structures, leading to the co-stabilization of the double-stranded configuration, the ultraviolet (UV) irradiation of *trans*-azobenzene yields *cis*-azobenzene that lacks binding affinity to the duplex DNA structure. The reverse visible light photoisomerization of *cis*-azobenzene to *trans*-azobenzene can then be used to re-stabilize the duplex DNA structure.

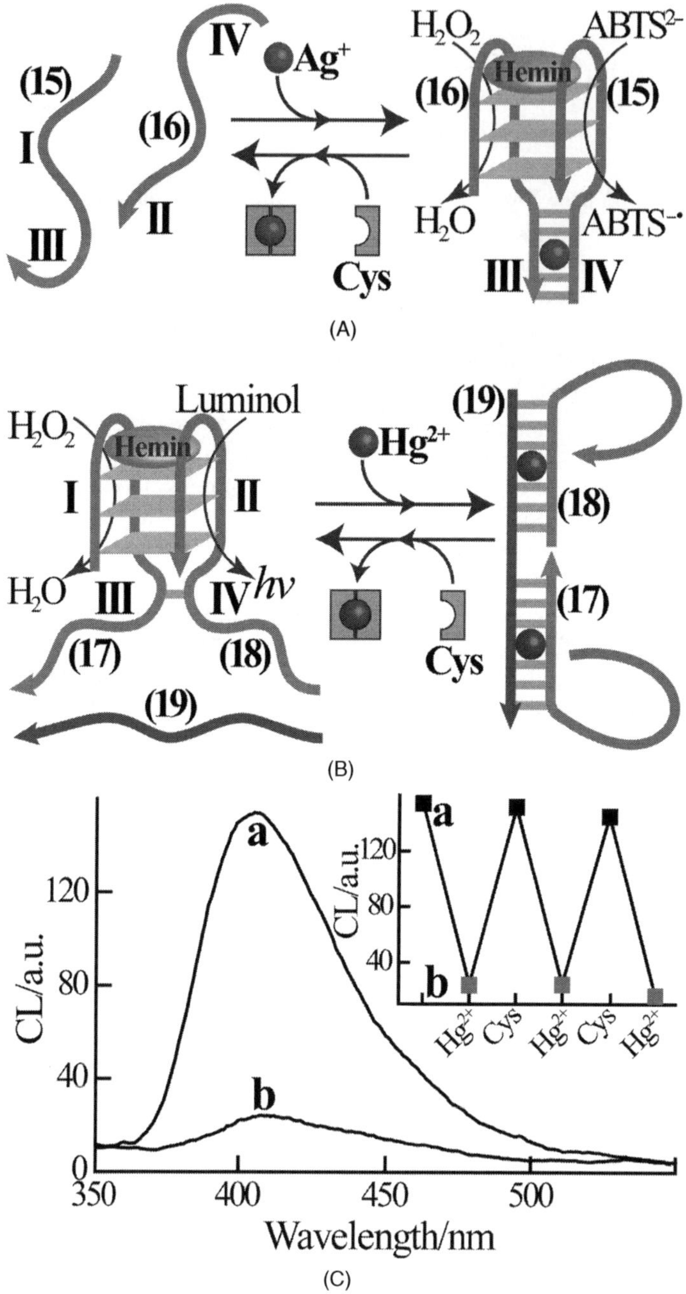

Figure 5.6.8 *(A) Switchable assembly of a catalytic hemin–G-quadruplex HRP-mimicking DNAzyme in the presence of Ag^+ ions, and its separation in the presence of cysteine (Cys). Reproduced with permission from [121]. Copyright © 2009, WILEY-VCH Verlag GmbH & Co. KGaA, Weinheim. (B) Cyclic Hg^{2+}–Cys-stimulated, switchable, reconfiguration of a duplex structure to a hemin–G-quadruplex DNAzyme and the reverse process. The reconfiguration is imaged via probing the biocatalyzed generation of chemiluminescence by the hemin–G-quadruplex catalyzed oxidation of luminol by H_2O_2. (C) Chemiluminescence spectra corresponding to: (a) the system in the hemin–G-quadruplex configuration and (b) the Hg^{2+} ion-stabilized reconfigured duplex assembly. Inset: Cyclic chemiluminescence intensities upon switching the system between the hemin–G-quadruplex structure (a) and the Hg^{2+}-stabilized duplex (b). Reproduced with permission from [122]. Copyright 2013, American Chemical Society*

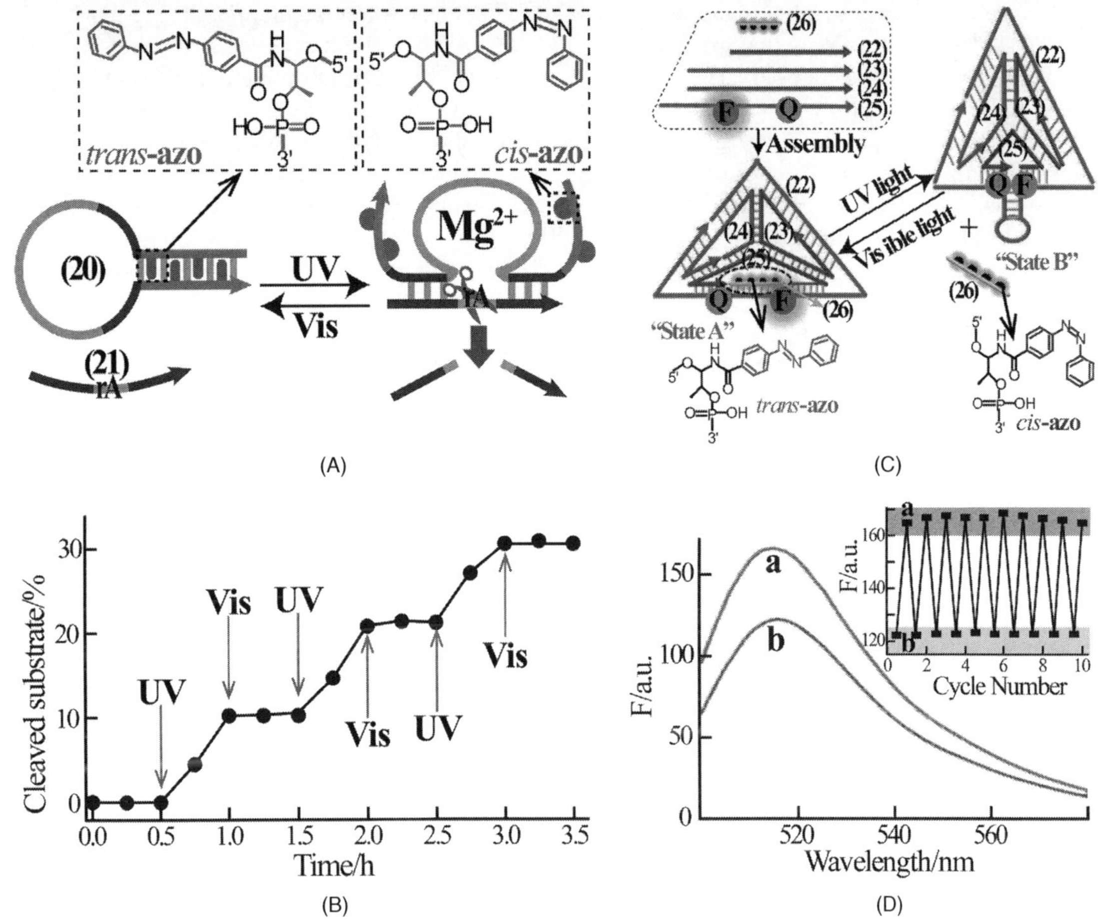

Figure 5.6.9 *(A) Photoinduced cyclic switching of an azobenzene-modified hairpin structure into a Mg^{2+}-dependent DNAzyme and the reverse photo-triggered process. (B) The light-induced "ON/OFF" activation of the Mg^{2+}-dependent DNAzyme followed by the cleavage of the DNAzyme substrate and applying gel electrophoresis to identify the fragmented products. Reproduced with permission from [124]. Copyright © 2010, WILEY-VCH Verlag GmbH & Co. KGaA, Weinheim. (C) Photoinduced reconfiguration of a DNA tetrahedron structure between an extended configuration, state A, and a compressed structure, state B. (D) Cyclic fluorescence monitoring of the light-induced transitions between states A and B, respectively. Inset: Cyclic fluorescence intensity changes at $\lambda = 520$ nm, corresponding to: (a) extended state A and (b) compressed state B. Reproduced with permission from [125] The Royal Society of Chemistry*

This photo-controlled stabilization–destabilization of duplex DNA structures has been implemented to switch the functional and structural properties of supramolecular DNA systems. Figure 5.6.9(A) depicts the light-controlled switchable "ON/OFF" activation and deactivation of the Mg^{2+}-dependent DNAzyme [124]. The hairpin structure (**20**) is co-stabilized by *trans*-azobenzene units, tethered as internal photoisomerizable components. The loop region includes the Mg^{2+}-dependent DNAzyme sequence in a caged catalytically inactive configuration. UV-stimulated photoisomerization of the *trans*-azobenzene units to the *cis*-state weakens the stability of the stem region, resulting in the reconfiguration of the hairpin structure into the Mg^{2+}-dependent DNAzyme structure, which is stabilized by hybridization with the ribonucleobase-containing DNAzyme substrate (**21**). The DNAzyme-catalyzed cleavage of substrate (**21**) yielded fragmented nucleic acid products that were followed electrophoretically. The reverse visible light photoisomerization of the *cis*-azobenzene units to *trans*-azobenzene regenerated the catalytically inactive *trans*-azobenzene-stabilized hairpin structure. Through the cyclic irradiation of the system with UV light and visible light, the catalytic functions of the system were switched between "ON" and "OFF" states, respectively, Figure 5.6.9(B). The cyclic photo-stimulated reconfiguration of a DNA nanostructure has been demonstrated by the reversible transitions of a DNA tetrahedron across an extended configuration (state A) and a compressed structure (state B) [125], Figure 5.6.9(C). The strands (**22**), (**23**), (**24**), and (**25**) include complementarities that enable the formation of a DNA tetrahedron. In the presence of the co-added *trans*-azobenzene-modified strand (**26**), the tetrahedron adapts an extended configuration, state A, where the *trans*-azobenzene units of (**26**) co-stabilize the duplex between the auxiliary strand (**26**) and the complementary domain associated with strand (**25**). UV-stimulated photoisomerization of the *trans*-azobenzene units to *cis*-azobenzene released the strand (**26**) and the resulting free single-stranded domain in (**25**) folded into a hairpin structure, leading to the compression of the tetrahedron, state B. By the internal modification of strand (**25**) with a fluorophore–quencher (F–Q) pair, the structural transitions of the DNA tetrahedron between the extended configuration, state A, and the compressed structure, state B, could be followed by fluorescence spectroscopy, Figure 5.6.9(D). In the extended structure, the system revealed higher fluorescence while the compressed system showed lower fluorescence, due to the enhanced quenching of the fluorophore. Through the cyclic photoisomerization of the system between the *trans*- and *cis*-azobenzene isomers, the cyclic, switchable transitions of the DNA tetrahedron between states A and B were demonstrated, Figure 5.6.9(D) inset.

The photochemical stabilization of duplex nucleic acid structures by means of *trans*-azobenzene photoisomerizable units was further implemented to switch the cyclic aggregation/de-aggregation of Au nanoparticles (Au NPs) [126], Figure 5.6.10(A). Two types of Au NPs, I and II, were prepared, where Au NPs of type I were modified with the nucleic acid (**27**) and the Au NPs of type II were functionalized with the azobenzene-modified nucleic acid (**28**). The nucleic acids (**27**) and (**28**) include partial complementarity, but this is insufficient to stabilize duplex structures between the NPs. In the presence of the *trans*-azobenzene units the cooperative stabilization of the duplex structures between (**27**) and (**28**) proceeds, leading to the aggregation of the NPs. The UV photoinduced isomerization of *trans*-azobenzene units to the *cis*- azobenzene state separates the bridging DNA duplexes that crosslinked the NPs, leading to de-aggregation of Au NPs. By the cyclic photoisomerization of the system with UV light and visible light, the Au NPs system was switched between de-aggregated and aggregated states, respectively, and the transitions were followed spectroscopically, Figure 5.6.10(B).

DNA switches may be similarly triggered by electrical signals. This is exemplified with the electrically triggered transitions between a Pb^{2+} ion-stabilized G-quadruplex and a random-coil structure [127], Figure 5.6.11(A). In the presence of Pb^{2+} ions, the nucleic acid (**29**) is stabilized as a G-quadruplex. The electrochemical reduction of Pb^{2+} ions and the deposition of metal Pb^0 on the electrode support lead to the separation of the G-quadruplex, and to the random coil structure. Incorporation of crystal violet (CV) into the system leads to low fluorescence of CV in the presence of the G-quadruplex state and to high fluorescence upon interaction with the random coil configuration. Thus, by the electrochemical reduction of the Pb^{2+} ions to Pb^0 and the reverse oxidation of Pb^0 to Pb^{2+} ions, the system was cycled between the random-coil structure and the Pb^{2+} ion-stabilized G-quadruplex, respectively, and the switching events were followed by the fluorescence intensities of CV, Figure 5.6.11(B).

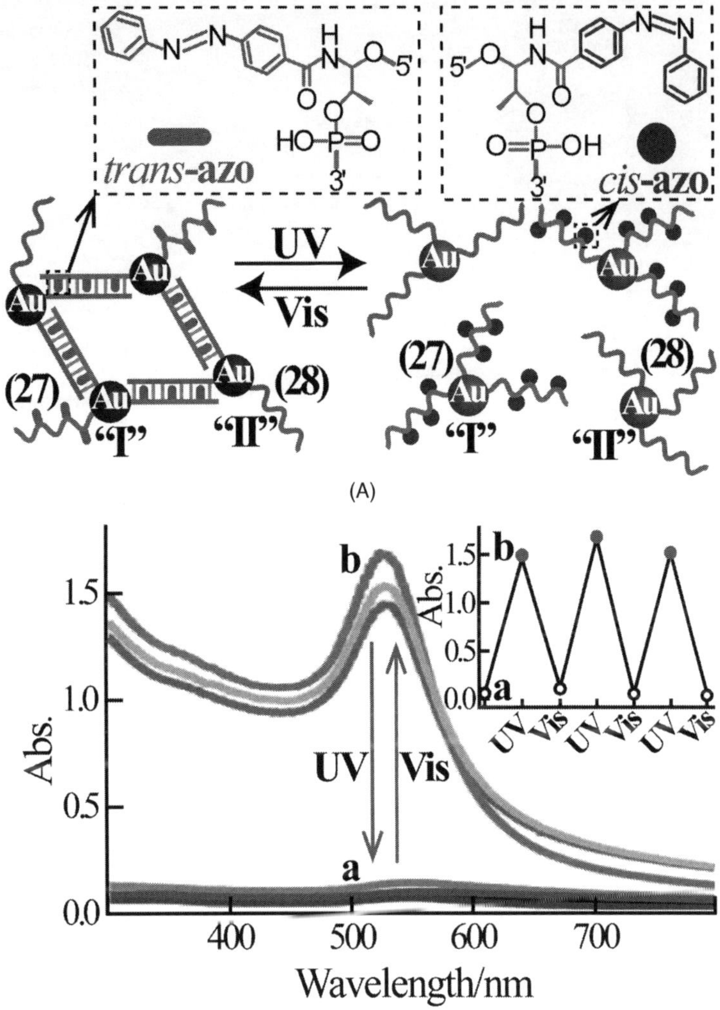

Figure 5.6.10 *(A) Photochemically controlled aggregation and de-aggregation of Au NPs using azobenzene-modified nucleic acid bridges. (B) Absorption spectra corresponding to: (a) the trans-azobenzene stimulated aggregated Au NPs and (b) the cis-azobenzene-induced de-aggregated Au NPs. Inset: Cyclic absorbance changes, at $\lambda = 530$ nm, upon the aggregation (a) and de-aggregation (b) of the Au NPs. Reproduced with permission from [126]. Copyright 2012, American Chemical Society*

5.6.3 Switchable DNA machines

The different stimuli used for the cyclic switching of the structures or functions of DNA assemblies can be also implemented to drive switchable devices that perform machine-like functions, such as tweezers [119, 128–130], walkers [122, 131, 132], gears [133], rotors [134], and more [135, 136]. The mechanical functions of DNA nanostructures may be driven in cyclic switchable directions, and can proceed across two

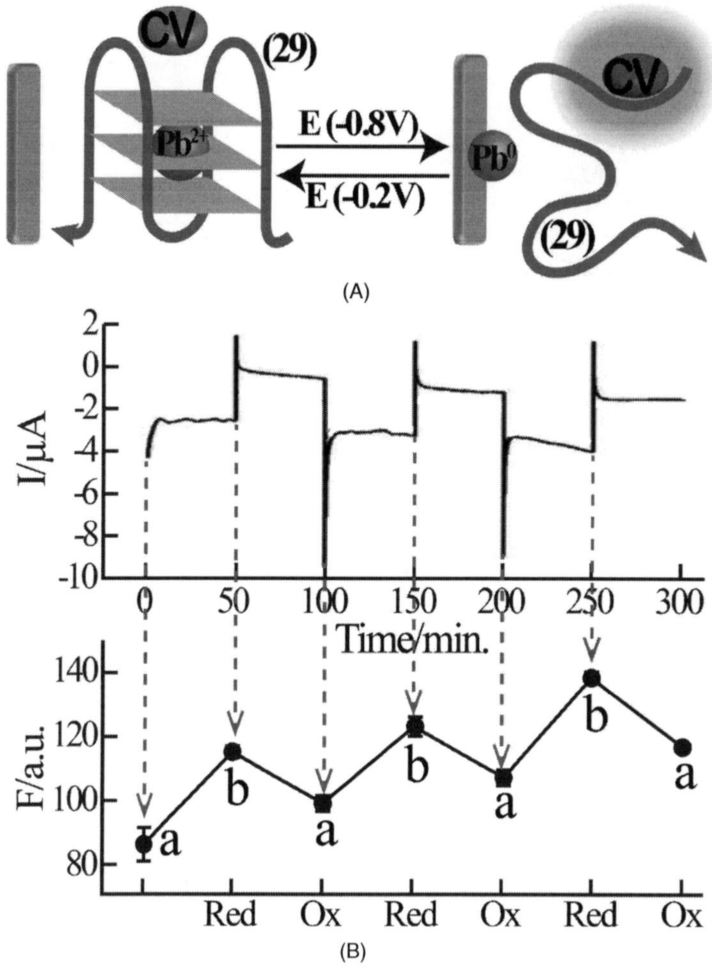

Figure 5.6.11 *(A) Electrochemically induced formation and dissociation of the Pb^{2+} ion-stabilized G-quadruplex. The voltammetric reduction of Pb^{2+} ions to Pb^0 separates the G-quadruplex while the oxidation of Pb^0 to Pb^{2+} ions regenerated the Pb^{2+}-stabilized G-quadruplex. (B) Voltammetric step corresponding to the oxidation of Pb^0 and the reduction of Pb^{2+} ions to Pb^0 and the concomitant fluorescence changes upon the dissociation of crystal violet from the G-quadruplex and its interaction with the separated G-rich random-coil strand. Reproduced with permission from [127] Elsevier*

distinct states or along several different states. Furthermore, the DNA machines can operate on surfaces, thereby allowing the transductions of the switchable molecular functions by the surfaces, for example, electrochemical transductions of the different states of the DNA machines [122, 135]. In this sub-section, the operation of cyclic and switchable DNA machines in solution and on surfaces will be discussed.

5.6.3.1 Two-state switchable DNA machines

The opening and closing of supramolecular DNA structures mimicking the functions of macroscopic tweezers have been demonstrated using different fuels–anti-fuels as triggering stimuli. Figure 5.6.12(A) depicts the operation of DNA tweezers using nucleic acids as fuel and anti-fuel triggers [128]. The tweezers consist of

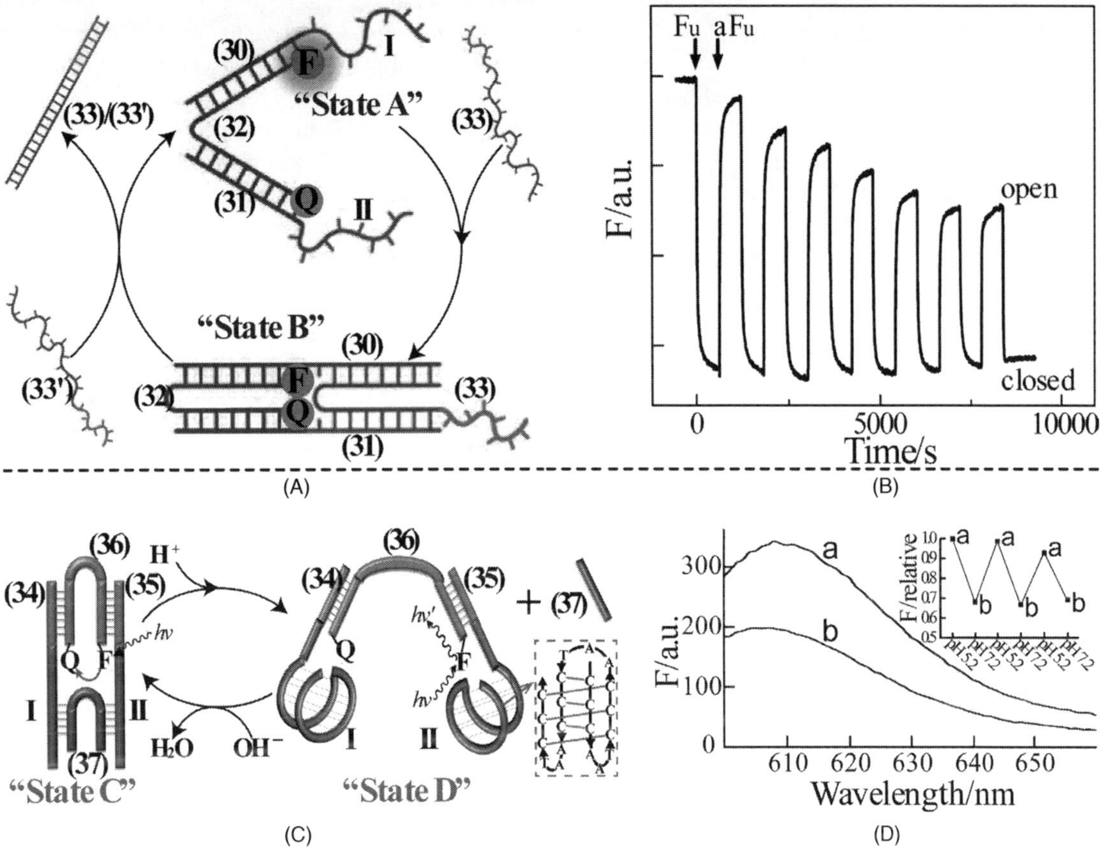

Figure 5.6.12 (A) Switchable transitions of a DNA tweezers structure using the strand-displacement mechanism. (B) Probing the "mechanical" functions of the DNA tweezers by fluorescence. Reproduced with permission from [128]. Copyright © 2000, Rights Managed by Nature Publishing. (C) pH-stimulated reversible opening and closure of the DNA tweezers. (D) Fluorescence properties of the open (a) and closed (b) states of the tweezers. Inset: Cyclic fluorescence changes upon the pH-stimulated opening and closure of the tweezers. Reproduced with permission from [129]. Copyright 2009, American Chemical Society

two "arms" (**30**) and (**31**) bridged together by the strands (**32**) and (**33**). In the presence of the fuel strand (**33′**), the bridging strand (**33**) is displaced, through the formation of the energetically-favored duplex (**33**)–(**33′**), resulting in the opening of the tweezers, state A. Re-addition of strand (**33**), results in the bridging of the "arms" and the closure of the tweezers, state B. The labeling of the bridging strand (**32**) with a fluorophore–quencher (F–Q) pair provides a fluorescence probe for the opening and closing of the tweezers. The close proximity between the fluorophore and quencher in the closed tweezers (state B) leads to effective quenching of the fluorophore, while the spatial separation of the fluorophore from the quencher in the open tweezers (state A) leads to less efficient quenching of the fluorophore. The cyclic switching of the tweezers structures between closed and open states was then followed by the fluorescence intensities of the fluorophore probe, Figure 5.6.12(B). A related approach has implemented pH as fuel–anti-fuel triggers that switch the tweezers device [129], Figure 5.6.12(C). The nucleic acids (**34**) and (**35**) act as the tweezers "arms", and these are bridged by the nucleic acid strands (**36**) and (**37**). The strands (**34**) and (**35**) include C-rich sequences that self-assemble into i-motif structures at acidic pH (pH = 5.2) in domains I and II, respectively. The binding of the "arm" units (**34**) and (**35**) with the bridging strands (**36**) and (**37**) at neutral pH (pH = 7.2), yields the closed configuration of the tweezers, state C. At acidic pH (pH = 5.2), the arms reconfigure into the energetically favored i-motif structures that result in the release of the bridging strand (**37**) and the opening of the tweezers, state D. Neutralization of the system (pH = 7.2) dissociates the i-motif structures, resulting in the rebinding of strand (**37**) to the "arms" and the closure of the tweezers, state C. By labeling of the bridging nucleic acid (**36**) with a fluorophore–quencher (F–Q) pair, the cyclic mechanical "transitions" of the tweezers system between the open and closed states were followed by the fluorescence intensities of the fluorophore probe, Figure 5.6.12(D).

Similarly, light stimuli, and photoisomerizable azobenzene intercalators, have been applied to switch tweezer structures between open and closed states [130], Figure 5.6.13(A). The tweezers consisted of two "arm" strands (**38**) and (**39**) bridged by the nucleic acid strands (**40**) and (**41**), resulting in the closed tweezers, state A. The bridging strand (**41**) included internally modified *trans*-azobenzene photoisomerizable units. The *trans*-azobenzene units cooperatively stabilized the duplex structures between domains I and II, associated with the "arms" (**38**) and (**39**), and the complementary sequences of bridging unit (**41**), through intercalation into the double-stranded regions. UV-light irradiation of the system (330–350 nm) photoisomerized the *trans*-azobenzene units to the *cis*-azobenzene state, which lacks affinity for the duplex structures. This weakened the duplex regions between the "arms" and bridging strand (**41**), resulting in the separation of strand (**41**), and the opening of the tweezers, state B. The reverse irradiation of the open tweezers and the separated strand (**41**) with visible light (440–460 nm) photoisomerized the *cis*-azobenzene units to the *trans*-azobenzene state, leading to the re-hybridization of the *trans*-azobenzene-modified strand (**41**) with the "arms" of the tweezers, and to the closure of the molecular devices. By the reversible light-induced switching of the azobenzene units between the *trans*- and *cis*-azobenzene states, the tweezers were cycled between closed (state A) and open (state B) configurations. As the bridging strand (**40**) linking the tweezer "arms" (**38**) and (**39**) is functionalized with a fluorophore–quencher (F–Q) pair, the fluorescence of the fluorophore label provided a means to follow the mechanical functions of the DNA device, Figure 5.6.13(B).

Furthermore, the formation of K^+-stabilized G-quadruplexes, and the dissociation of the quadruplexes by the elimination of the K^+ ions, have been applied to reversibly switch the opening and closing of the DNA tweezers [119], Figure 5.6.13(C). The tweezers consist of two "arm" strands (**42**) and (**43**) bridged together by the bridging strands (**44**) and (**45**), resulting in the closed tweezers, state C. The domains I and II in the "arms" include G-rich sequences, which, under appropriate conditions, can self-assemble into the G-quadruplexes. Subjecting the closed tweezers to the K^+ ions results in the self-assembly of domains I and II into K^+-stabilized G-quadruplexes, the separation of linker strand (**45**) from the supramolecular structure, and the opening of the tweezers, state D. The subsequent addition of the kryptofix [2.2.2] receptor eliminates the K^+ ions from the G-quadruplexes, resulting in their separation. This leads to the re-hybridization of strand (**45**) with the "arms", and to the closure of the tweezers. As the bridging unit (**44**) is modified by a fluorophore–quencher (F–Q) pair, the opening and closing of the tweezers are probed by the fluorescence intensities of the fluorophore. Figure 5.6.13(D) shows the cyclic fluorescence changes upon switching the tweezers across states C and D using K^+ ion and kryptofix [2.2.2] as triggers.

A different mechanical DNA device includes the reversible and switchable stimuli-triggered walk-over of a DNA strand between two footholds associated with a nucleic acid scaffold [131]. This is exemplified in Figure 5.6.14(A) with the cyclic, photoinduced transitions of a nucleic acid strand L (**46**) between two nucleic acid footholds (**47**) and (**48**) hybridized with a nucleic acid scaffold (**49**). The nucleic acids (**47**) and (**48**) include protruding single-stranded tethers I and II, which provide the binding sites for walker unit L (**46**). The protruding tether II is internally modified with azobenzene photoisomerizable units, and the two tethers I and II exhibit partial complementarities to the "moving" element L. When the azobenzene

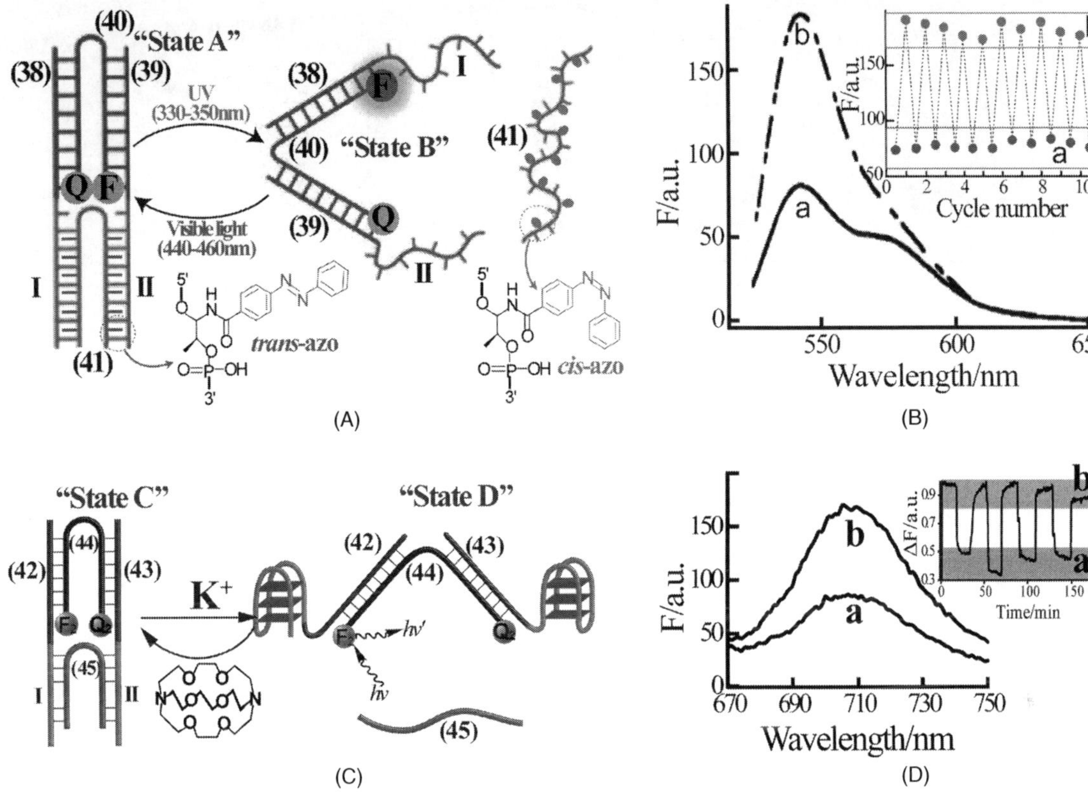

Figure 5.6.13 *(A) Photoinduced opening and closure of DNA tweezers modified with photoisomerizable azobenzene units. In the trans-azobenzene configuration the tweezers are closed, whereas in the cis-azobenzene state the tweezers are opened. The opening and closing of the tweezers is followed by labeling the bridging strand with a fluorophore–quencher pair and following the fluorescence of the fluorophores in the different states. (B) Fluorescence spectra of the system: (a) in the closed state (b) in the open state. Inset: Cyclic fluorescence changes upon switching the tweezers between the open (a) and closed (b) states. Reproduced with permission from [130]. Copyright © 2008, WILEY-VCH Verlag GmbH & Co. KGaA, Weinheim. (C) K^+–cryptate stimulated opening and closing of DNA tweezers. The open and closed states of the tweezers are transduced by the fluorescence intensities generated by a fluorophore–quencher modified-bridging unit: closed state – low fluorescence; open state – high fluorescence. (D) Fluorescence spectra corresponding to: (a) the closed tweezers and (b) the open tweezers. Inset: Cyclic fluorescence intensities upon the reversible opening and closure of the tweezers in the presence of K+ ions and cryptate (CP), respectively. Reproduced with permission from [119]. Copyright © 2014, WILEY-VCH Verlag GmbH & Co. KGaA, Weinheim*

photoisomerizable units, associated with tether II, exist in the *cis*-configuration, which lack affinity towards duplex nucleic acid structure, the "walker" L forms the energetically stabilized duplex with tether I, state A. Photoisomerization of the *cis*-azobenzene units to the *trans*-configuration results in a strand-displacement process where the walker element L steps-over to tether II, to yield the energetically favored duplex between walker element L and tether II, state B, originating from the cooperative stabilization of the duplex as a result of intercalation of the *trans*-azobenzene units into the hybrid. Subjecting state B to the UV-irradiation isomerizes the *trans*-azobenzene components into the *cis*-configuration. This weakens the binding interactions between walker L and tether II, resulting in the reverse strand-displacement process, and the

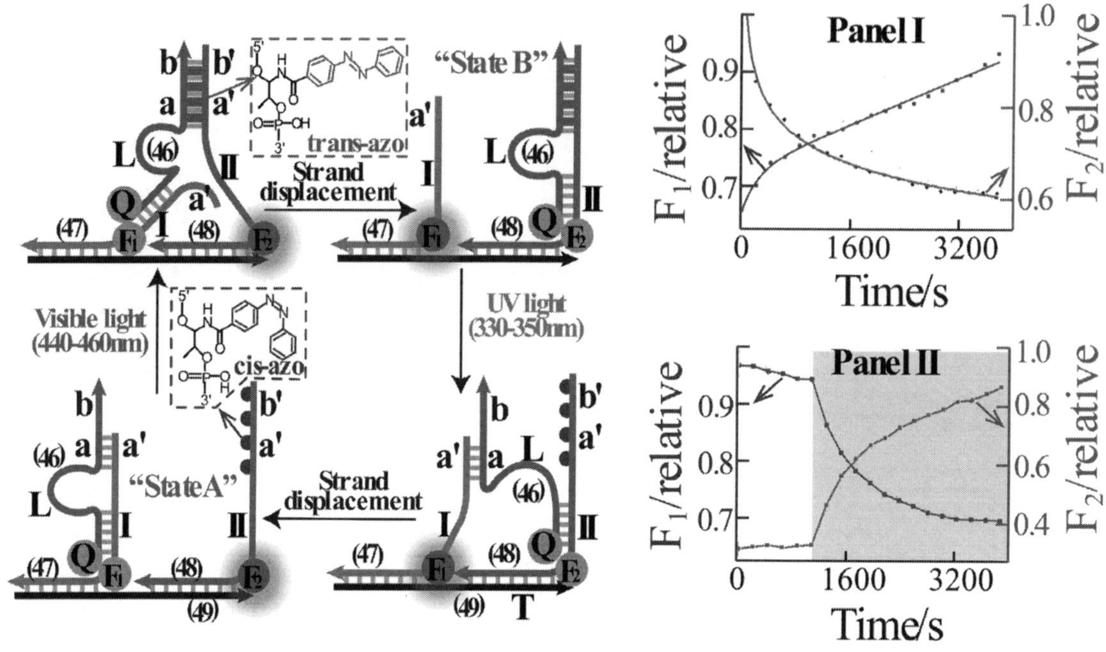

Figure 5.6.14 (A) A light-induced two-state DNA walker device driven by the reversible photoisomerization of cis- and trans-azobenzene units. (B) Fluorescence imaging of the two-state device using two fluorophores (F1 and F2) associated with the footholds and a quencher (Q) unit linked to the walker unit. Panel I: Time-dependent fluorescence changes upon translocation of the walker system from states A to B. Panel II: Time-dependent fluorescence changes upon translocation of the walker from state B to state A. Reproduced with permission from [131]. Copyright 2012, American Chemical Society

translocation of L to foothold I exhibiting energetically favored base-pairing hybridization. Through the cyclic irradiation of the system with visible light (440–460 nm) or UV light (300–350 nm), the walker unit L is reversibly switched between states B and A, respectively. By the internal modification of footholds (**47**) and (**48**) with two different fluorophores, F_1 and F_2, and via the labeling of the "walker" unit L (**46**) with a quencher unit, Q, the time-dependent transitions of the "walker" across the two different states, and the switchable mechanical functions of the DNA device could be probed by the fluorescence features of F_1 and F_2, Figure 5.6.14(B). While in state A, F_1 revealed low fluorescence and F_2 showed high fluorescence, state B showed high fluorescence of F_1 and low fluorescence of F_2.

The operation of two-state DNA machines was further examined on surfaces, particularly on electrode supports. This enables to probe and transduce the mechanical properties of the molecular structures by electrochemical means. Figure 5.6.15(A) depicts the cyclic operation of a "DNA arm" on an Au electrode surface [135]. The thiolated nucleic acid (**50**) was assembled on an Au electrode and it acted as a scaffold for the construction of the "arm" components on it. The β-cyclodextrin (β-CD), receptor was functionalized with nucleic acid (**51**), and the conjugate was hybridized with domain I of the scaffold (**50**). The nucleic acid (**52**) was functionalized at its 5′-end with the redox-active ferrocene (Fc) label and it folded into a stable hairpin structure, where the free tether II′ was hybridized with domain II of the scaffold. The hybridization of the (**51**)-modified β-CD receptor and of the ferrocene-functionalized hairpin structure (**52**) with the scaffold

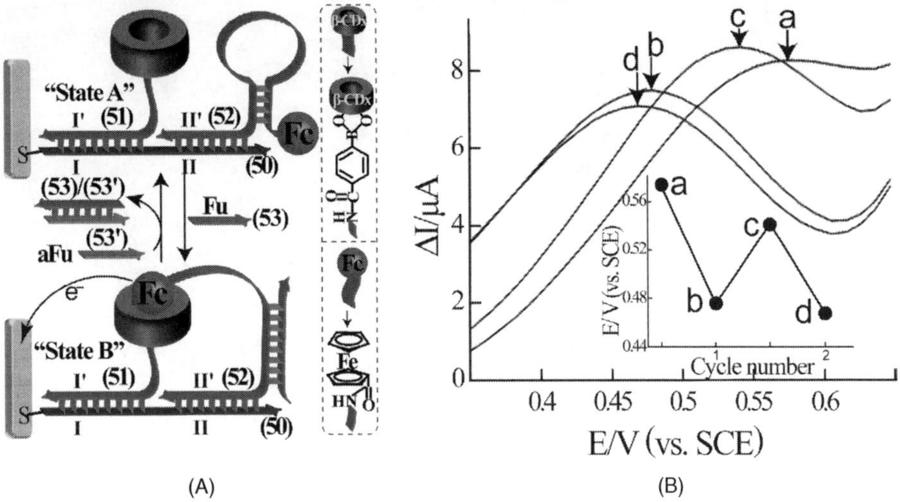

Figure 5.6.15 (A) Activation of a DNA "arm" on a nucleic acid scaffold associated with an electrode. The arm device consists of a nucleic acid foothold (**51**) functionalized with a β-cyclodextrin (β-CD) receptor, and a second foothold composed of the "arm" sequence (**52**) modified by a ferrocene redox-label, caged in a hairpin structure, associated with the scaffold (**50**). The opening of the hairpin by the fuel (Fu) strand releases the ferrocene-labeled "arm" that binds to the β-CD receptor. The removal of the fuel strand by the anti-fuel (aFu) strand results in the regeneration of the cagedarm configuration in the energetically stabilized hairpin structure. The cyclic transitions of the "arm" between states A and B are followed by the voltammetric responses of the ferrocene redox-label. (B) Voltammetric responses of the ferrocene redox-label (a) and (c) correspond to state A, (b) and (d) correspond to state B. Inset: Switchable voltammetric responses of the DNA "arm" upon switching the system between states A and B, respectively. Reproduced with permission from [135] The Royal Society of Chemistry

generated state A. Subjecting the system to the fuel strand (**53**) opens the hairpin structure, leading to a flexible Fc-labeled arm that binds the ferrocene probe to the β-CD receptor, to yield state B of the device. Treatment of state B with the anti-fuel strand (**53**′) displaces the fuel strand (**53**) while generating the energetically stabilized duplex (**53**)–(**53**′) as waste product. The release of strand (**53**) from the hybrid structure reconfigures the "arm" in the caged configuration of the energetically favored hairpin structure. As the spatial positions of the ferrocene redox-active label vary in states A and B, the voltammetric responses of the redox probe transduce the positions of the arm. Through the cyclic treatment of the system with the fuel (**53**) and anti-fuel (**53**′) strands, the position of the arm is controlled between states B and A, respectively, Figure 5.6.15(B).

The cyclic walk-over of a nucleic acid strand between two footholds associated with a DNA scaffold associated with an electrode was similarly transduced by electrochemical means [122], Figure 5.6.16(A). The thiolated nucleic acid (**54**) was assembled on an Au electrode. The nucleic acid (**55**) was hybridized with domain I of the scaffold and its single-stranded protruding tether provided foothold ***a*** of the device. The nucleic acid (**56**) was hybridized with domain II of the scaffold, and its single-stranded protruding tether ***b*** acted as the second foothold of the "walker" device. The walker unit (**57**) includes a G-rich region, capable of forming, under appropriate conditions, the G-quadruplex, and it is hybridized with foothold ***b*** of strand (**56**), where the G-rich sequence is caged in a duplex configuration, state A. Although the "walker" strand includes partial complementarity to foothold ***a*** of strand (**55**) it rests as duplex on foothold ***b*** due to favored energetic base pairing. In the presence of the fuel strand (**58**) the strand displacement of walker unit (**57**) from foothold ***b***.

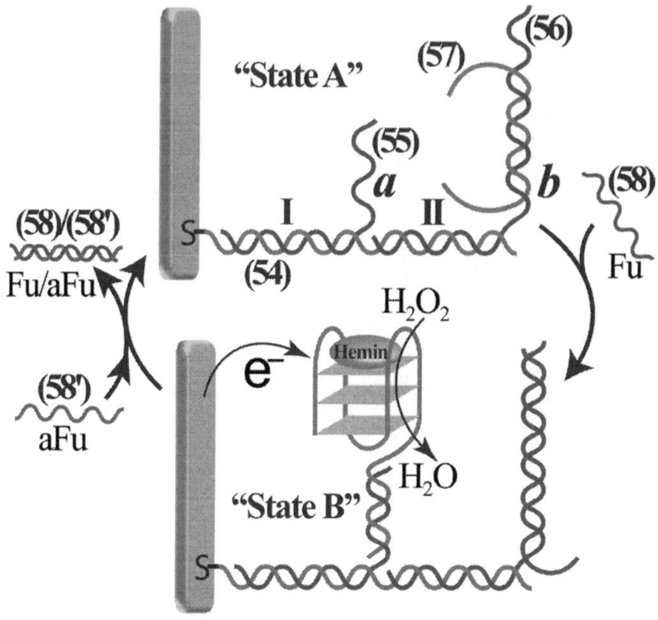

(A)

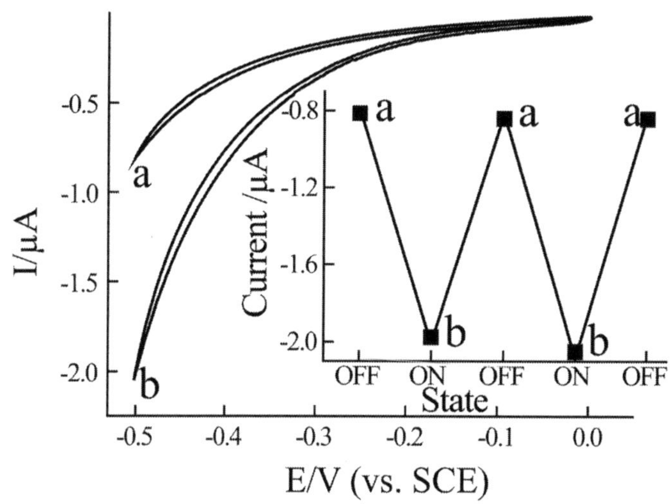

(B)

Figure 5.6.16 *(A) A two-step walker on an electrode surface driven by fuel and anti-fuel strands. The transition of the walker from state A to state B is accompanied by the generation of the hemin–G-quadruplex DNAzyme electrocatalyst. The reverse transition of the walker from states B to A is stimulated by the removal of the fuel strand, the separation of the G-quadruplex and the translocation of the walker back to the original position on the scaffold. The position of the walker is transduced by the electrocatalyzed cathodic currents generated by the hemin–G-quadruplex DNAzyme-catalyzed reduction of* H_2O_2*. (B) Cyclic voltammograms corresponding to the electrocatalyzed cathodic currents corresponding to: (a) system in state A and (b) system in state B. Inset: Cyclic current responses of the walker system upon the reversible switching of the walker between state A (a) and state B (b). Reproduced with permission from [122]. Copyright 2013, American Chemical Society* proceeds due to the favored duplex stability between foothold ***b*** and the fuel strand (**58**). The displaced walker strand (**57**) then translocates to the less-favored foothold ***a*** by its hybridization with domain ***a*** of strand (**55**), state B. The displaced walker strand (**57**) results in, however, the uncaging of the G-rich region that self-assembles into the G-quadruplex. The association of hemin to the resulting G-quadruplex yields an active electrocatalyst for the electrochemical reduction of H_2O_2 to water. Subjecting the system in state B to the anti-fuel strand (**58′**) results in the displacement of the fuel strand (**58**) from foothold ***b***, while forming the stable duplex structure (**58**)–(**58′**) as a waste product. The release of the fuel strand from foothold ***b*** triggers the reverse walk-over process, where the walker unit (**57**) is translocated back to foothold ***a*** of strand (**55**), state A, forming the energetically stabilized duplex (**57**)–(**55**). The separation of the hemin–G-quadruplex depletes the electrocatalytic functions of the system. Through the cyclic treatment of the system with the fuel and anti-fuel strands, the reversible switchable transitions of the "walker" between states B and A were transduced by the electrocatalytic "ON/OFF" functions of the hemin–G-quadruplex structure, Figure 5.6.16(B).

5.6.3.2 Multi-state DNA machines

Enhancing the complexity of the supramolecular DNA machines allows the design of multi-state devices. The dictated reversible transitions across different states can proceed by two alternative methods: (i) the implementation of the strand-displacement principle using different fuel strands that follow a programmed gradient of energies of the duplexes corresponding to the different states, and the application of anti-fuel strands to control the reverse processes – this method is feasible, yet the number of states is limited due to the equilibrium population of the different states; (ii) the implementation of different triggering stimuli such as pH, ions/ligands, photochemical stimuli, and so on, for the generation of the different states.

Figure 5.6.17(A) depicts the construction of a "DNA gear" operating across three states using DNA fuel and anti-fuel strands and the strand displacement principle [133]. The gear is composed of two DNA rings α to which the strands (**59**)–(**61**) and (**62**)–(**64**) were hybridized, to form the ring structures R_A and R_B, respectively. The nucleic acid strands (**59**)–(**61**) of ring structure R_A include protruding single-stranded tethers T_1^A, T_2^A, and T_3^A, and similarly, the nucleic acids (**62**)–(**64**) of the ring structure R_B include protruding single-stranded tethers T_1^B, T_2^B, and T_3^B. Subjecting the rings R_A and R_B to the fuel strand L_1 bridges the tethers T_1^A and T_1^B to yield the gear element consisting of the interconnected rings R_A and R_B, state I. Treatment of the system with the second fuel strand L_2 leads to the bridging of the tethers T_2^A and T_2^B to yield the doubly-bridged rings R_A and R_B, state II. The further treatment of state II with the anti-fuel strand L_1' eliminates the primary bridging fuel strand L_1 through the formation of the energetically stabilized duplex L_1–L_1', to yield state III. The transition from state I to state III represents a one-step rotation of ring R_A with respect to ring R_B. By the stepwise treatment of the nanostructure with additional fuel and anti-fuel strands the connected circles (gear) were rotated one against the other in anti-clockwise or clockwise directions.

Different triggers have been implemented to drive a cyclic switchable "bipedal-walker" device operating across three different states [132], Figure 5.6.17(B). The system consists of a DNA scaffold composed of four inter-hybridized DNA units (**65**)–(**66**), (**66**)–(**67**), (**67**)–(**68**), and (**68**)–(**69**). The scaffold includes four protruding single-stranded tethers I, II, III, and IV, to which the strands (**70**), (**71**), (**72**), and (**73**) are hybridized as footholds for the triggered binding of the bipedal walker unit. The walker element consists of two "pedals" (**74**) and (**75**) hybridized with a "bar" unit (**76**). The single-stranded tethers associated with footholds (**70**), (**71**), (**72**), and (**73**) were labeled with four different fluorophores F_1, F_2, F_3, and F_4, and the bar unit (**76**) was functionalized at its 5′- and 3′-ends with the quenchers Q_1 and Q_2, respectively. In the rest position of the system the energetically favored hybridization of the bipedal walker element occurred on footholds I and II of strands (**70**) and (**71**), state A. The pedal (**74**) includes, however, partial complementarity to the

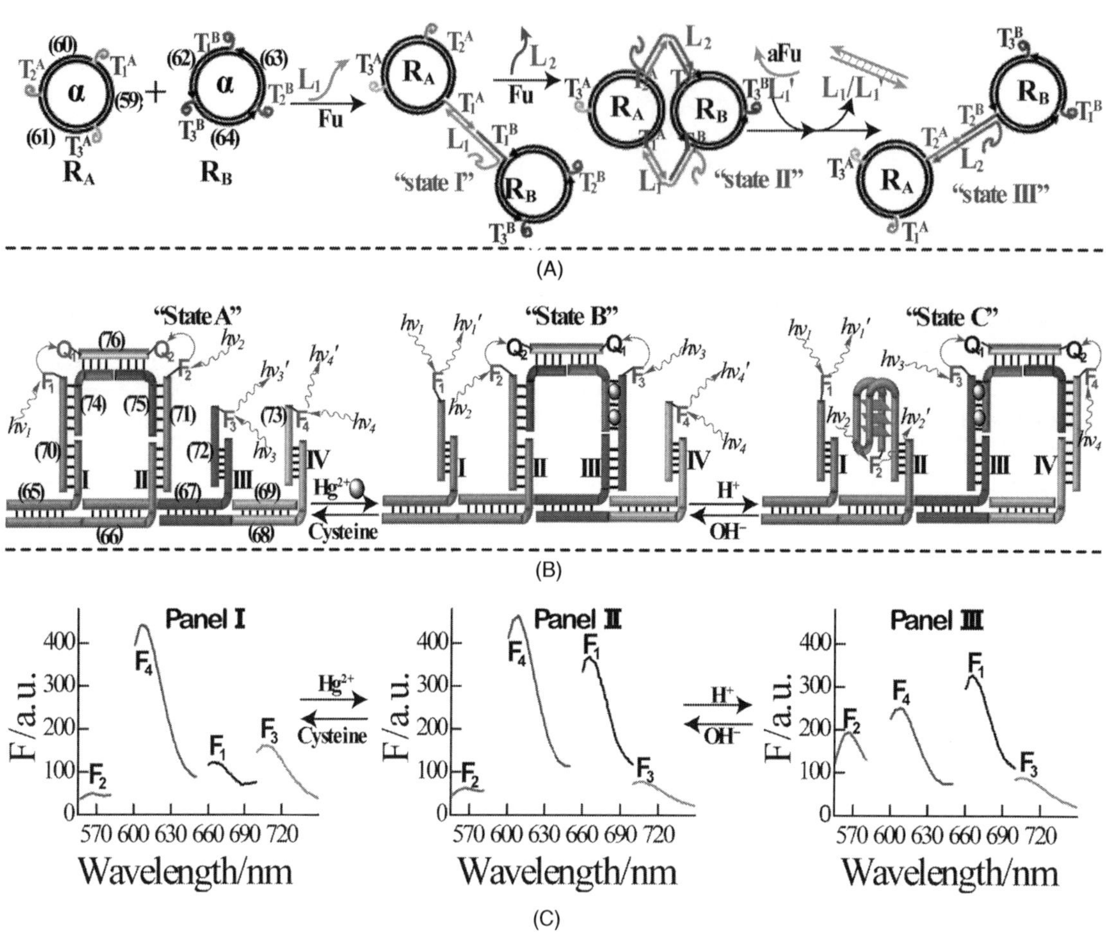

Figure 5.6.17 (A) A DNA "gear" stimulating dictated rotation in the presence of fuel–anti-fuel strands. Reproduced with permission from [133]. Copyright 2004, American Chemical Society. (B) A reversible bipedal "walker" system driven by Hg^{2+}-ion–cysteine and variable pH stimuli H^+/OH^-. The "walker" system consists of four inter-hybridized footholds I to IV. The walker element (**76**) is hybridized in each of the states (I) to (III) to the respective footholds using (**74**) and (**75**) as bridging "pedal" units. (C) Imaging of the walking process and walker positions by labeling of footholds I, II, III, and IV with the fluorophores F_1, F_2, F_3, and F_4, respectively, and the walker units with the quenchers Q_1 and Q_2. The position of the walker is determined by the fluorescence intensities of the respective fluorophores: state A – Panel I; state B – Panel II; state C – Panel III. Reproduced with permission from [132]. Copyright 2011, American Chemical Society

single-stranded tether III of (**72**) and T–T mismatches with the foothold III. In the presence of Hg^{2+} ions, the pedal (**74**) forms an energetically favored duplex with the foothold III of strand (**72**) due to the cooperative stabilization of the duplex by complementary base pairing and T–Hg^{2+}–T bridges. This leads to the translocation of the pedal (**74**) from foothold I of strand (**70**) to foothold III of strand (**72**), to yield state B. Treatment of state B with cysteine (Cys), eliminates the Hg^{2+} ions from the duplex (**72**)–(**74**) through the formulation of the $Hg^{2+}(Cys)_2$ complexes, and this results in the reverse translocation of pedal (**74**) to the primary foothold I of strand (**70**), to yield state A. The single-stranded domain II of (**71**) consists of a

C-rich region that can self-assemble into an i-motif nanostructure. Accordingly, treatment of state B under acidic pH conditions (pH = 5.2) leads to the folding of foothold II of strand (**71**) into the i-motif structure, and to the release of pedal (**75**) of the walker unit that is translocated to foothold IV of strand (**73**), which yields the stable walker nanostructure, state C. Neutralization of the system dissociated the i-motif structure and this resulted in the reverse translocation of the pedal (**75**) to foothold II of strand (**71**). Through the cyclic application of Hg^{2+}–Cys or H^+–OH^- triggers the cyclic switchable, bi-directional translocations of the "bipedal walker" were stimulated. The fluorescence features of the fluorophores associated with the four footholds allowed the probing of the positions of the walker element. In state A, the fluorescence of F_1 and F_2 are quenched while the fluorescence of F_3 and F_4 are high, Figure 5.6.17(C), panel I. In state B, the fluorophores F_2 and F_3 are quenched, while the fluorescence of F_1 and F_4 are high, Figure 5.6.17(C), panel II, and in state C, the fluorescence intensities of F_3 and F_4 are quenched, and the fluorophores F_1 and F_2 are high, Figure 5.6.17(C), panel III. By following the time-dependent fluorescence changes of the fluorophores, the dynamics of the transitions of the walker across the different states were followed.

Interlocked circular DNA structures (catenanes) attract growing interest as functional DNA nanostructures that perform programmed mechanical operations [136]. Figure 5.6.18(A) depicts the synthesis of a three-ring DNA catenane system and the cyclic operations of the device, undergoing transitions across three different states, using fuel–anti-fuel strands and strand displacement as the driving mechanism. The synthesis of the three-ring catenane involved the primary preparation of two rings α and γ through the ligation of strand, α_L (**77**) and γ_L (**78**) in the presence of the respective caps α_C and γ_C. The inter-threading of the strand β_L (**79**) into rings α and γ through the formation of duplex domains with the respective rings, followed by capping of the threaded strand with β_C and its ligation, leads, after the removal of the capping units, to the linear interlocked three-ring catenane L, consisting of interlocked three-ring α-β-γ. The interlocked catenane includes, in ring α, two identical regions I and II, complementary to domain III associated with ring γ. The hybridization of ring γ with ring β is, however, energetically favored. Subjecting the linear structure L to the fuel (Fu) strand displaces ring γ from ring β, resulting in the "mechanical" transitions of ring γ above and below the rims of ring β to hybridize with the energetically less-favored sites I and II associated with ring α, to yield the structures P_1 and P_2, respectively, Figure 5.6.18(B). As the sites I and II consist of similar sequences, the configurations P_1 and P_2 are formed at a 1:1 ratio. Treatment of the P_1 and P_2 mixture with the anti-fuel (aFu) strand, displaces the fuel (Fu) strand hybridized with ring β, while releasing the "waste" product Fu–aFu, path a. This releases the binding site on ring β, and ring γ is translocated back to ring β, to yield the energetically favored structure L. Directional transitions of ring γ above or below the rims of ring β was demonstrated by the blocking of the rims of ring β with appropriate blocker units. The hybridization of the nucleic acid blocker B_1 with the upper rim domain of ring β yielded a duplex blocking site on ring β, structure L_1. The subsequent treatment of the linear structure L_1 with the fuel (Fu) strand displaced ring γ, and the released ring γ was selectively translocated along the lower rim of ring β to form a hybrid between domains II and III, structure P_3. The reverse treatment of structure P_3 with the anti-fuel (aFu) strand and the anti-blocker (aB_1) strand restored the linear structure L. Similarly, subjecting the linear structure L with strand B_2, acting as blocker for the lower rim domain of ring β, yields the blocked linear structure L_2. The subsequent treatment of structure L_2 with the fuel (Fu) strand releases ring γ from ring β, resulting in the translocation of ring γ along the upper rim of ring β to form a hybrid between regions I and III, structure P_4. The reverse treatment of structure P_4 with the anti-fuel (aFu) strand and the anti-blocker (aB_2) strand restored the original structure L, Figure 5.6.18(B). The cyclic and switchable transitions of the three-ring catenane across the structure P_1–P_4 were probed by the labeling of rings α and γ with the fluorophore–quencher pairs F_1–Q_2 and F_2–Q_1, respectively The time-dependent fluorescence changes of fluorophores F_1 and F_2, upon the treatment of the three-ring catenane with the respective Fu–aFu strands and B–aB (B_1–aB_1 or B_2–aB_2) strands allowed the characterization of the dynamic transitions of the device across the different states, Figure 5.6.18(C). In state P_3, the fluorophore F_2

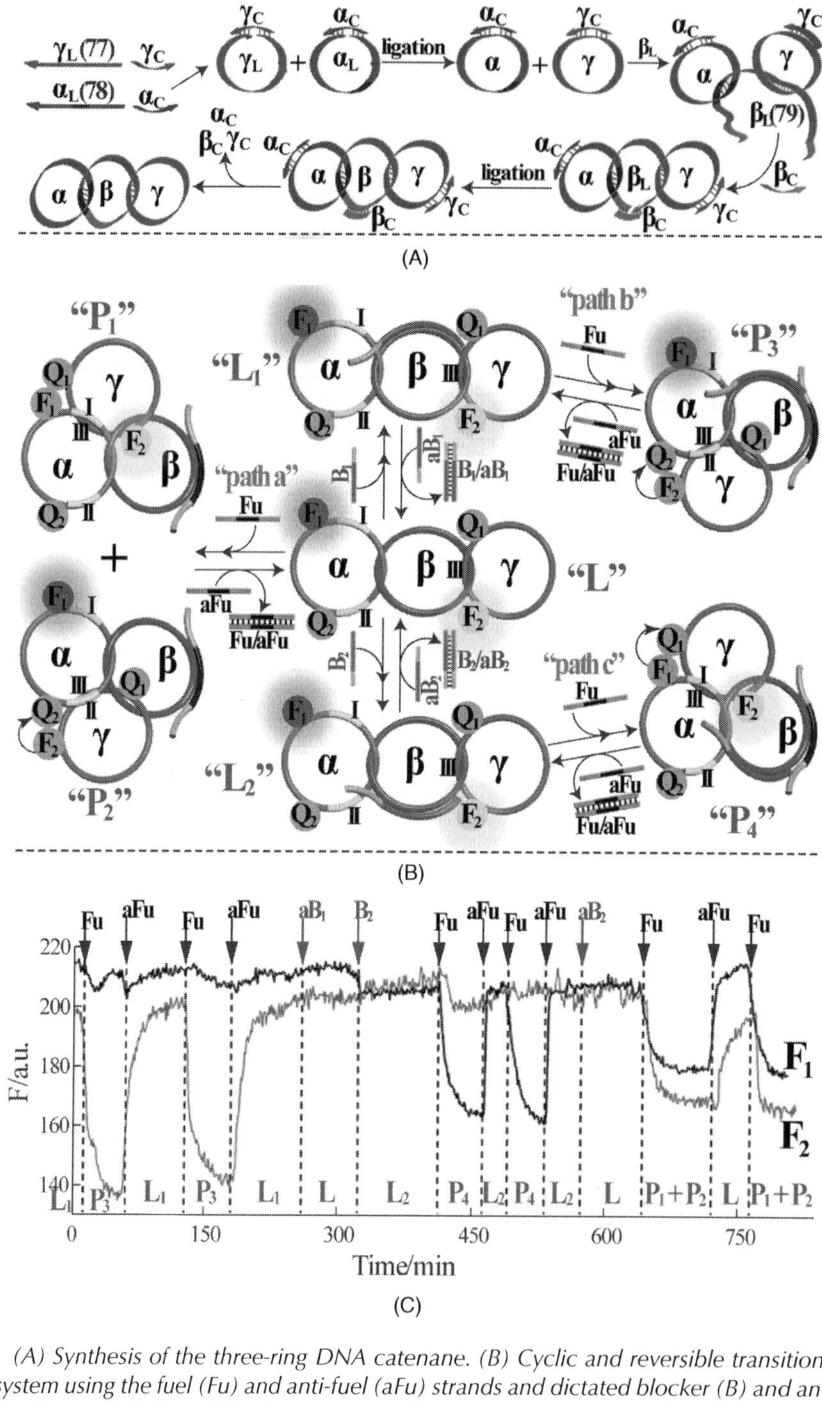

Figure 5.6.18 *(A) Synthesis of the three-ring DNA catenane. (B) Cyclic and reversible transitions of the three-ring catenane system using the fuel (Fu) and anti-fuel (aFu) strands and dictated blocker (B) and anti-blocker (aB) strands. The transitions are followed by the respective fluorescence intensities of F_1 and F_2 in the presence of the quencher units Q_1 and Q_2 positioned on the respective domain of the catenane nanostructure. (C) Time-dependent fluorescence intensity changes of F_1 and F_2 upon triggering the transitions between the different states of the device using the respective Fu–aFu and B–aB strands. Reproduced with permission from [136]. Copyright © 2012, WILEY-VCH Verlag GmbH & Co. KGaA, Weinheim*

is effectively quenched while the fluorescence of F_1 is unaffected. In state P_4, F_1 is effectively quenched and F_2 is unaffected. The formation of the equi-energetic structures P_1 and P_2, at a ratio of 1:1, yields to the concomitant quenching of F_1 and F_2 to an extent of about 50% of the fluorescence observed for the respective fluorophores in states P_4 and P_3, respectively.

An interlocked two-ring system that operates as a three-state rotor was synthesized and activated by pH and Hg^{2+}–Cys as triggers [134], Figure 5.6.19(A). The system consists of the two interlocked rings α and β, that rest in state I where the hybridization between domains ***a*** and ***b*** yields the energetically favored configuration I. Domain ***c*** of ring β and the region ***a*** in ring α include partial complementarity, and two T–T mismatches. In the presence of Hg^{2+} ions, the formation of the energetically improved T–Hg^{2+}–T bridged duplex between domains ***a*** and ***c*** stimulates the anti-clockwise rotation of ring α to form state II. Subjecting the system of state II to cysteine, eliminates the Hg^{2+} ions from the ***a–c*** duplex domain resulting in the reverse clockwise transition of ring α to form state I. Similarly, the sequence ***d*** in ring β exhibits partial complementarity to domain ***a*** associated with ring α, and it is somewhat blocked by strand (**80**), which prohibits the hybridization of ring α with domain ***d***. The strand (**80**) includes, however, a C-rich sequence that at pH = 5.2 self organizes into the i-motif structure. Accordingly, subjecting the system in state I to acidic conditions, pH = 5.2, leads to the self-assembly of strand (**80**) into an i-motif structure that releases from ring β. The unblocking of domain ***d*** yields an energetically favored site ***d*** for the hybridization with domain ***a*** of ring α, to form the stable duplex ***a–d***. As a result, ring α rotates clockwise to form state III. The reverse neutralization of state III to pH = 7.2 dissociated the i-motif structure, and the strand (**80**) displaced ring α from domain ***d*** of ring β to yield the stable duplex between strand (**80**) and ring β, while rotating ring α to site ***b*** of ring β and generating state I. Similarly, subjecting state III, to pH = 7.2 and Hg^{2+} ions resulted in the dissociation of the i-motif structure and the clockwise rotation of ring α to domain ***c*** of ring β to form the T–Hg^{2+}–T bridged energetically stabilized state II. The reverse treatment of state II with cysteine under acidic conditions, pH = 5.2, restored state III. That is, by the application of the appropriate fuel–anti-fuel elements, the ring α could be rotated selectively in clockwise or anti-clockwise directions on ring β. By the labeling of ring β with internal fluorophores (Cy3 and Cy5), and the functionalization of the rotating ring α with a quencher unit (BHQ-2), using a hybridized quencher-labeled nucleic acid (**81**), the position of the rotor in the different states could be probed. For example, Figure 5.6.19(B) depicts the time-dependent fluorescence changes of the two fluorophores upon the anti-clockwise rotation of ring α from states I→II→III and the reverse clockwise rotation of states III→II→I. By the detailed kinetic analysis of the time-dependent fluorescence changes of the two fluorophores upon subjecting the system to the respective fuels–anti-fuels, it was concluded that the rotation of ring α proceeds with controlled directionality (clockwise or anti-clockwise). That is, the direction of rotation is dictated by the shortest path for occupying the respective sites.

5.6.4 Applications of DNA switches and machines

The metal-ion stimulated switchable transitions of DNA nanostructures have been used to control the transport of ions through pores [137]. For example, nanopores were functionalized with a nucleic acid (**82**), and the flexibility of the polymer chains blocked the passage of ions through the pores, Figure 5.6.20(A). The single-stranded chain (**82**) included partial complementarities and T–T mismatches, and, thus, subjecting the system to Hg^{2+} ions led to the reconfiguration of strand (**82**) into rigid hairpin structures stabilized in their stem region by cooperative base pairing and T–Hg^{2+}–T bridges. The formation of the rigid hairpins opened the pore, thus enhancing the current flow through the pores. Treatment of the hairpin-functionalized pores with cysteine eliminated the stabilizing Hg^{2+} ions, a process that re-closed the pores. By the cyclic treatment

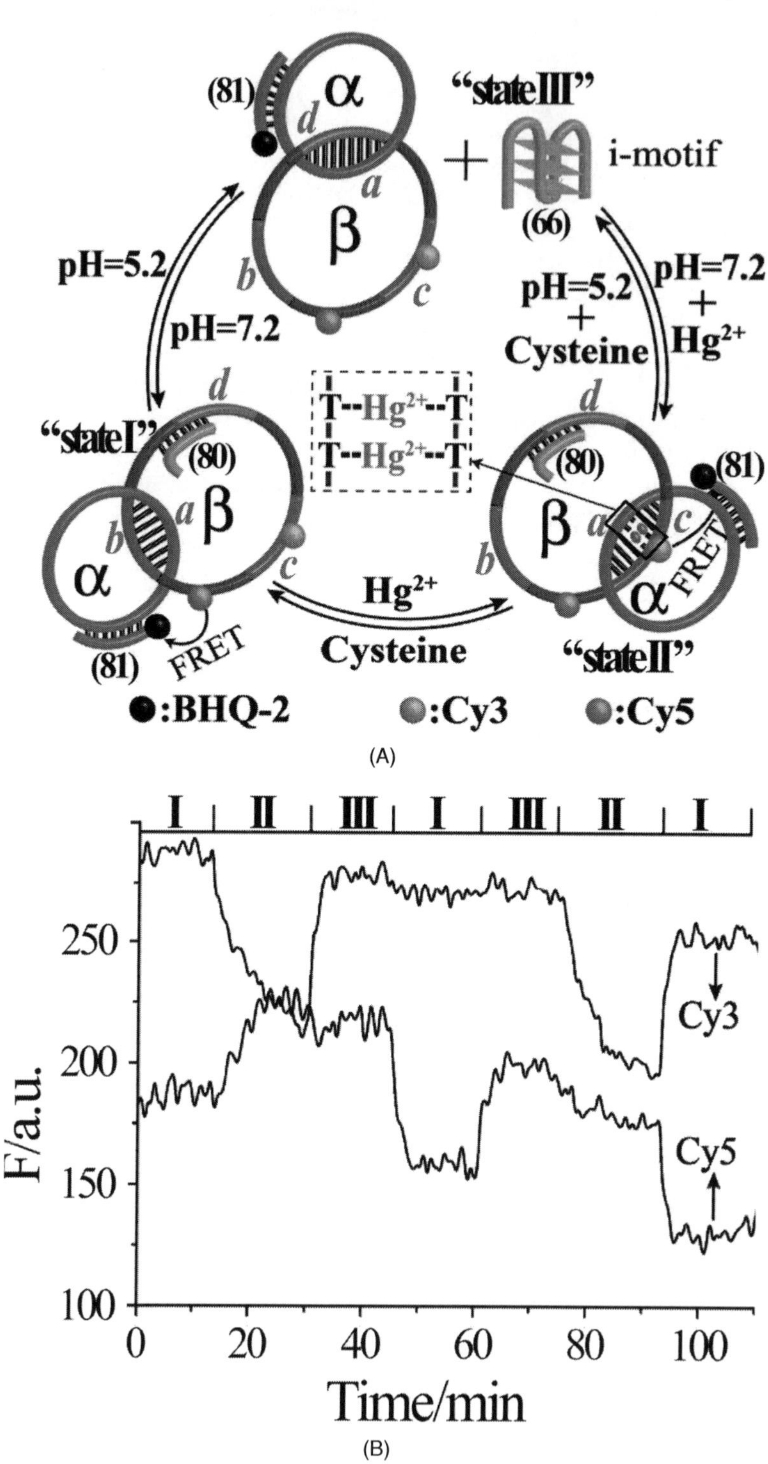

Figure 5.6.19 (A) A DNA rotor performing dictated clockwise and anti-clockwise rotations in the presence of variable pH values and/or Hg^{2+} ion–cysteine as triggering stimuli. The rotor is cycled across three different states, I, II, and III. The transitions of the rotor across the different states are imaged by following the fluorescence changes of two fluorophore labels (F_1 and F_2) associated with ring β, and a quencher unit (Q) associated with ring α. (B) Time-dependent fluorescence changes of the two fluorophores upon the anti-clockwise rotation I→II→III→I and the subsequent clockwise rotation I→III→II→I. Reproduced with permission from [134]. Copyright 2013, American Chemical Society

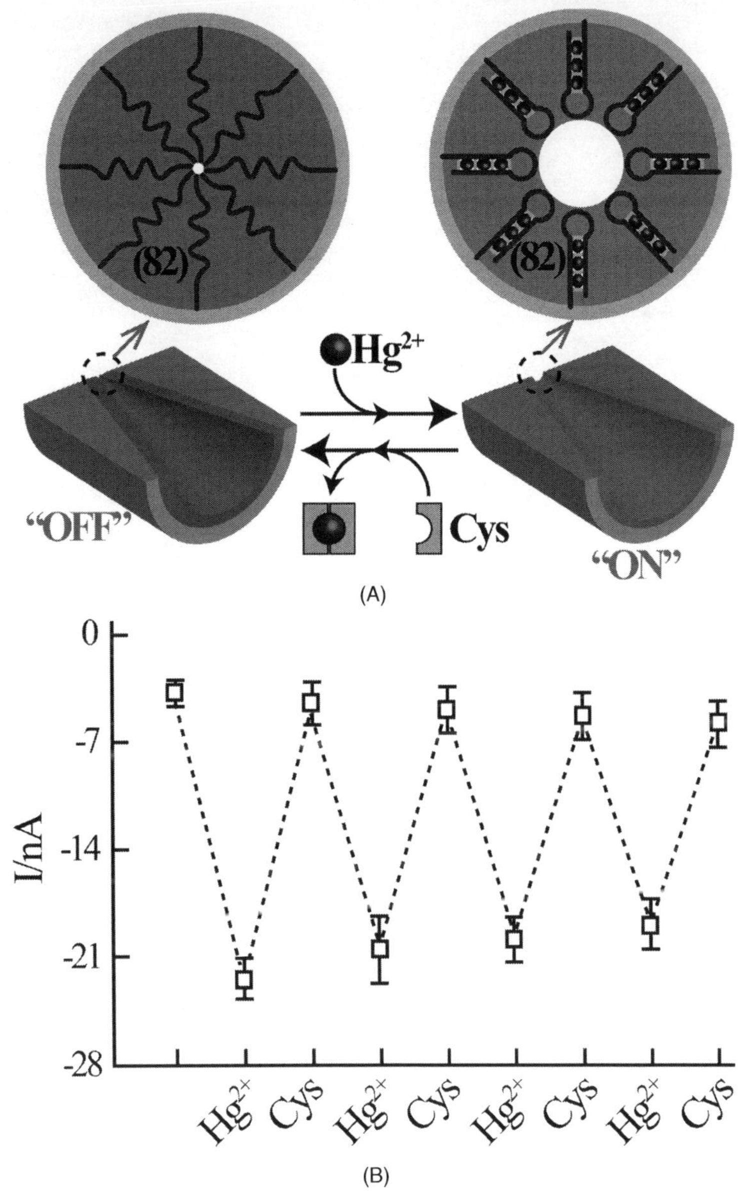

Figure 5.6.20 *(A) Hg^{2+}-ion–cysteine-stimulated switchable opening and blocking of nanopores. (B) Switchable ion currents stimulated in the presence of closed nanopores (low currents) or open nanopores (high currents), in the presence of Hg^{2+} ions and cysteine, respectively. Reproduced with permission from [137] The Royal Society of Chemistry*

of the (**82**)-modified pores with Hg^{2+} ions and cysteine, the ion current flow through the pores was switched between low and high values, Figure 5.6.20(B), implying that the DNA structures might act as switchable valves for transporting ions or substrates through pores.

Similarly, the metal-ion-stimulated switching of DNA-based polymers between hydrogels and polymer solutions has been demonstrated [138]. A supramolecular Y-shaped DNA nanostructure, consisting of three

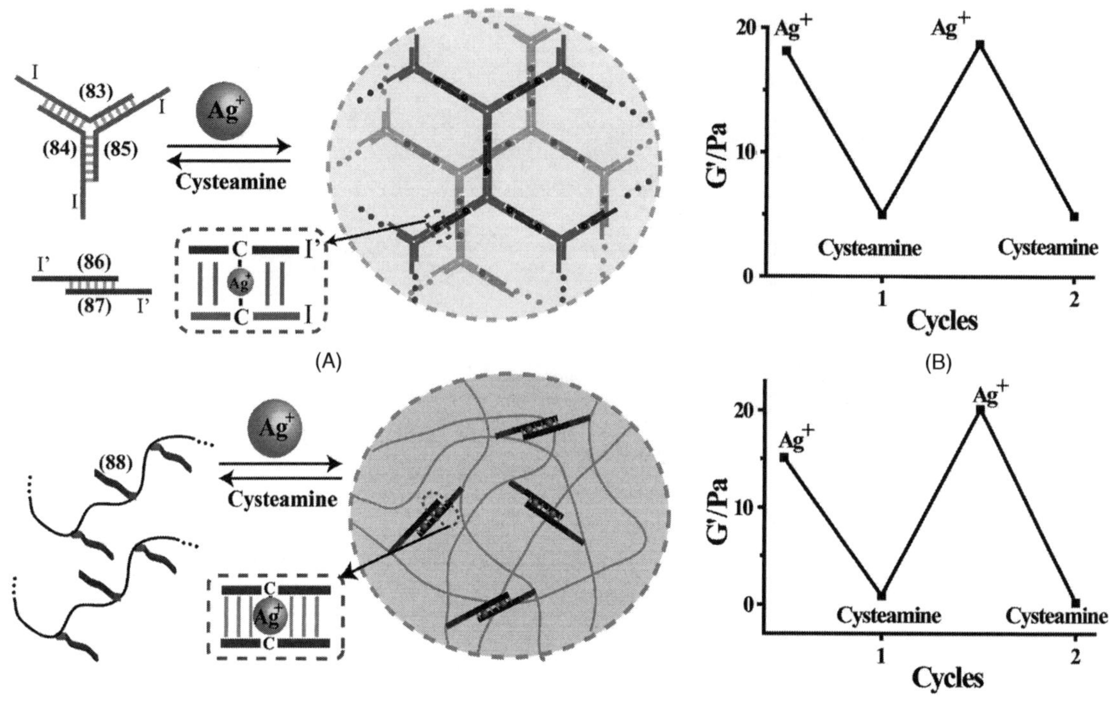

Figure 5.6.21 *(A) Switchable all-DNA based hydrogel formation by the Ag^+-ion-stimulated crosslinking of a Y-shaped DNA nanostructure with a duplex DNA crosslinker, using C–Ag^+–C bridges as cooperative stabilizers of the crosslinking units. The hydrogel is separated into the solution phase of the DNA subunits by the cysteamine-mediated elimination of the Ag^+-ions from the crosslinking units. (B) Cyclic changes of the storage modulus (G′) of the system upon the Ag^+-ion-stimulated formation of the hydrogel and the cysteamine-induced separation of the hydrogel into a solution phase. (C) Switchable Ag^+-stimulated crosslinking of acrylamide/acrylamide nucleic acid copolymer chains by the bridging of the nucleic acid tethers by means of C–Ag^+–C bridges, and the separation of the hydrogel by the cysteamine-induced elimination of the Ag^+ ions from the bridging units. (D) Cyclic changes of the storage modulus (G′) of the system upon the treatment of the copolymer chains with Ag^+ ions to form the hydrogel and the separation of the hydrogel into a solution mixture upon the addition of cysteamine. Reproduced with permission from [138] The Royal Society of Chemistry*

nucleic acid strands (**83**), (**84**), and (**85**) was cross-linked, in the presence of Ag^+ ions, by a duplex structure (**86**)–(**87**) that included toehold domains complementary to the single-stranded tethers, associated with the Y-shaped units, to yield the branched hybrid hydrogel structure, Figure 5.6.21(A). Treatment of the hydrogel with cysteamine removed the Ag^+ ions, resulting in the dissolution of the hydrogel into a homogenous polymer solution. The cyclic switchable transitions of the system between hydrogel and solution phases were characterized by rehometric experiments and the phase transitions were probed by following the cyclic storage modulus (G′) changes, Figure 5.6.21(B). A further reversible Ag^+-ion-stimulated polymer solution-to-hydrogel was demonstrated using a copolymer consisting of an acrylamide unit and the nucleic acid (**88**)-modified acrylamide unit (ratio 140:1), Figure 5.6.21(C). The nucleic acid (**88**) exhibits partial self-complementarity and C–C mismatches. In the presence of Ag^+ ions, the (**88**) tethers are crosslinked by cooperative base pairing and C–Ag^+–C bridges, leading to the formation of the hydrogel. Treatment of the hydrogel with cysteamine eliminated the Ag^+ ions from the crosslinking units, resulting in the dissociation of the hydrogel into a polymer solution. The cyclic and reversible polymer solution-to-hydrogel transitions were probed by rehometry experiments that followed the cyclic changes of the storage modulus values, Figure 5.6.21(D).

Analogous cyclic stimuli-triggered polymer solution-to-hydrogel solutions were demonstrated by using pH as the triggering signal for the formation or dissociation of i-motif bridges [139]. Also, acrylamide copolymers composed of acrylamide units and G-quadruplex subunit-modified acrylamide units were used as functional polymer chains for the reversible transitions between a polymer-solution and a hydrogel [140]. The K^+ ion stimulated formation of the G-quadruplexes, by the polymer tethering subunits, crosslinked the polymer chains, leading to the formation of the hydrogel. The subsequent treatment of the hydrogel with 18-crown-6-ether (CE), eliminated the K^+ ions from the crosslinked G-quadruplexes, resulting in the dissociation of the hydrogel into a polymer solution. Interestingly, the binding of hemin to the K^+-ion stabilized G-quadruplexes-crosslinked hydrogel, led to a catalytically active horseradish peroxidase (HRP)-mimicking DNAzyme hydrogel. The 18-crown-6-ether (CE) elimination of K^+ ions from the G-quadruplexes, and the dissociation of the hydrogel blocked the catalytic functions of the system.

The mechanical switching of DNA devices has been implemented to reconfigure metallic nanoparticle structures. This is exemplified in Figure 5.6.22 with the use of a DNA tweezers system to switch the structures composed of three Au nanoparticles (NPs) [108]. The molecular device consisted of two nucleic acid "arms"

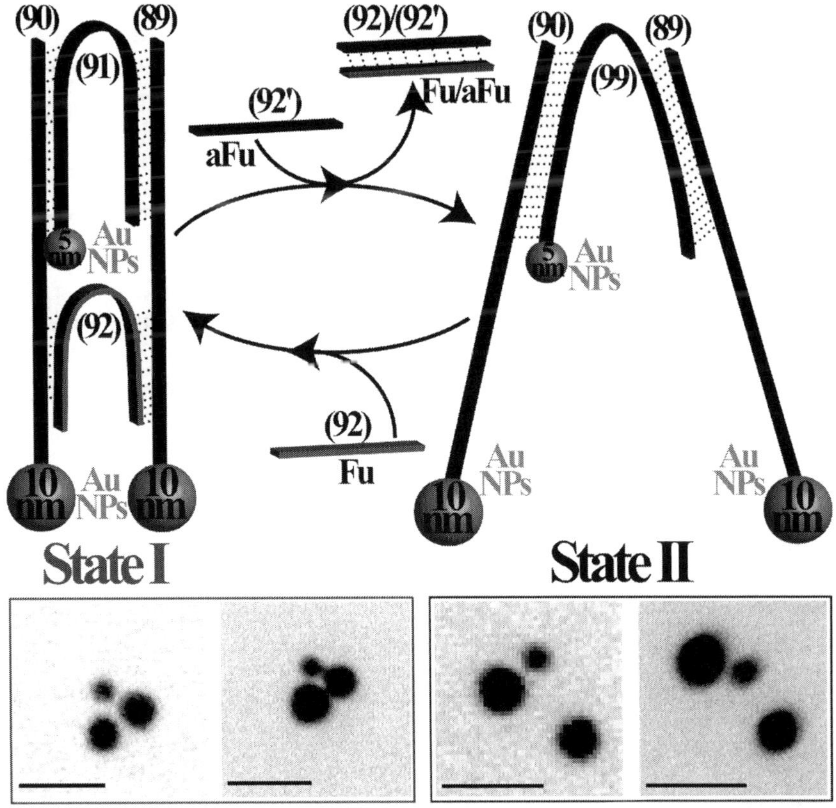

Figure 5.6.22 Programmed assembly of Au NPs by the closure and opening of a DNA tweezers device using fuel and anti-fuel strands. Bottom: TEM images of the Au NPs in the presence of closed tweezers scaffold (state I, left); and in the presence of open tweezers scaffold (state II, right). Reproduced with permission from [108]. Copyright 2013, American Chemical Society

(**89**) and (**90**), each tethered to a 10 nm sized Au NPs. The two arms were bridged by the nucleic acid (**91**)-functionalized 5 nm sized Au NPs and by the nucleic acid (**92**), exhibiting appropriate base-pairing complementarities to the "arms". Bridging the "arms" by the (**91**)-modified Au NPs and by the linker strand (**92**) yields the closed tweezers, which results in a compressed configuration of the three NPs, state I. Treatment of the closed tweezers with the anti-fuel strand (**92′**) displaces the bridging unit (**92**) by forming the energetically favored duplex (**92**)–(**92′**) as "waste" product. This leads to the opening of the tweezers and to the formation of an extended configuration of the three-NPs cluster, state II. The reverse treatment of the open tweezers with the fuel strand (**92**) closed the tweezers and the associated NPs to form state I. Analogous reconfiguration of Au NPs was demonstrated by using the mechanical transitions of interlocked catenated systems as the driving mechanism [109].

The DNA machines were further applied as functional mechanical devices for the switchable control of the plasmonic interactions between fluorophores and Au NPs. Metallic NPs usually quench the fluorophore. At certain distances that separate the fluorophores from the plasmonic particles, surface-enhanced fluorescence (SEF) occurs. This distance is controlled by the quantum yield of the fluorophores, the size of the NPs, the orientation of the chromophores relative to the NPs, and the refractive index of the medium. Since the distance separating the fluorophores from Au NPs can be controlled by a mechanical DNA scaffold, the cyclic transitions between states exhibiting surface-quenched and surface-enhanced fluorescence can be envisaged. Indeed, it has been demonstrated that mechanical switchable transitions of DNA devices may control the fluorescence properties of fluorophore–Au NPs conjugates. This is exemplified in Figure 5.6.23 with the assembly of two tweezers T_1 and T_2, which include in their structures the Cy3 fluorophore at different spatial positions relative to a 10 nm sized Au NP [108]. In tweezers T_1, the fluorophore and the Au NP are positioned at the 3′-end of the arm strand (**90a**) and 5′-end of the arm strand (**89**), respectively. The "arms" were bridged with the strand (**91a**) to yield the open tweezers, state I, Figure 5.6.23(A). In turn, tweezers T_2 includes an internally Cy3-modified arm (**90b**) and the nucleic acid (**89**)-modified Au NP as the second arm. The bridging of the two arms with (**91a**) yielded the open tweezers T_2, state III, Figure 5.6.23(B).

Treatment of the two tweezers with the fuel (Fu) strand (**92**) results in the closure of the two tweezers, states II and IV, respectively. Subjecting the tweezers to the anti-fuel (aFu) strand (**92′**) resulted in the elimination of the fuel strand (**92**) in the form of an Fu–aFu duplex (**92**)–(**92′**), and the opening of the tweezers structures. By the cyclic treatment of the tweezers with the fuel and anti-fuel strands, the tweezers were switched between closed and open states. The closure of tweezers T_1 led to quenching of the fluorophore, Figure 5.6.23(C), consistent with the close proximity between the fluorophore and the Au NP. In turn, closure of tweezers T_2 leads to an enhancement of the fluorescence, despite the shorter distance between the fluorophore and Au NP, as compared with the open tweezers, Figure 5.6.23(D). These experimental results were theoretically explained. Figure 5.6.23(E) and (F) show the theoretically calculated distance-dependent fluorescence yields of Cy3 and 10 nm sized Au NP in the tweezers structures T_1 and T_2, respectively. On the particular curves the estimated distances separating the fluorophore (Cy3) from the 10 nm sized Au NP are plotted, respectively. For tweezers T_1, the distances separating the fluorophore–Au NP pair were estimated to be d_o = 10–18 nm for the open tweezers, state I, and d_c = 1–2 nm for the closed tweezers, state II. Evidently, the theoretical curve suggests that at the intimate proximity between the fluorophore and Au NP in state II, an effective quenching proceeds, as observed experimentally. In turn, for tweezers T_2, the separation distances between the fluorophore and Au NP were estimated to be d_o = 9–14 nm for the open tweezers, state III, and d_c = 3.8–4 nm for the closed tweezers, state IV. The distance separating the fluorophore and Au NP in state IV is in the surface-enhanced fluorescence region, suggesting an intensified fluorescence as compared with the open tweezers, as, indeed, is observed experimentally. Similar surface-enhanced fluorescence and surface-quenched fluorescence phenomena were observed upon the mechanical transitions of a three-ring interlocked DNA catenane machine, decorated with a fluorophore–Au NP pair, above and below the central ring [109].

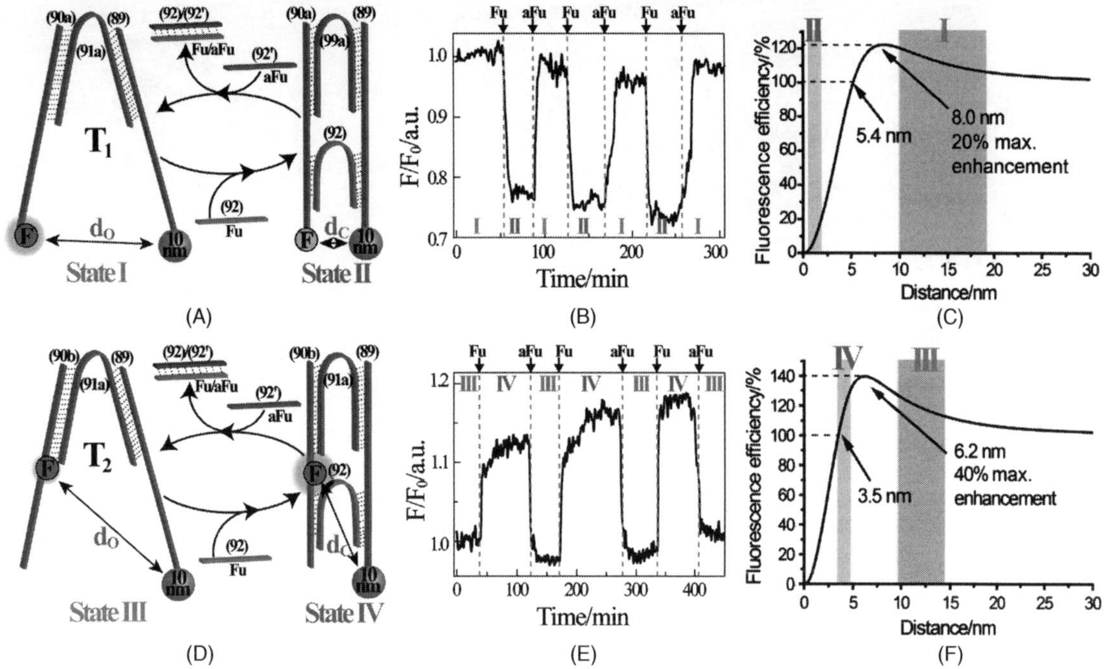

Figure 5.6.23 *Controlling the fluorescence properties of fluorophore–Au NPs pairs by the switchable spatial separation of the fluorophore–Au NPs pairs using the mechanical opening and closure of DNA tweezers. (A) Tweezers in configuration T_1. (B) Switchable fluorescence changes upon the fueled closure of state I to state II and re-opening by the anti-fuel (results imply fluorescence quenching upon closure of the tweezers). (C) Theoretical modeling of the distance-dependent fluorescence yields of the fluorophore (Cy3) in the presence of 10 nm sized Au NPs in tweezers system T_1. (D) Tweezers in configuration T_2. (E) Switchable fluorescence changes upon the fueled closure of state III to state IV and re-opening by the anti-fuel (results imply fluorescence enhancement upon closure of the tweezers). (F) Theoretical modeling of the distance-dependent fluorescence yields of the fluorophore (Cy3) in the presence of 10 nm sized Au NPs in tweezers system T_2. Reproduced with permission from [108]. Copyright 2013, American Chemical Society*

The switchable mechanical functions of supramolecular DNA nanostructures can be implemented to control chemical transformations, and specifically biotransformations. This is exemplified in Figure 5.6.24 with the switchable "ON" and "OFF" activation of a bi-enzyme cascade using DNA tweezers [141]. The tweezers are composed of two inter-hybridized "arms" that are bridged by a hairpin structure. The "arms" of the tweezers were each functionalized with the enzymes glucose oxidase (GOx) and horseradish peroxidase (HRP), respectively. Under these conditions, the tweezers exist in a closed configuration, state A. In this structure, the close proximity between the two enzymes allows the activation of the bi-enzyme cascade where the GOx-catalyzed oxidation of glucose yields gluconic acid and H_2O_2, and the resulting high localized concentrated H_2O_2 acts as a substrate for HRP that catalyzes the oxidation of $ABTS^{2-}$ to the colored product, $ABTS^{\bullet-}$. Subjecting the closed tweezers to the fuel (Fu) hairpin structure III′ leads to the opening of the bridging hairpin unit III and to the open tweezers structure, state B. The spatial separation between the enzymes inhibited the communication between the two enzymes. The H_2O_2 generated by GOx diffuses to the bulk solution, leading to a lower concentration at the HRP site, resulting in an inefficient bi-enzyme cascade. The reverse interaction of state B with the anti-fuel (aFu) displaces the fuel strand Fu by forming the energetically stabilized Fu–aFu duplex, leading to the closure of the tweezers, state A, and to the

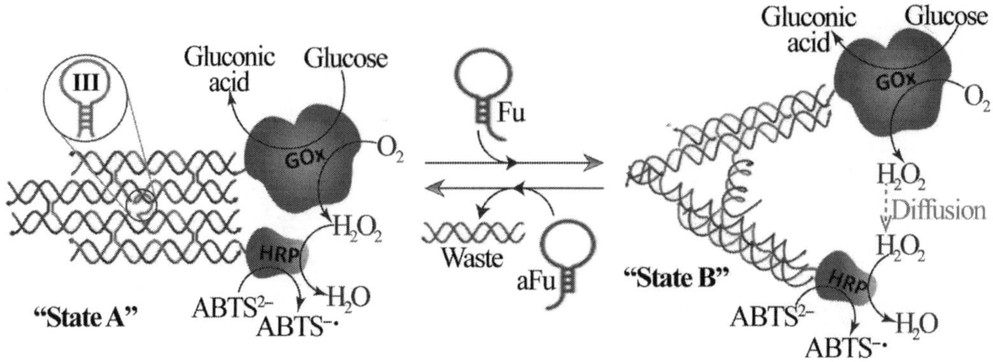

Figure 5.6.24 Cyclic activation and deactivation of the bi-enzyme cascade consisting of glucose oxidase–horseradish peroxidase (GOx–HRP) by the reversible closure, switched "ON", and opening, switched "OFF", of a DNA tweezers structure. Reproduced with permission from [141]. Copyright © 2013, WILEY-VCH Verlag GmbH & Co. KGaA, Weinheim

activation of the bi-enzyme cascade. By the cyclic treatment of the system with the fuel Fu and anti-fuel aFu strands, the bi-enzyme cascades were cycled between "OFF" and "ON" states, respectively. Related tweezers structures were used to switch the activity of the NAD^+-dependent glucose-6-phosphate dehydrogenase (G6pDH) [142].

5.6.5 Conclusions and perspectives

The design of dynamically switchable DNA systems and devices turned, in the last decade, into a central topic of DNA nanotechnology. The possibility to encode structural and functional information in the sequences of oligonucleotides enabled the design of supramolecular DNA nanostructures using simple rules. This has been exemplified in the present chapter by introducing the principles to construct DNA molecular devices acting as switches and machines. Different fuel–anti-fuel stimuli were implemented to drive the cyclic DNA switches and machines including pH (H^+/OH^-), strands/anti-strands, ions, photonic, and electrical signals. The switchable DNA devices were operated in solutions and on surfaces. A variety of methods were used for the readout of the mechanical functions of the DNA systems, and these included fluorescence resonance energy transfer (FRET), chemiluminescence resonance energy transfer (CRET), electrochemical transduction (voltammetric, Faradaic impedance spectroscopy), catalytic functions of DNAzymes, and more.

Although substantial progress in designing DNA switches and machines has been demonstrated, significant fundamental challenges are ahead of us. It is important to develop new fuels–anti-fuels to drive multi-state switchable devices of enhanced complexities. For example, the development of redox-active DNA intercalators, undergoing electrically controlled binding/dissociation to duplex DNAs, the incorporation of new synthetic purine–pyrinidine ligands into DNA chains as selective ion-binding sites (ligandosides) [143], and the development of new photochromic intercalators (beyond azobenzene derivatives) could be interesting paths to follow.

Also, the advances in scanning probe microscopies could allow us to follow the activities and functions of molecular machines at the single-molecule level. Indeed, several recent studies have been followed by atomic force microscopy (AFM): the imaging of K^+-ion and crown-ether-stimulated formation and dissociation of a G-quadruplex on an origami frame [144], or the stepwise motion of a DNA walker along a dictated path

[145, 146]. Thus, the implementation of this technique for imaging the dynamic functions of DNA machines of enhanced complexities is an important future goal.

The potential applications of DNA switches and machines represent, however, a major challenge to consider. Various applications of DNA machines have been exemplified, including the use of switchable nucleic acid modified pores for selective transport [137], the dynamic programmable organization of nanoparticle structures [108, 109], and the dynamic control of the plasmonic features of fluorophore–Au NPs conjugates associated with structurally switchable DNA scaffolds [108–110]. The control of biocatalytic cascades by means of switchable DNA systems is an interesting approach to control chemical reactivity, and the design of multi-state devices that regulate biocatalytic cascades in different dictated directions could provide a means to mimic cellular signal-triggered catalytic networks.

Albeit substantial progress in the applications of DNA switches and machines has been demonstrated, many new applications in different fields may be envisaged. The use of these systems in nano-medicine (diagnostics or therapy) has sparked substantial interest and several preliminary reports highlighting the future potential of such systems. For example, pH-triggered DNA tweezers were introduced into cells and used for imaging temporal and spatial pH changes in these cells [147]. Also, DNA machines and switchable DNA-based materials could be used for controlling drug delivery. For example, stimuli-triggered hydrogels undergoing gel-to-solution transitions could release entrapped drugs, in response to pH, ions or the biomarker triggers [138–140].

Additionally, DNA switchable machineries conjugated to mesoporous SiO_2 nanoparticles were recently demonstrated as functional materials for the stimuli-controlled release of drugs. For example, the anti-cancer drug doxorubicin was entrapped in the pores of mesoporous SiO_2 nanoparticles, and locked in the pores by their capping with programmed aptamer–DNAzyme sequences [148–150]. The unlocking of the pores by the ATP-biomarker induced activation of the DNAzyme, through the formation of the ATP–aptamer complexes, and released the anti-cancer drug. Impressive selectivity in the release of the chemotherapeutic doxorubicin drug in breast cancer cells versus normal breast cells has been observed, and the effective targeted death of breast cancer cells demonstrated. Other "smart" DNA caps that lock drugs in the pores of mesoporous SiO_2, and lead to the release of the drug by the stimuli-controlled unlocking of the pores by various other biomarkers (e.g., metal ions, pH, G-quadruplexes) may be envisaged as versatile sense-and-treat theranostics materials.

The dynamic and switchable control of the organization of plasmonic particles or fluorophore–plasmonic-nanoparticle conjugates holds great promise in material science. New switchable optical plasmonic functions may be envisaged, such as switchable surface-enhanced Raman spectroscopy or chiroplasmonic switches, are anticipated to be observed. Finally, DNA switches and machines are expected to provide effective scaffolds for programmed synthesis by the dictated stimuli-triggered interactions of chemical reactants [104–107, 151].

The impressive progress in developing DNA switches and machines suggests that these systems will have a significant impact in material science, physics, medicine, and chemistry.

References

[1] T. Schnitzler and A. Herrmann, DNA block copolymers: functional materials for nanoscience and biomedicine, *Acc. Chem. Res.*, **45**, 1419–1430 (2012).

[2] S. Modi, D. Bhatia, F. C. Simmel, and Y. Krishnan, Structural DNA nanotechnology: from bases to bricks, from structure to function, *J. Phys. Chem. Lett.*, **1**, 199–2005 (2010).

[3] C. Lin, Y. Liu, S. Rinker, and H. Yan, DNA tile based self-assembly: building complex nanoarchitectures, *ChemPhysChem*, **7**, 1641–1647 (2006).

[4] O. I. Wilner and I. Willner, Functionalized DNA nanostructures, *Chem. Rev.*, **112**, 2528–2556 (2012).

[5] F. Wang, C. H. Lu, and I. Willner, From cascaded catalytic nucleic acids to enzyme-DNA nanostructures: controlling reactivity, sensing, logic operations, and assembly of complex structures, *Chem. Rev.*, **114**, 2881–2941 (2014).

[6] S. Burge, G. N. Parkinson, P. Hazel, A. K. Todd, and S. Neidle, Quadruplex DNA: sequence, topology and structure, *Nucleic Acids Res.*, **34**, 5402–5415 (2006).

[7] D. Sen and W. Gilbert, Formation of parallel four-stranded complexes by guanine-rich motifs in DNA and its implications for meiosis, *Nature*, **334**, 364–366 (1988).

[8] J. T. Davis and G. P. Spada, Supramolecular architectures generated by self-assembly of guanosine derivatives, *Chem. Soc. Rev.*, **36**, 296–313 (2007).

[9] T. Simonsson, G-quadruplex DNA structures-variations on a theme, *Biol. Chem.*, **382**, 621–628 (2001).

[10] K. Gehring, J. L. Leroy, and M. Guéron, A tetrameric DNA structure with protonated cytosine•cytosine base pairs, *Nature*, **363**, 561–565 (1993).

[11] D. Collin and K. Gehring, Stability of chimeric DNA/RNA cytosine tetrads: implications for i-motif formation by RNA, *J. Am. Chem. Soc.*, **120**, 4069–4072 (1998).

[12] S. Nonin and J. L. Leroy, Structure and conversion kinetics of a bi-stable DNA i-motif: broken symmetry in the [d(5mCCTCC)]4 tetramer, *J. Mol. Biol.*, **261**, 399–414 (1996).

[13] J. L. Leroy, M. Gueron, J. L. Mergny, and C. Helene, Intramolecular folding of a fragment of the cytosine-rich strand of telomeric DNA into an i-motif, *Nucleic Acids Res.*, **22**, 1600–1606 (1994).

[14] Y. Miyake, H. Togashi, M. Tashiro, H. Yamaguchi, S. Oda, M. Kudo, Y. Tanaka, Y. Kondo, R. Sawa, T. Fujimoto, T. Machinami, and A. Ono, MercuryII-mediated formation of thymine-HgII-thymine base pairs in DNA duplexes, *J. Am. Chem. Soc.*, **128**, 2172–2173 (2006).

[15] Y. Tanaka, S. Oda, H. Yamaguchi, Y. Kondo, C. Kojima, and A. Ono, 15N–15N J-coupling across Hg(II): direct observation of Hg(II)-mediated T-T base pairs in a DNA duplex, *J. Am. Chem. Soc.*, **129**, 244–245 (2007).

[16] T. Li, S. Dong, and E. Wang, Label-free colorimetric detection of aqueous mercury ion (Hg^{2+}) using Hg^{2+}-modulated G-quadruplex-based DNAzymes, *Anal. Chem.*, **81**, 2144–2149 (2009).

[17] A. Ono, S. Q. Cao, H. Togashi, M. Tashiro, T. Fujimoto, T. Machinami, S. Oda, Y. Miyake, I. Okamoto, and Y. Tanaka, Specific interactions between silver(I) ions and cytosine-cytosine pairs in DNA duplexes, *Chem. Commun.*, 4825–4827 (2008).

[18] Y. Wen, F. Xing, S. He, S. Song, L. Wang, Y. Long, D. Li, and C. Fan, A graphene-based fluorescent nanoprobe for silver(I) ions detection by using graphene oxide and a silver-specific oligonucleotide, *Chem. Commun.*, **46**, 2596–2598 (2010).

[19] R. Freeman, T. Finder, and I. Willner, Multiplexed analysis of Hg^{2+} and Ag^{+} ions by nucleic acid functionalized CdSe/ZnS quantum dots and their use for logic gate operations, *Angew. Chem., Int. Ed.*, **48**, 7818–7821 (2009).

[20] P. J. Mitchell and R. Tjian, Transcriptional regulation in mammalian cells by sequence-specific DNA binding proteins, *Science*, **245**, 371–378 (1989).

[21] J. Boch, H. Scholze, S. Schornack, A. Landgraf, S. Hahn, S. Kay, T. Lahaye, A. Nickstadt, and U. Bonas, Breaking the code of DNA binding specificity of TAL-type III effectors, *Science*, **326**, 1509–1512 (2009).

[22] T. I. Lee and R. A. Young, Transcription of eukaryotic protein-coding genes, *Annu. Rev. Genet.*, **34**, 77–137 (2000).

[23] F. C. Wefald, B. H. Devlin, and R. S. Williams, Functional heterogeneity of mammalian TATA-box sequences revealed by interaction with a cell-specific enhancer, *Nature*, **344**, 260–262 (1990).

[24] K. Calvin and H. Li, RNA-splicing endonuclease structure and function, *Cell. Mol. Life Sci.*, **65**, 1176–1185 (2008).

[25] B. L. Stoddard, Homing endonuclease structure and function, *Q. Rev. Biophys.*, **38**, 49–95 (2005).

[26] R. Gupta, N. Capalash, and P. Sharma, Restriction endonucleases: natural and directed evolution, *Appl. Microbiol. Biotechnol.*, **94**, 583–599 (2012).

[27] S. West, N. Gromak, and N. J.Proudfoot, Human 5′→3′ exonuclease Xrn2 promotes transcription termination at co-transcriptional cleavage sites, *Nature*, **432**, 522–525 (2004).

[28] S. E. Osborne, I. Matsumura, and A. D. Ellington, Aptamers as therapeutic and diagnostic reagents: problems and prospects, *Curr. Opin. Chem. Biol.*, **1**, 5–9 (1997).

[29] T. Hermann and D. J. Patel, Adaptive recognition by nucleic acid aptamers, *Science*, **287**, 820–825 (2000).

[30] M. Famulok and G. Mayer, Aptamer modules as sensors and detectors, *Acc. Chem. Res.*, **44**, 1349–1358 (2011).

[31] I. Willner and M. Zayats, Electronic aptamer-based sensors, *Angew. Chem., Int. Ed.*, **46**, 6408–6418 (2007).

[32] M. Famulok, J. S. Hartig, and G. Mayer, Functional aptamers and aptazymes in biotechnology, diagnostics, and therapy, *Chem. Rev.*, **107**, 3715–3743 (2007).

[33] J. C. Achenbach, W. Chiuman, R. P. G Cruz, and Y. Li, DNAzymes: from creation in vitro to application in vivo, *Curr. Pharm. Biotechnol.*, **5**, 321–336 (2004).

[34] I. Willner, B. Shlyahovsky, M. Zayats, and B. Willner, DNAzymes for sensing, nanobiotechnology and logic gate applications, *Chem. Soc. Rev.*, **37**, 1153–1165 (2008).

[35] G. F. Joyce, Directed evolution of nucleic acid enzymes, *Annu. Rev. Biochem.*, **73**, 791–836 (2004).

[36] W. Pan and G. A. Clawson, Catalytic DNAzymes: derivations and functions, *Expert Opin. Biol. Ther.*, **8**, 1071–1085 (2008).

[37] S. K. Silverman, Catalytic DNA (deoxyribozymes) for synthetic applications-current abilities and future prospects, *Chem. Commun.*, 3467–3485 (2008).

[38] M. N. Stojanovic, P. de Prada, and D. W. Landry, Aptamer-based folding fluorescent sensor for cocaine, *J. Am. Chem. Soc.*, **123**, 4928–4931 (2001).

[39] G. Yang, J. Chun, H. Arakawa-Uramoto, M. A. Gawinowicz, K. Zhao, and D. W. Landry, Anti-cocaine catalytic antibodies: a synthetic solution to improved diversity, *J. Am. Chem. Soc.*, **118**, 5881–5890 (1996).

[40] L. C. Bock, L. C. Griffin, J. A. Latham, E. H. Vermaas, and J. J. Toole, Selection of single-stranded DNA molecules that bind and inhibit human thrombin, *Nature*, **355**, 564–566 (1992).

[41] R. F. Macaya, J. A. Waldron, B. A. Beutel, H. Gao, M. E. Joesten, M. Yang, R. Patel, A. H. Bertelsen, and A. F. Cook, Structural and functional characterization of potent antithrombotic oligonucleotides possessing both quadruplex and duplex motifs, *Biochemistry*, **34**, 4478–4492 (1995).

[42] Y. Li, C. R. Geyer, and D. Sen, Recognition of anionic porphyrins by DNA aptamers, *Biochemistry*, **35**, 6911–6922 (1996).

[43] P. Travascio, Y. Li, and D. Sen, DNA-enhanced peroxidase activity of a DNA-aptamerhemin complex. *Chem. Biol.*, **5**, 505–517 (1998).

[44] P. Travascio, A. J. Bennet, D. Y. Wang, and D. Sen, A ribozyme and a catalytic DNA with peroxidase activity: active sites versus cofactor-binding sites, *Chem. Biol.*, **6**, 779–787 (1999).

[45] P. Travascio, P. K. Witting, A. G. Mauk, and D. Sen, The peroxidase activity of a hemin-DNA oligonucleotide complex: free radical damage to specific guanine bases of the DNA, *J. Am. Chem. Soc.*, **123**, 1337–1348 (2001).

[46] V. Pavlov, Y. Xiao, R. Gill, A. Dishon, M. Kotler, and I. Willner, Amplified chemiluminescence surface detection of DNA and telomerase activity using catalytic nucleic acid labels, *Anal. Chem.*, **76**, 2152–2156 (2004).

[47] R. R. Breaker and G. F. Joyce, A DNA enzyme that cleaves RNA, *Chem. Biol.*, **1**, 223–229 (1994).

[48] T. Pan and O. C. Uhlenbeck, In vitro selection of RNAs that undergo autolytic cleavage with Pb^{2+}, *Biochemistry*, **31**, 3887–3895 (1992).

[49] T. Pan, O. C. Uhlenbeck, A small metalloribozyme with a two step mechanism. *Nature*, **358**, 560–563 (1992).

[50] D. Y. Zhang and E. Winfree, Control of DNA strand displacement kinetics using toehold exchange, *J. Am. Chem. Soc.*, **131**, 17303–17314 (2009).

[51] F. Wang, B. Willner, and I. Willner, DNA nanotechnology with one-dimensional selfassembled nanostructures, *Curr. Opin. Biotechnol.*, **24**, 562–574 (2013).

[52] Z. G. Wang, O. I. Wilner, and I. Willner, Self-assembly of aptamer-circular DNA nanostructures for controlled biocatalysis, *Nano Lett.*, **9**, 4098–4102 (2009).

[53] Y. Weizmann, A. B. Braunschweig, O. I. Wilner, Z. Cheglakov, and I. Willner, A polycatenated DNA scaffold for the one-step assembly of hierarchical nanostructures, *Proc. Natl. Acad. Sci., U.S.A.* **105**, 5289–5294 (2008).

[54] Y. He, Y. Chen, H. Liu, A. E. Ribbe, and C. Mao, Self-assembly of hexagonal DNA two-dimensional (2D) arrays, *J. Am. Chem. Soc.*, **127**, 12202–12203 (2005).

[55] S. H. Park, R. Barish, H. Li, J. H. Reif, G. Finkelstein, H. Yan, and T. H. Labean, Three-helix bundle DNA tiles self-assemble into 2D lattice or 1D templates for silver nanowires, *Nano Lett.*, **5**, 693–696 (2005).

[56] E. Winfree, F. Liu, L. A. Wenzler, and N. C. Seeman, Design and self-assembly of two-dimensional DNA crystals, *Nature*, **394**, 539–544 (1998).

[57] Y. Ke, N. V. Voigt, K. V. Gothelf, and W. M. Shih, Multilayer DNA origami packed on hexagonal and hybrid lattices, *J. Am. Chem. Soc.*, **134**, 1770–1774 (2012).

[58] U. Majumder, A. Rangnekar, K. V. Gothelf, J. H. Reif, and T. H. LaBean, Design and construction of doubledecker tile as a route to three-dimensional periodic assembly of DNA, *J. Am. Chem. Soc.*, **133**, 3843–3845 (2011).

[59] Y. Ke, L. L. Ong, W. M. Shih, and P. Yin, Three-dimensional structures self-assembled from DNA bricks, *Science*, **338**, 1177–1183 (2012).

[60] C. Teller and I. Willner, Organizing protein-DNA hybrids as nanostructures with programmed functionalities, *Trends Biotechnol.*, **28**, 619–628 (2010).

[61] C. M. Niemeyer, Semisynthetic DNA-protein conjugates for biosensing and nanofabrication, *Angew. Chem., Int. Ed.*, **49**, 1200–1216 (2010).

[62] Y. Liu, C. Lin, H. Li, and H. Yan, Aptamer-directed self-assembly of protein arrays on a DNA nanostructure. *Angew. Chem., Int. Ed.*, **44**, 4333–4338 (2005).

[63] Z. Cheglakov, Y. Weizmann, A. B. Braunschweig, O. I. Wilner, and I. Willner, Increasing the complexity of periodic protein nanostructures by the rolling-circle-amplified synthesis of aptamers, *Angew. Chem., Int. Ed.*, **47**, 126–130 (2008).

[64] B. Saccà and C. M. Niemeyer, Functionalization of DNA nanostructures with proteins, *Chem. Soc. Rev.*, **40**, 5910–5921 (2011).

[65] Z. Deng, Y. Tian, S. H. Lee, A. E. Ribbe, and C. Mao, DNA-encoded self-assembly of gold nanoparticles into one-dimensional arrays, *Angew. Chem., Int. Ed.*, **44**, 3582–3585 (2005).

[66] J. Sharma, R. Chhabra, Y. Liu, Y. Ke, and H. Yan, DNA-templated self-assembly of two-dimensional and periodical gold nanoparticle arrays, *Angew. Chem., Int. Ed.*, **45**, 730–735 (2006).

[67] H. Li, S. H. Park, J. H. Reif, T. H. LaBean, and H. Yan, DNA-templated self-assembly of protein and nanoparticle linear arrays, *J. Am. Chem. Soc.*, **126**, 418–419 (2004).

[68] J. Sharma, R. Chhabra, A. Cheng, J. Brownell, Y. Liu, and H. Yan, Control of self-assembly of DNA tubules through integration of gold nanoparticles, *Science*, **323**, 112–116 (2009).

[69] Y. Ke, S. Lindsay, Y. Chang, Y. Liu, and H. Yan, Self-assembled water-soluble nucleic acid probe tiles for label-free RNA hybridization assays, *Science*, **319**, 180–183 (2008).

[70] H. Pei, N. Lu, Y. Wen, S. Song, Y. Liu, H. Yan, and C. Fan, A DNA nanostructure-based biomolecular probe carrier platform for electrochemical biosensing, *Adv. Mater.*, **22**, 4754–4758 (2010).

[71] Y. Weizmann, M. K. Beissenhirtz, Z. Cheglakov, R. Nowarski, M. Kotler, and I. Willner, A virus spotlighted by an autonomous DNA machine, *Angew. Chem., Int. Ed.*, **45**, 7384–7388 (2006).

[72] F. Wang, J. Elbaz, R. Orbach, N. Magen, and I. Willner, Amplified analysis of DNA by the autonomous assembly of polymers consisting of DNAzyme wires, *J. Am. Chem. Soc.*, **133,** 17149–17151 (2011).

[73] C. C. Chen, Y. P. Lin, C. W. Wang, H. C. Tzeng, C. H. Wu, Y. C. Chen, C. P. Chen, L. C. Chen, and Y. C. Wu, DNA-gold nanorod conjugates for remote control of localized gene expression by near infrared irradiation, *J. Am. Chem. Soc.*, **128**, 3709–3715 (2006).

[74] W. Li, J. Wang, J. Ren, and X. Qu, Near-infrared- and pH-responsive system for reversible cell adhesion using graphene/gold nanorods functionalized with i-motif DNA, *Angew. Chem., Int. Ed.*, **52**, 6726–6730 (2013).

[75] Z. Xiao, C. Ji, J. Shi, E. M. Pridgen, J. Frieder, J. Wu, and O. C. Farokhzad, DNA self-assembly of targeted near-infrared-responsive gold nanoparticles for cancer thermo-chemotherapy, *Angew. Chem., Int. Ed.*, **51,** 11853–11857 (2012).

[76] R. Huschka, A. Barhoumi, Q. Liu, J. A. Roth, L. Ji, and N. J. Halas, Gene silencing by gold nanoshell-mediated delivery and laser-triggered release of antisense oligonucleotide and siRNA, *ACS Nano*, **6**, 7681–7691 (2012).

[77] S. Yamashita, H. Fukushima, Y. Akiyama, Y. Niidome, T. Mori, Y. Katayama and T. Niidome, Controlled-release system of single-stranded DNA triggered by the photothermal effect of gold nanorods and its in vivo application, *Bioorg. Med. Chem.*, **19**, 2130–2135 (2011).

[78] J. Elbaz, O. Lioubashevski, F. Wang, F. Remacle, R. D. Levine, and I. Willner, DNA computing circuits using libraries of DNAzyme subunits, *Nat. Nanotechnol.*, **5**, 417–422 (2010).

[79] L. Qian and E. Winfree, Scaling up digital circuit computation with DNA strand displacement cascades, *Science*, **332**, 1196–1201 (2011).

[80] D. Miyoshi, M. Inoue, and N. Sugimoto, DNA logic gates based on structural polymorphism of telomere DNA molecules responding to chemical input signals, *Angew. Chem., Int. Ed.*, **45**, 7716–7719 (2006).

[81] J. Zhu, L. Zhang, T. Li, S. Dong and E. Wang, Enzyme-free unlabeled DNA logic circuits based on toehold-mediated strand displacement and split G-quadruplex enhanced fluorescence, *Adv. Mater.*, **25**, 2440–2444 (2013).

[82] M. N. Stojanovic, S. Semova, D. Kolpashchikov, J. Macdonald, C. Morgan, and D. Stefanovic, Deoxyribozyme-based ligase logic gates and their initial circuits. *J. Am. Chem. Soc.*, **127**, 6914–6915 (2005).

[83] H. T. Maune, S. P. Han, R. D. Barish, M. Bockrath, Goddard, W. A. III. Goddard, P. W. K. Rothemund, and E. Winfree, Self-assembly of carbon nanotubes into two-dimensional geometries using DNA origami templates, *Nat. Nanotechnol.*, **5**, 61–66 (2010).

[84] N. A. Bell, C. R. Engst, M. Ablay, G. Divitini, C. Ducati, T. Liedl, and U. F. Keyser, DNA origami nanopores, *Nano Lett.*, **12**, 512–517 (2012).

[85] T. Yamazaki, Y. Aiba, K. Yasuda, Y. Sakai, Y. Yamanaka, A. Kuzuya, Y. Ohya, and M. Komiyama, Clear-cut observation of PNA invasion using nanomechanical DNA origami devices, *Chem. Commun.*, **48**, 11361–11363 (2012).

[86] R. Wei, T. G. Martin, U. Rant, and H. Dietz, DNA origami gatekeepers for solid-state nanopores, *Angew. Chem., Int. Ed.*, **51**, 4864–4867 (2012).

[87] D. Liu, S. H. Park, J. H. Reif, and T. H. LaBean, DNA nanotubes self-assembled from triple-crossover tiles as templates for conductive nanowires, *Proc. Natl. Acad. Sci. U.S.A.*, **101**, 717–722 (2004).

[88] O. I. Wilner, S. Shimron, Y. Weizmann, Z. G. Wang, and I. Willner, Self-assembly of enzymes on DNA scaffolds: en route to biocatalytic cascades and the synthesis of metallic nanowires, *Nano Lett.*, **9**, 2040–2043 (2009).

[89] F. Patolsky, Y. Weizmann, O. Lioubashevski, and I. Willner, Au-nanoparticle nanowires based on DNA and polylysine templates, *Angew. Chem., Int. Ed.*, **41**, 2323–2327 (2002).

[90] M. K. Beissenhirtz and I. Willner, DNA-based machines, *Org. Biomol. Chem.*, **4**, 3392–3401 (2006).

[91] J. Bath and A. J. Turberfield, DNA nanomachines, *Nat. Nanotechnol.*, **2**, 275–284 (2007).

[92] Y. Krishnan and F. C. Simmel, Nucleic acid based molecular devices, *Angew. Chem., Int. Ed.*, **50**, 3124–3156 (2011).

[93] C. Teller and I. Willner, Functional nucleic acid nanostructures and DNA machines, *Curr. Opin. Biotechnol.*, **21**, 376–391 (2010).

[94] C. M. Niemeyer and M. Adler, Nanomechanical devices based on DNA, *Angew. Chem., Int. Ed.*, **41**, 3779–3783 (2002).

[95] Z. G. Wang, J Elbaz, and I. Willner, A dynamically programmed DNA transporter, *Angew. Chem., Int. Ed.*, **51**, 4322–4326 (2012).

[96] Z. G. Wang, J. Elbaz, F. Remacle, R. D. Levine, and I. Willner, All-DNA finite-state automata with finite memory, *Proc. Natl. Acad. Sci., U. S. A.*, **107**, 21996–22001 (2010).

[97] J. Elbaz, M. Moshe, and I. Willner, Coherent activation of DNA tweezers: a "SET-RESET" logic system, *Angew. Chem., Int. Ed.*, **48**, 3834–3837 (2009).

[98] H. Pei, L. Liang, G. Yao, J. Li, Q. Huang, and C. Fan, Reconfigurable three-dimensional DNA nanostructures for the construction of intracellular logic sensors, *Angew. Chem., Int. Ed.*, **51**, 9020–9024 (2012).

[99] S. M. Douglas, I. Bachelet, and G. M. Church, A logic-gated nanorobot for targeted transport of molecular payloads, *Science*, **335**, 831–834 (2012).

[100] Q. Jiang, C. Song, J. Nangreave, X. Liu, L. Lin, D. Qiu, Z. G. Wang, G. Zou, X. Liang, H. Yan, and B. Ding, DNA origami as a carrier for circumvention of drug resistance, *J. Am. Chem. Soc.*, **134**, 13396–13403 (2012).

[101] Y. X. Zhao, A. Shaw, X. Zeng, E. Benson, A. M. Nyström, and B. Högberg, DNA origami delivery system for cancer therapy with tunable release properties, *ACS Nano*, **6**, 8684–8691 (2012).

[102] J. Elbaz and I. Willner, DNA origami: Nanorobots grab cellular control, *Nat. Mater.*, **11**, 276–277 (2012).

[103] J. Fu and H.Yan, Controlled drug release by a nanorobot, *Nat. Biotechnol.*, **30**, 407–408 (2012).

[104] D. M. Rosenbaum and D. R. Liu, Efficient and sequence-specific DNA-templated polymerization of peptide nucleic acid aldehydes, *J. Am. Chem. Soc.*, **125**, 13924–13925 (2003).

[105] F. E. Alemdaroglu, K. Ding, R. Berger, and A. Herrmann, A-templated synthesis in three dimensions: Introducing a micellar scaffold for organic reactions, *Angew. Chem., Int. Ed.*, **45**, 4206–4210 (2006).

[106] M. L. McKee, P. J. Milnes, J. Bath, E. Stulz, A. J. Turberfield, and R. K. O'Reilly, Multistep DNA-templated reactions for the synthesis of functional sequence controlled oligomers, *Angew. Chem., Int. Ed.*, **49**, 7948–7951 (2010).

[107] J. B. Ravnsbæk, M. F. Jacobsen, C. B. Rosen, N. V. Voigt, and K. V. Gothelf, DNA-programmed Glaser-Eglinton reactions for the synthesis of conjugated molecular wires, *Angew. Chem., Int. Ed.*, **50**, 10851–10854 (2011).

[108] S. Shimron, A. Cecconello, C. H. Lu, and I. Willner, Metal nanoparticle-functionalized DNA tweezers: from mechanically programmed nanostructures to switchable fluorescence properties, *Nano Lett.*, **13**, 3791–3795 (2013).

[109] J. Elbaz, A. Cecconello, Z. Fan, A. O. Govorov, and I. Willner, Powering the programmed nanostructure and function of gold nanoparticles with catenated DNA machines, *Nat. Commun.* **4**, 2000 (2013).

[110] A. Cecconello, C. H. Lu, J. Elbaz, and I. Willner, Au nanoparticle/DNA rotaxane hybrid nanostructures exhibiting switchable fluorescence properties, *Nano Lett.*, **13**, 6275–6280 (2013).

[111] H. Asanuma, T. Ito, T. Yoshida, X. G. Liang, and M. Komiyama, Photoregulation of the Formation and Dissociation of a DNA Duplex by Using the cis-trans Isomerization of Azobenzene, *Angew. Chem., Int. Ed.*, **38**, 2393–2395 (1999).

[112] H. Asanuma, T. Takarada, T. Yoshida, X. Liang, and M. Komiyama, Enantioselective incorporation of azobenzenes into oligodeoxyribonucleotide for effective photoregulation of duplex formation, *Angew. Chem., Int. Ed.*, **40**, 2671–2673 (2001).

[113] H. Asanuma, X. G. Liang, T. Yoshida, and M. Komiyama, Photocontrol of DNA duplex formation by using azobenzene-bearing oligonucleotides, *ChemBioChem.*, **2**, 39–44 (2001).

[114] S. Shimron, N. Magen, J. Elbaz, and I. Willner, pH-programmable DNAzyme nanostructures, *Chem. Comm.*, 8787–8789 (2011).

[115] J. Elbaz, S. Shimron, and I. Willner, pH-triggered switchable Mg^{2+}-dependent DNAzymes, *Chem. Comm.*, **46**, 1209–1211 (2010).

[116] A. Arola and R. Vilar, Stabilisation of G-quadruplex DNA by small molecules, *Curr. Top. Med. Chem.*, **8**, 1405–1415 (2008).

[117] N. H. Campbell and S. Neidle, G-Quadruplexes and Metal Ions, *in Interplay between Metal Ions and Nucleic Acids*, A. Sigel, H. Sigel and R. K. O. Sigel (Eds), Metal Ions in Life Sciences Series 10, Springer Netherlands (2012).

[118] T. Li, S. Dong, and E.Wang, A lead(II)-driven DNA molecular device for turn-on fluorescence detection of lead(II) ion with high selectivity and sensitivity, *J. Am. Chem. Soc.*, **132**, 13156–13157 (2010).

[119] M. A. Aleman-Garcia, R. Orbach, and I. Willner, Ion-responsive hemin-G-quadruplexes for switchable DNAzyme and enzyme functions, *Chem. Eur. J.* **20**, 5619–5624 (2014).

[120] B. Ge, Y. C. Huang, D. Sen, and H. Z. Yu, A robust electronic switch made of immobilized duplex/quadruplex DNA, *Angew. Chem., Int. Ed.*, **49**, 9965–9967 (2010).

[121] T. Li, L. Shi, E. Wang, and S. Dong, Silver-ion-mediated DNAzyme switch for the ultrasensitive and selective colorimetric detection of aqueous Ag^+ and cysteine, *Chem. Eur. J.*, **15**, 3347–3350 (2009).

[122] X. Liu, A. Niazov-Elkan, F. Wang, and I. Willner, Switching photonic and electrochemical functions of a DNAzyme by DNA machines, *Nano Lett.*, **13**, 219–225 (2013).

[123] C. Dohno and K. Nakatani, Control of DNA hybridization by photoswitchable molecular glue, *Chem. Soc. Rev.*, **40**, 5718–5729 (2011).

[124] M. Zhou, X. Liang, T. Mochizuki, and H. Asanuma, A light-driven DNA nanomachine for the efficient photoswitching of RNA digestion, *Angew. Chem., Int. Ed.*, **49**, 2167–2170 (2010).

[125] D. Han, J. Huang, Z. Zhu, Q. Yuan, M. You, Y. Chen, and W. Tan, Molecular engineering of photoresponsive three-dimensional DNA nanostructures, *Chem. Commun.*, **47**, 4670–4672 (2011).

[126] Y. Yan, J. I. Chen, and D. S. Ginger, Photoswitchable oligonucleotide-modified gold nanoparticles: controlling hybridization stringency with photon dose, *Nano Lett.*, **12**, 2530–2536 (2012).

[127] B. Y. Won, C. Jung, K. S. Park, and H. G. Park, An electrochemically reversible DNA switch, *Electrochem. Commun.* **27**, 100–103 (2013).

[128] B. Yurke, A. J. Turberfield, A. P. Jr Mills, F. C. Simmel, and J. L. Neumann, A DNA-fuelled molecular machine made of DNA, *Nature*, **406**, 605–608 (2000).

[129] J. Elbaz, Z. G. Wang, R. Orbach, and I. Willner, pH-stimulated concurrent mechanical activation of two DNA "tweezers". A "SET-RESET" logic gate system, *Nano Lett.*, **9**, 4510–4514 (2009).

[130] X. Liang, H. Nishioka, N. Takenaka, and H. Asanuma, A DNA nanomachine powered by light irradiation. *ChemBioChem*, **9**, 702–705 (2008).

[131] M. You, F. Huang, Z. Chen, R. W. Wang, and W. Tan, Building a nanostructure with reversible motions using photonic energy, *ACS Nano*, **6**, 7935–7941 (2012).

[132] Z. G. Wang, J. Elbaz, and I. Willner, DNA machines: bipedal walker and stepper. *Nano Lett.*, **11**, 304–309 (2011).

[133] Y. Tian and C. Mao, Molecular gears: a pair of DNA circles continuously rolls against each other, *J. Am. Chem. Soc.*, **126**, 11410–11411 (2004).

[134] C. H. Lu, A. Cecconello, J. Elbaz, A. Credi, and I. Willner, A three-station DNA catenane rotary motor with controlled directionality, *Nano Lett.*, **13**, 2303–2308 (2013).

[135] G. Pelossof, R. Tel-Vered, X. Liu, and I. Willner, Switchable mechanical DNA "arms" operating on nucleic acid scaffolds associated with electrodes or semiconductor quantum dots, *Nanoscale*, **5**, 8977–8981 (2013).

[136] J. Elbaz, Z.-G. Wang, F. Wang, and I. Willner, Programmed dynamic topologies in DNA catenanes. *Angew. Chem., Int. Ed.*, **51**, 2349–2353 (2012).

[137] Y. Tian, Z. Zhang, L. Wen, J. Ma, Y .Zhang, W. Liu, J. Zhai, and L. Jiang, A biomimetic mercury(II)-gated single nanochannel, *Chem. Commun.,* **49**, 10679–10681 (2013).

[138] W. Guo, X. J. Qi, R. Orbach, C. H. Lu, L. Freage, I. Mironi-Harpaz, D. Seliktar, H. H. Yang, and I. Willner, Reversible Ag^{+}-crosslinked DNA hydrogels, *Chem. Commun.,* **50**, 4065–4068 (2014).

[139] E. Cheng, Y. Xing, P. Chen, Y. Yang, Y. Sun, D. Zhou, L. Xu, Q. Fan, and D. Liu, A pH-triggered, fast-responding DNA hydrogel, *Angew. Chem., Int. Ed.*, **48**, 7660–7663 (2009).

[140] C. H. Lu, X. J. Qi, R. Orbach, H.-H. Yang, I. Mironi-Harpaz, D. Seliktar, and I. Willner, Switchable catalytic acrylamide hydrogels cross-linked by hemin/G-quadruplexes, *Nano Lett.*, **13**, 1298–1302 (2013).

[141] L. Xin, C. Zhou, Z. Yang, and D. Liu, Regulation of an enzyme cascade reaction by a DNA machine, *Small*, **9**, 3088–3091 (2013).

[142] M. Liu, J. Fu, C. Hejesen, Y. Yang, N. W. Woodbury, K. Gothelf, Y. Liu, and H. Yan, A DNA tweezer-actuated enzyme nanoreactor. *Nat. Commun.*, **4**, 2127 (2013).

[143] G. H. Clever, C. Kaul, and T. Carell, DNA–metal base pairs, *Angew. Chem., Int. Ed.*, **46**, 6226–6236 (2007).

[144] Y. Sannohe, M. Endo, Y. Katsuda, K. Hidaka, and H. Sugiyama, Visualization of dynamic conformational switching of the G-quadruplex in a DNA nanostructure, *J. Am. Chem. Soc.*, **132**, 16311–16313 (2010).

[145] K. Lund, A. J. Manzo, N. Dabby, N. Michelotti, A. Johnson-Buck, J. Nangreave, S. Taylor, R. Pei, M. N. Stojanovic, N. G. Walter, E. Winfree, and H. Yan, Molecular robots guided by prescriptive landscapes, *Nature*, **465**, 206–210 (2010).

[146] S. F. Wickham, M. Endo, Y. Katsuda, K. Hidaka, J. Bath, H. Sugiyama, and A. J. Turberfield, Direct observation of stepwise movement of a synthetic molecular transporter, *Nat. Nanotechnol.* **6**, 166–169 (2011).

[147] S. Modi, S. M. G, D. Goswami, G. D. Gupta, S. Mayor, and Y. Krishnan, A DNA nanomachine that maps spatial and temporal pH changes inside living cells, *Nat. Nanotechnol.*, **4**, 325–330 (2009).

[148] Z. Zhang, D. Balogh, F. Wang, and I. Willner, Smart mesoporous SiO_2 nanoparticles for the DNAzyme-induced multiplexed release of substrates, *J. Am. Chem. Soc.*, **135**, 1934–1940 (2012).

[149] Z. Zhang, D. Balogh, F. Wang, S. Y. Sung, R. Nechushtai, and I. Willner, Biocatalytic release of an anti-cancer drug from nucleic acids-capped mesoporous SiO_2 using DNA or molecular biomarkers as triggering stimuli, *ACS Nano*, **7**, 8455–8468 (2013).

[150] C. H. Lu, B. Willner, and I. Willner, DNA nanotechnology: from sensing and DNA machines to drug-delivery systems, *ACS Nano*, **7**, 8320–8332 (2013).

[151] Y. He and D. R. Liu, Autonomous multistep organic synthesis in a single isothermal solution mediated by a DNA walker. *Nat. Nanotechnol.*, **5**, 778–782 (2010).

5.7

DNA-Based Asymmetric Catalysis

Ana Rioz-Martínez and Gerard Roelfes

Stratingh Institute for Chemistry, University of Groningen, Groningen, The Netherlands

5.7.1 Introduction

The discovery of the DNA double helical structure by James Watson and Francis Crick in 1953 was one of the major achievements in science [1]. Since then, its structure has been the focus of research in many different fields ranging from biology and medicine to materials science [2, 3]. In the past ten years, DNA has also emerged as a powerful tool for synthesis, as this molecule has several properties that make it a very promising chiral scaffold for the design of hybrid catalysts. It is chemically stable, commercially available and natural DNA is inexpensive. Furthermore, its solubility and biodegradability make DNA an attractive material for the development of asymmetric catalysis in aqueous medium, giving it 'green' credentials [4]. Finally, its well defined chemical structure, the iconic right-handed double helix of B-DNA, is a highly attractive source of chirality for asymmetric catalysis.

In this chapter we will describe the use of DNA as chiral bio-scaffold in the design of hybrid catalysts and their application in asymmetric catalysis [5]. Some current and relevant examples will be discussed, followed by an overview of mechanistic studies.

5.7.2 Concept of DNA-based asymmetric catalysis

DNA-based asymmetric catalysis is an exponent of the general concept of hybrid catalysts, which aims to merge the attractive properties of homogeneous ('chemical') and bio-catalysis. A DNA-based catalyst comprises a transition metal complex, that is, a metal ion coordinated to a non-chiral ligand, which is able to bind

---

DNA in Supramolecular Chemistry and Nanotechnology, First Edition.
Edited by Eugen Stulz and Guido H. Clever.

to DNA. In this way, the chiral microenvironment provided by the DNA, also referred to as 'the second coordination sphere', directs the catalysed reaction towards the selective formation of one of the enantiomers of the product, resulting in an enantiomeric excess.

5.7.3 Design approaches in DNA-based asymmetric catalysis

There are two main approaches to the anchoring of a transition metal complex to DNA (Scheme 5.7.1). In covalent anchoring, the ligand for the metal is attached to the DNA via a chemical bond. The main advantage of covalent anchoring is that the position of the metal complex within the DNA is known exactly. Thus, there

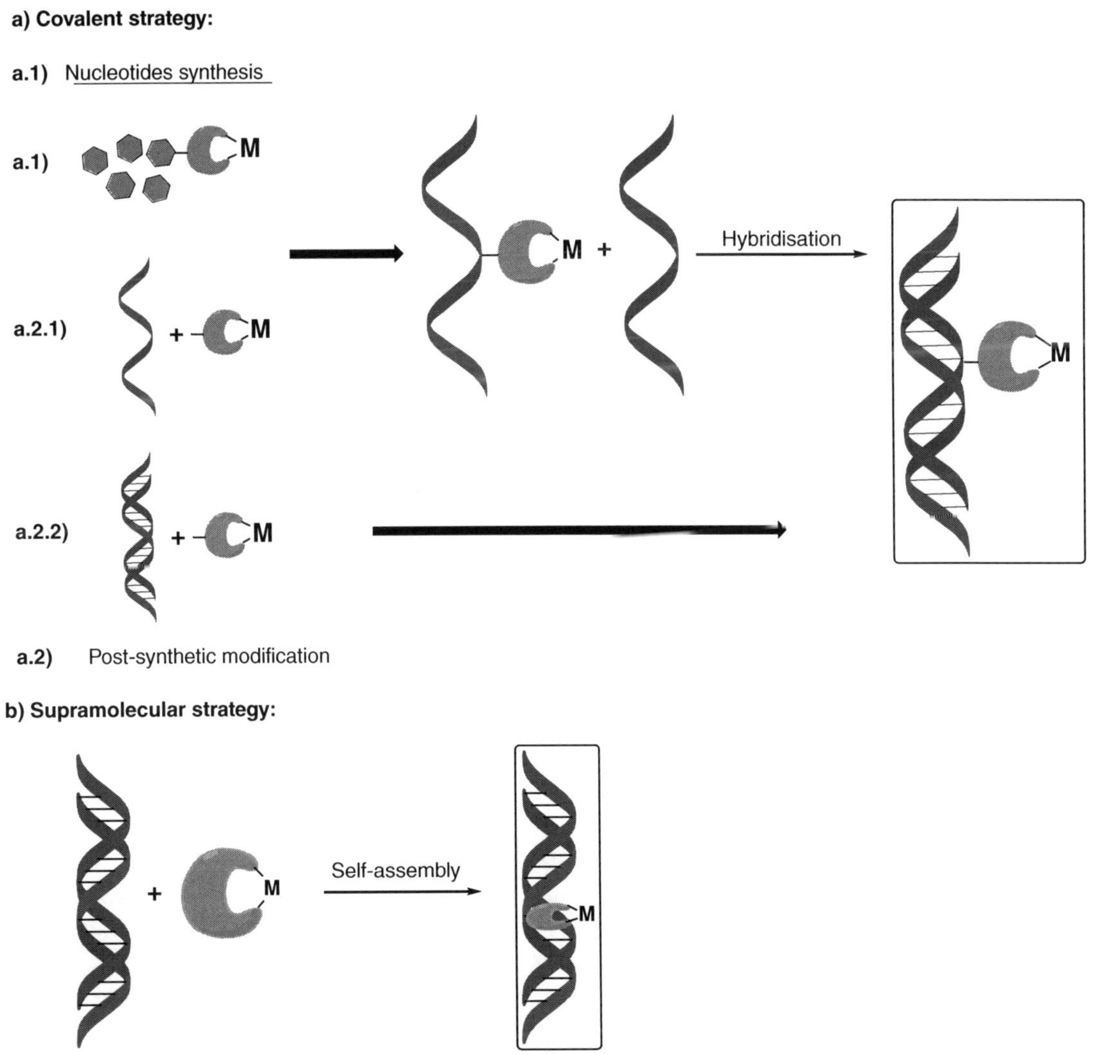

Scheme 5.7.1 Schematic representation of DNA-based catalysts using a covalent (a) or supramolecular (b) anchoring strategy

is a high degree of control over the second coordination sphere interactions. However, the main limitation of this methodology is the laborious synthesis and purification of chemically modified oligonucleotides. This potentially complicates the optimization of DNA-based catalysts for a given reaction. The second method is supramolecular anchoring, which involves the ligand being bound to the DNA molecule via intercalation or groove binding using non-covalent interactions. This approach is based on the ability of some small molecules to bind to the DNA by π-stacking, hydrophobic, electrostatic and/or hydrogen bonding interactions.

A particularly attractive aspect of supramolecular anchoring is the easy formation of the catalyst since it involves spontaneous self-assembly of the transition metal complex with DNA. Moreover, in this approach it is usually DNA from natural sources, such as calf thymus or salmon testes DNA, that is used. However, an apparent limitation of this strategy is that generally there is very little control over where the metal catalysts bind to the DNA and, hence, over the second-generation sphere. This may give rise to a very heterogeneous mixture of different catalysts, which could be difficult to control.

In the next section, examples of both classes and their application in asymmetric catalysis will be described. This is then followed by a discussion of the mechanistic aspects of these reactions.

5.7.4 Covalent anchoring

The covalent anchoring approaches reported to date have been based mainly on the derivatization of nucleobases with a ligand that is able to chelate transition metals. The first research on this topic was reported by Jäschke and coworkers. They described the modification of a series of nucleotides by reacting a phosphine-containing activated ester with aminoalkyl-modified oligonucleotides (yields from 38 to 78%) [6]. Unfortunately, no catalysis has been reported with these catalysts. Later, Kamer's group described the palladium catalysed synthesis of nucleosides functionalized with phosphine moieties and their application in a Pd(II) catalysed allylic amination. High enantioselectivities of up to 82% were obtained in THF, using a single modified nucleotide as the ligand. Extension to longer oligonucleotides did not give rise to good results in catalysis [7].

Later, Fournier *et al.* reported the allylic amination of phenyl allyl acetate with morpholine, catalysed by a hybrid catalyst, which consists of an iridium complex covalently attached to DNA (Scheme 5.7.2) [8]. The enantioselectivity obtained was low (24% *ee*), but it was observed that the stereochemical outcome of the reaction depends on the nucleic acid structure and nature, that is, when a complementary RNA strand was employed, the opposite enantiomer of the product was obtained.

In the same year, the synthesis and application of a hybrid catalyst composed of a Cu(II) chelating polyaza crown ether incorporated in the backbone of single-stranded DNA was reported [9]. After hybridization with a complementary DNA strand, the catalyst was evaluated in the asymmetric Diels–Alder reaction of azachalcone with cyclopentadiene. Unfortunately, only a low enantioselectivity of 10% was achieved.

Better results were achieved in the same reaction using a DNA-based catalyst assembled using a modular approach (Scheme 5.7.3) [10]. In this case, a 2,2′-bipyridine Cu(II) complex was covalently linked to the terminus of an oligonucleotide. As shown in Scheme 5.7.3, the catalytic system was formed by an oligonucleotide that is functionalized with the 2,2′-bipyridine ligand (on1), an unfunctionalized oligonucleotide (on2) and a template strand complementary to both on1 and on2. An attractive feature of this design is that the second coordination sphere is relatively easily optimized by exchange of one of the oligonucleotide modules: no chemical synthesis of novel modified DNAs is required. The highest enantiomeric excess reported was 93% and it was found that the *ee* mainly depended on the 3′-terminal nucleotides of on2 and the corresponding nucleotides in the template. Changes, such as GTA (3′–5′) and CAT (5′–3′), respectively, in these sequences increased the conversion and the enantioselectivity. Also, shorter linkers between the

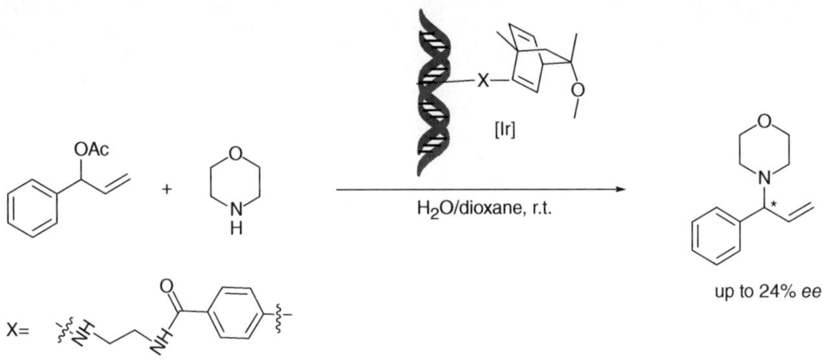

Scheme 5.7.2 *Schematic representation of the DNA-based iridium-catalysed asymmetric allylic amination of phenyl allyl acetate*

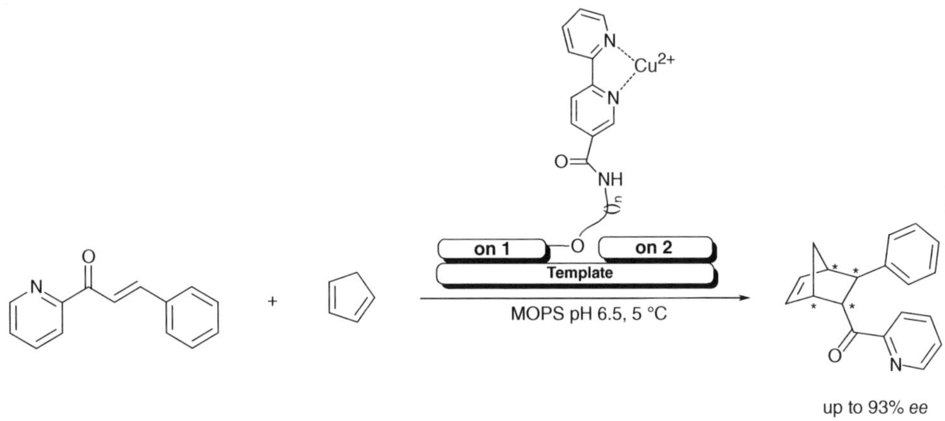

Scheme 5.7.3 *Schematic representation of the DNA-based copper-catalysed asymmetric Diels–Alder reaction using a modular approach to catalyst assembly*

oligonucleotide and the 2,2′-bipyrdine ligand were found to be beneficial for selectivity, demonstrating that the transfer of chirality is more efficient when the metal complex is in close proximity to the DNA molecule.

Recently, a new covalent anchoring strategy has been developed, which is based on the affinity of cisplatin complexes to DNA [11]. These novel heteronuclear complexes (Cu–Pt) contain two key structural features: a bipyridine moiety that acts as a ligand for the Cu(II), and a cisplatinum complex that binds to the DNA. The corresponding hybrid catalyst was tested in an asymmetric Friedel–Crafts alkylation and a Diels–Alder cycloaddition in water (Scheme 5.7.4). The hybrid system (8 mol% loading) catalysed the addition of 5-methoxy indole to the corresponding α,β-unsaturated 2-acyl imidazole with complete conversion and 64% *ee* after one day of reaction. This catalyst was also efficient in the Diels–Alder reaction between azachalcone and cyclopentadiene: a quantitative conversion and 73% *ee* of the *endo* product were achieved with the same catalyst loading (8 mol%). Finally, the DNA-based catalyst was recycled ten times without significant loss of activity or selectivity. This methodology represents a starting point for the covalent binding of metal complexes in a straightforward way to natural DNA.

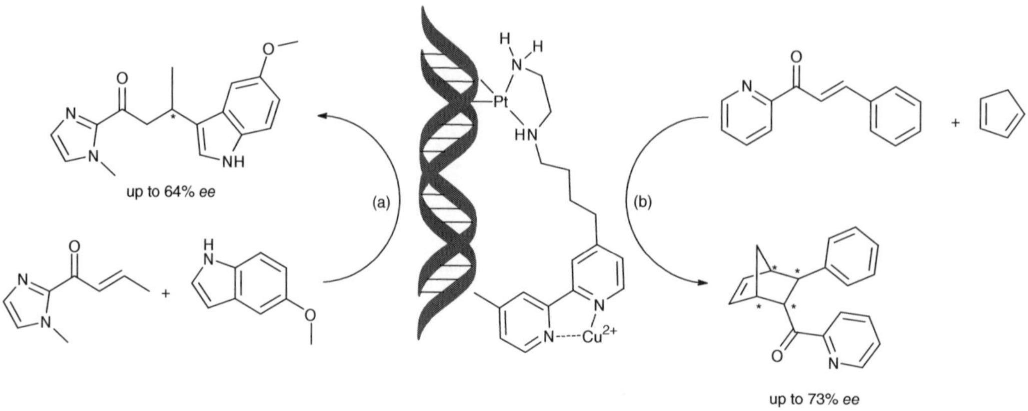

Scheme 5.7.4 *Schematic representation of the DNA-based catalysts assembled using cisplatin to bind the catalytically active Cu(II) complex to DNA and application in the asymmetric Friedel–Crafts alkylation and Diels–Alder cycloaddition reaction*

5.7.5 Supramolecular anchoring

5.7.5.1 First generation DNA-based catalysts

The design of the first DNA-based catalyst was reported in 2005 by Roelfes and Feringa [12]. This hybrid catalyst was formed *in situ* from a copper(II) salt, ligand **1** and salmon testes DNA (st-DNA) (Scheme 5.7.5). Ligand **1** consists of three essential parts: a DNA-intercalating moiety, such as 9-aminoacridine, a spacer and a metal-binding group such as aminomethyl pyridine. This class of ligands is known as first generation ligands. They are optimized readily via variation of the substituent (R) and the length of the spacer (*n*). The benchmark reaction that was used for proof of concept was the Cu(II) catalysed Diels–Alder reaction between azachalcone and cyclopentadiene in water (Scheme 5.7.5). This reaction was chosen for several reasons: (a) an aqueous environment favours this reaction [13], (b) it is a reaction that, due to the large structural changes occurring during the reaction, is sensitive to its environment, as is demonstrated by the fact that Diels–Alder reactions have also been catalysed by RNA-, DNA-zymes and catalytic antibodies [14–16], and (c) there are no changes in the oxidation state of the metal involved, avoiding possible damage to the DNA resulting from the undesired formation of reactive oxygen species [17].

The Diels–Alder product was obtained as a mixture of the *endo* (major) and *exo* (minor) isomers. The *endo*:*exo* ratio as well as the enantiomeric excess was found to be ligand-dependent. The best results were obtained with R = 1-napththylmethyl (**L1a**), for which the ratio *endo* to *exo* was 98:2 and 49% *ee* was measured for the *endo*(–) enantiomer. This suggests the importance of π-stacking interactions between the ligand and the bound substrate. An enantiomeric excess of 37% of the opposite enantiomer was achieved with R = 3,5-dimethoxybenzyl (**L1b**). Thus, although only right-handed helical DNA was employed in these studies, both enantiomers could be accessed by judicious choice of the achiral ligand. Another important issue that affects the enantiomeric excess is the spacer length: short spacers, that is, *n*-propyl and ethyl, resulted in the highest selectivities. This point underscores how the proximity of the Cu(II) ion to the DNA is key to achieving selectivity in the reaction.

5.7.5.2 Improvement of the catalytic system – second generation catalysts

The fact that using short spacers gave rise to high selectivities led to a design in which the Cu(II) ion is in direct contact with the double helix, that is, a ligand without a spacer to connect the DNA and metal-binding moieties. These new, 'second-generation', catalysts involve metal complexes that bind directly to DNA,

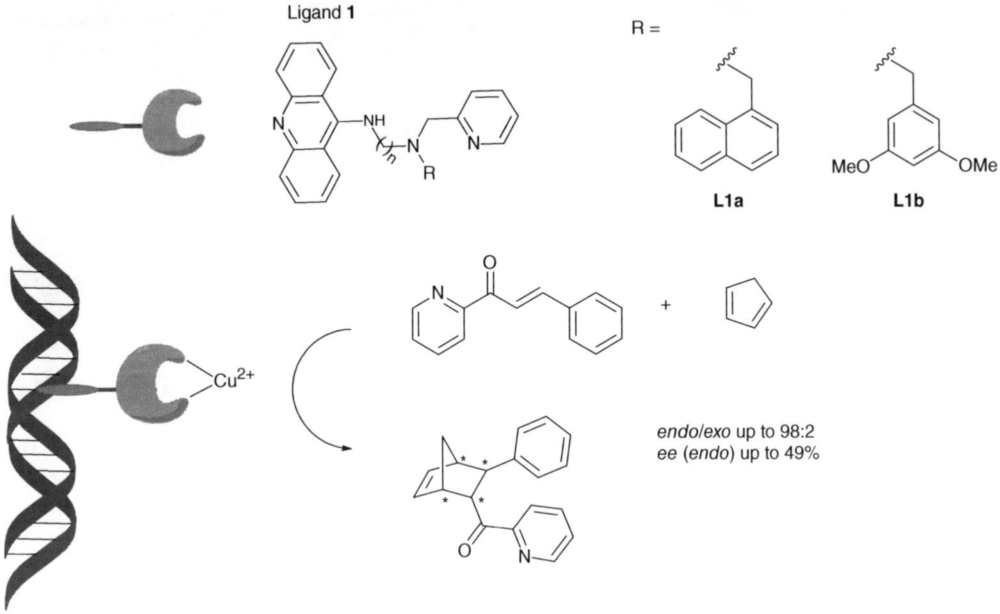

Scheme 5.7.5 *Schematic representation of the asymmetric Diels–Alder reaction of aza-chalcone with cyclopentadiene, catalysed by copper complexes of ligand **1** in the presence of DNA. Typical conditions: 1.3 mg mL^{-1} st-DNA, 30 mol% catalyst, 1 mM azachalcone, 5 mM cyclopentadiene, 20 mM 3-(N-morpholino)-propanesulfonic acid (MOPS buffer, pH 6.5), 5 °C, 3 days*

without needing a separate DNA binding moiety (Figure 5.7.1). Indeed, the second-generation DNA-based catalysts gave rise to a substantial increase in both the enantio- and the *endo*-selectivity of the Diels–Alder reaction compared with the first generation [18]. Depending on the ligand, enantioselectivities ranged from 49% for [Cu(dppz)(NO$_3$)$_2$] to up to 99% for [Cu(dmbipy)(NO$_3$)$_2$]. Furthermore, an improvement in the *endo*:*exo* selectivity was achieved, of from 96:4 to 99:1, respectively. Interestingly, the weaker DNA binding metal complex in these series, [Cu(dmbipy)(NO$_3$)$_2$], gave rise to the highest selectivity. However, all these catalysts produced the same enantiomer of the Diels–Alder product in excess. This apparent limitation was overcome by using tridentate ligands for the Cu(II) ion [19]. Several terpyridine ligands (**L9**) were evaluated and surprisingly the opposite enantiomers of the Diels–Alder product were obtained. This was attributed to a different binding geometry of the substrate bound Cu(II) complex in the DNA, resulting in a preferential attack of the diene from the opposite prochiral face of the enone.

α,β-Unsaturated 2-acyl imidazoles were introduced as alternative class of substrates for DNA-based catalysis (Scheme 5.7.6) [20]. These substrates, which, similar to aza-chalcone, can bind in a bidentate fashion to the Cu(II) ion, were introduced by Evans *et al.* [21]. The advantage of these substrates is that the imidazole moiety is readily removed after the catalysis, allowing for further synthetic application. The enantioselectivities achieved, which range from 80 to 98% *ee*, were similar to the values obtained with aza-chalcone. The best results were obtained using [Cu(dmbipy)(NO$_3$)$_2$] as the metal complex in the presence of st-DNA, obtaining the Diels–Alder products with excellent diastereoselectivity and enantioselectivity.

5.7.5.3 Catalytic scope of DNA-based catalysts

The DNA-based catalysis concept has been applied in many other Lewis acid catalysed reactions. The majority of these are conjugate addition reactions, such as the (oxa-) Michael addition, the vinologous Friedel–Crafts alkylation reaction, and the conjugate addition of water [22–24]. Additionally, other reactions

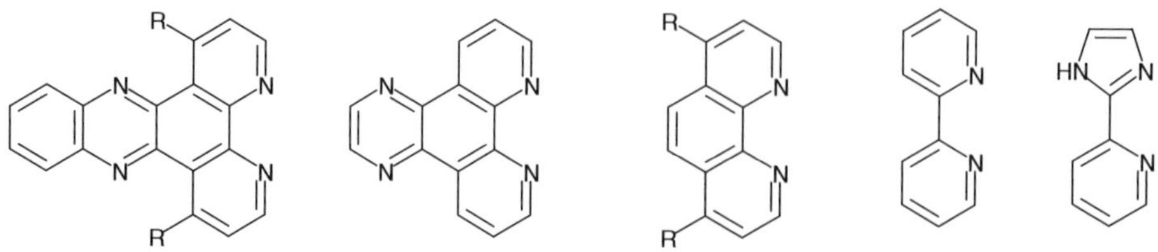

L2a R = H (dppz)
L2b R = Me **L3 (dpq)** **L4a R = H (phen)**
L4b R = Me **L5 (bipy)** **L6**

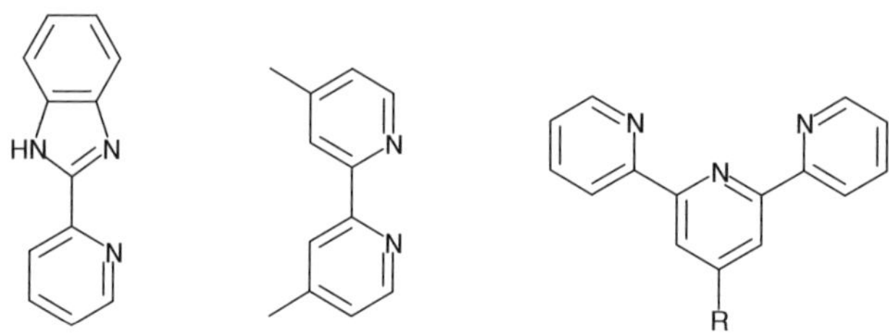

L9 (terpy)
R = H, MeO, *p*-tolyl, Cl

L7 **L8 (dmbipy)**

Figure 5.7.1 *Second-generation ligands used in DNA-based asymmetric reactions*

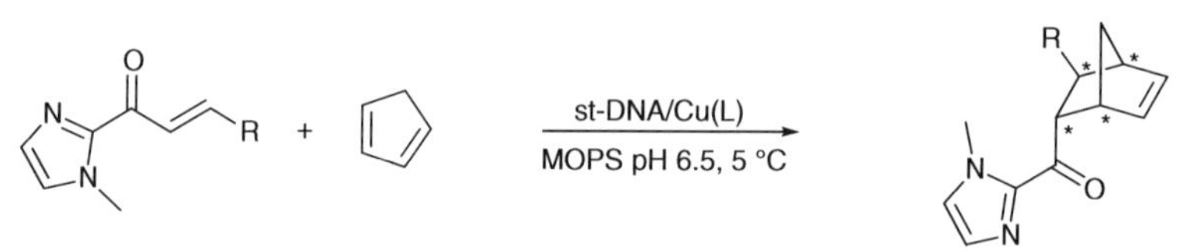

R = Ph, Me, H, 4-MeOPh, 4-ClPh, 2-BrPh, 2-furanyl, cyclohexyl

up to 98% *ee*

Scheme 5.7.6 *Schematic representation of the DNA-based copper-catalysed asymmetric Diels–Alder reaction of α, β-unsaturated 2-acyl imidazoles and cyclopentadiene*

such as fluorinations and hydrolytic epoxide ring opening reactions have also been reported [25, 26]. Recently, the catalytic scope of DNA-based asymmetric catalysis has been expanded beyond Lewis acid catalysis when it was applied successfully in a Cu(I) catalysed intramolecular cyclopropanation of α-diazo-β-ketosulfones [27].

5.7.5.3.1 Conjugated addition reactions

The DNA-based catalytic asymmetric Michael additions were evaluated using dimethyl malonate and nitromethane as nucleophiles and α,β-unsaturated 2-acyl imidazoles as Michael acceptors, in the presence of copper complexes and st-DNA (Scheme 5.7.7). In order to achieve complete conversion in both reactions, an excess of nucleophile was required (100 equivalents for dimethyl malonate and 1000 equivalents for nitromethane). The best results were obtained using a [Cu(dmbipy)$(NO_3)_2$] complex in combination with st-DNA: full conversion and high enantiomeric excess of up to 99 and 84% using dimethyl malonate and nitromethane, respectively.

Recently, Li *et al.* further broadened the substrate scope by including malononitrile and cyanoacetates as nucleophiles [22c]. High conversions were achieved in almost all cases and, again, the highest enantioselectivities (*ee* up to 84%) were obtained when the reaction was catalysed by [Cu(dmbipy)$(NO_3)_2$] in the presence of st-DNA.

The catalytic enantioselective oxa-Michael addition of several alcohols to the corresponding α,β-unsaturated 2-acyl imidazoles was also performed, as shown in Scheme 5.7.8 [22b]. The best yields with the highest ratio of alcohol addition product were achieved when 40% v/v of the corresponding alcohol was employed. Interestingly, the highest enantioselectivities were obtained using Cu(II) complexes derived from **L1b**. *Ee* values of up to 81 and 86% were achieved for the addition of methanol and *n*-propanol to enones, respectively.

Finally, the same substrates were used as electrophiles in the enantioselective Friedel–Crafts alkylation of a set of indoles in water (Scheme 5.7.9). Again, the hybrid catalyst consisted of st-DNA and [Cu(dmbipy)$(NO_3)_2$]. A broad range of indoles was studied, achieving complete conversion and enantioselectivities from 69 to 83%, in less than 10 hours. Note that these reactions exhibit the highest reaction rates of all DNA-based catalytic conjugate additions. This can tentatively be explained by the fact that indoles as π-nucleophiles are neutral. Thus, no electrostatic repulsion between the nucleophile and the negatively charged DNA occurs.

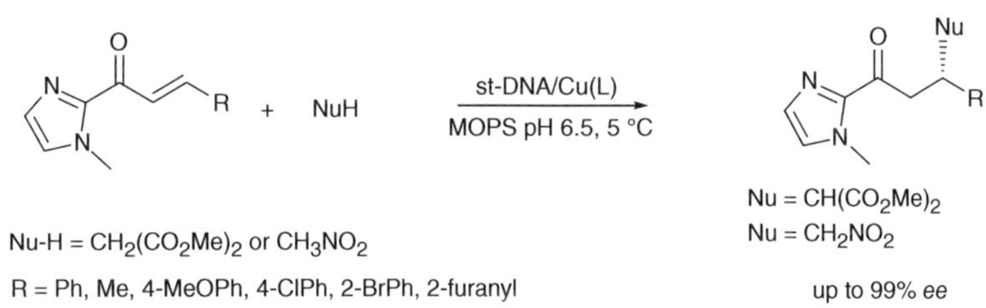

Scheme 5.7.7 *The asymmetric DNA-based copper catalysed Michael reaction of α,β-unsaturated 2-acyl imidazoles and dimethyl malonate or nitromethane as nucleophiles*

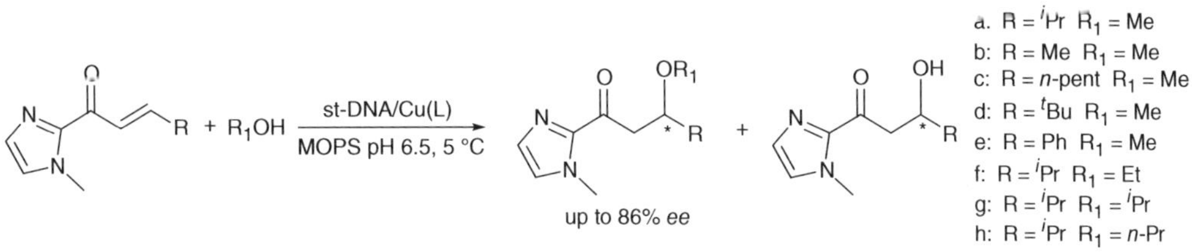

Scheme 5.7.8 *The asymmetric DNA-based copper catalysed oxa-Michael reaction of α,β-unsaturated 2-acyl imidazoles in water*

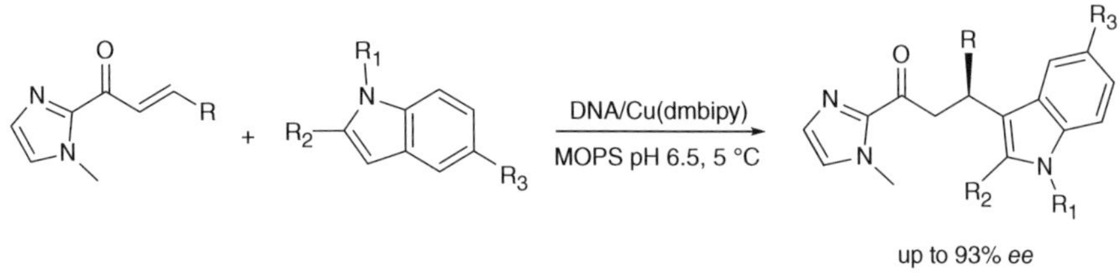

Scheme 5.7.9 *The asymmetric DNA-based copper catalysed Friedel–Crafts reaction of α,β-unsaturated 2-acyl imidazoles in water*

The highest enantiomeric excess (83%) was found using 5-methoxyindole as nucleophile. This was further increased up to 93% by employing the self-complementary oligonucleotide d(TCAGGGCCCTGA)$_2$. It is noteworthy that these results were obtained with a very low catalyst loading of 0.15 mol%.

An asymmetric intramolecular Friedel–Crafts alkylation has been described by Sugiyama and coworkers (Scheme 5.7.10) [23b]. A yield of 64% and a 71% enantiomeric excess were obtained with the hybrid catalyst based on st-DNA and the complex [Cu(5,6-dmp)(NO_3)$_2$], derived from the ligand **L4**. This value was improved up to 77% *ee* by employing the self-complementary oligonucleotide d(TGTGTGCACACA)$_2$. These results demonstrated that the sequence selectivity of the catalytic reaction differs from the intermolecular version [22a]. In these studies a model for the binding of the copper complex to the DNA was also proposed.

The Michael addition, as well as the Friedel–Crafts alkylation reaction has been performed in the presence of water-miscible organic co-solvents, such as alcohols, DMF, DMSO and 1,4-dioxane [28]. For both reactions it was found to have a positive effect on the reactivity. Furthermore, the use of co-solvents in DNA-based catalysis allowed the increase of the reaction scale and the decrease of the reaction temperature, which is of interest for the application of the DNA-based asymmetric catalysis in organic synthesis.

Recently, the group working with Smietana and Arseniyadis reported the first example of mirror-image DNA-based asymmetric catalysis, applied in two different reaction-types: (i) the Michael addition of nitromethane and dimethyl malonate and (ii) the Friedel–Crafts addition, to α,β-unsaturated 2-acyl imidazoles (Scheme 5.7.11) [29]. The L-oligonucleotides chosen for the experiments were the mirror image of the D-DNA sequences, which gave the best results in the corresponding reactions [L-d(TCAGGGCCCTGA)$_2$] [22a, 23a]. Reactions were performed using a hybrid catalyst consisting of these sequences and the complex [Cu(dmbpy)(NO_3)$_2$] in the optimal conditions previously reported for these reactions. In all cases, using the

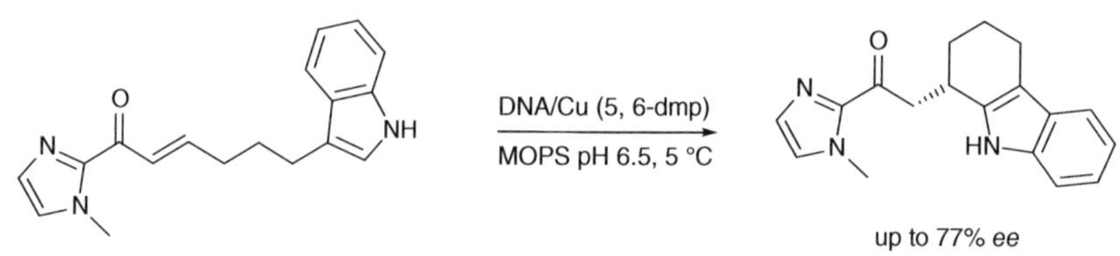

Scheme 5.7.10 *The asymmetric DNA-based copper catalysed intramolecular Friedel–Crafts reaction in water*

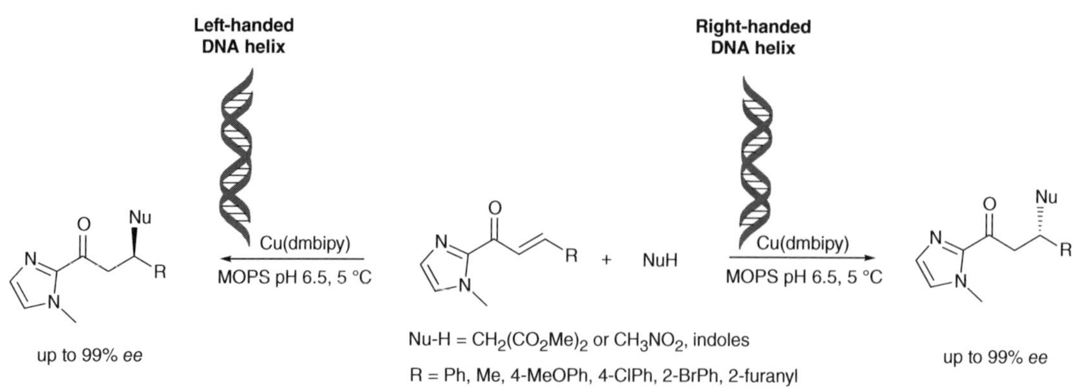

Scheme 5.7.11 *The asymmetric L- and D-DNA-based copper catalysed Friedel–Crafts and Michael reaction*

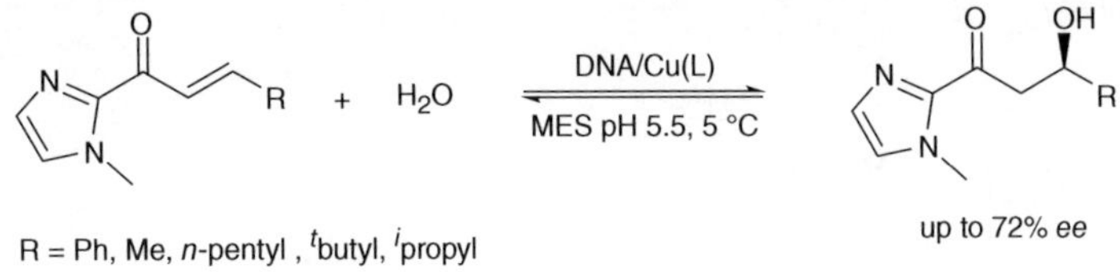

Scheme 5.7.12 *The DNA-based copper catalysed asymmetric hydration of α,β-unsaturated 2-acyl-(1-methyl) imidazoles*

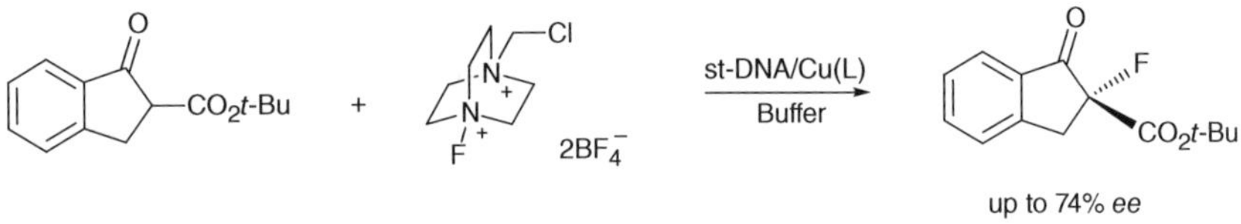

Scheme 5.7.13 *The DNA-based copper catalysed asymmetric fluorination of β-keto esters*

left-handed double helical structure of L-DNA resulted in formation of the opposite enantiomer compared with when using D-DNA.

A particularly exciting class of reactions achieved with DNA-based catalysis is the catalytic enantioselective and diastereospecific *syn* hydration of α,β-unsaturated ketones (Scheme 5.7.12) [24]. The product of these reaction is a β-hydroxy ketone, which is a key structural motif in many natural products. This is a reaction that has no equivalent using conventional homogeneous catalysis. The main challenges associated with this reaction are the small size of water, the reversible nature of the reaction and that water is a actually poor nucleophile under neutral conditions. Using the combination of st-DNA and [Cu(dmbipy)$(NO_3)_2$] as the catalytic system gave the final β-hydroxy ketone with low *ee* (19%). Surprisingly, in this case the use of first generation ligands gave better results. The highest enantiomeric excess was achieved using **L1b** with a spacer ($n = 2$) in the presence of st-DNA, with 72% *ee*. This value was increased up to 82% when D_2O was used instead of water. Furthermore, a study of the substrate scope of the reaction demonstrated that with an increase of the steric bulk of the substituent at the β-position of the enone, a higher enantiomeric excess was achieved. With a phenyl group at the β-position no conversion was obtained, which is most likely caused by the fact that, for this more conjugated enone, the hydration reaction is thermodynamically unfavourable.

5.7.5.3.2 Miscellaneous reactions

The DNA-based catalytic asymmetric fluorination reaction was described by Toru and coworkers [25]. In this case, different Cu(II) complexes in the presence of st-DNA catalysed the enolization of several β-ketoesters and the subsequent nucleophilic attack on an electrophilic fluoride source. The best results were obtained when [Cu(dmbipy)$(NO_3)_2$]/st-DNA was applied together with Selectfluor®, achieving up to 75% yield and 74% *ee* in the fluorination of *tert*-butylindanone carboxylate (Scheme 5.7.13).

The same catalyst was applied in the hydrolytic kinetic resolution of 2-pyridyloxiranes (Scheme 5.7.14) [26]. The highest selectivity factor (*S*) found was 2.7 for the resolution of *trans*-β-phenyl pyridyloxirane.

Recently, the first DNA-based intramolecular cyclopropanation of α-diazo-β-ketosulfones in water was reported (Scheme 5.7.15) [27]. In this case, the role of the copper is to assist formation of a carbenoid species, followed by an intramolecular cyclopropanation reaction. It was observed that the reaction could be

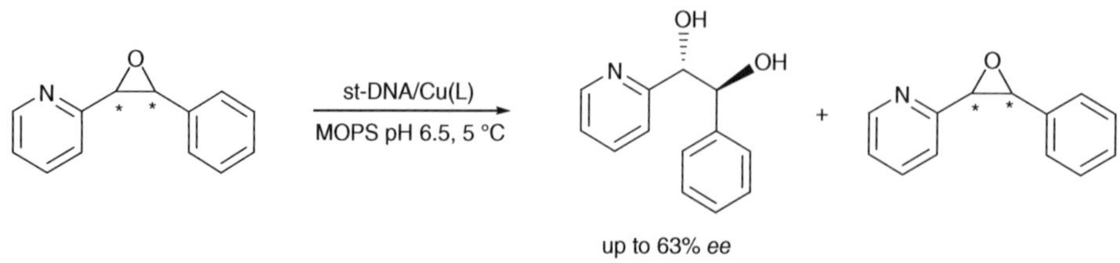

Scheme 5.7.14 *The asymmetric DNA-based hydrolytic kinetic resolution of 2-pyridyloxiranes*

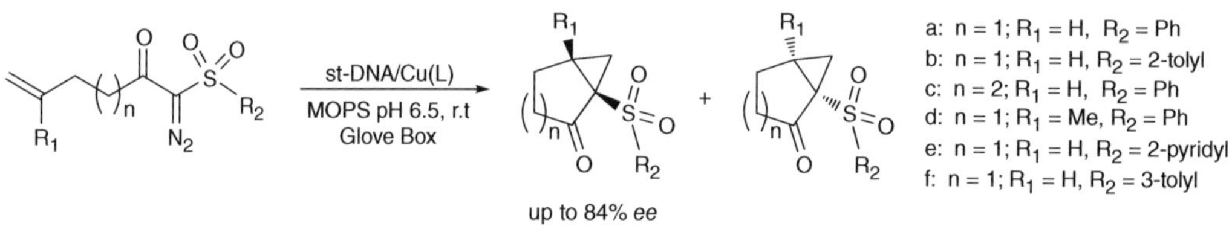

Scheme 5.7.15 *The DNA-based copper catalysed asymmetric cyclopropanation of α-diazo-β-ketosulfones*

performed using a Cu(II) complex, whereby Cu(II) is reduced *in situ* to the catalytically active Cu(I) state. Copper complexes derived from **L1** and **L8** in combination with st-DNA gave moderate conversions, since in addition to the desired product, a significant amount of a side product resulting from the insertion of the carbene into the O–H bond of water was obtained. Interestingly, the conversion and the enantiomeric excess increased when the aromatic system of the ligand was larger, that is, when derivatives of ligand **L2** were employed. After ligand optimization, the best results were achieved with **L2b** (60% *ee*). The addition of two equivalents of this ligand with respect to copper increased this value up to 84% *ee*. The substrate scope of this system was analysed for a series of diazosulfones. In general, low yields and moderate enantioselectivities were achieved.

It is significant that in this case, in contrast to the Lewis acid catalysed reactions described earlier, a strongly intercalating ligand, such as **L2b**, is necessary in order to obtain the cyclopropanation product in high enantiomeric excess.

5.7.5.4 Mechanistic studies and role of DNA in catalysis

Mechanistic studies were carried out to understand and explain the results of DNA-based catalysis. To date, mechanistic studies have been performed only for the supramolecular DNA-based catalysis approach. Since there is no sequence selectivity in the binding of the second-generation Cu(II) complexes, they will bind at random positions to the DNA. This means that there is a different environment for each complex. Nevertheless, in spite of this 'stochasticity', excellent activities and selectivities were obtained. Kinetic and DNA sequence dependence studies were performed in order to explain this fact, as well as the role of DNA in DNA-based catalytic reactions. To date, almost all the mechanistic studies have been carried out for the Diels–Alder reaction. However, there is also some information available about other reactions, such as the Michael addition, the Friedel–Crafts alkylation and the hydration of enones.

5.7.5.4.1 Diels–Alder cycloaddition

The proposed catalytic cycle of the Diels–Alder reaction catalysed by Cu(II) is depicted in Scheme 5.7.16. The cycle starts with a reversible bidentate chelation of the dienophile, which is azachalcone in this particular case, to give the Cu(II) complex. The following step consists of the irreversible Diels–Alder reaction between the activated dienophile and the cyclopentadiene. Finally, the cycloaddition product dissociates from the copper ion, and the catalyst is ready to start a new catalytic cycle [30]. There are three parameters that contribute to the overall rate of the reaction: K_a, the equilibrium constant for the reversible binding of the dienophile to the Cu(II) complex; k_{cat}, the rate of the Diels–Alder reaction between the activated dienophile and cyclopentadiene; and K_{dis}, the dissociation constant for the release of the product from the copper complex.

A kinetic study of this reaction revealed a different kinetic behaviour depending on the ligand used. With the first generation of catalysts, the reaction was actually slower in the presence of DNA compared to the metal complex alone. The opposite behaviour was observed when second generation ligands were used in combination with DNA. This acceleration depended on the structure of the ligand employed. Cu–**L4** and Cu–**L5**, together with DNA, displayed only a modest increase in the rate of the reaction. Interestingly, using Cu–**L8** and DNA resulted in a 58-fold acceleration with respect to Cu–**L8** in the absence of DNA. A further investigation of the kinetics of the reaction revealed that the k_{cat} increased two orders of magnitude, whereas K_a remained constant. This means that the rate acceleration is predominantly a kinetic effect.

A series of different synthetic oligonucleotides having self-complementary sequences were tested to establish the role of the DNA sequence in catalysis. Different trends were found for first- and second-generation ligands [31]. **L1b** and **L8** were chosen as model ligands for the studies. In both cases a strong correlation

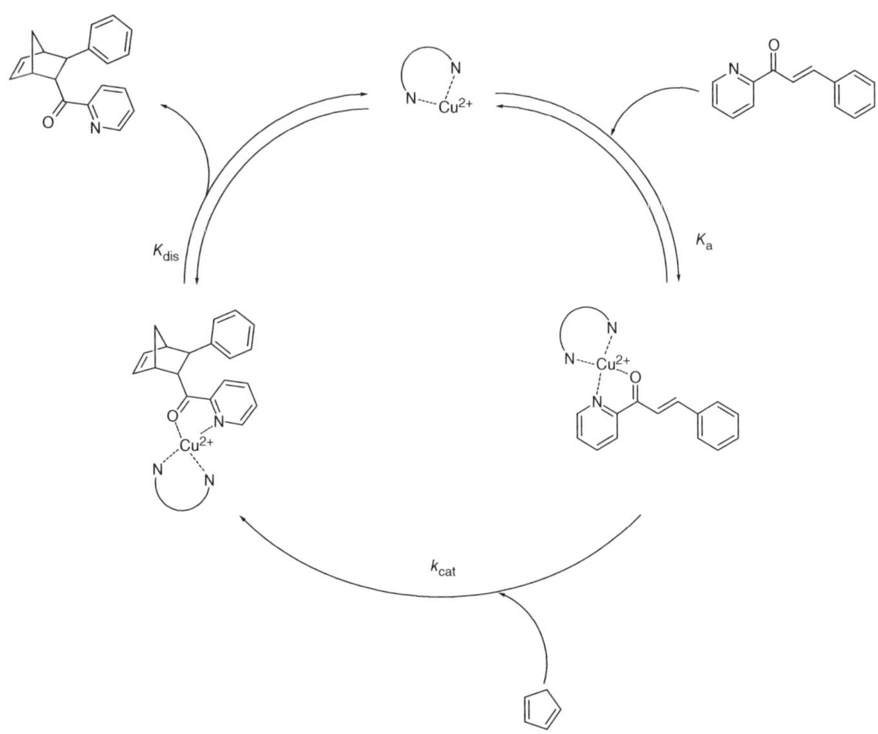

Scheme 5.7.16 *Schematic representation of the proposed catalytic cycle of Diels–Alder reaction catalysed by Cu(II)–**L***

between the enantioselectivity and DNA sequence was observed. The value of the enantioselectivity increased significantly (from 37 to 62%) when GC sequences were used instead of st-DNA, in combination with [Cu(**L1b**)(NO_3)$_2$]. In contrast, when alternating GC sequences/[Cu(dmbipy)(NO_3)$_2$] were employed as the hybrid, the enantiomeric excess dropped from 99 to 78% using the same oligonucleotide as the scaffold. In this case, excellent results were achieved using oligonucleotides containing short G tracts, such as, d(TCAGGGCCCTGA)$_2$.

The binding affinities (K_b) for all the oligonucleotides were measured and no apparent differences between the various sequences and the copper complexes were found. The k_{app} (apparent second-order rate constant) of the catalytic reaction in the presence of these oligonucleotides were also analysed. This study demonstrated that those sequences that gave higher enantiomeric excesses also led to higher rate acceleration. Combined, these results explain why such a heterogeneous system as st-DNA/[Cu(dmbipy)(NO_3)$_2$] can lead to such high enantiomeric excesses (99% *ee*), even though not all of the complexes are bound to the DNA: the reaction is highly accelerated by DNA, and those Cu(II) complexes that reside in DNA sequences which give rise to the highest *ee* values also cause the largest rate accelerations. Thus, these complexes dominate the outcome of the reaction. This represents an interesting situation where catalytic activity and selectivity are actually coupled, in contrast to what is often observed in conventional transition metal catalysis.

In addition to duplex DNA, alternative DNA structures can also be used. Moses and coworkers described the first asymmetric DNA catalytic system where duplex DNA has been substituted for G-quadruplexes [32]. This hybrid catalyst, which comprises G-quadruplex binding Cu(II) complexes based on second-generation ligands (Figure 5.7.1), was again tested in the benchmark Diels–Alder cycloaddition. The best results were achieved using Cu-**L4a**–*h*-Tel (*h*-Tel: human telomeric G-quadruplex) and Cu-**L6**–*h*-Tel as catalysts (conversion >85 and modest *ee* values of up to 51%). More recently, another G-quadruplex based-catalyst has been reported and applied in a Diels–Alder reaction [33]. This hybrid system consists of a Cu(II) cationic porphyrin (TMpyP4.Cu) bound to different oligonucleotides with G-quadruplex-forming sequences. In general, high conversions, but low to moderate *ee* values were achieved in all the experiments. It is worth noting that small changes to the structure of G-quadruplexes can significantly influence the activity of the hybrid catalyst.

5.7.5.4.2 Friedel–Crafts reaction

A 30-fold increase in reaction rate was observed with the st-DNA/[Cu(dmbipy)(NO_3)$_2$] in the Friedel–Crafts reaction. A series of enones and indoles with different substituents were analysed. In all the examples st-DNA showed a positive effect on the rate acceleration (9–27-fold). The faster reactions and the higher rate accelerations were observed for α,β-unsaturated 2-acyl imidazoles carrying an aromatic group with an electron withdrawing substituent at the β-position. In general, no relation between the DNA acceleration factor and the selectivity of the reaction was observed. In contrast to the Diels–Alder cycloaddition, the sequences that gave higher enantiomeric excesses did not significantly more accelerate the reaction. In addition, it was observed that the optimal sequence in terms of optical purity strongly depend on the substrate employed. The best result obtained (93% *ee*) was with the sequence d(TCAGGGCCCTGA)$_2$ and the α,β-unsaturated 2-acyl imidazole carrying a methyl group at the β-position.

5.7.5.4.3 Michael addition

Also for this reaction, it was observed that the reaction rate was affected by the presence of DNA. A rate increase from three- to six-fold was obtained when dimethyl malonate was used as a nucleophile in the st-DNA/[Cu(dmbipy)(NO_3)$_2$] catalysed Michael addition. In contrast, a smaller two- to four-fold rate decrease was measured when nitromethane was employed. This difference is at present not understood.

The sequence d(TCAGGGCCCTGA)$_2$ gave as high enantiomeric excess as st-DNA (93 and 90%, respectively) and increased the reaction rate 2.5 times.

5.7.5.4.4 *Hydration of enones*

Finally, the *syn*-hydration of α,β-unsaturated ketones catalysed by st-DNA/[Cu(**L1b**)(NO_3)$_2$] was studied. As described previously, this reaction is reversible. This fact was corroborated by monitoring the reaction over time. It was found that after 24 hours reaction, the conversion increased to 65%, but at the same time a decrease in the enantiomeric excess of the product was observed (from (*R*)-72% *ee* to (*S*)-23% *ee*). This means that at longer reactions times, the dehydration reaction becomes significant and consequently the *ee* starts to decrease, a hallmark of a reversible catalytic enantioselective reaction. This reaction was also performed in D_2O. Although when using this solvent the process was slower, a higher conversion and a higher selectivity than in water were achieved. This suggests the presence of both a kinetic and an equilibrium isotope effect. Apparently, this final product is more stable than that obtained in water and, consequently, its dehydration is less favourable, resulting in a higher *ee*. Using D_2O also allowed the diastereoselectivity of the reaction to be determined. It was concluded that the hydration is diastereospecific and occurs in a *syn* fashion. However, it was found that this latter aspect is not due to the DNA, because in the absence of DNA the reaction also proved to be *syn* specific.

Finally, regarding the DNA sequences, a different pattern was observed when compared with the catalysed C–C bond forming reactions. The highest enantioselectivities were found with oligonucleotides that contain AT base pairs in the middle of a self-complementary sequence [34]. The best results were obtained with d(CAAAAATTTTTG)$_2$ and d(GCGCTATAGCGC)$_2$, achieving up to 82% *ee* when D_2O was used as nucleophile.

5.7.5.4.5 *Stereochemistry in DNA-based catalytic asymmetric reactions*

Unfortunately, to date the molecular structure of the catalytic system is still unknown. However, structural knowledge is required to be able to propose a model that could explain how DNA is able to transfer its chirality to the catalysed reaction. Yet, some interesting conclusions about the mechanism of the reaction can be drawn based on the stereochemistry of the final products that are obtained. The absolute configuration of the Diels–Alder cycloaddition and the conjugated addition products were obtained by direct comparison of their optical rotations with those previously reported in the literature or via derivatization. Comparison of the absolute configuration of the products from all classes of DNA-based Lewis acid catalysed C–C bond-forming reactions revealed that the stereochemistry is predictable: the diene or the nucleophile always attacks from the same π-face of the enone moiety (Scheme 5.7.17). Thus, apparently the Diels–Alder cycloaddition, the Michael addition and the Friedel–Crafts reactions follow the same stereochemical path where the DNA directs the nucleophile and the diene to the preferred face of the enone.

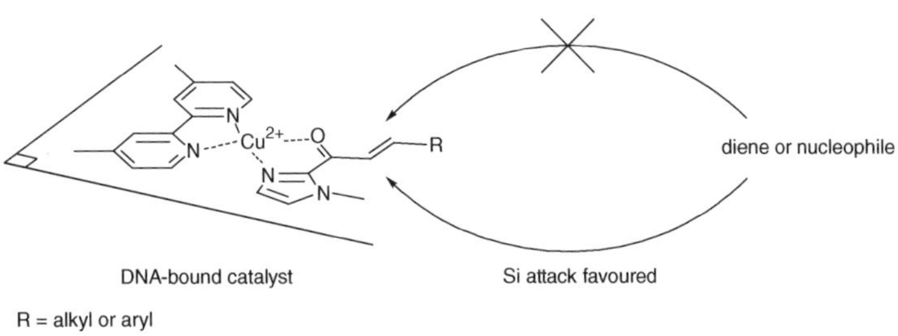

Scheme 5.7.17 *Stereochemistry of the approach of the diene or nucleophile*

A different behaviour was observed for the Cu-**L1b**–st-DNA catalysed conjugate addition of water. In this case, the nucleophile attacks from the *Re* face of the enone [24]. Note that when the reaction was performed in the absence of a ligand (only st-DNA and $Cu(NO_3)_2$ present), the reaction was also enantioselective and the opposite enantiomer was obtained. This means that the sterochemical outcome of the reaction is determined by the combination of ligand/st-DNA. It is proposed that in this case the induction of the enantioselectivity is due to hydrogen-bonding interactions of the nucleobases in the groove, that is, the second coordination sphere, wherein a spine of hydration is formed with highly localized water molecules [35]. It is likely that this complex fluid network plays a role in assisting and directing the approach of the water nucleophile to the enone substrate.

5.7.6 Conclusions and perspectives

This chapter describes the evolution of the DNA-based catalysis concept over the last few years. Although it is still a very young idea, some impressive results have already been achieved. As a result, DNA-based catalysis may represent a feasible alternative to conventional synthetic methodologies. It is, sometimes, even the only available way to perform some challenging reactions, as is illustrated by the DNA-based enantioselective conjugate addition of water in aqueous medium [24]. The DNA-based catalysis approach described herein is generally competitive with conventional asymmetric catalytic methodologies for the same reactions, as a result of the excellent activity and selectivity in water as solvent. Furthermore, this system has several benefits that makes it a potential synthetic tool for organic synthesis:

- St-DNA is an abundant and inexpensive source of chirality. This means that it can be an excellent alternative to many chiral ligands for asymmetric catalysis. Its solubility in water allows for the reactions in this solvent, adhering to one of the main principles of 'green' chemistry [4]. Additionally, the majority of the reactions are performed at 5 °C and almost neutral pH. The recyclability of the DNA-based catalyst is another attractive property [22a,23a]. Related to this, Sugiyama and coworkers recently described the synthesis and the application of a catalytic system formed by silica-supported st-DNA and [Cu(dmbipy)$(NO_3)_2$], in the asymmetric Diels–Alder reaction [36]. Herein, the DNA can be isolated and reused for up to ten cycles without loss of activity or selectivity.
- DNA-based catalysed reactions can be scaled up to the mmol scale. Moreover, although an auxiliary group is required for the coordination of the copper, it can easily be replaced after the catalytic reaction, expanding the potential synthetic application of the catalyst. Water miscible organic cosolvents can also be used, which increases the solubility of the substrates and allows for scaling up of the reactions, as well as the decrease of the reaction temperature. This improves the enantio- and chemoselectivity of the reactions.

However, there are also issues that still need to be addressed. Firstly, the reaction scope has to be increased further. As has previously been discussed, almost all the examples of DNA-catalysis involve Lewis acid catalysis. Thus, it would be of great interest to broaden this scope to other organometallic reactions in addition to the cyclopropanation of α-diazo-β-ketosulfones described in this chapter [27].

Finally, another important goal for DNA-based asymmetric catalysis is to understand how the chiral second coordination sphere provided by the DNA contributes to achieving the high rate of accelerations and enantioselectivites that are observed. This knowledge will inspire the design of novel DNA-based catalytic systems. Nevertheless, the lessons learned from DNA-based catalysis will also undoubtedly contribute to further advancing our understanding of how to design an active and selective catalyst in general.

References

[1] J. D. Watson and F. H. Crick, Molecular structure of nucleic acids: A structure for Deoxyribose Nucleic Acid, *Nature*, **171**, 737–738 (1953).

[2] K. M. Wang, Z. W. Tang, C. Y. J. Yang, Y. M. Kim, X. H. Fang, W. Li, Y. R. Wu, C. D. Medley, Z. H. Cao, J. Li, P. Colon, H. Lin, and W. H. Tan, Molecular engineering of DNA: Molecular beacons, *Angew. Chem., Int. Ed.*, **48**, 856–870 (2009).

[3] S. D. Patil, D. G. Rhodes, and D. J. Burgess, DNA-based therapeutics and DNA delivery systems: A comprehensive review, *AAPS J*, **7**, E61–E77 (2005).

[4] P. Anastas and N. Eghbali, Green chemistry: Principles and practice, *Chem. Soc. Rev.*, **39**, 301–312 (2010).

[5] (a) S. Park and H. Sugiyama, DNA-based hybrid catalysts for asymmetric organic synthesis, *Angew. Chem., Int. Ed.*, **49**, 3870–3878 (2010); (b) S. Park and H. Sugiyama, DNA as a chiral scaffold for asymmetric synthesis, *Molecules*, **17**, 12792–12803 (2012).

[6] M. Caprioara, R. Fiammengo, M. Engeser, and A. Jäschke, DNA-based phosphane ligands, *Chem. - Eur. J.*, **13**, 2089–2095 (2007).

[7] L. Ropartz, N. J. Meeuwenoord, G. A. van der Marel, P. W. N. M. van Leeuwen, A. M. Z. Slawin, and P. C. J. Kamer, Phosphine containing oligonucleotides for the development of metallodeoxyribozymes, *Chem. Commun.*, 1556–1558 (2007).

[8] P. Fournier, R. Fiammengo, and A. Jäschke, Allylic amination by a DNA-diene-iridium(I) hybrid catalyst, *Angew. Chem., Int. Ed.*, **48**, 4426–4429 (2009).

[9] U. Jakobsen, K. Rohr, and S. Vogel, Toward a catalytic site in DNA: Polyaza crown ether as non-nucleosidic building blocks in DNA conjugates, *Nucleosides, Nucleotides and Nucleic Acids*, **26**, 1419–1422 (2007).

[10] N. Sancho Oltra and G. Roelfes, Modular assembly of novel DNA-based catalysts, *Chem. Commun.*, 6039–6041 (2008).

[11] L. Gjonaj and G. Roelfes, Novel catalyst design by using cisplatin to covalently anchor catalytically active copper complexes to DNA, *ChemCatChem*, **5**, 1718–1721 (2013).

[12] G. Roelfes and B. L. Feringa, DNA-based asymmetric catalysis, *Angew. Chem., Int. Ed.*, **44**, 3230–3232 (2005).

[13] D. C. Rideout and R. Breslow, Hydrophobic acceleration of Diels-Alder reactions, *J. Am. Chem. Soc.*, **102**, 7816–7817 (1980).

[14] M. Helm, M. Petermeier, B.X. Ge, R. Fiammengo and A. Jäschke, Allostericallly activated Diels-Alder catalysis by a ribozyme, *J. Am. Chem. Soc.*, **127**, 10492–10493 (2005).

[15] M. Chandra and S. K. Silverman, DNA and RNA can be equally efficient catalysts for carbon-carbon bond formation, *J. Am. Chem. Soc.*, **130**, 2936–2937 (2008).

[16] (a) D. Hilvert, K. W. Hill, K. D. Nared and M.-T. M. Auditor, antibody catalysis of a Diels-Alder reaction, *J. Am. Chem. Soc.*,**111**, 9621–9623 (1989); (b) A. C. Braisted and P. G. Schultz, An antibody-catalyzed bimolecular Diels-Alder reaction, *J. Am. Chem. Soc.*, **112**, 7430–7433 (1990).

[17] M. Pitié and G. Pratviel, Photoenhanced oxidative DNA cleavage with non-heme iron(II) complexes. *Chem. Rev.*, **110**, 1018–1059 (2010).

[18] G. Roelfes, A. J. Boersma, and B. L. Feringa, Highly enantioselective DNA-based catalysis, *Chem. Commun.*, 635–637 (2006).

[19] A. J. Boersma, B. de Bruin, B. L. Feringa, and G. Roelfes, Ligand denticity controls enantiomeric preference in DNA-based asymmetric catalysis, *Chem. Commun.*, 2394–2396 (2012).

[20] A. J. Boersma, B. L. Feringa, and G. Roelfes, α,β-Unsaturated 2-acyl imidazoles as a practical class of dienophiles for the DNA-based catalytic asymmetric Diels-Alder reaction in water, *Org. Lett.*, **9**, 3647–3650 (2007).

[21] D. A. Evans, K. R. Fandrick, and H. J. Song, *J. Am. Chem. Soc.*, **127**, 8942–8943 (2005).

[22] (a) D. Coquière, B.L. Feringa and G. Roelfes, DNA-based catalytic enantioselective Michael reactions in water, *Angew. Chem., Int. Ed.*, **46**, 9308–9311 (2007); (b) R. P. Megens and G. Roelfes, DNA-based catalytic enantioselective intermolecular oxa-Michael addition reactions, *Chem. Commun.*, 6366–6368 (2012); (c) Y. Li, C. Wang, G. Jia, S. Lu, and C. Li, Enantioselective addition reactions in water using a DNA-based catalyst, *Tetrahedron*, **69**, 6585–6589 (2013).

[23] (a) A. J. Boersma, B. L. Feringa, and G. Roelfes, Enantioselective Friedel-Crafts reactions in water using a DNA-based catalyst, *Angew. Chem,. Int. Ed.*, **48**, 3346–3348 (2009); (b) S. Park, K. Ikehata, R. Watabe, Y. Hidaka, A. Rajendran, and H. Sugiyama, Deciphering DNA-based asymmetric catalysis through intramolecular Friedel–Crafts alkylations, *Chem. Commun.*, 10398–10400 (2012).

[24] A. J. Boersma, D. Coquière, D. Geerdink, F. Rosati, B. L. Feringa, and G. Roelfes, Catalytic enantioselective syn-hydration of enones in water using a DNA-based catalyst, *Nature Chem.*, **2**, 991–995 (2010).

[25] N. Shibata, H. Yasui, S. Nakamura, and T. Toru, DNA-mediated enantioselective carbon-fluorine bond formation, *Synlett*, 1153–1157 (2007).

[26] E. W. Dijk, B. L. Feringa, and G. Roelfes, DNA-based hydrolytic kinetic resolution of epoxides, *Tetrahedron: Asymmetry*, **19**, 2374–2377 (2008).

[27] J. Oelerich and G. Roelfes, DNA-based asymmetric organometallic catalysis in water, *Chem. Sci.*, **4**, 2013–2017 (2013).

[28] R. P. Megens and G. Roelfes, Organic co-solvents in aqueous DNA-based asymmetric catalysis, *Org. Biomol. Chem.* **8**, 1387–1393 (2010).

[29] J. Wang, E. Benedetti, L. Bethge, S. Vonhoff, S. Klussmann, J.-J. Vasseur, J. Cossy, M. Smietana, and S. Arseniyadis, DNA vs. mirror-image DNA: a universal approach to tune the absolute configuration in DNA-based asymmetric catalysis, *Angew. Chem., Int. Ed.*, **52**, 11546–11549 (2013).

[30] S. Otto, F. Bertoncin, and J. B. F. N. Engberts, Lewis-acid catalysis of a Diels-Alder reaction in water. *J. Am. Chem. Soc.*, **118**, 7702–7707 (1996).

[31] (a) F. Rosati, A. J. Boersma, J. E. Klijn, A. Meetsma, B. L. Feringa, and G. Roelfes, A kinetic and structural investigation of DNA-based asymmetric catalysis using first-generation ligands, *Chem.-Eur. J.*, **15**, 9596–9605 (2009); (b) A. J. Boersma, J. E. Klijn, B. L. Feringa, and G. Roelfes, DNA-based asymmetric catalysis: sequence-dependent rate acceleration and enantioselectivity, *J. Am. Chem. Soc.*, **130**, 11783–11790 (2008).

[32] S. Roe, D. J. Ritson, T. Garner, M. Searle, and J. E. Moses, Tuneable DNA-based asymmetric catalysis using a G-quadruplex, *Chem. Commun.*, **46**, 4309–4311 (2010).

[33] M. Wilking and U. Hennecke, The influence of G-quadruplex structure on DNA-based asymmetric catalysis using the G-quadruplex-bound cationic porphyrin TMPyP4.Cu, *Org. Biomol. Chem.*, **11**, 6940–6945 (2013).

[34] F. Rosati and G. Roelfes, A ligand structure-activity study of DNA-based catalytic asymmetric hydration and Diels-Alder reactions, *ChemCatChem*, **3**, 973–977 (2011).

[35] H. M. Berman and B. Schneider, *Oxford Handbook of Nucleic Acid Structure*, S. Neidle (Ed.), Oxford University Press, Oxford (1999).

[36] S. Park, K. Ikehata, and H. Sugiyama, Solid-supported DNA for asymmetric synthesis: A stepping-stone toward practical applications. *Biomater. Sci.*, **1**, 1034–1036 (2013).

Index

DNA in Supramolecular Chemistry and Nanotechnology, First Edition.
Edited by Eugen Stulz and Guido H. Clever.